Computing and Systems Technology Division 2017

Core Programming Area at the 2017 AIChE Annual Meeting

Minneapolis, Minnesota, USA
29 October - 3 November 2017

ISBN: 978-1-5108-5799-5

Printed from e-media with permission by:

Curran Associates, Inc.
57 Morehouse Lane
Red Hook, NY 12571

Some format issues inherent in the e-media version may also appear in this print version.

Printed by Curran Associates, Inc. (2018)

AIChE
120 Wall Street, FL 23
New York, NY 10005-4020

Phone: (800) 242-4363
Fax: (203) 775-5177

www.aiche.org

Additional copies of this publication are available from:

Curran Associates, Inc.
57 Morehouse Lane
Red Hook, NY 12571 USA
Phone: 845-758-0400
Fax: 845-758-2633
Email: curran@proceedings.com
Web: www.proceedings.com

TABLE OF CONTENTS

Some abstracts in this book refer to images that were not made available by the publisher

493368 Accounting for Safety System Activation within Economic Model Predictive Control

Abstract:

Chemical plants utilize a set of barriers against process accidents consisting of the control system and the safety system (e.g., the alarm, emergency shut-down, and safety relief systems) [1]. Recent advances in chemical process safety utilizing a systems engineering perspective [2] have sought to incorporate safety considerations within economic model predictive control (EMPC) [3] through safety-based constraints (either Lyapunov level set-based constraints [4] or constraints based on a safety metric termed the Safeness Index [5]). These designs have assumed that the process model does not change throughout the prediction horizon, meaning that no actions of the safety system (which are typically on/off-type actions such as bringing a valve from its fully open to fully closed position) are considered to be taken during process operation. However, the activation of the safety system should be accounted for within EMPC design because the actions of the safety system change the process dynamics and potentially the input availability (e.g., in the case that a valve actuated by the safety system is in series with a valve actuated by the control system), and thus the state predictions from the EMPC should account for activation of the safety system to avoid significant plant-model mismatch throughout the prediction horizon that may cause the EMPC to choose less suitable control actions (from both an economics viewpoint and a safety viewpoint) than it would choose if it was aware of the change in the plant due to safety system activation.

Motivated by the above considerations, we develop an EMPC design that accounts for the activation conditions of the safety system and changes the process model when the safety system is predicted to be activated at any time throughout the prediction horizon to reflect the dynamics of the process after safety system activation. We consider modeling of the automated actions of both the emergency shutdown system and the safety relief system within the EMPC. We investigate the impact that the foreknowledge of the safety system activation has on the state and input trajectories of a chemical process under such a control design compared to the case that the EMPC is unaware of the safety system activation. We discuss the impact of the plant-model mismatch that may occur due to disturbances if the state predictions utilized for choosing the control actions indicate activation of the safety system but it does not activate for the process with disturbances (similarly, we discuss the impact of the safety system activating when it was not predicted within the EMPC to do so). We investigate the addition of constraints to the EMPC as the state predictions approach the safety system activation levels to encourage it to choose control actions that will prevent the safety system activation. The closed-loop modeling strategy developed that includes the safety system activation conditions may also provide a systematic methodology for identifying safety system triggers and thus for coordinating the control and safety systems.

[1] Marlin T. Operability in Process Design: Achieving Safe, Profitable, and Robust Process Operations. McMaster University in Ontario, Canada, 2012.

[2] Leveson NG, Stephanopoulos G. A system-theoretic, control-inspired view and approach to process safety. AIChE Journal. 2014;60:2-14.

[3] Ellis M, Durand H, Christofides PD. A tutorial review of economic model predictive control methods. Journal of Process Control. 2014;24:1156-1178.

[4] Albalawi F, Alanqar A, Durand H, Christofides PD. A feedback control framework for safe and economically-optimal operation of nonlinear processes. AIChE Journal. 2016;62:2391-2409.

[5] Albalawi F, Durand H, Christofides PD. Process operational safety using model predictive control based on a process Safeness Index. Computers & Chemical Engineering. in press.

502118 A Novel Biomimetic Approach to Process Control By Exploiting Memory and Cognition

Abstract:

Typical learning algorithms used in the area of process control utilize the 'experiences' in the immediate past. However, for uncertain cyclic systems or for systems that 'revisit' the same or similar transients due to inherent evolution of the system, occurrence of certain disturbances, and/or desired control trajectory, experiences learnt in the 'distant' past can be exploited to achieve superior control performance. The main challenges in utilizing past experiences are in identifying the similarity of the current problem with one or more previous control problem(s), and then in calculating the optimal control moves based on the evaluation of 'dissimilarity' measure. In this work, we propose a novel biomimetic control system that mimics human memory and cognition system by drawing from the past experiences to achieve superior performance for the current control problem.

In this work, a dynamic sparse coding (DSC) algorithm is developed to identify the features of a non-linear time-varying system for the purpose of estimating the similarity to previous control scenarios. Each experience is stored in a compact cluster of "memories" by using a subtractive clustering technique for unsupervised learning of unique, or "core", control problems. Cores are automatically updated to improve future performance based on the new 'experience' gathered while solving the current control problem. When there is an acceptable similarity to a past control problem, then the experience from the past controller is delivered in the form of updated controller parameters. Overall, the control system automatically evolves to an optimal system with continued improvement as new 'experiences' are gathered.

Recognizing that it is unlikely for a controller to experience all possible forms of control scenarios upfront, or even after long exposure, an algorithm is developed that mimics the human cognitive decision support system. By integrating the memory base with a fuzzy cognitive map (FCM), informed decisions can be made on how to optimally proceed against a large array of un-encountered control scenarios. Overall, it is desired that a seamless independent operation of the control system be achieved.

The algorithms and approaches are validated on an acid gas removal unit as part of an integrated gasification combined cycle plant. This unit consists of large number of equipment items, is highly nonlinear, and involves significant mass and heat integration along with considerable time delay. It is observed that even if the controller performance is poor to start with, repeated use of past experiences leads to superior control performance. Performance of the control system is also studied under various un-encountered control scenarios.

497642 Control with Soft Feedback in Social Systems: Mathematical Principles, Empirical Evidence, and Applications

Abstract:

Feedback control has long been regarded as essential in engineering, biological, and ecological systems. Control in social systems, a recent addition, reflects further advances of this discipline. Feedback here includes opinions, consensus, polling results, market signals, rumors, product ratings, etc. Compared with the same concept in process control, feedback in social systems is "soft" and need not be followed. This unique property makes control in social systems challenging and interesting. To quantify the extent to which individuals adopt feedback, we introduce a new quantity—degree of social influence: It critically determines the closed-loop dynamics of a social control system. In this presentation, we will first describe a social system with multiple intelligent agents. Each agent has an optimization problem to solve. That problem, for instance, can be improving one's health or finding the optimal state-wise taxation depending on the context. Our objective is to understand whether soft feedback makes the problem-solving process more efficient for the multi-agent system. If so, what is the optimal degree of social influence to achieve the most improvement? How can we make intelligent crowds even smarter? We will show via mathematical proofs and empirical evidence that it is indeed possible to make crowds smarter. Our work has a diverse set of potential applications in areas such as healthcare, revenue management, and policy-making.

A Novel, Biomimetic Approach to Self-Organizing, Optimal Control Structure Design

Temitayo Bankole[1], Debangsu Bhattarcharrya[1], Berhane Gebreslassie[2]
Urmila Diwekar[2]

[1]Department of Chemical and Biomedical Engineering
West Virginia University

[2]Viswamitra Research Institute
Center For Uncertain Systems: Tools For Optimization & Management

AIChE , Minneapolis
Oct. 2017

Outline

1. Control Structure Design
   - Previous Work

     Exhaustive search

     Motivation
2. Dynamic Causal Model

   Dynamic Causal Model Architecture

   EM Algorithm
3. Framework

   Acid Gas Removal Unit
4. Results

   Sets

   Loss

Controlled variable selection (CVS)

Approach to CVS?

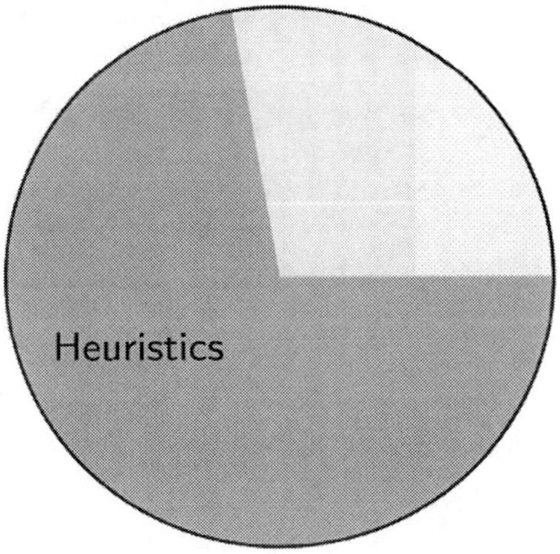

- Heuristics
  - Process knowledge
    Suspected key variables
- Systematic Approaches[1,2]
  - Optimization
    - Economic Impact
    - Controllability

Controlled variable selection (CVS)

Approach to CVS?

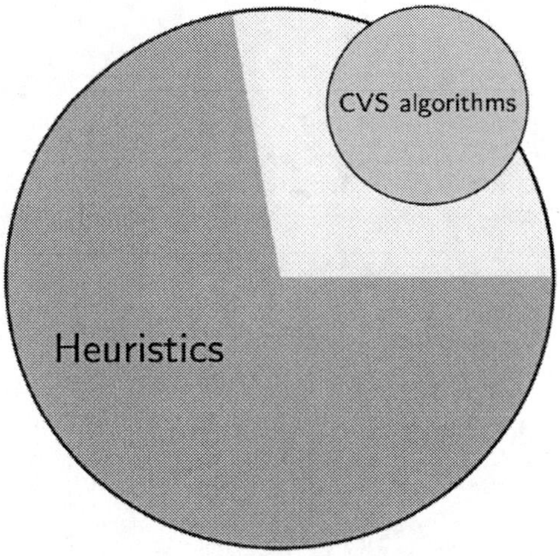

- Heuristics
  - Process knowledge
    Suspected key variables

Systematic Approaches[1,2]
- Optimization
  - Economic Impact
  - Controllability

[1] Jones D. et al Plant-wide control system design: Secondary controlled variable selection. Computers & Chemical Engineering

[2] Jones D. et al Plant-wide control system design: Primary controlled variable selection. Computers & Chemical Engineering

MultiObjective search

- APriori

$$y = G_p u + G_d d + W_n e$$
$$\left\|(G_p)_i\right\|_\infty \geq 1 \; \forall \, i$$
$$\left\|(G_p)_i\right\|_\infty \geq \left\|(G_d)_i\right\|_\infty \; \forall \, i$$
$$\tau\left(u_i, y_j\right) \leq \xi_j$$

Prescreen poor candidates

Control Structure Design | Exhaustive search

MultiObjective search

APriori

$$y = G_p u + G_d d + W_n e$$
$$\left\|(G_p)_i\right\|_\infty \geq 1 \; \forall \, i$$
$$\left\|(G_p)_i\right\|_\infty \geq \left\|(G_d)_i\right\|_\infty \; \forall \, i$$
$$\tau\left(u_i, y_j\right) \leq \xi_j$$

Prescreen poor candidates

Optimization:

$$\min_{P_n} \left(L(P_n), J_c(P_n)\right)$$
$$J_c(P_n) = \underline{\sigma}^{-1}\left(\hat{G}(P_n)\right)$$
$$L = \frac{1}{6(n_y + n_d)} \left\| \left(G_p J_{uu}^{\frac{1}{2}}\right)^{-1} \left(G_p J_{uu}^{-1} J_{ud} - G_d\right) W_d \, W_n \right\|_F^2$$

Temitayo Bankole | Biomimetic Control Structure Design | AIChE Conference | 4 / 21

MultiObjective search

- APriori

$$y = G_p u + G_d d + W_n e$$
$$\left\|(G_p)_i\right\|_\infty \geq 1 \; \forall \; i$$
$$\left\|(G_p)_i\right\|_\infty \geq \left\|(G_d)_i\right\|_\infty \; \forall \; i$$
$$\tau\left(u_i, y_j\right) \leq \xi_j$$

Prescreen poor candidates

Optimization:

$$\min_{P_n} \left(L(P_n), J_c(P_n)\right)$$
$$J_c(P_n) = \underline{\sigma}^{-1}\left(\hat{G}(P_n)\right)$$
$$L = \frac{1}{6(n_y + n_d)} \left\| \left(G_p J_{uu}^{\frac{1}{2}}\right)^{-1} \left(G_p J_{uu}^{-1} J_{ud} - G_d\right) W_d \, W_n \right\|_F^2$$

Posteriori Analysis:

Evaluate alternatives

Bidirectional Branch and Bound (Exhaustive)

Divide and Conquer

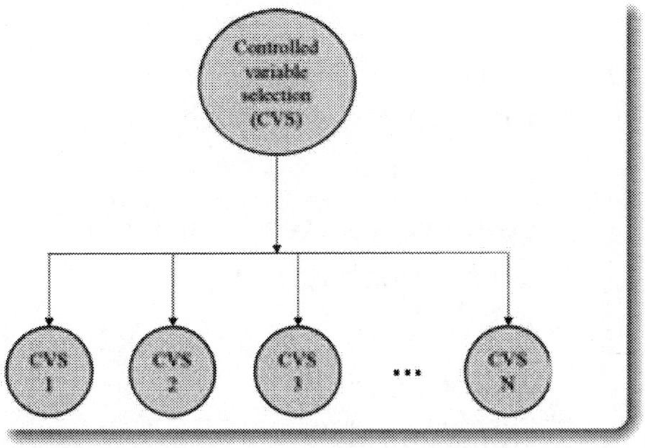

- Numerous candidates impossible to solve real time.

  Operational Objective can change dynamically

Divide and Conquer

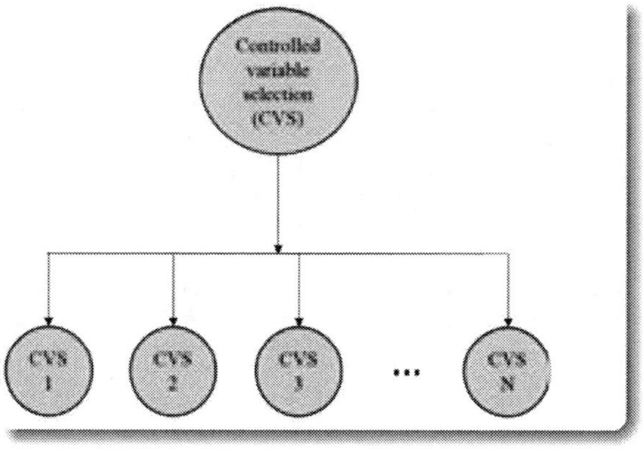

- Numerous candidates impossible to solve real time.
  Operational Objective can change dynamically

  Establish connectivity/dependence pattern.
  Decompose primary problem into subproblems.
  Solve subproblems independently.

Neuroscience

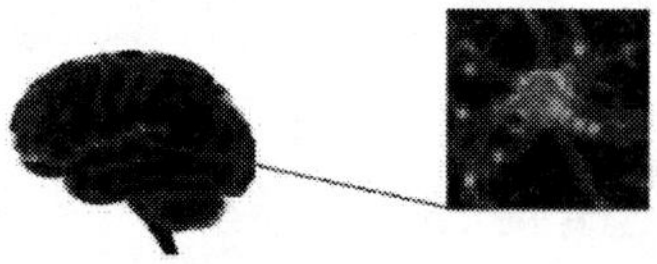

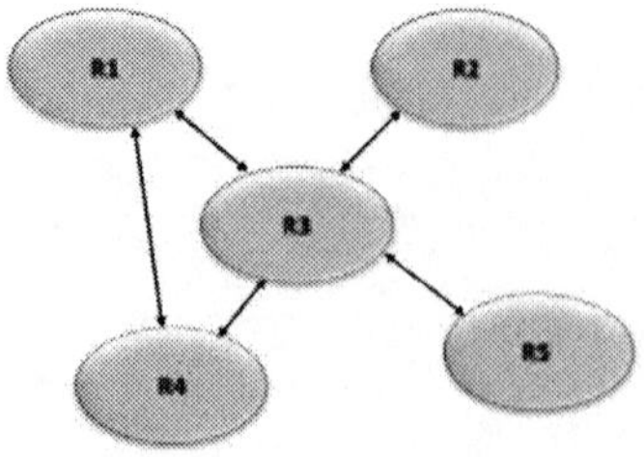

- Cortical regions of the brain
  Connectivity strengthens/weakens depending on external stimuli
  Estimation of this connectivity is performed with a dynamic causal model

Approach

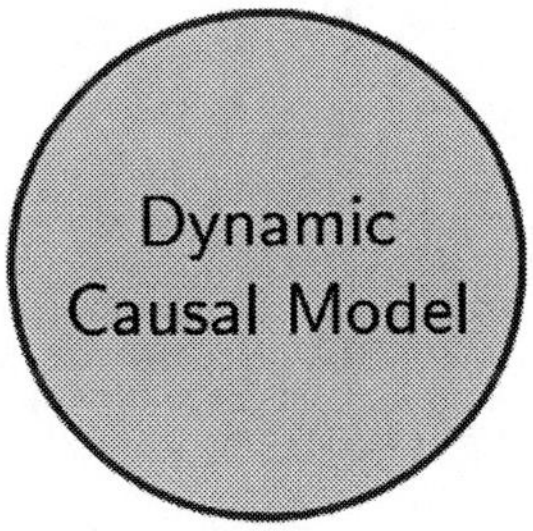

Approach

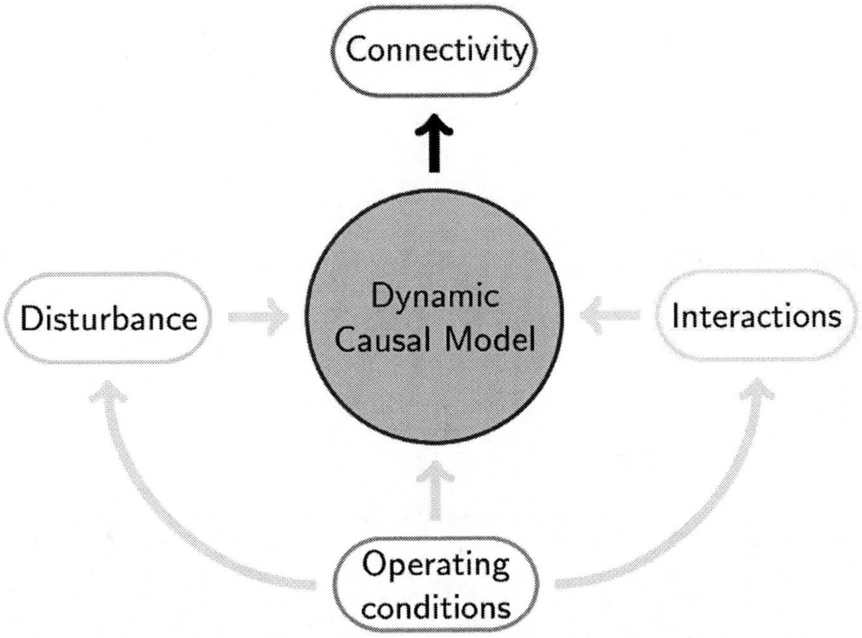

Model Architecture

- Approach
  - Represent process with nodes

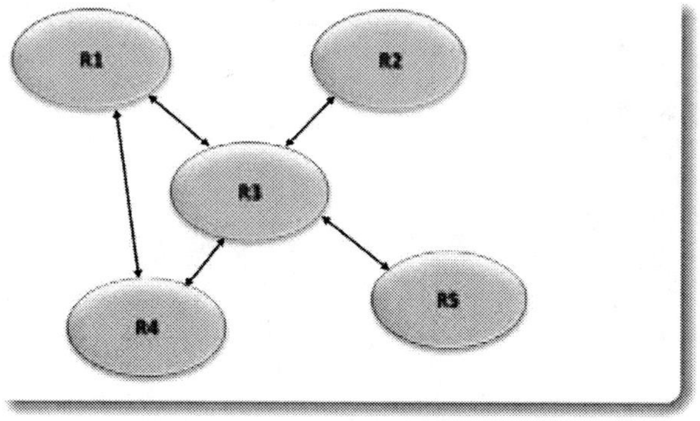

Dynamic Causal Model | Dynamic Causal Model Architecture

Model Architecture

- Approach
  - Represent process with nodes
    Categorize coupling effects as
    - Latent,
    - Induced.

$$\dot{y} = \left(A + \sum_j u_j B^j \right) y + Cu$$

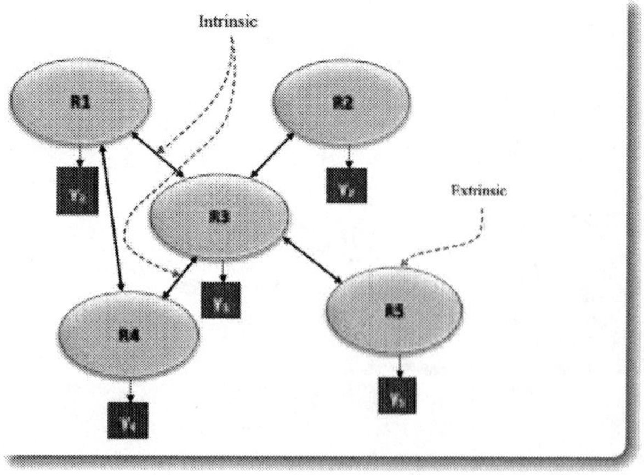

Model Architecture

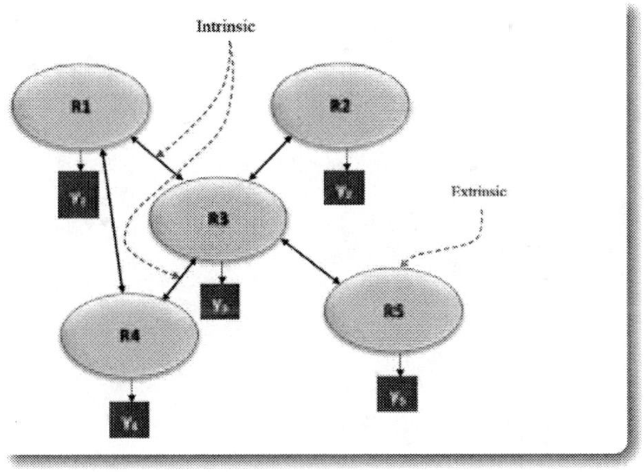

- Approach
  - Represent process with nodes
    Categorize coupling effects as
    - Latent,
    - Induced.

$$\dot{y} = \left(A + \sum_j u_j B^j \right) y + Cu$$

Model Parameters

Latent connectivity
Induced Connectivity
Extrinsic influence of inputs

Expectation Maximization

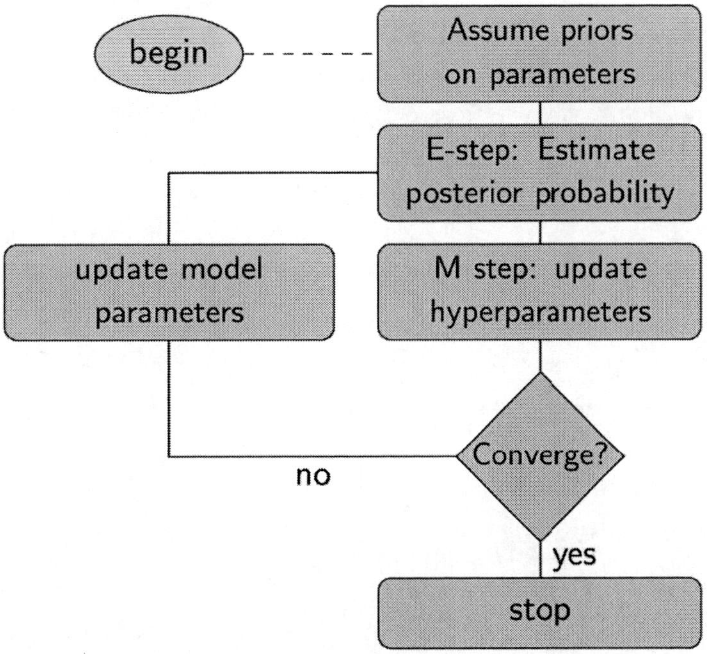

- Expectation

$$y = h(u,\theta), C_\epsilon = \sum_j \lambda_j Q_j$$

- Expectation

$$y = h(u,\theta), C_\epsilon = \sum_j \lambda_j Q_j$$
$$l = \ln p(\theta|y,\lambda;u|)$$
$$l = \ln p(y|\theta,\lambda;u) + \ln p(\theta;u)$$

- Expectation

$$y = h(u,\theta), C_\epsilon = \sum_j \lambda_j Q_j$$
$$l = \ln p(\theta|y, \lambda; u|)$$
$$l = \ln p(y|\theta, \lambda; u) + \ln p(\theta; u)$$
$$l = -\frac{1}{2}\left(y - h(u,\theta)\right)^T C_\epsilon^{-1}\left(y - h(u,\theta)\right) - \frac{1}{2}\left(\eta_\theta - \theta\right)^T C_\theta^{-1}\left(\eta_\theta - \theta\right)$$

Dynamic Causal Model | EM Algorithm

- Expectation

$$y = h(u,\theta), C_\epsilon = \sum_j \lambda_j Q_j$$
$$l = \ln p(\theta|y, \lambda; u|)$$
$$l = \ln p(y|\theta, \lambda; u) + \ln p(\theta; u)$$
$$l = -\frac{1}{2}\left(y - h(u,\theta)\right)^T C_\epsilon^{-1}\left(y - h(u,\theta)\right) - \frac{1}{2}\left(\eta_\theta - \theta\right)^T C_\theta^{-1}\left(\eta_\theta - \theta\right)$$
$$\eta_{\theta|y} \leftarrow \eta_{\theta|y} - \left(\frac{\partial^2 l}{\partial \theta^2}\right)^{-1} \frac{\partial l}{\partial \theta}$$

Maximization

Temitayo Bankole | Biomimetic Control Structure Design | AIChE Conference | 11 / 21

- Expectation

$$y = h(u,\theta), C_\epsilon = \sum_j \lambda_j Q_j$$
$$l = \ln p(\theta|y,\lambda;u|)$$
$$l = \ln p(y|\theta,\lambda;u) + \ln p(\theta;u)$$
$$l = -\frac{1}{2}\left(y - h(u,\theta)\right)^T C_\epsilon^{-1}\left(y - h(u,\theta)\right) - \frac{1}{2}\left(\eta_\theta - \theta\right)^T C_\theta^{-1}\left(\eta_\theta - \theta\right)$$
$$\eta_{\theta|y} \leftarrow \eta_{\theta|y} - \left(\frac{\partial^2 l}{\partial \theta^2}\right)^{-1}\frac{\partial l}{\partial \theta}$$

Maximization

$$\ln p(y|\lambda;u) = \ln \int q(\theta)\frac{p(\theta,y|\lambda;u)}{q(\theta)}d\theta \geq$$
$$F = \int q(\theta)\ln\frac{p(\theta,y|\lambda;u)}{q(\theta)}d\theta$$

Dynamic Causal Model EM Algorithm

- Expectation

$$y = h(u,\theta), C_\epsilon = \sum_j \lambda_j Q_j$$
$$l = \ln p(\theta|y,\lambda;u|)$$
$$l = \ln p(y|\theta,\lambda;u) + \ln p(\theta;u)$$
$$l = -\frac{1}{2}\left(y - h(u,\theta)\right)^T C_\epsilon^{-1}\left(y - h(u,\theta)\right) - \frac{1}{2}\left(\eta_\theta - \theta\right)^T C_\theta^{-1}\left(\eta_\theta - \theta\right)$$
$$\eta_{\theta|y} \leftarrow \eta_{\theta|y} - \left(\frac{\partial^2 l}{\partial \theta^2}\right)^{-1}\frac{\partial l}{\partial \theta}$$

Maximization

$$\ln p(y|\lambda;u) = \ln \int q(\theta)\frac{p(\theta,y|\lambda;u)}{q(\theta)}d\theta \geq$$
$$F = \int q(\theta)\ln\frac{p(\theta,y|\lambda;u)}{q(\theta)}d\theta$$
$$\lambda \leftarrow \lambda - \left(\frac{\partial^2 F}{\partial \lambda^2}\right)^{-1}\frac{\partial F}{\partial \lambda}$$

Acid Gas Removal Unit

- Apply connectivity estimation algorithm
  Integrated mass and heat exchange and high nonlinearity
  Subsequently perform controlled variable selection.

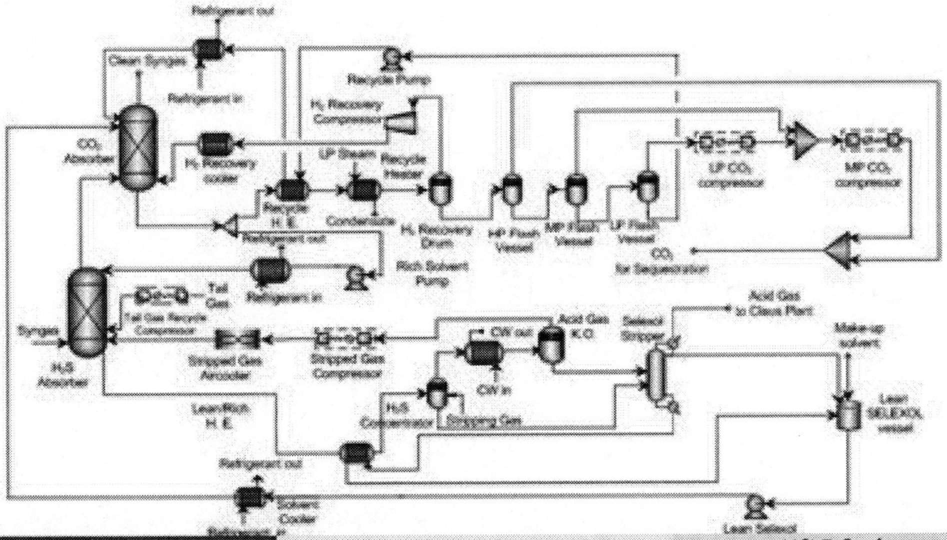

Connectivity

- CO_2 absorber
  H_2S absorber
  H_2 recovery drum
  MP flash
  H_2S concentrator

Define a threshold for connectivity ζ Islanding:

$\zeta = \infty$, 5 islands $\rightarrow$ [1], [2], [3], [4], [5]

Connectivity

- CO_2 absorber
  H_2S absorber
  H_2 recovery drum
  MP flash
  H_2S concentrator

Define a threshold for connectivity ζ Islanding:

$\zeta = \infty$, 5 islands → [1], [2], [3], [4], [5]

$\zeta = 100$, 4 islands → [1,2], [3], [4], [5]

Framework Acid Gas Removal Unit

Connectivity

- CO_2 absorber
  H_2S absorber
  H_2 recovery drum
  MP flash
  H_2S concentrator

Define a threshold for connectivity ζ Islanding:

$\zeta = \infty$, 5 islands → [1], [2], [3], [4], [5]

$\zeta = 100$, 4 islands → [1,2], [3], [4], [5]

$\zeta = 10$, 3 islands → [1,2,5], [3], [4]

$\zeta = 1$, 2 islands → [1,2,3,5], [4]

$\zeta = 0.1$, 1 island → [1,2,3,4,5]

Biomimetic approach

- Alternative to branch and bound

  Exploration and Exploitation

Multiagent approach employs agents in parallel:

  Simulated Annealing

  Ant Colony

  Genetic Algorithm

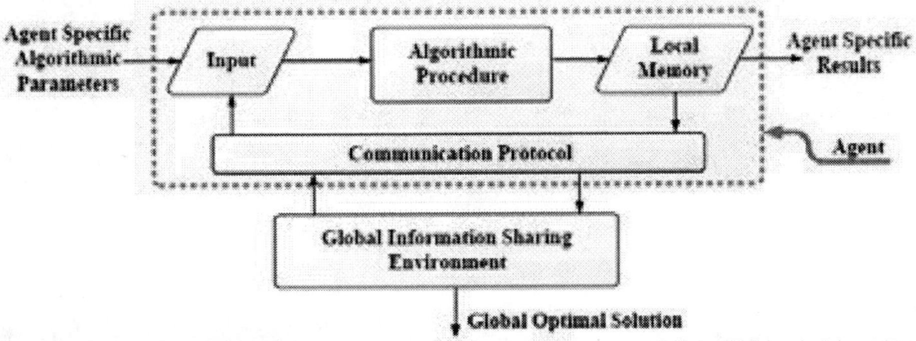

Results

- Similar results obtained with Branch and bound

  Minor variations from island to island

  66, 67 $= T_5, T_6$ of CO_2 absorber

ζ	islands	CV sets				
0.1	1	67	124	198	202	211
1	2	66	130	193	203	228
10	3	66	130	193	203	228
100	4	61	93	194	199	217
∞	5	10	185	194	199	217

Results

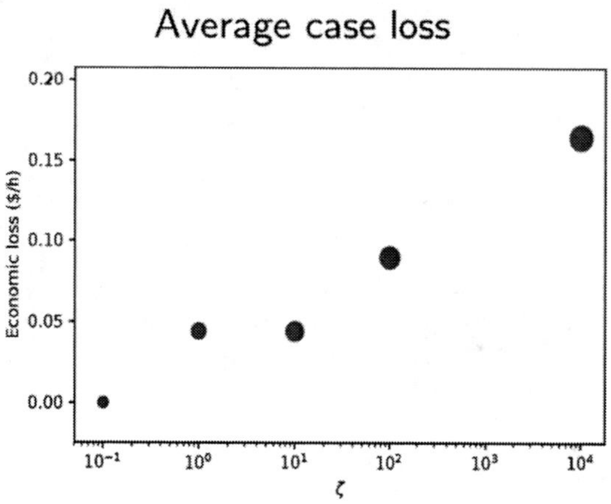

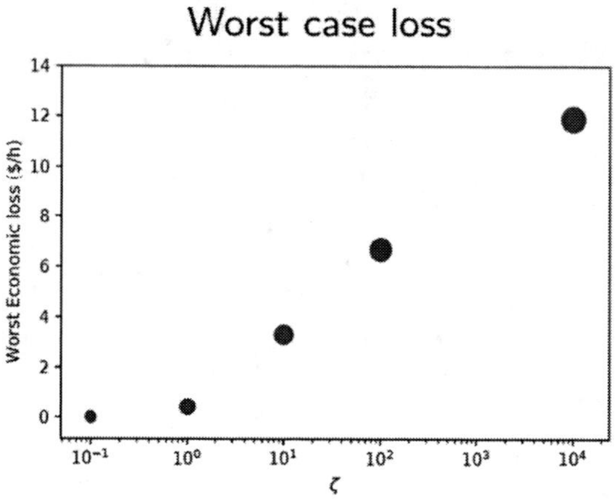

Conclusion

- Dynamic causal model allows decomposition of process into islands

  Decomposed islands allow for parallel processing of Controlled variable selection

  Multiagent optimization techniques provide coordination

- Acknowledgement
  The authors gratefully acknowledge support from NETL DOE through grant no. DE-FE0012451 titled AOI 1: Development of Integrated Biomimetic Framework with Intelligent Monitoring, Cognition and Decision Capabilities for Control of Advanced Energy Plants

References I

Jones, D., Bhattacharyya, D., Turton, R., Zitney, S. E.
Plant-wide control system design: Primary controlled variable selection.
Computers Chemical Engineering, 71 (2014):220-234.

Jones, D., Bhattacharyya, D., Turton, R., Zitney, S. E.
Plant-wide control system design: Secondary controlled variable selection.
Computers Chemical Engineering, 71 (2014):253-262.

Friston, Karl J., Lee Harrison, and Will Penny.
Dynamic causal modelling.
Neuroimage,19.4 (2003): 1273-1302.

Questions

Control Structure Configuration

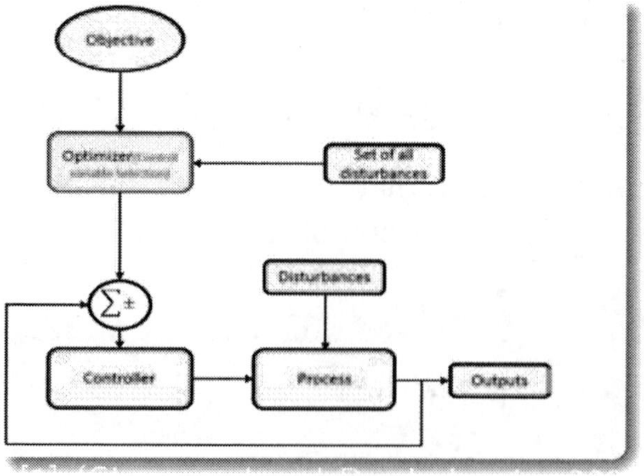

- Preselect a number of controlled variables
  - Optimally selected

Control Structure Configuration

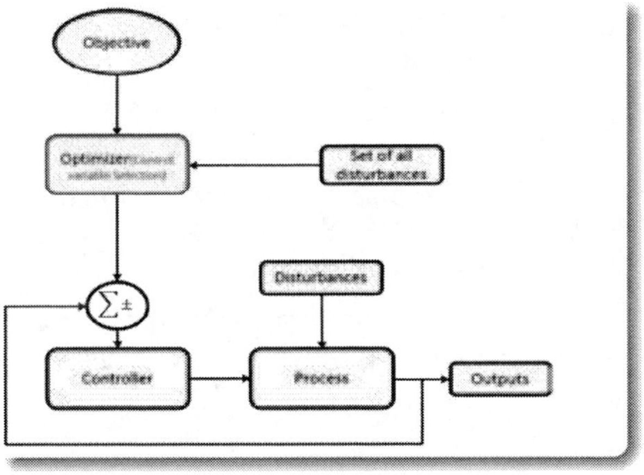

- Preselect a number of controlled variables
  - Optimally selected
    Constant set point policy

[1](Skogestad and Postlethwaite 2007)
[2](Dustin, et al. 2014)

Control Structure Configuration

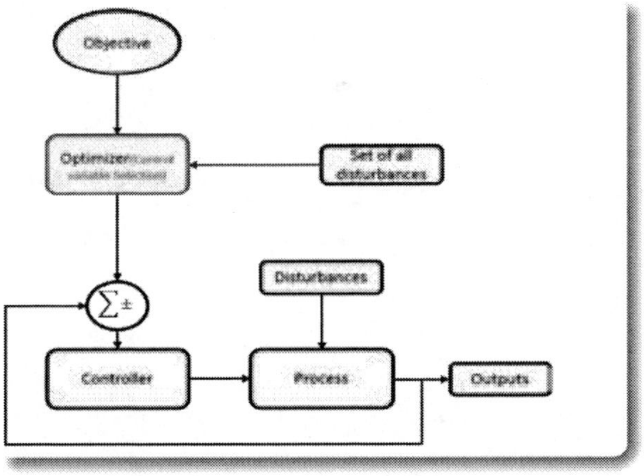

- Preselect a number of controlled variables
  - Optimally selected
    Constant set point policy
    Replace real time optimizer.

[1](Skogestad and Postlethwaite 2007)
[2](Dustin, et al. 2014)

Control Structure Configuration

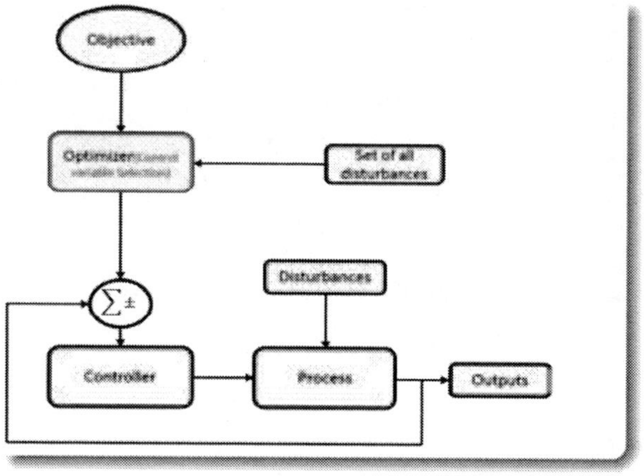

- Preselect a number of controlled variables
  - Optimally selected
  - Constant set point policy
  - Replace real time optimizer.
- Good candidates?
  - Economic impact of control variables [1]
  - Controllability [2]

[1](Skogestad and Postlethwaite 2007)
[2](Dustin, et al. 2014)

495206 Closed-Loop Active Fault Diagnosis for Uncertain Nonlinear Systems

Abstract:

Fault detection and diagnosis allows for maintaining stable, reliable, and profitable operation of complex systems in the presence of component malfunctions, drifting system parameters, and other abnormal events [1]. However, nonlinear dynamics, system uncertainty, disturbances, and measurement noise often render fault diagnosis a challenging task in many applications. The problem of fault diagnosis is further complicated by the increasingly stringent requirements on the high-performance operation of practical systems, which necessitate retaining the system within admissible operational constraints during fault diagnosis. To address these challenges, a wide range of methods have been developed including residual- and observer-based [2], set-based [3], and data-based [1] approaches. The vast majority of these methods are *passive*, where the status of the system is deduced by comparing measurements with model predictions or historical data. However, system uncertainties and/or corrective actions of the control system can often obscure faulty behavior to an extent that impairs reliable diagnosis of faults. In these cases, auxiliary input signals can be applied to the system to enhance the detectability and isolability of faults in the system measurements [4].

This notion has led to the development of *active fault diagnosis* (AFD) methods in which an input sequence is designed to diagnose the system faults within a given time window in the presence of uncertainty. Worst-case "robust" AFD methods have been reported for designing input sequences that account for the worst-case values of the uncertainty. These approaches, largely developed for linear systems, look to minimize the energy of the input sequence while guaranteeing separation of the reachable output sets for all model hypotheses [4,5]. When probabilistic descriptions of the system uncertainty are available, a more natural approach is to account for probability distribution of the model outputs in place of deterministic sets. By accounting for distributional information, the input sequences can be designed in a (possibly) less conservative manner when compared to worst-case approaches. Recently, a model predictive control approach with active model discrimination is presented to account for not only the uncertainty but also the presence of control in active fault diagnosis [6].

This talk addresses the problem of AFD for nonlinear systems subject to stochastic disturbances and measurement noise as well as probabilistic model uncertainty in parameters and initial conditions (representing our incomplete knowledge of the system). The talk further discusses the challenges of integrating the AFD and control problems for improved diagnosability without adversely affecting the average control performance. The natural choice of the AFD objective function is the *Bayes risk*, which represents the probability of misclassifying the model; however, the Bayes risk in known to be an intractable optimization criterion as discussed in [7]. Therefore, we instead minimize an upper bound on the Bayes risk that can be defined in terms of the Bhattacharyya coefficient, which has a closed-form solution for Gaussian random variables [8]. Input and state constraints are enforced to guarantee safe system operation that is minimally intrusive during the fault diagnosis period. State constraints are enforced as *chance constraints*to provide flexibility and minimal conservatism. A key challenge in solving this stochastic

AFD problem is propagation of uncertainties through the nonlinear system dynamics since there is no closed-form expression for the probability distribution of the predicted state evolution. This work explores two methods for efficient uncertainty propagation (including linearization and the unscented transform) to determine the evolution of the moments of the predicted states.

The proposed AFD method can be used to compute an input trajectory once offline or it can be implemented in an online (i.e., receding-horizon) fashion. Online AFD allows for re-planning the inputs (i.e., fault diagnosis experiment) at each measurement sampling time based on new system observations/measurements. This online procedure involves solving a nonlinear state estimation problem to recursively update the model probabilities using Bayes' rule. The effectiveness of the proposed tractable approach for offline and online AFD is demonstrated on a continuous bioreactor with multiple fault scenarios including operational and structural changes to the process. Simulation results indicate that the designed input trajectories achieve a correct fault diagnosis in all considered cases despite the presence of uncertainty. For this case, online AFD shows superior performance compared to an offline implementation and the AFD optimization problem can be solved sufficiently fast for real-time implementation. When integrating control and AFD, the simulation results show that AFD can be performed with minimal loss of control performance and that adding AFD to the predictive control algorithm improves average control performance for certain fault scenarios.

References

[1] L. H. Chiang, E. L. Russell, R. D. Braatz. *Fault Detection and Diagnosis in Industrial Systems*. Springer, 2000.

[2] R. J. Patton, J. Chen. Observer-based fault detection and isolation: Robustness and applications. *Control Engineering Practice*, 5, 671–682, 1997.

[3] R. Nikoukhah. Guaranteed active failure detection and isolation for linear dynamical systems. *Automatica*, 34, 1345–1358, 1998.

[4] A. E. Ashari, R. Nikoukhah, S. L. Campbell. Auxiliary signal design for robust active fault detection of linear discrete-time systems. Automatica, 47, 1887–1895, 2011.

[5] J. K. Scott, R. Findeisen, R. D. Braatz, D. M. Raimondo. Input design for guaranteed fault diagnosis using zonotopes. *Automatica*, 50, 1580–1589, 2014.

[6] Heirung, T. A. N., Mesbah, A. Stochastic nonlinear model predictive control with active model discrimination: a closed-loop fault diagnosis application. In *Proceedings of the IFAC World Congress*, 2017, Toulouse.

[7] T. Cover, P. Hart. Nearest neighbor pattern classification. *IEEE Transactions on InformationTheory*, 13, 21–27, 1967.

[8] L. Blackmore, B. Williams. Finite horizon control design for optimal discrimination between several models. In *Proceedings of the 45th IEEE Conference on Decision and Control*, pp 1147–1152, 2006, San Diego.

493488 Development of a Biologically-Inspired Approach for Advanced Adaptive Control of Clean Energy Systems

Abstract:

Over the last decades, biologically-inspired control techniques have emerged as attractive alternatives for process control applications. For example, a Biologically-Inspired Optimal Control Strategy (BIO-CS) has been recently proposed and illustrated via examples including a nonlinear fermentation process[1-3] and a hybrid energy system that integrates different process components[4]. Also, a low-dimensional system associated with the Acid Gas Removal (AGR) unit of an Integrated Gasification Combined Cycle (IGCC) plant was controlled using BIO-CS considering setpoint tracking cases[5]. Specifically, BIO-CS is an agent-based control strategy that is inspired by the ant's rule of pursuit idea in combination with gradient-based optimal control concepts[1-5]. In previous applications of BIO-CS, the multi-agent aspects of the algorithm were studied in detail and different termination criteria were analyzed to accommodate computational time limitations. However, the existence of plant-model mismatches as well as multiple disturbances in the control of high-dimensional systems has not yet been addressed using BIO-CS. To fill these gaps, in this presentation, BIO-CS is further extended to incorporate an adaptive component based on an Artificial Neural Network (ANN) method into the controller formulation. The proposed framework is then implemented for tackling high-dimensional control structures of the IGCC-AGR process simulation considering more realistic scenarios that could be encountered in practice.

For this study, a Multi-Input-Multi-Output (MIMO) system from the IGCC-AGR process in DYNSIM (software used for dynamic simulations of chemical processes) is chosen. In particular, CO_2/H_2S absorption units of the IGCC-AGR process are analyzed for the control structure selection. Simplified dynamic models are derived for BIO-CS employing system identification techniques, such as the autoregressive model with exogenous inputs (ARX). The optimal control trajectories are then computed by BIO-CS using the simplified model to maintain multiple outputs at their desired setpoints. The controlled/output variables considered in this implementation are the compositions of the outgoing stream and the temperature of the incoming solvent related to the CO_2 absorption unit. The manipulated/input variables are the flowrates of the recycled solvent and the refrigerant associated with the same unit. Similar variables are selected for the H_2S absorption system, thus resulting in an overall high-dimensional system with multiple control islands. The proposed controller framework is designed in MATLAB and the control laws are communicated to the IGCC-AGR process simulation in DYNSIM by employing a MATLAB-DYNSIM link. This application represents the realistic scenario of the inherent mismatch between the plant and the model used by the controller. To mitigate this mismatch, an ANN-based adaptive component that has online learning capabilities is incorporated into the BIO-CS formulation. Using the information of the tracking errors, outputs, and available states, the adaptive BIO-CS brings the system back to the desired operating point. Preliminary results on the implementation of the proposed framework for the CO_2 capture island show promising capabilities in terms of maintaining the system at the required level of carbon capture. In addition, these results demonstrate the potential of the BIO-CS with adaptive component framework to tackle multiple

challenges, such as nonlinearities, high dimensionality and plant-model mismatches that are commonly encountered in the process control industries.

References:

1. Lima F. V., Li S., Mirlekar G. V., Sridhar L. N. and Ruiz-Mercado G. J., "Modeling and advanced control for sustainable process systems". Sustainability in the Analysis, Synthesis and Design of Chemical Engineering Processes, G. Ruiz-Mercado and H. Cabezas (eds.), Elsevier, 2016.
2. Li S., Mirlekar G. V., Ruiz-Mercado G. J. and Lima F. V., "Development of chemical process design and control for sustainability". Processes, 4(3):23, 2016.
3. Mirlekar G. V., Li S. and Lima F. V., "Design and implementation of a Biologically-Inspired Optimal Control Strategy (BIO-CS) for chemical process control". Submitted for publication.
4. Mirlekar G. V., Pezzini P., Bryden M., Tucker D. and Lima F. V., "A Biologically-Inspired Optimal Control Strategy (BIO-CS) for hybrid energy systems". To appear in Proceedings of 2017 American Control Conference.
5. Mirlekar G. V. and Lima F. V., "Design and implementation of a Biologically-Inspired Optimal Control Strategy (BIO-CS) for advanced energy systems". In AIChE Annual Meeting, San Francisco, CA, 2016.

488222 A New Dynamic Response Surface Methodology for Modeling the Dynamics of Nonlinear Processes

Abstract:

In the era of Big Data, there is an ever-increasing amount of measured process variables that have been collected and stored. However, such data have not yet resulted in accurate models for process optimization and control purposes. Often the information that might be extracted from them, a challenging task, might not be sufficient[1]. In addition, most of the proposed modeling methodologies for extracting such information, such as PCA and PLS, are linear and static. However, the fact remains that both the input and output variables are time-varying and their true interrelationship is nonlinear.

Our research group has recently introduced two data-driven modeling methodologies that account for the nonlinearities and process dynamics. They generalize the classical Design of Experiments (DoE)[2,3] framework by considering dynamic input as well as dynamic output variables. They also accurately model nonlinear dependencies among the variables. The Design of Dynamic Experiments (DoDE)[4] approach addresses the issue of dynamic inputs and the Dynamic Response Surface Methodology (DRSM)[5] models time-varying outputs. These novel data-driven methodologies have been successfully applied to the modeling, optimization and control many batch processes[6-10].

In this paper, we present a new and more promising version of the DRSM methodology in order to accurately model dynamic data over the semi-infinite time horizon. The novelty of the new method rests on a nonlinear transformation of time, the independent variable. The new method has the following advantages in comparison with the previous DRSM approach. It is capable of:

1) Modeling both continuous as well as batch processes, handling semi-infinite as easily as finite time domains

2) Using data that are not equidistant in time

3) Using data segments that are of varied durations due to possible strong nonlinearities in dynamics

We have examined the accuracy of the new approach in two representative processes: a complex continuous polymerization process[11] and a semi-batch penicillin fermentation[12]. In both cases, statistical measures prove the obtained DRSM model to be very accurate. We will also discuss how such DRSM models can be used for control and optimization purposes.

Reference

1. Reis MS, Braatz RD, Chiang LH. Big Data: Challenges and Future Research Directions. *Chemical Engineering Progress.* 2016;112(3):46-50.

2. Box GEP, Wilson KB. ON THE EXPERIMENTAL ATTAINMENT OF OPTIMUM CONDITIONS. *Journal of the Royal Statistical Society Series B-Statistical Methodology.* 1951;13(1):1-45.

3. Montgomery DC. *Design and Analysis of Experiments.*8th ed. New York: Wiley; 2013.

4. Georgakis C. Design of Dynamic Experiments: A Data-Driven Methodology for the Optimization of Time-Varying Processes. *Industrial & Engineering Chemistry Research.* 2013;52(35):12369-12382.

5. Klebanov N, Georgakis C. Dynamic Response Surface Models: A Data-Driven Approach for the Analysis of Time-Varying Process Outputs. *Industrial & Engineering Chemistry Research.* 2016;55(14):4022-4034.

6. Fiordalis A, Georgakis C. Data-driven, Using Design of Dynamic Experiments, versus Model-driven Optimization of Batch Crystallization Processes. *Journal Of Process Control.* 2013;23(2):179-188.

7. Georgakis C, Chin S, Hayot P, Wassick J, Chiang LH. Optimizing an Industrial Batch Process Using the Design of Dynamic Experiments Methodology. Paper presented at: AICHE Spring Meeting; April 10-14, 2016; Houston,TX.

8. Wang Z, Klebanov N, Georgakis C. DRSM Model for the Optimization and Control of Batch Processes. Paper presented at: Dynamics and Control of Process Systems, including Biosystems/IFAC2016; Trondhelm,Norway.

9. Wang Z, Georgakis C. Data-Driven Optimization Using an Evolutionary Design of Dynamic Experiments for Biopharmaceutical Processes. Paper presented at: AICHE Annual Meeting; Nov 13-18, 2016; San Francisco, CA.

10. Troup GM, Georgakis C. Process systems engineering tools in the pharmaceutical industry. *Computers & Chemical Engineering.* 2013;51:157-171.

11. Zacca JJ, Ray WH. Modeling of the liquid-phase polymerization of olefins in loop reactors. *Chem. Eng. Sci.* 1993;48(22):3743-3765.

12. Bajpai RK, Reuss M. A Mechanistic Model for Penicillin Production. *J. Chem. Technol. Biotechnol.* 1980;30(6):332-344.

486920 The Stabilization of Input-Constrained Nonlinear Systems with Maximal Region of Attraction

Abstract:

All physical control systems exhibit input constraints: all real motors have finite torque and valves can only open so wide. However, many control system fail to incorporate this limitation into their design, resulting in potential loss of stability. Regardless of the control law, input constraints often limit the set of initial conditions from where stabilization is achievable to the so- called null controllable region (NCR).

For linear control systems, it was shown in [2] that the boundary of the NCR is formed from certain trajectories of the time-reversed system. These trajectories can be shown to be solutions to a time-optimal control problem [3]. There currently do not exist any methods to determine the NCR for general nonlinear systems.

In this work, we will introduce a new, computationally efficient algorithm to determine the NCR for any nonlinear system. The method consists of expanding a quadrative Lyapunov-function based estimate of the NCR with successive estimations of the NCR. By simulating trajectories just outside estimated NCR, the set of 'stable' states grows by adding states that can reach the current region. This can be seen as exploiting the positive invariance property of the NCR.

We next use knowledge of the NCR to design a stabilizing control law. This design is the first to be demonstrated which can guarantee stabilization from everywhere in the NCR (it is impossible to stabilize the system outside of the NCR) for nonlinear systems, and is similar to the 'Constrained control Lyapunov function'-based design which was first reported in [5] for linear systems.

The design can be summarized as driving the states away from the boundary of the NCR. Before online implementation, we compute several NCR "shells" corresponding to various input constraints. Then, we implement a model predictive controller that drives the states into successively smaller shells, which results in asymptotic stabilization to the origin. The versatility and efficacy of our results are illustrated with several linear and nonlinear examples in various dimensions.

References

[1] P. Mhaskar, N. El-Farra, P. Christofides (2006). Stabilization of nonlinear systems with state and control constraints using Lyapunov-based predictive control. Systems & Control Letters, 55, 650-659.

[2] T. Hu, Z. L. and Qiu, L. (2002). An explicit description of null controllable regions of linear systems with saturating actuators. Systems & Control Letters, 47, 65-78.

[3] Lewis, A. (2006). The Maximum Principle of Pontryagin in Control and in Optimal Control. Department of Mathematics and Statistics, Queen's University, Kingston, Canada.

[4] Mahmood, M. and Mhaskar, P. (2014). Constrained control Lyapunov function based model predictive control design. Int. J. Robust Nonlinear Control, 24, 374-388

496569 MILP Formulation and Nested Decomposition Algorithm for Electric PowerInfrastructure Planning

Abstract:

Energy systems planning models allow the evaluation of alternate scenarios for future growth, providing information to support the decision-making process and the selection of technologies in the power sector. Generation and transmission expansion models can vary widely in scope (local versus regional) as well as in resolution of time and space. These models can be used to study the impact of new technology developments, resource cost trends, and policy shifts on the projected generation mix in order to meet future demand. The increasing contribution of intermittent renewable generation to the power grid poses new challenges that are not fully covered by the traditional power systems planning models, such as the multi-scale integration of detailed operating decisions in the hourly (or sub-hourly) level with investment decisions over a few decades.

In this paper, we address the long-term planning of electric power infrastructures considering high renewable penetration. To capture the intermittency of those sources, we propose a deterministic MILP planning model that includes operating decisions on an hourly basis. This model optimizes the generation expansion required to meet the projected electricity demand over the next few decades while taking into account detailed operational constraints (i.e., unit commitment), the variability and intermittency of renewable generation sources, and the power flow between regions. The modeling framework takes the viewpoint of a central planning entity whose goal is to identify the energy source (nuclear, coal, natural gas, wind and solar), generation technology (e.g., steam, combustion and wind turbines, photovoltaic and concentrated solar panels), location (regions), and generation capacity that can meet the projected electricity demand, while minimizing the amortized capital investment of all new generating units, the retrofit cost of old units that reached their expected lifetime, the operation and maintenance costs of both new and existing units, and potential environmental costs (e.g. carbon tax and renewable generation quota).

The major challenge lies in the multi-scale integration of detailed operating decisions at the hourly level with investment planning decisions over a few decades, which directly impacts its computational burden. Therefore, we adopt judicious approximations and aggregations, such as time sampling [1, 2] and generator clustering [3] to reduce the size of the model. Although there is novelty in the formulation of the problem, the main contribution of this work is in the solution strategy. We propose a decomposition algorithm based on Nested Benders Decomposition [4] for mixed-integer multi-period problems. This framework was originally developed for stochastic programming [5, 6] but we have adapted it to deterministic multi-period problems, and extended it to handle integer and continuous state variables. We introduce three valid cuts to be used in this decomposition, and propose acceleration techniques to improve the overall performance of the algorithm [7]. This algorithm targets large-scale multi-period problems and allows us to evaluate the impact of the number of representative days per year in the planning strategy over a long time horizon.

The proposed formulation and algorithm are applied to a case study in the region managed by the Electric Reliability Council of Texas (ERCOT) for a 30 year planning horizon. We first solve the proposed model for 1 to 12 representative days per year to assess the impact of the number of day samplings in the planning strategy. We show that for the proposed model and case study, it is sufficient to have 4 representative days per year in order to adequately represent the variability in load and capacity factor for the different regions. Furthermore, we solve our model as a full-space MILP, as well as using the Nested Decomposition with the proposed cuts for the 4 representative days variant of the model, and applied the acceleration technique to the cut with the best performance. The two best versions of the algorithm are also tested for the 12 representative day. For both the 4-days and 12-days representations, we demonstrate massive computational savings with our decomposition.

References

[1] A. Pina, C. Silva, and P. Ferrão, (2011). Modeling hourly electricity dynamics for policy making in long-term scenarios, Energy Policy, vol. 39, no. 9, pp. 4692 - 4702.

[2] P. Nahmmacher, E. Schmid, L. Hirth, and B. Knopf (2016), Carpe diem: A novel approach to select representative days for long-term power system modeling, Energy, vol. 112, pp. 430 - 442.

[3] Palmintier, B. S., & Webster, M. D. (2014). Heterogeneous Unit Clustering for Efficient Operational Flexibility Modeling. IEEE Trans. Power Syst. IEEE Transactions on Power Systems, 29, 1089-1098.

[4] Zou, J., Ahmed, S., & Sun, X. A. (2016). Stochastic Dual Dynamic Integer Programming, Submitted for publication, url: http://www.optimization-online.org/DB_FILE/2016/05/5436.pdf

[5] J. R. Birge, (1985). Decomposition and partitioning methods for multistage stochastic linear programs, Operations Research, vol. 33, no. 5, pp. 989-1007.

[6] M. V. F. Pereira and L. M. V. G. Pinto, (1991). Multi-stage stochastic optimization applied to energy planning, Mathematical Programming, vol. 52, no. 1, pp. 359-375.

[7] Lara, C. L., Mallapragada, D., Papageorgiou, D., Venkatesh, D., & Grossmann, I.E. (2017) MILP Formulation and Nested Decomposition Algorithm for Electric Power Infrastructure Planning, Submitted for publication.

500291 SPICE: A Computer-Aided Framework for Systematic Process Intensification of Chemical Enterprises

Abstract:

Process intensification facilitates the improvement of process performance and leads to substantially smaller, cleaner, safer, and more energy-efficient technologies [1-2]. Several methods (e.g., [3-6]) and computer-aided tools and software frameworks (e.g., [7-11]) exist for the intensification of chemical processes. In this work, we describe a computational framework SPICE which stands for Systematic Process Intensification of Chemical Enterprises. SPICE identifies optimal intensified process designs for user-specified raw materials, product specifications, chemical species, plausible chemistry (reactions), and enabling materials for separation, conversion and/or storage. We have recently used a new superstructure using fundamental building blocks for the systematic design and intensification of chemical processes [6]. A mixed-integer integer nonlinear optimization (MINLP) model describing the building block superstructure has been also developed. This MINLP model constitutes the core of SPICE. Specifically, the framework (i) contains the interface for collecting the model input data on feeds, products and materials, (ii) generates optimal block configurations by solving the MINLP model to optimality using a commercial solver, and (iii) converts the resultant block configurations into classical process flow diagrams. A key feature of SPICE is that the users do not need to specify any superstructure *a priori*. The MINLP-based model incorporates general mass and energy conservations, and descriptions of reaction, separation and material selection towards designing an intensified chemical process. SPICE also provides options for incorporating new reaction models and materials properties. Additional constraints can be also set on product demands and waste generation to meet certain industrial and/or environmental specifications.

Using several intensification examples, we will demonstrate the capabilities of the software framework in effective collection and transferring of input data through a user-friendly interface, and automatic generation of process flowsheet. In addition, we will discuss a systematic method, which has been implemented in SPICE, to automatize the conversion of building block-based diagrams to classical process flow diagrams and vice-versa. This method allows to systematically identify the primary and auxiliary units from the corresponding solutions of the building-block based MINLP model. Evaluation on flowrate variables between blocks gives the information on connectivity for construction of an adjacency matrix with nodes as identified units and edges as connectivity [12]. This adjacency matrix is used to interpret the connectivity within a process flowsheet.

References:

[1] Stankiewicz, A.I., Moulijn, J.A., 2000. Process intensification: transforming chemical engineering. *Chemical Engineering Process*, 2000, 96(1), 22-34.

[2] Reay, D.; Ramshaw, C.; Harvey, A. Process intensification: engineering for efficiency, sustainability and flexibility. Waltham, MA: Butterworth-Heinemann, 2013.

[3] Arizmendi-Sanchez, J. A., and Sharratt, P. Phenomena-based modularisation of chemical process models to approach intensive options. *Chemical Engineering Journal*, 2008, 135(1):83-94.

[4] Lutze P., Babi, D. K., Woodley, J. M., Gani, R. Phenomena based methodology for process synthesis incorporating process intensification. *Industrial & Engineering Chemistry Research*, 2013, 52(22):7127–44.

[5] Babi, D. K., Holtbruegge, J., Lutze, P., Gorak, A., Woodley, J. M., and Gani, R. Sustainable process synthesis–intensification. *Computers & Chemical Engineering* 2015, doi:10.1016/j. compchemeng.2015.04.030.

[6] Demirel, S.E., Li, J., Hasan, M.M.F. Systematic Process Intensification using Building Blocks. *Computers & Chemical Engineering*, http://dx.doi.org/10.1016/j.compchemeng.2017.01.044.

[7] Siirola, J. J., Powers, G. J., Rudd, D. F. Synthesis of system designs: III. Toward a process concept generator. AIChE J., 17 (3) (1971), pp. 677–682.

[8] Mahalec, V., Motard, R. L. Procedures for the initial design of chemical processing systems. Comput. Chem. Eng., 1 (1) (1977), pp. 57–68.

[9] Kravanja, Z. and I.E. Grossmann. PROSYN: An MINLP Process Synthesizer. *Computers & Chemical Engineering*, 1990, 14, 1363-1378.

[10] Tula, A. K., Eden, M. R., Gani, R. Process synthesis, design and analysis using a process-group contribution method. Comput. Chem. Eng., 81 (2015), pp. 245–259.

[11] Bertran, M.-O., Frauzem, R., Sanchez-Arcilla, A.-S., Zhang, L., Woodley, J. M., Gani, R. A generic methodology for processing route synthesis and design based on superstructure optimization, Computers & Chemical Engineering, 2017, http://dx.doi.org/10.1016/j.compchemeng.2017.01.030.

[12] Biggs, Norman. Algebraic graph theory. Cambridge university press, 1993.

495212 Distributed Output-Feedback Fault Detection and Isolation of Cascade Process Networks

Abstract:

Due to the demands for improved efficiency and profitability, modern process networks that are comprised of tightly interconnected units have attracted substantial research attention within the chemical industry. These process networks are normally featured by complex and highly integrated structures, which pose significant challenges to the design of satisfactory automated control systems. For these process networks, centralized approaches are not favorable due to the issues of computational complexity, fault tolerance, etc. As one common architecture in process control, decentralized control systems may not provide desired control performance. Specifically, the subsystem interaction is completely neglected or treated in a relatively conservative way, which may lead to much degraded control performance compared to the centralized counterpart. Alternatively, distributed control frameworks can give overall improved performance by enabling information exchange among local regulators, and serve as a very promising candidate for automatic control of modern processes.

However, the control performance of a distributed control system is subject to potential faults which are commonly encountered in complex and integrated process networks. Particularly, faults (e.g., actuator faults and sensor faults) in some key components may result in control performance degradation, a higher risk of system failures, reduction in economic profits, or even catastrophic impacts on operation safety as well as the environment. Since these undesired consequences may propagate from one single subsystem to the entire plant very quickly in modern process networks, fault detection and isolation (FDI) is especially important for these plants. During the past two decades, we have witnessed rapid progress in FDI. Model-based Fault detection, diagnosis, isolation and fault tolerant control issues have been studied within distributed frameworks to account for spatially distributed nonlinear processes with medium to large sizes. However, most of these results are based on the availability of measurements of the entire system states, which may not be satisfied in many applications. Therefore, state estimation based distributed FDI approaches are highly desirable from an application perspective.

In this work, we focus on cascade nonlinear processes that have been widely encountered in various manufacturing processes. We propose a distributed output-feedback FDI approach accounting for both actuator and sensor faults. First, a distributed state estimation network is designed for the cascade process based on an approach developed in [1]. It is assumed that for each subsystem, there already exists a decentralized estimator of exponential convergence when its subsystem interaction is at steady-state values. A compensator is designed for each subsystem (except the first subsystem) to better handle subsystem interaction. Each developed compensator is then connected to the associated decentralized estimator to form an augmented estimator. The estimators for the subsystems are required to communicate continuously to exchange subsystem state estimates and the output measurements. In a fault-free case, we show that the estimation error of the developed distributed state estimation scheme converges to zero using the integral input-to-state stability (iISS). Subsequently, we design a state predictor to generate two sets of reference state predictions for each subsystem. Each predictor is re-initialized with the subsystem estimate given by the distributed state estimation system after every prediction horizon in two paralleled manners. For each subsystem, we also design a subsystem residual generator, which gives two residual signal sequences for FDI. The thresholds on the residuals are also characterized. A distributed output feedback FDI mechanism is established. A fault can be detected and further isolated by examining all the residuals of the subsystems and the associated thresholds. It is demonstrated that the proposed approach can address FDI problems for medium-sized systems. The proposed method also has the potential to handle processes with larger sizes.

[1] X. Yin, J. Zeng, J. Liu. From decentralized to distributed state estimation. In proceedings of American Control Conference, Seattle, WA, May 2017.

494587 A Branch-and-Bound Scheme for K-Adaptability Problems in Robust Optimization

Abstract:

Multi-stage decision-making problems, in which a sequence of decisions must be made over time, both in anticipation of and in response to the realization of uncertain problem parameters, arise in a wide variety of engineering and management applications. Besides faithfully modeling the dynamic nature of decision-making processes in practice, multi-stage problems are essential to mitigate the conservatism of traditional single-stage problems. However, despite intensive research efforts, multi-stage optimization problems continue to pose several theoretical and computational challenges.

Amongst the various methodologies that have been proposed to address such problems, *robust optimization* [1] has received a lot of attention over the last decade since it avoids many of the shortcomings associated with traditional approaches. In particular, it replaces probability distributions with *uncertainty sets* as the fundamental models of uncertainty and seeks to determine decisions that remain feasible for *any* realization of the uncertainty from within the modeled uncertainty set.

While multi-stage robust optimization problems involving *continuous recourse decisions* have been well-studied [1, 2], the literature on problems involving *discrete recourse decisions* is relatively sparse. In the special case of two-stage problems in which the uncertainty appears only in the constraint "right-hand sides," a nested *column-and-constraint generation* solution approach that converges in finite time is developed in [3]. Other approaches are concerned with the design of conservative approximations of the true multi-stage problem and fall into one of three categories: (i) *decision rule* approaches, in which the recourse decisions are modeled as explicit functions of the uncertainty [4, 5]; (ii) *uncertainty set partitioning* approaches, in which the recourse nature of discrete decisions is simulated by designing separate sets of decisions for different, pre-specified subsets of the uncertainty set [6, 7]; and (iii) *K-adaptability* approaches, in which K sets of discrete recourse decisions are designed before the uncertainty is resolved and the best decision is implemented after observing the actual realization of the uncertainty [8, 9].

In this work, we focus on the K-adaptability approach to address two-stage robust optimization problems. In contrast to the other two approaches, K-adaptability has been shown to provide a remarkably good approximation of the true problem, both in theory and numerical experiments [8, 9]. Furthermore, rather than designing the optimal recourse decision for every possible realization of the uncertainty [3], it aims to design a small number of contingency plans, which may be preferable for operational or organizational reasons.

Existing solution approaches for K-adaptability problems are either heuristic [8] or consist of reformulating the problem as a finite-dimensional mixed-integer linear program under the additional assumption that both the first- and second-stage decisions are binary [9]. In this work, we alleviate both of these shortcomings and devise a solution scheme that is both exact and applicable for problems featuring mixed-integer decisions and constraints with uncertainty affecting their "left-hand sides." By viewing the K-adaptability problem as a semi-infinite disjunctive program, our solution scheme uses a sampling-based disjunctive branch-and-bound search procedure to asymptotically converge to an optimal solution. For problems with uncertainty sets of finite cardinality, our solution scheme converges in finite time. In addition to our convergence proof, we also analyze the solvability, convexity and tractability properties of general K-adaptability problems, and present conditions under which they reduce to single-stage (also known as *static*) robust optimization problems, and under which their optimal values coincide with that of the true two-stage problem.

Extensive numerical experiments conducted on benchmark data from a number of applications demonstrate that: (a) we can significantly improve upon decisions determined via static robust optimization within small computing times, and (b) our solution scheme for K-adaptability problems is a practically tractable approach for addressing two-stage robust optimization problems that improves upon the current state-of-the-art.

References:

[1] D. Bertsimas, D.B. Brown, and C. Caramanis (2011). Theory and applications of robust optimization. *SIAM Review*, 53(3):464—501.

[2] A. Ben-Tal, A. Goryashko, E. Guslitzer, and A. Nemirovski (2004). Adjustable robust solutions of uncertain linear programs. *Mathematical Programming A*, 99(2):351—376.

[3] L. Zhao and B. Zeng (2012). An exact algorithm for two-stage robust optimization with mixed integer recourse problems. Technical report, University of South Florida.

[4] D. Bertsimas and A. Georghiou (2015). Design of near optimal decision rules in multistage adaptive mixed-integer optimization. *Operations Research*, 63(3):610—617.

[5] D. Bertsimas and A. Georghiou (2014). Binary Decision Rules for Multistage Adaptive Mixed-Integer Optimization. Available on Optimization Online.

[6] D. Bertsimas and I. Dunning (2016). Multistage Robust Mixed-Integer Optimization with Adaptive Partitions. *Operations Research*, 64(4):980-998.

[7] K. Postek, D. den Hertog (2016). Multistage Adjustable Robust Mixed-Integer Optimization via Iterative Splitting of the Uncertainty Set. *INFORMS Journal on Computing*, 28(3):553—574.

[8] D. Bertsimas and C. Caramanis (2010). Finite adaptability in multistage linear optimization. *IEEE Transactions on Automatic Control*, 55(12):2751—2766.

[9] G.A. Hanasusanto, D. Kuhn, W. Wiesemann (2015). K-Adaptability in Two-Stage Robust Binary Programming. *Operations Research*, 63(4):877—891.

501419 A Novel Noncooperative Modeling Framework for Economic and Environmental Life Cycle Optimization of Supply Chains and Product Systems: Miblp Model and Efficient Solution Algorithm

Abstract:

Sustainable design and operations of supply chains and product systems considering both economic and environmental performance have become increasingly important in industries [1-4]. Inspired by the life cycle analysis (LCA) approach and multiobjective optimization method, life cycle optimization methods are proposed for sustainability optimization from a life cycle perspective [5-7]. However, existing life cycle optimization studies rely on centralized models, assuming all the components in a supply chain or a product system operate in a cooperative way towards a universal objective [8, 9]. Supply chains and product systems normally involve multiple stakeholders, and different life cycle stages of a certain product are managed in a decentralized way. Consequently, the optimal strategies obtained from a centralized life cycle optimization model may be overly optimistic or even infeasible under a noncooperative environment [10-13]. Existing studies on optimization of noncooperative supply chains and product systems solely focus on the economic performance and fail to consider the corresponding environmental performance, especially from a life cycle perspective [14, 15]. To investigate the impact of decentralized feature on life cycle optimization and to better capture the corresponding life cycle performance, it is important to explicitly model and analyze the noncooperative feature in the life cycle optimizations of supply chains and product systems.

To fill this knowledge gap, we propose a general life cycle optimization framework for noncooperative supply chains and product systems. In this holistic model, we couple the sophisticated Stackelberg game theory model with the state-of-the-art life cycle optimization approach, which enables us to simultaneously address the trade-offs between conflicting objectives as well as the interactions between different stakeholders. Following the Stackelberg game, two types of stakeholders, namely a leader and a follower, are identified in the optimization problem [16]. The leader enjoys the priority of making decisions and has the knowledge of potential reactions of the follower. Due to the essential position and information advantage, the leader senses more responsibility of proposing sustainable strategies in terms of its economic and life cycle environmental performance. Once the leader's decisions are made, the follower reacts rationally to optimize its own decisions. Due to the limited information, the follower is mainly driven by its economic objective. The conflicting objectives from different stakeholders will eventually lead to a Stackelberg equilibrium [17]. This modeling framework is general enough to allow for the consideration of both design and operational decisions for the leader and the follower. The resulting problem can be formulated as a mixed-integer bilevel linear fractional program (MIBLFP), which cannot be solved directly using any off-the-shelf optimization solvers [18]. We present a tailored solution strategy integrating the parametric algorithm with a projection-based reformulation and decomposition algorithm to tackle this computational challenge. To illustrate the application of proposed modeling framework and solution algorithm, a "well-to-wire" Marcellus shale gas supply chain is considered. Two major decision makers are identified as the operator of power plants and shale gas producer. Multiple decisions with respect to drilling, production, processing, transmission, and power generation are considered. The optimal levelized cost of electricity ranges from $72/MWh to $128/MWh, and the corresponding life cycle GHG emissions are 477 kg CO2-eq/MWh and 107 kg CO2-eq/MWh, respectively. Through a detailed comparison among the noncooperative models and their centralized counterparts, we conclude that noncooperative environment affects the optimal results of life cycle optimization, especially for the upstream of a shale gas supply chain. Although centralized models are easier to solve and lead to better results in most cases, the life cycle economic and environmental performance is over-optimistic, and the corresponding optimal strategy can be infeasible in a noncooperative shale gas supply chain.

References

[1] B. Mota, M. I. Gomes, A. Carvalho, and A. P. Barbosa-Povoa, "Towards supply chain sustainability: economic, environmental and social design and planning," *Journal of Cleaner Production,* vol. 105, pp. 14-27, 2015.

[2] A. Hugo and E. N. Pistikopoulos, "Environmentally conscious long-range planning and design of supply chain networks," *Journal of Cleaner Production,* vol. 13, pp. 1471-1491, 2005.

[3] J. Gong and F. You, "Sustainable design and synthesis of energy systems," *Current Opinion in Chemical Engineering,* vol. 10, pp. 77-86, 2015.

[4] C. He and F. You, "Toward more cost-effective and greener chemicals production from shale gas by integrating with bioethanol dehydration: Novel process design and simulation-based optimization," *AIChE Journal,* vol. 61, pp. 1209-1232, 2015.

[5] F. You, L. Tao, D. J. Graziano, and S. W. Snyder, "Optimal Design of Sustainable Cellulosic Biofuel Supply Chains: Multiobjective Optimization Coupled with Life Cycle Assessment and Input–Output Analysis," *AIChE Journal,* vol. 58, pp. 1157-1180, 2012.

[6] D. Yue, M. A. Kim, and F. You, "Design of Sustainable Product Systems and Supply Chains with Life Cycle Optimization Based on Functional Unit: General Modeling Framework, Mixed-Integer Nonlinear Programming Algorithms and Case Study on Hydrocarbon Biofuels," *ACS Sustainable Chemistry & Engineering,* vol. 1, pp. 1003-1014, 2013.

[7] J. Gao and F. You, "Shale gas supply chain design and operations toward better economic and life cycle environmental performance: MINLP model and global optimization algorithm," *ACS Sustainable Chemistry & Engineering,* vol. 3, pp. 1282-1291, 2015.

[8] J. E. Santibañez-Aguilar, J. B. González-Campos, J. M. Ponce-Ortega, M. Serna-González, and M. M. El-Halwagi, "Optimal planning of a biomass conversion system considering economic and environmental aspects," *Industrial & Engineering Chemistry Research,* vol. 50, pp. 8558-8570, 2011.

[9] T. V. Bartholomew and M. S. Mauter, "Multiobjective optimization model for minimizing cost and environmental impact in shale gas water and wastewater management," *ACS Sustainable Chemistry & Engineering,* vol. 4, pp. 3728-3735, 2016.

[10] D. J. Garcia and F. Q. You, "Supply chain design and optimization: Challenges and opportunities," *Computers & Chemical Engineering,* vol. 81, pp. 153-170, 2015.

[11] J. Gao and F. You, "Game theory approach to optimal design of shale gas supply chains with consideration of economics and life cycle greenhouse gas emissions," *AIChE Journal.* DOI: 10.1002/aic.15605, 2017.

[12] J. Gao and F. You, "Design and optimization of shale gas energy systems: Overview, research challenges, and future directions," *Computers & Chemical Engineering.* DOI: http://doi.org/10.1016/j.compchemeng.2017.01.032.

[13] J. Gjerdrum, N. Shah, and L. G. Papageorgiou, "Fair transfer price and inventory holding policies in two-enterprise supply chains," *European Journal of Operational Research,* vol. 143, pp. 582-599, 2002.

[14] K. Hjaila, J. M. Laínez-Aguirre, L. Puigjaner, and A. Espuña, "Scenario-based dynamic negotiation for the coordination of multi-enterprise supply chains under uncertainty," *Computers & Chemical Engineering,* vol. 91, pp. 445-470, 2016.

[15] D. Yue and F. You, "Game-theoretic modeling and optimization of multi-echelon supply chain design and operation under Stackelberg game and market equilibrium," *Computers & Chemical Engineering,* vol. 71, pp. 347-361, 2014.

[16] D. Yue and F. You, "Stackelberg-game-based modeling and optimization for supply chain design and operations: A mixed integer bilevel programming framework," *Computers & Chemical Engineering.* DOI: 10.1016/j.compchemeng.2016.07.026, 2016.

[17] H. Von Stackelberg, D. Bazin, R. Hill, and L. Urch, *Market Structure and Equilibrium.* Springer Berlin Heidelberg, 2010.

[18] J. T. Moore and J. F. Bard, "The mixed integer linear bilevel programming problem," *Operations Research,* vol. 38, pp. 911-921, 1990.

493617 A Sequential Approach to Global Flowsheet Optimization Using Mccormick Relaxations

Abstract:

Local optimization methods are widely employed for flowsheet optimization, using formulations based on either the equation-oriented mode or the sequential modular mode of flowsheet simulation, or some intermediate version [5]. While in the former case, all model variables are considered as optimization variables and all model equations as equality constraints, the latter approaches reduce the problem size by operating on a reduced set of variables that typically correspond to the degrees of freedom in designing the process, augmented by the variables associated with tear streams (in case of infeasible path methods), while the majority of the model equations are `hidden' from the optimizer in external modules [1]. Deterministic global methods for flowsheet optimization have so far almost exclusively relied on an equation-oriented approach (cf., e.g., [11]), because the branch-and-bound-based (B&B) solvers used require explicit access to the model equations in order to supply convex relaxations [7, 10]. As the flowsheet size and model complexity increases, this approach can get challenging because of the exponential worst-case runtime of B&B solvers, and because the user generally has to supply sufficiently tight bounds on a large number of variables. From this perspective, approaches that operate in a reduced space seem particularly attractive for deterministic global optimization.

In this work, we consider a method for global optimization of flowsheets that is similar to the sequential modular infeasible path method employed in local optimization in the sense that the majority of the model variables and equations are moved to external modules, while only few are left to the optimizer. While a similar approach was originally introduced using interval methods [3], a recent modification based on McCormick relaxations has been shown to enable significant reductions in computational time compared to equation-oriented formulations [2]. This approach makes use of the automatic propagation of McCormick relaxations [6] and their subgradients through computer codes [8] that allow to supply tight relaxations for functions that are not explicitly handled by the optimizer, but rather act as a black box from which function values or relaxations and their subgradients can be queried as needed. Through an appropriate selection of the equations to be left to the optimizer, the applicability of the relaxation propagation is ensured. A possible alternative would be to optimize only in the degrees of freedom, which would require the relaxation of implicit functions [9]. A simple B&B solver has been implemented that can treat external module definitions. The modules are implemented as template functions, and the propagation of relaxations is conducted using MC++ [4]. The application of this reduced-space formulation is demonstrated for simple process flowsheets. We compare solution times of the reduced-space formulation with those of the conventional equation-oriented formulation and discuss the influence of the selection of variables and constraints left to the optimizer.

References:

[1] L.T. Biegler, R.R. Hughes. AIChE J., 28(6):994–1002, 1982.
[2] D. Bongartz, A. Mitsos. Deterministic global optimization of process flowsheets in a reduced space using McCormick relaxations. Submitted to Journal of Global Optimization.

[3] R.P. Byrne, I.D.L. Bogle. Ind. Eng. Chem. Res. 39(11):4296–4301, 2000.
[4] B. Chachuat. MC++ (version 2.0): Toolkit for Construction, Manipulation and Bounding of Factorable Functions, 2014. Available at https://omega-icl.bitbucket.io/mcpp/?. Last accessed 11 April 2017.
[5] T.F. Edgar, D.M. Himmelblau, L. Lasdon. Optimization of Chemical Processes. McGraw-Hill, New York, NY, 2001.
[6] G.P. McCormick. Math. Prog., 10:147-175, 1976.
[7] R. Misener, C.A. Floudas. J. Global Optim., 59:503–526, 2014.
[8] A. Mitsos, B. Chachuat, P.I. Barton. SIAM J. Optim., 20(2):573–601, 2009.
[9] M.D. Stuber, J.K. Scott, P.I. Barton. Optim. Method. Softw., 30:424–460, 2015.
[10] M. Tawarmalani, N.V. Sahinidis. Math. Prog., 99(3):563–591, 2004.
[11] J.M. Zamora, I.E. Grossmann. Comput. Chem. Eng., 22(12):1749–1770, 1998.

493086 Semi-Supervised Anomaly Detection for Production Oil Wells

Abstract:

Anomaly detection and diagnosis (ADD) plays an important role in preventing process faults and failures. ADD is a challenging area of process monitoring due to lack of *a priori* information regarding anomalies. Traditionally, either fault class information is required or an unsupervised (i.e., one-class [1]) approach is adopted. Process monitoring as applied to modern chemical systems is further complicated due to processes that operate continuously, perhaps in several modes, with feedback control. To tackle these issues, we extend the semi-supervised approach of Blanchard et al. [2] for static data to accommodate time series data. Motivated by the work of Fulcher and Jones [3], an expert feature space is proposed as the input data. A likelihood ratio test is employed with the nominal and anomalous distributions modeled using kernel density estimates with Gaussian kernels. Early operations are leveraged as semi-supervised training data under the premise that nominal operating days are more easily labeled than anomalous days. Results are visualized using a monitoring plot similar to a contribution plot [4-6]. The success of the approach is demonstrated on operational data collected from real production oil wells.

[1] D. M. J. Tax (2001) One-class classification. Ph.D. Thesis, Delft University of Technology.

[2] G. Blanchard, G. Lee, and C. Scott (2010) Semi-supervised novelty detection. Journal of Machine Learning Research, 11, 2973-3009.

[3] B. D. Fulcher and N. S. Jones (2014) Highly comparative feature-based time-series classification. IEEE Transactions on Knowledge and Data Engineering, 26, 3026-3037.

[4] J. A. Westerhuis, S. P. Gurden, and A. K. Smilde (2000) Generalized contribution plots in multivariate statistical process monitoring. Chemometrics & Intelligent Laboratory Systems, 51(1), 95-114.

[5] X. Zhu and R. D. Braatz (2014) 2D contribution map for fault identification. IEEE Control Systems, 33(5):72-77.

[6] B. Jiang, D. Huang, X. Zhu, F. Yang, and R. D. Braatz (2015) Canonical variate analysis-based contributions for fault identification. Journal of Process Control, 26, 17-25.

499750 A Critical Comparison of Stochastic and Worst-Case Robust Approaches to Optimal Experiment Design

Abstract:

Dynamic models are indispensable tools for analysis, optimization, and control of complex technical systems. Construction of accurate models relies on experiments, which should provide as much system information as possible while being safe and least intrusive to the nominal system operation. However, generating informative system data can be cost, time, and labor intensive. To this end, model-based methods have been developed for optimal design of experiments, generally aiming at systematically maximizing the information content of experiments while fulfilling the operational limitations of the system and possibly reducing the cost of experiment [1].

This work considers optimal experiment design (OED) for parameter estimation, where experiments are designed to generate informative data for the adequate estimation of model parameters. However, as OED is based on the *current best estimate*of the model parameters and initial conditions, two problems are likely to arise due to parametric uncertainty: (i) the predicted information content of the experiment can differ from that of the real experiment and (ii) the system constraints may be violated during the experiment, leading to potentially unsafe and ineffective experimental conditions. Thus, ensuring the robustness of the designed experiments to system uncertainties is critical to both effectively enhancing the information content of the experiment as well as enforcing the system constraints [2].

In this work we will discuss two formulations for OED that can explicitly account for the system uncertainty â€" a stochastic OED formulation based on probabilistic descriptions of uncertainties and a worst-case robust OED formulation. The stochastic OED formulation relies on generalized polynomial chaos [3], for which an efficient sample-based implementation is presented. The worst-case robust OED formulation, on the other hand, requires second-order sensitivities in the OED objective function [4], necessitating the use of efficient automatic differentiation tools for the third and fourth order derivatives. The OED approaches are implemented in CasADi, a dedicated framework for efficient algorithmic differentiation and numerical optimization [5], and are illustrated on a benchmark biological system. The performance of the OED approaches is critically evaluated in terms of their theoretical formulation, computational complexity, and extensibility to larger scale systems.

References

[1] Franceschini, G. and Macchietto, S. (2008). Model-based design of experiments for parameter precision: State of the art. *Chemical Engineering Science*, 63, 4846-4872.

[2] Asprey, S. and Macchietto, S. (2002). Designing robust optimal dynamic experiments. *Journal of Process Control*, 12(4), 545-556.

[3] Mesbah, A. and Streif, S. (2015). A probabilistic approach to robust optimal experiment design with chance constraints. In: *International Symposium on Advanced Control of Chemical Processes*(ADCHEM), 100 -105.

[4] Telen, D., Logist, F., Van Derlinden, E. and J. Van Impe J. (2012). Robust Optimal Experiment Design: A Multi-Objective Approach. In Proceedings of the 7th Vienna International Conference on Mathematical Modelling â€" MATHMOD.

[5] Andersson, J. (2013). A General-Purpose Software Frame-work for Dynamic Optimization. PhD thesis, KU Leuven, Department of Electrical Engineering

491150 Supply Chain Logistics for Personalized Medicines

Abstract:

Traditional supply chains emphasize the need to be efficient, fast and continuously stable. Strategic supply chain optimization for the pharmaceutical industries is always investigated through product management, capacity management, and trading structure to select a product development and introduction strategy and a capacity planning and investment strategy [1]. However, in the targeted therapy supply as the foundation of precision medicine which is becoming a trend, the individually treated end-users and precise manufacturing process have increased the complexity of optimization. The product demand should be classified into multiple sub-categories, instead of being modelled as random parameters that follow normal distribution with a mean value and standard deviation range, according to the central limit theorem, in normal supply chain design [2].

Modern supply chains are oriented to responsiveness, flexible and configurable to custom needs. When applied to cell and gene therapies, novel concepts are required that streamline and de-risk manufacturing and delivery to patient activities in a personalized model in economic fashions [3]. In this paper we demonstrates typical processes included in a supply chain for cell therapies, encompassing end-to-end flow of information, materials, and money. Multi-objective stochastic supply chain optimization models are developed can fit into the scope to obtain a pareto-optimal curve that considers criteria including both the supply and demand side.

The results depict how the optimized supply chain logistics guarantee personalized medicines realizable in real-world bio-manufacturing and delivery to patients. Furthermore, the cognitive technology is able to gather and correlate information from various resources such as historical clinic data, environmental variables forecast and social media information, to track and predict supply chain disruptions and tread. Machine-learning technology can also be applied to new product introduction into the supply chain through studying user profiles and behaviors. Future work is also discussed about how to enhance the supply chain theories with lessons from these emerging healthcare domains.

[1] Papageorgiou, L., Rotstein, G., & Shah, N. (2001). Strategic supply chain optimization for the pharmaceutical industries. *Industrial Engineering Chemical Research*, 40(1), 275â€“286.

[2] You, F., & Grossmann, E. (2008). Design of responsive supply chain under demand uncertainty. *Computers & Chemical Engineering*, 32, 3090â€“3111.

[3] Waldman, S. A., & Terzic, A. (2013). Information Hierarchies Optimize Patient-Centered Solutions. *Clinical Pharmacology & Therapeutics*, 93(1), 3â€“7.

491307 A Dynamic Game Theoretic Framework for Strategic Production Planning

Abstract:

Chemical industries operate on the competitive edge of their profit margins; production planning algorithms must generate competitively viable plans in order to be useful[1]. Production planning methods which do not consider market behaviour and the presence of antagonistic competitors may not generate competitively viable production plans[2]. Game theory provides the tools necessary to model competitive behaviour in production planning problems. We present a dynamic game theoretic framework for production planning. In this framework, prices change dynamically in time based on historical market supply of product; such behaviour is modelled using a modified dynamic Cournot oligopoly model with sticky prices formulated as a dynamic potential game[3,4]. The production plans resulting from this framework are obtained for all competitors, and are dynamic Nash equilibrium trajectories, meaning they are optimal in their interpretation as mutual best responses to all other competitorsâ€™ plans[5].

This framework is used to investigate patterns in the petroleum refining industry in Western Canada, namely the persistence of small refineries in the presence of much larger refineries (owned by different parent companies) serving the same markets. The large refiners possess the market power to force the small refiner out of business, thus obtaining larger market shares for themselves[6,7]. Yet this does not appear to be the typical outcome, and the small refiner persists through time. We model this scenario as a dynamic game theoretic production planning problem, and present our results and interpretations of this phenomenon, including under what conditions the small refiner should be concerned with the threat of closure, what actions are available to the small refiner to prevent closure (if the threat is legitimate) and why there exist situations in which there is no legitimate threat of closure.

These results have an important impact on organization level planning and the integration of production planning into strategic organizational planning. At the production planning level, we are able to capture market-dependent phenomena, and at the organizational decision level, production capabilities are a driving force of strategy.

1. Sahebi H, Nickel S, Ashayeri J. Strategic and tactical mathematical programming models within the crude oil supply chain context â€“ a review. *Comput Chem Eng*. 2014; 68: 56-77.
2. Tominac P, Mahalec V. A game theoretic framework for petroleum refinery strategic production planning. *AICHE J*. 2017; DOI 10.1002/aic/15644.
3. Fershtman C, Kamien MI. Dynamic duopolistic competition with sticky prices. *Econometrica*. 1987; 55(5): 1151-1164.
4. Monderer D, Shapley L. Potential Games. *Game Econ Behav*. 1996; 14(1): 124-143.
5. Webb JN. *Game Theory: Decisions, Interaction and Evolution*. London, UK: Springer London; 2007.
6. Government of Canada, National Energy Board. Feature article: understanding the production, transport, and refining of crude oil in Canada. https://www.neb-one.gc.ca/nrg/ntgrtd/mrkt/prcstrdrtcl/qrtrlprcpdts/ftrrtcl/2016-04-29ndrstndngcrl-eng.html. Updated May 2, 2016. Accessed August 17, 2016.
7. Johnston D. Refining report complexity index indicates refinery capability, value. *Oil gas J*. 1996; 94(12).

499411 Mixed-Integer Models for Simultaneous Optimization of Safety Stock and Inventory Policies in Supply Chain Planning

Abstract:

Demand forecasts are prone to error. Holding safety stock is necessary to serve demands larger than forecasted while waiting on replenishment to arrive. Inventory policies provide a simple and effective strategy to manage inventories while taking into account the uncertainties of daily operation. In this paper we analyze four mathematical programming formulations for incorporating safety stock in supply chain planning models and extend them to consider dynamic demand and multi-echelon systems. Additionally, novel formulations to represent inventory policies are derived using disjunctive programming. The safety stock formulations considered are: 1) proportional to throughput, 2) proportional to throughput with risk-pooling, 3) explicit risk-pooling, and 4) guaranteed service time. The last one is especially suited for multi-echelon systems. Formulations 1 and 2 give rise to LP or MILP models, while formulations 3 and 4 yield NLP or MINLP models. The inventory policies considered are the (r,Q) and (s,S) policies. In the (r,Q) inventory policy an order for Q units is placed every time the inventory level reaches level r, whereas in the (s,S) policy inventory is reviewed in predefined intervals. If the inventory is found to be below level s, an order is placed to bring the level back to level S. Traditionally, these parameters are obtained from single echelon analysis, while in this paper they are optimization variables in supply chain models. The models proposed allow simultaneous optimization of safety, reserve, and base stock levels together with material flows in supply chain planning. The service level achieved with each of the formulations was evaluated using discrete event simulation. Real-world examples are included to illustrate the proposed approaches.

References

- Shen, Z. J. M., Coullard, C., & Daskin, M. S. (2003). A joint location-inventory model. *Transportation science*, *37*(1), 40-55.
- You, F., & Grossmann, I. E. (2008). Mixed-integer nonlinear programming models and algorithms for large-scale supply chain design with stochastic inventory management. *Industrial & Engineering Chemistry Research*, *47*(20), 7802-7817.
- Graves, S. C., & Willems, S. P. (2000). Optimizing strategic safety stock placement in supply chains. *Manufacturing & Service Operations Management*, *2*(1), 68-83.
- Garcia-Herreros, P., Agarwal, A., Wassick, J. M., & Grossmann, I. E. (2016). Optimizing Inventory Policies in Process Networks under Uncertainty. *Computers & Chemical Engineering,* 92, 256-272.

494543 Modeling and Optimization of Multi-Product Supply Chains

Abstract:

Multi-product network models are routinely used to evaluate the economic and environmental performance of supply chains and infrastructure networks [1, 2]. Such models are also used to identify optimal resource allocation strategies for strategic planning and to perform optimal network design [3-6]. Several models exist in the literature to deal with different types of systems, constraints, and study areas [7]. Unfortunately, these models are usually developed on a case-by-case basis. The existing lack of coherence between models limits systematic comparisons (benchmarks), collaboration and sharing of data sets, analysis of solution properties, and development of solution algorithms and software implementations.

In this talk, we present optimization formulations for multi-product supply chain networks. The formulations use a general graph representation that considers a set of technologies placed at a set of nodes and under which a set of products undergo transformations. Interactions between products are captured by using a hierarchical graph that maps products at each node using a transformation matrix and that maps network nodes using transportation paths (arcs). The proposed graph abstraction combines modeling concepts from supply chain and infrastructure network modeling. With this, we seek to provide a formal and general representation that can cover a wide range of settings existing in the literature.

We show how to use the framework to compute compromise solutions that resolve geographical and stakeholder conflicts. In particular, we present case studies in which we seek to design supply chains to collect and process organic waste from a large number of farms in the State of Wisconsin to mitigate point phosphorus and methane emissions while minimizing investment and transportation costs.

References:

[1] G. Guillen-Gosalbez and I. E. Grossmann, "Optimal design and planning of sustainable chemical supply chains under uncertainty," AIChE Journal, vol. 55, no. 1, pp. 99–121, 2009.

[2] F. You, L. Tao, D. J. Graziano, and S. W. Snyder, "Optimal design of sustainable cellulosic biofuel supply chains: Multiobjective optimization coupled with life cycle assessment and input-output analysis," AIChE Journal, 2012.

[3] Y. Kim, C. Yun, S. B. Park, S. Park, and L. T. Fan, "An integrated model of supply network and production planning for multiple fuel products of multi-site refineries," Computers and Chemical Engineering, vol. 32, no. 11, pp. 2529–2535, 2008.

[4] S. M. Neiro and J. M. Pinto, "A general modeling framework for the operational planning of petroleum supply chains," Computers and Chemical Engineering, vol. 28, no. 6-7, pp. 871–896, 2004.

[5] L. G. Papageorgiou, G. E. Rotstein, and N. Shah, "Strategic supply chain optimization for the pharmaceutical industries," Industrial Engineering Chemical Research, vol. 40, no. 1, pp. 275–286, 2001.

[6] F. You, J. M. Wassick, and I. E. Grossmann, "Risk Management for a Global Supply Chain Planning Under Uncertainty: Models and Algorithms," AIChE J, vol. 55, pp. 931–946, 2009.

[7] I. E. Grossmann, "Challenges in the new millennium: Product discovery and design, enterprise and supply chain optimization, global life cycle assessment," Computers and Chemical Engineering, vol. 29, no. 1, pp. 29–39, 2004.

490832 A Branch-Price-and-Cut Approach for Robust Optimization in Vehicle Routing

Abstract:

The Vehicle Routing Problem (VRP) is one of the most highly studied combinatorial optimization problems [1-2] in the area of Supply Chains and Logistics. It aims to determine a set of minimum-cost routes for vehicles to serve customers such that all known demands are satisfied and in a way such that all vehicle capacities are respected. Recently, VRP under demand uncertainty [3-4] has attracted considerable attention. It is motivated by the fact that customer demands are often not readily available at the time the distributor needs to commit to some routing plan. In this work, we combine Robust Optimization (RO) [5] and the Branch-Price-and-Cut (BP&C) optimization method for mixed-integer linear programming, in order to develop a BP&C approach for robust VRP under demand uncertainty.

Traditionally, the Branch-and-Cut (B&C) approach was the most popular method for solving deterministic VRPs; however, BP&C has been gradually accepted as the state-of-the-art algorithm to do so, and is especially advantageous for large size and complex, "rich" VRP variants [6]. The BP&C algorithm is a sophisticated combination of Column Generation (CG) and cut generation in the context of Branch-and-Bound (B&B). The linear relaxation of every node in the whole B&B tree is solved by CG and cutting planes are added to strengthen the relaxation. In our VRP case, columns denote feasible routes, and they are generated by solving the Shortest Path Problems with Resource Constraints (SPPRC) using a Dynamic Programming (DP) technique [7-8]. It has been shown that Rounded Capacity Cuts (RCC) [9] and Limited-Memory Subset Row Cuts [10] can efficiently tighten the linear model in the deterministic setting.

In the context of VRPs, RO can be a promising approach to deal with uncertainty in customer demands, guaranteeing the immunity of the optimal solution from all demand realization in a given uncertainty set. Though some progress has been made in the past via a B&C approach for robust VRP under demand uncertainty [4], these algorithms cannot address VRP instances of a large size and cannot be readily extended to VRP variants such as the Heterogeneous Vehicle Routing Problem, among others.

In this study, we propose for the first time a BP&C approach for robust VRP under demand uncertainty. In our approach, the columns that are dynamically added denote robust feasible routes under a postulated demand uncertainty set. Since our computational experience suggested that solving SPPRC under uncertainty to guaranteed optimality is very expensive, and in order to alleviate the burden in the DP phase, we propose a Tabu Search method [11] to heuristically generate robust feasible columns, and we resort to the exact solution of SPPRC instances only when our heuristic fails. We also add robust RCC [4] to tighten the relaxations at each node of the B&B tree. To the best of our knowledge, BP&C and RO are incorporated together for the first time, enabling us to tackle robust VRPs under uncertainty from a totally new perspective.

[1]. Laporte, G. (2009). Fifty years of vehicle routing. *Transportation Science*, 43(4), 408-416.

[2]. Baldacci, R., Toth, P., & Vigo, D. (2007). Recent advances in vehicle routing exact algorithms. *4OR: A Quarterly Journal of Operations Research*, 5(4), 269-298.

[3]. Sungur, I., Ordónez, F., & Dessouky, M. (2008). A robust optimization approach for the capacitated vehicle routing problem with demand uncertainty. *IIE Transactions*, 40(5), 509-523.

[4]. Gounaris, C. E., Wiesemann, W., & Floudas, C. A. (2013). Demand Uncertainty The Robust Capacitated Vehicle Routing Problem Under Demand Uncertainty. *Operations Research*, *61*(3), 677–693.

[5]. Ben-Tal, A., El Ghaoui, L., & Nemirovski, A. (2009). Robust optimization. *Princeton University Press.*

[6]. Toth, P., & Vigo, D. (Eds.). (2014). Vehicle routing: problems, methods, and applications. *Society for Industrial and Applied Mathematics.*

[7]. Irnich, S., Desaulniers, G., & others. (2005). Shortest path problems with resource constraints. *Column Generation*, 33–65.

[8]. Pecin, D. G. (2014). Exact Algorithms for the Capacitated Vehicle Routing Problem (*Doctoral dissertation*, PUC-Rio).

[9]. Fukasawa, R., Lysgaard, J., Poggi de Aragão, M., Reis, M., Uchoa, E., & Werneck, R. (2004). Robust branch-and-cut-and-price for the capacitated vehicle routing problem. *Integer programming and combinatorial optimization*, 11-17.

[10]. Pecin, D., Pessoa, A., Poggi, M., & Uchoa, E. (2014, June). Improved branch-cut-and-price for capacitated vehicle routing. In *International Conference on Integer Programming and Combinatorial Optimization* (pp. 393-403). Springer International Publishing.

[11]. Desaulniers, G., Lessard, F., & Hadjar, A. (2008). Tabu Search, Partial Elementarity, and Generalized k -Path Inequalities for the Vehicle Routing Problem with Time Windows. *Transportation Science*, *42*(3), 387–404.

497487 Allocation and Transportation Optimization of Products to Contracts Under Multiple up to Second Order Statistics Conditions on a Variable List of Product Characteristics

Abstract:

Given a general product with many characteristics and a list of contracts where given quantities of the product need to be delivered at certain destinations, and where contracts could freely impose conditions on some or all of the characteristics such as the average, or coefficient of variation of a dimension to be within certain limits for the overall contract, and assuming that various quantities of the product are spread geographically in many warehouses, the problem is to optimize the assignment of which individual product item from which warehouse should be given to which contract, such that the overall cost of both the intrinsic value of the products and their transportation be minimized.

Many products, particularly in the agricultural field can be described by the methodology presented here. To simplify things, a "product" will be assumed to be already packaged in a container which is economically unfeasible to break down. Therefore, when referring to the characteristics of a product, we will actually refer to the average values of the constitutive elements of that product in the package. As an example, let's assume that one of the characteristics of the product is length. If the product is a grain of some type, we can measure every single grain in a given container and determine the average length of the grains in that container. In short, we will refer to this as the length of the product (in that container). Details on the sampling techniques and methodology on determining, for every characteristic the corresponding value in each container, are beyond the scope of this work.

Customers are limited in their need of products in several ways. They depend on the particular end product they manufacture which may require specific ranges of the input product. They are also restricted by the machinery they have available and which may not be able to cope with a wide variability of one or another characteristic. For example, while the end product requirements might be satisfied by a group of products with an average length between 1 and 4 units, the machines available may not be able to process the product effectively unless the length range is reduce to between 2 and 3 units.

Characteristics of a product are assumed to be open ended and generic. Some taken in consideration are: length, height, weight, colors (3 numbers), as well as very specific ones. Some characteristics might be of a nominal type, such as the name of the location where the product was harvested. The list of products is open ended because, even if so far a particular one was not taken in consideration, such as humidity, the methodology should be flexible to allow the use of this new variable without change.

The contracts can specify for a given characteristic only a limited set of condition types. As an example, a contract can specify the **length** of the product to be between a min and a max, the **average length** to be between a min and a max, and **the coefficient of variation of the length** of the product to be within certain limits. Not all types of conditions are present in a given contract for every single characteristic, and in that regard, the contracts can be assume to be of "free form".

The actual example for which an optimization solution was built deals with approximately 3,000,000 products, each product has about 15 characteristics, and the products are distributed over several hundred warehouses. The number of contracts were also in the hundreds. For practical purposes, every single product can be considered to be unique because of the individual resolution of each of the characteristics measurements. Similarly, no two contracts are alike, even from the same customer – probably because even if they manufacture the same end product as before, their internal optimization needs require more restrictive ranges of the product.

Additional restrictions in the assignment process apply because as products are assigned to a contract, deliveries in batches start to happen and once on the move, deliveries cannot be stopped. Even more, at a given moment in time assignments can be made to a contract based on the available products at that time. However, at a later time, when some of the previously assigned products have not yet shipped, and are therefore still in play, and furthermore, if products better suited for that contract (meeting the conditions and also be cheaper), a reassignment can take place.

Another restriction is dictated by transportation needs to group products in containers or truck loads.
The solution proposed for the optimization starts with the intuitive observation that this is an iterative process, where at every individual step additional products need to be assigned to a contract such that at the end, the contract conditions are met, and also the value of the assigned products together with the transportation cost to destination is minimized. To take the example of length as one of the characteristics, we can assume a contract condition specifying the **average length** to be between min_length and MAX_length. If we face a shortage of long products which are also cheap at the beginning of the assignment process we will naturally be forced to assign short products, some of which might even be delivered. However, after a while, because we assigned many short products, we can still satisfy the overall contract condition intuitively by assigning, from this point on, even longer products, which are even cheaper. In this way, we can regard the process of assigning products to a contract as a dynamic process, where, after each assignment of a product or group of products to the contract the individual conditions on each of the characteristics may change and still guarantee that in the end, the overall contract conditions will be met.

The paper presents in detail the recurrent mathematical relationships for a given characteristic linking the summary of what was assigned so far (average, coefficient of variation) with the overall contract conditions to determine the relative conditions to be imposed on the products still to be assigned. It is like at each step of the assignment we are dealing with a dynamic contract rather than an absolute one, where the conditions are adapted based on what was assigned before.

The overall algorithm is also presented, where the transportation cost is included. In brief, candidate groups of product are determined from each warehouse then the one with the overall minimum value after including the transportation cost is selected.

Additional practical IT considerations are also discussed. For example, the entire data repository was available in an Oracle database and rather than extracting it for processing in a more convenient computing platform, the optimization algorithm was developed in PL/SQL, which is the Oracle's procedural language, thus keeping the data in place and bringing the algorithm to it.

500223 Process Synthesis Using Grid Superstructure

Abstract:

Process synthesis recognizes the optimal process flowshseet among many feasible alternatives to minimize/maximize economic, environmental, social or other objectives [1-2]. A process superstructure is often used to represent all plausible process configurations within a large flowsheet. Superstructure optimization, a systematic method to address process synthesis problems, is then employed to find the optimal processing routes [3]. While superstructure-based process synthesis has been very successful in systematic design of chemical processes, a key remaining challenge is that whenever a new problem is addressed, a different superstructure needs to be postulated.

To this end, we have recently proposed a building-block based representation of chemical processes that allows us to dissect classical unit operations into different phenomena, represent these phenomena using fundamental building blocks, and then recombine the building blocks to generate new and intensified unit operations, equipment and process flowsheets [4]. In this work, we generalize the idea towards a comprehensive superstructure for process synthesis, in which all plausible unit operations are represented using finite blocks that are arranged in a two dimensional grid. This grid-based representation of a superstructure facilitates the automated selection of different unit operations as required. We have also formulated a mixed-integer nonlinear optimization (MINLP) problem for process synthesis model. Furthermore, the connectivity among different units is identified via the connectivity between different blocks, and the interaction of blocks is achieved through intra-block material and energy transfer. While reaction, mixing, heating and cooling operations are modeled within the block, separation operations are achieved via the block boundary between adjacent blocks. The systematic arrangement of building blocks results in as many process flowsheets as possible without *a priori* postulation of their connectivity.

In this presentation, we will also address the model complexity for large grid superstructure which may make it difficult for a global solver to find the optimal solution in reasonable time. To tackle this challenge, we have developed an iterative solution strategy based on the expansion of the grid superstructure to facilitate the solution of the MINLP model. During the initialization step, the initially-set grid superstructure is solved for a feasible solution. This feasible solution is further utilized as an initial guess for an expanded superstructure. Through a range of applications, we will demonstrate that the proposed method is capable of simplifying the procedure for superstructure-based process synthesis, and unifying the mathematical formulation for various synthesis problems.

References:

[1] Chen, Q. and Grossman, I.E., 2017. Recent Developments and Challenges in Optimization-Based Process Synthesis. *Annual Review of Chemical and Biomolecular Engineering*, 8(1).

[2] Babi, D.K., Holtbruegge, J., Lutze, P., Gorak, A., Woodley, J.M. and Gani, R., 2015. Sustainable process synthesisâ€"intensification. *Computers & Chemical Engineering*,Â 81, pp.218-244.

[3] Wu, W., Henao C.A, Maravelias C.T. A superstructure representation, generation, and modeling framework for chemical process synthesis. *AIChE Jounal*, 2016, 62(9): 3199-3214.

[4] Demirel, S.E., Li, J., Hasan, M.M.F. Systematic Process Intensification using Building Blocks. *Computers & Chemical Engineering*, http://dx.doi.org/10.1016/j.compchemeng.2017.01.044.

501188 Stability and Performance of Economic Model Predictive Control with Discrete Actuators

Abstract:

With ever-increasing computing power and availability of data, there exists a significant potential to use mathematical optimization to improve process operation. Model predictive control (MPC) is an advanced process control technique whereby inputs are chosen by solving (in real time) an optimization problem with an embedded process model. Tracking-oriented MPC (in which the objective function is artificially designed to drive the system to a predetermined operating point) has achieved success due to its ability to incorporate satisfaction of constraints with process dynamics. The downside of this approach is that the setpoint must be chosen outside of the control layer, and therefore its selection may not be informed by the best available disturbance forecasts. To ease this restriction, economic MPC has recently been developed to allow the objective function to represent a more concrete goal (e.g., minimizing energy use or maximizing profit) However, accompanying control theory for this class of problem has traditionally applied only to systems with continuous-valued actuators, which means discrete-valued decisions (e.g., on/off, either/or, etc.) must still be made in a different layer, often via heuristic methods. This separation requires MPC to react to the discrete choices rather than proactively make them, and thus it would be desirable to consider all of the decisions in a single optimization. Recent results have extended the theory for tracking MPC to cover systems in which actuators are both discrete and continuous. In this presentation, we take this idea a step further and extend results for stability and closed-loop performance of economic MPC to allow discrete-valued inputs.

By incorporating discrete actuators, economic MPC can be applied to a much wider variety of systems, for example switched systems whose dynamics can change based on the current operating mode, as well as scheduling environments in which discrete assignments must be made between units and specific tasks. Due to the geometry of the input set, steady-state operation is often not sufficient to achieve low economic cost while still satisfying all of the necessary constraints, and thus we show how a periodic reference trajectory can be used to bound closed-loop performance. In addition, we demonstrate how cost structures like demand charges, which penalize the peak value of a state rather than a time-varying sum, can be incorporated into the economic MPC framework. Finally, by means of of simulation, we will examine the gap between realized closed-loop cost and the optimal infinite-horizon cost. The end goal is to provide a means to utilize the ongoing advances in math programming methods to implement real-time dynamic optimization within a rigorous mathematical framework. By incorporating feedback at each timestep, these controllers can quickly respond to unexpected disturbances to ensure that the process continues to operate as close to optimally as possible. With the addition of discrete actuators, economic MPC can be used to make higher-level decisions that have previously been outside the realm of feedback control, leading to improved economic performance.

Works Cited

1. Rawlings, J.B., Risbeck, M.J., 2017. *Model predictive control with discrete actuators: Theory and application.* Automatica **78**, 258-265.

2. Angeli, D., Amrit, R., Rawlings, J.B., 2012. *On average performance and stability of economic model predictive control.* IEEE Trans. Auto. Cont. **57**, 1615-1626.

3. Allan, D.A., Risbeck, M.J., Rawlings, J.B., 2016. *Stability and robustness of model predictive control with discrete actuators.* Proceedings of the American Control Conference, 32-37.

496673 Global Optimization Algorithm for Capacitated Multi-Facility Continuous Location-Allocation Problems

Abstract:

In this paper, we address the design of networks that involves the selection and location of facilities in the continuous 2-dimensional space. The problem is formulated as a version of the continuous facility location and allocation problem with limited capacity, also known as the Capacitated Multi-facility Weber Problem (CMWP) [1]. The objective of this type of problem is to determine locations in continuous 2-dimensional space for opening new facilities based on their maximum capacity and the given coordinates of the suppliers or customers. Cooper [2] was the first to attempt solving this type of location-allocation problem in 1972. He proposed exact and approximate solution methods based on the property that an optimal allocation occurs at the extreme point of the transportation polytope, while the optimal set of locations lies on the convex hull of the locations of the existing facilities. Sherali and Shetty [3] develop a cutting plane algorithm for the rectilinear distance location-allocation problem. Sherali and Tuncbilek [4] propose a branch-and-bound algorithm for the squared-Euclidean distance location-allocation problem. Sherali, Al-Loughani, Subramanian [5] developed a branch-and-bound algorithm based on the partitioning of the allocation space that finitely converges to a global optimum within a tolerance.

We propose an extension of the original CMWP that considers fixed cost for opening new facilities, fixed transportation costs, and two sets of fixed-location points: suppliers *i* and customers *j* [6]. The latter goes back to the original Weber problem, in which the location of the facility had to be determined in relation to 2 suppliers and 1 customer. The model is a nonlinear Generalized Disjunctive Programming (GDP), reformulated as a nonconvex Mixed-Integer Nonlinear Programming (MINLP). This is, to the best of our knowledge, an original problem not reported before that has high practical applicability.

We develop a bilevel decomposition algorithm that consists of decomposing the problem into a master problem and a subproblem. The master problem is based on a relaxation of the nonconvex MINLP, which yields an MILP that predicts the selection of facilities and their links to suppliers and customers, as well as a lower bound on the cost of problem. The subproblem corresponds to a nonconvex NLP of reduced dimensionality that results from fixing the binary variables in the MINLP problem, according to the binary variables predicted in the MILP master problem. Based on the bounding properties of their subproblems, Îµ-convergence is proved for this algorithm.

We apply the proposed method to random instances varying from 2 suppliers and 2 customers to 40 suppliers and 40 customers, from one type of facility to 3 different types, and from 2 to 32 potential facilities. The results show that the algorithm is more effective at finding global optimal solutions than general purpose global optimization solvers tested.

[1] Brimberg, J., Hansen, P., Mladonovic, N., Salhi, S.: A survey of solution methods for the continuous location allocation problem. International Journal of Operational Research 5(1), 1â€“12 (2008).

[2] Cooper, L.: The transportation-location problem. Operations Research 20(1), 94-108 (1972).

[3] Sherali, A.D., Shetty, C.M.: The rectilinear distance location-allocation problem. A I I E Transactions 9(2), 136-143 (1977).

[4] Sherali, H.D., Tuncbilek, C.H.: A squared-euclidean distance location-allocation problem. Naval Research Logistics (NRL) 39(4), 447-469 (1992).

[5] Sherali, H.D., Al-Loughani, I., Subramanian, S.: Global optimization procedures for the capacitated Euclidean and l_p distance multifacility location-allocation problems. Operations Research 50(3), 433-448 (2002).

498204 Nonlinear System Identification: Finding Normal Forms By Iteratively Uncovering Informed Geometries

Abstract:

We present a data-driven technique with roots in manifold learning for recovering the underlying structure of nonlinear dynamical systems. The data from such a dynamical system can be organized along three axes: the many trials with different parameter settings, the many measurement channels associated with each trial, and the many time instances. We assume no knowledge of the dimensionality or organization of the parameter settings or the measurement channels; we only know an index identifying the parameter trial, an index identifying the measurement channel, and an index describing the measurement time.

Our algorithm analyzes the structure of the data along any two of the axes and uses the results to construct a more informed geometry for the third axis. This process can proceed iteratively, improving the geometry for all three axes. We demonstrate a version using multilevel clustering to structure the data, as well as a version using diffusion maps. Using an earth mover's distance in another version of the algorithm can even overcome scenarios where the channel indices for the different trials are scrambled. We also show that applying our algorithm to invertible functions of the data gives rise to homeomorphic (or even isometric) embeddings conveying the same prototypical behavior.

502629 Model-Predictive Safety System for Predictive Detection of Operation Hazards: Off-Line Calculation of Most Aggressive Control Actions and Worst-Case Uncertainties

Abstract:

In 2016 we introduced a method of model-predictive safety (MPS) system design [1]. The MPS system generates alarm signals that are predictive and systematically account for process nonlinearities and interactions, while typical existing functional safety systems generate reactive, non-interacting alarm signal(s) when a process variable exceeds a threshold, For the first time, the MPS system design method proposed a systematic utilization of the dynamic process models to generate predictive alarm signals (alerts) for the detection of present and future operation hazards (OHs) in real time. An MPS system uses a process model to project in real-time the process operability status and to generate alarm signal(s) indicating the presence of a present or future OH with reasonable accuracy; it generates predictive alarm signals that alert the process personnel to imminent and potential future OHs before the actual OHs occur. Although the method uses the concepts of moving-horizon, model-based prediction and state estimation, it does not implement any control; when an MPS system identifies a current or future hazard in the operation of a process, it indicates that no controller is or will be able to assure safe process operation. Compared to a first-principles MPC system, an MPS system is easier to implement, and its on-line computational cost is significantly less.

The implementation of an MPS system requires off-line calculations of: (a) the most aggressive control action that minimizes each operability-constraint index when uncertain model parameters take their nominal values, and (b) the most aggressive control action that minimizes each operability-constraint index when uncertain model parameters take their worst-case values. In this paper, after reviewing the MPS system design method, we present novel optimization methods that can perform the off-line calculations systematically. The performance and ease-of-use of the methods will be shown using chemical process examples.

References

[1] Mohseni Ahooyi, T., J.E. Arbogast, W.D. Seider, U.G. Oktem, and M. Soroush, "Model-Predictive Safety System for Proactive Detection of Operation Hazards," AIChE J., 62, 2024-2042 (2016).

499060 Bilevel Programs with Coupling Equality Constraints for Parameter Estimation of Thermodynamic Property Models

Abstract:

As early as 1982, bilevel programs and related semi-infinite programs have been considered in chemical engineering, e.g., in the context of process flexibility [1]. However, classical bilevel theory is limited to the solution of linear bilevel programs or at least programs with convex lower-level program.

In recent years, more general bilevel programs with nonconvex lower-level programs have seen considerable activity, both in terms of algorithms [2, 3, 4, 5, 6, 7] and engineering applications. Examples are parameter estimation problems for thermodynamic property models with sufficient thermodynamic stability criteria [8] and the worst case stability analysis of power grids [9]. While real-world problems have been solved, current algorithms still pose restrictions on the problems that can be solved. As such, bilevel programs with equality constraints coupling the lower and upper level pose a significant challenge, either due to presenting an adverse case in terms of computational effort spent by the algorithms, or due to violating assumptions that are required for convergence.

Indeed, discretization methods for bilevel programming derive single-level subproblems by discretizing the lower-level variable space of the bilevel program. Iterative population strategies for those discretizations have been shown to yield convergent bounding algorithms under appropriate assumptions. These usually include an assumption of a Slater-like condition for the lower-level program, which fails to be satisfied in the presence of lower-level equality constraints coupling the lower and upper level.

Based on similar assumptions as related work for semi-infinite programs [10, 11], we propose an extension of an existing discretization method for nonlinear mixed-integer bilevel programs [2, 3] to alleviate that problem and allow lower-level equality constraints coupling the lower and upper level. The method is guaranteed to converge to an ε-optimal solution in finite time under appropriate assumptions. The central assumption is concerned with a uniqueness property of the solution to the equality constraints. Beyond an incrementally increasing number of constraints typically associated with discretization methods, the subproblems of the proposed method exhibit an incremental increase in the number of variables.

The viability of the extended method is assessed based on the solution of parameter estimation problems for thermodynamic property models with the equality constraints stemming from cubic equations of state.

[1] I. E. Grossmann and K. P. Halemane. Decomposition strategy for designing flexible chemical plants. AIChE Journal, 28(4):686–694, 1982.
[2] A. Mitsos. Global solution of nonlinear mixed-integer bilevel programs. Journal of Global Optimization, 47(4):557–582, 2010.
[3] A. Mitsos, P. Lemonidis, and P. I. Barton. Global solution of bilevel programs with a nonconvex inner program. Journal of Global Optimization, 42(4):475–513, 2008.
[4] P.-M. Kleniati and C. S. Adjiman. Branch-and-sandwich: a deterministic global optimization algorithm for optimistic bilevel programming problems. part i: Theoretical development. Journal of Global Optimization, 60(3):425–458, 2014.
[5] P.-M. Kleniati and C. S. Adjiman. Branch-and-sandwich: a deterministic global optimization algorithm for optimistic bilevel programming problems. part ii: Convergence analysis and numerical results. Journal of Global Optimization, 60(3):459–481, 2014.
[6] A. Tsoukalas, B. Rustem, and E. N. Pistikopoulos. A Global Optimization Algorithm for Generalized Semi-infinite, Continuous Minimax and Bi-level Problems. Journal of Global Optimization 44(2):235–250, 2009.
[7] S. Avraamidou and E. N. Pistikopoulos. Multi-Parametric Programming Based Algorithms for the Exact and Global Solution of Multi-Level Mixed-Integer Linear and Quadratic Programming Problems. Global Optimization Conference, College Station, Texas, USA, 2017.
[8] G. M. Bollas, P. I. Barton, and A. Mitsos. Bilevel optimization formulation for parameter estimation in vapor-liquid(-liquid) phase equilibrium problems. Chemical Engineering Science, 64:1768–1783, 2009.
[9] S. Fliscounakis, P. Panciatici, F. Capitanescu, and L. Wehenkel. Contingency ranking with respect to overloads in very large power systems taking into account uncertainty, preventive, and corrective actions. IEEE Transactions on Power Systems, 28(4):4909–4917, 2013.
[10] M. D. Stuber and P. I. Barton. Robust simulation and design using semi-infinite programs with implicit functions. International Journal of Reliability and Safety 5(3-4):378-397, 2011.
[11] M. D. Stuber and P. I. Barton. Semi-Infinite Optimization with Implicit Functions. Industrial & Engineering Chemistry Research 54(1):307-317, 2015.

501816 A Branch and Bound Algorithm to Solve Large-Scale Multistage Stochastic Programs

Abstract:

Multistage stochastic programing is a scenario-based approach for optimization under uncertainty (Birge & Louveaux, 2011). In chemical engineering, MSSPs are used to find optimal decisions in oil field investment planning, clinical trial scheduling, and process synthesis under uncertainty. The MSSP problem structure consists of three parts. The first part is a set of constraints for each scenario sub-problem. The remaining two parts are blocks of non-anticipativity constraints, which relate the decision variables in each scenario sub-problem to ensure that the outcomes of uncertain parameter(s) are not anticipated. The ability to generate the MSSP formulation for large-scale problems becomes prohibitive due to the rate at which the number of equations and variables grows with problem size. Increasing the number of uncertain parameters or the number of realizations for each uncertain parameter causes exponential growth in the number of scenarios. As such, there is great opportunity to develop exact solution approaches, which reduce the size of the problem formulation. Some recent examples of this type of work in chemical engineering include Colvin and Maravelias (2010) where the clinical trial planning problem formulation size is reduced by gradually adding non-anticipativity constraints that are violated with the in progress solution. Gupta and Grossmann (2011) presented a Lagrangean decomposition strategy, which solved the Lagrangean problem with relaxed conditional non-anticipativity constraints and dualized first time-period constraints to bound the solution. The authors then use a heuristic approach to find a feasible solution.

In this work, we present a branch and bound algorithm to solve multistage stochastic programs. For the algorithm we compare two different relaxation approaches for generating infeasible bounds. The first approach utilizes a modified progressive hedging approach. Progressive hedging is an augmented-Lagrangean-inspired scenario decomposition strategy, which iteratively converge to the optimal solution of stochastic programs (Rockafellar & Wets, 1991). The approach is shown to be provably optimal for multistage stochastic programs with decision independent uncertainty and continuous variables (Watson & Woodruff, 2011). In order to apply the algorithm to MSSPs with endogenous uncertainty and integer variables, we relax all integrality constraints of the problem and relax the conditional non-anticipativity constraints which cannot be treated as exogenous. Applying the progressive hedging approach results in solving a set of individual modified scenario sub-problems until variance in the decision variables is sufficiently small. The advantage of using the progressive hedging approach is that scenario sub-problems are able to completely separated which allows for easy parallelization.

The second approach bounds the problem by solving the MSSP without enforcing any non-anticipativity constraints. This approach provides a weaker bound than the progressive hedging approach, however, the scenario sub-problems are completely independent of each other and do not require multiple iterations in order to converge. This approach may be particularly useful when the size of the scenario set is prohibitively large. The feasible bound is generated using a memory efficient knapsack problem based decomposition heuristic (Christian & Cremaschi, 2015).

Computational studies were performed on case studies of the pharmaceutical R&D pipeline management problem originally presented in Colvin and Maravelias (2008). Results from the work show that the algorithm generates feasible bounds and bounds these solutions efficiently for instances of the clinical trial planning problem using either the progressive hedging bounding approach or Optimally solving Scenario Sub-problems (OSS) without enforcing non-anticipativity constraints. In both cases, the memory requirements were lower than the memory needed to generate the MSSP formulation. In the case of the OSS approach, the algorithm solved a seven-product instance of the pharmaceutical clinical trial planning problem with more than 11 million binary variables and 249 million constraints to a 7% optimality gap. Memory savings in the algorithm was found to be highest in the largest case studies, where the effect of the problem scaling is most noticeable.

References

Birge, J. R., & Louveaux, F. (2011). *Introduction to Stochastic Programming*. Springer New York.

Christian, B., & Cremaschi, S. (2015). Heuristic solution approaches to the pharmaceutical R&D pipeline management problem. *Computers and Chemical Engineering*, *74*, 34–47.

Colvin, M., & Maravelias, C. T. (2008). A stochastic programming approach for clinical trial planning in new drug development. *Computers & Chemical Engineering*, *32*, 2626–2642. http://doi.org/10.1016/j.compchemeng.2007.11.010

Colvin, M., & Maravelias, C. T. (2010). Modeling methods and a branch and cut algorithm for pharmaceutical clinical trial planning using stochastic programming. *European Journal of Operational Research*, *203*(1), 205–215. http://doi.org/10.1016/j.ejor.2009.07.022

Gupta, V., & Grossmann, I. E. (2011). Solution strategies for multistage stochastic programming with endogenous uncertainties. *Computers and Chemical Engineering*, *35*(11), 2235–2247. http://doi.org/10.1016/j.compchemeng.2010.11.013

Rockafellar, R. T., & Wets, R. J.-B. (1991). Scenarios and Policy Aggregation in Optimization Under Uncertainty. *Mathematics of Operations Research*, *16*(1), 119–147.

Watson, J. P., & Woodruff, D. L. (2011). Progressive hedging innovations for a class of stochastic mixed-integer resource allocation problems. *Computational Management Science*, *8*(4), 355–370. http://doi.org/10.1007/s10287-010-0125-4

501588 A New Decomposition Framework for Solving Multi-Stage Stochastic Programs with Endogenous Uncertainty

Abstract:

Optimization problems with decision-dependent, i.e., endogenous, uncertainty [1] are commonly observed in process industry, e.g., project portfolio management or R&D pipeline management problems [2], investment and operational planning of gas field developments [3], synthesis of process networks with uncertain process yields [4], and biomass-to-commodity chemicals investment planning [5]. Most optimization problems with endogenous uncertainty, by nature, can be modeled as multi-period multi-stage stochastic programs (MSSP). Unfortunately, these stochastic programming formulations grow exponentially with the increases in the numbers of scenarios and periods, and quickly become computationally intractable for real-world sized problems. An excellent review of existing literature for solving multi-stage problems with endogenous uncertainty can be found in [6]. In general, the solution approaches for large-scale stochastic programs with endogenous uncertainty rely heuristic or approximation approaches, and it has been shown that moderate-size problems can be solved to optimality. For example, Colvin and Maravelias [7] proposed a branch and cut algorithms for solving the pharmaceutical R&D pipeline management problems. Colvin and Maravelias [7, 8] and Gupta and Grossmann [9] introduced several theoretical reduction properties with iterative schemes based on relaxation of non-anticipativity constraints. Recently, Gupta and Grossmann [10] developed a new Lagrangean decomposition algorithm by decomposing the problem into scenario sub-problems.

In this talk, we introduce a new framework for efficiently solving multi-stage stochastic programs with endogenous uncertain parameters. The framework builds upon our Knapsack-problem based decomposition approach (KDA) [11, 12], which decomposes the original multi-period multi-stage stochastic program into a series of Knapsack problems, and solves these problems at appropriate decision points of the planning horizon. Our computational results revealed that the KDA provides tight feasible bounds for the original problem, and generates these bounds several orders of magnitude faster than solving the original problem. For example, for the pharmaceutical R&D pipeline management problem, the solutions of KDA are within 0.65%, 0.99%, 1.87% and 1.88% of the true solution for two-, three-, five- and six-product three-clinical trial cases, respectively. The KDA yielded these optimality gaps in 0.001, 0.03, 0.36 and 1.82 CPU seconds, which are up to five orders of magnitude faster than solving the original multi-stage stochastic programs. The decomposition framework - Lagrangean KDA (LKDA) - introduced here, employs KDA with a modified Lagrangean decomposition. The algorithm starts by executing KDA, which quickly generates a feasible solution and a lower bound (LB, assuming minimization). This feasible solution is used to calculate the first set of Lagrange multipliers. The modified Lagrangean problems are generated by removing the initial and conditional NACs, and dualizing the initial NACs in the objective function; this decomposes the Lagrangean problem to individual scenario problems, which are solved efficiently. The solution of the Lagrangean problem yields an upper bound. Next, the KDA is directed using the solution of the Lagrangean problem, and a new iteration is stated. As the iterations progress, to tighten the upper bound, violated conditional NACs are dualized and gradually added to modified Lagrangean problems. The algorithm terminates when the optimality gap is below a pre-specified value or when the maximum number of iterations are reached.

We applied the LKDA to solve two multi-stage stochastic problems: (1) the artificial lift infrastructure-planning problem [13], and (2) the pharmaceutical clinical trial planning problem [2]. The results show that LKDA successfully reduced the optimality gap below 5% quickly.

References

1. Goel, V. and I.E. Grossmann, *A Class of stochastic programs with decision dependent uncertainty.* Mathematical Programming, 2006. **108**(2/3): p. 355-394.
2. Colvin, M. and C.T. Maravelias, *A stochastic programming approach for clinical trial planning in new drug development.* Computers & Chemical Engineering, 2008. **32**(11): p. 2626-2642.
3. Goel, V. and I.E. Grossmann, *A stochastic programming approach to planning of offshore gas field developments under uncertainty in reserves.* Computers & Chemical Engineering, 2004. **28**(8): p. 1409-1429.
4. Tarhan, B. and I.E. Grossmann, *A multistage stochastic programming approach with strategies for uncertainty reduction in the synthesis of process networks with uncertain yields.* Computers & Chemical Engineering, 2008. **32**(4-5): p. 766-788.
5. Fahmi, I. and S. Cremaschi, *A prototype simulation-based optimization approach to model feedstock development for chemical process industry.* Chemical Engineering Research and Design, 2013. **91**(8): p. 1499-1507.
6. Apap, R.M. and I.E. Grossmann, *Models and Computational Strategies for Multistage Stochastic Programming under Endogenous and Exogenous Uncertainties.* Computers & Chemical Engineering.
7. Colvin, M. and C.T. Maravelias, *A branch and cut framework for multi-stage stochastic programming problems under endogenous uncertainty*, in *Computer Aided Chemical Engineering*. 2009. p. 255-260.
8. Colvin, M. and C.T. Maravelias, *Scheduling of testing tasks and resource planning in new product development using stochastic programming.* Computers & Chemical Engineering, 2009. **33**(5): p. 964-976.
9. Gupta, V. and I.E. Grossmann, *Solution strategies for multistage stochastic programming with endogenous uncertainties.* Computers and Chemical Engineering, 2011. **35**(11): p. 2235-2247.
10. Gupta, V. and I.E. Grossmann, *A new decomposition algorithm for multistage stochastic programs with endogenous uncertainties.* Computers & Chemical Engineering, 2014. **62**(0): p. 62-79.
11. Christian, B. and S. Cremaschi, *Variants to a knapsack decomposition heuristic for solving R&D pipeline management problems.* Computers & Chemical Engineering, 2017. **96**: p. 18-32.
12. Christian, B. and S. Cremaschi, *Heuristic solution approaches to the pharmaceutical R&D pipeline management problem.* Computers & Chemical Engineering, 2015. **74**: p. 34-47.
13. Zeng, Z. and S. Cremaschi, *Artificial Lift Infrastructure Planning for Shale Gas Producing Horizontal Wells*, in *Foundations of Computer Aided Process Operations / Chemical Process Control*. 2017: Tucson, Arizona.

497627 A Framework for Modeling and Optimizing Complex, Structured Problems

Abstract:

Algebraic modeling languages (AMLs) have drastically simplified the implementation of algebraic optimization problems. However, there are still many classes of optimization problems that are not easily represented in most AMLs. For example, generalized disjunctive programming, stochastic programming, and dynamic optimization problems all incorporate high-level mathematical constructs (i.e. disjuncts/disjunctions, scenario trees, and differential equations) in order to clearly and concisely describe complex relationships between different parts of a model. However, most AMLs cannot represent these high-level mathematical constructs directly. Instead, these classes of problems are typically reformulated before implementation, which requires significant effort and time from the modeler and obscures the original problem structure or context. In this talk we demonstrate how the Pyomo AML can be used to represent complex optimization problems using high-level modeling constructs. We also describe the general model transformations available in Pyomo for automatically transforming these high-level constructs into a form that off-the-shelf optimization solvers understand. We focus on problems that span multiple classes of optimization problems and demonstrate the combination of Pyomo extensions for generalized disjunctive programming, stochastic programming, and dynamic optimization. Finally, we demonstrate the scalability of this framework on a large-scale process model of a bubbling fluidized-bed adsorber.

492759 Identification of Optimization Decomposition Structures Via Community Detection Algorithms

Abstract:

Many optimization problems of interest for chemical engineers scale poorly with size. This is due to the presence of nonlinear, nonconvex terms in many chemical models, including, for example, bilinear terms in mass balance constraints, nonconvex power functions of variables for capital costs, or reaction equilibria terms. To solve these complex optimization problems, a classical method widely applied in process systems engineering is to seek a decomposition, namely to separate the original problems into subproblems of smaller sizes with reduced computational complexity by first removing some complicating variables and constraints, and then coordinating the subproblems through a master problem or primal-dual approaches [1, 2].

As a motivating example, consider our previous work which determined the optimal design of a plant which couples biofuel and renewable energy production [3]. The original optimization formulation was a very large (~60,000 variables) and computationally intractable mixed integer nonlinear program (MINLP). However, the overall plant consisted of two subsystems, the biofuel producing subsystem and the energy producing subsystems, which shared a few variables. As such, the original problem was decomposed into a much smaller (~700 variables) nonlinear program (NLP) which determined the optimal biofuel plant design, and a large mixed integer *linear* program (MILP) which determined the optimal renewable energy plant design. A master problem decided values for the three variables shared by the two systems. This approach made the optimization problem tractable, solving the problem within a reasonable (<10%) optimality gap.

For a general optimization problem, one cannot determine an optimal decomposition intuitively as in the motivating example. As such, this work proposes a generic automated approach to finding subnetworks within optimization problems using community detection algorithms. Such an approach has been proven useful in the control-relevant decomposition of chemical plant dynamic networks for distributed model predictive control [4, 5]. For the purpose of optimization, the problem is converted to a novel bipartite graph with one category of nodes representing equations and the other category of nodes representing variables. Community detection algorithms, e.g. spectral clustering or fast unfolding, are then used to identify communities (i.e. clusters of equations and variables) which are connected by a minimum number of shared variables or constraints. The framework allows for the incorporation of other considerations; for example, ensuring that the communities are tightly interacting amongst themselves, or requiring that the most difficult nonconvex equations all appear in the same community which is made as small as possible. We apply the proposed method to various optimization problems from existing test libraries, and document its capability to systematically generate efficient decompositions whenever they are warranted by the structure of the problem.

[1] Conejo, A.J., Castillo, E., Minguez, R., and Garcia-Bertrand, R. Decomposition techniques in mathematical programming: Engineering and science applications (2006). Springer.
[2] Grossmann, I.E. Advances in mathematical programming models for enterprise-wide optimization. Comput. Chem. Eng., 47(2012), pp. 2-18.
[3] Allman, A., and Daoutidis, P. Optimal design of synergistic renewable fuel and power systems. Renew. Energy, 100(2017), pp. 78-89.
[4] Tang, W. and Daoutidis, P. Hybrid distributed/hierarchical control architecture design. 20th IFAC World Congress, (2017), accepted.
[5] Pourkargar, D.B., Almansoori, A., and Daoutidis, P. Distributed model predictive control of process networks: Impact of control architecture. 20th IFAC World Congress, (2017), accepted.

488914 Integrating Mixed-Integer Optimisation and Satisfiability Modulo Theories

Abstract:

A major challenge in process systems engineering is integrating: (i) long-term strategic planning decisions, (ii) medium-term tactical planning, and (iii) short-term scheduling [1]. One way to address these multi-scale optimisation problems is by integrating logic and optimisation [2]. For example, a scheduling problem may have two levels: (i) assigning orders to machines and (ii) sequencing orders on each machine [3]. In a minimum cost model, assigning orders to machines is a mixed-integer optimisation problem, sequencing orders is a constraint satisfaction problem. The entire problem may be reformulated as either an optimisation or logic problem, but this misses the chance to use optimisation and logic synergistically.

Hybrid optimisation/logic approaches have been developed combining mixed-integer linear programming (MILP) and constraint programming (CP) [3, 4, 5]. The hybrid formulations usually use logic-based Benders decomposition (LBBD) [2], a generalisation of Benders decomposition [6]. The principles of Benders decomposition remain: we have a master problem and a subproblem which generates cuts if the solution from the master problem is infeasible. Differently from classical Benders decomposition, LBBD requires a logic proof deriving an objective bound.

But CP requires specialised, bespoke constraints. We consider modifying the hybrid method by replacing CP with satisfiability modulo theories (SMT). SMT is a constraint satisfaction technique combining propositional satisfiability with a background theory [7]. A background theory is a set of axioms and symbols, e.g. the theory of arithmetic. An SMT solver consists of a satisfiability solver and a theory solver. The idea is to leverage the strength and robustness of modern satisfiability solvers to search for a feasible solution. The modelling framework exposed by SMT allows for Boolean variables to be used with background theory variables, e.g $z \rightarrow (x \geq 0)$ where x is continuous and z is Boolean, so SMT is a natural choice when logical decisions form a part of the system being modelled.

For a test set of 335 scheduling problems, we computationally compare pure CP, MILP, and SMT approaches to LBBD methods combining (i) MILP/CP and (ii) MILP/SMT. We find that the MILP/CP and MILP/SMT logic-based Benders decomposition approaches work very well on a minimum cost model for scheduling and perform significantly better than any of CP, MILP, or SMT alone. But the hybrid MILP/CP and MILP/SMT methods are weaker than CP, MILP, SMT on a minimum makespan model.

We further discuss the computational and modelling trade-offs between CP and SMT. In particular, we show how logic-based modelling can be used to manage symmetry and degeneracy in applications such as bin packing and scheduling. Our work is supported by extensive computational testing on large-scale instances.

References

1. Maravelias, C. T. and Sung, C. (2009). Integration of production planning and scheduling: Overview, challenges and opportunities. *Comput Chem Eng*, 33(12):1919 – 1930.
2. Hooker, J. N. and Ottoson, G. (2003). Logic-based Benders decomposition. *Math Progam*, 96(1):33–60.
3. Jain, V. and Grossmann, I. E. (2001). Algorithms for Hybrid MILP/CP Models for a Class of Optimization Problems. *INFORMS J Comput*, 13(4):258–276.
4. Li, H. and Womer, K. (2008). Scheduling projects with multi-skilled personnel by a hybrid MILP/CP Benders decomposition algorithm. *J Sched*, 12(3):281–298.
5. Sitek, P. (2014). A hybrid CP/MP approach to supply chain modelling, optimization and analysis. In Computer Science and Information Systems (FedCSIS), pages 1345–1352.
6. Benders, J. F. (1962). Partitioning procedures for solving mixed-variables programming problems. *Numerische Mathematik*, 4(1):238–252.
7. De Moura, L. and Bjørner, N. (2008). Z3: An Efficient SMT Solver, pages 337–340. Springer.
8. Hooker, J. N. (2007). Planning and scheduling by logic-based Benders decomposition. *Oper Res*, 55(3):588–602.

497440 Solution Methods for Multiperiod Blend Scheduling MINLP Models

Abstract:

Over the past few decades, the integrated scheduling-blending problem of oil refinery has been studied extensively, since significant economic benefits can be achieved by operating the refinery according to a schedule with efficient blending. Based on their structure, refinery operations can be classified into three groups, namely, (1) crude oil operations, (2) intermediate processing, and (3) final product blending. The majority of work in the area is based on a decomposition into problems considering the three aforementioned groups of operations.

In this work, we first propose an integrated model for the multiperiod blending and scheduling problem taking into account all three refinery subsystems. We employ a discrete representation of time and, for illustration, we use linear mixing rules resulting in bilinear terms. We show that accounting for all subsystems simultaneously can lead to better solutions.

Second, to solve the resulting MINLP model, we propose a series of preprocessing and tightening methods based on flow and property propagation. The preprocessing algorithms calculate parameters that are then used to generate new tightening constraints. The algorithm runs in seconds. Three types of constraints are generated:

(1) Constraints on property demands. Using the lower bounds of the property specifications of a product and its demand, we propose novel constraints on available streams for blending, which consider the product demand and property specifications simultaneously.

(2) Constraints on single stream flow. For a given product, by examining its property specifications, we determine if there exists a stream that is required to be blended in order to satisfy the property specifications. If such stream exists, we enforce its flowrate to satisfy certain lower bound, which is calculated from product demands.

(3) Constraints on multiple stream flows. For every property specification of a product, we determine a subset of streams, at least one of which is required to be blended. We propose constraints for streams that are involved in these subsets, to ensure both product demand and property specifications are satisfied.

The integrated MINLP model and proposed solution methods are tested on a number of instances of different horizon and process complexity. The addition of the generated constraints leads to significant improvement in computational performance. Some instances that originally cannot be solved to optimality in hours, are solved in a few minutes. To further show the effectiveness of the methods, we test them using two MILP models from the literature [1]. Using the proposed methods leads to one order of magnitude improvement in solution times for both models, thus allowing us to address previously intractable instances.

Finally, we study, theoretically, the tightness of these proposed constraints. We show that they are often tighter than the McCormick envelops for the relaxation of the bilinear terms appearing in the "property balance" constraints. The relaxation of the MILP models is also significantly reduced when adding the proposed tightening constraints.

References

1. Kolodziej SP, Grossmann IE, Furman KC, Sawaya NW. A discretization-based approach for the optimization of the multi period blend scheduling problem. *Comput Chem Eng.* 2013; 53: 122–42.

Servo-Control of Selective Catalytic Reduction in Diesel-Powered Vehicles

Xiaodong Xu, Stevan Dubljevic

Department of Chemical and Materials Engineering
University of Alberta, Edmonton, Canada

October 29 - November 3, 2017

2017 AIChE Annual Meeting
Minneapolis, MN, USA

Outline

A. Process Description

B. Distributed Parameter SCR Model

D. Steady State

E. Simulation Study

F. Conclusion

Process Description

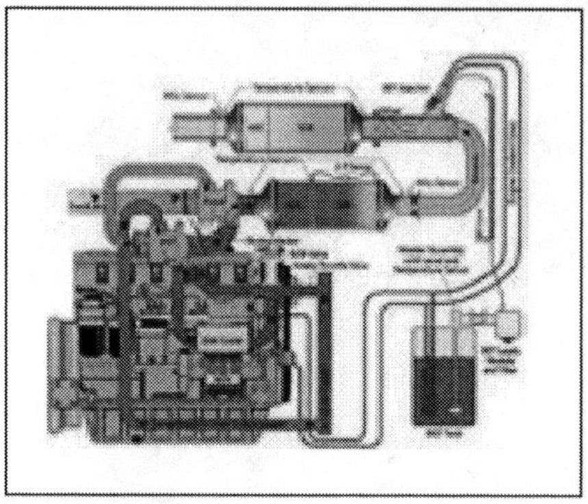

- Better fuel economy and higher power
- Emission of particulate matter and oxides of nitrogen (NO_x)
- Emission control technologies

  - Diesel Oxidation Catalyst (DOC)
  - Diesel Particulate Filter (DPF)
  - Selective Catalytic Reduction (SCR)

Selective Catalytic Reduction (SCR)

- hydrocarbon-based SCR

  NO_x reduction 40%

- Ammonia-based SCR

  NO_x reduction 95%

  Ammonia or urea solution

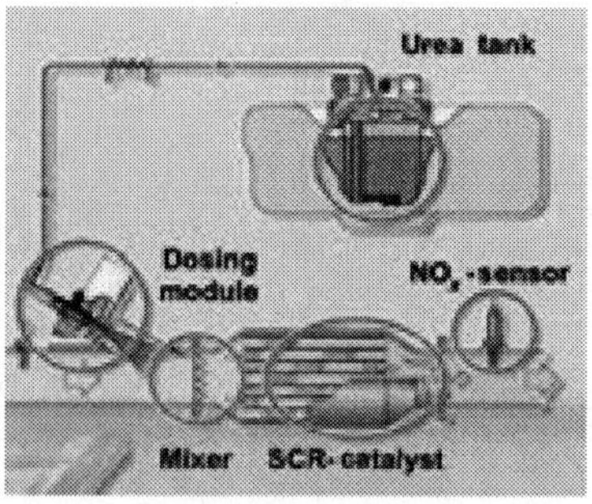

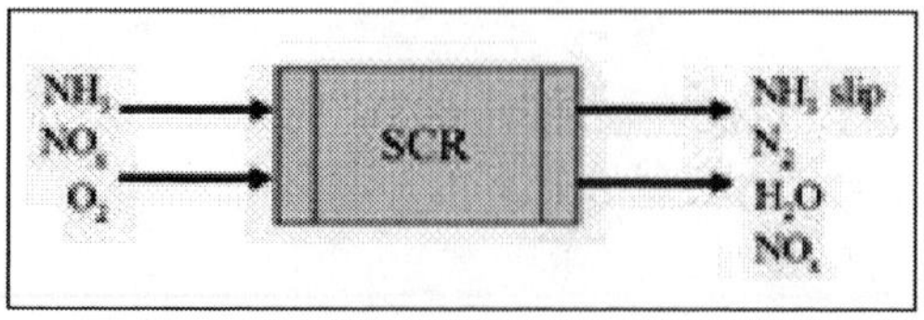

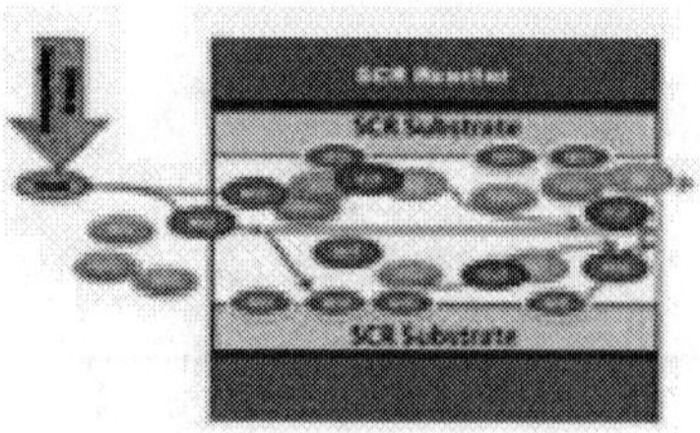

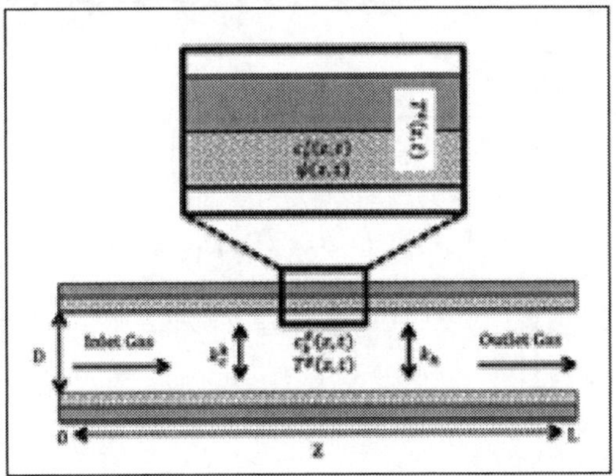

Parameter	Value	Unit
a	2660	$[m^2\,m^{-3}]$
a_{ext}	1	$[m^2\,m^{-3}]$
k_{ext}	15	$[W\,m^{-2}\,K^{-1}]$
ε^g	0.68	[-]
ρ^s	1770	$[kg\,m^{-3}]$
c_p^s	900	$[J\,kg^{-1}\,K^{-1}]$
λ_s	1	$[W\,m^{-1}\,K^{-1}]$
$\Psi_{NH_3}^{cap}$	200	$[mol/m^3]$
ΔH_1	−378.514	$[kJ\,mol^{-1}]$
ΔH_2	−407.129	$[kJ\,mol^{-1}]$
ΔH_3	−341.664	$[kJ\,mol^{-1}]$
ΔH_4	−116.839	$[kJ\,mol^{-1}]$
ΔH_5	−226.549	$[kJ\,mol^{-1}]$

NH_3 formation process

$H_4N_2CO \rightarrow NH_3 + HNCO$
$HNCO + H_2O \rightarrow NH_3 + CO_2$

NO_x conversion process

$4NH_3 + 2NO + 2NO_2 \rightarrow 4N_2 + 6H_2O$
$4NH_3 + 4NO + O_2 \rightarrow N_2 + 6H_2O$
$8NH_3 + 6NO_2 \rightarrow 7N_2 + 12H_2O$

Parameter	Equation	Unit
k_h	$= 19 + 0.1748T^g - 18.118 \times 10^{-5}(T^g)^2$	$[W\,m^{-2}\,K^{-1}]$
k_c^{NO}	$= 2.745 \times 10^{-5}(T^g)^{1.75}$	$[m\,s^{-1}]$
$k_c^{NO_2}$	$= 2.212 \times 10^{-5}(T^g)^{1.75}$	$[m\,s^{-1}]$
$k_c^{NH_3}$	$= 2.950 \times 10^{-5}(T^g)^{1.75}$	$[m\,s^{-1}]$
$k_c^{O_2}$	$= 2.390 \times 10^{-5}(T^g)^{1.75}$	$[m\,s^{-1}]$

Distributed Parameter SCR Model

Mass balance (flowing gas)

$$\frac{\partial c_k^g}{\partial t} = -v\frac{\partial c_k^g}{\partial z} + \frac{k_c^k a}{\varepsilon^g}\left(c_k^s - c_k^g\right),$$
$$k = NO, NO_2, NH_3, O_2$$

Mass balance (catalytic surface)

$$\frac{d\psi_{NH_3}(z,t)}{dt} = \frac{1}{\Psi_{NH_3}^{cap}}\left(R_{ad} - R_{des} - aR_4\right)$$

Enthalpy balance (flowing gas)

$$\frac{\partial T^g}{\partial t} = -v\frac{\partial T^g}{\partial z} + \frac{k_h^g a}{\rho^g C_p^g \varepsilon^g}\left(T^s - T^g\right)$$

Enthalpy balance (solid phase)

$$\frac{\partial T^s}{\partial t} = \frac{\lambda^s}{\rho^s C_p^s}\frac{\partial^2 T^s}{\partial z^2} - \frac{k_h^g a}{\rho^s C_p^s(1-\varepsilon^g)}\left(T^s - T^g\right)$$
$$-\frac{k_{ext} a_{ext}}{\rho^s C_p^s(1-\varepsilon^g)}\left(T^s - T^{ext}\right) - \frac{a\sum_{j=1}^{5}\Delta H_j R_j}{\rho^s C_p^s(1-\varepsilon^g)}$$

Boundary conditions

$$T^g(0) = T_{in}^g,\ c_k^g(0) = c_{k,in}^g,$$
$$\frac{\partial T^s(0)}{\partial z} = 0 = \frac{\partial T^s(L)}{\partial z}$$

Transfer coefficients

$k_c^k = f_k(T^g)$

$k = NO, NO_2, NH_3, O_2$

Reaction rates

$R_j = f_j(c_k^s),$

$k = NO, NO_2, NH_3, O_2$

$j = 1, 2, 3, 4, 5$

$R_{ad} = 0.82c_{NH_3}^s\left(1 - \psi_{NH_3}\right)$

$R_{des} = 3.67 \times 10^6 \times$
$\exp\left(\frac{-12992(1-0.31\psi_{NH_3})}{T^s}\right)\psi_{NH_3}$

Optimal Steady State

Mass balance (flowing gas)

$$v\frac{\partial \bar{c}_k^g}{\partial z} = \frac{\bar{k}_c^k a}{\varepsilon^g}\left(c_k^s - \bar{c}_k^g\right)$$
$$k = \mathrm{NO}, \mathrm{NO_2}, \mathrm{NH_3}, \mathrm{O_2}$$

Enthalpy balance (flowing gas)

$$v\frac{\partial \bar{T}^g}{\partial z} = \frac{\bar{k}_h^g a}{\rho^g C_p^g \varepsilon^g}\left(\bar{T}^s - \bar{T}^g\right)$$

Mass balance (catalytic surface)

$$R_{ad} - R_{des} - aR_4 = 0$$

Enthalpy balance (solid phase)

$$\frac{\partial^2 \bar{T}^s}{\partial z^2} = \frac{\bar{k}_h^g a}{\lambda^s(1-\varepsilon^g)}\left(\bar{T}^s - \bar{T}^g\right)$$
$$+\frac{k_{ext}a_{ext}}{\lambda^s(1-\varepsilon^g)}\left(\bar{T}^s - T^{ext}\right) + \frac{a\sum_{j=1}^{5}\Delta H_j \bar{R}_j}{\lambda^s(1-\varepsilon^g)}$$

Boundary Conditions:

$$\bar{T}^g(0) = T_{in}^g, \bar{c}_k^g(0) = c_{k,in}^g$$
$$\frac{\partial \bar{T}^s(0)}{\partial z} = 0 = \frac{\partial \bar{T}^s(L)}{\partial z}$$

$$\min_{c_{\mathrm{NH_3,in}}} \int_0^L y(z)Qy(z)^T dz + Ru^2$$

Subject to: System steady state

$$0 \le c_{\mathrm{NH_3,in}} \le c_{\mathrm{NH_3}}^*$$
$$y(z) = \left[\bar{c}_{\mathrm{NO}}(z), \bar{c}_{\mathrm{NO_2}}(z), \bar{c}_{\mathrm{NH_3}}(z)\right]$$
$$u = c_{\mathrm{NH_3,in}}$$

Optimal Steady State Profile

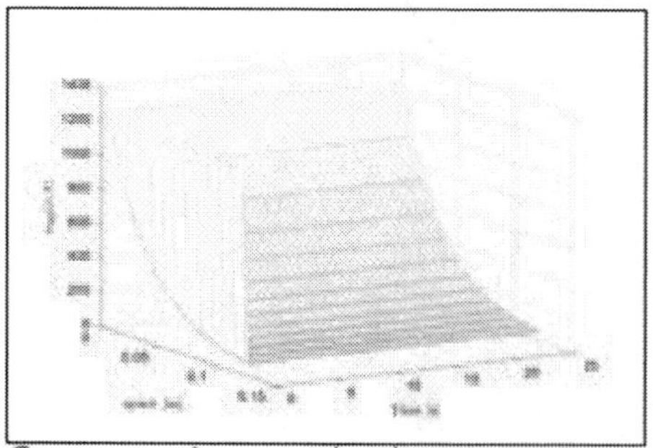

Concentration $c_{\mathrm{NO}}(z,t)$

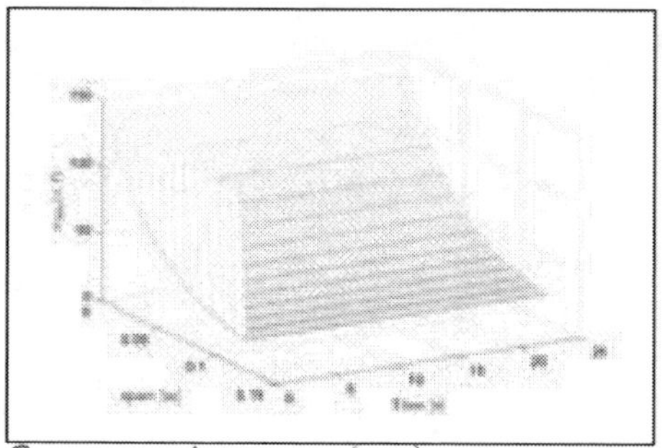

Concentration $c_{\mathrm{NO_2}}(z,t)$

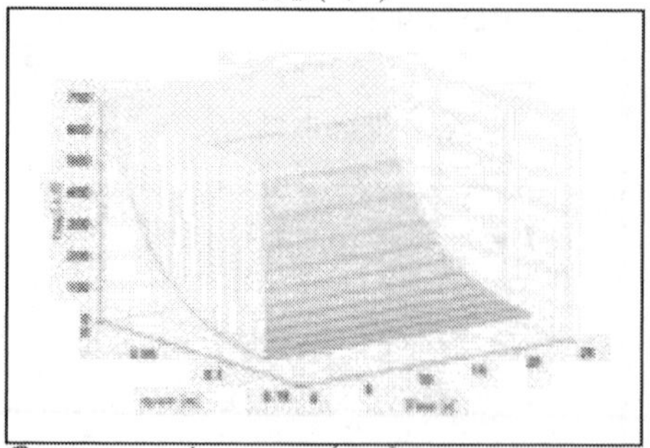

Concentration $c_{\mathrm{NO}}(z,t)$

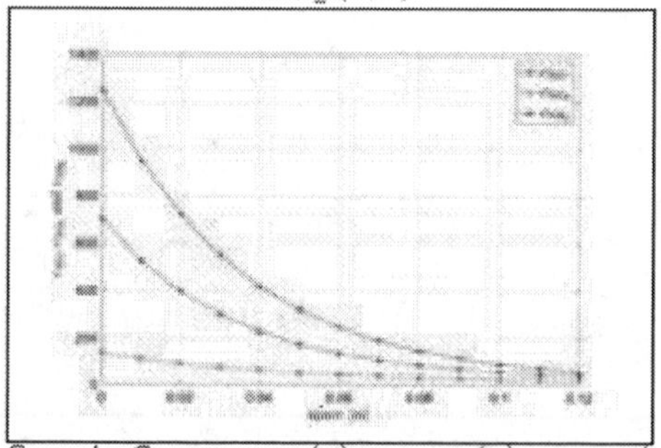

Steady State $c_{\mathrm{NO}}(z)$, $c_{\mathrm{NO_2}}(z)$, $c_{\mathrm{NH_2}}(z)$

Linearized Model

Components Concentration

$$\frac{\partial \tilde{c}_k^g}{\partial t} = -v\frac{\partial \tilde{c}_k^g}{\partial z} - \frac{\bar{k}_c^k a}{\varepsilon^g}\tilde{c}_k^g + \left.\frac{\partial k_c^k}{\partial T^g}\right|_{\bar{T}^g} \Delta c_k \tilde{T}^g$$

$k = NO, NO_2, NH_3, O_2$

$\Delta c_k = c_k^s - \bar{c}_k^g$

Temperature (flowing gas)

$$\frac{\partial \tilde{T}^g}{\partial t} = -v\frac{\partial \tilde{T}^g}{\partial z} + \frac{\bar{k}_h^g a}{\rho^g C_p^g \varepsilon^g}\left(\tilde{T}^s - \tilde{T}^g\right)$$

Enthalpy balance (solid phase)

$$\frac{\partial \tilde{T}^s}{\partial t} = \frac{\lambda^s}{\rho^s C_p^s}\frac{\partial^2 \tilde{T}^s}{\partial z^2} - \frac{\bar{k}_h^g a}{\rho^s C_p^s(1-\varepsilon^g)}\left(\tilde{T}^s - \tilde{T}^g\right) - \frac{k_{ext} a_{ext}}{\rho^s C_p^s(1-\varepsilon^g)}\tilde{T}^s$$

Boundary Conditions:

$$c_{NO_x}^g(0,t) = 0, c_{O_2}^g(0,t) = 0,$$
$$c_{NH_3}^g(0,t) = u(t)$$
$$\tilde{T}^g(0,t) = 0$$
$$\frac{\partial \tilde{T}^s(L,t)}{\partial z} = 0 = \frac{\partial \tilde{T}^s(0,t)}{\partial z}$$

Output:

$$y(t) = c_{NO_x}^g(L,t)$$

Output regulation

- Stabilization problem
- Tracking problem

Stability Analysis

$$\frac{\partial x_k}{\partial t} = -v\frac{\partial x_k}{\partial z} - \alpha_k x_k + \lambda_k T_1$$

$$\frac{\partial}{\partial t}\begin{bmatrix} T_1 \\ T_2 \end{bmatrix} = A\begin{bmatrix} T_1 \\ T_2 \end{bmatrix}$$

with boundary conditions:

$$x_k(0,t) = 0, k = NO, NO_2, O_2$$
$$x_{NH_3}(0,t) = u(t), T_1(0,t) = 0,$$
$$\frac{\partial T_2(0,t)}{\partial z} = 0 = \frac{\partial T_2(L,t)}{\partial z}$$

$$A = \underbrace{\begin{bmatrix} -v\frac{d}{dz} - \beta_1 & 0 \\ 0 & d_0\frac{d^2}{dz^2} - \gamma_2 \end{bmatrix}}_{A_0} + \underbrace{\begin{bmatrix} 0 & \beta_1 \\ \gamma_1 & 0 \end{bmatrix}}_{D}$$

Theorem

The linearized system is exponentially stable.

Output Regulator Configurations:

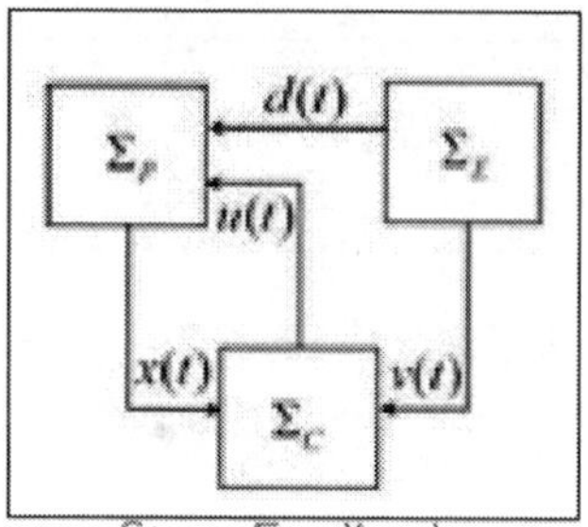

State Feedback

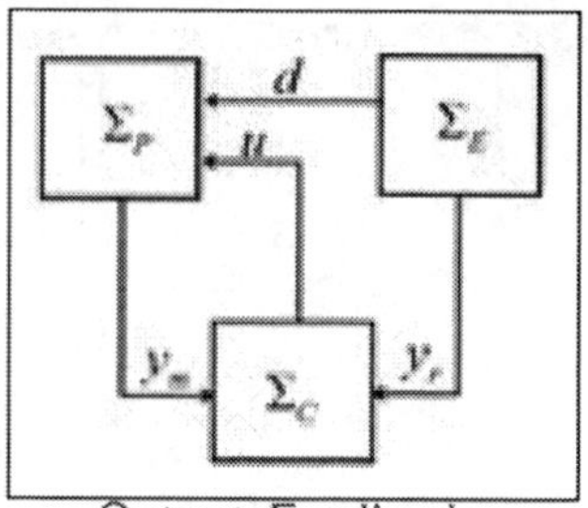

Output Feedback

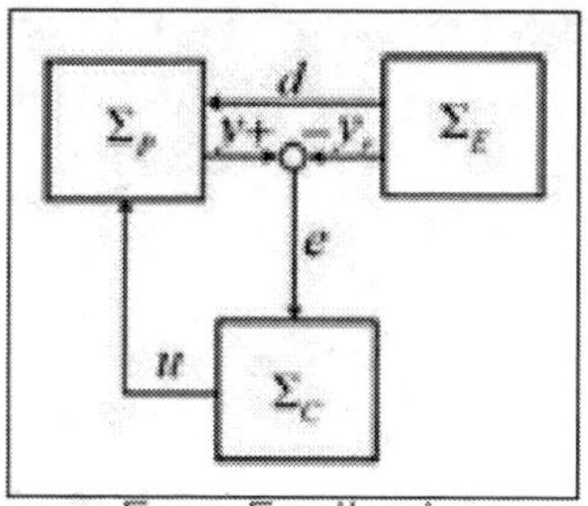

Error Feedback

- Plant Σ_P: $x(t)$ full state, $y(t)$ controlled output, $y_m(t)$ measured output;
- Exosystem Σ_E: $v(t)$ state, $y_r(t)$ reference signal, $d(t)$ disturbance;
- Controller (regulator) Σ_C: $u(t)$ control signal;

- State Feedback Regulator: ① $x(t)$ and $v(t)$ available ② Reference tracking;
- Output Feedback Regulator: ① $y_m(t)$ and $y_r(t)$ available ② Disturbance rejection and signal tracking.
- Error Feedback Regulator: ① Tracking error $e(t)$ available ② Disturbance rejection and signal tracking.

Boundary Output Regulation

Exosystem:

Unknown disturbances

$$\dot{w}(t) = Sw(t), w(0) = w_0 \in \mathbb{C}^{n_v}$$
$$d_i(t) = p_{d_i}^T w(t), i = 1, 2, \cdots$$
$$y_r(t) = q^T w(t), t \in R^+$$

Reference signals

Assumption 1:

- The matrix S is diagonalizable and $\sigma(S) \subset j\mathbb{R}$. All eigenvalues in $\sigma(S)$ are distinct.

Assumption 2:

- The matrix S is block diagonalizable: $S = \text{bdiag}(S_d, S_r) = \text{bdiag}(S_m, S_n)$. S_n is a block of S_r. $\sigma(S_m) \subset j\mathbb{R}$ are distinct.

- Assumption 1: The generation of harmonic and step-like signals.
- Assumption 2: The generation of polynomial signals.

Boundary Regulator Design

Tracking Errors (change of variables)

$$x_k = e_k + m_k(z)w(t)$$
$$T_1 = e_1 + m_1(z)w(t)$$
$$T_2 = e_2 + m_2(z)w(t)$$

$$\frac{\partial e_k}{\partial t} = -v\frac{\partial(e_k)}{\partial z} - \alpha_k e_k + \lambda_k e_1$$
$$+\Big(-v\frac{dm_k(z)}{dz} - m_k(z)S - \alpha_k m_k(z) + \lambda_k m_1(z)\Big)\, w(t)$$
$$\frac{\partial e_1}{\partial t} = -v\frac{\partial e_1}{\partial z} - \beta_1 e_1 + \beta_1 e_2$$
$$+\Big(-v\frac{dm_1(z)}{dz} - m_1(z)S - \beta_1 m_1(z) + \beta_1 m_2(z)\Big)\, w(t)$$
$$\frac{\partial e_2}{\partial t} = d_0\frac{\partial^2 e_2}{\partial z^2} + \gamma_1 e_1 - \gamma_2 e_2$$
$$+\Big(d_0\frac{d^2 m_2(z)}{dz^2} - Sm_2(z) - \gamma_2 m_2(z) + \gamma_1 m_1(z)\Big)\, w(t)$$

$$e_k(0,t) = -m_k(0)w(t)$$
$$e_{NH_3}(0,t) = u(t) - m_{NH_3}(0)w(t)$$
$$e_1(0,t) = -m_1(0)w(t)$$
$$\frac{\partial e_2(0,t)}{\partial z} + \frac{dm_2(0)}{dz}w(t) = 0$$
$$\frac{\partial e_2(L,t)}{\partial z} + \frac{dm_2(L)}{dz}w(t) = 0$$

Extended Regulator Equation

$$v\frac{dm_k(z)}{dz} = m_k(z)S - \alpha_k m_k(z) + \lambda_k m_1(z)$$
$$v\frac{dm_1(z)}{dz} = m_1(z)S - \beta_1 m_1(z) + \beta_1 m_2(z)$$
$$d_0\frac{d^2 m_2(z)}{dz^2} = m_2(z)S + \gamma_2 m_2(z) - \gamma_1 m_1(z)$$
$$m_k(0) = 0, m_1(0) = 0, \frac{dm_2(0)}{dz} = 0 = \frac{dm_2(L)}{dz}$$

Theorem

Suppose that $G(s)$ is the transfer function of linearized model. Then the regulator equations have a unique solution if and only if $G(\lambda) \neq 0\ \forall \lambda \in \sigma(S)$.

The feedforward control law:

$$u(t) = m_{NH_3}(0)w(t)$$

Simulation Study

Working Scenario

Mode	Temperature [°C]	Velocity [m·s⁻¹]	NO_x [ppm]
1	120	0.068	150
2	303	0.148	749
3	425	0.197	1146
4	491	0.236	1294
5	515	0.318	1250
6	290	0.210	996
7	368	0.282	1045
8	416	0.363	1176
9	470	0.435	1215
10	270	0.262	497
11	334	0.350	778
12	376	0.439	1003
13	439	0.493	1113

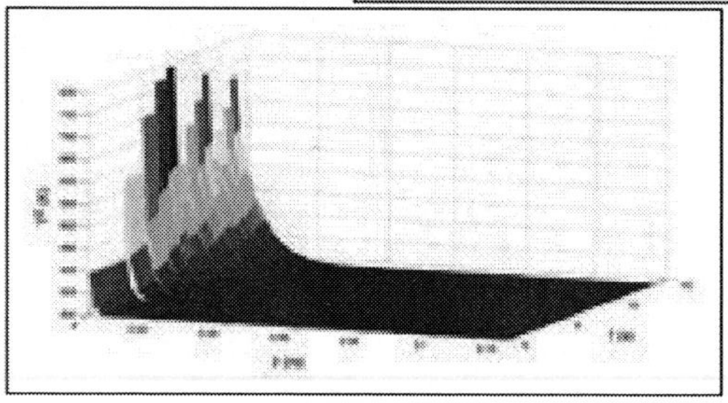

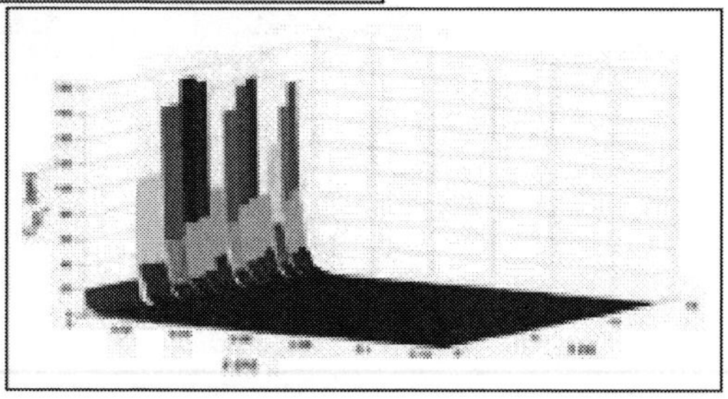

Simulation Study

Concentration of NO_x

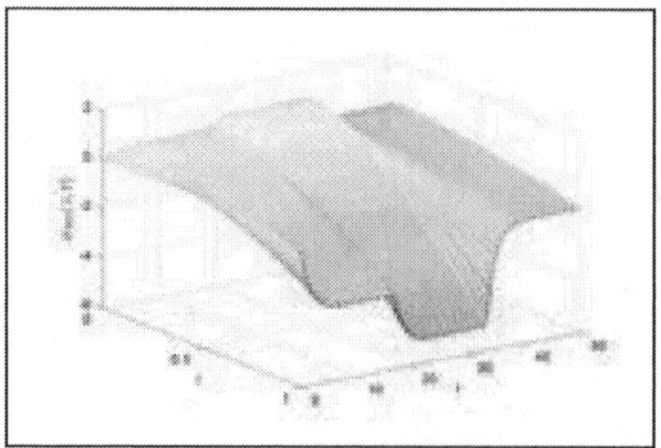

Dimensionless Concentration $x_{NO}(z, t)$

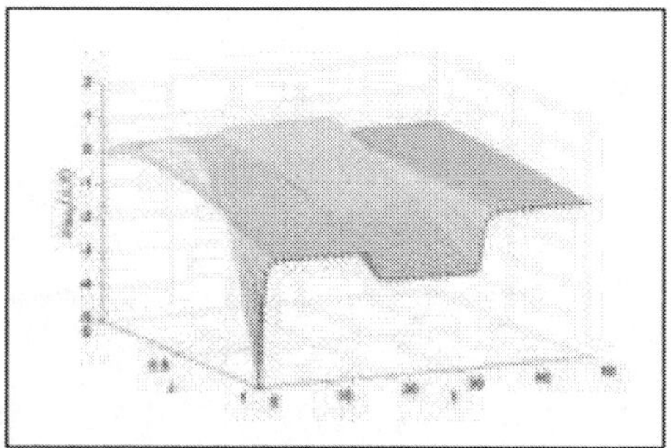

Dimensionless Concentration $x_{NO_2}(z, t)$

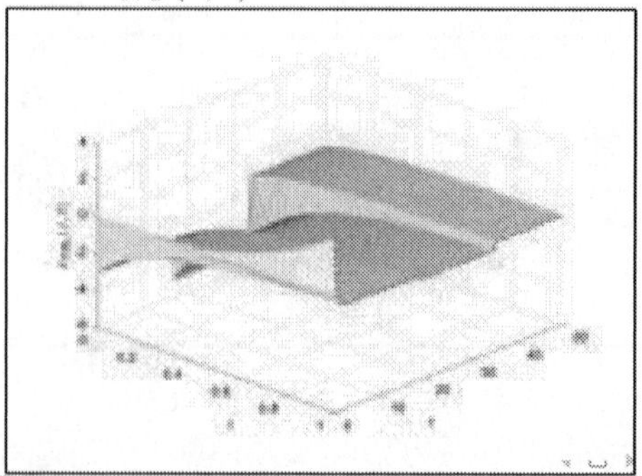

Dimensionless Concentration $x_{NH_3}(z, t)$

Simulation Study

Concentration of NO at $z = 1$

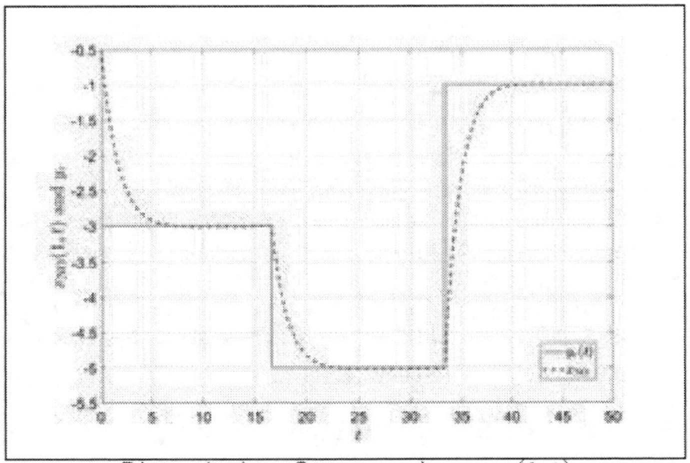

Dimensionless Concentration $x_{NO}(1, t)$

Conclusion:

- Mathematic Model for SCR Process
  - Mass Balance (flow gas)
  - Enthalpy Balance (flow gas)
  - Enthalpy Balance (Solid phase)
- Optimal Steady State
- Servo-control of Linearized System
  - Stability Analysis
  - Output Regulation

491790 A Parallel Fluid Solid Coupling Model with Lattice Boltzmann Fluid Solver and Molecular Dynamics Simulator Using the Immersed Boundary Method

Abstract:

The study of viscous fluid flow coupled with rigid or deformable solids has many applications in biological and engineering problems, e.g., blood cell transport, drug delivery, and particulate flow. We developed a partitioned approach to solve this coupled Multiphysics problem. The fluid motion was solved by Palabos (Parallel Lattice Boltzmann Solver), while the solid displacement and deformation was simulated by LAMMPS (Large-scale Atomic/Molecular Massively Parallel Simulator). The coupling was achieved through the immersed boundary method (IBM). The code modeled both rigid and deformable solids exposed to flow. The code was validated with the classic problem of rigid ellipsoid particle orbit in shear flow and demonstrated essentially linear scaling over multiple cores. The results showed a 1st order convergence rate. An example of the fluid-solid coupling was given for flexible filaments (drug carriers) transport in a flowing red blood cell suspensions, highlighting the advantages and capabilities of the developed code.

501333 Efficient Linear Underestimators for Dynamic Process Systems

Abstract:

Established methods for deterministic global optimization typically compute useful global bounding information by constructing and minimizing convex underestimators of process models. However, generating useful convex underestimators for dynamic systems is difficult; dynamics obscure our intuition about the dependence of a model on its parameters, and any discrepancy between a dynamic model and its convex underestimator may well grow exponentially with system time. Nevertheless, there is room for improvement; supplying tighter convex underestimators for dynamic models would broaden the class of dynamic optimization problems that can be solved to global optimality with given computational resources, as would reducing the computational effort required for subgradient evaluations.

This presentation combines and extends several recent results to obtain an efficient technique for constructing useful affine underestimators for optimization problems with embedded systems of parametric ordinary differential equations (ODEs). Differentiable relaxations are constructed for the ODE right-hand side using a method [1] developed in the â €œmultivariate McCormickâ€ framework [3]. These are then substituted into a description [2] of efficient ODE relaxations as an auxiliary hybrid discrete/continuous system. It may be shown that various properties of the relaxations of [1] prevent the discrete behavior of this hybrid system from manifesting, yielding an auxiliary ODE system that describes convex relaxations. Standard adjoint sensitivity analysis techniques may be applied to this system to compute affine underestimators for the overarching optimization problem efficiently using standard ODE solvers. Implications and examples are discussed.

References

[1] KA Khan, HAJ Watson, and PI Barton, Differentiable McCormick relaxations, J. Glob. Optim., 67:687-729, 2017.

[2] JK Scott and PI Barton, Improved relaxations for the parametric solutions of ODEs using differential inequalities, J. Glob. Optim., 57:143-176, 2013.

[3] A Tsoukalas and A Mitsos, Multivariate McCormick relaxations, J. Glob. Optim., 59:633-662, 2014.

496644 Model Misspecifications in Metabolic Flux Analysis: Biases, Tests and Fixes

Abstract:

Model Misspecifications in Metabolic Flux Analysis: Biases, Tests and Fixes

Rudiyanto Gunawan*,** , Sandro Hutter*,**

*Institute for Chemical and Bioengineering, ETH Zurich, Zurich, Switzerland

**Swiss Institute of Bioinformatics, Lausanne, Switzerland

Metabolic flux analysis (MFA) has become an indispensible tool in metabolic engineering, not only for analyzing metabolic phenotypes (flux distribution), but also for optimizing metabolic pathways to improve productivity and/or to gain new capabilities. The MFA is based on a stoichiometric model of the metabolic reaction network under the pseudo-steady state assumption. The simplest variant of the MFA, hereupon referred to as the overdetermined MFA, uses a reduced stoichiometric model of the cellÕs metabolism such that the flux estimation constitutes an over- (or fully) determined linear regression problem. Thanks to its simple formulation and numerical implementation, the overdetermined MFA is still widely used in metabolic engineering applications [1-2].

Despite the long history of MFA, the impact of misspecifications of the stoichiometric model, particularly the issue of missing or omitting reactions, has not received much attention. A recent study demonstrated that modeling errors could lead to gross uncertainty in the estimated intracellular flux values [3]. In this study, we employed the framework of linear least square regressions to quantify the effects of missing reactions on the flux estimates, and evaluated several statistical tests for detecting omitted reactions in the stoichiometric model. Finally, we proposed an iterative algorithm for resolving the issue of missing reactions in the overdetermined MFA.

The MFA can be formulated as a linear least square regression problem as follow:

where **S** denotes the stoichiometric matrix, **v** denotes the fluxes, the subscripts **I** and **E** refer to the intracellular and exchange (uptake) fluxes, respectively, and **e** denotes the vector of measurement errors. In the overdetermined MFA, the matrix $\mathbf{S_I}$ has a full column rank and the exchange fluxes $\mathbf{v_E}$ are measured (or could be computed from the measurements). Assuming that the error **e** has a zero mean and a constant variance, the ordinary least square (OLS) solution provide the minimum variance unbiased estimates (MVUE) for the intracellular fluxes $\mathbf{v_I}$ as follow:

When the variance of the error is not constant but known, one can use the generalized least square estimation to provide the MVUE for $\mathbf{v_I}$. When the true model is in fact given by:

one can show that the OLS solution may no longer the MVUE because

has a possible specification bias of

In the above derivation, the subscript **O** refers to the omitted reactions. Thus, the specification bias is non-zero when the omitted fluxes $\mathbf{v_O}$ are non-zero and the missing reaction stoichiometry is not orthogonal to $\mathbf{S_I}$.

We demonstrated using a Chinese hamster ovary (CHO) metabolic model and cell culture data that the omission of even a single reaction from the network could lead to disproportionally large specification biases in the intracellular flux estimation. Importantly, while a poor statistical significance of the linear regression is a good indication of large specification biases, an acceptable statistical significance can still result from the omission of a reaction that causes significant flux biases.

We further evaluated three tests for detecting missing reactions, including RamseyÕs RESET test, F-test for model selection, and the Lagrange multiplier (LM) test [4]. Using 1000 randomly generated biochemical reaction networks [5] and *in silico* flux values, we computed the true positive, false positive, true negative and false negative rates of the above tests. The results showed that the F-test robustly outperforms the RESET and LM tests across different network sizes, numbers of omitted reactions, numbers of exchange fluxes and levels of errors.

Based on the above findings, we then proposed an iterative strategy to resolve the issue of missing reactions using the F-test. Briefly, in each iteration, we employed the F-test to detect whether the addition of a candidate reaction would lead to a statistically significant improvement in the linear regression. We then incorporated the reactions that passed the statistical threshold into the model, and repeated the F-tests for the remaining reactions against the updated model. Using the CHO metabolic model above and *in silico* generated data, we demonstrated that the proposed iterative strategy could recover the majority of missing reactions (~80%), and leave out reactions that were not part of the reaction network.

The findings of this study are available in a pre-print publication [6].

References:

1. Naderi, S.; Meshram, M.; Wei, C.; Mcconkey, B.; Ingalls, B.; Budman, H.; Scharer, J. Development of a mathematical model for evaluating the dynamics of normal and apoptotic Chinese hamster ovary cells. *Biotechnol. Prog.* **2011**, *27*, 1197–1205.

2. Nolan, R. P.; Lee, K. Dynamic model of CHO cell metabolism. *Metab. Eng.* **2011**, *13*, 108–24.

3. Sokolenko, S.; Quattrociocchi, M.; Aucoin, M. G. Identifying model error in metabolic flux analysis – a generalized least squares approach. *BMC Syst. Biol.* **2016**, *10*, 1–14.

4. Long, J. S.; Trivedi, P. K. Some Specification Tests for the Linear Regression Model. *Sociol. Methods Res.* **1992**, *21*, 161–204.

5. Aho, T.; Smolander, O.-P.; Niemi, J.; Yli-Harja, O. RMBNToolbox: random models for biochemical networks. *BMC Syst. Biol.* **2007**, *1*, 22.

6. R. Gunawan; S. Hutter. Assessing and Resolving Model Misspecifications in Metabolic Flux Analysis. *Preprints.* **2017**, doi:10.20944/preprints201703.0124.v1.

498569 Optimal Flow Control for Oil Production Under Gas Coning Conditions in Oil-Rim Reservoirs

Abstract:

In oil-rim reservoirs, a thin oil layer is trapped between an aquifer and a gas cap. Oil can be produced from the thin oil layer by drilling horizontal wells just above the aquifer. During the oil production, the Gas-Oil Contact (GOC) near the wellbore lowers and this phenomenon is called gas coning [1]. The gas coning may lead to gas breakthrough, which will significantly decrease the oil flow rate and in turn increase the gas production rate. Since the gas price is much cheaper than the oil price, we typically operate the oil extraction process only until the gas breakthrough to prevent the gas production. However, in practice, we can extract the oil even after the gas breakthrough at the expense of a reduced oil production rate. In this case, we have to impose a constraint on the gas-oil ratio (GOR) to restrict the amount of gas extracted from the well. Recently, there have been some efforts to design feedback control systems for the oil extraction from the oil-rim reservoirs, but they did not consider the oil extraction after the gas breakthrough, which is called the super-critical phase [2, 3, 4].

Motivated by these considerations, we propose a model-based control framework to maximize the oil production from the oil-rim reservoir during the super-critical phase. For this purpose, we first develop a first principles model to describe the gas-coning behavior, particularly the spatiotemporal evolution of GOR and GOC, which is determined by the oil extraction rate. Second, the data generated from the high-fidelity model is used to construct a reduced-order model (ROM). The ROM is used to design a Kalman filter that will utilize the real-time measurements at the wellbore to estimate GOR and GOC at the locations where the measurements are not available. Third, model predictive control (MPC) theory is applied in the receding horizon fashion for the design of a feedback control system that manipulates the oil extraction rate at the wellbore to maximize the net present value (NPV) of the produced oil in the super-critical phase. In the controller, we will explicitly consider practical constraints such as the actuator limitation and a limit on GOR for economically-feasible operations. Lastly, the developed feedback control system will be compared with the traditional back-stepping methods during the super-critical phase.

References:

[1] Mjaavatten, A., Aasheim, R., Saelid, S., & Gronning, O. (2006, January). A Model for Gas Coning and Rate-Dependent Gas/Oil Ratio in an Oil-Rim Reservoir (Russian). In *SPE Russian Oil and Gas Technical Conference and Exhibition*. Society of Petroleum Engineers

[2] Hasan, Agus Ismail, et al. "Modeling, simulation, and optimal control of oil production under gas coning conditions." SPE EUROPEC/EAGE Annual Conference and Exhibition. Society of Petroleum Engineers, 2011.

[3] Hasan, A., Foss, B., & Sagatun, S. (2012). Flow control of fluids through porous media. Applied Mathematics and Computation, 219(7), 3323-3335.

[4] Maree, J. P., & Imsland, L. (2014, October). Optimal control strategies for oil production under gas coning conditions. In Control Applications (CCA), 2014 IEEE Conference on (pp. 572-578). IEEE.

503670 Digital Alchemy for Assembly Engineeringc

Abstract:

Back in the day, Stuart Churchill was a pioneer in the use of computers for modeling chemical and thermal processes. Since that time, the role of computing in chemical engineering has become indispensable. In this talk, we present new computational methods we call "digital alchemy" that allow for predictive modeling of thermodynamic assembly, and permit the design of building blocks optimized for assembly into target structures with desired properties. In this approach, standard thermodynamic ensembles are generalized to include alchemical variables and their thermodynamic conjugates, allowing for rapid, real-time computational investigation of large, complex design spaces.

503671 Improved Algebraic, Numerical, and Graphical Representations in Fluid Mechanics

Abstract:

The concept of equivalents is applied to fully developed flows through straight channels, and two applications are introduced: the equivalents that produce equal rates of flow through channels of different cross-section, and the diminishment that occurs between the regimes of laminar and turbulent flow. The validity of the eddy diffusivity and the invalidity of the mixing-length as measures of the shear stress due to turbulence have perhaps never before been demonstrated so clearly. The possible choices of primary variables and the consequences are also examined. The objective of these explorations is an improved understanding of the basic fluid mechanics by teachers, students, and industrial practitioners.

503672 The Scaling of Turbulence Near the Wall and the Churchill Turbulent Flux Correlation: Insights with Lagrangian Simulations

Abstract:

The evaluation of mean velocity and mean temperature in wall-bounded turbulent flows has been based on the "classical" theory of turbulence and the law of the wall. The viscous wall parameters, namely the friction velocity, the friction temperature, and the kinematic viscosity of the fluid have been used to scale not only the mean values of velocity, but also other mean turbulent quantities close to the wall, including the Reynolds stresses and the turbulent velocity and temperature fluctuations. Stuart Churchill, as he questioned the validity of the Reynolds analogy between momentum and heat or mass transfer in his works [1], he realized that extending the scaling of momentum to the scaling of scalar quantities in turbulence was not a sound approach. He proposed a scaling of turbulence that depends on the turbulent momentum flux and the turbulent heat flux. In his AIChE Institute Lecture (in Miami, 1998), Churchill presented a summary of his work on this topic, and closed form solutions for the mean velocity and temperature [2]. The turbulent Prandtl number, the most significant parameter for the modeling of turbulent transport not only in theory but also in modern CFD commercial software packages, was also interpreted physically based on these ideas [3]. We have used direct numerical simulation results and results from Lagrangian simulations [4] for the transport of passive scalars in turbulent channel flows to explore Churchill's ideas on turbulent flow scaling and to develop an updated theory for the scaling of heat transport in turbulent flows [5-7]. His passion to remove empiricisms and approximations from the development of predictive correlations in turbulent transport has been an inspiration. In this talk, we will revisit the Churchill scaling in view of our simulation results, and will discuss the counterintuitive finding that wall turbulence can be used to separate rather than mix substances under special circumstances [8].

1. Churchill, S. W., 1996, "A Critique of Predictive and Correlative Models for Turbulent Flow and Convection," Ind. Eng. Chem. Res., 35, pp. 3122–3140.
2. Churchill, S.W., 2000, "Progress in the Thermal Sciences: AICHE Institute Lecture," AIChE Journal, 46(9), pp. 1704-1722.
3. Churchill, S. W., 2002, "A reinterpretation of the Turbulent Prandtl Number," Ind. Eng. Chem. Res., 41, pp. 6393–6401.
4. Papavassiliou, D.V, 2002, "Turbulent transport from continuous sources at the wall of a channel," Int. J. Heat Mass Transfer, 45(17), pp. 3571-3583.
5. Le, P.M., and Papavassiliou, D.V., 2008, "On the scaling of heat transfer using thermal flux gradients for fully developed turbulent channel and Couette flows," Int. Commun. Heat Mass Transfer, 35(4), pp. 404-412.
6. Srinivasan, C., and Papavassiliou, D. V., 2011, "Prediction of Turbulent Prandtl Number in Wall Flows With Lagrangian Simulations," Ind. Eng. Chem. Res., 50(15), pp. 8881–8891.
7. Srinivasan, C., and Papavassiliou, D.V., 2013, "Heat transfer scaling for wall bounded turbulent flows," Applied Mechanics Review 65(3), Art. 031002.

8. Nguyen, Q., Srinivasan, C., and Papavassiliou, D.V., 2015, "Flow induced separation in wall turbulence", Phys Rev E, 91, Art. 033019.

503674 Flow Boiling Using a Piranha Pin Fin Heat Sink

Abstract:

Flow Boiling using a Piranha Pin Fin Heat Sink

Corey Woodcock, Xiangfei Yu, Yoav Peles[1] and Joel Plawsky

Rensselaer Polytechnic Institute; [1]University of Central Florida

The *Piranha Pin Fin* (*PPF*) microdevice is a heat exchanger engineered to dissipate extreme thermal loads using either dielectric fluids such as HFE7000 or water. Fabrication of the device is CMOS compatible and the first generations of the device have been fashioned from silicon substrates. The PPF operates on the principle that heat transfer would be more efficient if heated fluids could be removed nearly as fast as they are created. This process would keep the driving force as high as possible. Thus the heat exchanger is built as two layers; a primary substrate that dissipates the bulk of the heat and shunts the heated fluids to a booster section that continues to heat the fluid and extends the total surface area of the device. Three generations of devices were fabricated. The first two generations were single-layer systems designed to help understand the operation of the primary layer and optimize the shape and location of the pin fins. The third generation was the first, two-layer system to be fabricated and tested.

Figure 1 Representative PPF array

Correlations for the heat transfer coefficient, h_{tp}, based on the Boiling Number and the Weber Number in the PPF design was developed and validated with Gen II. The discrepancy of heat transfer coefficient between predicted and measured values was within 20%. The ratio of two-phase to single-phase pressure drop was correlated with the Boiling Number (Bo) and the Jakob Number (Ja). The discrepancy of the two-phase multiplier between prediction and experimental measurements were within 12%.

The third generation *PPF* heat sinks were the first series to provide both a primary substrate populated with *PPF*s, as well as a booster substrate populated with solid micro pin fins. The substrates are micro-aligned and fusion-bonded to provide a full three dimensional (3D) architecture with bulk silicon properties. Experimental results from the Gen III *PPF* microdevices provide a proof of concept of the true *PPF* heat sink concept with fluid venting into multiple substrates. At low flowrates (G = 670 [kg/s/m^2]) the Gen III *PPF* heat sink has been shown to dissipate heat loads approaching 1.1 [kW/cm^2] at an exit quality of 95%, while at higher flowrates (G = 3,300 [kg/s/m^2]) the thermal limit of the heat sink remains unknown due to exceeding the available power supply at >1.4 [kW/cm^2].

503673 Transport Problems in the Spirit of Stuart Churchill for Teaching and Research at the University of Michigan

Abstract:

Stuart Churchillâ€™s solutions to important heat transport problems have greatly influence both pedagogy and research. The correlation of Churchill and Chu for the natural convection heat transfer to or from a vertical plate, over an astonishing range of Rayleigh numbers is based partly on a similarity solution of a type common in heat, mass, and momentum transfer for problems in chemical engineering. One such problem is asphaltene or wax transport and deposition onto the wall of an oil pipeline, which can be solved using similar methods, as will be discussed. In addition, such problems, and the wealth of their applications make for excellent motivations and examples in the teaching of graduate or undergraduate transport, as will be discussed in the talk.

491768 Nonlinear Distributed Model Predictive Control of Gas Sweetening Processes

Abstract:

Gas sweetening is one of the key processes in the production train of liquefied natural gas and in the post-combustion flue gas treatment, where acid gases like CO_2 and H_2S are removed through selective absorption in solvents (e.g., diethanolamine, potassium carbonate, etc.) to reduce their release into the environment [1, 2]. A typical gas sweetening plant comprises single or multi-stage absorber columns for the absorption of acid gases into the solvent, and stripper units to regenerate the spent solvent for recycle. Material recycles and heat integration in the plant are essential for minimization of energy and utility waste. Many efforts have been made to design, model and optimize the gas sweetening plants, but fewer research efforts have been dedicated to controlling the process. Single stage absorption-regeneration has been widely studied from the control viewpoint, e.g. control under feed flow rate fluctuations [3], decentralized control of the absorber and stripper column under disturbances [4, 5], self-optimizing control for determining the control configuration [6], and single-input-single-output PID control [7, 8]. In these studies, the interaction effects due to integration between process units have not been considered [9]. These become important in the context of transient operation of such plants. In addition, the gas sweetening plants are frequently subject to disturbances in the form of fluctuations in the mass flow rate of the acid gas or the acid content of the feed. These disturbances can propagate through the integrated plant, making its control challenging. Motivated by these, this work addresses the nonlinear optimization-based control of such plants.

The optimization-based control of gas sweetening process units and their interconnections, which can be described by a set of algebraic and differential equations (ODEs and PDEs), requires solving a large-scale constrained nonlinear dynamic optimization in real-time [10], which is computationally expensive. In this work, we implement a distributed model predictive control (DMPC) architecture comprising local controllers with some level of cooperation and communication [11, 12]. A prerequisite for the application of such a strategy is the decomposition of the large-scale system into the corresponding subsystems. To this end, we adopt a recently developed framework that determines optimal decompositions based on community detection methods [13-15]. The resulting decompositions maximize the modularity of the corresponding subsystems, thereby minimizing the interactions among them.

Specifically we implement an iterative DMPC architecture to address the output regulation problem. The sweetening process consists of two absorbers, two regenerator columns, and two heat exchangers which are tightly integrated. The optimal decomposition separates the first-stage absorption and regeneration columns from those in the second stage, recommending a distributed control structure of two local controllers which can communicate over the network. The local controllers are developed based on the corresponding ODE systems and coarse discretizations of the PDEs describing the absorber/stripper columns. The closed-loop performance and the average computation time are evaluated using the detailed process model implemented in gPROMS. We additionally compare the results with those of centralized and fully decentralized MPC synthesis.

[1] A. Bedelbayev, T. Greer, B. Lie, Model based control of absorption tower for CO_2 capturing, Proceedings, 49th International Conference of Scandinavian Simulation Society (SIMS 2008), October 7th-8th 2008, Oslo University College, Oslo, Norway. ISBN-13: 978-82-579-4632-6, 2008.

[2] M. Abu-Zahra, J. Niederer, P. Feron, G. Versteeg, CO_2 capture from power plants Part II: A parametric study of the economical performance based on mono-ethanolamine, International Journal of Greenhouse Gas Control, 1, 135–142, 2007.

[3] A. Lawal, M. Wang, P. Stephenson, G. Koumpouras, H. Yeung, Dynamic modelling and analysis of post-combustion CO_2 chemical absorption process for coal-fired power plants, Fuel, 89(10), 2791–2801, 2010.

[4] S. Ziaii, G. Rochelle, T. Edgar, Dynamic modeling to minimize energy use for CO_2 capture in power plants by aqueous monoethanolamine, Industrial & Engineering Chemistry Research, 48 (13), 6105–6111, 2009.

[5] H. Kvamsdal, J. Jakobsen, K. Hoff, Dynamic modeling and simulation of a CO_2 absorber column for post-combustion CO2 capture, Chemical Engineering Process, 48 (1), 135–144, 2009.

[6] M. Schach, R. Schneider, H. Schramm and J. Repke, Control structure design for CO_2 absorption processes with large operating ranges, Energy Technology, 1, 233 – 244, 2013.

[7] C. Biliyoka, A. Lawal, M. Wanga, F. Seibert, Dynamic modelling, validation and analysis of post-combustion chemical absorption CO_2 capture plant, International Journal of Greenhouse Gas Control, 9, 428–445, 2012.

[8] Y. Lin, T. Pan, D. Wong, S. Jang, Y. Chi, C. Yeh, Plantwide control of CO_2 capture by absorption and stripping using monoethanolamine solution, Industrial & Engineering Chemistry Research, 50, 1338 – 1345, 2011.

[9] M. Wang, A. Lawal, P. Stephenson, J. Sidders, C. Ramshawa, Post-combustion CO_2 capture with chemical absorption: A state-of-the-art review, Chemical Engineering Research and Design, 89, 1609–1624, 2011.

[10] R. Lopez-Negrete, F.J. Damato, L.T. Biegler, and A. Kumar. Fast nonlinear model predictive control: Formulation and industrial process applications, Computers & Chemical Engineering, 51, 55–64, 2013.

[11] D.Q. Mayne, Model predictive control: Recent developments and future promise, Automatica, 50(12), 2967–2986, 2014.

[12] P.D. Christofides, R. Scattolini, D. Munoz de la Pena, and J. Liu. Distributed model predictive control: A tutorial review and future research directions, Computers & Chemical Engineering, 51, 21–41, 2013.

[13] S. Heo and P. Daoutidis. Control-relevant decomposition of process networks via optimization-based hierarchical clustering, AIChE Journal, 62(9), 3177–3188, 2016.

[14] S. Heo, W.A. Marvin, and P. Daoutidis. Automated synthesis of control configurations for process networks based on structural coupling, Chemical Engineering Science, 136, 76–87, 2015.

[15] M. Moharir, L. Kang, A. Almansoori, and P. Daoutidis. Control relevant network decomposition of distributed parameter system networks, Computers & Chemical Engineering, 2017, submitted.

493356 Dynamic Modeling and Explicit Model Predictive Control of Absorption of Hydrogen in a $LaNi_5$ Bed

Abstract:

Hydrogen is an important energy carrier that is pivotal to the evolution of the energy sector. Compounds containing hydrogen absorption energy from the environment to produce hydrogen gas, which has higher gravimetric energy density. Hydrogen is stored in solids by adsorption - where hydrogen physically or chemically adheres to the surface of a material with high affinity and surface area (ex. Nanomaterial like carbon nanotubes) or by absorption – when hydrogen percolates into the lattice of a material consequently forming a different material (ex. Metal hydrides). Technologies have been developed for storing compressed hydrogen in specialized cylinder at high pressure (~ 10,000 psi) onboard a light duty fuel Cell Vehicle (FCV). However, Metal Hydride (MH) hydrogen storage remain an inherently safer storage medium for onboard storage. Two of the performance target areas listed by DOE [1] is the cooling capacity and thermal management of the MH. The temperature surge and plummet during the hydriding and dehydriding processe respectively are inherent properties of the MH. Design and control are basic considerations for performance improvement of the operation of MH systems. Here we design an explicit model predictive controller for the absorption of hydrogen into the MH bed using the PAROC framework. The schematic of the proposed framework can be found in [2]. First, we present a detailed high-fidelity dynamic model of the processes based on first principles. Several studies have investigated the various aspect of metal hydride processes [3-7]. Optimization and controls can play an important role in the thermal management of the MH process[7]. Several studies have appeared in the open literature describing the optimization and design but only a few studies address the control of adsorption and desorption processes of the metal hydride. Panos et al [8] presented a robust explicit model predictive control of hydrogen storage in MH tanks focusing on the desorption of hydrogen.

The high-fidelity dynamic model described in this work is based on first principles according to [7, 9] and accounts for the spatial variations of some thermodynamic properties. The model makes the following assumptions; the ideal gas law holds in the gas phase, the solid and fluid temperatures are essentially equal, axial and radial dispersion are included in the mass balances for the interstitial fluid, the axial and radial pressure drops in the bed depend linearly on the superficial velocity through Darcy's law. The model is developed in gPROMS software and it is used to simulate the absorption process. It is subsequently validated with experimental data. An optimization procedure is implemented on the model to determine optimal operation condition and results of the optimization will be used to determine appropriate constraints for the control design. The optimization strategy minimizes the storing time and the output pressure while satisfying, a number of operating and safety constraints. In the second part of this work, we develop a reduced order approximate model suitable for multi-parametric model predictive control (mp-MPC) based on prior analysis. The reduced order state-space (SS) model (usually linear) is developed using model identification. The SS model is developed such that it can approximate the high fidelity model. The main contribution is the design of a multi-parametric/explicit model predictive controller to regulate the absorption of hydrogen in the $LaNi_5$bed. Unlike the MPC the mp-MPC[10] MPC avoids the online

optimization procedure. The optimization is done offline to determine the control action at every state obtained while simultaneously accounting for physical and operational constraints. For system with faster dynamics (like the MH), mp-MPC is ideal. The explicit controller is designed in MATLAB and it is linked to gPROMS through gO:MATLAB.

Reference

1. Satyapal, S., J. Petrovic, C. Read, G. Thomas, and G. Ordaz, *The U.S. Department of Energy's National Hydrogen Storage Project: Progress towards meeting hydrogen-powered vehicle requirements.* Catalysis Today, 2007. **120**(3–4): p. 246-256.

2. Pistikopoulos, E.N., N.A. Diangelakis, R. Oberdieck, M.M. Papathanasiou, I. Nascu, and M. Sun, *PAROC—An integrated framework and software platform for the optimisation and advanced model-based control of process systems.* Chemical Engineering Science, 2015. **136**: p. 115-138.

3. Silin, N. and M. Melnichuk, *A case study of a hydride container performance applying non dimensional parameters.* International Journal of Hydrogen Energy, 2014. **39**(36): p. 21060-21067.

4. Choi, H. and A.F. Mills, *Heat and mass transfer in metal hydride beds for heat pump applications.* International Journal of Heat and Mass Transfer, 1990. **33**(6): p. 1281-1288.

5. Askri, F., A. Jemni, and S. Ben Nasrallah, *Study of two-dimensional and dynamic heat and mass transfer in a metal–hydrogen reactor.* International Journal of Hydrogen Energy, 2003. **28**(5): p. 537-557.

6. Demircan, A., M. Demiralp, Y. Kaplan, M.D. Mat, and T.N. Veziroglu, *Experimental and theoretical analysis of hydrogen absorption in – reactors.* International Journal of Hydrogen Energy, 2005. **30**(13–14): p. 1437-1446.

7. Kikkinides, E.S., M.C. Georgiadis, and A.K. Stubos, *On the optimization of hydrogen storage in metal hydride beds.* International Journal of Hydrogen Energy, 2006. **31**(6): p. 737-751.

8. Panos, C., K.I. Kouramas, M.C. Georgiadis, and E.N. Pistikopoulos, *Dynamic optimization and robust explicit model predictive control of hydrogen storage tank.* Computers & Chemical Engineering, 2010. **34**(9): p. 1341-1347.

9. Jemni, A. and S.B. Nasrallah, *Study of two-dimensional heat and mass transfer during absorption in a metal-hydrogen reactor.* International Journal of Hydrogen Energy, 1995. **20**(1): p. 43-52.

10. Pistikopoulos, E.N., V. Dua, N.A. Bozinis, A. Bemporad, and M. Morari, *On-line optimization via off-line parametric optimization tools.* Computers & Chemical Engineering, 2002. **26**(2): p. 175-185.

495141 Effective Dose Delivery in Plasma Medicine Using a Robust MPC Approach for Mixed Stochastic and Deterministic Uncertainty

Abstract:

Cold atmospheric plasma (CAP) is a weakly ionized gas generated in ambient conditions via application of modulated electric field to gases such as helium, argon, and air. CAPs have recently gained increasing attention for treatment of heat and pressure sensitive (bio)materials in surface etching/functionalization, environmental, and biomedical applications [1,2]. An attractive feature of CAPs is that they are non-thermal plasma discharges, that is, the electrons have a high temperature (~10,000 K) while the temperature of the background gas remains near to room temperature. The high-temperature, energetic electrons in a CAP discharge can generate reactive chemical species through a plethora of excitation and chemical reactions with the background gas and species diffusing in from the ambient air. The reactive chemical species, along with other plasma effects such as UV photons, thermal energy, and electric fields have been postulated to induce therapeutic effects such as tissue disinfection, healing of chronic wounds, shrinkage of cancerous tumors, and regulation of multidrug resistant bacteria [2].

The notion of plasma medicine generally relies on evoking a desired response in a living system through the delivery of a multitude of plasma effects. The highly nonlinear, spatially distributed dynamics of CAP discharges, the cumulative (i.e., "non-retractable") response of a living system to plasma effects, and the development of instabilities and plasma mode transitions to arc or spark due to variations in plasma properties are some of the key challenges to safe, consistent, and effective dose delivery in plasma medicine. Recently, we have demonstrated the promise of model predictive control (MPC) for effective dose delivery in atmospheric pressure plasma jets (APPJs) [3,4]. APPJs have been identified as a promising class of CAP discharges for plasma medicine due to their operational flexibility and versatile discharge chemistry [2]. A key challenge in applying MPC to APPJs arises from uncertain knowledge of the plasma behavior, which is intrinsically stochastic. Incomplete representations of the complex plasma dynamics can lead to inadequate system models for the use in MPC, possibly affecting the MPC performance for dose delivery. In addition, unmeasured system disturbances, which are typically ubiquitous in APPJ applications, cannot be systematically handled in the standard MPC framework, possibly compromising the safety, consistency, and efficacy of the plasma dose delivery.

In this work, we will first present a robust MPC method for offset-free tracking of piece-wise constant references in the presence of bounded deterministic and stochastic uncertainty [5]. The deterministic uncertainty source accounts for the unknown structural/parametric plant-model mismatch (i.e., incomplete knowledge of the plasma dynamics), whereas the stochastic uncertainty source represents exogenous system disturbances (e.g., random variations in properties of the target substrate in an APPJ application). Concepts from robust tube-based and stochastic tube-based MPC [6,7] will be used to derive a tractable MPC formulation, with computational complexity comparable to that of nominal MPC. We will then adopt the proposed robust MPC method for plasma dose delivery in a kilo-hertz APPJ in helium, which is a prototypical

medical therapy device. The performance of the robust MPC method for consistent and effective dose delivery in the presence of plant-model mismatch and disturbances will be demonstrated using extensive simulation results performed on a nonlinear APPJ simulator developed in COMSOL.

References

[1] P. Chu, J. Chen, L. Wang, and N. Huang, Plasma-surface modification of biomaterials, *Materials Science and Engineering: Reports*, 36,143-206, 2002.

[2] M. Laroussi, M. Kong, G. Morfill, and W. Stolz, Plasma Medicine. New York, NY: Cambridge University Press, 2012.

[3] D. Gidon, D.B. Graves, and A. Mesbah, Model predictive control of thermal effects of an atmospheric pressure plasma jet for biomedical applications, In *Proc. ACC*, 4889-4894, 2016.

[4] D. Gidon, D.B. Graves, and A. Mesbah, Effective dose delivery in atmospheric pressure plasma jets for plasma medicine: A model predictive control approach, *Plasma Sources Science and Technology*, Under review, 2017.

[5] J.A. Paulson, L. Xie, and A. Mesbah, Robust offset-free MPC of systems with mixed stochastic and deterministic uncertainty. In *Proceedings of the IFAC World Congress*, 2017, Toulouse.

[6] D.Q. Mayne, M.M. Seron, and S.V. Rakovic, Robust MPC of constrained linear systems with bounded disturbances. *Automatica*, 41, 219–224, 2005.

[7] B. Kouvaritakis, M. Cannon, S.V. Rakovic, and Q. Cheng, Explicit use of probabilistic distributions in linear predictive control. *Automatica*, 46, 1719–1724, 2010.

495172 A Model Reduction and Decomposition Approach for Economic MPC of Wastewater Treatment Plants

Abstract:

To minimize the environmental impacts of wastewater and to recycle the water, complex wastewater treatment plants (WWTPs) have been widely used for the treatment of wastewater. A typical WWTP is a large-scale nonlinear process comprised of several interconnected biological reactors and a settler. All the components involved in a WWTP are required to be well monitored and regulated to meet the strict environmental regulations. On the other hand, significant variability of the inlet flow rates to WWTPs and unexpectable perturbations to the wastewater composition lead to complex dynamics of the plant models. Therefore, great challenges are posed to the design of advanced control/monitoring schemes for WWTPs.

In the literature, several control approaches have been proposed for WWTPs. One commonly adopted solution is to design a proportional-integral (PI) regulator. Another regulating approach is based on the combination of feedforward control and PI control. Despite the simple structures, the above approaches may not handle system constraints and provide optimal control actions. To address these shortcomings, more and more research attention has been given to model predictive control (MPC) algorithms during the last decade. However, the computational complexity of these algorithms is non-negligible. In our previous work [1], an economic MPC algorithm was developed for the control of WWTPs. It was demonstrated that economic MPC can improve the effluent quality while reducing the operational costs compared with the classical set-point tracking MPC. While economic MPC has the potential to improve the control of WWTPs, one critical issue is its high computational complexity. Note that in the implementation of set-point tracking MPC, the computational issue may be overcome by using identified linear input-output models around the set-point. However, these input-output models may not be suitable for economic MPC if the objective function in the economic MPC is a function of the entire process state (instead of the outputs) and the process exhibits time-varying (instead of steady-state) operation like in WWTPs.

In this work, we present an objective-oriented model reduction and subsystem decomposition approach for the control of WWTPs. A benchmark simulation model for a WWTP of five activated sludge reactors and a secondary reactor is used. In the proposed approach, the nonlinear model is reduced and simplified through a few steps. First, the model is reduced by considering the time-scale multiplicity in the process dynamics; second, the dynamic sensitivities of the objective function (effluent quality and operational cost) with respect to all the process states are evaluated along typical operating trajectories and the most important states are determined according to the dynamic sensitivities; third, a hybrid model is developed based on linearizing those not so important states around a steady-state and piecewise linearizing those important states (with high dynamic sensitivities). Based on the simplified hybrid model, we also illustrate how recent results on input-output pairing [2] and subsystem decomposition [3] may be applied to the WWTP. Simulation results are also provided to demonstrate the effectiveness of the proposed approach.

[1] J. Zeng and J. Liu. Economic model predictive control of wastewater treatment processes. Industrial & Engineering Chemistry Research, 54:5710-5721, 2015.

[2] X. Yin and J. Liu. Input-output paring accounting for both structure and strength in coupling. AIChE Journal, 63:1226-1235, 2017.

[3] X. Yin, K. Arulmaran, J. Liu and J. Zeng. Subsystem decomposition and configuration for distributed state estimation. AIChE Journal, 62:1995-2003, 2016.

499498 Real-Time Energy Management for Electric Arc Furnace Operation

Abstract:

Electric arc furnaces (EAFs) are extensively employed by steel production companies to produce steel from scrap metal. The energy intensive batch operation involves melting of the scrap using electrical and chemical energy, and modifying its chemistry. Electrical energy is added via electrodes, while combustion reactions provide chemical energy. The melting continues and a flat bath of molten metal is achieved. Oxygen lancing leads to the formation of metal oxides due to oxidation, which floats on the top as slag. The slag chemistry is altered by flux additions (carbon, lime and dolomite), carbon and oxygen lancing. The amount and timing of input additions are determined mainly by the operators. Since EAFs are operated under very harsh conditions with limited measurements available during the heat, process automation is a challenging task. The composition of offgas is the only set of measurements generally available at a regular frequency. With electricity contributing approximately 60% of the energy requirement [1], external variations in electricity price needs to be considered for optimal energy usage. An energy-efficient control strategy is envisaged to generate significant savings due to an optimal use of electrical power, natural gas, oxygen, carbon and the fluxes in response to electricity price changes.

Control strategies have been developed in the past for the EAF electrode system [2], with linear model predictive control (MPC) for the offgas system [3]. The use of a first-principles model for nonlinear MPC assuming full state feedback was reported in [4]. Economic model predictive control, on the other hand, has become popular due to its straightforward handling of an economic target. The traditional set-point tracking and input move suppression objectives are either replaced by, or used in conjunction with, an economic objective [5, 6]. A time-varying stage cost in EMPC [7] is suitable for incorporating a changing electricity price profile. EMPC for continuous processes has been extended to include demand response (DR) [8]. Literature in which DR is considered for batch operation of EAFs is sparse. Real-time control applications require state estimation to work in tandem with MPC. Multi-rate moving horizon estimation (MHE) has been used for EAFs to reconstruct the states using both the fast and slow measurements [9]. A decision support tool for EAF operators based on a combined application of shrinking horizon dynamic optimization and multi-rate MHE was proposed in [10]. The study is extended here to employ a coupled EMPC-MHE implementation for EAFs to utilize time varying electricity cost for optimal control.

In this work, we aim to develop an EAF control strategy that effectively reduces the energy costs in real-time while exploiting the electricity price profile. A shrinking horizon EMPC is coupled with a multi-rate MHE to develop an integrated decision-making framework. The EMPC objective function comprises time-varying cost coefficients to handle a volatile wholesale electricity market. The control structure employs a modified first-principles differential-algebraic equation (DAE) model proposed in [11, 12]. The model divides the furnace into four zones (solid scrap, molten metal, slag-metal, gas) with mass transfer limitations across the zone interfaces. Chemical equilibrium is considered for the slag-metal and the gas zone through an embedded Gibbs free energy minimization. The complex DAE model is discretized and used within a simultaneous solution framework for solving the dynamic optimization problems. Online solve times of EMPC-MHE problems are critical for a real-time implementation of the control strategy. We introduce a novel initialization scheme to obtain fast solutions of EMPC-MHE optimization problems.

To evaluate the proposed strategy, we conduct case studies to investigate the performance of EMPC under various Day Ahead Market conditions, in which the electricity price changes every hour. The case studies are chosen based on a detailed analysis of actual and forecasted 2016-2017 prices for the Ontario (Canada) wholesale market. The EMPC is shown to balance the electrical and chemical energy consumption in an optimal way to generate cost savings. Moreover, the peak electricity demand during the batch operation is reduced when a time-varying electricity price profile is used. The multi-rate MHE demonstrates a strong tracking ability to support the EMPC. Finally, the computational results show a significant reduction in EMPC-MHE solve times when the proposed initialization scheme is used. In summary, an EMPC-MHE based strategy is presented for curtailing the energy costs in real-time. Current work includes formulations for emission reduction through constraints or a carbon credit-based approach. As a future work, we intend to study stochastic formulations for differentiating forecasts from actual prices. Additionally, the strategy will be tested and compared for 15 minute and 5 minute electricity markets. References

[1] R. J. Fruehan, The Making, Shaping, and Treating of Steel: Ironmaking volume, vol. 2, AISE Steel Foundation, 1999.

[2] B. Boulet, G. Lalli and M. Ajersch, "Modeling and control of an electric arc furnace," in *American Control Conference (ACC)*, 2003.

[3] J. G. Bekker, I. K. Craig and P. C. Pistorius, "Model predictive control of an electric arc furnace off-gas process," *Control Engineering Practice,* vol. 8, no. 4, pp. 445-455, 2000.

[4] R. D. M. MacRosty and C. L. E. Swartz, "Nonlinear predictive control of an electric arc furnace," *IFAC Proceedings Volumes,* vol. 40, no. 11, pp. 285-290, 2007.

[5] R. Amrit, J. B. Rawlings and D. Angeli, "Economic optimization using model predictive control with a terminal cost," *Annual Reviews in Control,* vol. 35, no. 2, pp. 178-186, 2011.

[6] M. Ellis, H. Durand and P. D. Christofides, "A tutorial review of economic model predictive control methods," *Journal of Process Control,* vol. 24, no. 8, pp. 1156-1178, 2014.

[7] D. Angeli, A. Casavola and F. Tedesco, "Theoretical advances on Economic Model Predictive Control with time-varying costs," *Annual Reviews in Control,* vol. 41, pp. 218-224, 2016.

[8] J. Feng, A. Brown, D. O'Brien and D. J. Chmielewski, "Smart grid coordination of a chemical processing plant," *Chemical Engineering Science,* vol. 136, pp. 168-176, 2015.

[9] S. Shyamal and C. L. E. Swartz, "A Multi-rate Moving Horizon Estimation Framework for Electric Arc Furnace Operation," *IFAC-PapersOnLine,* vol. 49, no. 7, pp. 1175-1180, 2016.

[10] S. Shyamal and C. L. E. Swartz, "Optimization-based Online Decision Support Tool for Electric Arc Furnace Operation," in *IFAC World Congress*, Toulouse, France, 2017 (accepted).

[11] R. D. MacRosty and C. L. E. Swartz, "Dynamic modeling of an industrial electric arc furnace," *Ind.Eng.Chem.Res.,* vol. 44, pp. 8067-8083, 2005.

[12] Y. Ghobara, "Modeling, Optimization and Estimation in Electric Arc Furnace," *Master's thesis,* no. McMaster University, 2013.

498219 Multi Variable Model Predictive Control to Improve Oil Production for Steam-Assisted Gravity Drainage (SAGD)

Abstract:

Steam-Assisted Gravity Drainage (SAGD) is a major thermal recovery technique in Western Canada. It involves injection of superheated steam through an upper injection well to mobilize and drain the highly viscous bitumen to form a liquid pool above a bottom producer well before its subsequent production. Existing research has shown that controlling the temperature difference between the injected steam and the produced oil-water emulsion, which is also called the subcool temperature difference, can regulate the height of the liquid pool such that it is favorable for production. This technique, called steam trap control, prevents the production of live steam and also protects the formation from fracturing, by keeping the steam injection rates within allowable limits. However, major drawbacks of this technique are the absence of a method to optimally control the production of the fluids from the liquid pool and the lack of guidelines to determine the ideal subcool temperature difference setpoint.

In this study, we address these issues through the development of a multi input multi output (MIMO) model predictive controller (MPC) that incorporates adaptive oil rate and steam trap controls. The addition of the adaptive oil rate control helps improve the oil production rate from the liquid pool and drainage rate of oil into the liquid pool. Furthermore, the steam trap control setpoint was turned into an optimization variable, such that the optimizer determined the optimal subcool setpoint to satisfy the adaptive oil rate constraints. It should be noted that oil rate control by itself leads to formation damage due to high injection rates of steam and live steam production due to high production rates of the oil-water emulsion from the liquid pool. Therefore, it was used in conjunction with steam trap control, which not only reduces the live steam production, but also protects the formation from fracturing. The novel MIMO control strategy was compared with a multi input single output (MISO) MPC steam trap control strategy. The real time control study for both strategies were enabled using a bidirectional communication that was established between CMG STARS (virtual oil reservoir) and MATLAB (onsite controller) software. The results show a 171% improvement of oil recovery for the novel MIMO controller as compared to the MISO controller over a period of 6 simulated years. Moreover, there was a 35% improvement in the cumulative steam-to-oil ratio (a performance metric for SAGD) for the MIMO controller over the MISO controller.

494399 Real-Time Adaptive Model Predictive Control Framework of Plasma Process

Abstract:

Plasma process is one of the most critical process in semiconductor manufacturing. Since it determines the dimension of entire processes, it has a decisive effect on the performance of devices. With the continually evolving process technologies that require high quality constraint, semiconductor industry has been imperatively considering real-time control over current run-to-run based control.

They have been to use *recipe* that is terminology indicating optimal process inputs which produce maximal yields. The *recipe* is obtained from several experiments by trial-and-error and is applied to consecutive processes with run-to-run adjustment. However, even with this *recipe*, the process may not reproduce the desired results especially when the required dimension gets smaller. This is because the process is essentially driven by the plasma formed by the *recipe* and can be disturbed by external factors. Thus, it is necessary to implement a real-time cascade control to include the plasma as a control variable [1] and to reject the disturbance fast.

The biggest obstacle to real-time control is the nature of the plasma. Plasma is generated by applying power at nearly vacuum pressure of 1.3 Pa (0.00001 bar). Hence this sparse system of multi-physics can be intrinsically affected even if there are small disturbances. The key considerations in controlling such systems are: the use of sensors that do not affect the system, the control of highly nonlinear system, and the way to dealt with inter- and intra- system variability.

In this work, we used optical emission spectroscope (OES), since it is not only having no concern about invasiveness toward process, but also a default plasma diagnostic tool for every plasma etcher in production line. We calibrated it with Langmuir probe and calculated the core plasma variables such as electron density and electron temperature from the intensity data [2,3].

To cope with nonlinearities in system gain, the response surface was piecewise linearized and structured as a linear parameter varying (LPV) system [4,5]. The other system dynamic parameters related to time were identified through the design of experiment, since the dynamics is very fast nearly as the minimal sampling rate.

Inter- and intra- system variability are known to be caused by the difference in equipment/recipes, discrete events between unit processes, and history of precedent processes. Due to the nature of the semi-batch process, the process is more likely to be exposed to these effects than other conventional processes. Therefore, deterioration of control performance coming from the model-plant-mismatch is inevitable. Even in the case of process drift, it is difficult to model it because the information is distributed over multiple nonlinear signals. Therefore, we utilized recurrent neural network (RNN) to track process state variables and drawn initial values for operating condition.

Consequently, an adaptive model predictive control (AMPC) is constructed. The objective function was formulated with a mixed-integer quadratic programming (MIQP) since only integer inputs are applicable to the equipment as a practical restriction. The developed controller was implemented on the field equipment with 100ms of sampling rate and proved it can enhance the reproducibility of process results. This novel framework is expected to deal with the major challenges in real-time control of plasma systems.

[1] Ringwood, J. V., Lynn, S., Bacelli, G., Ma, B., Ragnoli, E., & McLoone, S. (2010). Estimation and control in semiconductor etch: Practice and possibilities. IEEE Transactions on Semiconductor Manufacturing, 23(1), 87-98.

[2] Boffard, J. B., Lin, C. C., & DeJoseph Jr., C. A. (2004). Application of excitation cross sections to optical plasma diagnostics. J. Phys. D: Appl. Phys. 37, R143-R161.

[3] Zhu, X.-M., Pu, Y.-K., (2010). Optical emission spectroscopy in low-temperature plasmas containing argon and nitrogen: determination of the electron temperature and density by the line-ratio method. J. Phys. D: Appl. Phys. 43, 403001.

[4] Keville, B., Zhang, Y., Gaman, C., Holohan, A. M., Daniels, S., & Turner, M. M. (2013). Real-time control of electron density in a capacitively coupled plasma. Journal of Vacuum Science & Technology A, 31(3), 031302.

[5] White, A. P., Zhu, G., & Choi, J. (2013). Linear parameter-varying control for engineering applications. New York: Springer.

502123 Load-Following Control for a 10 MWe Supercritical CO_2 Recompression Brayton Power Cycle

Abstract:

Supercritical carbon dioxide (sCO_2) Brayton power cycles have recently gained attention as an attractive alternative to the more conventional Brayton and Rankine cycles. The high density of the sCO_2 working fluid allows for equipment with a more compact design and a power cycle with potentially higher thermal efficiencies and lower overall cost of electricity. Due to these advantages over traditional systems, the U.S. Department of Energy, under its Supercritical Transformational Electric Power Program, is currently in the process of building a nominal 10 MWe (net) sCO_2 Brayton cycle test facility to accelerate scale-up, development, and commercial deployment. In this work, a previously designed dynamic process model of a 10MWe indirect recuperated sCO_2 recompression Brayton cycle is used to develop traditional as well as advanced controllers for efficient load-following performance.

Temperature control at various strategic locations is critical for achieving high efficiency in the sCO_2 cycle while ensuring stable operation during load-following. Critical locations where temperature must be controlled within a narrow window at the face of disturbances include: turbine, main compressor, main and bypass fluid mixing point, and the sCO_2 cooler. Temperature control at the turbine, main compressor, and mixing point is integral to efficient cycle operation, while the main compressor inlet temperature is also critical to avoid transition to the sub-critical phase. Since the main compressor inlet temperature remains very close to the critical point, only a narrow temperature margin is affordable to avoid phase transition. Since the cooling water temperature in the sCO_2 cooler at the main compressor inlet should not exceed certain value to avoid scale formation and the compressor surge must be avoided under all circumstances, controlling the main compressor inlet temperature becomes challenging. Furthermore, critical temperatures mentioned earlier are tightly coupled leading to a highly interacting system. Configurational changes along with novel control structures are developed to address this control problem.

Performance of the control system is evaluated under nominal and multiple off-design scenarios under various load ramp rates. For the turbine inlet temperature control, two approaches are considered. The first approach uses inventory control, whereby CO_2 is removed from the closed cycle to maintain the turbine inlet temperature during cycle heat input disturbances. The second approach uses heat input adjustments (natural gas and/or air flow to an external combustor) to maintain this temperature. Studies on load following performance will also be presented when using the first control approach for variations in cycle heat input and the second control approach for fluctuations in inventory. For the main compressor inlet temperature control, configurational changes include recycling a portion of the cooling water from the outlet of the main compressor inlet cooler. The presentation will include critical analysis of trade-offs between systems efficiency and control system performance under the constraints of process operating window at the face of various load ramp rates.

497299 Optimization-Based Process Synthesis of Processes with Seasonal and Daily Variability: Application on Concentrating Solar Power Plants with Thermochemical Energy Storage

Abstract:

The majority of research in the area of chemical process synthesis has focused on processes which operate on a single "mode" during a day and also have reasonably small variability in terms of availability and quality of resources. A "nominal" operation is typically assumed and variability is considered through "deviations" from the nominal operation. This treatment is insufficient however when the variability is so large that large subsystems of the facility are basically operating in completely different "modes" during, for example, the course of a day; and, in addition, the changes in quality and resource availability are very large and "persistent". These features arise, for example, in solar-based technologies. Accordingly, the goal of this work is to develop methods that allow us to account for daily and seasonal variability in an accurate yet tractable manner, so that the optimization model can still employ relatively accurate unit models.

The system we use as a testbed is concentrating solar power plants with thermochemical energy storage (TCES). Energy storage is crucial for solar power to avert the grid instability caused by the periodic nature and uncertainty of the solar insolation. Concentrating solar power (CSP) is a technology that converts concentrated sunlight to heat, which then drives a turbine to generate electricity. Using heat as an intermediate provides CSP the opportunity to incorporate efficient and low cost thermal energy storage (TES) [1]. Among all TES technologies, thermochemical energy storage (TCES), that is, the reversible conversion of solar-thermal energy to chemical energy, has high energy density and low heat loss over long periods. Although many reaction types have been studied as TCES candidates [2], past research has mostly focused on experimental studies and reactor design. The limited system-level studies are for specific reactions [3] and consider TCES in isolation, neglecting its interactions with the solar field and power generation system [4]. In addition, a common assumption for previous design model is that the process operates at a single steady state. However, CSP plants have daily and seasonal variations in operation due to the periodic nature of solar irradiation.

In this work, we develop a general optimization process model for CSP plants integrated with TCES systems [5] which allows us to synthesize a process under daily and seasonal solar variations, while considering the interactions between TCES and the power generation system. The model is built upon unit mass and energy balances in TCES process and performance models of the solar field, receiver and turbine. It includes a reactor and heat exchanger network which enables heat integration (the ability to design the coupling between exothermic reactor and steam generation is crucial to target high reaction conversion and turbine efficiency). In addition, we impose constraints tailored to prevent violation of the second law of thermodynamics during steam generation.

To model the "day" and "nigh" modes, a physical unit is represented as two units in the model to refer to its day and night operational modes. Special considerations are taken for unit sizing and cost calculations. To address the daily variation in storage compression, we derive the analytic relation between compression work and pressure variables by integrating the daily storage tank pressure profile. Furthermore, to account for the seasonal variation in solar resource, we calculate annual capacity factors for plant subsystems from the annual probability distribution of daily irradiation. These factors then serve as correction factors for the actual plant power output.

The objective function is to minimize the plant levelized cost of electricity (LCOE) by manipulating stream and unit operational variables, as well as plant sizing variables. The model can be applied to different TCES systems given their corresponding reaction properties (e.g., reaction enthalpy and entropy) and unit technologies.

To enhance the solution of the resulting model, we (1) develop algorithms for variable bound generation; (2) we propose case specific model simplifications based on physical insights; and (3) we employ scaling methods for modeling reaction equilibrium.

We apply the process model to a 100 MW CSP plant with methane TCES system, which shows an LCOE reduction potential over current CSP plants using two-tank molten salt storage.

References:

[1] J. Stekli, L. Irwin and R. Pitchumani, *J. Therm. Sci. Eng. Appl.*, 2013, 5, 21011.

[2] P. Pardo, A. Deydier, Z. Anxionnaz-Minvielle, S. Rougé, M. Cabassud and P. Cognet, *Renew. Sustain. Energy Rev.*, 2014, **32**, 591–610.

[3] A. Luzzi, K. Lovegrove, E. Filippi, H. Fricker, M. Schmitz-goeb, M. Chandapillai and S. Kaneff, *Sol. Energy*, 1999, **66**, 91–101.

[4] C. Corgnale, B. Hardy, T. Motyka, R. Zidan, J. Teprovich and B. Peters, *Renew. Sustain. Energy Rev.*, 2014, 38, 821–833.

[5] X. Peng, T. W. Root and C.T. Maravelias, *Green. Chem*, 2017, DOI: 10.1039/c7gc00023e.

487802 Anchor-Tenant Models for the Synthesis of Eco-Industrial Parks through Carbon-Hydrogen-Oxygen Symbiosis Networks

Abstract:

An eco-industrial park (EIP) is a network of entities who choose to interact by exchanging byproducts, wastes and/or energy to gain benefits greater than unintegrated operations. The concept is particularly attractive for developing industrial clusters and integrating processing facilities within an industrial city.

Recently, the concept of Carbon-Hydrogen-Oxygen Symbiosis Network (CHOSYN) has been introduced [1] for the integration of hydrocarbon processing facilities. The key idea of CHOSYN is to use atomic-level information to set targets for the integration opportunities and to employ a multi-scale systems approach to devising macroscopic strategies to attain the atomic-based benchmarks. CHOSYN takes advantage of the common ground of the chemical industry, utilizing streams containing carbon, oxygen and hydrogen atoms, to create synergism among multiple entities. Optimization approaches and shortcut methods have been developed to synthesize implementation alternatives of a mass integrated network that reduces raw material usage and waste disposal while offering significant economic benefits [1, 2].

In synthesizing new CHOSYNs or retrofitting an existing system, it is important to account for the various relationships among the EIP participants. The objective of this paper is to adopt the Anchor-Tenant model in the synthesis of CHOSYNs. Anchors are first invited as the key participants in the EIP. "Tenants" are potential plants that could be developed and integrated with the existing "Anchor(s)" thus creating a genesis of an EIP. A multi-scale optimization approach is developed to identify and screen the tenants and to determine performance benchmarks for individual plants and for the whole EIP. Different cooperative and non-cooperative scenarios are examined. The scope is of this study is extended to include nitrogen-based chemistries. A case study is developed and solved to demonstrate the key theoretical concepts and the applicability of the new approach.

References

[1] Noureldin, M.M.B. and M.M. El-Halwagi, Synthesis of C-H-O Symbiosis Networks. AIChE Journal, 2015. 61(4), 1242-1262.

[2] El-Halwagi, M.M., "A Shortcut Approach to the Multi-Scale Atomic Targeting and Design of C-H-O Symbiosis Networks", Process Integration and Optimization for Sustainability (DOI: 10.1007/s41660-016-0001-y, 2017)

492292 Optimal Renewable Production of Ammonia from Water and Air

Abstract:

Wind and solar energy are plentiful natural resources; however they are intermittent and difficult to handle. Chemical processes are capable of storing this energy in the form of useful products. Lately methane and methanol have been produced from CO_2 and electrolytic hydrogen [1,2]. Both are easy to transport and to deal with within the chemical industry. Ammonia is typically produced from fossil sources. However, it can also be produced using air and water. It is not entire new idea; a process was discussed in the early 1925's [3]. Nowadays, we can evaluate the use of solar and wind energy to produce the power required to produce the basic species for the fertilizer industry.

In this work we have proposed the optimal design and operation of a production facility of ammonia from air and water. The separation of air as well as water electrolysis are energy intense processes. Air is filtered and compressed before dehydration. Next, it is further compressed and cooled down before entering Linde's double column. The product streams are used in the cooling stage of compressed air. In parallel, water is splitted into hydrogen and oxygen. Both streams must be dehydrated and compressed. Finally, hydrogen is mixed with nitrogen and fed to the reactor. Two reactor designs are considered. Both are three packed bed reactors. Each bed operates adiabatically. The first design is a three packed bed reactor that uses indirect cooling between beds. The second alternative is also a three packed bed reactor. However, in this case the streams exiting each bed are cooled down by mixing with fresh syngas before being fed to the next. All the units are modelled using mass and energy balances, thermodynamic principles such as chemical equilibrium and surrogate models based on rigorous simulation to predict the liquefaction performance, the distillation column operation and the ammonia recovery. In particular, ammonia converters are modeled in detail in MATLAB based on industrial case studies [4,5] to determine the conversion reached at each bed since typically equilibrium is not reached.

The optimization of the process selects the use of solar energy. Furthermore, a direct cooling reactor results in better performance. However, the production cost for ammonia is high, 0.9€/kg, as well as the investment, 1050MM€ for an average production of 320t/day. Half of the investment cost is due to the solar panels and the rest is the chemical plant as such. Further analysis is required to include the operation of these facilities with variable solar and wind availability under uncertainty.

References

1.-Davis, W., Martín, M (2014) Optimal year-round operation for methane production from CO_2 and Water using wind and/or Solar energy. J. Cleaner Prod. 80, 252-261.

2.- Martin, M. Grossmann I.E: (2017) Optimal integration of a self sustained algae based facility with solar and/or wind energy J Clean Prod. 10.1016/j.jclepro.2017.01.051

3.- Ernest, F.A., Reed, F.C:, Edwards, W.L., (1925) A direct synthetic ammonia Plant. Ind.Eng. Chem. 17 (8) 775--788

4.-Elnashaie, S.S., Abash, M.E:, Al-Ubaid, A.S. (1988) Simulation and optimization of an industrial ammonia reactor. Ind. Eng. Chem., Res, 27, 2015-2022

5.-Galnes L.D., 1977 Optimal Temperatures for Ammonia synthesis converters. Ind. Eng. Chem., Process Des. Dev., Vol. 16, No. 3, 381-389

499853 Practical Bounds on Reaction Selectivity

Abstract:

Knowledge of the maximum attainable selectivities for a given chemical reaction network would be very useful during process design and operation. However, until recently a methodology to obtain such information for chemistries with more than two or three reactions was not readily available. Horn first introduced the concept of attainable regions (AR) in 1964 in an attempt to find optimal process designs for a given chemistry [1]. Since then, much creativity and effort has resulted in methodologies to enumerate ARs using geometric arguments for systems involving reaction and mixing [2-8] as well as systems involving reactions, separations, and mixing [9]. Unfortunately, all of these geometric methods are limited to a small number of components and reactions, which limits their utility.

The Continuous Flow Stirred Tank Reactor (CFSTR) Equivalence Principle, developed by Feinberg and Ellison [10], abandons the geometric approach altogether. This principle allows one to decompose any and every reaction/mixing/separation process into an equivalent process comprising only R+1 CFSTRs and a perfect mixer-separator (a â€œFeinberg Decompositionâ€), where R is the number of linearly independent reactions. Feinberg and coworkers [10, 11] showed that one can use the CFSTR Equivalence Principle to find the maximum attainable production flowrate of some desired product from a given chemistry for a given feed and reaction volume. This is done by maximizing the flowrate of the Feinberg Decomposition (with a given feed and total reactor volume) specified by the CFSTR Equivalence Principle. The utility of this method results from the fact that one can optimize an objective function over the AR without having to first identify the AR or its boundary. Frumkin and Doherty [12] showed that one can use the CFSTR Equivalence Principle together with a global optimization routine to find the maximum selectivity of a chemistry entirely independent of process design using kinetic models. They also show that constraints on overall reaction conversion and molar flowrates on the Feinberg Decomposition can be enforced to obtain bounds on selectivity that are more practical than the unconstrained case.

In this work, the problem is reformulated as a mixed-integer nonlinear program to solve the optimization problem. The problem is nonlinear and non-convex, and we implement a more robust, deterministic global optimization using a spatial branch-and-bound algorithm (BARON). We use this optimization routine to investigate the selectivity limits of the production of acrolein, the chemistry for which has thirteen components ant seventeen reactions. We find upper bounds on selectivity that are less than the stoichiometric selectivity limit, and thus provide valuable process information which may otherwise be unknown. Although the Feinberg Decomposition given by the CFSTR Equivalence Principle may not be practical, the selectivity limits it yields provide a useful target or benchmark for process design and operation. Furthermore, an optimization of the Feinberg Decomposition can provide practical hints for improved process designs. Finally, by using the CFSTR Equivalence Principle to compare synthesis routes, we can gain insight into which reaction path amongst a number of options is superior. In conclusion, the CFSTR Equivalence Principle is a useful engineering tool that can be applied to obtain important and practical information about a chemistry of interest.

[1] Horn, FJM. Attainable and Non-attainable Regions in Chemical Reaction Technique. In: *Chemical Reaction Engineering (Proceedings of the Third European Symposium).* 1964.

[2] Glasser, D; Hildebrandt, D; Crowe, CA. *Ind. Eng. Chem. Res.* **1987**, 46, 1803-1810.

[3] Hildebrandt, D; Glasser, D. *Chem. Eng. Sci.* **1990**, 45(8), 2161-2168.

[4] Hildebrandt, D; Glasser, D; Crowe, CM. *Ind. Eng. Chem. Res.* **1990**, 29(1), 49-58.

[5] Feinberg, M; Hildebrandt, D. *Chem. Eng. Sci.* **1997**, 52(10), 1637-1665.

[6] Feinberg, M. *Chem. Eng. Sci.* **2000**, 55(13), 2455-2479.

[7] Feinberg, M. *Chem. Eng. Sci.* **2000**, 55(17), 3553-3565.

[8] Abraham, T; Feinberg, M. *Ind. Eng. Chem. Res.* **2004**, 43(2), 449-457.

[9] Nisoli, A. Malone, M; Doherty, M. *AIChE Journal.* **1997**, 43(2), 374-387.

[10] Feinberg, M.; Ellison, P. *Ind. Eng. Chem. Res.* **2001**, *40*, 3181-2194.

[11] Tang, Y.; Feinberg, M. *Ind. Eng. Chem. Res.* **2007**, *46*, 5624-5630.

[12] Frumkin, J. Doherty, M. *Submitted to AIChE Journal.* **2016**.

494199 Pattern Recognition in Chemical Process Flowsheets

Abstract:

Process design is a nontrivial and complex task that has motivated the development of many systematic search and optimization techniques over the past four decades. As a result, a practical designed flowsheet will follow certain design patterns. In this work, we are interested in the development of a methodology for discovering patterns in existing chemical process flowsheets. There are two primary motivations for this work. First, a systematic comparison of a large number of successful chemical processes may help us identify design rules and operational characteristics that contribute to the success of these processes. Second, comparing the flowsheet of a potentially new process against flowsheets of existing processes may help assess the costs associated with the detailed design of the new process. In particular, potential new designs with a high degree of similarity with existing processes are likely to be easier to analyze as well as less expensive to build. A tool capable of performing such comparisons can be used to screen potential new projects and optimize the selection of product portfolios.

There is no prior literature on tools and methodologies for mining process flowsheets. To close this gap, we have developed a systematic methodology for discovering process patterns in existing chemical process flowsheets. The proposed methodology proceeds in three major steps to: (1) generate graph representations to denote general process flowsheets; (2) convert process flowsheets to string notations [1]; and (3) identify process patterns with sequence alignment algorithms [2, 3]. This three-step methodology has polynomial complexity at each step and can be applied automatically to compare general process flowsheets. We demonstrate the effectiveness of the proposed method through several case studies. By comparing flowsheets that produce the same chemical product, we assess the similarity of process flowsheets and discover structure features that contribute to successful processes. While comparing flowsheets that produce different chemical products, we identified many general structural patterns that reflect engineering practice or heuristic rules in literature [4].

References cited

[1] Tula, A. K., M. R. Eden, and R. Gani. Process synthesis, design and analysis using a process-group contribution method. Computers & Chemical Engineering 81 (2015): 245-259.

[2] Smith, T. F., and M. S. Waterman. Identification of common molecular subsequences. Journal of molecular biology 147.1 (1981): 195-197.

[3] Needleman, S. B., and C. D. Wunsch. A general method applicable to the search for similarities in the amino acid sequence of two proteins. Journal of molecular biology 48.3 (1970): 443-453.

[4] Biegler, L. T., I. E. Grossmann, and A. W. Westerberg. Systematic methods for chemical process design. Prentice Hall (1997).

494760 Robust Process Flowsheeting through Nonsmooth Models and Generalized Derivatives

Abstract:

The use of analytical derivatives in process simulation and optimization routines is well known to be beneficial for achieving rapid convergence and high accuracy (e.g. [1]). These advantages can be directly extended to flowsheets containing nondifferentiable process models through the use of generalized derivatives and implicit function sensitivity results for nonsmooth functions [2, 3]. Importantly, this allows process simulation and optimization formulations to be augmented with nonsmooth algorithms for non-ideal vapor-liquid equilibrium ("flash") calculations for greater reliability when the phase regimes at the results of these calculations are not known or fixed *a priori* [4]. This significantly improves the robustness of flash subroutines underlying many key process units, e.g. flash drums, throttle valves, compressors and turbines, without increasing the size or complexity of the problem compared to the case of using classical flash formulations. Furthermore, process models for inherently nonsmooth unit operations such as multistream heat exchangers may be seamlessly integrated into process flowsheets, so long as sensitivities of these models' outputs in the form of generalized derivatives with respect to their inputs are computed correctly and communicated to the simulation or optimization algorithm. These techniques may be used in sequential-modular formulations, as well as hybrid methods in which only the most difficult to converge submodels are treated in a modular manner, while the remaining flowsheet equations are handled simultaneously as in an equation-oriented framework. This new nonsmooth flowsheeting strategy is capable of solving process simulation and optimization problems more reliably and efficiently than the algorithms implemented in existing software, and, in some cases, even allows for the solution of problems that are beyond the capabilities of classical approaches. As examples of the latter, it will be shown that the nonsmooth approach is particularly well-suited for highly accurate simulation and optimization of cryogenic liquefaction processes, in which many or all of the aforementioned nonsmooth modeling elements are present simultaneously in combination with non-ideal thermodynamic behavior and complex heat transfer considerations.

[1] D. Wolbert, X. Joulia, B. Koehret, L. T. Biegler, 1994. Flowsheet optimization and optimal sensitivity analysis using analytical derivatives. Computers & Chemical Engineering, 18(11/12): 1083–1095.

[2] K. A. Khan, P. I. Barton, 2015. A vector forward mode of automatic differentiation for generalized derivative evaluation. Optimization Methods and Software. 30(6):1185-1212.

[3] K. A. Khan, P. I. Barton, In Press. Generalized Derivatives for Hybrid Systems. IEEE Transactions on Automatic Control.

[4] H. A. J. Watson, M. Vikse, T. Gundersen, P. I. Barton, 2017. Reliable Flash Calculations: Part 1. Nonsmooth Inside-Out Algorithms. Industrial & Engineering Chemistry Research, 56(4): 960-973.

492387 A Tightly Constrained MINLP-Based Formulation for the Identification of Energy Efficient Distillation Configurations

Abstract:

Distillation is one of the widely used unit operation to carryout multicomponent separation. It is well known that the number of configurations available to perform the separation increases to millions with increase in number of components in the feed. Identifying the optimal solution from the search space is challenging as the governing equations, Underwood constraints, are nonconvex. Moreover, a sub-optimal solution often results in a severe energy penalty[1].

We propose a novel Mixed-Integer Nonlinear Program (MINLP) based formulation to identify distillation configurations that are energy efficient and cost effective. We will present a novel formulation to describe the search space of basic configurations in a way that is simultaneously tighter and uses fewer binary variables than those available in the literature. The problem is formulated under the assumption of constant overflows and relative volatilities. We use ideas from Nallasivam *et al.*[2] and Caballero and Grossmann[3], and couple them with convexification techniques, such as the Reformulation-Linearization Technique (RLT), to construct tighter constraints for mass balance and Underwood equations. The formulation can be solved using standard global optimization solvers such as BARON[4].

The optimal solution to the MINLP may not be easily implementable when the underlying assumptions are relaxed. Hence, we present a method to identify a handful configurations, K, of the top solutions. Rigorous tray-by-tray calculations can then be performed only on these K configurations to determine the appropriate configuration. Towards the end, we present a few five and six-component cases that were solved to global optimality using our formulation.

[1]. Shah, Vishesh H., and Rakesh Agrawal. "A matrix method for multicomponent distillation sequences." *AIChE journal* 56.7 (2010): 1759-1775.

[2]. Nallasivam, Ulaganathan, *et al*. "Global optimization of multicomponent distillation configurations: 2. Enumeration based global minimization algorithm." *AIChE Journal* 62.6 (2016): 2071-2086.

[3]. Caballero, José A., and Ignacio E. Grossmann. "Structural considerations and modeling in the synthesis of heat-integrated– thermally coupled distillation sequences." *Industrial & Engineering Chemistry Research* 45.25 (2006): 8454-8474.

[4]. Tawarmalani, Mohit, and Nikolaos V. Sahinidis. "A polyhedral branch-and-cut approach to global optimization." *Mathematical Programming* 103.2 (2005): 225-249.

498020 A Superstructure-Based Model for Multistream Heat Exchanger Design within Flow Sheet Optimization

Abstract:

A multistream heat exchanger (MHEX) is a special heat-transfer equipment that allows multiple streams to exchange heat simultaneously. Its typical applications include air separation, natural gas (NG) processing, liquid hydrogen, petrochemicals, and liquefied natural gas (LNG).[1] Most processes employing MHEXs are capital- and energy intensive, and often involve phase changes of multicomponent mixtures. Consequently, modeling and optimization of such processes require efficient models for MHEXs that can ensure the feasibility of all heat exchanges, accommodate arbitrary phase changes, use phase-dependent nonlinear stream properties, and enable seamless and simultaneous optimization of the overall process. The design targets (e.g. flows and exchange areas) for MHEXs and operational (e.g. flows, temperatures, pressures, etc.) targets for the flowsheet are obtained from such optimization.[2]

While commercial process simulators have MHEX models, these models are generally proprietary, and do not inherently impose ensure minimum driving force for heat transfer.[3,4] Considering these problems, Kamath et al.[3] proposed a pinch-based equation-oriented formulation for MHEXs that ensures minimum driving force during flowsheet optimization. Their model is a special case of simultaneous optimization and heat integration of chemical processes without utilities. They presented a disjunctive formulation to handle phase changes, and approximated the non-differentiable max function, inherent in a pinch-based model, by a smoothing function. Watson et al.[4] revised the equation describing the pinch locator from Kamath et al.[3], and suggested a non-smooth solution strategy for the design and simulation of MHEXs. However, their study primarily focused on the simulation of MHEXs outside of an optimization framework. On the other hand, Pattison and Baldea[2] introduced a pseudo-transient framework for the equation-oriented modeling of MHEX avoiding Boolean variables. Their formulation involves ordinary differential and algebraic equations (DAEs), ensures feasible heat exchange, handles phase changes, and is suitable for flowsheet optimization applications. MHEXs can also be modeled as a network of two-stream exchangers as opposed to the composite-curve approach. Yee et al.[5]proposed a stage-wise superstructure for simultaneous heat integration along with area and energy targeting, and presented its application to model an MHEX. However, they assumed constant heat capacities, and did not account for phase changes. Furthermore, they assumed isothermal mixing and neglected pressure drops.

In this work, we propose an equation-based model for MHEX design that guarantees the feasibility of heat exchanges, and allows seamless and simultaneous overall process optimization. This model accommodates arbitrary and unknown phase changes of streams during the optimization process without the need to use any Boolean variables. Similar to Yee et al.[5], we treat the MHEX as a network of simple two-stream exchangers, but with a novel single-stage superstructure. Since we address overall process optimization and MHEX design, the state variables (flows, compositions, inlet/outlet temperatures, and inlet/outlet pressures) of MHEX streams may be unknown optimization variables. Our formulation is novel, uses exact physical properties from a process simulator, and offers three advantages for MHEX design. First, it provides heat transfer areas for each pair of heat-exchanging streams in the MHEX. Second, it allows the user to impose constant-phase two-stream heat exchanges, if desired. This enables better estimation of heat transfer coefficients (HTCs). Third, the use of precise HTCs and simulator-based physical properties result in accurate heat exchange areas. Lastly, our formulation uses a new form of the LMTD constraints to eliminate the numerical problems arising due to non-positive arguments in a logarithmic function.

We use the two most complex case studies in the literature[2-4] to illustrate the capabilities of our model to optimize processes employing MHEXs. The first study (Liquefied Energy Chain or LEC) is a transport chain to produce LNG from stranded natural gas using liquid inert nitrogen (LIN) and liquid carbon dioxide (LCO2). It was first suggested by Wechsung et al.[6], and subsequently simulated by Watson et al.[4]. The second study (PRICO®) is an industrial natural gas liquefaction process discussed by Kamath et al.[3] and Pattison and Baldea.[2] For the first case study, we consider more variables for optimization than the literature[6], minimize LIN flow and report a comparable solution (0.8979 kg/s vs 0.898 kg/s[6]). For the second study on the highly non-linear PRICO® process, we reduce the compressor power by 5.16% over the best-reported solution in the literature.[2]

References:

1. Hasan MMF, Karimi IA, Alfadala HE, Grootjans H. Operational modeling of multistream heat exchangers with phase changes. *AIChE Journal.* 2009;55(1):150-171.

2. Pattison RC, Baldea M. Multistream heat exchangers: Equation-oriented modeling and flowsheet optimization. *AIChE Journal.* 2015;61(6):1856-1866.

3. Kamath RS, Biegler LT, Grossmann IE. Modeling multistream heat exchangers with and without phase changes for simultaneous optimization and heat integration. *AIChE Journal.* 2012;58(1):190-204.

4. Watson HAJ, Khan KA, Barton PI. Multistream heat exchanger modeling and design. *AIChE Journal.* 2015;61(10):3390-3403.

5. Yee TF, Grossmann IE, Kravanja Z. Simultaneous optimization models for heat integrationâ€"I. Area and energy targeting and modeling of multi-stream exchangers. *Computers & Chemical Engineering.* 1990/10/01 1990;14(10):1151-1164.

6. Wechsung A, Aspelund A, Gundersen T, Barton PI. Synthesis of heat exchanger networks at subambient conditions with compression and expansion of process streams. *AIChE Journal.* 2011;57(8):2090-2108.

509054 Solve this! Fundamental Approach to Problem Solving in Industrial Processes I (Invited Talks)

Abstract:

Solving complex plant problems is an everyday exercise that affects all projects and budgets, and it is very rarely introduced to students at any level. In fact, by the time young engineers are few years into working in industry, most have lost the engineering skills that they developed in school. We hope to encourage future engineers to retain those skills and learn the importance of plant problem-solving.

497393 From Atom Groups to Molecules and Mixtures/Formulations: A Comprehensive Design Methodology with Generalized Disjunctive Programming

Abstract:

In the chemical industry, the design of mixtures and blends that meet a specified set of physical and chemical properties is an important and challenging activity across a wide range of applications. Several chemical examples with different functions and qualities, such as agrochemicals (pesticides, insecticides, etc.), functional chemicals (solvents, refrigerants, lubricants, etc.) and household products (cosmetics & personal care products, pharmaceuticals & drugs, etc.), come as formulations/mixtures and are used in most aspects of human life. Computer-Aided Molecular (CAMD) and Mixture/blend (CAM^{b}D) Design[1,2] is a promising approach for efficient and reliable design of promising molecules, mixtures or blends that meet predefined target properties and optimise a given performance measure. When formulating a mixture problem via optimisation, the main purpose is to identify innovative solutions, which can be a new molecular structure, an existing molecule for a new property function, or a new mixture of existing/new molecules. Mixture design, however, can be very challenging because it requires finding the optimal number, identities and compositions of mixture components, and using nonlinear property models, which results in complex nonlinear problems with a large combinatorial space. Thus, a reduced version of the general CAM^{b}D problem is usually posed: the number of mixture ingredients is fixed *a priori* and the identity of a compound (or of all compounds) that can participate in a mixture is selected from a restricted list of candidate compounds, leading to suboptimal solutions.

In view of these challenges, our work focuses on developing a comprehensive methodology within the computer-aided mixture/blend design framework by combining mathematical modelling, optimisation and chemical engineering insights to formulate the general mixture problem. Within this approach, the optimal number of components in a mixture, their identities and compositions are determined simultaneously[3,4] and the desired molecules are designed from a large set of atom groups (UNIFAC groups). A logic-based methodology, Generalised Disjunctive Programming[5,6] (GDP), is used to express the general mixture problem within a mathematical framework and formulate the discrete choices inherent in the problem (i.e., how many components are designed, which specific molecules should be used - what atom groups are required). In order to exploit existing MINLP algorithms, Big-M approach[7] is employed to transform the disjunctive constraints into mixed-integer form.

The proposed framework has been applied successfully to two case studies where the design of solvent mixtures for separation processes is presented. The first case study involves the design of optimal solvent and antisolvent mixtures for cooling and drowning out crystallization, respectively. In the second case study, optimal solvent mixtures are determined to separate acetic acid from water in a single stage liquid extraction process. Finally, integer cuts are introduced to the general mixture formulations and a list of optimal solutions (i.e., list of mixtures with different number, identity and compositions of ingredients) is

obtained for each problem. Significant benefits can accrue by employing the general framework (number of mixture constituents not fixed a priori and molecules designed from functional groups) in mixture and product design: avoid evaluating explicitly every choice of the number of components, which can be computationally or experimentally costly and time-consuming, especially as the number of desirable ingredients increases; consider larger design spaces where many molecules and mixtures are designed and not selected from a limited set of choices.

1. Achenie LEK, Gani R, Venkatasubramanian V. *Computer aided molecular design: theory and practice.* Amsterdam, The Netherlands: Elsevier. 2003.
2. Gani R. Chemical product design: challenges and opportunities. *Computers & Chemical Engineering.* 2004;28:2441–2457.
3. Jonuzaj S, Akula PT, Kleniati PM, Adjiman CS. The formulation of optimal mixtures with Generalized Disjunctive Programming: A solvent design case study. *AIChE Journal.* 2016;62:1616–1633.
4. Jonuzaj S, Adjiman CS. Designing optimal mixtures using Generalized Disjunctive Programming: Hull relaxations. *Chemical Engineering Science.* 2017;159:106–130.
5. Balas E. Disjunctive Programming. In: *Discrete Optimization II*, edited by Hammer PLK, Johnson EL, Korte BH, vol. 6 of *Annals of Discrete Mathematics*, chap. 6, pp. 3–51. Amsterdam, The Netherlands: Elsevier. 1979.
6. Raman R, Grossmann IE. Modelling and computational techniques for logic based integer programming. *Computers & Chemical Engineering.* 1994;18:563–578.
7. Nemhauser GL, Wolsey LA. *Integer and combinatorial optimization*, vol. 41. New York: Wiley & Sons, Inc. 1999.

497375 A Novel Computer-Aided Model-Based Tool for Chemical Product Design

Abstract:

In chemical product design, the application of model-based methodologies can help to design/improve products so as to reach the market faster by reducing costly and time-consuming experiments at the early stages of design [1]. That is, experiments are only performed during the last stage as a verification step. Since product candidates involve thousands of mixtures/blends of chemicals that need to be evaluated, a huge amount of data on physico-chemical properties of chemicals and their mixtures and/or models that can reliably predict them are needed for tailor-made design of products. This is a challenging task requiring data acquisition, data testing, model development, model-based design method development, etc, that needs to be integrated within a computer-aided framework so that chemicals based products can be designed, analyzed and verified in a fast, efficient and systematic manner [2].

In this work, the development of a systematic model-based framework for product design and evaluation, implemented in a new product design simulator called ProCAPD is presented. The chemical product simulator works in the same way as a chemical process simulator, but has additional features. Like a process simulator, the chemical product simulator is able to perform calculations on the specified product. Instead of the process flowsheet, the product molecular structure (in the case of a single molecule) or a mixture of molecules is given as input. Analogous to the process conditions, the product application conditions are specified and similar to the mass and energy balance calculations in the process simulator, the properties-functions of the product are calculated. Like the contents of the process simulator, the product simulator needs a database of chemicals and properties, a library of models, numerical routines to solve mathematical problems as well as various calculation options. Also, like the process simulator, the product simulator comes with a user-interface to describe the problems and to obtain the simulation results.

In order to make the chemical product simulator versatile with a wide range of applications, it includes a suite of a built-in knowledge base that has a suit of databases containing properties of different classes of chemicals (lipids, solvents, aroma, polymers, cosmetics, etc.) and blended products (gasoline, jet-fuels, lubricants, hair spray, etc.). It can calculate 55 pure component properties (such as critical properties, acentric factor and solubility parameter) and 10 functional properties (such as vapor pressure, liquid viscosity, and surface tension) and includes databases covering more than 24,000 compounds. Also, it has models for phase equilibria predictions such as UNIQUAC, UNIFAC, NRTL and PC-SAFT. It has special model parameters and data for lipids and ionic liquids. Transport related properties and new interaction parameters for a UNIFAC model for lipids systems have also been incorporated. These features together allow the design of a very wide range chemical-based products. The large amount of data, models and calculation-algorithms are managed through a specially developed ontology. All these capabilities are based on the prototype tool developed by Kalakul et al. (2017) [3].

The presentation will highlight the software architecture, the implemented computer aided methods-tools and the scope-significance will be illustrated through new chemical product design-evaluation applications.

References

[1] M. Hill, 2009. Chemical Product Engineering â€“ The third paradigm, Computer & Chemical Engineering, 33(5), 947-953.

[2] R. Gani, 2004, Chemical Product Design: Challenges & Opportunities, Computers and Chemical Engineering, 28, 2441-2457.

[3] S. Kalakul, S. Cignitti, L. Zhang, R. Gani, 2017, Chapter 3 â€“ VPPD-Lab: The Chemical Product Simulator, Tools For Chemical Product Design, Computer Aided Chemical Engineering, 39, 61-94

501375 Antibiotics Molecular Design Using Artificial Bee Colony Optimization

Abstract:

Antibiotics Molecular Design Using Artificial Bee Colony Optimization

Authors: Shweta Mapari, Matthew Hartenstein, Rex Gaumer, Kyle Camarda

In a recently published WHO report, a list of bacteria which pose the greatest threat to human health was given. The ESKAPE list also defines pathogens which have become resistant to multiple drugs and for which there exists a critical need for alternate therapies. Acinetobacter *baumannii,* which is on both of these lists, is a pathogen causing infections such as pneumonia and bacteremia. The goal of this research is to apply Computer Aided Molecular Design (CAMD) techniques to develop a novel antibiotic agent to treat this bacterial infection.

In order to treat A. *baumannii*, two molecular design problems were formulated, one for a beta lactam antibiotic and the other for a fluoroquinolone compound. CAMD is used to find analogous structures of small molecules which have reduced values of the Minimum Inhibitory Concentration (MIC) and toxicity for beta lactam and flouroquinolone drugs. We used piperacillin, amoxicillin, and ticarcillin as starting points for the beta lactam drugs and paufloxacin as a starting point for the fluoroquinolone drugs.

This work applies topological molecular descriptors to develop Quantitative Structure Activity Relationships (QSAR) which are employed to predict MIC values and toxicity. These expressions are combined with structural feasibility constraints to formulate an optimization problem. Because of the large number of potential chemical structures and the uncertainty in the structure-property correlations, stochastic algorithms are preferred to solve the resulting MINLP. One stochastic method which has shown promise to solve these problems is the Artificial Bee Colony algorithm, which relies on principles of swarm intelligence to find near-optimal solutions efficiently. The Artificial Bee Colony algorithm described in this paper is used to derive solutions which serve as lead compounds for a narrowed search for novel antibiotics, which would employ molecular simulation and experimental validation.

500584 An Integrated Methodology for Chemical Product Design: Application to Hair-and Skin-Care Emulsions

Abstract:

Due to the large number of ingredients commercially available for the formulation of consumer products, and because their multi-functionality, emulsified cosmetic product design is a challenging task. The traditional design approach often relies on costly and time consuming trial-and-error experimental procedures. In recent years, systematic approaches to reduce time and resources using computer-aided mixture/blend design (CAM^bD) have been proposed [1-2]. Based on predictive models for product properties and mixed-integer optimization, these design methodologies are able to explore a large domain of possibilities. They are however of limited applicability to few formulated products due to the lack of reliable models for properties prediction or for synergic effects.

In particular industrial sectors such as the cosmetics one, there is a considerable amount of heuristic knowledge, namely regarding qualitative function of ingredients, their incompatibilities and positive synergies, as well as their impact on sensorial attributes, which is a critical aspect for product acceptance [3]. This heuristic knowledge has been recently incorporated into a systematic CAM^bD methodology [4] using mixed-integer optimization with logical constraints [5] and in this work is further developed for cosmetic emulsions. The method considers a list of available ingredients classified according to their main function (e.g., emollients, emulsifiers, rheology modifiers). Then, heuristics regarding ingredientsâ€™ selection, their concentrations in final product, and the manufacturing procedures are formulated as logical conditions, and further translated into algebraic constraints. These constraints, together with other known restrictions (technical, toxicological and/or legal, etc.), define a reduced design space. The search in this space is guided by property models relating product composition to key physicochemical properties or sensorial attributes. The multi-objective function is composed of a weighted sum of Taguchi functions with a target value for each cost/performance requirement (e.g. softness, greasiness and thickness of the product, etc.), which could be adapted by the designer. Through integer cuts, the entire set of feasible solutions within the reduced design space or a subset of interest may be generated.

To illustrate the proposed methodology, two case studies of a rinse-off conditioner and a body lotion are presented. From an initial list of 24 emollients, 4 rheology modifiers and 8 emulsifiers, a set of alternative formulations and respective manufacturing procedures is listed. The first 5% of these resu-lting alternative formulations were produced and then analyzed using instrumental and sensorial methods. The predicted values of the performance were similar for most of the formulations, confirming the potential of this methodology to model emulsified products.

REFERENCES

[1] L. E. K. Achenie, R. Gani, V. Venkatasubramanian, Eds., *Computer Aided Molecular Design: Theory and Practice*. Amsterdam: Elsevier, 2002.

[2] M. Mattei, G. M. Kontogeorgis, R. Gani, «A Systematic Methodology for Design of Emulsion Based Chemical Products», In: Karimi IA, Srinivasan R, eds., *Computer-Aided Chemical Engineering*, vol 31. Amsterdam: Elsevier, 2012, pp. 220-224.

[3] A. M. Pensé-Lhéritier, «Recent developments in the sensorial assessment of cosmetic products: A review», *Int. J. Cosmet. Sci.*, 2015, vol. 37, pp. 465-473.

[4] J. Arrieta-Escobar, F. Bernardo, A. Orjuela, M. Camargo, L. Morel, « An Integrated Methodology for Emulsified Cosmetic Product Formulation Using Integer Programming with Logical Constraints», (Provisionally accepted paper for ESCAPE 27)

[5] H. P. Williams, *Model building in mathematical programming*, 5th ed. John Wiley & Sons, 2013.

491188 Product Design: A Systematic Pricing Framework for Maximizing Company Profits

Abstract:

The importance of chemical product design has been well-recognized in the chemical engineering community. Many articles have been published to consider various aspects in product design, including material selection, product conceptualization, and process design to meet the desired product attributes. However, pricing strategy and pricing decisions receive little attention in the chemical engineering community. This is a serious omission as product pricing is important for a company to set the optimal price for getting a reasonable return and at the same time has a good market penetration.

A systematic pricing framework has been developed for pricing a chemical product. First, a pricing strategy is selected based on product characteristics, company strategy, and market situation. Market information and company data are collected next. The collected data are used to predict the parameters in a pricing model. For example, advertising data are used to estimate the product awareness in the market, whereas the consumer preference data between our product and the competing product is used to estimate whether the consumer will purchase our product or the competing product. With all these information, a relation between product price and product demand can be obtained from a pricing model. Finally, the optimal product price is obtained by maximizing the companyâ€™s profit. This framework is illustrated with an energy drinks example. With this framework, product design can be optimized not only from a technical perspective, but also from a consumer perspective and a business perspective.

487526 A Minimalist Model for Rapid Simulation Enabling Optimization of the Uniformity of Multiple Simultaneous Hydraulic Fracture Growth

Abstract:

Hydraulic fracturing is vital for stimulating production of oil and gas, especially from low permeability reservoirs. There has been an increasing drive to optimize hydraulic fracturing treatments. One of the senses in which there is a growing desire for optimization is to maximize the uniformity of the stimulation obtained when attempting to simultaneously generate multiple hydraulic fractures growing from several entry points distributed along a horizontal. Simulations show that maximizing uniformity can be equivalent to maximizing the surface area generated by stimulation which, in turn, is expected to maximize recovery. Hence, a typical desire is to choose the controllable parameters (spacing between fractures, fluid viscosity and pumping rate, etc.) to generate a maximally-uniform hydraulic fracture growth pattern. However, a pre-requisite to optimization is a forward model that both captures the relevant physical behaviors and that executes rapidly enough to be called upon hundreds or thousands of times by an optimization scheme. Fully coupled models can take days or even weeks for a single simulation, hence optimization is often impractical or impossible with current numerical models.

Motivated, then, by the need for rapidly computing models, we present a minimalist approach for modeling simultaneous growth of multiple, radially-symmetric hydraulic fractures. Our approach leads to a drastic reduction in simulation time wherein simulations that take 1 week to solve with a fully coupled simulator can be simulated in seconds to minutes. This reduced model nonetheless maintains sufficient fidelity to track even complex dynamics of interacting hydraulic fracture growth that are predicted by a fully coupled simulator that is used for benchmarking purposes. The first key to the minimalist model is to avoid the need to mesh the hydraulic fractures or the domain around them through a judicious choice of functional approximation of the internal fluid pressure. The second key is to make use of a global energy balance relationship to enforce partitioning of the fluid among the multiple fractures in a manner compatible with energy minimization principles. The simulator is then used to demonstrate how optimization can substantially increase the predicted uniformity and generated surface area associated with stimulation of a given interval of a horizontal well at different stage lengths, entry point locations, injection rates, and fracturing times.

491223 Development of a Multi-Objective Optimization-Tool for Simulation-Based Chemical Process Synthesis and Design Tasks

Abstract:

For the synthesis and design of chemical processes numerous methods have been developed. Generally speaking, they can be divided into two main approaches. The first and traditional approach follows the so-called â€œonion logicâ€. In this approach first the reactor followed by the separation system etc. are designed. Due to this procedure, every decision is based on the information available at the particular stage of the process design and therefore on an incomplete picture [1]. As a consequence, a series of best local decisions are made [1]. These processes satisfy the chemical and technical requirements very well, but no guarantee on global optimality with respect to for instance its cost can be given. Moreover, the ability to look ahead to the completed design might have led to different decisions [1]. The second approach overcomes this disadvantage by creating a so-called superstructure. Superstructures contain every feasible variant of the process structure. In a simulation-based design approach, the process is modeled in one of the various flowsheet-simulators available on the market. Subsequently, the design problem is formulated as a mathematical optimization problem. A set of objective functions, e.g. operating and investment costs, as well as constraints, e.g. the chemical purity of the product streams, are chosen. In addition, decision variables are defined. In most cases, they are operative or constructive parameters of the process. The completed optimization problem is then solved by a solution algorithm. Yet the suitability of different solution algorithms depends highly on the optimization problem. In order to provide a universally applicable method to process optimization, an innovative tool, the Advanced Process Optimizer (AdvPO), was developed and intensively tested at the Institute of Process and Plant Engineering.

In this presentation the AdvPO as well as selected optimization methods will be introduced. The AdvPO is capable of performing single- or multi-objective structural and parameter (SOSP-/MOSP-) optimizations automatically. As a result, constructive and operative parameters are optimized as well as the process structure itself. This was realized by a bidirectional interface between the AdvPO and various flowsheet simulators, for example Aspen Plus. Due to the modular concept of the AdvPO basically every process can be optimized. The solution algorithm is a specially tailored evolutionary algorithm. The basic structure and functionality of the AdvPO will is part of the presentation. In case of a MOSP-optimization the obtained Pareto-optimal solutions, visualized as the Pareto-front, represent the best trade-offs between the multiple and often opposed objectives. As the optimization problem is highly complex, different innovative concepts and methods had to be developed and extensively tested. One example for such a method is the Two-Step-Optimization which will be part of the presentation as well. The Two-Step-Optimization combines SOSP and MOSP approaches to enlarge the obtained Pareto-Front. As a result, the decision maker can choose his personal optimal design on the basis of a complete picture of the best trade-offs. The exemplary optimization problem deals with a dimethyl ether process. The objective functions are the investment and operation costs of the process. The superstructure is composed of three different reactor types and two different separation systems. The optimization results will be compared with reference solutions achieved by conventional approaches [2]. The results show a significant reduction of both the investment and operating costs compared to the reference solutions.

References

[1] R. Smith, *Chemical process design and integration*, Wiley, Chichester, West Sussex, England, Hoboken, NJ **2005**.

[2] W.L. Luyben, *Principles and case studies of simultaneous design*, Wiley, Hoboken, N.J. **2011**.

Radial Basis Function Interpolation Using Quenched Periodic and Fourier Extensions

Rafael G. Henriquez and Ludwig C. Nitsche

University of Illinois at Chicago

rafaelgh@uic.edu

Interpolation

Interpolation of regular data

- Digital image processing
- Particle computer simulations

Goal

- Interpolate a divergence-free function on a finite domain

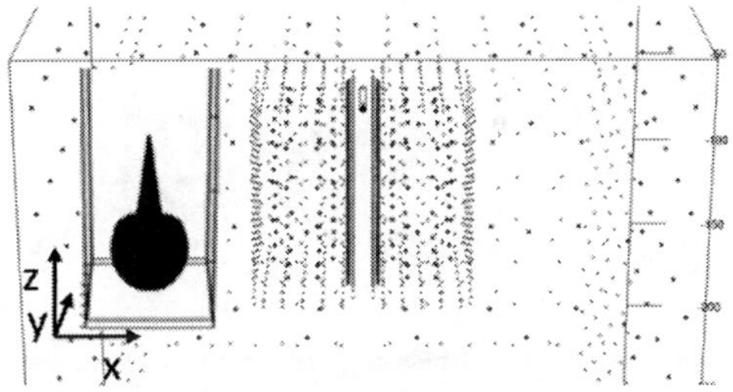

Data for interpolation

- Map particle forces onto grid
- Fast Stokes solver on grid
- Interpolate grid velocities back to particles

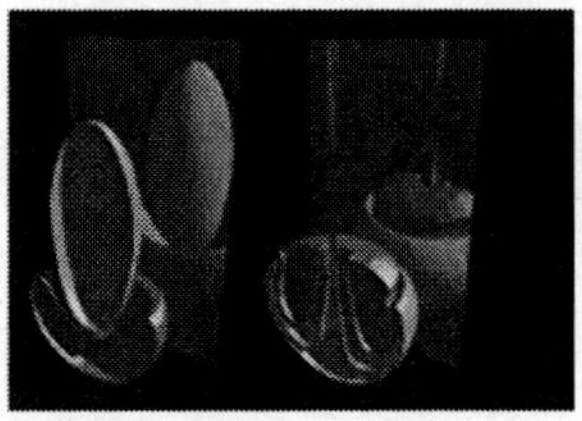

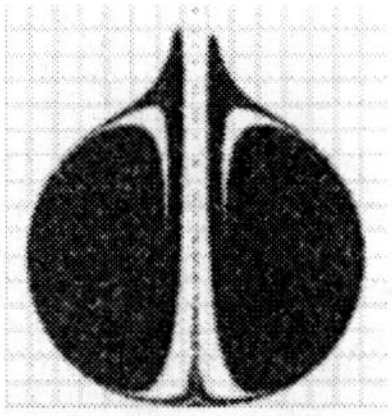

Radial basis functions (RBF)

RBF $\quad f_{\text{interp}}(\mathbf{x}) = \sum_{i=1}^{n} c_i \, \phi(||\mathbf{x} - \mathbf{x}_i||)$

Solving PDEs

Additional constraints (*e.g.* divergence free)

Interpolation using RBF

Solve sparse linear system for RBF coefficients

Linear equations are typically ill conditioned and difficult to solve by ordinary methods

If periodically continued, convolution theorem and FFT

If not periodic (finite domain), what to do?

Scalar RBF

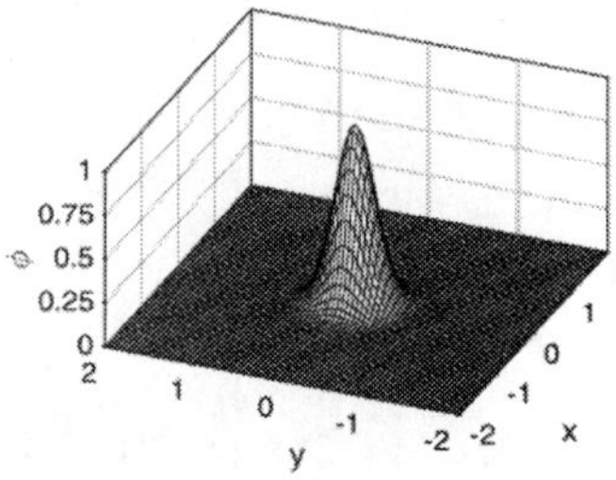

Divergence-free RBF

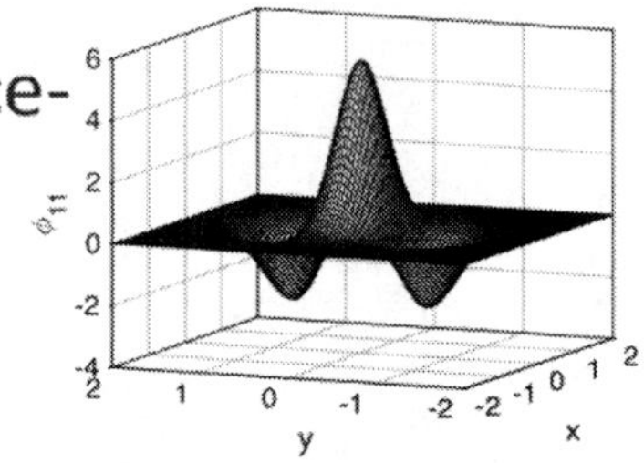

Periodically continued function

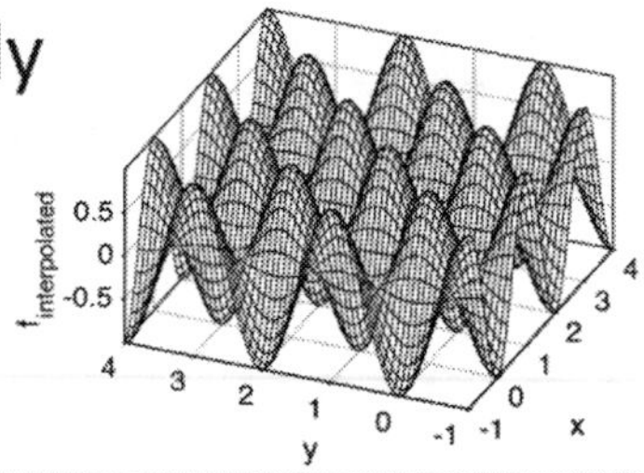

Quenched periodic and Fourier extensions

Original function

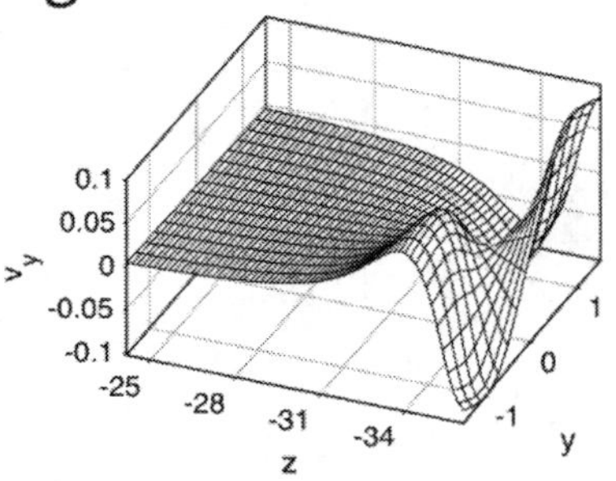

Quenched periodic extension

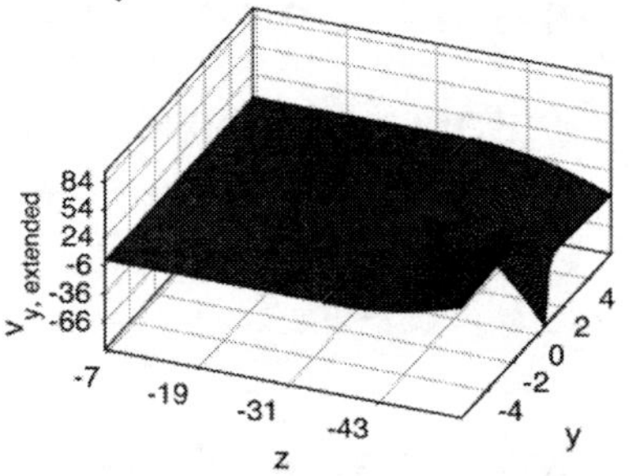

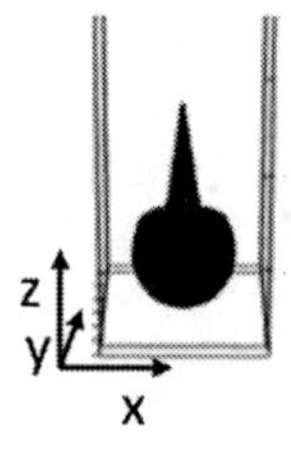

Fourier extension

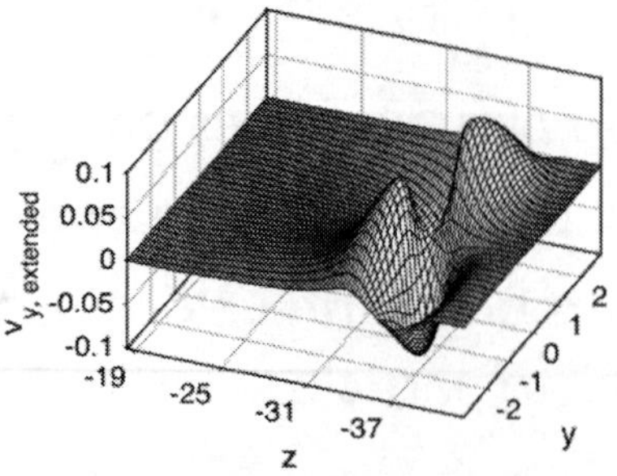

Interpolated function

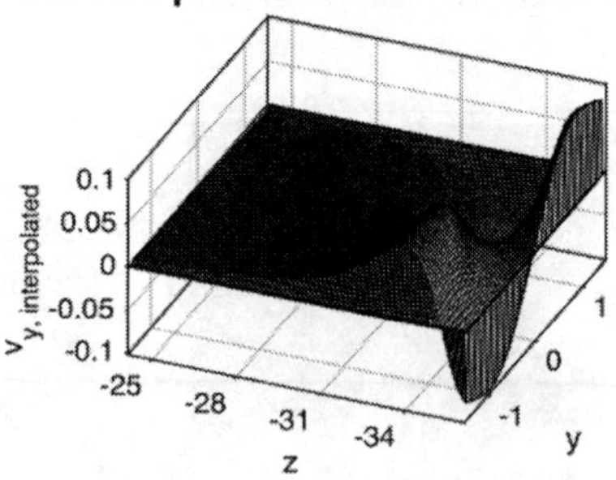

A Two-Phase Imbibition-Drainage Model for Soils Amended with Biochars

Yi Chen and Kyriacos Zygourakis
Department of Chemical and Biomolecular Engineering
Rice University
YC47@rice.edu

Wetting and drying dynamics of soils amended with biochar

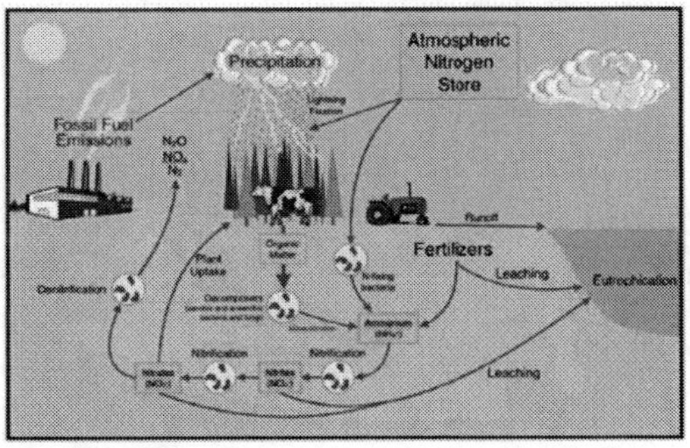

Biochar addition slows down N transport in top soil

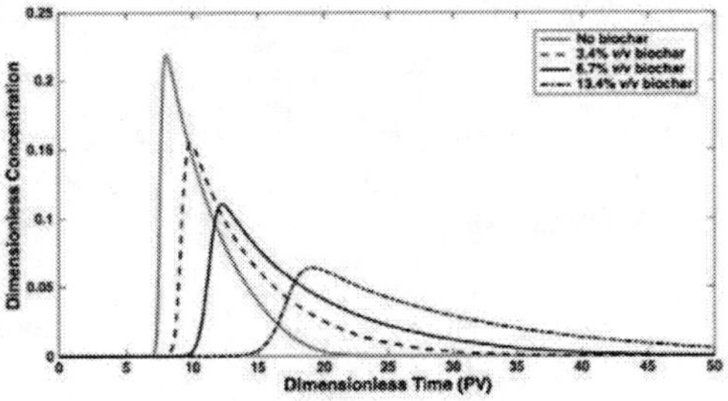

From H. Sun et al., *Ind. Eng. Chem. Research*, **54** (2015)

Key questions:

- How fast will biochar-amended soil saturate when it starts raining?
- How fast will it dry when it stops raining?
- Does biochar affect the hydraulic conductivity of soil or its water retention capacity?

Two-phase model for structured soils amended with biochar

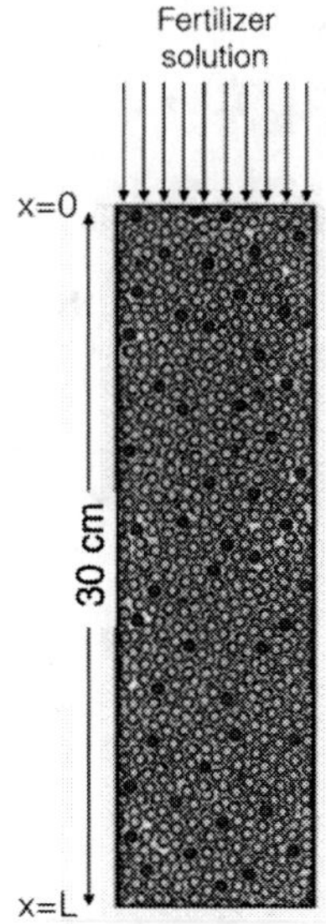

Structured soil with 3 pore populations:

#1: Interstitial voids

#2: Intra-pores in soil aggregates

#3: Intra-pores in biochar particles

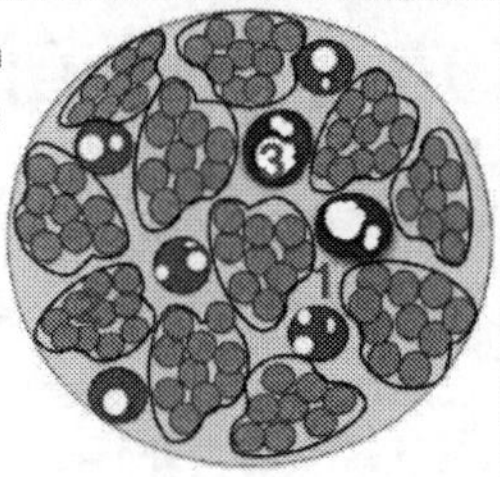

Procedure:

- Measure water retention curve experimentally
- Fit data with a linear combo of lognormal CDFs

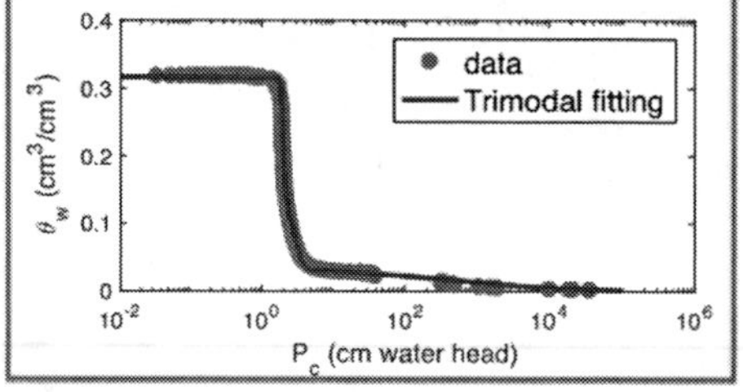

Two-phase model:

Assumptions:

- Local equilibrium
- Constant densities and viscosities
- Negligible gravity effects

$$\varepsilon_{\text{tot}} \frac{\partial S_{\beta,\text{tot}}}{\partial t} = \frac{\partial}{\partial x}\left[\frac{k_\beta k_{r,\beta}}{\mu_\beta}\left(\frac{\partial P_\beta}{\partial x}\right)\right]$$

$$\varepsilon_{\text{tot}} \frac{\partial S_{\gamma,\text{tot}}}{\partial t} = \frac{\partial}{\partial x}\left[\frac{k_\gamma k_{r,\gamma}}{\mu_\gamma}\left(\frac{\partial P_\gamma}{\partial x}\right)\right]$$

$$S_{\beta,\text{tot}} + S_{\gamma,\text{tot}} = 1$$

$$P_\beta - P_\gamma = P_C\left(S_{\beta,\text{tot}}\right)$$

Use IMPES algorithm to solve these PDEs subject to the appropriate initial and boundary conditions for the **wetting** and **drying** phases.

Biochar-amended soil gets saturated a bit more slowly, but holds more water

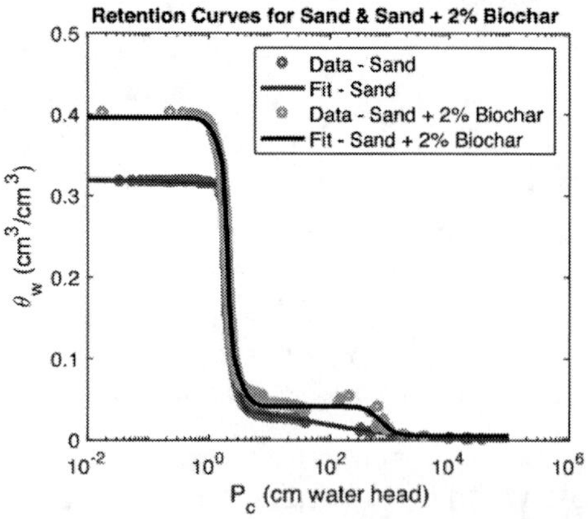

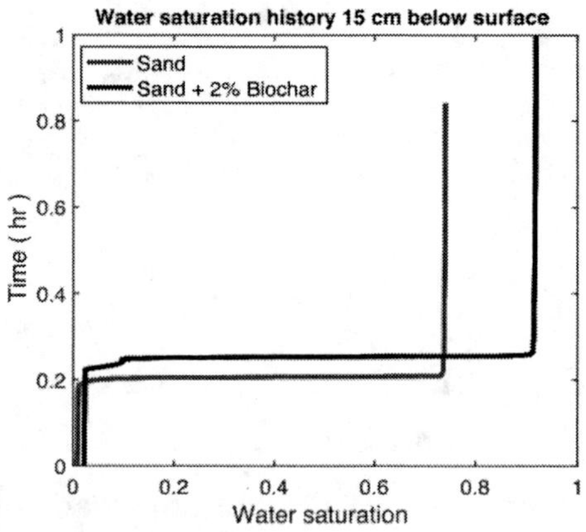

- Even 2% of biochar can increase the water holding capacity of soils.
- Soil hydraulic conductivity remains largely unaffected.
- Top soil can become fully saturated in about an hour.
- But, drying takes an order of magnitude longer.

493036 The Benefits of Resource Partitioning and Division of Labor in Biofilm Based Microbial Consortia

Abstract:

Microbial consortia are commonly observed in natural and synthetic systems, and these consortia frequently result in higher biomass production relative to monocultures. The focus here is on the impact of initial spatial localization and substrate diffusivity on the growth of a model microbial consortium consisting of two different producer strains that consumes glucose and produce acetate or lactate and a scavenger strain that consumes the acetate and lactate. The mathematical model is based on an individual cell model where growth is described by Monod kinetics, and substrate transport is described by a continuum-based, non-equilibrium reaction-diffusion model where convective transport is negligible (e.g., in a biofilm). The impact of the initial population density and substrate diffusivity is examined in different systems. Depending on the geometry of the system, transitions are observed from the highest initial density resulting in the greatest cell growth to cell growth being independent of initial density. A high initial density minimizes diffusive transport time and is typically expected to result in the highest growth, but this expected behavior is not predicted in environments with lower diffusivity or larger length scales. For example, when the producer cells are placed on the bottom of the domain with the scavenger cells above in a layered biofilm arrangement, a critical transition is observed from a thin, dense initial scavenger layer is optimal for cell growth at the highest diffusivities to a thicker, less dense initial scavenger layer providing maximal growth at lower diffusivities. The overall conclusion is that high density clustering of members of a food chain is optimal under most common transport conditions, but under some slow transport conditions, high density clustering may not be optimal for microbial growth.

497029 Non-Linear Behavior of Coupled Autocatalytic Reaction Systems

Abstract:

In an autocatalytic reaction one of the products catalyzes its own formation. In this work, we analyze a system consisting of two species, each participating in two autocatalytic reactions. Each species is also responsible for the autocatalytic generation of the other. Such reactions can be used to model systems which show interconversion between two social or religious groups or systems where there is interconversion between two isomers. We analyze how the kinetics of the reaction determines the non-linear behavior of the system such as multiple steady-states. Two cases are considered depending on the kinetics of the two reactions. In the first case both autocatalytic reactions are assumed to be quadratic in nature, while in the second case both autocatalytic reactions are assumed to be cubic. Singularity theory and bifurcation theory are employed to comprehensively understand the multiplicity features of the steady states and to classify the bifurcation behavior of this system. Regions in parameter space are identified in the system where a specific kind of bifurcation diagram is obtained. The analysis in this work is focused on determining operating conditions under which a pure enantiomer can be produced from a racemic mixture or one social group completely gets converted to another. The model can also be used to address evolution of species in an eco-system.

Theoretical Analysis and Process Design for Micromixer Cooling Crystallization

Mo Jiang, J. Carl Pirkle, Jr., Richard D. Braatz
mojiang@mit.edu

VCU
VIRGINIA COMMONWEALTH UNIVERSITY

Massachusetts Institute of Technology

Unexpected discovery of "cooling" micromixers

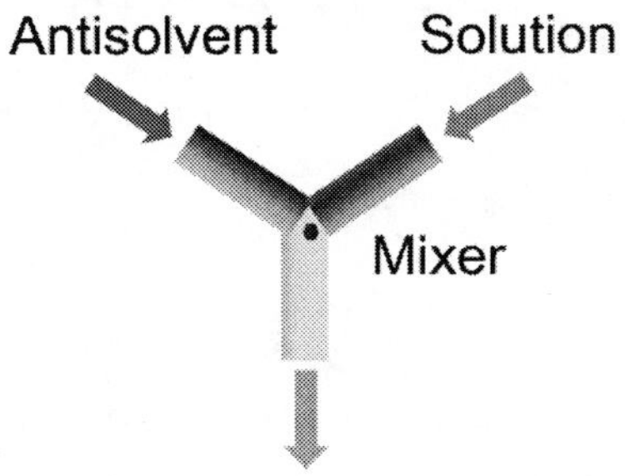

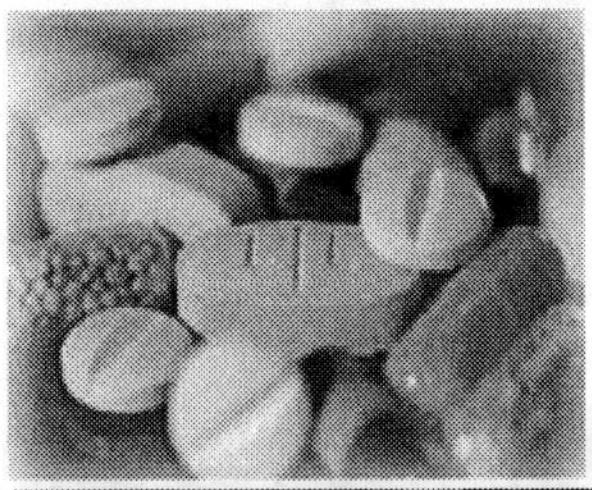

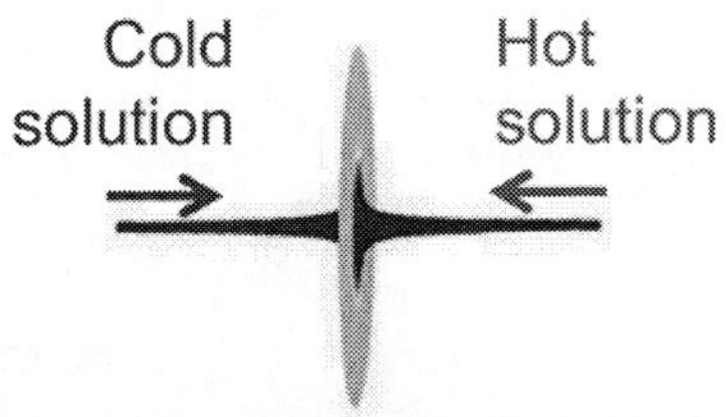

Dual-impinging-jet mixe

- Can continuously generat small crystals (remove mi
- Allow reaction or antisolv crystallizations

pharmasyte.co.uk

M. Jiang et al., *Chem. Eng. Sci.*, 77, 2–9, 2012.

Heat/mass transfer modeling for analytical solutions

Heat transfer: $v_r \frac{\partial T}{\partial r} + v_z \frac{\partial T}{\partial z} = \alpha \left[\frac{1}{r} \frac{\partial}{\partial r} \left(r \frac{\partial T}{\partial r} \right) + \frac{\partial^2 T}{\partial z^2} \right]$

Mass transfer: $v_r \frac{\partial c}{\partial r} + v_z \frac{\partial c}{\partial z} = D \left[\frac{1}{r} \frac{\partial}{\partial r} \left(r \frac{\partial c}{\partial r} \right) + \frac{\partial^2 c}{\partial z^2} \right]$

($v_r \frac{\partial T}{\partial r} \cong 0$, $\frac{1}{r}\frac{\partial}{\partial r}\left(r\frac{\partial T}{\partial r}\right) \cong 0$)

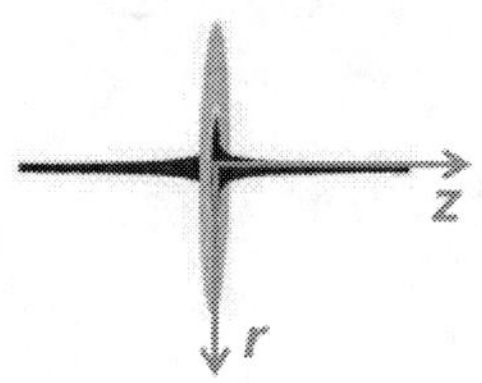

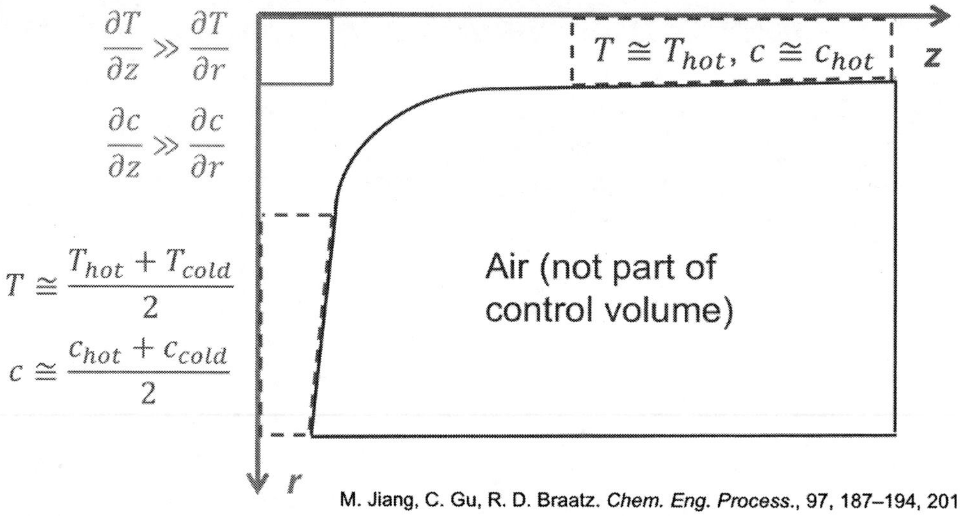

M. Jiang, C. Gu, R. D. Braatz. *Chem. Eng. Process.*, 97, 187–194, 2015.

High supersaturation can occur near the interface

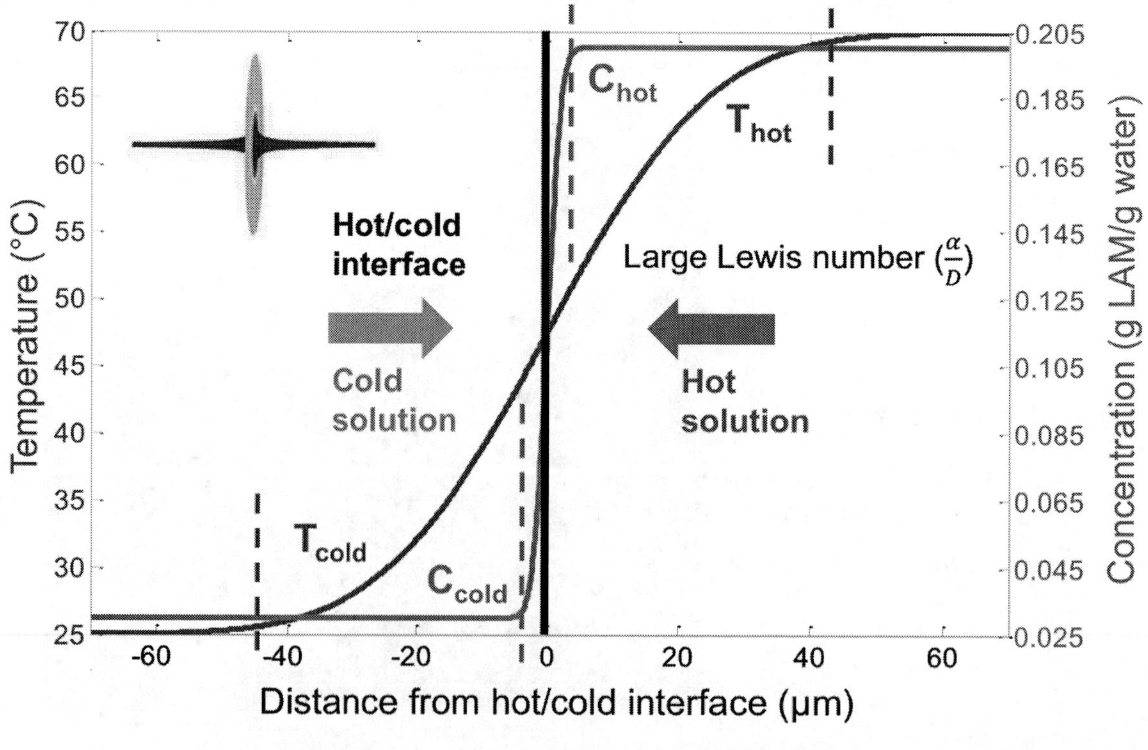

490050 Analytical Solution of the Period of Belousov-Zhabotinsky Reaction Using a Feedback Mechanism

Abstract:

Analytical solution of the period of Belousov-Zhabotinsky reaction using a feedback mechanism

Chi Zhai[a,b], Wei Sun[a], Ahmet Palazoglu[b*]

[a] Beijing Key Lab of Membrane Science and Technology, College of Chemical Engineering, Beijing University of Chemical Technology, 100029 Beijing, China

[b] Department of Chemical Engineering, University of California, Davis,

CA 95616, USA

Abstract

One of the most important models for the development of Prigogine¡¯s dissipative structure theory is the ¡°Bursselator¡±,[1] whose prototype is the *Belousov–Zhabotinsky*[2] (BZ) reaction. Since the BZ reaction is far-from-equilibrium, the dynamics of the system does not obey the Onsager reciprocal relations and symmetry breaking bifurcation may cause the system to generate self-organized patterns. Figure 1 depicts the generation of the self-oscillatory waveform observed in experiments.

Figure 1. The snapshot (and amplification) of the diffusive BZ reaction, which can generate waveform of color change from blue to red periodically.[3]

By the dissipative structure theory, the BZ reaction is viewed as an open system with constant negative entropy consumption, and a portion of the overall reaction entropy-change is consumed for the maintenance of the periodic color-change structure. Figure 2 is the schematic of the reaction kinetics, indicating a periodically dissipative system. From the viewpoint of a dynamic system, the nonlinearity of the intermediate terms (X, Y, Z) bring about a Hopf bifurcation where increasing one of the parameters beyond the critical point may cause a periodical color-change waveform to emerge.

It is clear that the period of the self-oscillatory pattern is relevant to the input parameter, i.e., the input entropy flux. Knowing how the period is related to the parameter change would be critically beneficial in identifying the characteristics of the self-oscillatory structure. Our goal in this study is to develop mathematical methods to compute the period of the limit cycles as a function of parametric changes.

Figure 2. The Oregonator[4] model kinetics and the abstracted structure. Species identification with respect to the FKN mechanism[5] of the BZ reaction are X=$HBrO^2$, Y=Br^-, Z=Ce(IV), A=BrO^3, B=Organic species, P=HOBr. The reactant and product species A, B and P are normally present in much higher concentrations than the dynamic intermediate species X, Y and Z and are assumed to be constant on the time scale of a few oscillations. The oscillatory exchange of the intermediates causes Z to vary between Ce(IV) and Ce(III) back and forth, and with the presence of the ferroin indicator, the media would change color between blue to red repeatedly.

Often one can use a numerical continuation method[6] or a shooting method[7] to obtain the period of the limit cycles as a function of changes in the parameters, but it would be much more effective to find analytical relationships as they would reveal the specific characteristics of the self-oscillatory system, i.e., the maintenance entropy flux of the self-oscillatory structure.

Figure 3. The closed-loop equivalence of the dynamic system and the criterion of self-oscillation on the Nyquist diagram.

Mees and Chua[8] have proposed a method to solve the period of the limit cycle analytically based on the frequency domain Hopf bifurcation theory. A self-oscillatory dynamic system can be reformulated as a feedback system as shown in Figure 3. Here, $G(s)$ in Figure 3 is the Laplace transform of the linear part and f is the memoryless nonlinear part. The generalized Nyquist criterion provides the necessary condition for the closed-loop system with $v = d = 0$ to generate oscillatory outputs, which is clear in the right hand side of Figure 3. The intersection point satisfies the following condition: *y* = *G*(*i*¦Ø)*u* ¡Ö *G*(*i*¦Ø)*N*(*A*)*y*, which leads to *G*(*i*¦Ø) = -1/*N*(*A*). Here, $N(A)$ is the amplitude correlated approximation of the nonlinear term f, and $N(A)$ is identified by the harmonic balance method. For example, if the system is approximated by the 2nd-order harmonics, $N(A)$ could be written as 1+*A*²¦*Î*(¦Ø). [9]

However, the BZ reaction is a multivariable system, and, when the method above is applied, tensor operations[8] will render the calculations tedious and impractical. Since the BZ reaction is a highly nonlinear system, higher order harmonics[10] may also be needed to approximate the oscillatory behavior accurately, which would complicate the calculation process further.

Figure 4. Block diagram for the numerical computation of functional expansions.

In the current study, we propose to use the Laplace-Borel (LB) transfer function representation to express the feedback system. The LB transform is an extension of the Laplace transform to the nonlinear polynomial terms by an infinite series of iterated integrals, and the transformed system obeys the shuffle algebraic operation. [11, 12] The functional expansion (FEx) method approximates the analytical solution of the system which can be presented graphically as in Figure 4. However, for a specific order of expansion $X_p(t)$, the approximated solution by the FEx method does not promise the solution to be a closed cycle as time approaches infinity.

Similar to Figure 3, we propose to close the system by adding dash lines to Figure 4, and set $u(t) = 0$. By utilizing the shuffle algebra, each variable of the nonlinear part f_i (i¡¯s expansion) could be decoupled if the order of the harmonics is specified, which is especially attractive for multivariable systems. The approximation procedure is progressive and the residuals of each expansion are used to identify the parameters of the harmonics. Because (1). the nonlinear blocks in Figure 4 may introduce higher order harmonics than the input ones; (2). the output orders of the harmonics is predictable if the structure of the system is given, then, the formula of $N(A)$ is known and the parameter of $N(A)$ can be identified by setting higher order residuals as zero. This method makes $N(A)$ being flexible for different systems and expendable to higher order approximations.

Keywords: dissipative structure; self-organization; Laplace-Borel transform; harmonic balance method; frequency domain Hopf bifurcation.

References:

1. Prigogine, I., 1978. Time, structure, and fluctuations. Science, 201(4358), 777-785.

2. Winfree, A. T., 1984. The prehistory of the Belousov-Zhabotinsky oscillator. J. Chem. Educ., 61(8), 661.

3. http://hopf.chem.brandeis.edu/anatol.htm

4. Field, R. J., Noyes, R. M., 1974. Oscillations in chemical systems. IV. Limit cycle behavior in a model of a real chemical reaction. The Journal of Chemical Physics, 60(5), 1877-1884.

5. Field, R. J., Koros, E., Noyes, R. M., 1972. Oscillations in chemical systems. II. Thorough analysis of temporal oscillation in the bromate-cerium-malonic acid system. Journal of the American Chemical Society, 94(25), 8649-8664.

6. Zhai, C., Palazoglu, A., Wang, S., et al. 2017. Strategies for the Analysis of Continuous Bio-ethanol Fermentation under Periodical Forcing. I&ECR, in press.

7. Doedel, E. J., 1981. AUTO: A program for the automatic bifurcation analysis of autonomous systems. Congr. Numer, 30, 265-284.

8. Mees, A., Chua, L., 1979. The Hopf bifurcation theorem and its applications to nonlinear oscillations in circuits and systems. IEEE Transactions on Circuits and Systems, 26(4), 235-254.

9. Allwright, D. J., 1977. Harmonic balance and the Hopf bifurcation. Mathematical Proceedings of the Cambridge Philosophical Society. Cambridge University Press, 82(03), 453-467.

10. Moiola, J., Chen, G., 1993. Computations of limit cycles via higher-order harmonic balance approximation. IEEE Transactions on Automatic Control, 38(5), 782-790.

11. Harris, K. R., Palazoglu, A., 1997. Studies on the analysis of nonlinear processes via functional expansions-II. Forced dynamic responses. Chem. Eng. Sci. 52, 3197-3207.

12. Batigun, A, Harris, K. R., Palazoglu, A., 1997. Studies on the analysis of nonlinear processes via functional expansions-I. Solution of nonlinear ODEs. Chem. Eng. Sci. 1997, 52(18): 3183-3195.

Physically based modeling for predictive simulation of a net-zero home

Alan Uy
Raymond A. Adomaitis

Institute for Systems Research
Chemical & Biomolecular Engineering
University of Maryland

Poster 189m
Poster location: 647

Overview

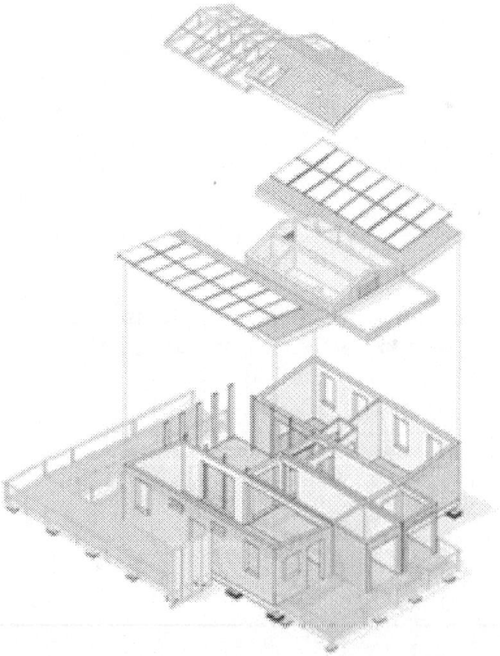

- Poster describes physically based modeling effort to support Team Maryland's entry to the 2017 Solar Decathlon

  Work was motivated by
  - the need to do model-based systems integration of the electrical/mechanical subsystems
  - provide predictive support of resource allocation decision making during operation

Model elements

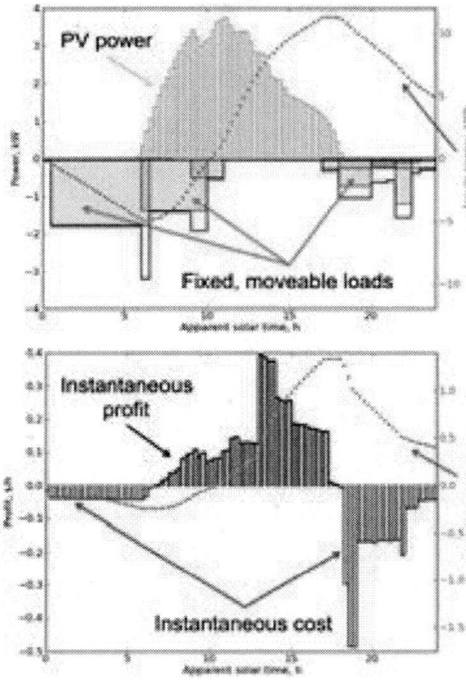

1. First-principles, time-resolved solar irradiance and projection calculations for PV array input and exterior wall heat transfer
2. PV array *I* versus *V* prediction with model parameters identified from manufacturer performance data
3. Scheduled events consuming/producing energy, water, or CO_2 coded in XML
4. Dynamic modeling of internal heat transfer, greywater balance; CO_2 balance, and more

Conclusion: net-zero success

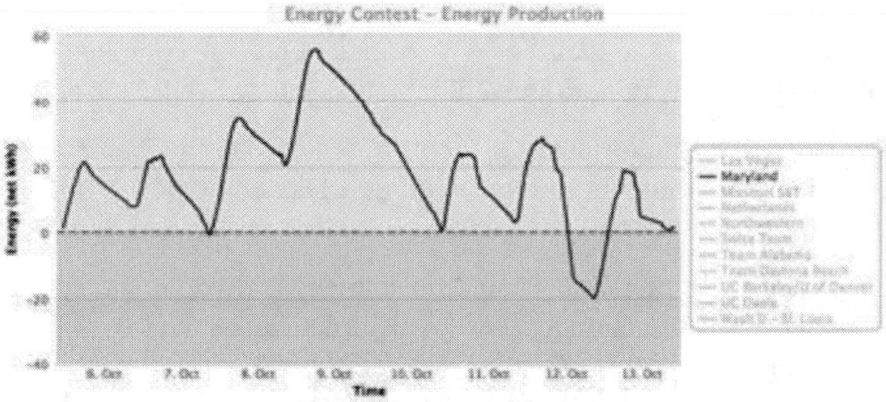

Final Scores

Rank	Team	Score
1.	Swiss Team	872.910
2.	Maryland	822.683
3.	UC Berkeley/U of Denver	807.875
4.	Missouri S&T	758.315
5.	Team Alabama	757.873
6.	Northwestern	750.758
7.	Netherlands	750.275

Comparison of Various Techniques for Solving Complex Chemical Equilibrium Problems

Mordechai Shacham
Ben-Gurion University
Beer-Sheva, Israel
shacham@bgu.ac.il

Neima Brauner
Tel-Aviv University
Tel-Aviv, Israel

Motivation

- Computation of the equilibrium composition in complex reactions, formulated as systems of non-linear algebraic equations (NLE) is considered.
- Such systems are notoriously difficult to solve - the mathematical model often contains discontinuities and/or regions where some of the functions are undefined.
- Consequently the iterations may not converge, diverge, or converge to physically infeasible complex or real solutions with some negative concentration values

Objectives

- Initial estimates - a crucial issue – a method based on atom balances and Gibbs energy values of the compounds is suggested to find initial estimates close to the physical solution.
- Solution technique - Compare various problem formulations and solution algorithms to select the one/s that have the highest probability to converge to the physical solution (i.e., all composition values are real & positive).

The Example Problem - Combustion of Propane (C_3H_8) in Air*

➢ Reaction: $C_3H_8 + (R/2)(O_2 + 4N_2) \rightarrow$ Products

R is the relative air to fuel ratio

➢ 10 compounds are assumed in the equilibrium mixture:

CO_2, H_2O, N_2, CO, H_2, H, OH, O, NO, O_2

No.	Equation	Variable	Initial Guess
1	f(CO2) = CO2+CO-3 # Carbon dioxide (Carbon balance)	CO2	2.9
2	f(H2O) = 2*CO2+H2O+CO+OH+O+NO+2*O2-R # Water (Oxygen	H2O	4
3	f(N2) = 2*H2O+2*H2+H+OH-8 # Nitrogen (Hydrogen balance)	N2	20
4	f(CO) = 2*N2+NO-4*R # Carbon monoxide (Nitrogen balance)	CO	0.1
5	f(H2) = CO2*H2-0.193*H2O*CO # Hydrogen	H2	0
6	f(H) = H*sqrt(CO2)-0.002597*sqrt(H2O*CO*xs/p) # Hydrogen atom	H	0
7	f(OH) = OH*sqrt(CO)-0.003448*sqrt(CO2*H2O*xs/p) # Hydroxyl	OH	0
8	f(O) = O*CO-1.799e-5*CO2*xs/p # Oxygen atom	O	0
9	f(NO) = NO*CO-0.0002155*CO2*sqrt(N2*xs/p) # Nitric oxide	NO	0
10	f(O2) = O2*CO^2-3.846e-5*CO2^2*xs/p # Oxygen	O2	0
11	R = 10 # Relative amounts of air and fuel (propane)	R	
12	p=40 # Pressure	p	
13	xs = CO2+H2O+N2+CO+H2+H+OH+O+NO+O2 # Total moles,	xs	

The sqrt terms may yield error messages or physically infeasible complex solutions if the associated molar amount becomes negative during the iterations

Example Problem
Used to demonstrate results of the following activities:

Initial Estimates for the equilibrium concentrations are based on Gibbs energy values and atom balances.

Alternative problem formulations:

(1) A system of Nonlinear Algebraic Equations (NLE) using Equilibrium Constants.

(2) Gibbs energy minimization using Lagrange Multipliers.

Alleviation of the solution process:

- Selection of the solution algorithm
- Eliminating/modifying expressions that are undefined for variable values < 0, or ≤ 0
- Replacing some implicit equations by explicit ones

496527 Development and Proving of an Information System

Abstract:

The process and manufacturing industries are currently in a dynamic competitive situation. On an increasingly global market, at least the medium-term growth rates of past decades are no longer to be expected. On the other hand, the classical core technologies of the two industries have now achieved a degree of maturity, which allows only limited rationalization. Companies in both industries must tackle this pressure through appropriate strategies. These strategies consist, among other things, to make operational processes faster, more efficient and more cost-effective. In addition, access to "hidden" capital in the form of skills, knowledge and experience of the employees creates additional competitive advantages. As a result, the problem of expert (employee) knowledge as well as of the transfer and management of knowledge within the framework of smart manufacturing is of great importance. This greatly favors the need to develop appropriate information systems.

In the presentation, a comparison between the process and the manufacturing industry is developed using relevant criteria. This is followed by a critical assessment of modeling strategies employed in both industries with regard to the identification, preservation, distribution, development and collection of implicit and explicit knowledge of employees in the context of knowledge management. The consistent use of knowledge greatly supports the planning of the production and the achievement of stable operating conditions. As a result, the costs and time required for the production of the products are greatly reduced, thereby considerably increasing the competitiveness of the company. This is achieved by a high degree of automation. The presentation will be closed with the introduction and the proving of a developed information system in the manufacturing industry.

EVALUATING HOSPITAL PERFORMANCE USING PROCESS SYSTEM ENGINEERING TOOLS:

INVESTIGATING EFFECT OF SPECIALIZATION ON HOSPITAL PERFORMANCE

Jangwon Lee[1] (jzl0164@auburn.edu)
Ravi Chinta[2]
Q. Peter He[1]

[1]Department of Chemical Engineering
Auburn University
[2]Department of Business Management
Auburn University at Montgomery

AUBURN UNIVERSITY

MOTIVATION & HYPOTHESIS

Motivation

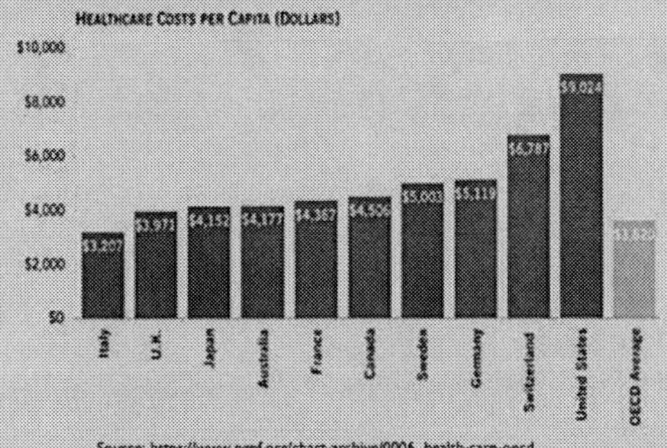

Source: https://www.pgpf.org/chart-archive/0006_health-care-oecd

Greater Healthcare Spending

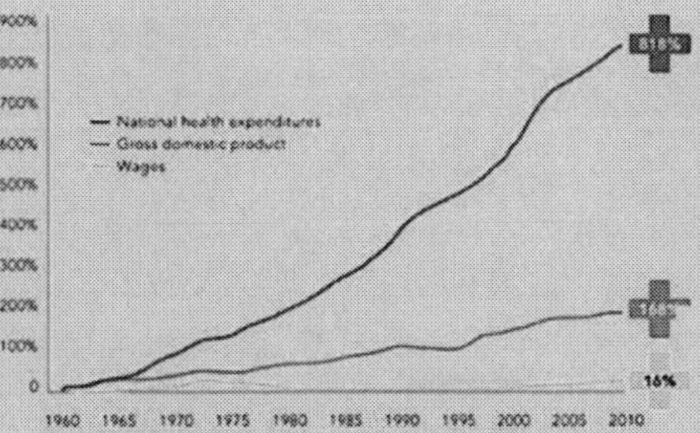

Source: https://www.huffingtonpost.com/2013/10/03/health-care-costs-_n_3998425.html

Faster Growth of Healthcare Spending

Hypothesis

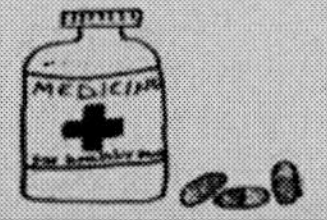

High Expenditure ⟶ Solution ? ⟶ SPECIALIZATION

Focused Factory Theory: factory that concentrates on narrow range of service or operation can provide better product at low costs

⇒ **Hypothesis:** focused factory theory can be applied to hospital industry.

DATA SOURCE & TERMINOLOGY

Data Source

https://www.hcup-us.ahrq.gov/

- Healthcare Cost and Utilization Project (HCUP) includes the largest collection of longitudinal hospital care data with all-payer level information in the United States.
- The HCUP data (2012) was used in this study.

Terminology

EXPENSIVE DRG CASE

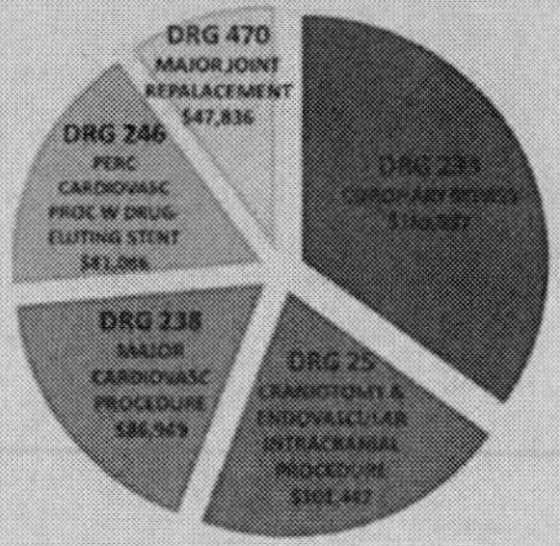

INEXPENSIVE DRG CASE

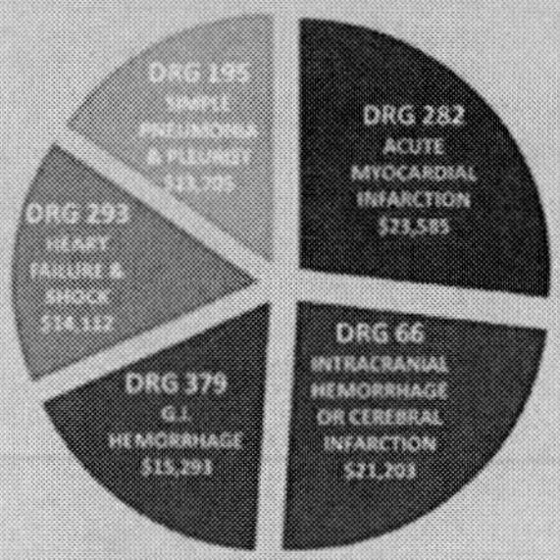

- DRG: Diagnosis-Related Group
- DRG is code assigned by diagnosis or procedure patients take

RESULTS & CONCLUSION

Results

Effect of Specialization on Total Charge		Effect of Specialization on Death of Patient	
Expensive DRG cases	Inexpensive DRG cases	Expensive DRG cases	Inexpensive DRG cases
⊖	⊖	⊖	Insignificant

Conclusion

- Total charges are lower in specialized hospitals than in non-specialized hospitals
- Clinical outcomes are better (i.e., lower mortality rates) in specialized hospitals than in non-specialized hospitals for expensive DRG cases

⇒ **Focused factory theory is applicable to hospital industry for expensive DRGs**

- For more details, please visit our poster (# 650)

496392 Deepmetabolism: A Deep Learning Algorithm to Predict Phenotype from Genome Sequencing

Abstract:

High-throughput sequencing technology has brought life science into a "big data" era with an unrivaled explosion in the amount of genomic and transcriptomic data. The falling cost (<$1,000 per human genome) and increasing speed (<1 day per human genome) of high-throughput sequencing lead to the snowballing data at petabyte level. However, it is still difficult to transfigure such "Big Data" to valuable biological insights such as cell growth rate and metabolic pathway activities. The gap between genome sequencing and cell phenotypes is one of the biggest challenges that need to be solved to achieve "Data-to-Insight". In recent five years, the rapid development of artificial intelligence, especially deep learning, provides a novel option to overcome this challenge. Deep learning is found to be extremely effective in learning and modeling complex systems based on the graphic processing unit computation. In this study, we developed DeepMetabolism, a deep learning algorithm that predicts cell phenotypes from genome sequencing data such as transcriptomics data. DeepMetabolism uses biological knowledge to design a neural network model and integrates unsupervised learning with supervised learning for predicting multiple phenotypes. In a prototypic application on *E. coli*, DeepMetabolism is able to predict phenotypes with high accuracy (PCC>0.92), high speed (<30 min for >100 GB data using a single GPU), and high robustness (tolerate up to 75% noise). We envision DeepMetabolism to bridge the gap between genotype and phenotype and serve as a springboard for applications in synthetic biology and precision medicine.

2017 AIChE Annual Meeting, Minneapolis, MN

Computing and Systems Technology Division

CAST Rapid Fire Session IV

Multiscale Dynamics System Identification of Time Series of Riser Reactor Temperature in FCC process Based on HHT

Daofan Cao

State Key Laboratory of Heavy Oil Processing
China University of Petroleum-Beijing

2017.10.30

Content

1. Introduction
2. Methods
3. Results & Discussion
4. Wrap-up

Introduction

Overview of Fluid Catalytic Cracking (FCC)

- Contributes: 75% to the gasoline pool
  35% to the diesel pool
  40% to the propene pool

Safe, stable and environment-friendly running is vital to enterprises

nonlinear effects and multiscale structure

- ☐ Reactor temperature is a critical variable that can illustrate the reaction status in FCC riser reactor.
- ☐ In order to develop system model, the prerequisite is to know the system structure clearly. So multiscale method and nonlinear analysis must be implemented.

3

Methods

Multiscale Method Comparison

	Fourier	Wavelet	Hilbert-Huang ✓
Basis	a priori	a priori	Adaptive
Frequency	Integral transform: Global	Integral transform: Regional	Differentiation: Local
Presentation	Energy-frequency	Energy-time-frequency	Energy-time-frequency
Nonlinear	no	no	yes
Non-stationary	no	yes	yes
Uncertainty	yes	yes	no
Harmonics	yes	yes	no

4

Methods

Flowchart of HHT method

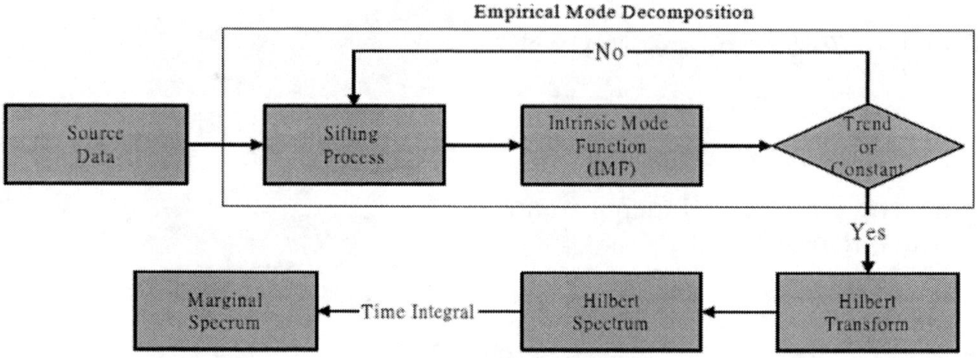

$$h(\omega) = \int_0^T H(\omega, t)dt \qquad X(t) = \sum_{j=1}^{n} a_j(t)e^{i2\pi\int f_j(t)dt} \qquad Y(t) = \frac{1}{\pi}PV\int_{-\infty}^{\infty}\frac{X(t')}{t-t'}dt'$$

5

Methods

Nonlinear Analysis

Indexes used in the dynamic analysis framework

Index	Brief description	Significance
D2, Correlation dimension	Measure of chaos, by exploiting the fact that chaotic systems have fractal dimensions	If the correlation exponent increases and then saturates as the embedding dimension increases, signal is chaotic
K2, Kolmogrov entropy	Measure of the rate of information loss along the attractor	For a chaotic signal, K should be positive and finite
LLE, Largest Lyapunov exponent	A measure of chaos, whose sign signifies chaos and whose value indicates the extent of chaos	A chaotic process would have a positive first LE

6

Results & Discussion

Data acquiring

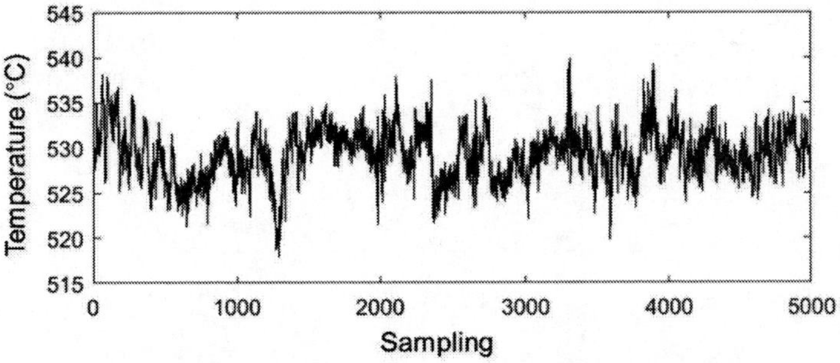

Fig. 1. Time series for temperature of riser reactor

- Sampling type: Reactor temperature data for FCC
- Sampling time interval: 5 minutes
- 5000 data points adopted

7

Results & Discussion

Decomposition

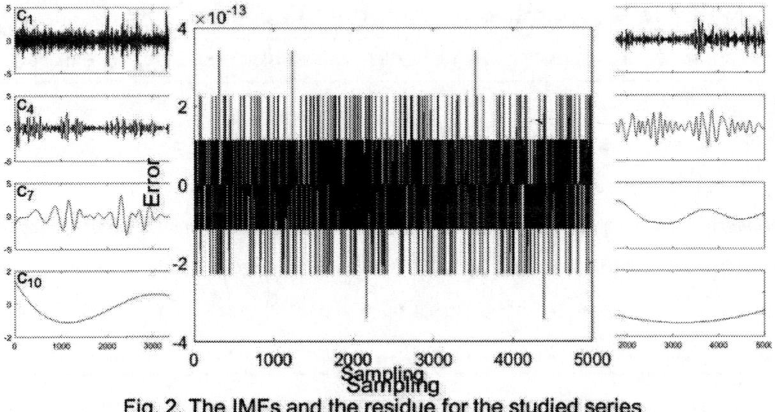

Fig. 2. The IMFs and the residue for the studied series

Fig. 3. The error from EMD decomposition

- Original series is consisted of 11 IMFs and a trend
- Reconstruction loss is very small

8

Results & Discussion

Decomposition reliability

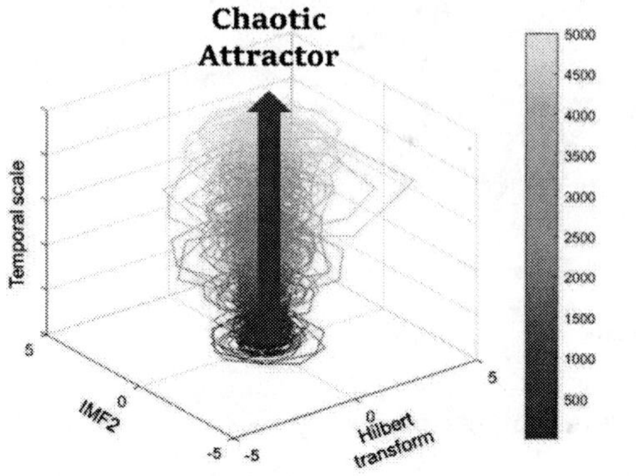

Fig. 4. Trajectories for the first IMF of the time series in the complex plane

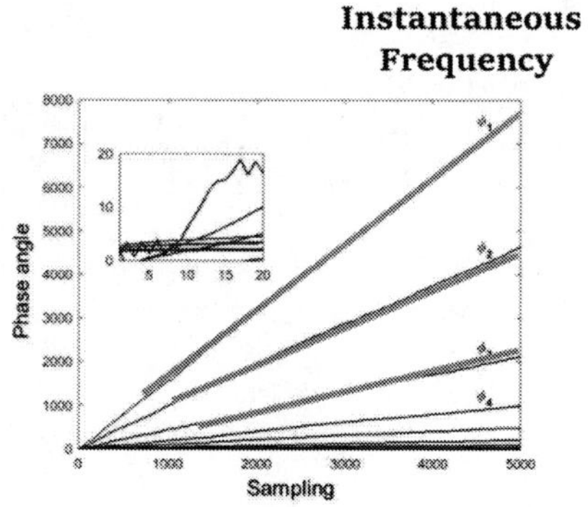

Fig. 5. Phase function φ of each IMF

- The trajectory in Fig. 4 moves in a certain direction (clockwise or anticlockwise) with a center, which means this system has a attractor.
- The slopes in Fig. 5 of IMF(1) has the largest one among all subscales and gr adually decrease with the scales order increasing.

Results & Discussion

Nonlinear analysis for each subseries

Multiscale dynamics system identification

	OTS	IMF(1)	IMF(2)	IMF(3)	IMF(4)	IMF(5)	IMF(6)	IMF(7)	IMF(8)	IMF(9)	IMF(10)	IMF(11)	R(11)
ω	-	1.56	0.94	0.43	0.20	0.10	0.04	0.02	0.01	0.00	0.00	0.00	0.00
τ	3	7	2	4	8	15	30	35	30	26	14	9	12
m	17	∞	16	16	13	12	19	12	6	5	4	3	3
D_2	7.945	∞	7.115	7.504	5.803	5.425	8.763	5.027	2.407	1.566	1.127	0.9961	0.9869
K_2	0.0797	∞	0.0906	0.0535	0.0141	0.0092	0.0035	0.0061	0.0033	0.0037	≈0	≈0	≈0
LLE	0.0478	∞	0.064	0.0487	0.0386	0.0241	0.0035	0.0066	0.0329	0.0123	≈0	≈0	≈0
D.S.I	Chaos	Noise	Chaos	Chaos	Chaos	Chaos	Chaos	Chaos	Chaos	Chaos	-	-	-

- Original temperature shows the chaotic behaviors with low time lag (3) and hig h embedding dimension (17).
- IMF(1) is a noise; IMF(2)-(9) are chaos series; others move as cycles (K2 = 0).
- IMF(2) and IMF(3) have similar properties with OTS in these indexes.

Wrap-up

1. Hilbert-Huang Transform method can be used to analysis the temperature series acquired from FCC process database. The temperature of riser reactor can be decomposed into 11 subscales and a residue by EMD.
2. Several subscales exist in the temperature series representing different instinct elements in this system, which shown the multiscale structure of this series.
3. For future perspective, more effective control model for riser temperature can be developed based on multiscale view revealed by HHT.

11

AIChE The Global Home of Chemical Engineers | DOING A WORLD OF GOOD

2017 AIChE Annual Meeting, Minneapolis, MN

Computing and Systems Technology Division

CAST Rapid Fire Session IV

Thank you!

Daofan Cao

Acknowledgement:
Prof. Xingying Lan & Dr. Xiaogang Shi
State Key Laboratory of Heavy Oil Processing

2017.10.30

AUBURN UNIVERSITY

STATISTICAL PROCESS MONITORING FOR SMART MANUFACTURING

Q. Peter He (qhe@auburn.edu)
Jin Wang
Kerul Suthar

Department of Chemical Engineering
Auburn University

2017 AIChE Annual Meeting, Minneapolis, MN

Statistics Pattern Analysis (SPA)

AUBURN UNIVERSITY
SAMUEL GINN COLLEGE OF ENGINEERING

- SPA: use various **statistics** to quantify process characteristics, and monitoring these statistics **instead of process variables** themselves to perform process monitoring

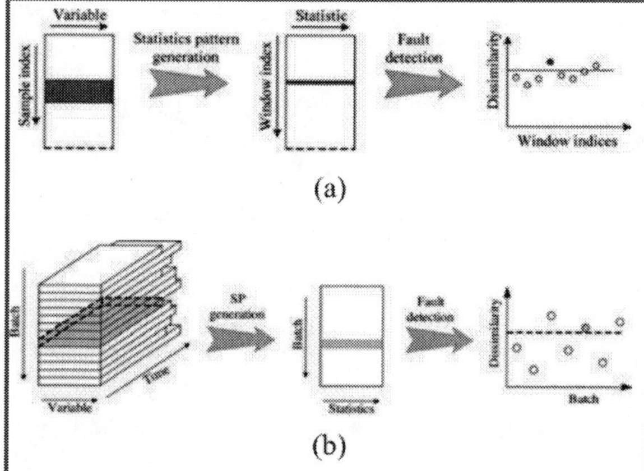

Fig. 1: SPA based fault detection for continuous (a) and batch (b) processes.

SPA: capture process characteristics

AUBURN UNIVERSITY
SAMUEL GINN COLLEGE OF ENGINEERING

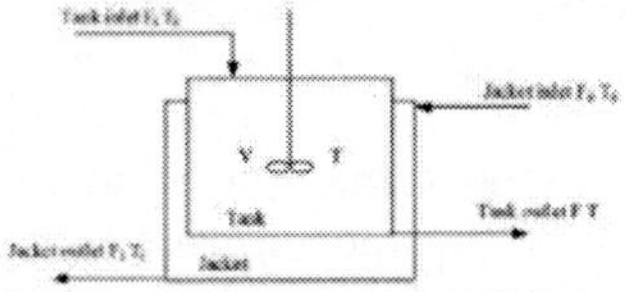

Fig. 2 Jacketed stirred tank heater: the tank inlet stream is heat up through heating jacket.

Process model:

$$\frac{dT}{dt} = \frac{F}{V}(T_i - T) + \frac{UA(T_j - T)}{V\rho c_p}$$

$$\frac{dT_j}{dt} = \frac{F_j}{V_j}(T_{ji} - T_j) + \frac{UA(T_j - T)}{V_j \rho_j c_{pj}}$$

Disturbance:

$$F_j = F_{js} + 0.5\sin\left(\frac{t\cdot\pi}{10}\right) + N(0, 0.01)$$

Fault:

$$F_{jf} = F_j - 0.3$$

- SPA removes the dynamics, nonlinearity and non-Gaussianity.
- None of the other methods able to achieve the same.

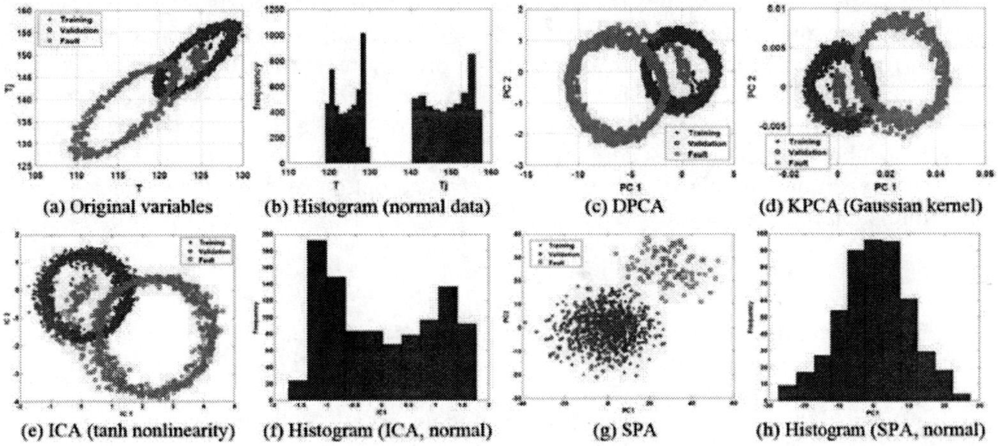

SPA: address big data

AUBURN UNIVERSITY
SAMUEL GINN COLLEGE OF ENGINEERING

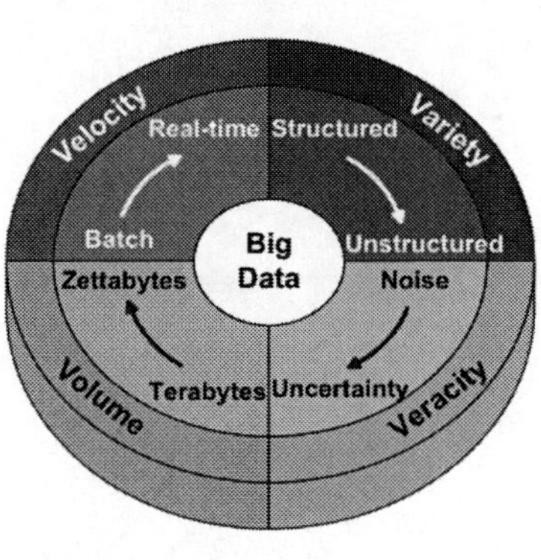

Fig. 3 The 4 V's of big data

- ***For volume,*** since statistics can be computed sample wise and variable wise, SPA can significantly reduce the volume;
- ***For variety,*** the easy integration of variable statistics can help address data variety;
- ***For velocity,*** since SPA reduces volume, it can handle real-time data;
- ***For veracity,*** statistics are inherently less sensitive to noises and missing data as multiple measures are used to compute statistics.

493083 Reaction Identification and Parameter Estimation from Chemical Process Data

Abstract:

Mixed-integer non-linear programming (MINLP) has been utilized to facilitate model reduction in chemical reaction networks in the literature [1, 2]. Inspired by recent success in the application of mixed-integer programming to the domain of surrogate model development [3], we propose a MINLP methodology to identify the reactions, mechanisms, and estimate the associated kinetic rate parameters of a reaction network from chemical process data. Application of this approach to dynamic process data requires the use of algebraic surrogate models for the concentration or conversion profiles [4]. This approach utilizes a superset of chemical reactions, which can contain different mechanistic pathways for the same reaction. A model-building procedure, similar to one described previously for subset selection in multiple linear regression [3], is then used to minimize an objective that balances the complexity of the model with its fit in order to identify a probable subset of reactions and estimate the associated kinetic rate parameters. The utility of this methodology is demonstrated on a number of steady-state and continuous computational problems, as well as an experimental data set that affords the consideration of different oxidation mechanisms in a chemical looping combustion reactor.

References cited

[1] Edwards, Keith, T. F. Edgar, and V. I. Manousiouthakis. "Reaction mechanism simplification using mixed-integer nonlinear programming." *Computers & Chemical Engineering* 24.1 (2000): 67-79.

[2] Bhattacharjee, Binita, et al. "Optimally-reduced kinetic models: reaction elimination in large-scale kinetic mechanisms." *Combustion and Flame* 135.3 (2003): 191-208.

[3] Cozad, A., N. V. Sahinidis, and D. C. Miller, Automatic learning of algebraic models for optimization, AIChE Journal, **60**, 2211-2227, 2014.

[4] Wilson, Zachary T., and Nikolaos V. Sahinidis. "The ALAMO approach to machine learning." *Computers & Chemical Engineering,* accepted (2017).

\Orchestrating a brighter world NEC

Plant-Wide Visualization for Situation Awareness Using Ising Model Based Clustering of Vanishing Correlations

Masanao Natsumeda*
Security Research Laboratories, NEC Corporation, Japan
*Email: m-natsumeda@ah.jp.nec.com

Research Motivation

Visualization for situation awareness is important for correct and timely decision making. However, existing visualization **doesn't effectively bring out the potential of operators**.

Contribution map

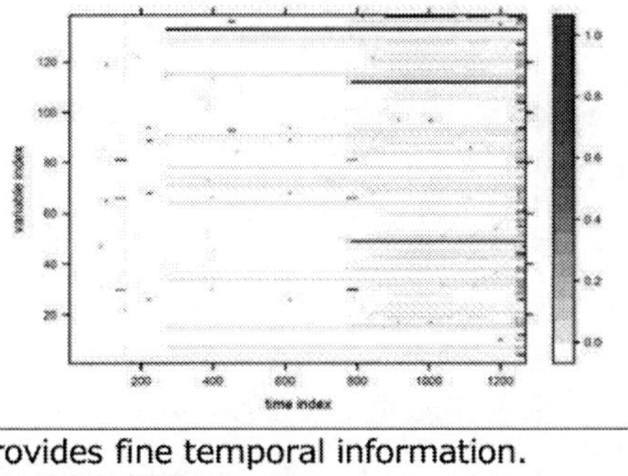

- provides fine temporal information.
- is not intuitive.
- burdens operators to interpret the results.

Fault propagation path diagram

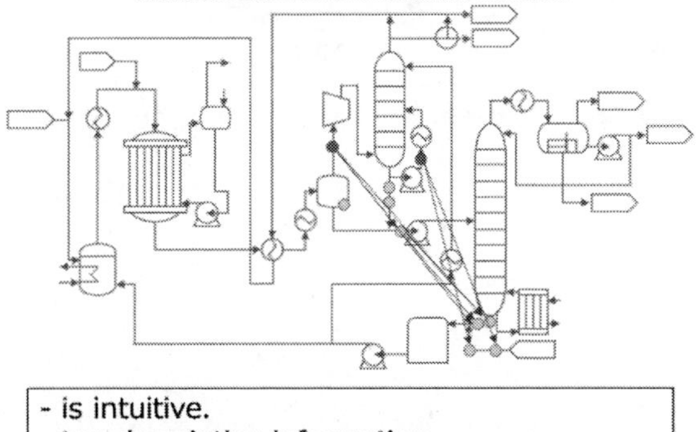

- is intuitive.
- too descriptive information.
- disturbs operators to utilize their knowledge

  \Orchestrating a brighter world NEC

Our proposal

Visualize distribution of related faulty process variables accompanied by their transition onto the process flow diagram.

Proposed visualization

Ising model based clustering of vanishing correlations

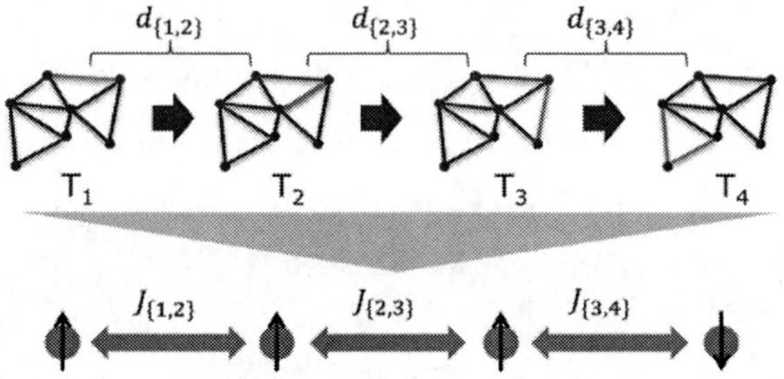

- is intuitive and comprehensive.
- is indirect.
- encourages operators to utilize their knowledge.

- consider affinity of their transitions
- can naturally consider temporal order.
- gives their transition chunk and faulty variables.

Results with a vinyl acetate monomer production plant simulator

Supposed three consecutive but individual faults and Our method gave three clusters. Faulty variables in each cluster were located around their corresponding faulty subsystem.

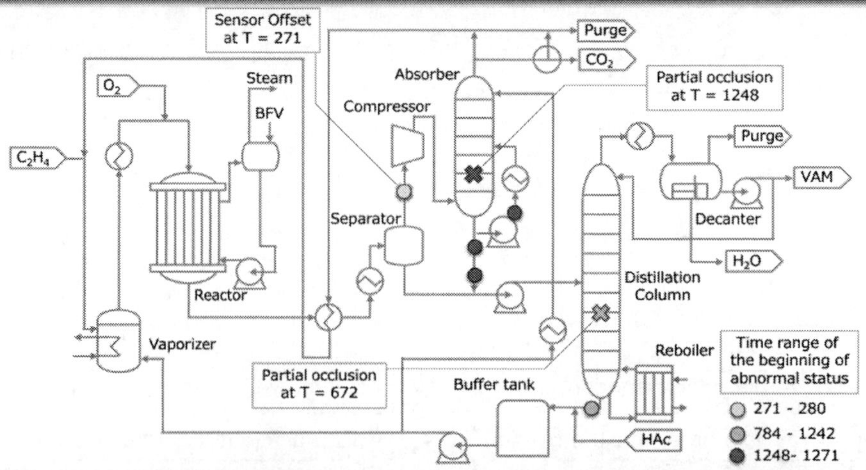

497606 Process Monitoring Using a PCA-Based Exponentially Weighted Generalized Likelihood Ratio Chart

Abstract:

Most industrial processes are multivariate in nature and can be highly dimensional and complex. These processes require continuous monitoring in order to ensure that they run safely, while maintaining consistent product quality and minimizing pollution to the environment. A critical part of process monitoring methods is fault detection. Process history or data based methods are frequently used to carry out process monitoring, as they do not do require the process model to be explicitly defined. A popular input-based process monitoring technique is the principal component analysis (PCA) method [1].

PCA is a linear dimensionality reduction technique that is widely utilized to monitor a broad range of industrial processes. PCA functions by applying an orthogonal transformation to a set of highly correlated variables in order to produce a set of uncorrelated values, where the largest principal component is in the direction that accounts for the greatest degree of correlation between two variables. The dimensionality of a process is then reduced by only retaining the principal components that account for a majority of correlation between variables in the process (95-99%). Numerous authors have produced extensions of the conventional PCA method. However, many of these extensions have not been implemented in practice, as most techniques only offer marginal improvement in results and may increase computational time, which is an important factor since most industrial processes require quick detection.

The hypothesis testing method, the generalized likelihood ratio (GLR) method has recently been incorporated with the PCA model in order to further enhance its performance [2]. However, the conventional GLR method only utilizes a single observation to compute the detection statistic. Results have demonstrated that the utilization of a moving window when implementing the GLR statistic can improve the performance of the technique even further [3], [4]. However, since the moving window GLR method only utilizes a mean filter when computing the detection statistic, it is possible to improve the results further. The exponentially weighted moving average (EWMA) is a method contains process memory, and is known to show improved performance over a simple moving average technique. Therefore, a control chart that integrates the advantages of the GLR test with those of the EWMA method should provide improved performance, as it utilizes exponential weights with decaying memory in order to compute the GLR statistic [5].

This work focuses on extending the conventional PCA model, to one that utilizes an exponentially weighted GLR chart, in order to provide enhanced results. The fault detection performance of PCA-based EW-GLR chart will be demonstrated through illustrative examples using real-world applications, including chemical processes, and will be evaluated using the three main fault detection criteria: missed detection rate, false alarm rate, and out-of-control average run length (ARL_1). This work will also demonstrate that the PCA-based EW-GLR algorithm is relatively easy to implement, and therefore encourages its application in practice for process monitoring purposes.

[1] D. C. Montgomery, *Introduction to Statistical Quality Control*, 7th ed. Hoboken, NJ: John Wiley & Sons, 2013.

[2] F. Harrou, M. N. Nounou, H. N. Nounou, and M. Madakyaru, "Statistical fault detection using PCA-based GLR hypothesis testing," *J. Loss Prev. Process Ind.*, vol. 26, no. 1, pp. 129–139, 2013.

[3] M. R. Reynolds and J. Y. Lou, "An Evaluation of a GLR Control Chart for Monitoring the Process Mean," *J. Qual. Technol.*, vol. 42, no. 3, pp. 287–310, 2010.

[4] M. Z. Sheriff, M. Mansouri, M. N. Karim, H. Nounou, and M. Nounou, "Fault detection using multiscale PCA-based moving window GLRT," *J. Process Control*, vol. 54, pp. 47–64, Jun. 2017.

[5] J. Zhang, C. Zou, and Z. Wang, "A control chart based on likelihood ratio test for monitoring process mean and variability," *Qual. Reliab. Eng. Int.*, vol. 26, no. 1, pp. 63–73, 2010.

COLLEGE OF ENGINEERING
LAMAR UNIVERSITY®
MEMBER THE TEXAS STATE UNIVERSITY SYSTEM™

TEXAS STATE OF MIND

187I: An Inverse-Model-Based Methodology for Real-Time Fault Diagnosis in Non-Square Multivariate Dynamic Systems

Liwen Chen, Qiang Xu
lchen1@lamar.edu
qiang.xu@lamar.edu

COLLEGE OF ENGINEERING

Inverse-model-based FDI

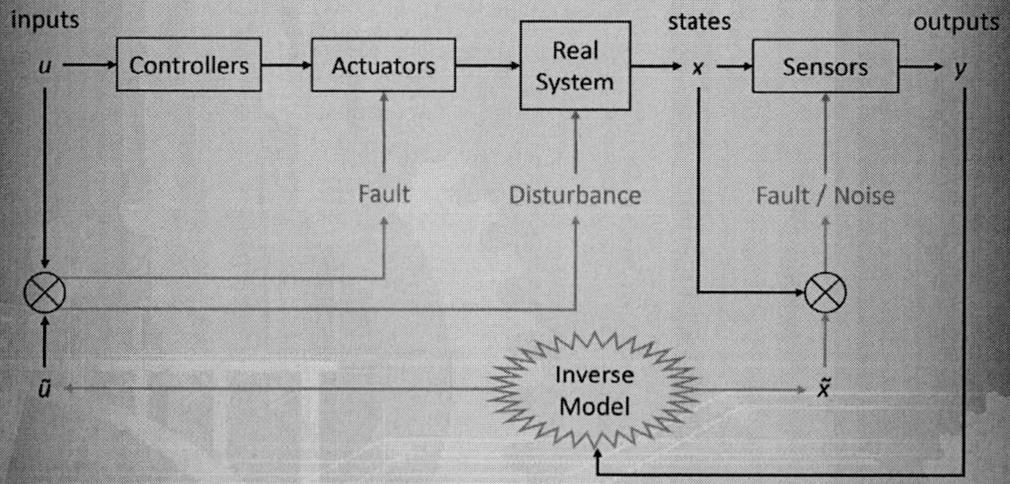

LAMAR UNIVERSITY

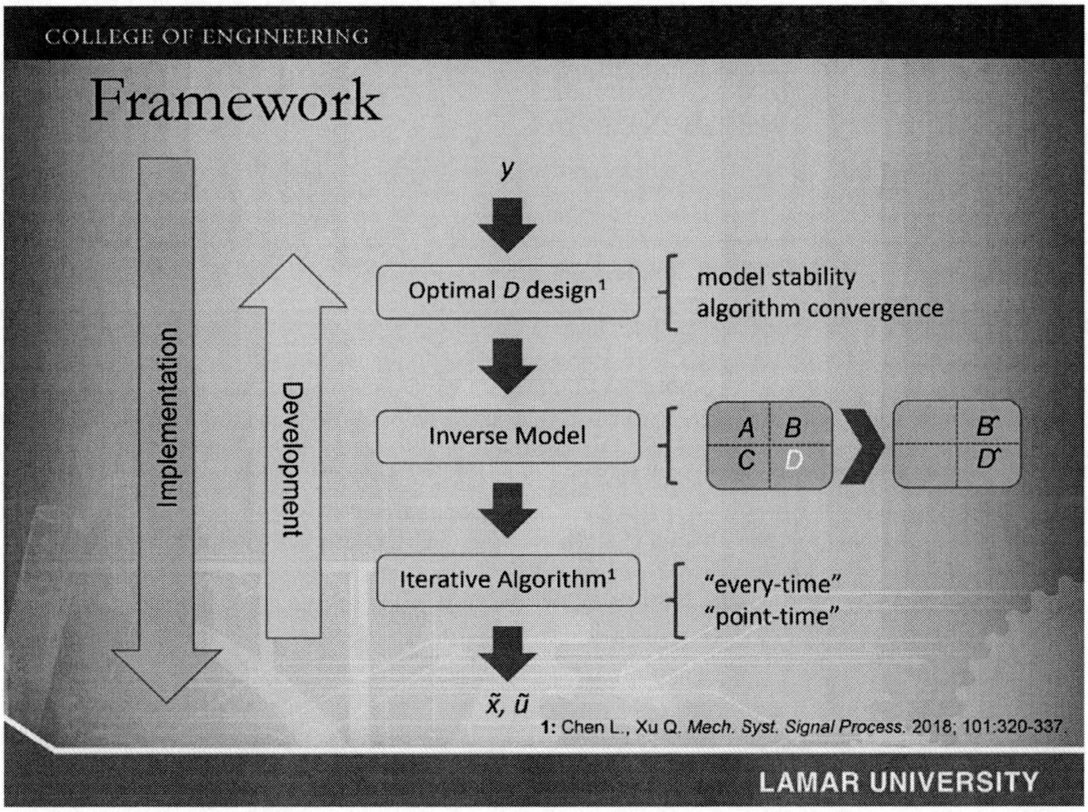
COLLEGE OF ENGINEERING
Framework
Implementation
Development
y
Optimal D design1
model stability
algorithm convergence
Inverse Model
A B
C D
B*
D*
Iterative Algorithm1
"every-time"
"point-time"
x̃, ũ
1: Chen L., Xu Q. Mech. Syst. Signal Process. 2018; 101:320-337.
LAMAR UNIVERSITY

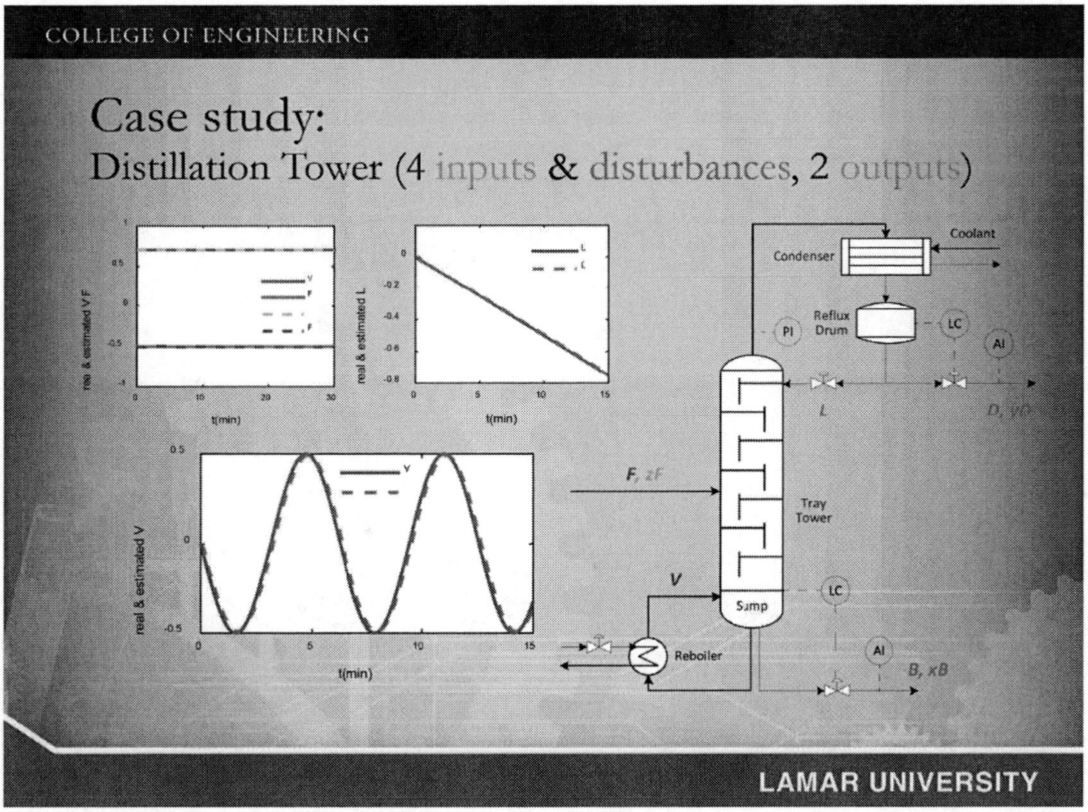
COLLEGE OF ENGINEERING
Case study:
Distillation Tower (4 inputs & disturbances, 2 outputs)
t(min)
t(min)
t(min)
real & estimated L
real & estimated V
Coolant
Condenser
Reflux Drum
PI
LC
AI
L
D, yD
F, zF
Tray Tower
V
Sump
Reboiler
B, xB
LAMAR UNIVERSITY

494245 Root Cause Diagnosis of Process Fault Using Modified Convergent Cross Mapping and Bayesian Network

Abstract:

Modern plantsâ€™ measurement devices and control systems are cross-linked with each other. A disturbance in major equipment may lead to plant-wide disturbances, which makes it difficult to find the root cause. This paper develops a method to combine data-driven methods and process knowledge to diagnose the root cause of an abnormal status in industrial process. Convergent cross mapping(CCM) is an algorithm to detect causality of two time series variables. CCM is based on the theory of Takensâ€™ embedding theorem. Compared to other causal algorithms, CCM can be applied to systems where causal variables have synergistic effects. The causality detected by the CCM between process variables and process knowledge are combined to construct the causality network. The process data is used to estimate the conditional probabilities of the Bayesian network. After an alarm is detected, the Bayesian network is updated by changing the status of the node. The proposed method enable the operator to diagnose the fault. To test the effectiveness of the proposed method, the method is applied to Tennessee Eastman Chemical Process.

487409 Uniting Lyapunov-Based MPC with Closed-Loop Subspace Identification

Abstract:

Model Predictive Control (MPC) is a well established control method to handle process systems with constraints, nonlinearity, and uncertainty while recognizing performance objectives. One tool that imbues MPC approaches with excellent stability properties is use of control Lyapunov function in the control design [1, 2] to design Lyapunov-based MPC (LMPC). The success of LMPC (like other MPC designs) require a reasonable predictive model of the system. Often, a good first principles model is unavailable, however, sufficient historical data exists to possibly build a data driven model for the purpose of control.

The subspace identification methods (SIM) approach uses a set of input and output data, to estimate linear time-invariant models in a state space form [3]. These methods are based on the geometrical projections and numerical linear algebra, and have recently been adapted to handle data from batch processes [4]. Numerical robustness, fewer user parameters, MIMO systems identification, model order reduction make SIMs more favorable as compared to other identification methods like prediction error methods for the industrial applications. SIMs are usually non-iterative methods, which makes their application, computationally affordable, and, this simplicity would make the model update, affordable.

In the present work, we develop a framework to integrate LMPC with the subspace models. To make the model adaptive, an appropriate model re-identification trigger is included in the control design. Furthermore, the complexities arising out of the closed-loop nature of the data is also accounted for. Simulations on a nonlinear CSTR are used to illustrate the method.

Key Words: Lyapunov-Based Model Predictive Control, Subspace Identification Methods Continuous Stirred-Tank Reactor

[1] Mahmood, M. and Mhaskar, P. (2012). Lyapunov-based model predictive control of stochastic nonlinear systems. Automatica, 48(9), 2271-2276.

[2] Mahmood, M. and P. Mhaskar, Constrained Control Lyapunov Function Based Model Predictive Control Design, Int. J. Rob. & Nonl. Contr., 24, 374â€"388, 2014.

[3] Qin, S. J. (2006). An overview of subspace identification. Computers & chemical engineering, 30(10), 1502-1513.

[4] Rashid, M. M., Mhaskar, P., & Swartz, C. L. (2017). Handling multiâ€rate and missing data in variable duration economic model predictive control of batch processes. AIChE Journal.

Development of Integrated Biomimetic Framework with Intelligent Monitoring, Cognition, and Decision Capabilities for Control of Advanced Energy Plants

Temitayo Bankole[1], Gaurav Mirlekar[1], Ghassan Al-sinbol[2], Berhane Gebreslassie[3], Fernando Lima[1], Mario Perhinschi[2], Urmila Diwekar[3], Richard Turton[1], Debangsu Bhattacharyya[1]

[1]Department of Chemical & Biomedical Engg., West Virginia University, Morgantown, WV 26506,

[2]Department of Mechanical & Aerospace Engg., West Virginia University, Morgantown, WV 26506,

[3]Vishwamitra Research Institute, Clarendon Hills, IL 60154.

tsbankole@mix.wvu.edu
Mineapolis, MN
October, 2017

Biomimetic Self Organization

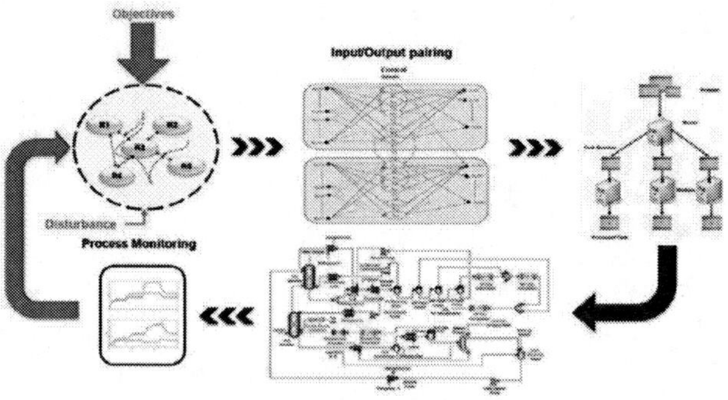

Motivation:
- Flexibility
- Adaptability
- Cognition

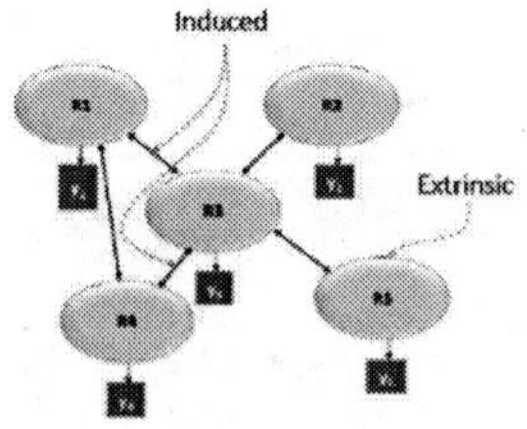

- Controlled variable selection* : Central Nervous system
- Controller design: Rule of pursuit of ants
- Detection and monitoring: Immune system

*Bankole, T. & Bhattacharyya, D. Algorithmic modelling of structural connectivity for process plants. 2016 American Control Conference (ACC), 2016. IEEE, 5038-5043.

Controller Design

Bio-inspired Optimal Control Strategy (BIO-CS)*:

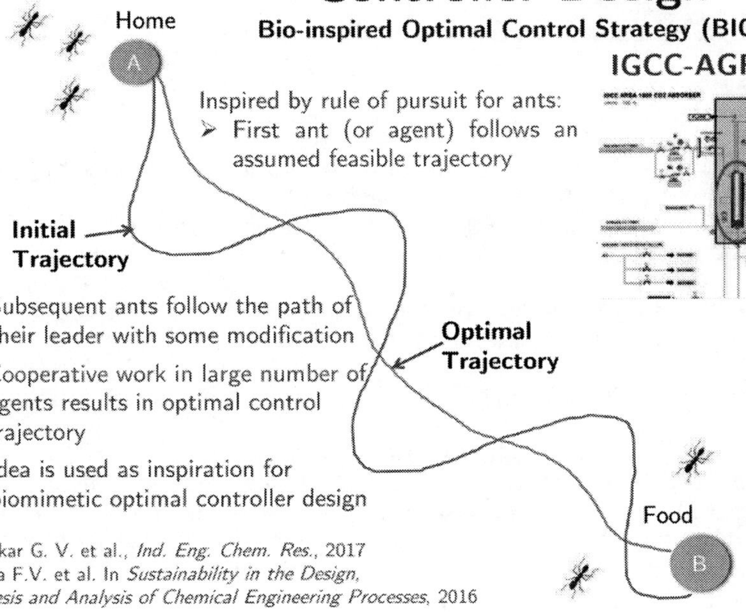

Inspired by rule of pursuit for ants:

- ➢ First ant (or agent) follows an assumed feasible trajectory
- ➢ Subsequent ants follow the path of their leader with some modification
- ➢ Cooperative work in large number of agents results in optimal control trajectory
- ➢ Idea is used as inspiration for biomimetic optimal controller design

IGCC-AGR Process Implementation

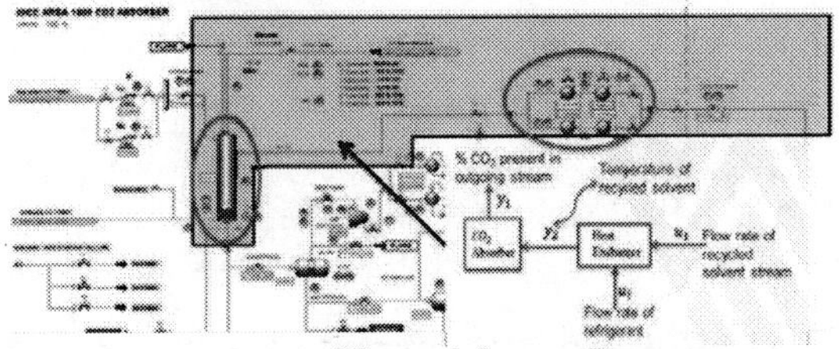

Closed-loop Responses

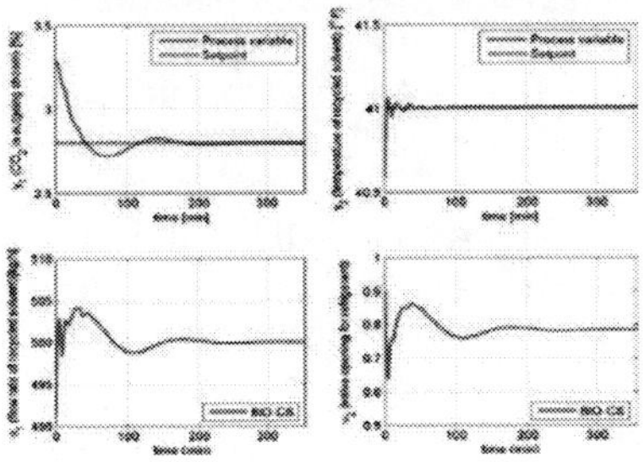

*Mirlekar G. V. et al., *Ind. Eng. Chem. Res.*, 2017
**Lima F.V. et al. In *Sustainability in the Design, Synthesis and Analysis of Chemical Engineering Processes*, 2016

AIS Monitoring & Detection

- Objective: Development of an intelligent, comprehensive, and integrated framework for advanced power plant monitoring and control
- Approach: mimic biological immune system mechanisms
- Abnormal condition (AC) detection, identification, and evaluation
- 100% detection rate
- Minimum 91% identification rate

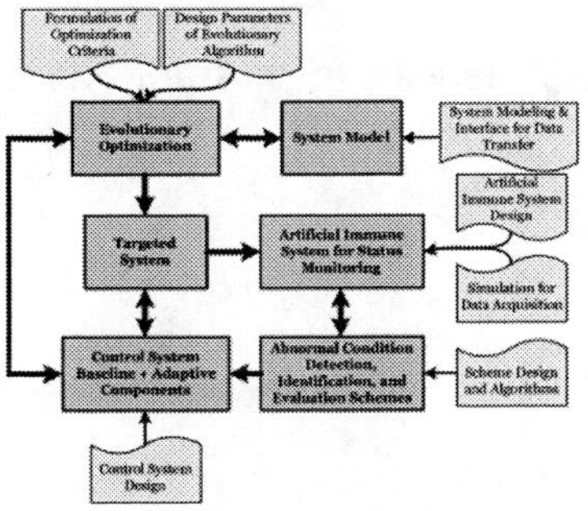

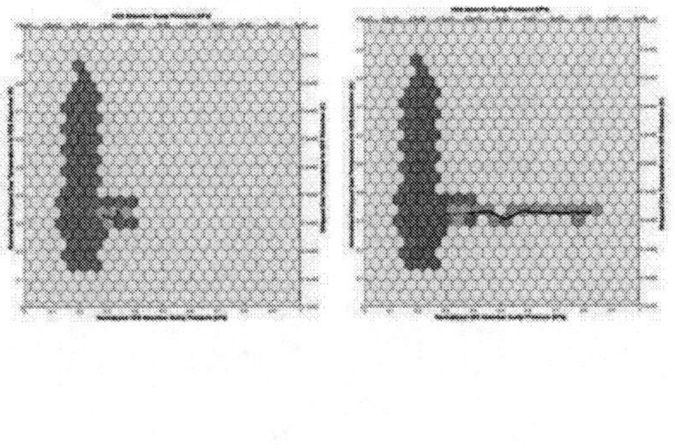

497513 Controller Design for CSTR Process Output Using a Combination of GA, PSO, Fuzzy and PID Algorithms for Quick Rejection of Process Disturbances

Abstract:

Research in methods for process control design has been reinvigorated due to the advancement of computational capabilities and development of optimization and machine learning algorithms e.g. Genetic Algorithm, Neural network, Particle Swarm Optimization, Fuzzy Logic. As a result, a significant increase in the development of new control design algorithms has been observed.

Efficient controller design is a vital component of chemical process design. The aim of the control system is to address any process upset as quickly as possible to maintain the process within the specified operating parameters. As such, fast response time for the controller is an important design criteria. One option for achieving this is to develop an adaptive controller that exhibits the lowest possible variable dependence.

In this work, a CSTR system output has been used to study and compare responses of controllers developed using GA-PID, GA-Fuzzy, and PSO-PID algorithms. The aim of this study is to evaluate which combination of algorithm is the quickest in rejecting disturbances in the system. In the case of GA-PID and PSO-PID, GA and PSO were used to determine best control weighting factor () and rise-time parameter (). However, in the case of GA-Fuzzy, GA was used to determine the best applicable Fuzzy membership functions for the operating parameters obtained from literature.

Selected References:

1. Rejane de B. Araújo, Antonio A.R. Coelho, Filtered predictive control design usingmulti-objective optimization based on geneticalgorithm for handling offset in chemical processes, Chemical engineering research and design, Vol. 117, 2017
2. Turker Tekin Erguzel, Fuzzy Controller Parameter optimization Using Genetic Algorithm for a Real Time Controlled System, World Congress on Engineering, Vol II, 2013
3. Yasue Mitsukura, Toru Yamamoto and Masahiro Kaneda, A Design of Self-Tuning PID Controllers Using a Genetic Algorithm, American Control Conference, 1999
4. Zwe-Lee Gaing, A Particle Swarm Optimization Approach for Optimum Design of PID Controller in AVR System, IEEE Transactions on Energy Conversation, Vol. 19, No. 2, 2004

498245 Strategies for Minimum Variance ALS Estimation of Noise Covariance Matrices

Abstract:

In process systems, engineers generally have a deterministic model describing the dynamics of the plant. Uncertainty manifests as noises and disturbances that are not included in this model, so it is common to augment the deterministic model with a stochastic model. Disturbances may occur within the process itself, affecting the dynamic response of the system, or they may occur at the output (i.e., sensor noise). Motivated by the central limit theorem, a typical choice is to model the process and measurement noises as independent zero mean Gaussian random variables. In this case the stochastic model is fully defined by the process and measurement noise covariance matrices, denoted Q and R respectively. Determining good values for Q and R is necessary for applications such as state estimation and controller performance monitoring.

The variances are generally not known a priori, so they must be estimated from plant data. The autocovariance least squares (ALS) algorithm casts the estimation as a linear regression problem (Odelson et al., 2006; Rajamani and Rawlings, 2009). To get the minimum variance ALS estimates of Q and R, the appropriate weighting matrix W* is required. However, W* is itself a function of the unknown Q and R. We will discuss the following strategies for obtaining minimum variance ALS estimates:

1. Pick an initial Q_0 and R_0 , calculate $W^*(Q_0, R_0)$, and solve the ALS problem with this weight to obtain Q_1 and R_1. Repeat until convergence.
2. Estimate W* itself directly from the plant data.

Strategy 1 was originally suggested by Rajamani and Rawlings (2009), but has not been widely adopted because calculation of W* from Q and R was thought to be intractable in most practical cases. We will present a new method of calculation that reduces the computational burden to more manageable levels.

Strategy 2 was originally suggested by Zagrobelny and Rawlings (2015). We will present improvements and new insights related to this strategy.

B. J. Odelson, M. R. Rajamani, and J. B. Rawlings. A new autocovariance least-squares method for estimating noise covariances. Automatica, 42(2):303â€"308, February 2006.

M. R. Rajamani and J. B. Rawlings. Estimation of the disturbance structure from data using semidefinite programming and optimal weighting. Automatica, 45(1):142â€"148, 2009.

M. A. Zagrobelny and J. B. Rawlings. Practical improvements to autocovariance least-squares. AIChE J., 61:1840â €"1855, 2015.

501768 Robust Economic Linear Optimal Control

Abstract:

Economic Linear Optimal Control (ELOC) has recently been established as a strategy to design linear feedback controllers with economic objectives (i.e., minimize operating cost). In this work, the ELOC method is extended to address processes with model uncertainty. In particular, a robust formulation of the ELOC problem will be presented. In addition to optimizing with respect economic objectives the resulting controller will be guaranteed stable for all realizations of the uncertain parameters. The Sum Of Squares (SOS) methodology in conjunction with semi-definite programming is employed to arrive at a computationally tractable (but not overly conservative) solutions to the proposed robust ELOC problem.

Optimal operation of heat exchanger networks (HENs) through heat duty redistribution using energy flow graphs

Sujit S. Jogwar

Department of Chemical Engineering
Indian Institute of Technology Bombay, India
jogwar@iitb.ac.in

October 30, 2017

Heat Exchanger Networks

- Practical manifestation of energy integration

  Energy minimization at the nominal operating conditions

  Disturbances
  - input fluctuations (fast)
    performance deterioration (slow)

  Regulatory control can lead to
    sub-optimal operation
    controller saturation

  Need for optimal operation

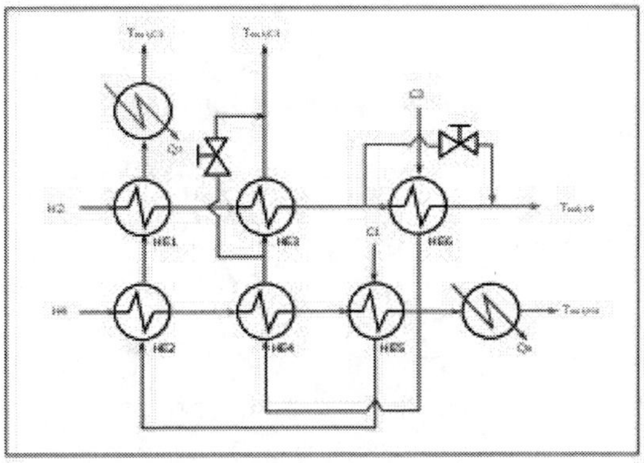

Optimal operation through heat duty redistribution

Heat Duty Redistribution Framework

1. Energy flow graph representation
2. Identification of closed loops
3. Circulation of loop loads (e_i)
4. LP optimization problem
   - Objective: minimize $e_{external}$
   - Variables: Loop loads
   - Constraints: exchanger duty, thermal driving force
5. Convert e_i into Q_i
6. Implementation
   - Intermediate temperature controllers
   - Additional duty controllers

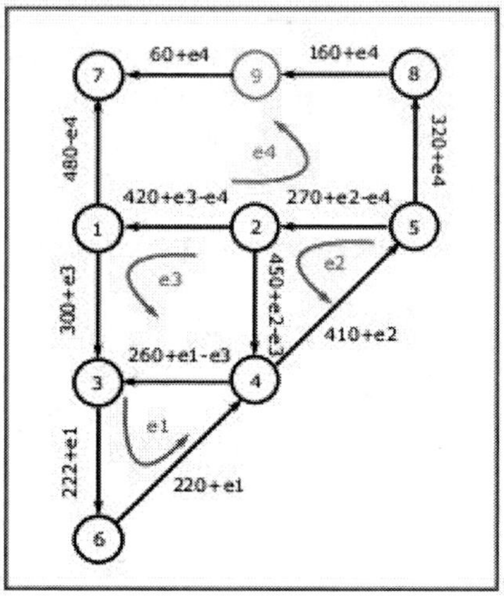

Fast and scalable solution

Case Study

System details:

Stream	Target	Stream	Target	Utility	Value
1	180 C	2	40 C	Hot	60 kW
3	130 C	4	40 C	Cold	160 kW

Disturbance: Stream 2 inlet temperature

	$\delta T_{in,4} = +5C$		$\delta T_{in,4} = -5C$	
	base control	hybrid control	base control	hybrid control
$T_{1,out}$	180.0 C	180.0 C	180.0 C	180.0 C
$T_{2,out}$	40.0 C	40.0 C	40.0 C	40.0 C
$T_{3,out}$	130.0 C	130.0	127.4 C	130.0 C
$T_{4,out}$	40.0 C	40.0	40.0 C	40.0 C
Q_{hot}	52.6 kW	48.1 kW	66.9 kW	90.9 kW
Q_{cold}	172.6 kW	168.1 kW	153.6 kW	170.9 kW

Acknowledgment: DST INSPIRE grant IFA 13 ENG 61

Thank you

496097 Steady State Real Time Optimization of a Reactor-Separator-Recycle Process

Abstract:

Real time optimization (RTO) for economic (near) optimum steady operation of a reactor-separator-recycle process module is evaluated. The RTO obtains the optimum value of the dominant unconstrained regulatory layer setpoint, namely, the reactor limiting reactant composition, by adjusting unknown parameters of a simple reaction kinetic model to best fit the available overall plant material balance data, and also the separator tray efficiency to best fit the measured column tray temperature profile. Rigorous non-linear dynamic simulation results demonstrate that for slow long term changes in the plant characteristics such as catalyst deactivation, column fouling and throughput / fresh feed composition changes, RTO delivers 4-5% higher energy savings and maximum achievable throughput compared to process operation at constant setpoint. Considerations of model adequacy are highlighted.

Imperial College London

A multi-parametric bi-level optimization strategy for hierarchical model predictive control

Styliani Avraamidou[1,2], Nikolaos A. Diangelakis[2*]
Efstratios N. Pistikopoulos[2]

1. Centre for Process Systems Engineering, Department of Chemical Engineering, Imperial College London, UK
2. Texas A&M Energy Institute and Artie McFerrin Department of Chemical Engineering, Texas A&M University, College Station, Texas, 77840, USA

*nikos@tamu.edu

Hierarchical MPC as a Bi-level Optimization Problem

Solution via B-POP and PAROC

Hierarchical Model Predictive Control → **Quadratic Bi-level Optimization Problem**

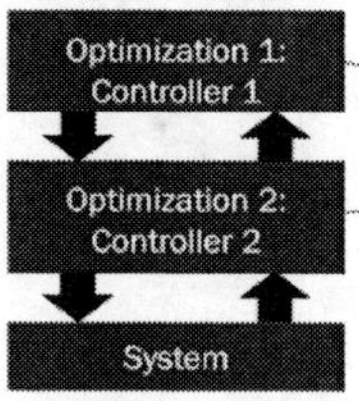

$$\begin{cases} \min\limits_{x_1} & (Q_1{}^T x + H_1\theta + c_1)^T x + (Q_{t1}\theta + c_{t1})^T\theta + c_{c1} \\ s.t. & A_1 x \le b_1 + F_1\theta \\ \begin{cases} \min\limits_{x_2} & (Q_2{}^T x + H_2\theta + c_2)^T x + (Q_{t2}\theta + c_{t2})^T\theta + c_{c2} \\ s.t. & A_2 x \le b_2 + F_2\theta \end{cases} \end{cases}$$

$$x \in \Re^n, \quad x = [x_1^T x_2^T]^T$$

- Each control level corresponds to an optimization level
- Both levels share the same variables
- Each level optimizes a subset of the optimization variables

Challenges:

- **Bi-level problems are NP-hard even in the linear case**
  - Need for Global optimization
- **MPC strategies require the solution of the optimization problem at every control time step**

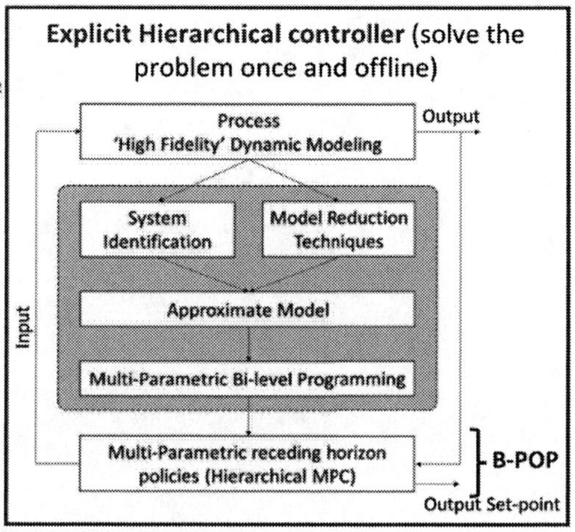

Pistikopoulos, E., Diangelakis, N., Oberdieck, R., Papathanasiou, M., Nascu, I., Sun, M., 2015. PAROC - An integrated framework and software platform for the optimisation and advanced model-based control of process systems. Chemical Engineering Science 136, 115–138.
Faísca, N.P., Dua, V., Rustem, B., Saraiva, P.M., Pistikopoulos, E.N. Parametric global optimisation for bilevel programming (2007) Journal of Global Optimization, 38 (4), pp. 609-623.

Bi-level (MI)P via Multi-Parametric Programming

B-POP (Bi-level Parametric OPtimization) toolbox

- A MATLAB toolbox for bi-level programming
- An extension to our already developed **POP toolbox**

- **Features Solvers for:**
  1. Bi-level linear and quadratic (B-LP and B-QP) problems
  2. Bi-level **mixed-integer** linear and quadratic (B-MILP and B-MIQP) problems
  3. Tri-level problems (T-LP, T-QP, T-MILP, T-MIQP)

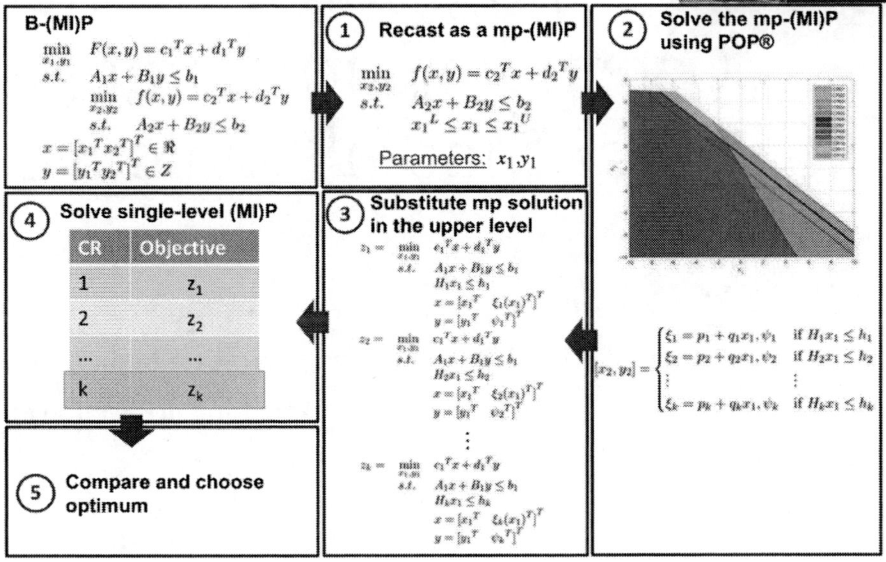

Case Study

Hierarchical Control of an Exothermic CSTR

Objectives:
1. Minimize cost
2. Keep temperature to 300 K

Manipulated variables:
1. Inlet mass flowrate (m)
2. Coolant Temperature (T_c)

Disturbances:
1. Inlet Concentration (C_{A0}: 9-10 mole/L)
2. Inlet Temperature (T_0: 200-350 K)

Control Variables:
1. Outlet Temperature (T)

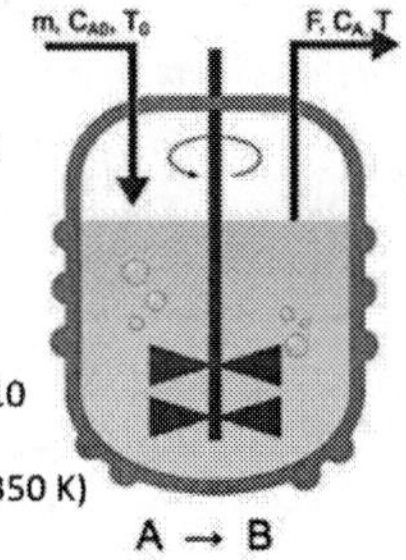

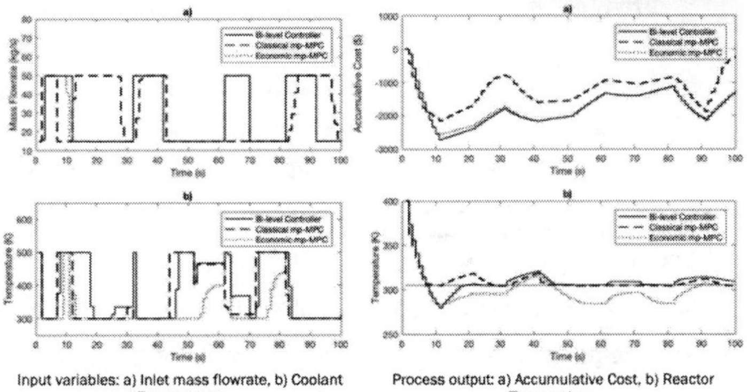

Input variables: a) Inlet mass flowrate, b) Coolant Temperature

Process output: a) Accumulative Cost, b) Reactor Temperature

<u>The resulting hierarchical controller:</u>
- Effectively rejects disturbances and economically maintains the system at the given set-points
- Performed more economically than the mp-MPC without significant set-point violations as the mp-eMPC.

	mp-MPC	mp-eMPC	B-eMPC
Accumulative Profit ($)	151.78	1194.54	1165.89
$\sum_{k=1}^{100} \lvert T_k - T_k^R \rvert$ (K)	499.40	1283.28	740.19
Average $\lvert T_k - T_k^R \rvert$ (K)	4.94	12.71	7.33

Closed loop validation results

Computation of Terminal Constraints for Large-scale Nonlinear Model Predictive Control

Devin Griffith[*], Sachin Patwardhan[**], and Lorenz Biegler[*]
[*]Dept. of Chemical Engineering, Carnegie Mellon University
[**]Dept. of Electrical Engineering, IIT Bombay

Quasi-Infinite Horizon NMPC

Focus of this work

$$V_N(x_k) = \min_{z_i, v_i} \sum_{i=0}^{N-1} l(z_i, v_i) + V_f(z_N)$$ *Sum of stage costs & terminal cost*

$s.t.\ z_{i+1} = f(z_i, v_i, 0)\ \forall\, i = 0 \ldots N-1$ *Dynamic system*

$z_0 = x_k$ *Initial condition*

$z_i \in \mathbb{X}\ \forall\, i = 1 \ldots N-1$ *State constraints*

$v_i \in \mathbb{U}\ \forall\, i = 0 \ldots N-1$ *Control constraints*

$z_N \in \mathbb{X}_f$ *Terminal constraint*

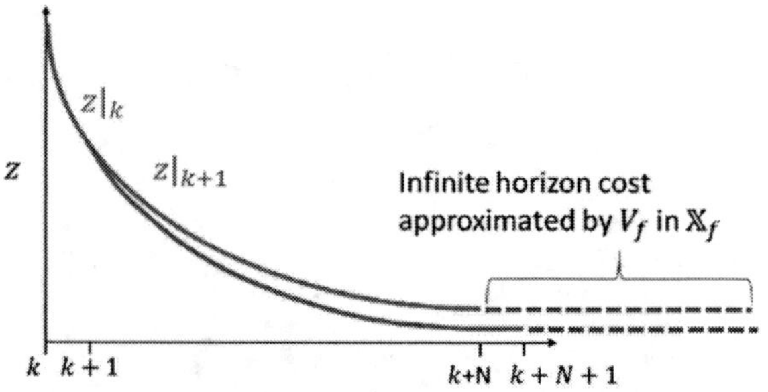

- **Nonlinear model predictive control** (NMPC) utilizes control actions computed by a dynamic optimization problem
- Selection of predictive horizon length N is a major difficulty in implementation
- Main requirement of N is the reachability of the terminal constraint
- **Quasi-Inifinite horizon NMPC**[1] approximates infinite horizon problem by choice of terminal conditions
- Use to derive terminal conditions that lead to better choice of N
- Terminal cost V_f represents a stabilizing controller (LQR) valid in the terminal region $\mathbb{X}_f$
- Choosing the terminal region $\mathbb{X}_f$ is then equivalent to finding a region for which the LQR is stabilizing for the full nonlinear system

[1] H. Chen and F. Allgöwer. A quasi-infinite horizon nonlinear model predictive control scheme with guaranteed stability. *Automatica*, 4:1205–1218, 1998.

Bounding Nonlinearities

- Stabilizing region for LQR can be computed given a bound on system nonlinearities – gives $\mathbb{X}_f$ for NMPC

$$|f(x, -Kx) - Ax + BKx| = |\phi(x)| \leq c|x|^q$$

- Nonlinear effects bounded by analyzing many one step simulations under LQR control – scalable
- Applied to distillation system[2] with 246 states
- Terminal regions for NMPC derived for multiple setpoints

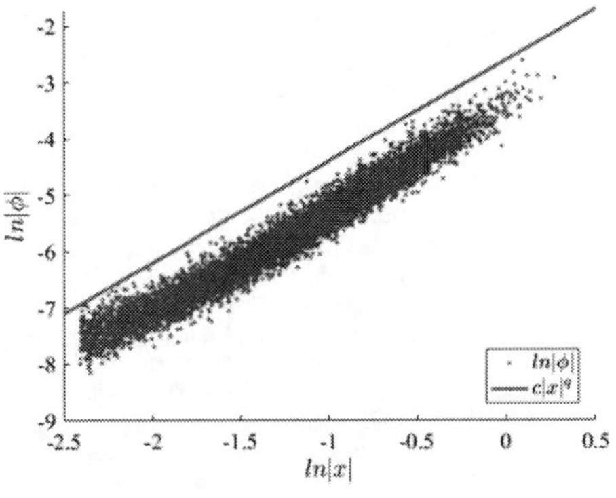

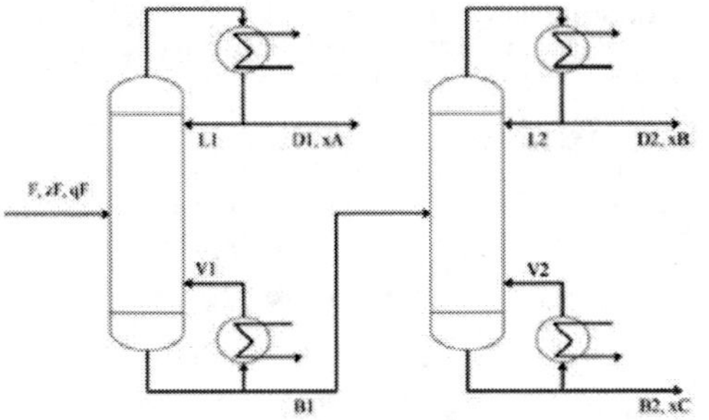

zF	$\mathbb{X}_f$, $\|z_N\| \leq$
0.4,0.2,0.4	0.41
0.4,0.4,0.2	0.36
0.2,0.4,0.4	0.38

[2] R.B. Leer. Self-optimizing control structures for active constraint regions of a sequence of distillation columns. Master's thesis, Norwegian University of Science and Technology, 2012.

Application: Adaptive Horizon NMPC

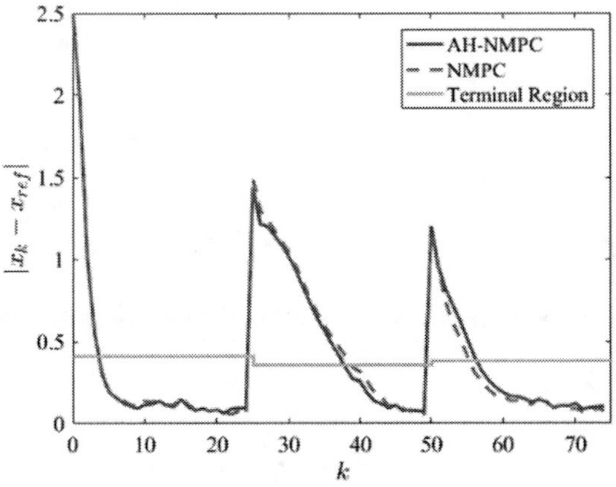

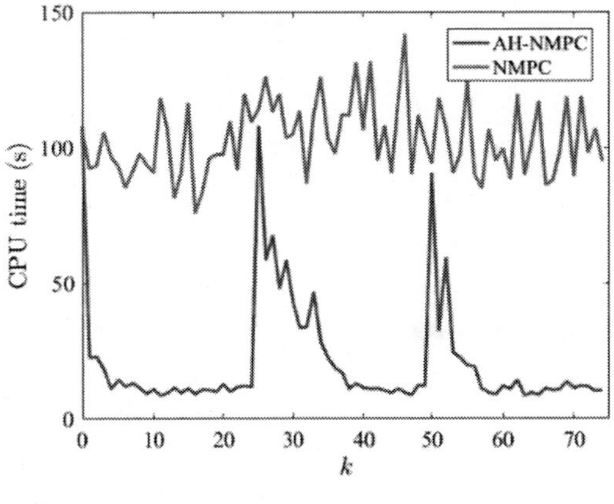

- Distillation system controlled through noise and setpoint changes
- Infinite horizon sufficiently approximated – horizon length N can change as long as reachability conditions are met
- Horizon lengths updated in real time via sensitivity updates from sIpopt
- **80% reduction in average computational time by better choice of N**

493546 Approximate Dynamic Programming for Nonlinear Process Control Under Uncertainty

Abstract:

Even though model predictive control (MPC) has become a dominated methodology for process control both in industry and academia, it still suffers from several drawbacks. First, it is difficult to find an optimal solution for large-scale nonlinear process with long control horizon. Second, the control performance cannot be guaranteed if uncertainties or model matches are not taken into account properly. To overcome these shortcomings, several strategies are proposed to handle online computational demands and uncertainties, such as fast MPC, robust MPC and stochastic MPC.

In this work, we focus on the approximate dynamic programming (ADP) [1] to improve MPC and develop a simple control strategy with lower computational complexities and better performance. ADP, or reinforcement learning in computer science community, is a data-driven method to solve multistage scheduling problem and overcome the curse of dimensionality for optimization in high dimensional space. The essential idea is extracting and generalizing useful information from sampled states trajectory to build a value function, which represents the control cost of current control policy. Then a better control action can be selected by minimizing the value function of process state in next step based on the model prediction. The key advantage of ADP is that only one-step prediction and online optimization is required such that computational burden is reduced significantly. Moreover, the model uncertainty can be compensated if the value function is constructed based on the real operational data. Different from conventional optimal control theory, ADP offers a reliable and practical way to improve the current control performance.

Clearly, the quality of proposed control approach is highly dependent on the value function, which is build on the post-state space to avoid considering uncertainty explicitly in the formulation. In this study, we utilize MPC to generate a number of states trajectories and then employ three different strategies to approximate value function for MPC. The first and also the easiest way is using polynomial function to fit the data, which requires to solve a least square problem. Second, convex regression is employed to build value function such that the desired steady state becomes the global minimum of the value function. Third, the hinge function, in a piecewise linear form, is used to approximate operational data with arbitrary accuracy. This will result in a mixed-integer linear programming (MILP) problem and an efficient computational strategy is proposed to obtain a suboptimal solution.

Each strategy mentioned above has pros and cons. In the numerical study, we design and compare three controllers based on these strategies for two nonlinear process with additive noises.

Reference

[1] Lee JM, Lee JH, Approximate dynamic programming-based approaches for inputâ€“output data-driven control of nonlinear processes. Automatica, 2005; 1281-1288.

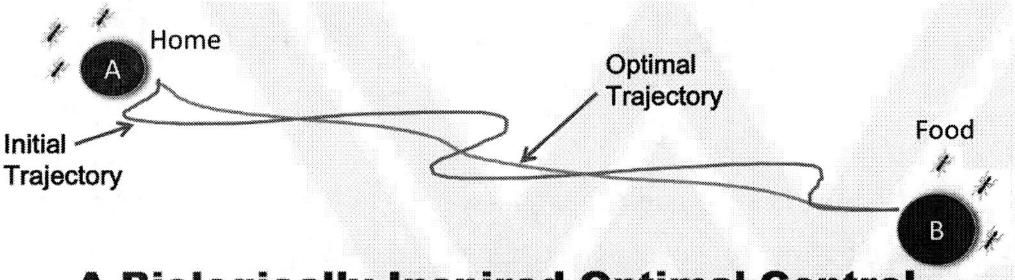

A Biologically-Inspired Optimal Control Framework: Application to the Hybrid Performance (Hyper) System

Gaurav Mirlekar[1], Paolo Pezzini[2], Kenneth M. Bryden[2], David Tucker[3] and Fernando V. Lima[1]
[1]Department of Chemical and Biomedical Engineering, West Virginia University, Morgantown, WV, USA
[2]Ames Laboratory, Ames, IA, USA
[3]National Energy Technology Laboratory (NETL), Morgantown, WV, USA

Presenting author's email: gvmirlekar@mix.wvu.edu

2017 AIChE Annual Meeting
Minneapolis, MN
October 30, 2017

BACKGROUND

- Hybrid process motivation and challenges:
  - ✓ Hybrid system integration increases efficiency by using less fuel when compared to stand-alone system components
  - ✓ Coupling effects among different system components
  - ✓ Nonlinearities present in the process model
  - ✓ Multivariable control and operational constraints

Gas turbine*

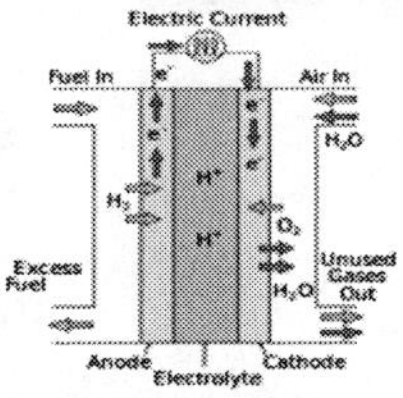

Fuel cell*

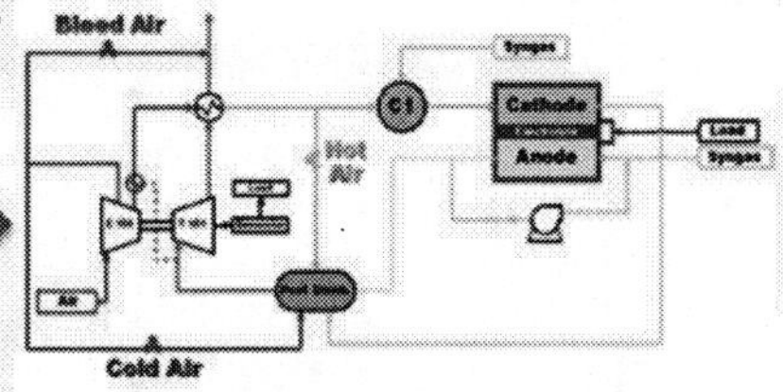

Hybrid process (Hyper from NETL)

- Objectives:
  - ✓ Design an advanced biomimetic control strategy inspired by biological systems
  - ✓ Implement strategy for hybrid energy processes

*https://www.wikipedia.org/

WEST VIRGINIA UNIVERSITY
DEPARTMENT OF CHEMICAL AND BIOMEDICAL ENGINEERING
LIMA RESEARCH GROUP
Control, Optimization and Design of Energy Systems (CODES)

BIO-INSPIRED OPTIMAL CONTROL STRATEGY (BIO-CS)*

- Inspired by rule of pursuit for ants
- Cooperative work in large number of ants (or agents) results in optimized path
- Intermediate optimal control problems are solved employing gradient-based solvers
- Unique features of BIO-CS for improved computational performance:
  - ✓ Algorithm termination at a suboptimal solution corresponding to a specific agent
  - ✓ Possibility of parallelizing optimal control problems associated with different agents

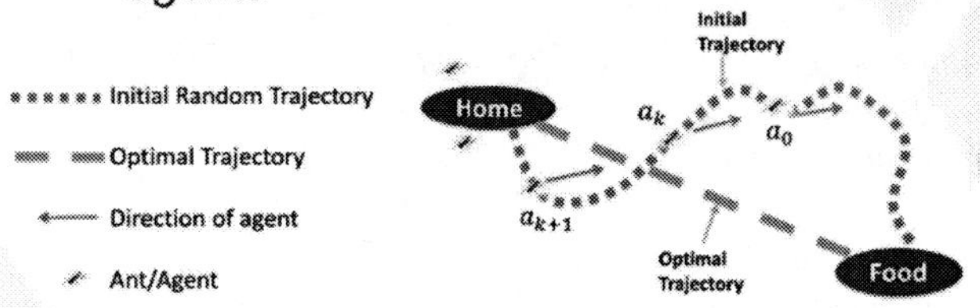

Ant's rule of pursuit

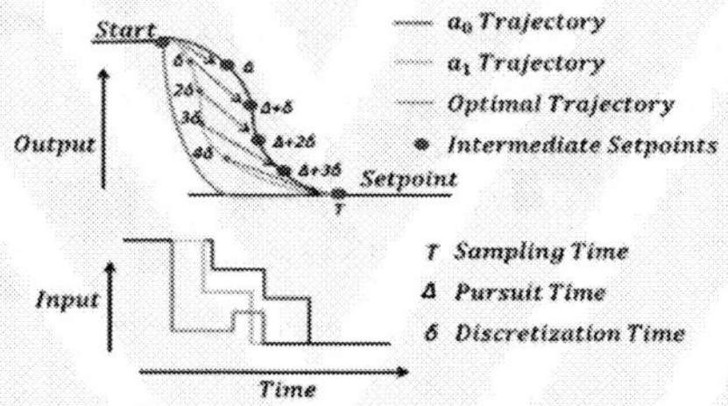

Agent-based framework of BIO-CS

*Mirlekar G., Li S. and Lima F. V., "Design and Implementation of a Biologically-Inspired Optimal Control Strategy (BIO-CS) for Chemical Process Control". Ind. Eng. Chem. Res., 56, 6468–6479, 2017.

WEST VIRGINIA UNIVERSITY
DEPARTMENT OF CHEMICAL AND BIOMEDICAL ENGINEERING
LIMA RESEARCH GROUP
Control, Optimization and Design of Energy Systems (CODES)

RESULTS

Application to Hyper process* from NETL:

- BIO-CS is implemented for MIMO (2×2) transfer function sub-system
- Multiple outputs need to be simultaneously controlled at their desired setpoints

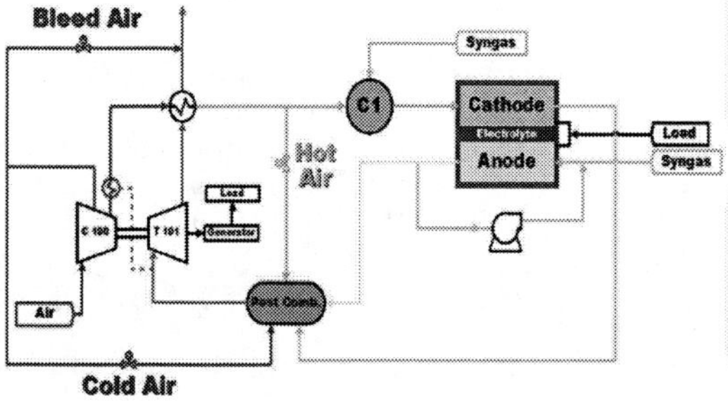

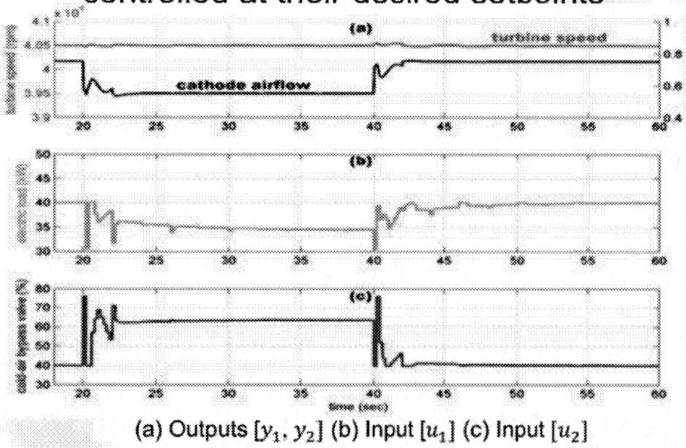

(a) Outputs $[y_1, y_2]$ (b) Input $[u_1]$ (c) Input $[u_2]$

BIO-CS implementation results:

- Effect on turbine speed due to coupling was limited to 0.2%
- Future plans include implementation of BIO-CS for Hyper process at NETL

*Mirlekar G., Pezzini P., Bryden M., Tucker D. and Lima F. V., "A Biologically-Inspired Optimal Control Strategy (BIO-CS) for Hybrid Energy Systems". In Proceedings of the 2017 American Control Conference (ACC), Seattle, WA.

WEST VIRGINIA UNIVERSITY
DEPARTMENT OF CHEMICAL AND BIOMEDICAL ENGINEERING
LIMA RESEARCH GROUP
Control, Optimization and Design of Energy Systems (CODES)

MARY KAY O'CONNOR PROCESS SAFETY CENTER
TEXAS A&M ENGINEERING EXPERIMENT STATION

CFD simulation of charged aerosol combustion under high electrical field

Shuai Yuan, Joseph Sang-Il Kwon, M. Sam Mannan
Mary Kay O'Connor Process Safety Center, Artie McFerrin Department of Chemical Engineering, Texas A&M University, College Station, Texas 77843-3122, USA

TEXAS A&M UNIVERSITY Texas A&M Energy Institute | TEXAS A&M ENGINEERING EXPERIMENT STATION

ıckground

- ɔid of fine solid particles or liquid droplets in air or other gas
- The liquid with lower flash point (less than 93°C) is ıable liquid
- ignited below their flash point because the fine droplets increase the surface area which promotes the mass and heat transfer
- Incidents, such as Buncefield explosion and Miliken Carpet factory fire, involved aerosol explosion
- The same volume of aerosol combustion releases ı the same volume of gas

1ethodology

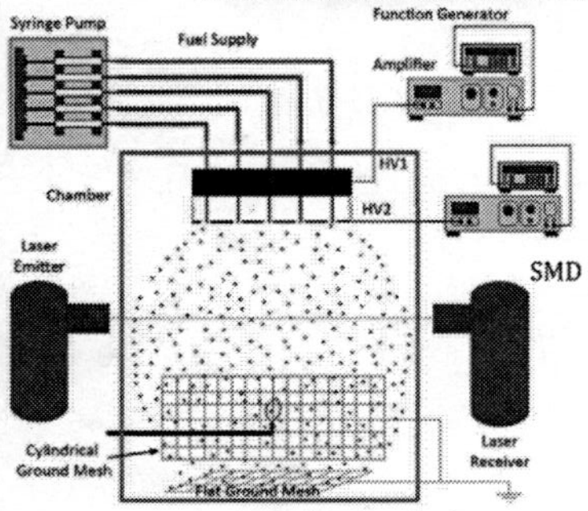

ectrospray device

$$\text{SMD } (D_{[3][2]}) = \frac{\sum n_i d_i^3}{\sum n_i d_i^2}$$

Experiment

Aerosol generation: Aerosol is generated by electrospray and it applied high voltage to atomize liquid. The droplets will carry like charges and move under high electrical field.

Ignition source: ropane flame is applied to ignite the aerosol cloud.

ırement: The droplet size is Tec Malvern Laser and use Sauter ʌID) to represent.

Methodology

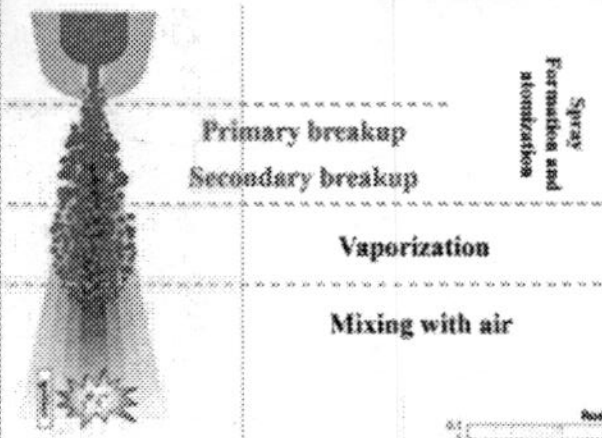

Simulation

Droplets: Three stages of aerosol generation include atomization, primary breakup and secondary breakup. The movement of droplets are described by Lagrangian manner.

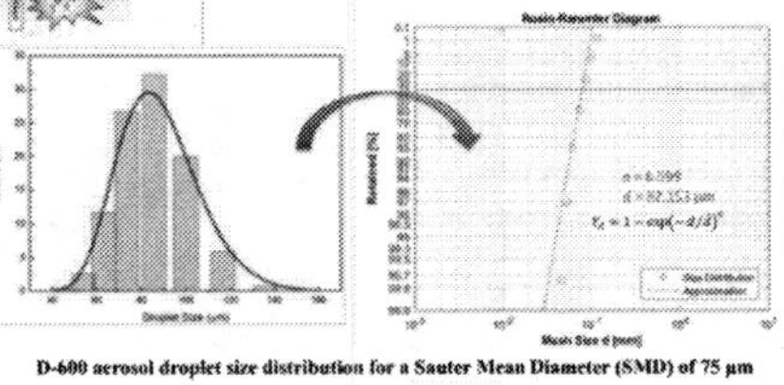

Initial spray: Fitting the size distribution measured from experiment with Rosin-Rammler Distribution.

D-600 aerosol droplet size distribution for a Sauter Mean Diameter (SMD) of 75 μm

Lagrange manner: The movement of droplets is tracked by Lagrange manner

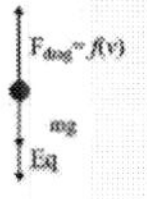

$$\sum \vec{F} = \vec{F}_{drag} + \vec{F}_g + \vec{F}_E$$

$$\vec{F}_{drag} = m_d \frac{18\mu}{\rho_d d_d^2} \frac{C_D Re}{24} (\vec{u} - \vec{u}_d)$$

$$\vec{F}_g = m_d g \qquad E = -\nabla V \qquad \vec{F}_E = Eq$$

Where

m_d is the mass of single droplet
ρ_d is the droplet density
d_d is the droplet diameter
u_d is the velocity of single droplet
C_D is the drag force
Re is the Reynolds number
g is gravitational constant

Eulerian manner: The gas phase are described through Eulerian manner. The mass, momentum, energy and species balance are shown below.

Mass balance: $\frac{\partial \rho}{\partial t} + \frac{\partial \rho u_j}{\partial x_j} = S_{ev}$

Momentum balance: $\frac{\partial \rho u_i}{\partial t} + \frac{\partial}{\partial x_j}(\rho u_i u_j - \tau_{ij}) = -\frac{\partial p}{\partial x_i} + S_{i,mo}$

Energy balance: $\frac{\partial \rho h}{\partial t} + \frac{\partial}{\partial x_j}\left(\rho u_j h - \frac{\rho v}{Pr}\frac{\partial h}{\partial x_j}\right) = \frac{\partial p}{\partial t} + \frac{\partial p u_j}{\partial x_j} + S_{he}$

Species balance: $\frac{\partial \rho X_f}{\partial t} + \frac{\partial}{\partial x_j}\left(\rho u_j X_f - \rho D \frac{\partial X_f}{\partial x_j}\right) = S_{ev}$

Where

ρ is the density of gas phase mixture ρ_d is the droplet density
u is the velocity of gas phase mixture
p is the gas phase pressure
τ is the stress tensor
v is the kinematic viscosity
h is the specific enthalpy
X_f is the mass fraction of the hydrocarbon in gas phase
D is the diffusion coefficient of the hydrocarbon vapor
$S_{ev}, S_{i,mo}$ and S_{he} are source terms contributed from evaporation in mass conservation, momentum conservation and energy conservation respectively.

Results

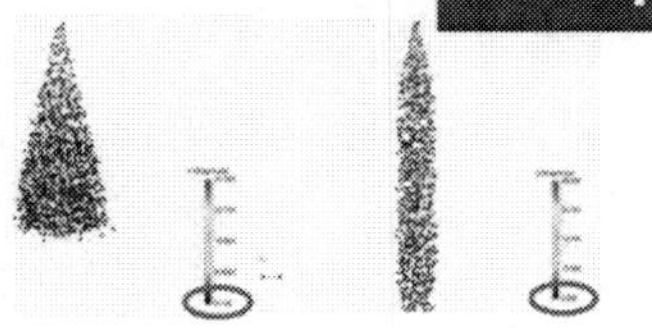

(a) Balance velocity (b) Balance velocity

(a) droplets movement without considering electric force
(b) droplets movement with considering electric force

The drag force is a function of velocity. When there is a electric force, it requires greater drag force to balance the electric force. Therefore, the balance velocity is much larger when considering the electric force.

Results

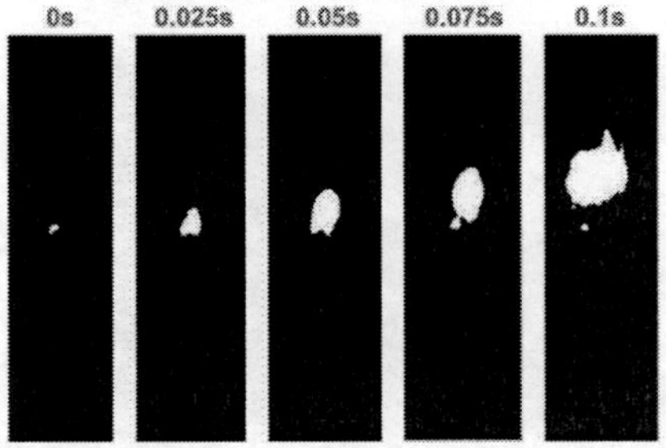

Experimental flame propagation under lean concentration (equivalence ratio $\phi = 0.01$) for droplɐ μm

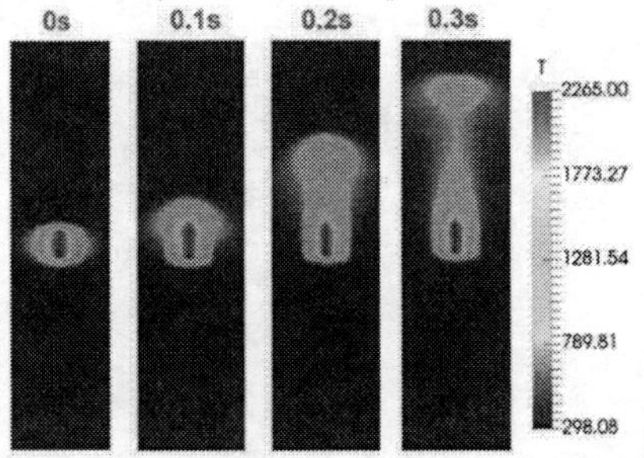

Flame propagation simulation under lean concentration (equivalence ratio $\phi = 0.01$) for droplet with SMD of 75 μm

Conclusic

This study implemented the electric field force in the OpenFOAM solver and proposed a procedure of simulating aerosol combustion under lean concentration condition. The simulation results show a slower flame than the experimental results, which is attributed to the insufficient high ignition energy and less turbulence in the simulation. Nevertheless, this study opens up the possibility to use simulation to study aerosol combustion.

References

[1] NFPA 30, Flammable and Combustible Liquids Code, 2015 edition, National Fire Protection Association: Quincy, MA.
[2] Globally Harmonized System of Classification and Labelling of Chemicals(G Nations
[3] Eichhorn, J., Carefull Mist can explode. Petroleum Refiner, 1955. 34(11): p. 194-196.
[4] Huang, Chen, and Andrei Lipatnikov. Modelling of gasoline and ethanol hollow- 2011-01-1896. SAE Technical Paper, 2011.
[5] D. R. Ballal and A. H. Lefebvre. Ignition and flame quenching of flowing heterogeneous fuel-air mixtures. Combust. Flame, 35:155-168, 1979
[6] Globally Harmonized System of Classification and Labelling of Chemicals(G Nations
[7] Burgoyne, J. and L. Cohen. The effect of drop size on flame propagation liquid Society of London A: Mathematical, Physical and Engineering Sciences, 1954. 225(1162): p. 375-392.

500448 A Case Study on Semi-Batch Endpoint Control

Abstract:

The endpoint control problem appears in many chemical processes in which neither of the two starting reactants can remain at the end of the reaction. Having unreacted materials in the final product above some threshold may ruin the batch due to a number of reasons, including difficultly of separation and progression of side reactions after limiting reagent has vanished. Modeling the process is vital in determining how much of each reactant to add in the presence of disturbances, side reactions, and measurement uncertainties. In this talk, a comparison study is performed between four modeling techniques for endpoint control of a semi-batch process: parameter estimation, principal component regression, projection onto latent structures (also known as partial least squares), canonical correlation analysis, and subspace identification.

The case study considers a semi-batch process taken from a collaboration with Eastman Kodak (Rawlings, Jerome, Hamer, & Bruemmer, 1989) (Rawlings & Ekerdt, 2012, Section 9.3). A dihalogenated precursor is dehalogenated using an organic base to form the divinyl product, which is used in photographic film production. Historically, this process has been manually controlled by operators. The batch reactor is charged initially with some approximately known amount of the precursor. Then, a fixed amount of the base B is added to the batch that is sure not to overshoot the end point. A measurement is obtained to check how close the batch is to the specification. Then, a series of injections of B are performed slowly with measurements being taken after each until the target product specification is met. Batch times can be long and vary based on operator experience. Since the conventional method involves conservative trial and error, production is slow and operating costs are high. The primary objective is to automate control and speed up the addition of B to reduce batch times and to increase the profitability of the process. Model predictive control (MPC) is one such advanced control system that can provide these benefits. Since a system model is required for MPC, several modeling techniques are explored in this document.

Both regression-based models of the time-unfolded data and dynamic state-space models are considered. Projection onto latent structures (PLS) and canonical correlation analysis (CCA) are used to develop the regression-based models (Wold, 1966; Hotelling, 1936). Subspace identification and parameter estimation techniques are used for the dynamic state-space models. Subspace identification techniques have been extended to batch processes using (Corbett & Mhaskar, 2016). After validation of these models, the MPC problem is formulated. Using feedback with regression-based models in this context is discussed to ensure zero offset. Simulation results comparing the closed-loop performance of the resulting controllers are presented. We conclude with remarks on the effectiveness and ease of implementation for each modeling technique.

References

Corbett, B., & Mhaskar, P. (2016). Subspace identification for data-driven modeling and quality control of batch processes. AIChE J., 62(5), 1581–1601.
Hotelling, H. (1936). Relations between two sets of variates. Biometrika, 28(3-4), 321–377.
Rawlings, J. B., & Ekerdt, J. G. (2012). Chemical reactor analysis and design fundamentals (2nd ed.).

Madison, WI: Nob Hill Publishing. (664 pages, ISBN 978-0-9759377-2-3)
Rawlings, J. B., Jerome, N. F., Hamer, J. W., & Bruemmer, T. M. (1989). End-point control in semi-batch chemical reactors. In Proceedings of the IFAC symposium on dynamics and control of chemical reactors, distillation columns, and batch processes (pp. 323–328).
Wold, H. (1966). Estimation of principal components and related models by iterative least squares. In P. R. Krishnaiah (Ed.), Multivariate analysis (pp. 391–420). Academic Press.

PUPUK KALTIM UNIVERSITY OF WATERLOO 17 AIChE Annual Meeting, Minneapolis, MN

Modeling the Effect of Tube Replacement on the Operation of Primary Reformer in Ammonia Plant

Muhamad Fariz Failaka, Fildzah Hanifati, and Ali Elkamel

Objective of The Study

Modeling of Primary Reformer's radiant and convection section using Aspen Plus 8.8 software is expected to be used to evaluate the performance of radiant and convection section of the primary reformer before and after replacement of catalyst tubes.

The replacement was intended to achieve the company long-term goal; i.e., to reduce the energy consumptions.

Furthermore, the modelling in Aspen Plus can also be used to obtain the energy efficiency and heat recovery of the radiant and convection section of the primary reformer.

Heat Utilization in Primary Reformer

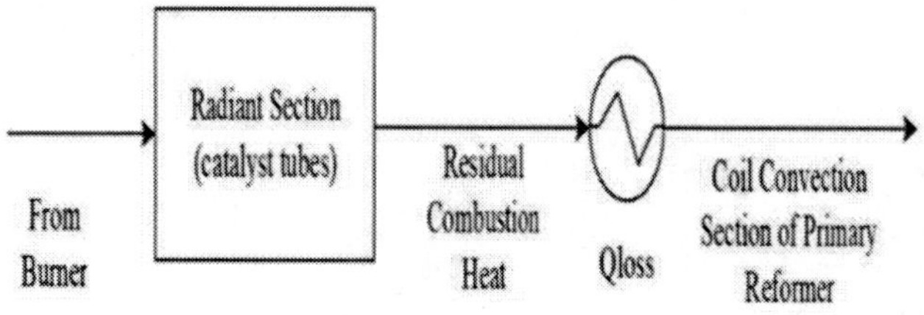

Technical Data of Catalyst Tube

Tube Material	Old	New
	IN-519 (24Cr/24Ni/Nb)	KHR-35CT (25Cr/35Ni/Nb-Ti)
OD Tube	152 mm	152 mm
ID Tube	120 mm	125 mm
Thickness	15,2 mm	12,7 mm
Length	13720 mm	13925 mm
Total Tube	168 (84 x 2 cell)	168 (84 x 2 cell)

The Algorithm

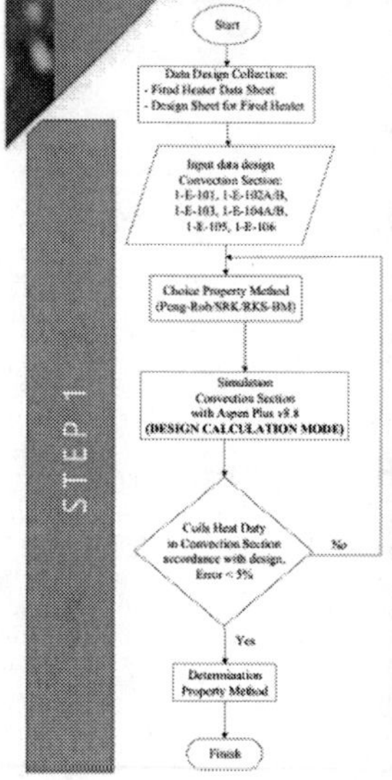

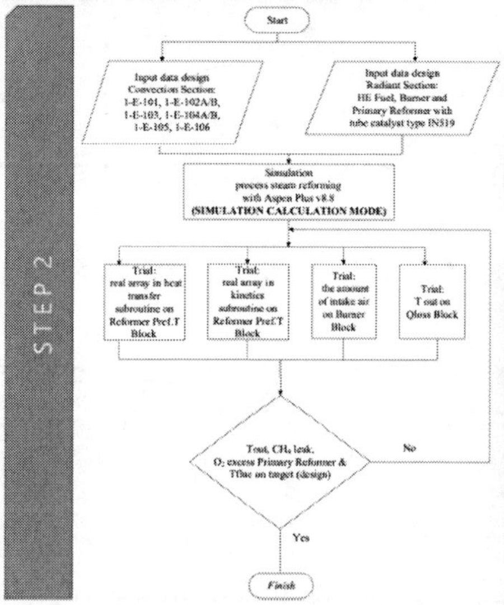

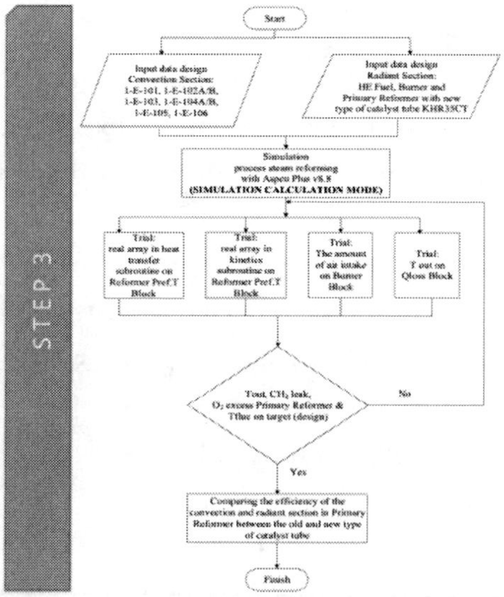

Results & Discussion

Simulation of Old Catalyst Tube

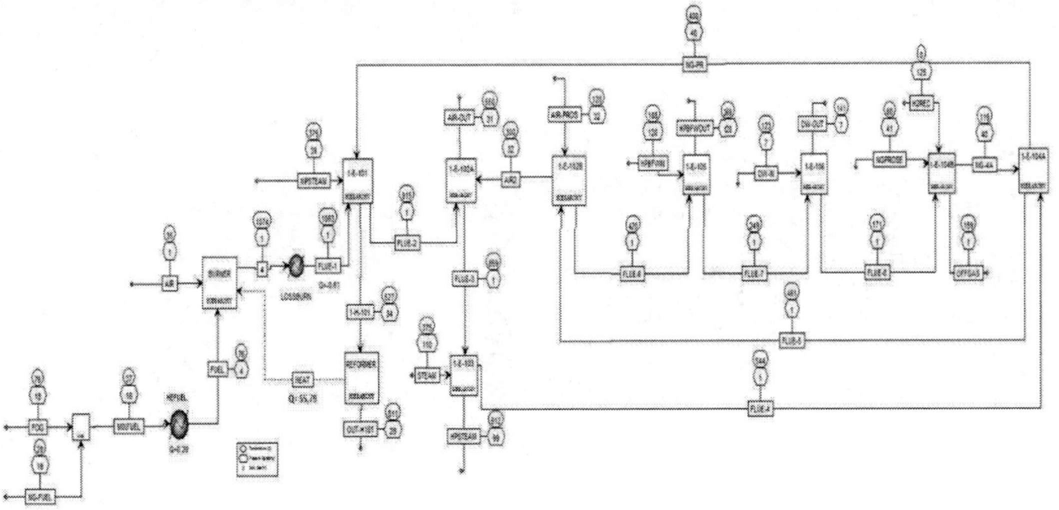

Results & Discussion

Simulation of New Catalyst Tube

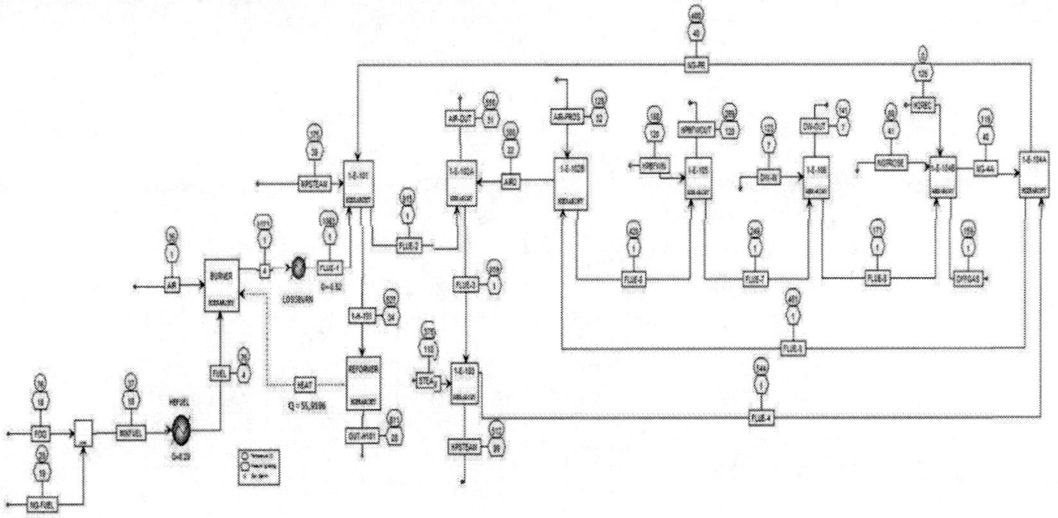

Results & Discussion

Comparison of Coils Temperature

Coil	Material	Temperature (°C)			
		Old Tube	New Tube	Diff.	Allowable
1-E-101	A335 P11	526.8612	526.8897	0.0285	592.5
1-E-102A	A213 T22	549.9463	549.9863	0.04	595
1-E-102B	A106 Gr.B	299.9008	299.9048	0.004	397
1-E-103	A213 T22	511.6304	511.6442	0.0138	548
	A213 T12	476.3054	476.3109	0.0055	512
1-E-104A	A335 P11	399.5086	399.5161	0.0075	447
1-E-104B	A106 Gr.B	119.2853	119.2923	0.007	200
1-E-105	A210 C	268.9413	268.9435	0.0022	298
1-E-106	A312 TP304	140.8268	140.8271	0.0003	200

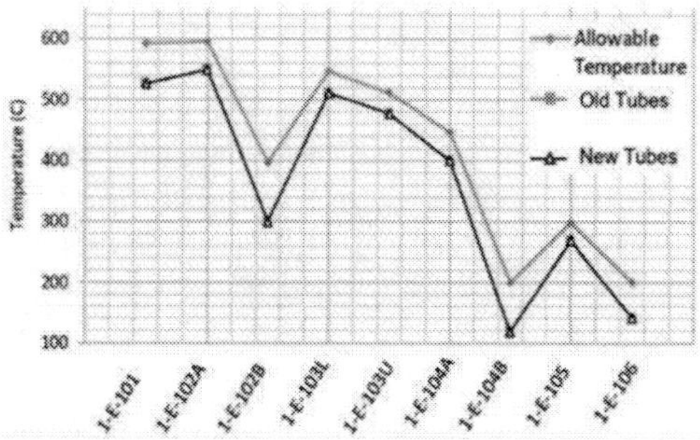

Results & Discussion

Comparison of Convection Section Temperature

	Old Tubes (ID = 120 mm)		New Tubes (ID = 125 mm)	
	T out, C	T flue, C	T out, C	T flue, C
Outlet Burner		1074.016		1072.506
Inlet Convection		1063.381		1063.489
1-E-101	526.8612	914.5493	526.8897	914.6319
1-E-102A	549.9463	858.5364	549.9863	858.6114
1-E-102B	299.9008	419.6672	299.9048	419.6739
1-E-103	511.6304	778.2357	511.6442	778.2953
	476.3054	544.319	476.3109	544.3272
1-E-104A	399.5086	460.8397	399.5161	460.8473
1-E-104B	119.2853	158.9992	119.2923	158.9997
1-E-105	268.9413	249.3252	268.9435	249.3269
1-E-106	140.8268	170.899	140.8271	170.8996

Results & Discussion

Heat Balance of Primary Reformer

	Heat in	Heat out			
	Burner (Gcal/hr)	Heat reaction in catalyst tubes (Gcal/hr)	Residual Combustion Heat (Gcal/hr)	Flue Gas Heat to Convection Section's Coil (Gcal/hr)	Diff. (Gcal/hr)
Old Tubes	121.8009	55.79	-66.0109	-66.6209	-0.61
New Tubes	121.8009	55.81	-65.98837	-66.50837	-0.52

Summary

1. The efficiency of the radiant section of primary reformer with old tubes is 56.12% and the efficiency of the convection section is 99.97%. Meanwhile, with the new catalyst tubes, the efficiency of radiant section is increased to 56.21% and the efficiency of the convection section is 99.99%.
2. The efficiency of the radiant section and the convection section were increased by 0.08% and 0.02%, respectively when using the new-type catalyst tubes.
3. The replacement of catalyst tubes in primary reformer has a potential issue about the increased workload of the coils in the convection section. However, the increase temperatures are still below the maximum allowable temperature of each coil.

Summary

4. With the new tubes type KHR-35CT, the heat recovery in the radiant section was increased up to 14.75% while maintaining the operating conditions as follows:
   - CH4 leak 10.14%,
   - excess Oxygen 1.05%,
   - the exit temperature of primary reformer 811 deg C,
   - the temperature of the flue gas 159 deg C.

Thank You

487567 Identification ofÂ PiecewiseÂ Autoregressive ExogenousÂ (PWARX) Model Using Efficient Optimization Algorithm

Abstract:

An efficient identification approach for piecewise autoregressive exogenous (PWARX) model is presented in this study. Specifically, we focus on a subclass, called hyperplane autoregressive exogenous (HHARX) model, which is able to approximate general nonlinear process as a sum of hinge function accurately and ensure the continuity of model even in a piecewise affine form. In fact, the ability and superiority of HHARX model have been shown in many literatures [1,2,3]. It receives great attentions from academia due to its simple structure, arbitrary accuracy and suitability for model predictive control (MPC). Hence, this empirical model is especially useful for the control and optimization of complex nonlinear process, for which first principle model is difficult to obtain.

However, the computational method for HHARX model identification has not been well-developed yet. For example, the widely used expectation maximization (EM) based algorithm is concerned with maximum likelihood estimation (MLE) but does not take the process continuity into account. Baesd on the same reason, the clustering-based method also cannot be used to build HHARX model. Among all existing methods, the mixed-integer programming (MIP) has been shown as a very promising technique for identifying all hinging function simultaneously to achieve a global optimal solution [1], but limited in the case where not too much data is available. In order to build HHARX model, the formulation proposed in Ref. [1] has to introduce many binary variables, proportional to the number of data and hyperplanes, such that even a state-of-the-art solver cannot handle it.

To build HHARX model for process control with big-data setting, a novel optimization approach is presented to identify hinging function sequentially. This scheme enables mixed-integer programming to solve an identification problem with relatively large-scale data set, and obtain a satisfied solution instead of the global optimum. In each step, only one hinging hyperplane is optimized to minimize the error between predictive and real outputs. Parameters for other hyperplanes are unchanged, such that only a few binary variables have to be considered. Using warm start, the modeling error is reduced continuously and finally a suboptimal solution will be founded promptly. Moreover, we also show that some practical issues in the general piecewise system identification problem, including switching frequency, missing output data and specified steady state, can be addressed efficiently within the same framework. Finally, the efficiency and accuracy of the proposed computational scheme for system identification are demonstrated by studying two nonlinear processes and comparing our results with conventional approaches.

Reference

[1] Roll J, Bemporad A, Ljung L. Identification of piecewise affine system via mixed-integer programming. Automatica. 2004;40:37-50.

[2] Breiman L. Hinging hyperplanes for regression, classification and function approximation. IEEE Transactions on Information Theory. 1993;39:999-1013.

[3] Pucar P, Sjoberg J. On the hinge-finding algorithm for hinging hyperplanes. IEEE Transactions on Information Theory. 1998;44:3310-3319.

495888 Closed-Loop Re-Identification of Multi-Rate System Using N4SID and Zone MPC

Abstract:

Closed-loop re-identification of multi-rate system using N4SID and zone MPC

ByungJun Park, Se-Kyu Oh and Jong Min Lee

School of Chemical and Biological Engineering, Institute of Chemical Processes,

Seoul National University, 1 Gwanak-ro, Gwanak-gu, Seoul 08826, South Korea

Changes in operating conditions or process units are common in industry and these changes result in the performance degradation of model predictive control (MPC). The quality of model plays a key part in maintaining the control performance of MPC, and it is desirable to re-identify the model under closed-loop rather than to perform open-loop plant tests. Multi-rate (MR) systems involve variables sampled at different rates, and they are difficult to identify under closed-loop without additional persistent excitation of the inputs, leading to poor regulation performance. In addition, most existing methods for closed-loop identification of MR systems involve polynomial transformation, which is not compatible with popular model forms for MPC such as step response and state space models.

This work presents a closed-loop re-identification method that provides a state-space model of multi-rate system suitable for MPC. The proposed approach consists of three major steps: First, a lifting technique is exploited to convert the multi-rate system to an augmented single-rate system [1]. Second, a modified MPC scheme, i.e., zone MPC, to strike a balance between input excitation and output regulation is introduced [2]. Finally, N4SID, the subspace identification algorithm, is employed to identify a state-space model [3]. This procedure, however, does not consider dynamic characteristics of systems like steady-state gain and settling time and structural characteristics due to using the lifting technique. This work modifies the N4SID algorithm to reflect steady-state, settling time and structural characteristic of the lifted model. First, the feedthrough matrix for the augmented state-space model is represented as a lower triangular form that is included as a linear constraint. Second, steady-state gain and settling time are explicitly included as constraints. Because the evaluation of steady-state gain involves matrix inversion, $C(sI-A)^{-1}B+D$, A and C are first identified without those two constraints and then B and D are identified with the constraints. This two-stage identification is computationally more robust and efficient than the standard N4SID algorithm that identifies A, B, C and D simultaneously.

Both single-input single-output linear time invariant (LTI) system and multiple-input multiple-output LTI system are illustrated as numerical examples. It is verified that the estimated lifted state-space model by the modified N4SID algorithm describe the dual-rate input-output data better than that of the standard N4SID. Results also show that the suggested scheme successfully improves the model accuracy while regulating the outputs within acceptable bounds.

References

[1] Z. Y. P. Qin, S. Kanae, "Identification of lifted models for general dual-rate sampled-data systems based on input-output data", *Proceedings of the IASTED Asian Conference Modelling and Simulation* (2007) 6–12.

[2] L. F. M. Oscar A.Z. Sotomayor, Darci Odloak, "Closed-loop model re-identification of processes under MPC with zone control", *Control Engineering Practice* 17 (2009) 551–563.

[3] B. D. M. P. Van Overschee, "N4sid: Subspace algorithms for the identification of combined deterministic-stochastic systems", *Automatica* 30 (1994) 75–93

Data-Driven Modeling and Optimization of an Ethane Steam Cracker

Burcu Beykal[1,2], Onur Onel[1,2,3], Efstratios N. Pistikopoulos[1,2]

[1]Artie McFerrin Department of Chemical Engineering, Texas A&M University, College Station, TX

[2]Texas A&M Energy Institute, Texas A&M University, College Station, TX

[3]Department of Chemical and Biological Engineering, Princeton University, Princeton, NJ

Email: burcubeykal@tamu.edu

Steam Cracking Motivation & Opportunity

- **The U.S. natural gas liquids production has been increasing for the past 10 years.**
  - ~110% from 2007 to 2017.
  - Projected increase of 29% by 2050.
  - **Increasing ethylene consumption** (from 43% to 51%).
  - <u>NGLs</u>: **High ethane content** (40% by vol.)
  - **Attractive feedstock for olefin production**

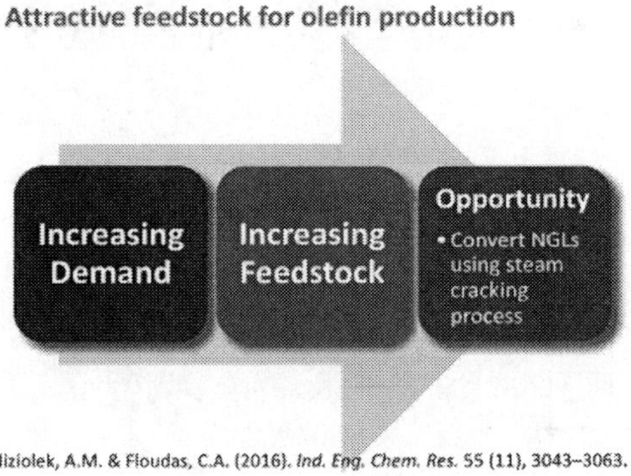

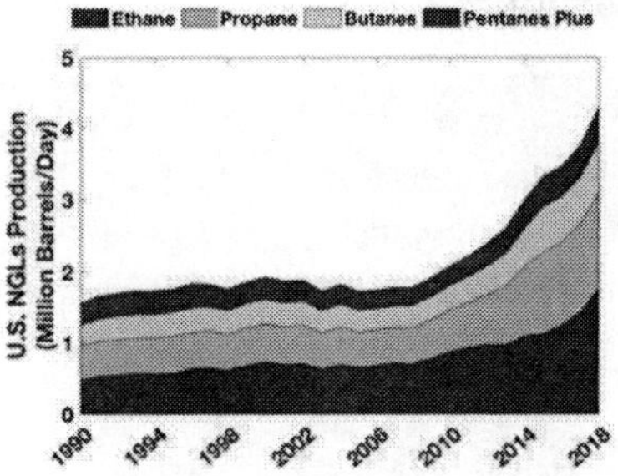

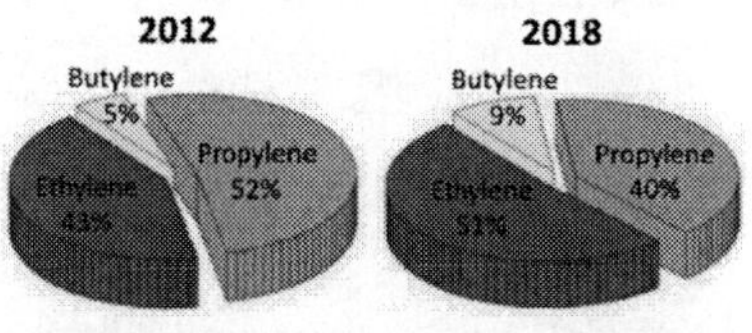

Ethylene Consumption is Increasing

Onel, O., Niziolek, A.M. & Floudas, C.A. (2016). *Ind. Eng. Chem. Res.* 55 (11), 3043–3063.

Methodology: Solving Steam Cracker Optimization Problem as a Grey-Box NLP

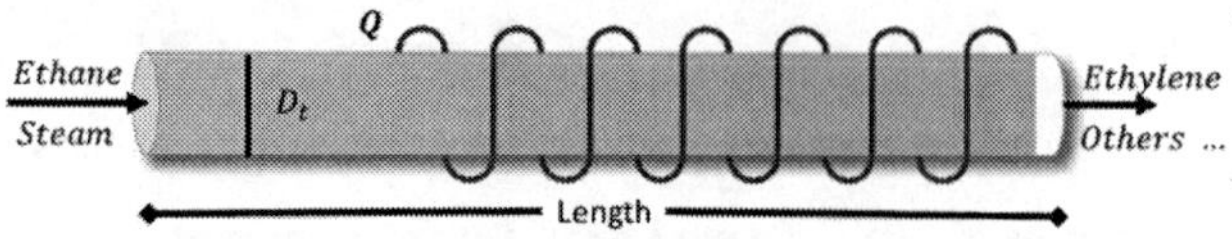

- **Previous Study: Orthogonal Collocation on Finite Elements**
  - Discretize the set of ODE's into nonlinear equality constraints
- **Current Study: Data- Driven Modeling & Optimization**
  - Via the ARGONAUT Algorithm

Objective: Maximize Profit = \$ Ethylene - \$ Costs

Steam Cracking Reactor Model:

Rate: $r = A_0 e^{-\frac{E_a}{RT}} . \prod_{i \in reactant} \left[\frac{F_i}{V}\right]^{v_i}$ **Coking**: $\Delta d_c = \frac{A_o \exp\left(-\frac{E_a}{RT}\right)\bar{C}\Delta t}{\rho_c . 10^6}$

Continuity: $\frac{dF_j}{dz} = -\left(\sum s_{ij} r_i\right)\frac{\pi D_t^2}{4} \quad \forall j$ **Energy**: $\frac{dT}{dz} = \frac{\left[Q(z)\pi D_t + \frac{\pi D_t^2}{4}\sum r_i(-\Delta H)_i\right]}{\sum F_j C_{pj}}$

Momentum: $\frac{dp_t}{dz} = \frac{\frac{d}{dz}\left(\frac{1}{M_m}\right) + \frac{1}{M_m}\left(\frac{1}{T}\frac{dT}{dz} + Fr\right)}{\frac{1}{M_m p_t} - \frac{p_t}{G^2 RT}}$

ODEs

Direct use of deterministic optimization methods are prohibitive!

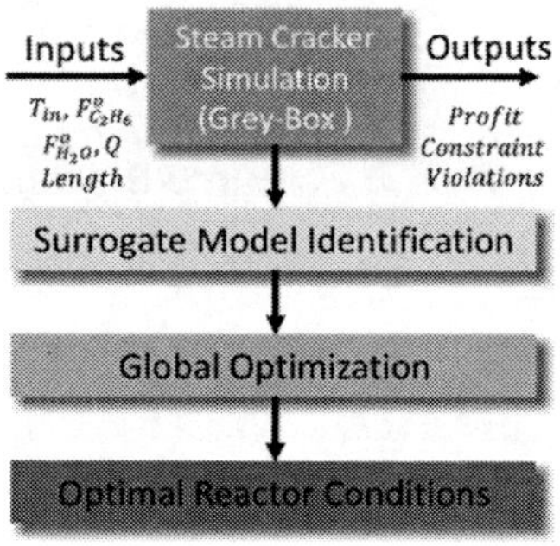

Boukouvala, F. & Floudas, C.A. (2017). *Optim. Lett.* 11, 895-913.

Key Results for the Steam Cracking Optimization

Results: No Coking

Reactor Coking (0%)
- **High Profit**
- Unfortunately not realistic!

Reactor Coking (0.139%)
- Slight coking **reduces profit by 54.2%**

Reactor Conditions	Optimal Results
Tube Length (m)	22.0287
Heat (MW)	7.47
Coke (%)	0
Profit (\$/s)	1.2413

Reactor Conditions	Optimal Results
Tube Length (m)	29.5469
Heat (MW)	5.93
Coke (%)	0.139
Profit (\$/s)	0.5675

Results: Coking

Orthogonal Collocation

Reactor Conditions	Optimal Results
Tube Length (m)	40
Heat (MW)	4.74
Coke (%)	0.21
Profit (\$/s)	0.4031
CPU Time (hr)	0.033

Data-Driven Optimization

Reactor Conditions	Optimal Results
Tube Length (m)	29.5469
Heat (MW)	5.93
Coke (%)	0.139
Profit (\$/s)	0.5675
CPU Time (hr)	26.513

Trade-off between methods

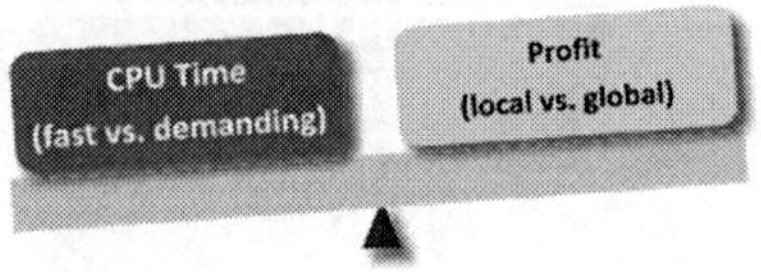

501920 Dual-Rate Approach for Data-Driven Modeling and Prediction of Behavior of Processes with Variations in Sampling Frequencies

Abstract:

Multi-rate systems, where inputs and/or outputs have different sampling rates, are encountered in many chemical and biological engineering applications. Typically, this occurs in systems where multiple variables are measured. Process variables such as temperature, pressure and pH are measured more frequently, at times nearly continuously, while other variables, such as species concentrations in a reaction mixture, are measured less frequently. An example of such processes is batch and fed-batch mammalian cell cultures, where sampling rates for viable hybridoma cells, nutrients glucose and glutamine, by-products lactate and ammonia, and the target product, a MAB, are significantly different. The simplest way to handle multi-rate systems is to neglect excess data from fast sampling signals and synchronize the signals with the slowest sampling rate. With this method, great amount of data will be discarded, the data that may contain crucial information regarding system dynamics. Therefore, some effort has been devoted to modeling, analysis and identification of multi-rate systems to take advantage of the rich information available in the experimental databases. In all the techniques proposed, multi-rate systems are generally simplified to multiple dual rate systems where the slower sampling rate is a positive integer () multiple of the faster sampling rate. One of the most successful techniques, polynomial transformation technique, can successfully transfer single rate models into dual rate models with ease. The dual rate model, after transformation, can utilize dual rate signals simultaneously. For such model structures, multiple identification techniques have been proposed with good estimation accuracy and convergence rate.

In this work, a second order discrete time system is considered first to demonstrate convergence of the proposed parameter estimation algorithm for a dual rate model. The convergence rates and estimation accuracy for various values are compared to examine efficacy of the dual rate parameter estimation technique. Illustrative examples are discussed. One of these pertains to application of a dual rate model with parameter estimation technique for modeling and prediction of mammalian cell cultures. The fastest sampling rate corresponds to viable hybridoma cell concentration, an intermediate sampling rate corresponds to glucose and glutamine concentrations, and the slowest sampling rate corresponds to monoclonal antibody (MAb) concentration. The sampling rate for monoclonal antibody is much slower than that for glucose and glutamine. Recursive time series models are developed for key culture variables. The model parameters are recursively estimated by using a least square estimation algorithm. Model stability is evaluated and confirmed by converting the time series model into a state space counterpart. The most rapidly measured variable, viable cell concentration in cell culture is expressed by a recursive ARMAX model. The performance of the dual rate model coupled with frequent parameter estimation in representation and prediction of mammalian cell culture producing a MAB is examined in considerable detail.

Monoclonal antibodies (MAbs) have extensive biomedical applications and are produced in mammalian cell bioreactors at a variety of scales, with glucose and glutamine being the principal carbon and nitrogen sources required for cellular metabolism. Fed-batch operation has certain inherent advantages over batch culture for MAb production. Design, optimization, scale-up, and control of bioreactors used for MAb production requires reliable predictive

empirical or mechanistic models for key cellular activities. The models used in prior studies have largely been the first principles-based models (FPMs), although data-driven models are receiving increasing attention due to their certain inherent benefits and increasing information available about biological processes. The simpler and much less rigid structure of data-driven models facilitates frequent updating of parameters and prediction of process trajectories, increasing their utility in representation, monitoring and control of these processes. Recursive time series models are developed in this work for representation of a mammalian cell culture, with process variables that are determinants of the performance of the culture being measured at different sampling frequencies. Appropriate parameter constraints have been imposed in parameter estimation algorithms and stability of these has been examined and is ensured. The data required for estimation of parameters are generated from simulated experiments using a well-tested FPM considering random variations in manipulated inputs, kinetic parameters in the FPM, and measurement error for an output. Glucose and glutamine being determinants of mammalian cell metabolism, their supply rates are considered to be the inputs. The predictive ability of the data-driven models is examined and demonstrated over a broad range of prediction horizon.

Nonlinear System Identification and Dynamic Real-Time Optimization of Postcombustion CO_2 Capture Processes for Cycling Applications

Rebecca Kim and Fernando V. Lima
Department of Chemical and Biomedical Engineering, West Virginia University, Morgantown, WV, USA

Presenting author's email: rhkim@mix.wvu.edu

2017 *AIChE Annual Meeting*

Minneapolis, MN
October 30, 2017

Background and Objective

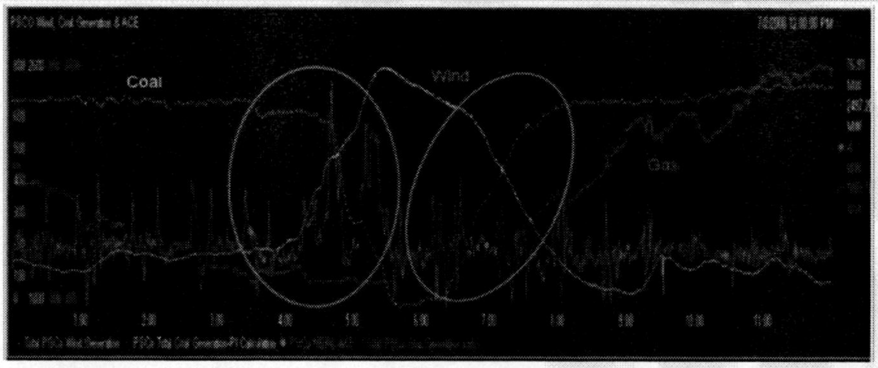

The increased focus on renewable energy sources is associated with the need for coal-fired power plants to cycle their load

Goal: Calculate optimal operating trajectories for a postcombustion amine-based CO_2 capture unit under cycling conditions

Figure from Bentek Energy LLC. "How less became more…. Wind power, and unintended consequences in the Colorado energy market." Evergreen, CO: Bentek Energy, LLC, 2010.

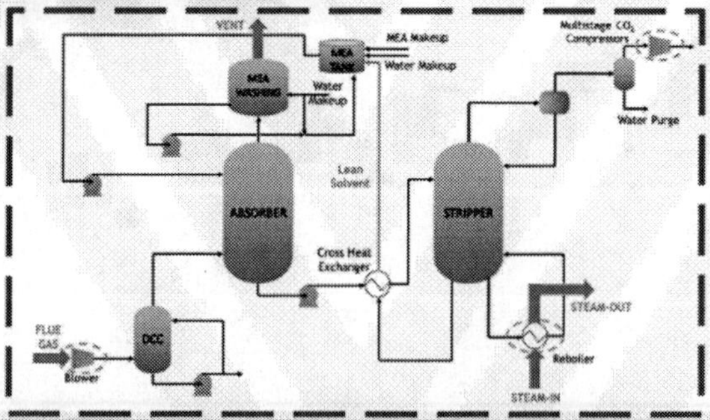

WEST VIRGINIA UNIVERSITY
DEPARTMENT OF CHEMICAL AND BIOMEDICAL ENGINEERING
LIMA RESEARCH GROUP
Control, Optimization and Design of Energy Systems (CODES)

Proposed Optimization Framework

Process Simulation
- **Dynamic Input-Output Data Generation**

System Identification
- **Nonlinear Auto-Regressive with eXogenous inputs (NARX) with wavelet networks model**

Optimization
- **Dynamic Real-Time Optimization (DRTO)**

<u>Disturbance Variable</u>: Flue gas flowrate during cycling

<u>Input/Manipulated Variables</u>:
- Steam flowrate
- Lean MEA solvent flowrate

<u>Output/Monitoring Variables</u>:
- Total power consumption
- Carbon capture rate
- Lean solvent CO_2 loading

FLUE GAS CYCLING

INPUT VARIABLES

MARKET PRICE

CARBON SCENARIOS
- Carbon taxation
- *Cap&Trade*

DYNAMIC REAL-TIME OPTIMIZATION

$$\min cost\,(\varphi) = \Delta t * \sum_t \begin{pmatrix} Electricity + Transportation + Storage - EOR^{*} \\ -Carbon\ credit + Carbon\ tax \end{pmatrix}$$

s.t.: Identified nonlinear **dynamic** NARX process model

(*) EOR = Enhanced Oil Recovery

Output variable optimal trajectories

WEST VIRGINIA UNIVERSITY
DEPARTMENT OF CHEMICAL AND BIOMEDICAL ENGINEERING
LIMA RESEARCH GROUP
Control, Optimization and Design of Energy Systems (CODES)

Results and Conclusion

Flue gas cycling & Cost parameters

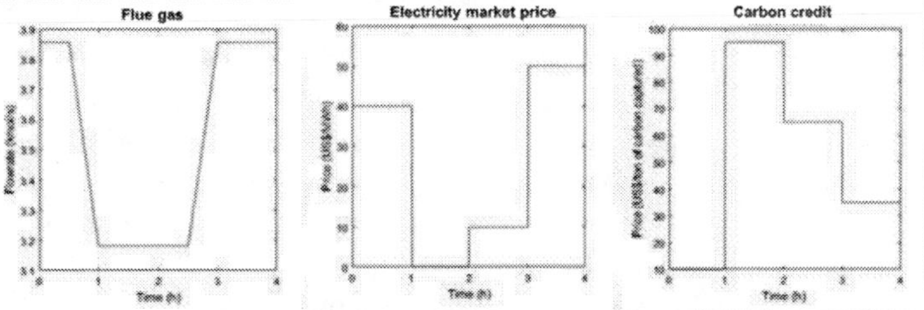

CASE 1

Transportation = 4.9
Storage = 13
EOR = 0
Carbon tax= 5
[parameter] = US$/ton CO_2

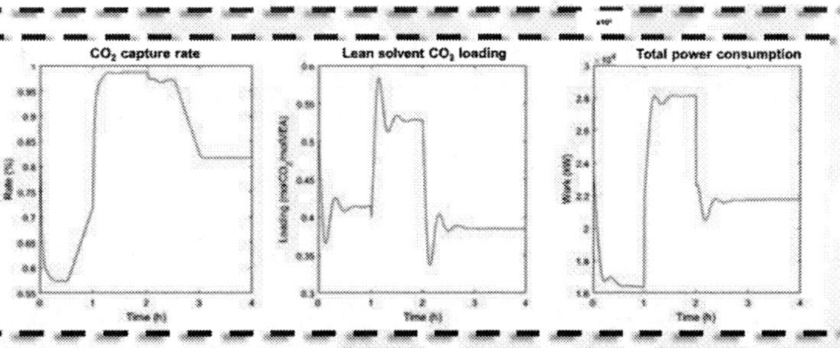

Results show that proposed **DRTO** approach can calculate the **optimal operating** conditions for the carbon capture unit according to **cost parameters**

WEST VIRGINIA UNIVERSITY
DEPARTMENT OF CHEMICAL AND BIOMEDICAL ENGINEERING
LIMA RESEARCH GROUP
Control, Optimization and Design of Energy Systems (CODES)

499785 Sparse Kernel Filtering Algorithms for Online Glucose Prediction in T1D

Abstract:

Development of predictive models for nonlinear systems with time-varying parameters is challenging. It is difficult to present a personalized physical model based on first principles, as the parameters of such models are usually time-variant or cannot be estimated easily. The alternative is data-driven modeling methods to estimate parameters and capture the relationships between inputs and outputs. But one of the challenges to achieve a real-time prediction based on empirical models is the dynamic changes related by many physical or environmental elements over time, which necessitate online updating of model parameters. Kernel-based modeling approaches with sparsification criteria, Approximate Linear Dependency (ALD) Criterion and Surprise Criterion (SC) are proposed in this work to enhance the accuracy of data-driven models. The approach is illustrated for real-time prediction of glucose concentration. With information from continuous glucose monitoring (CGM) systems, accurate online prediction of glucose concentration is becoming feasible for type 1 diabetes mellitus (T1D). In artificial pancreas (AP) systems, online glucose prediction is a key issue for proactive blood glucose regulation and for producing early alarms for hypo- and hyperglycemia. To obtain reliable and accurate glucose prediction values, various approaches based on physical or empirical models have been proposed.

Kernel-based modeling with sparsification criteria, ALD and SC is investigated for online glucose prediction for patients with T1D. The model is trained by current and historical CGM data and the kernel filtering algorithms are proposed to characterize the glycemic variability and serve as the online learning machine for glucose prediction. By combining sparsification criteria based on information theory, the learning dictionary online could be updated recursively by checking if the new kernel function is appropriate to be add into the subset. For ALD criterion, the distance of the new coming data to the linear span of the present dictionary in the reproducing kernel Hilbert spaces (RKHS) is indicated and the thresholds are set to classify if the data are learnable, redundant or abnormal. For SC criterion, the uncertainty of a new input-output pattern relate to the current knowledge of the learning system is quantified. It uses an information theoretic method that captures the surprise of the new exemplar and allows us to add or discard it in the previous learning system. Therefore, the dictionary growth of the online filter could be effectively curbed and the overfitting is avoided.

In AP systems, the time and space complexity of the glucose prediction model could be reduced by setting an appropriate threshold to ignore the redundant data without harming the prediction performance. Meanwhile, the abnormal readings of the CGM sensor could also be removed by setting the overfitting threshold. This is critical to compensate for CGM measurement noise and CGM inaccuracy. Based on above, the proposed real-time update scheme permits the prediction models to properly address time-varying characteristic of the glucose dynamics.

The validation of online glucose prediction is investigated from both *in silico* data (UVA/Padova metabolic simulator) and clinical data. Model predictions are compared with actual CGM measurements. The sampling time of the sensor is 5 minutes. Prediction accuracy is assessed on the basis of the metrics: mean absolute relative deviation [MARD (%)], root mean-square error [RMSE (mg/dL)], prediction-Error Grid Analysis (PRED-EGA) and network size. The proposed sparse kernel-based algorithms show better accuracy, especially in cases where random sensor noise is added and CGM sensor calibration procedure is performed.

Bounding Multistage Stochastic Programs Using Sample Average Approximation

Katie Martin, Brianna Christian, and Selen Cremaschi
Department of Chemical Engineering
Auburn University, Auburn, AL

AUBURN UNIVERSITY
SAMUEL GINN COLLEGE OF ENGINEERING

AIChE Annual Meeting October 30, 2017

Multistage Stochastic Programming (MSSP)

Scenario based modelling approach

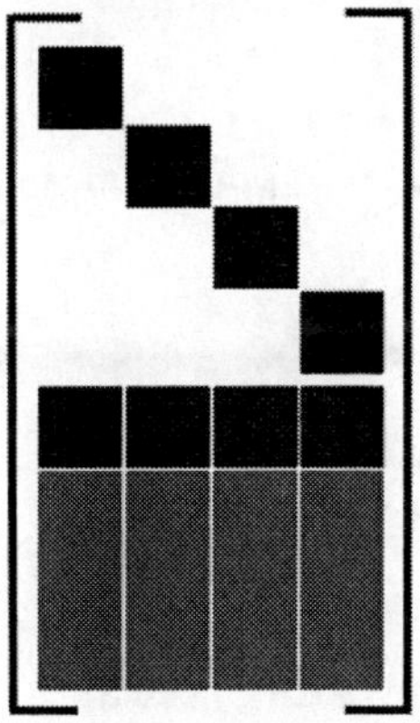

- Scenario specific constraints representing deterministic model
- First time period NACs
- Conditional NACs

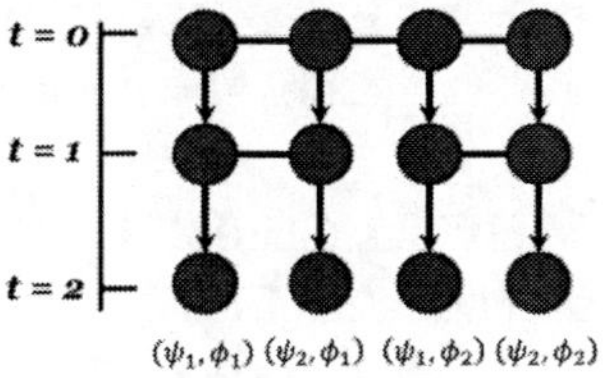

Discrete Planning Horizon

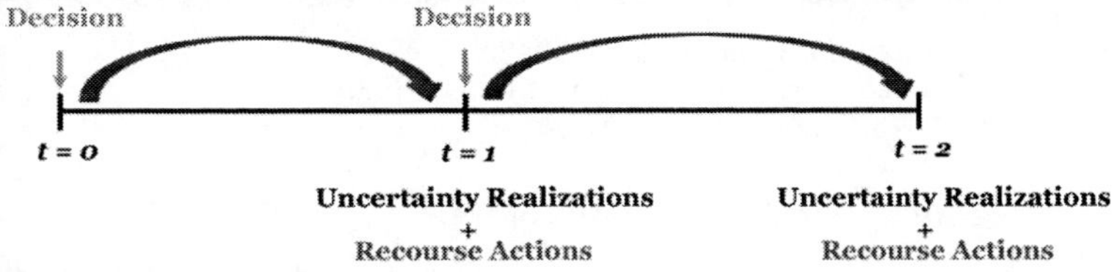

Scenario Generation

$\phi = \{\phi_1, \phi_2\}$

$\psi = \{\psi_1, \psi_2\}$

	ϕ_1	ϕ_2
ψ_1	(ψ_1,ϕ_1)	(ψ_1,ϕ_2)
ψ_2	(ψ_2,ϕ_1)	(ψ_2,ϕ_2)

$\Rightarrow$

$s_1 = (\psi_1,\phi_1)$
$s_2 = (\psi_2,\phi_1)$
$s_3 = (\psi_1,\phi_2)$
$s_4 = (\psi_2,\phi_2)$

Cartesian Product
$s = \phi \times \psi$

Pharmaceutical R&D Pipeline Management Problem

SIGNIFICANCE: Time from discovery to market has increased over last decade.
Limited patent life protection.
Increased need to boost pipeline efficiency.

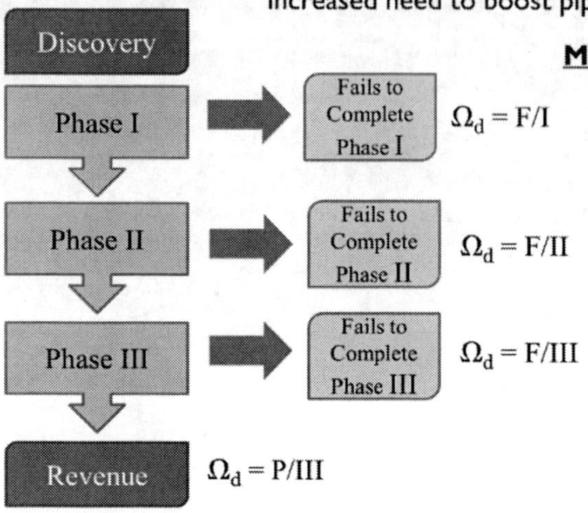

Mixed-Integer Linear Program (MILP)

- **Uncertainty-** Outcome of trial.
- **Constraints-** Available resources and non-anticipativity constraints (NAC)

COMPLEXITIES

Drugs	Trials	Scenarios
2	2	9
2	3	16
3	3	64
4	3	256
5	3	1,024

* problem becomes intractable

OBJECTIVE: $\max \textbf{\textit{Net Present Value}} = \sum_s p_s(Revenue_s + FreeRevenue_s - Cst_s)$

Sample Average Approximation (SAA) Provides Upper Bound to Solution

SAA: Uses a subset of the full scenario space to approximate a solution.
Pro- smaller scenario set translates to smaller problem size.
Con- NAC reductions aren't necessarily applicable.

SAMPLE NON-ANTICIPATIVITY CONSTRAINT (SNAC)*:

- Graph theory based polynomial time algorithm.
- Generates the minimum cardinality NAC set for any subset of the full scenario set generated by the Cartesian product.
- **Reduces number of constraints.**

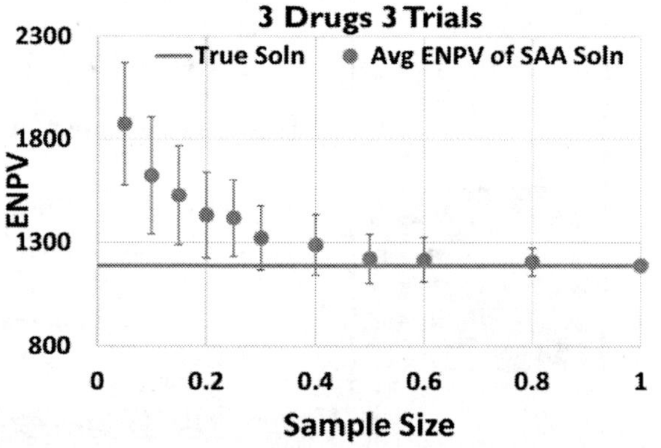

POSTER:

1. Impact of sample size and iterations on quality of upper bound generated from SAA.
2. Impact of SNAC* algorithm on SAA solution times versus using full NAC set.

***Christian, B. and Cremaschi, S.** *A Graph Theory Approach to Non-Anticipativity Constraint Generation in Multistage Stochastic Programs with Incomplete Scenario Sets.* AIChE Annual Meeting. San Francisco, CA.

499295 A New Proactive Methodology for Robust Berth Planning of Container Vessels

Abstract:

The rise in international trade is significantly observed over the years and one of the major transport mode used is the container ships, which also have potential impact on chemical industry. With many ships arriving at the port periodically, the operation schedule at container ports is becoming tighter. Over the past several years, a reduction in the productivity of port has been observed due to inefficient usage of the available resources. Amongst all the operations, the prime most important is an optimal berth allocation of the ship and yard allocation for the containers. The berth location of ships and the yard location of containers are inter-dependent on each other based on the transshipment of the containers. These operations are based on constraints like allowable time window for berthing, quay crane capacity and allowable process time for the ships. Thus, there is a mighty importance of scheduling of port operations and optimizing the available resources at the port terminal. Mathematical optimization has been used to improve the operational capability of port terminals using various realistic constraints.

Conventional berth allocation and planning models have difficulties to handle uncertainty of vessel arrival while constructing a nominal berth plan or nominal berth allocation. However, in reality, timetables become more and more vulnerable to disturbances. The delay of a single vessel might interrupt the entire schedule and make it very difficult or even impossible to recover. To deal with disturbances in transportation schedules, two approaches are getting more and more attention: (1) proactive robustness, which incorporates buffer times into strategic or tactical timetables to deal with disturbances and thus to prevent delay propagation through a well set schedule, and (2) reactive disruption management, which is concerned with operational recovery after a disruption. The research in this paper focuses on incorporating proactive robustness into the nominal berth planning.

Container vessels might arrive earlier or later than their nominal arrival time due to all kinds of uncertainties during port operations. Therefore, the customers and terminal operators agree upon an arrival window, which is placed around the nominal arrival time. The time interval between the nominal arrival and departure time is the nominal vessel process time. Existing publications are either focused on planning problem without considering realistic flexible arrival and departure time windows [1, 2], or using fixed maximal deviation from nominal arrival time that is still within the arrival window of vessels [3].

The problem is hence to develop a time window-based berth allocation planning model that minimizes the maximally required operation time for all scenarios where vessels arrive within their arrival time windows and depart within their departure time windows. The flexible arrival and departure time windows have been into account specifically as control variables. A robust mixed integer nonlinear programming (MINLP) model is developed, which explicitly incorporates the operation time agreements and minimizes the maximally required operation time. The general objective is to minimize the maximally required operation time while satisfy all the transportation demand and operation constrains. Commercial solvers such as BARON, DICOPT and ANTIGONE have been employed and

compared to obtain the optimal solution of the developed MINLP planning model. Computational results of the case study demonstrate the efficacy of the developed planning model, and a significant reduction in the maximal operation time can be obtained.

References

1. Wang, F. and A. Lim, *A stochastic beam search for the berth allocation problem.* Decis. Support Syst., 2007. **42**(4): p. 2186-2196.

2. Cordeau, J.-F., et al., *Models and Tabu Search Heuristics for the Berth-Allocation Problem.* Transportation Science, 2005. **39**(4): p. 526-538.

3. Hendriks, M., et al., *Robust cyclic berth planning of container vessels.* OR Spectrum, 2010. **32**(3): p. 501-517.

499882 Optimal Sampling Locations to Reduce Uncertainty in Contamination Extent in Water Distribution Systems

Abstract:

Water facilities rely on samples collected from the distribution system to provide assurance of water quality and meet regulatory requirements. If there is a water quality concern, additional sampling can be conducted to determine the extent of the water quality event and the source of the concern. Finding the location of the contaminant injection is important for long-term decontamination and recovery, while determining the extent of contamination is important for short-term response. Using models for this purpose poses a challenge because significant uncertainty exists in the system hydraulics, contaminant reaction dynamics, and incident characteristics. Determining the overall extent of contamination is important to allow utilities to prioritize their recovery efforts.

This presentation outlines an optimization formulation to identify strategic sampling locations in the water network that can quickly reduce uncertainty in extent of contamination. We present two mixed-integer linear programs (MILP) that seek to identify the best location or locations in the network to gather additional measurements to quickly determine the characteristics of the contamination incident. Similar approaches have been proposed for solving the source identification problem, however, the novelty of this work is that it incorporates probability metrics within a rigorous optimization formulation. In addition, the mixed integer programming formulations presented in this work have greatly improved computational complexity over the previous formulations. The new formulation solves within seconds, and is applicable for larger water network models.

Sandia National Laboratories is a multi-program laboratory managed and operated by Sandia Corporation, a wholly owned subsidiary of Lockheed Martin Corporation, for the U.S. Department of Energyâ€™s National Nuclear Security Administration under contract DE-AC04-94AL85000

501686 Computer Aided Analysis about the Flow Channel Structure Effect on the Vanadium Redox Battery

Abstract:

Modern civilization demands more energy ever than the past. However, the traditional energy source like fossil fuel cause pollution and global climate change. The renewable energies -for example, solar cell and wind turbine and biomass fuel- are suggested as alternative energy source. Electric power scale of renewable energy is usually smaller than ordinary energy source. Furthermore, Lots of renewable energy is influenced by circumstance and time. To achieve stable electric power supply, it is inevitable to develop high capacity energy storage system. Also, Smart grid is one of the technology to provide stable power supply. Redox flow battery (RFB) is an electrical battery which uses redox reaction of electrolyte ions to generate power. It is a prospective energy storage system (ESS) because of its flexible installation. Due to separation of generation and storage of VRB, it is easy to control capacity and power level of VRB by adjusting storage tank and cell size respectively. Those advantages allows VRB to be appeared at the various field. For example, VRB is widely used in applications including, but not limited to, remote area (e.g. island, Mountains) power supply, back-up power and power level and distribution optimization. The power scale of VRB is also wide. However, there are a lot of alternative device for ESS. VRB must be improved of its efficiency to be competitive

VRB, a kind of RFB, takes vanadium ions for its redox couple. Using vanadium redox couple had certain advantages compared to using other materials. Because cross-over loss was less than other redox flow batteries. Also, VRB side reaction is smaller than other redox flow batteries. Several researches has been carried out to improve VRB performance. Generally, porous carbon is used as VRB electrode. Proper heat and chemical treatment on porous carbon could assist electrochemical reaction of cell. Polysulfonic membrane had been used in early developed RFB until nafion was invented. Nowadays, almost all RFB uses nafion membrane for its high hydrogen ion delivery.

Â Researches has been progressed on VRB not only on the materials side, but also other aspects as well. Previous researches regarding ordinary electrochemical cell such as fuel cell show that well designed cell structure can improve cell performance. Cell structure can reduce activation loss by enhancing mass transportation. Similar research was conducted on VRB. The research revealed that the cell characteristics changed significantly following the type of channel structure. Pressure drop and limit current dramatically changed by cell type because of mass transportation.

Â Analytic modeling is widely used to analyze RFBâ€™s dynamic performance. Because analytic modeling require severe differential calculation, computer simulation is widely used for VRB research. Furthermore it is hard to measure inner side of cell experimentally. Therefore, Computer simulation can be a complementary tools to analyze and design RFB It should also be noted that parameters are easily changed in computer simulations. Recent computing power growth enable much more intensive simulation than before. Therefore, it is more convenient to analyze electric cell with computer simulation. Computer simulation is preferable way to analyze and designs cell structure. Recent studies shows that channel type influence cell performance characteristics such as current-potential curve. In this study, inner cell features such as concentration distribution, flow distributions, electrical potential and transfer current distribution were analyzed by channel type to investigate relationship between channel structure and performance. . Then, new VRB design which combined various advantages of each cell structure type are proposed. Modified cell performance was analyzed with computer simulation. Research was carried with COMSOL Multiphysics

487920 Reputation-Based Market on the Blockchain Platform: An Emission Trading Application

Abstract:

The application of blockchain in emission trading that encourages participant compliance and climate change conscious behaviour through trading advantages is presented in this paper. The designed scheme adds a reputation system to conventional cap-and-trade scheme implemented on the blockchain platform. The trading process in the application is heavily influenced by participant's reputation that indicates their business approach related to the emission abatement effort. The reputation system and the trading advantages distributed based on it act as an inducement for participant better business practice. The case study presented exhibit the trading process as implemented using blockchain while emphasizing the trading advantages some participant will have over the others on a free trading market. Multi-criteria analysis is employed to evaluate the viability of the scheme. It concludes that the proposed scheme that utilizes blockchain technology and reputation system can be used to improve the conventional emission trading scheme in reducing emission production and increasing the adoption of abatement technology.

490505 Systematic Process Design of a Styrene Production Plant Using a Hierarchical 12 Task Procedure: Waste Stream Utilization for Improved Sustainability

Abstract:

Monomeric styrene is consumed in large quantities for the production of polystyrene polymers. Improvements of the traditional process have been developed continually in order to improve product yield and reduce utility costs[1]. However, the process is still highly energy consuming due to the high temperatures and pressures required for the reactions, as well as the necessary purification of the crude styrene that is produced. As such, development of a more efficient and sustainable process is highly desirable in terms of economic savings and reduced environmental impact.

In this study, a process design for a styrene production plant is developed based on ethylene alkylation of benzene to ethylbenzene and a subsequent isothermal dehydrogenation of ethylbenzene to styrene. The developed process design has a capacity of approximately 150,000 tons of 99.92 wt% styrene per year, along with a recovery of approximately 4000 and 2500 tons per year of industrial grade toluene and hydrogen byproducts, respectively. Various options for utilizing the recovered hydrogen gas are considered. In order to reduce raw material consumption, polyethylbenzenes are recovered and reacted with benzene to form ethylbenzene in a transalkylation reaction. The recovery of ethylbenzene, polyethylbenzene, styrene, and benzene for recirculation, is performed using distillation columns.

The process design is developed according to a systematic 12 task procedure for process design[2 3]. The 12 tasks are solved hierarchically, and can be summarized as follows. Tasks 1-2: Investigation of the product(s), chemical process(es), and economic potential. Task 3: Process synthesis in order to obtain a process flowsheet. Task 4: Solution of the mass balance of the process flow sheet using simple models. Task 5: Specification of operating conditions. Task 6: Solution of the energy balance of the process flowsheet. Task 7: Simulation of the process using rigorous models for the various unit operations. Task 8-9: Equipment sizing and costing calculations along with an economic evaluation. Task 10-12: Heat and mass integration, environmental impact considerations, and process optimization. To facilitate the solution of the tasks, the commercial software PRO/II is used to simulate the process. Sustainable design alternatives for the process are evaluated using SustainPro, while a life cycle assessment is carried out using LCSoft.

By solving the 12 tasks, an efficient process design is obtained. This includes an investigation of more sustainable alternatives to the traditional styrene synthesis; the focus of this investigation being efficient recovery and utilization of the hydrogen produced by the dehydrogenation of ethylbenzene as well as optimized heat integration between unit operations. For the utilization of the hydrogen, two design cases are investigated: (i) Utilizing the hydrogen as fuel for plant burners, and (ii) recovery of the hydrogen to be sold on the global market.

References

1. James, D. H. & C. W. M. *et al.* Styrene. *Ullmannâ€™s Encycl. Chem. Technol.* **1,** 1â€“32 (2003).

2. Biegler, L. T., Grossmann, I. E. & Westerberg, A. W. *Systematic methods of chemical process design*. (Prentice-Hall, 1997).

3. Babi, D. K. *Teaching Sustainable Process Design Using 12 Systematic Computer-Aided Tasks. Computer Aided Chemical Engineering* **37,** (Elsevier, 2015).

492660 Sustainable Production of Dimethyl Carbonate and Ethylene Glycol Via a Systematic Process Design Framework

Abstract:

In the past several decades, there has been a growing awareness and an increasing concern in the public over the human impact on the environment. Governments are pushed toward enacting laws and policies to further safeguard the planet for the future. Consequently, the chemical industry has seen a drive to produce *greener* products and to reduce carbon dioxide emissions.

As a potential strategy to reduce carbon dioxide emissions into the atmosphere, CO
have gained much attention in recent times as means to produce fine chemicals. Dimethyl carbonate (DMC) is one such product that can be synthesized by utilizing CO_2.

Dimethyl carbonate is a non-toxic, readily biodegradable chemical that exhibits high reactivity [1]. DMC received its VOC exemption under the Clean Air Act on January 13, 2009 by the US EPA. While the traditional production from phosgene and methanol uses toxic and hazardous chemicals, DMC can now be produced via the transesterification of ethylene carbonate (EC) and methanol. EC is obtained from reacting ethylene oxide (EO) with CO_2. Ethylene glycol (EG) is a valuable product that is coproduced in this process.

A sustainable process is developed to produce DMC and EG using a systematic framework [2]—a twelve-step hierarchical decomposition method. Using this approach, a plant with an annual production capacity of about 100,000 tonnes of DMC is designed. EO, CO_2, and methanol are the raw materials. EO and CO_2 are first reacted at 120 ^{0}C and 20 bar to form EC. A flash is used to separate EC from impurities and unreacted components. The EC stream is sent into the second reactor to react with methanol at 60 ^{0}C and atmospheric pressure to produce DMC and EG. The outlet of this reactor contains DMC-methanol and EG-EC azeotropes that are separated in a distillation column at atmospheric pressure. The overhead stream of DMC and methanol is separated in a high-pressure distillation column at 40 bar while the bottoms stream of EC and EG are separated using a membrane. Once the sizing and costing are done, a base case design is obtained.

An economic analysis is performed on the base case design to identify process hotspots that are translated into design targets for process integration and optimization. The resulting process is then put through sustainability analysis [3] and life cycle assessment [4] to refine it further with respect to environmental impact. In this project, PRO/II is used as the simulation tool. The final objective is to obtain a more sustainable process design that is also able to achieve a negative net CO_2 emission.

[1] P. Tundo and M. Selva, "The Chemistry of Dimethyl Carbonate". *Acc Chem Res*, 2002, vol. 35 (9), pp 706–716.

[2] D. K. Babi, "Teaching Sustainable Process Design using 12 systematic Computer Aided

Tasks". *Computer Aided Chemical Engineering*, 2015, vol. 37, pp. 173-178.

[3] A. Carvalho, H. A. Matos, and R. Gani, "SustainPro—A tool for systematic process analysis, generation and evaluation of sustainable design alternatives". *Comput. Chem. Eng.*Mar. 2013, vol. 50, pp. 8–27.

[4] S. Kalakul, P. Malakul, K. Siemanond, and R. Gani, "Integration of life cycle assessment software with tools for economic and sustainability analyses and process simulation for sustainable process design". *J. Clean. Prod.*May 2014, vol. 71, pp. 98–109.

493676 Optimal Use of Water for Hydraulic Fracking of Gas Shale Production

Abstract:

The development of the gas shale industry by the continuous discovery of new shale reserves around the world, the improvement of extraction technologies, and the design of the new logistic process operations have produced the global interest in exploiting this resource to satisfy the current increasing energy demand based on the dependence on fossil energy sources. Although, shale gas is surrounded by several controversies, mainly environmental issues, due to the large amounts of water required for the extraction processes (hydraulic fracturing), as well as the large volumes of polluting effluents that are discharged to the environment (flowback fluid). In these sense, this work proposes an optimization approach to the design of a water network for gas shale production considering the recycle and reuse of water to reduce the fresh water consumption and the decrease the discharge of polluting effluents where it is included a storage system and a sophisticated treatment system.

The discharge of effluents produced by the shale gas extraction processes are composed by several substances that cause major problems, mainly high ecosystem damage and human health problems. The identified pollutants that cause major affectation and are presented in greater concentration are the TDS (Total Dissolved Solids), the treatment system is composed of specific separation units to obtain certain concentration of TDS according to the limits to be discharged to the final disposal or according to the specification to be recirculated in the water network. It is proposed the implementation of the eco-indicator 99 methodology to evaluate the efficiency for the treatment system, and quantify the total environmental impact of the water network operations. The model defines the number and type of treatment units in the water network according with the environment impact and cost. The multi-objective optimization model is defined by the economic objective function that aims to minimize the total annual cost for the water network that includes the storage system and the treatment system, and operations and capital costs. The environment objective is specified by the minimization of the environmental impact by the discharged concentration of TDS and the consumption of fresh water.

Finally, a case study for the northwest of Mexico is implemented to evaluate the proposed methodology, it should be noticed that there is a severe water scarcity problem in this region.

494466 A Systematic Process Design for Sustainable Dimethyl Carbonate Production through Carbon Dioxide Utilization

Abstract:

With greenhouse gas emissions becoming an increasing global concern and environmental laws becoming more stringent, processes that produce carbon dioxide should ensure minimal emissions. Dimethyl carbonate (DMC) is a profitable, specialty chemical, that could be produced using CO2 as a raw material. DMC is a biodegradable, non-toxic solvent with exceptional properties especially as an electrolyte in lithium ion batteries and other energy saving devices.

In this work, the systematic hierarchical decomposition design method consisting of twelve tasks as described by Babi et. al [2] is employed to design a process producing 100,000 metric tons of 99% pure DMC per year with an additional production of 73,000 metric tons of ethylene glycol (EG) of 91% purity. This method assists in constructing the process initially obtaining a base case and subsequently developing an optimized rigorous design. Data required to develop a plant of the desired capacity is assimilated in tasks 1-3 to facilitate preliminary design decisions. Those are then verified through a mass balance in task 4 employing the commercial design simulator PROII. The simulation is extended from the initially simple model to a rigorous one in tasks 5-6 leading to a full fledged rigorous model where mass and energy balances are confirmed in task 7. The process is then sized and an economic analysis is carried out with the software ECON in task 8-9 leading to a base case for the process. Through that, targets for improvement are identified and the process is improved through optimization and heat integration in tasks 10-12. This is followed by an environmental impact analysis after each improvement step, to asses the overall sustainability of the final design. Applying these last tasks enables the design of an economically viable and simultaneously sustainable process obtaining DMC.

DMC can be produced from methanol (MeOH) and CO2 through an intermediate alkylene carbonate production route. In this mechanism, CO2 reacts with EO to form ethylene carbonate (EC) in a reaction under elevated pressure. Subsequently, MeOH is added and a transalkylation of the cyclic intermediate occurs to form DMC and EG as a byproduct at near ambient conditions. Azeotrope formation between DMC-MeOH and EG-EC is a crucial part of the separation sequence in the process. DMC is obtained at high purity from the pressure sensitive DMC-MeOH azeotrope operating at 15 atm. The purity of byproduct EG coming from a regular distillation is further increased by hydrothermal conversion of the remaining EC and separated using a flash [3].

From an environmental standpoint, apart from the utilization of CO2, equivalent with a net reduction of CO2 emissions, a main focus is set to reducing waste. This is achieved by handling the azeotropes by minimizing the use of solvents which makes the separations environmentally benign. The process is made economically more attractive by mass integration through recycling EC and heat integration to use energy efficiently. Through this hierarchical decomposition method, a potential zero discharge closed loop system is designed through the recycle of CO2 produced from the hydrothermal decomposition of EG-EC azeotrope.

[1] Ullmann, Fritz, et al. Ullmann's encyclopedia of industrial chemistry. Vol. 7. Weinheim, Germany: Vch, 2012.

[2] D. K. Babi, "Teaching Sustainable Process Design using 12 systematic Computer Aided Tasks". Computer Aided Chemical Engineering, 2015, vol. 37, pp. 173-178.

[3] Seader, Junior D., and Ernest J. Henley. Separation process principles. Hoboken, NJ: Wiley, 2011.

494498 Evaluation of Carbon Monetization in Power Systems for Flaring Mitigation

Abstract:

One of the mayor problems that contributes to climate change is gas flaring. Flares are open flames used for disposing of waste gases during normal and abnormal operations as safety devices. The World Bank Group has a leadership role in the initiative for gas flaring reduction through the global gas flaring reduction partnership (GGFR), and some legislations have been taken to promote the minimization of emissions using economic incentives and penalties (monetization). Numerous projects have taken into account monetization to improve the economic feasibility of them; nevertheless, monetization of greenhouse gas emissions may not be the best way to mitigate environmental impact. This work aims to study the effect of monetization of emissions versus a multi-objective strategy when industrial flare streams are used as supplementary fuel in a cogeneration system to reduce fresh fuel consumption and at the same time to satisfy energy requirements. An optimization approach is developed to determine the optimal size of the main equipment of the cogeneration system while the profit is maximized and greenhouse gas emissions are minimized. There is presented a case study that represents an industrial complex with several flares to show the evaluation of three different scenarios; the first one looks for stimulate the use of flares through economic incentives by the reduction of emissions, the second one seeks to penalize the generation of CO_2 in order to do attractive the utilization of waste resources like flare streams. Finally, the third scenario proposes the creation of a multi-objective function that pursues to maximize the profit and minimize the emission simultaneously. The results shows that the monetization of emissions is not necessary the best method to incentive the investment in alternative to reduce emission by industries. It affects directly the economy of new projects; notwithstanding, it does not affect in the same proportion the quantity of emissions generated. On the other hand, when the multi-objective strategy is used, economic and environmental benefits are obtained simultaneously.

495817 A Study on Maximizing the Energy Utilization of Process Operation By Integrating Multiple Energy Sources

Abstract:

Process industries are under heavy pressure to minimize the operation cost by improving their energy costs. The external environment is continuously changing with the increasing price of fossil fuels and the deviating outputs of renewable sources. Therefore process industry should be prepared to get involved in multiple sources of different energy sources. Multiple sources can be organized into a hybrid system to have a synchronized effect. This paper investigates the technical and economic feasibility of the hybrid system. This paper is specifically concerned with a hybrid system consisting of a wind turbine, an electrolyzer, hydrogen storage, and a fuel cell. Excess energy produced by the wind turbine is transformed into hydrogen by electrolyzer and stored. The hydrogen may thus be used directly as a fuel or converted back to electricity by a fuel cell later. Therefore, the redundant wind power that could be curtained can be stored and used again. The feasibility study like the proposed one allows us to recognize the key bottleneck technologies for accelerating the introduction of hybrid systems and the associated electrolysis technologies.

495824 Development of Reaction Mechanism and Kinetics for the Production of Butadiene through Oxidative Dehydrogenation of Alkane or Alkene

Abstract:

One of the methods to produce butadiene product in industry is oxidative dehydrogenation (ODH) of alkane or alkene. In this study, reaction mechanism and kinetics were developed for the production of butadiene through ODH over an transition metal catalyst. For this, the kinetic data was obtained from a small scaled adiabatic reactor in the temperature range of 350 °C - 450 °C. Furthermore, deactivation behaviors of catalyst were studied under abnormal condition to investigate the impact of operating condition on catalyst life. Kinetic parameters were fitted against experiment data, using a linear method and computer curve-fitting technique. Integral experiment data was used to validate the proposed kinetic model. Lastly, various design and operating parameters were considred for the design of catalytic reactors (i.e. change of reaction temperature, feed composition, and residence time).

492439 Multiobjective Tabu Search for Plant Design Models

Abstract:

Multiobjective Tabu Search for Plant Design Models

Multiobjective optimization techniques are employed to visualize the relationship between operating parameters and efficiencies related to profit, emissions, and safety. The challenge in generating trade-off curves for these attributes comes from the nonlinearity and complexity of plant design models, so higher level optimization heuristics are needed to compute the Pareto optimality curves and surfaces. The purpose of this research is to investigate the efficacy and capabilities of the Tabu Search algorithm for multi-objective optimization specifically for plant design projects. A Tabu Search algorithm was developed to solve the 0-1 knapsack problem for multiple objectives. Multiple plant design optimization problems are then solved to determine Pareto surfaces. Results show how tuning the parameters of the algorithm leads to more efficient determination of the Pareto surface, for highly nonlinear plant design models.

492545 Stochastic Optimization to Reduce Cost of Energy for Parabolic Trough Solar Power Plant

Abstract:

The need for clean and cheap renewable energy is on the rise. Solar energy is one of the cleanest and readily available technologies with almost zero carbon emissions. Optimizing the resources to produce efficient power at low costs is the need of the day. However, solar energy power plants face a number of uncertainties like the weather. Since the technologies are new, cost uncertainties are common. In this work, we optimize a solar thermal power plant with two different capacities using the novel Better Optimization of Nonlinear Uncertain Systems (BONUS) algorithm. We use the System Advisory Model (SAM) system from NREL to model performance and economics of the power plant. Since this is a black box model, optimization and optimization under uncertainty becomes difficult. Unlike the deterministic optimization problem, in stochastic programming (or stochastic optimization), one has to consider the probabilistic functional of the objective function and constraints. The generalized treatment of such problems is to use computationally intensive probabilistic or stochastic models instead of the deterministic model, inside the optimization loop. BONUS can circumvent the problems associated with black box models, computational intensity of sampling, and perturbation derivative costs. Instead of the stochastic modeling loop, BONUS uses a statistical reweighting approach to obtain the probabilistic information. It has been shown that BONUS reduces computational intensity by 98 to 99% for large scale problems like this problem of optimization of a solar thermal power plant. We found that the solution of expected value cost minimization for the two different capacity solar plants is robust in the face of uncertainties.

Lingxun Kong and Christos T. Maravelias
University of Wisconsin – Madison
E-mail address: maravelias@wisc.edu

Simultaneous Utility and Heat Exchanger Area Targeting for Integrated Process Synthesis and Heat Integration

- Simultaneous process synthesis and heat integration
  - Optimization based on an economic metric
- Consider both utility usage and heat exchanger area in the heat integration
- Consider process streams with **variable** temperatures and flowrates

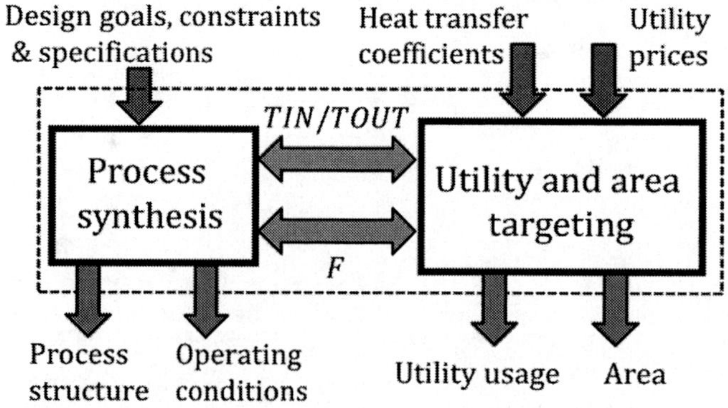

Additional features

- Multiple utilities and isothermal streams
- Unclassified streams (i.e., streams with unknown hot/cold identity)

Kong et al., *Industrial & Engineering Chemistry Research*. **2017**

Dynamic Enthalpy Grid

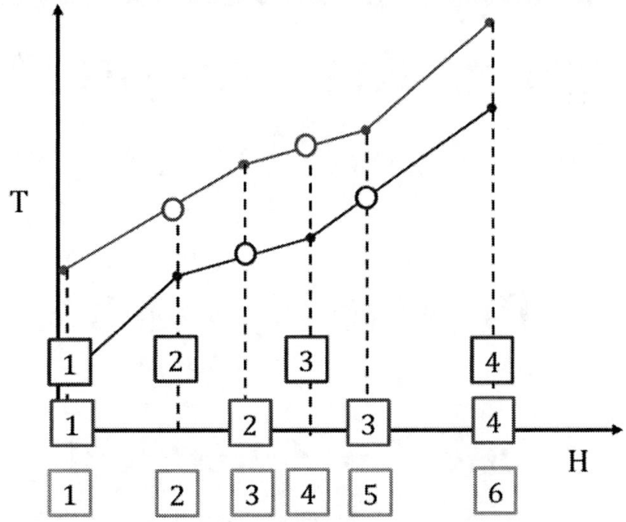

- Introduce a "dynamic" enthalpy grid
- Calculate enthalpy and temperatures at each grid point
  - Grid temperatures correspond to stream inlet/outlet temperatures
  - Implicit ordering of grid temperatures
- Estimate exchanger area at each interval

Illustrative Examples

Utility + area target

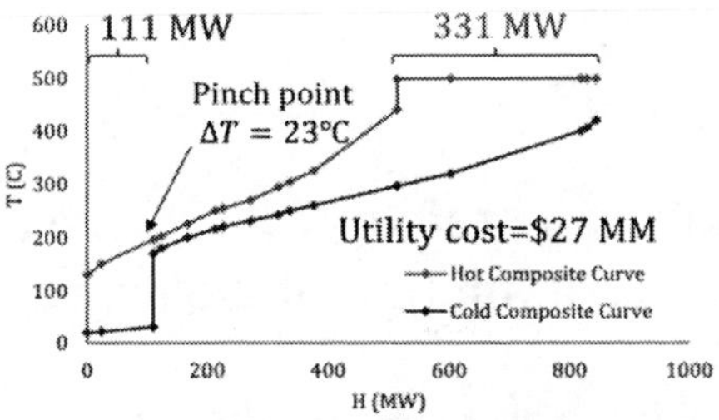

Area = 16473 m^2 Area cost=$9 MM

Total cost=$36 MM

Utility target only

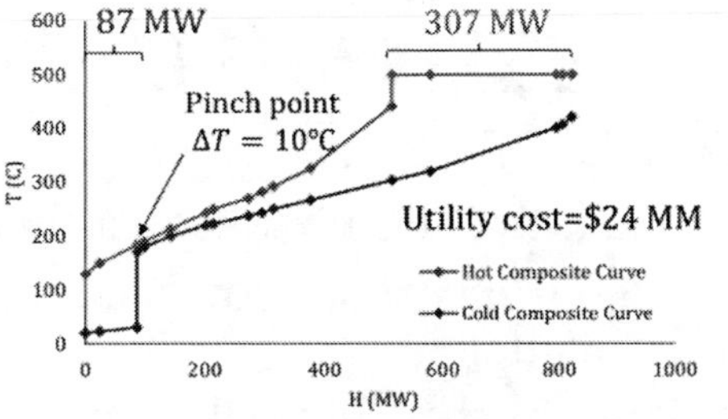

Area = 23926 m^2 Area cost=$13 MM

Total cost=$37 MM

Simultaneous process synthesis and heat integration

- Utility and area targets are simultaneously obtained
- Process structure and unit operating conditions are also determined

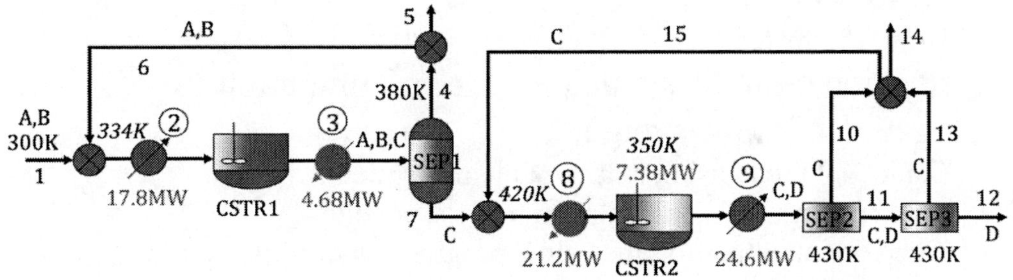

The Design of Beta Amino Acid Fragments to Inhibit Alpha Synuclein

Rex Gaumer (Presenter), Kyle Camarda

Department of Chemical & Petroleum Engineering

The University of Kansas, r430g812@ku.edu

Introduction

- **Objectives:**
  - Score the interaction between a ligand and a protein
  - Determine if this score correlates with the inhibition of the aggregation of alpha synuclein
  - Validate published Lipinski type rules for oral CNS drugs
- **Motivation:**
  - Parkinson's Disease
    - What is it
    - What does it do
    - How is it treated
  - Aggregation of alpha synuclein, and Inflammation
  - aggregation of alpha synuclein
    - Can be inhibited using beta synuclein fragments
    - May be inhibited using beta amino acid peptides
    - Aggregation inhibition may slow progression of Parkinson's

Approach

- Initial amino acid residue candidate selected from analysis of published work, (Petersen et al., 2010)
- New Scoring function based on the number of folds associated with an amino acid pair
- Successful scoring function will be used within a molecular design algorithm

Methodology

- **Requirements for CNS drugs**
- **Adherence to Lipinski type CNS Rules**
- **Novel scoring method**
- **Other Interactions between ligands and a proteins**

Future Work

- **Use protein fragments (Shaltiel-Kayo, et al., 2010), as ligands in Schroedinger and PASTA 2.0 software to develop a simple model for the interaction between these ligands and alpha synuclein.**
- **Eliminate the need for intensive molecular modeling**
- **Perform a search for optimized ligands within the 2400+ propectve quad-peptide and tri-peptide ligands composed of the selected beta amino acid residues.**
- **Use flow cytometry to study aggregation of alpha synuclein.**
- **Use cultured human tissue to observe if the designed ligands protect the tissue from aggregation of alpha synuclein.**

The University of Kansas

End of Rapid Fire Presentation

Rex Gaumer,
The University of Kansas,
r430g812@ku.edu

494416 Sustainable and Efficient CO_2Â Utilization:Â Production of Dimethyl Carbonate By an Indirect Route Using Ethylene Oxide and Methanol

Abstract:

Global warming is one of the most challenging phenomena to be faced by humankind in recent history. Anthropogenic greenhouse gases (GHG) are a major culprit in the phenomena. Around 60% of the anthropogenic GHG emissions between 1970 and 2010 are CO_2 from fossil fuel and industrial process[1]. However, the increasing demand for chemicals will consequently increase the amount of CO_2 emitted. This calls for additional means for CO_2 reduction. Such means could be the capture of CO_2 and using it as a new carbon source to produce a variety of chemicals. However, such process are few and not widespread.

The process considered consist of two parts: CO_2 capture plant and a CO_2 utilization plant that converts CO_2 to DiMethyl Carbonate (DMC). DMC is a versatile, non-volatile chemical compound used as reactive intermediate for methylation and carboxylation reactions, as fuel additive, as solvent for coating as well as in polymerization reactions [2, 3]. The market value for this green agent is rapidly on the rise; valued at US$ 390 mn in 2014 and projected to reach US$ 690 mn by 2023 [4]. The high demand for DMC represents the perfect platform for this sustainable design problem. The problem is mainly concerned with the sustainable design and optimization of the utilization plant. The design is preformed based on a 12-step hierarchical decomposition strategy. Starting from collecting information about the raw materials and the possible reaction pathways to generation of a preliminary flow-sheet based on which the process parameters are set. This is followed by mass and energy balance, which are then verified by performing a process simulation with a commercial simulator like PRO/II. Sizing and costing are then performed on all equipment used to enable a full economic evaluation of the process.

A brief description of the proposed process is as follows: CO_2 is reacted in excess with ethylene oxide (EO) to produce ethylene carbonate (EC), eventual impurities are removed by means of a distillation. EC is then reacted with excess methanol (MeOH) to produce ethylene glycol (EG) alongside the product of interest DMC. The recovery of the DMC and EG is done by: separating {DMC+MeOH} from {EG+EC} by regular distillation. The unreacted EC is hydrothermally converted into EG. The system {DMC+MeOH} form a pressure sensitive azeotrope and are therefore separated by pressurized distillation. The proposed process design is able to supply 100,000 metric tons DMC and similar quantity of EG, utilizing around around 53,000 metric tons of CO_2 and consuming around 75,000 metric tons EO and 74 metric tons MeOH on a yearly basis.

The base case is then subjected to environmental impact analysis as well as an investigation into further improvements and optimization e.g. mass and heat integration, enabling the setting of design targets for further improvement so that a more sustainable process design can be obtained. The original design is thus further developed to incorporate a more thorough sustainability analysis and a life-cycle assessment (LCA) to determine process bottlenecks that when removed through, for example, process intensification or alternative hybrid operations, leads to a more sustainable process design.

References

[1] Intergovermental Panel on Climate change (IPCC) IPCC fifth Assessment Report (AR5), Synthesis report, 2014.

[2] L. F. S. Souza, P. R. R. Ferreira, Jose Luiz de Medeiros, R. M. B. Alves and O. Q. F. Araujo. Production of DMC from CO2 via Indirect Route: Technical-Economical-Environmental Assessment and Analysis: ACS Sustainable Chem. Eng., 2014, vol. 2, page 62-69.

[3] H. Cui, T. Wang, F. Wang, C. Gu, P. Wang and Y. Dai. One-Pot Synthesis of Dimethyl Carbonate Using Ethylene Oxide, Methanol, and Carbon Dioxide under Supercritical Conditions: Industrial & Engineering Chemistry Research, July 2003, Issue. 42, 3865-3870.

[4] Transparencymarketresearch.com.
(2017). Global Dimethyl Carbonate Market to Exhibit 6.6% CAGR 2015-2023 owing to Rising Demand for Polycarbonates and Paint and Coatings. Available at: http://www.transparencymarketresearch.com/pressrelease/global-dimethyl-carbonate-market.htm [Accessed 29 Mar. 2017]

496039 Superstructure Formulation and Optimization of a Methane-Based Chemical Refinery for Co-Producing Olefins and Aromatics

Abstract:

Depletion of petroleum source along with the increased accessibility to cheap, abundant natural gas and shale gas reserves forms the ideal conditions to synthesize and design processes that upgrade these resources to petrochemicals. Methane as the primary component of natural gas and shale gas is becoming increasingly attractive as a feedstock. Up to date, various indirect and direct routes have been proposed for methane conversion. The most conventional route for converting methane sets syngas as intermediate. Direct conversion of natural gas without going through syngas route has been under investigation for several decades, but no commercial processes are practiced to date. As an effective complementary to experimental activity for methane conversion, superstructure formulation and optimization enable the identification of optimal technologies for potential methane conversion routes allowing more effective screening of catalysts and concepts, and identifying the types of devices and processes that should be considered for more detailed modeling and analysis.

In order to achieve a cost-effective and environmentally-benign methane-based chemical refinery producing C_2~C_4 olefins and C_6~C_8 aromatics simultaneously, we propose a process superstructure with alternatives regarding (i) methane reforming, (ii) syngas upgrading, (iii) methanol upgrading, (iv) direct methane to olefins/aromatics, (v) direct methane to methanol, and (vii) direct methanol to aromatics, among others. Simultaneous heat, power, and water integration is included to ensure minimum utility requirements and wastewater discharge. The objective function of the proposed mixed integer nonlinear programming model is to maximize the profit of the refinery. In addition to the input-output representations of complex reaction units via symbolic regression, the bilinear, trilinear, quadrilinear, and concave terms are reformulated to simplify the complexity associated with solving the MINLP model. In general, the proposed optimal methane-based chemical refinery effectively integrates the methane bi-reforming (steam reforming coupled with dry reforming), liquid-phase methanol synthesis, Dalian methanol to olefins, olefins to aromatics, and direct methanol to aromatics for maximum aromatics production to contribute highest net present value while achieving the lowest emission.

Qinglin Su

Research Associate
Davidson School of Chemical Engineering
Purdue University, West Lafayette

▪ Pharmaceutical continuous manufacturing

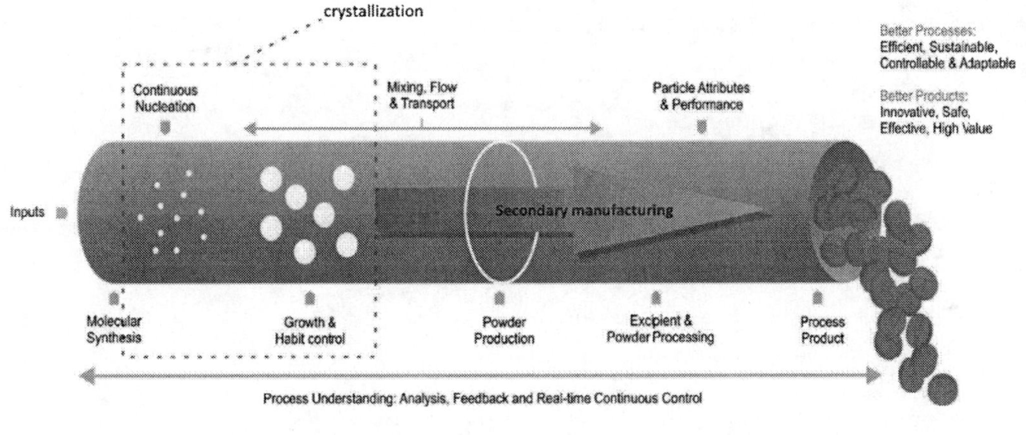

[1] UK EPSRC Center for Continuous Manufacturing and Crystallisation (CMAC)

PURDUE UNIVERSITY

▪ Batch vs Continuous crystallization process control

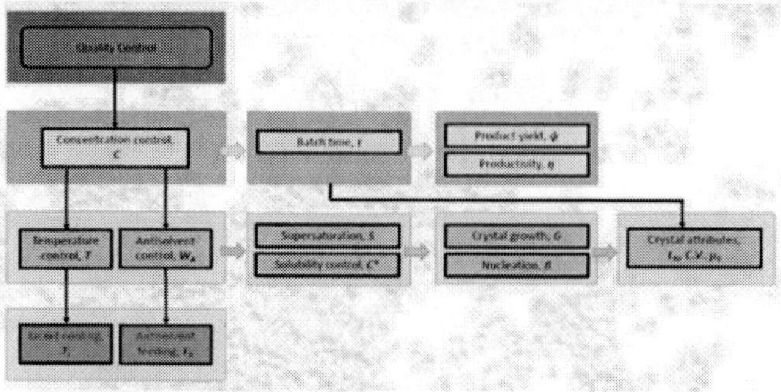

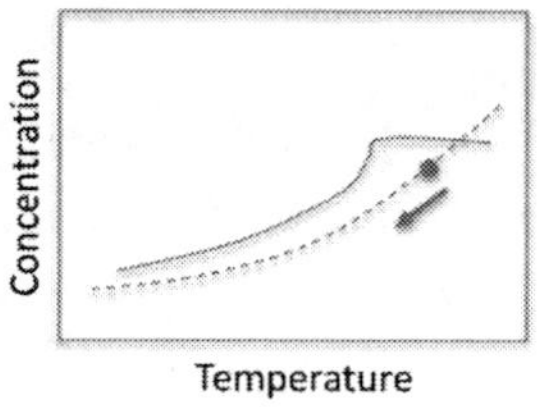

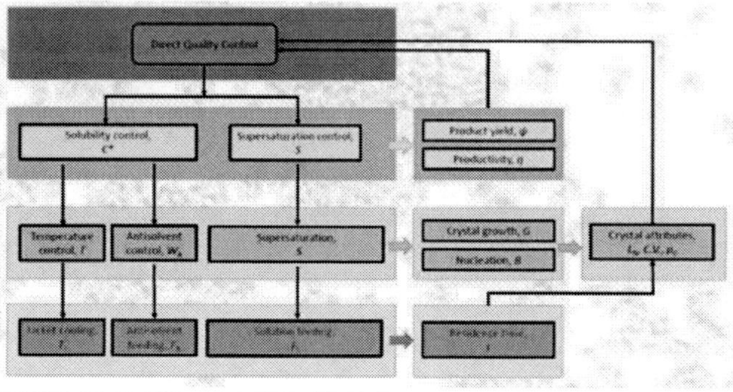

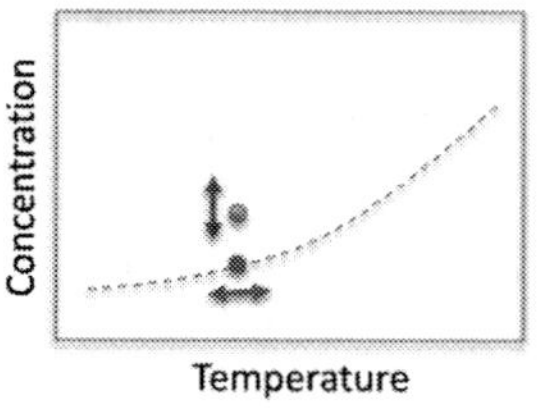

PURDUE
UNIVERSITY

▪ Applications of the mind-set change

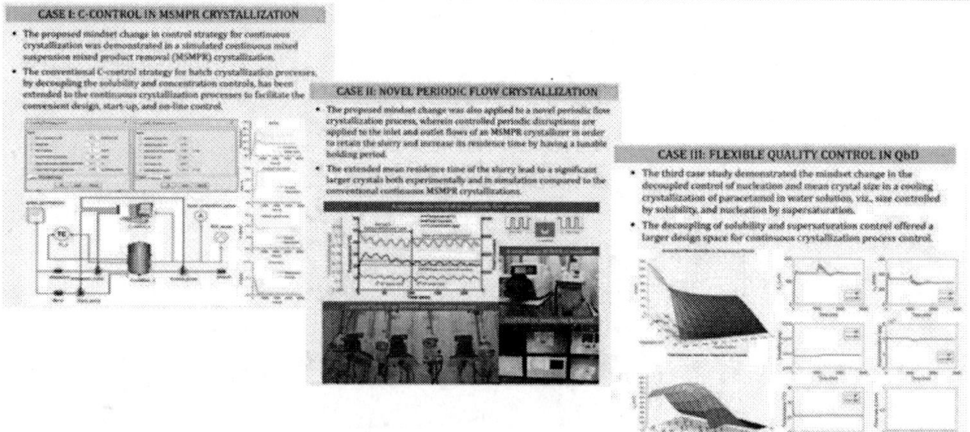

Publications

[1] *Su Q*, Rielly CD, Powell KA, Nagy ZK. Mathematical modelling and experimental validation of a novel periodic flow crystallisation processes using MSMPR crystallizers. AIChE Journal. 2017;63(4):1313-1327. **(The PSE Model-based Innovation Award 2017)**

[2] *Su Q*, Nagy ZK, Rielly CD. Pharmaceutical crystallisation processes from batch to continuous operation using MSMPR stages: modelling, design, and control. Chemical Engineering and Processing: Process Intensification. 2015; 89:41-53. **(The most downloaded paper in 2015)**

[3] *Su Q*, Benyahia B, Nagy ZK, Rielly CD. Mathematical modeling, design, and optimization of multisegment multiaddition plug-flow crystalliser for antisolvent crystallizations. Organic Process Research & Development. 2015; 19:1859-1870.

[4] *Su Q*, Braatz RD, Chiu MS. JITL-based concentration control for semi-batch pH-shift reactive crystallization of L-glutamic acid. Journal of Process Control. 2014; 24:415-421. **(The selected paper in ADCHEM 2012)**

PURDUE
UNIVERSITY

Biomass-to-Chemical Conversion Process: Production of 1-Butanol from Wheat Straw Feedstock

Authors: (MSc Students – Chemical & Biochemical Eng.)

- Andreas Nørgreen (Email: s130081@student.dtu.dk)
- Caroline Nørgreen
- Christina Etler
- Olivia Perederic

DTU Technical University of Denmark

Why Butanol?

- Interesting both as a biofuel and specialty chemical
- Utilization of cheap 2nd Generation feedstock has lead to a resurgence of interest in bio-based production
- Novel pretreatment and separation methods may pave the way for growth in the butanol market

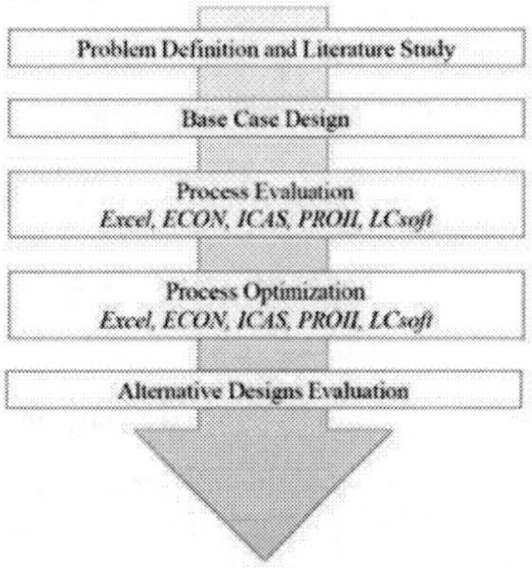

State-of-the-art Challenges:

- High pretreatment costs
- Metabolic process regulation
- Microbial product inhibition
- Dilute product streams and separation difficulties
- Market competitiveness with conventional production means
- Fluctuating commodity prices

Process Overview

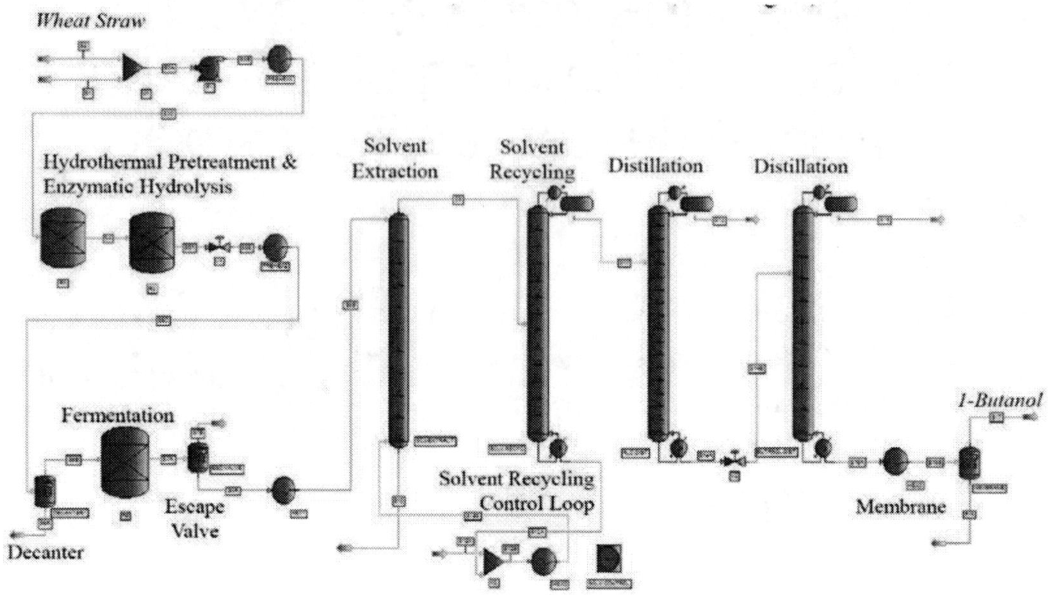

DTU

Economic Overview

- Post-Optimization / Heat integration production cost:
    - $1.18/kg butanol = $3.61/gallon
- Market Price Range: $3.5-4.0/gallon

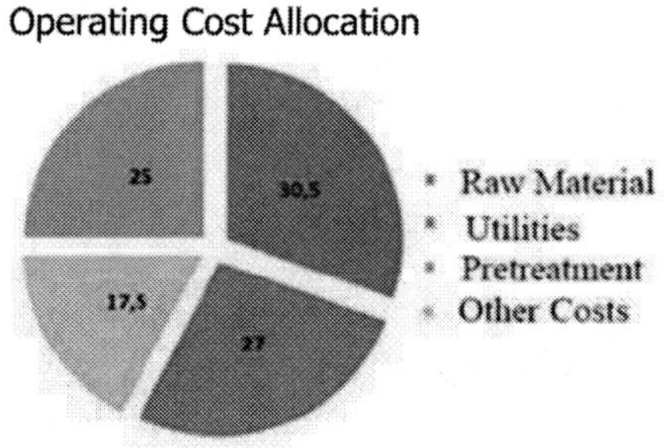

Concluding Remarks

- Pretreatment costs are a hinderance to economic viability
- Environmental hotspots
    - Extraction waste stream responsible for most of the environmental impact
    - Treatment/Recovery of waste stream for improved sustainability; alternatively use of environmentally-friendly solvents, i.e. ionic liquids
- Metabolic engineering of new strains is ongoing
    - Pilot- and industrial-scale testing of engineered strains is necessary

DTU

THE UNIVERSITY OF TEXAS AT AUSTIN

Probabilistic Process Design Under Uncertainty Via Dynamic Optimization

Calvin Tsay
calvintsay5@gmail.com
The Unviersity of Texas at Austin

CAST Rapid Fire

Optimal Design Under Parameter Uncertainty

- Solve scenarios in a "pseudo-time" domain
- Uncertain parameter is effectively a (continuous) disturbance acting on process model
- Reduced computational expense (only requires a single instance of the process)

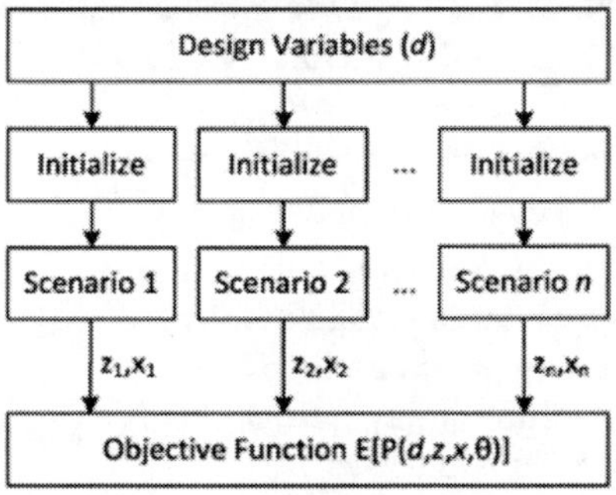

Multi-Scenario Approach

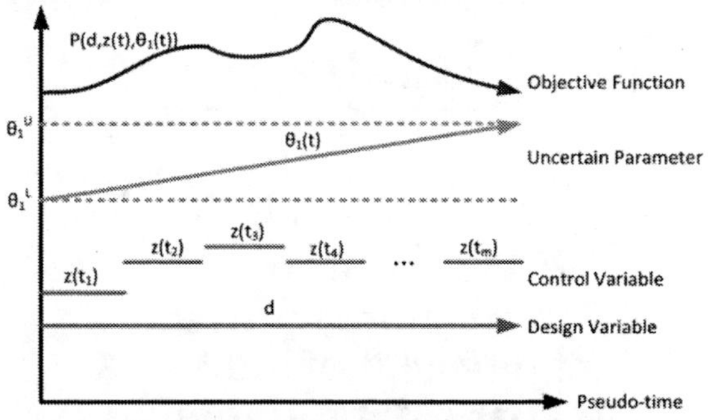

Dynamic Optimization Approach

Gutierrez, R.F.B., Pantelides, C.C., and Adjiman, C.S. *Comput.-Aided Chem. Eng.* **2006.** Tsay, Pattison, and Baldea. *Ind. Eng. Chem. Res.*. **2017.**

Case Study: Dimethyl Ether Production

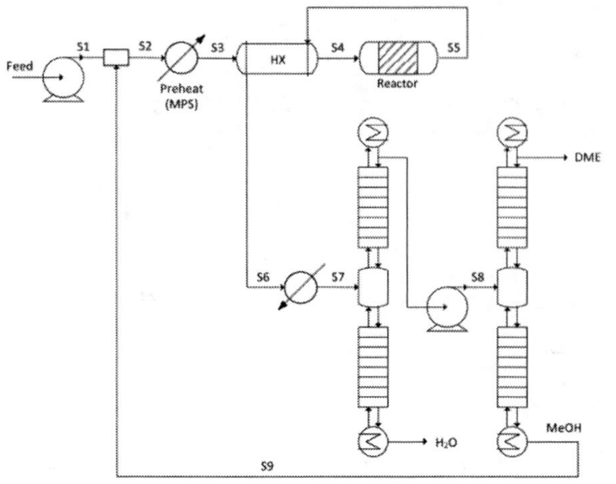

Uncertainty: Reaction rate not well-known

Special formulation used to avoid binary variables during column design optimization (see reference)

- Two cases:
  - $\delta \sim Uniform(-0.1, 0.1)$
  - $\delta \sim Normal(\mu = 0, \sigma = 0.5)$
- δ is bounded between -0.1 and 0.1 for both

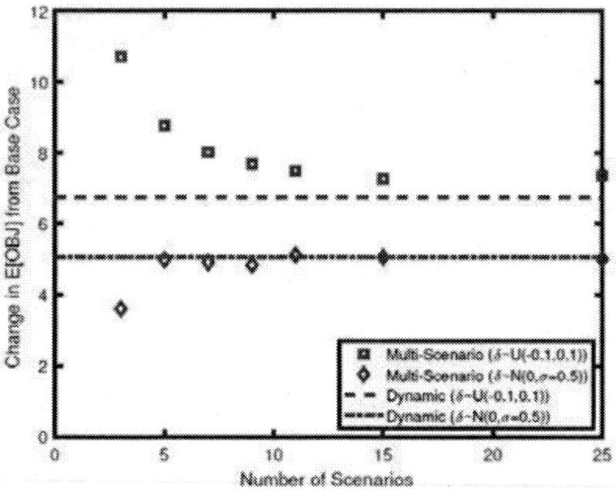

Tsay, Pattison, and Baldea. *Ind. Eng. Chem. Res.*. **2017.**

Significant Computational Benefits

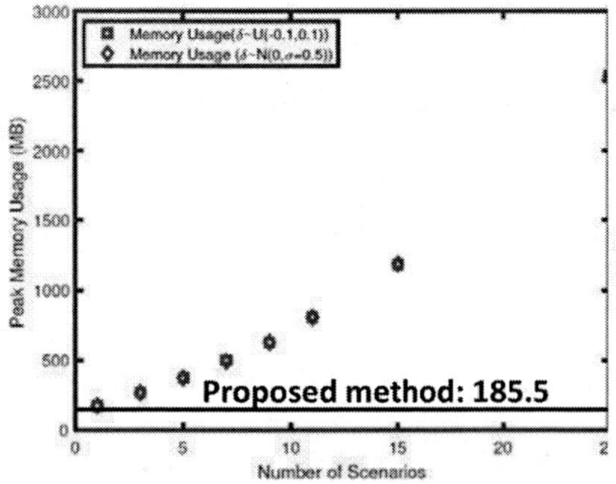

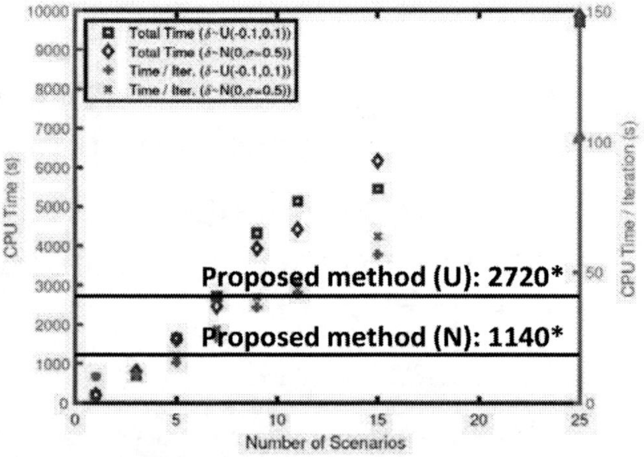

- Memory and CPU time are both reduced
- Formulation represents an infinite limit of multi-scenario optimization
  - See reference for proof

***estimated by replacing CPU time expended on initialization at every iteration with average pseudo-transient initialization; Actual CPU times: 21550s and 9350s**

Tsay, Pattison, and Baldea. *Ind. Eng. Chem. Res.*. **2017.**

Tony (Wenzhao) Wu

wwu67@wisc.edu

University of Wisconsin – Madison

Superstructure-Based Bio-Separation Process Synthesis

- **Problem**

  What's the 'optimal' separation process for bio-based chemicals X?

- **Solution**

  A general effluent from an upstream bio-conversion unit contains **cells**, the **product** (if EX), large amount of **water**, and other **impurities**.

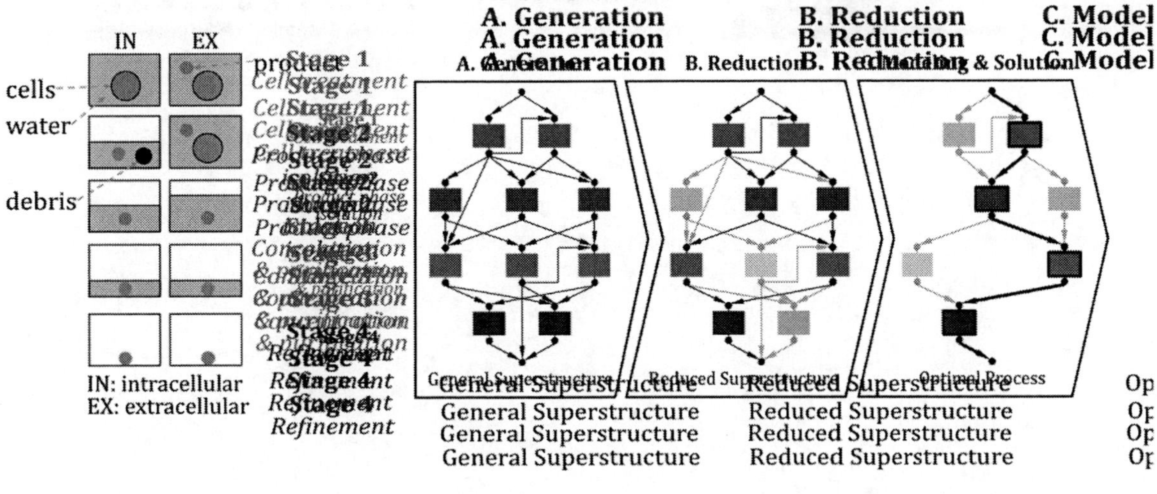

Superstructure-Based Bio-Separation Process Synthesis

❑ Example

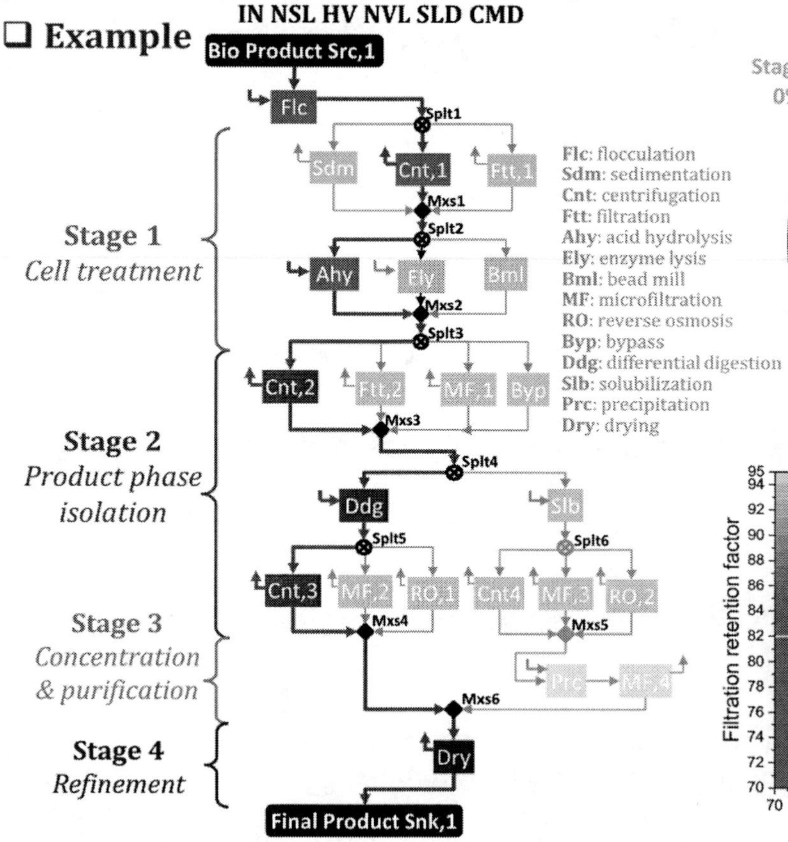

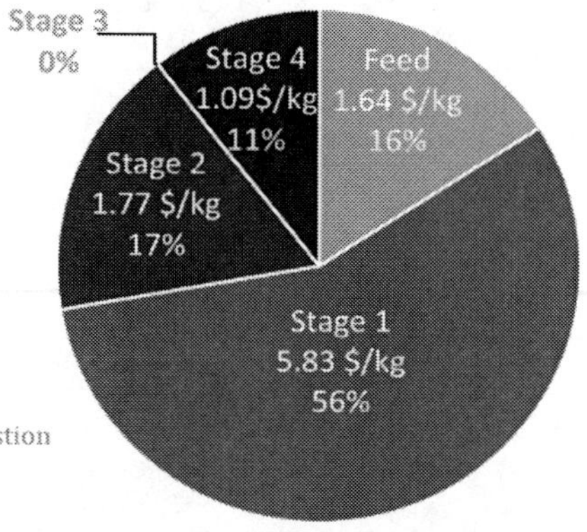

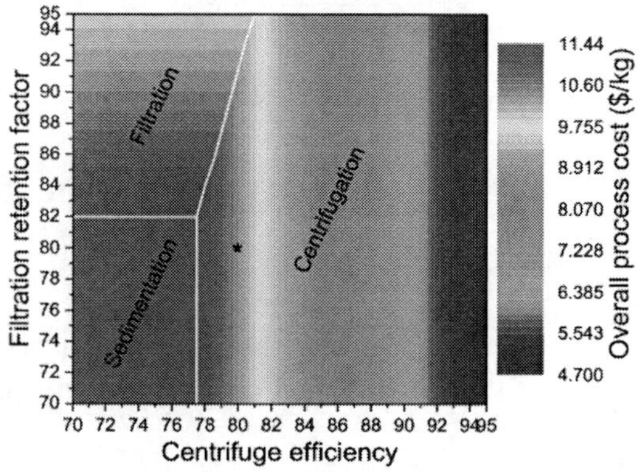

IN - intracellular; **NSL** - insoluble; **HV** - heavy; **NVL** - non-volatile; **SLD** - solid; **CMD** - commodity

CFD Modeling of Piston-Type Direct Work Exchangers

Aida Amini Rankouhi* and Yinlun Huang
Dept. of Chem. Eng. & Mat. Sci.
Wayne State University, Detroit, MI 48202
* aida.amini@wayne.edu

246 CAST Rapid Fire Session I
(Paper No. 246j)
AIChE Annual Meeting
Minneapolis, MN, Oct. 29 – Nov. 3, 2017

Energy Integration through WEN

- The energy loss in the manufacturing sectors is about 50% of the total energy consumption.
    - It includes 21% energy loss in Process Heating and 11% in Machine Drive System sections.
- Energy integration approach for thermal and mechanical energy recovery:

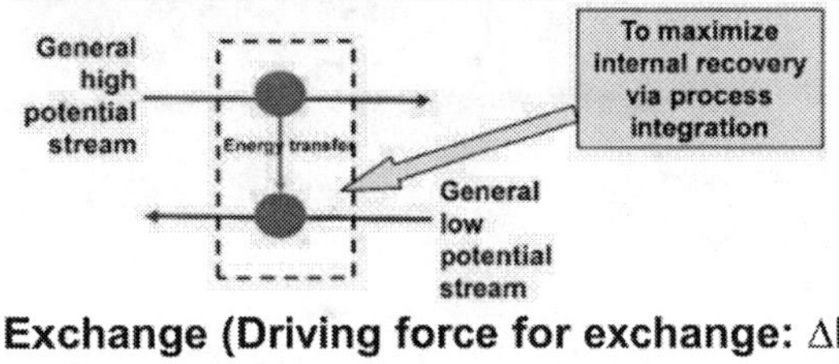

- A new class of exchange: Work Exchange (Driving force for exchange: ΔP)
- Various process analysis and design methodologies introduced to develop WEN systems.
- Direct and Indirect Devices for mechanical energy recovery
    - One-step: Flow Work Exchanger
    - Two-step: Single Shaft Turbine Compressor

Direct Work Exchanger

WAYNE STATE College of Engineering

- Commercialized in desalination process
- Energy recovery efficiency: 90-95%.

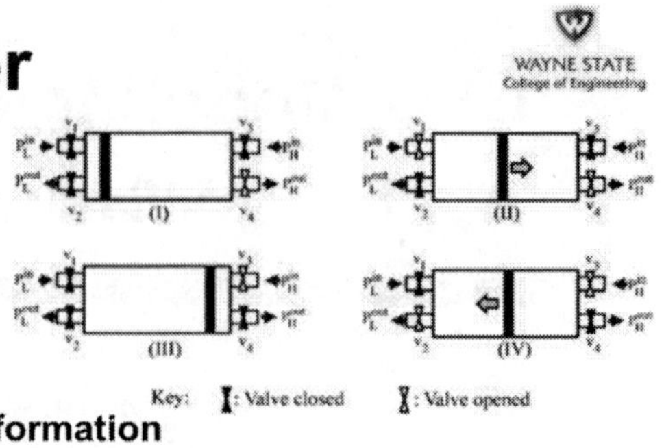

Challenges and Opportunities:

- Feasibility of using WE's in any chemical process
- Lack of experimental scale device and information about WE in different operating conditions
- Goal: <u>Performance of Piston type work exchangers-CFD Modeling</u>

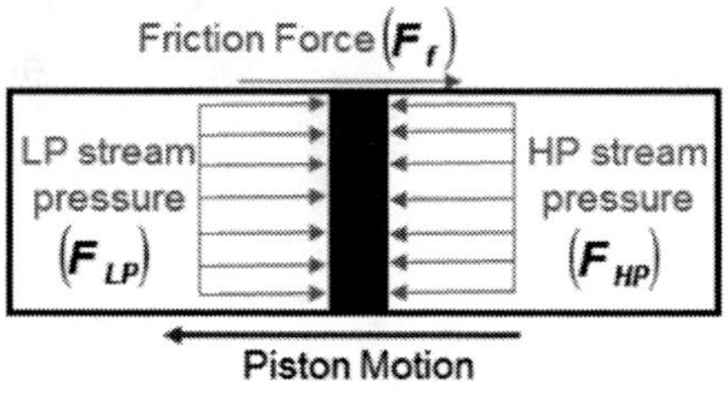

Balance of forces on Piston

- Motion of the Piston:

$$m_p \cdot \frac{d}{dt} v = F_P + F_f$$

$$F_P = p \cdot A \quad , \quad F_f = F_{f_h} + F_{f_c}$$

$$\frac{dv}{dt} = \frac{v_{new} - v_{old}}{t_{step}} \quad , \quad v_{new} = \frac{d_{p_{new}} - d_{p_{old}}}{t_{step}}$$

Then, $d_{p_{new}} = \frac{F_P + F_f}{m_p} \times (t_{step})^2 + v_{old} \times t_{step} + d_{p_{old}}$ ⟸ **Piston new position**

Modeling and Results Analysis

WAYNE STATE College of Engineering

- Transient model, HP at 6 atm, LP at 1 atm, air ideal gas.

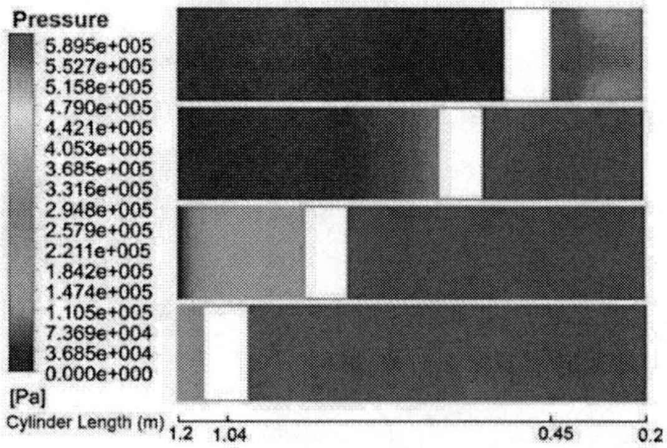

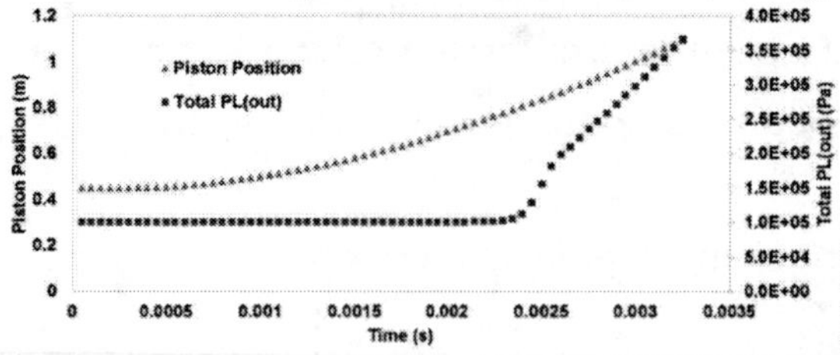

Piston position vs. Time under different operating condition

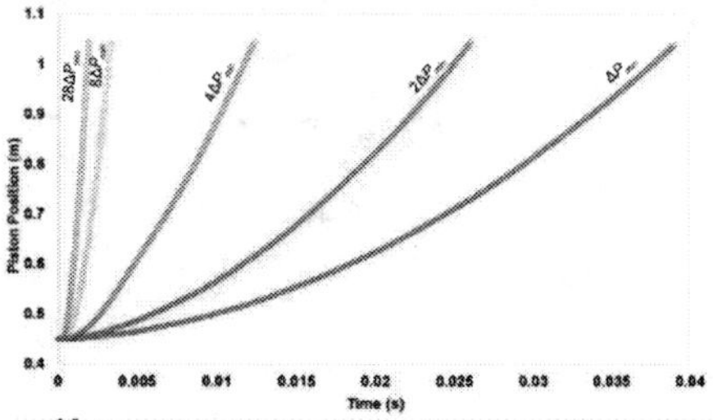

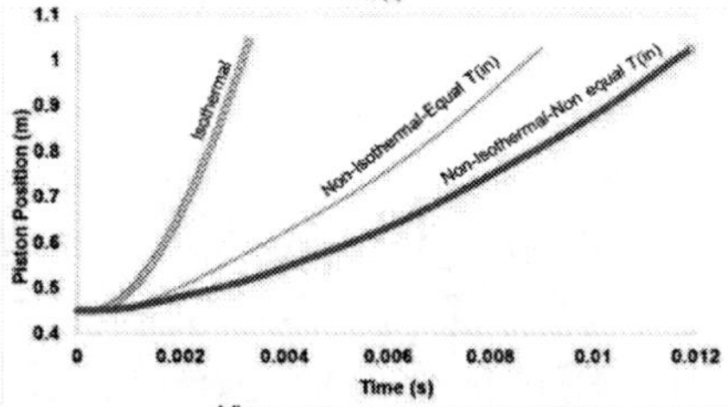

Compressibility ratio of LP stream for a maximum displacement of piston

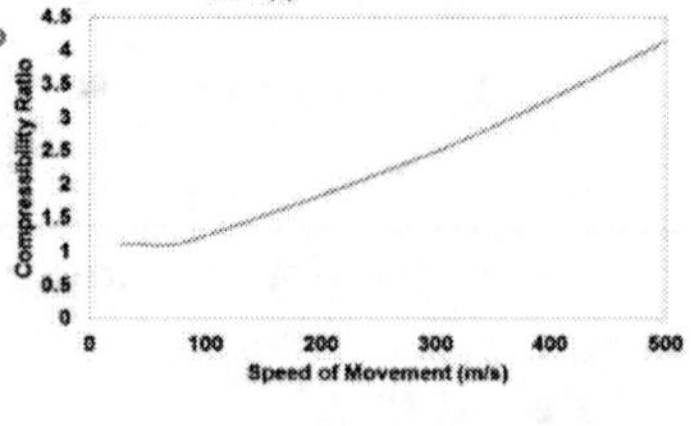

- Better insight into the unit for further design in experimental and industrial scales.

Simultaneous Process Synthesis and Heat Integration Using a Single Superstructure

Salih Emre Demirel, Jianping Li and M. M. Faruque Hasan

Artie McFerrin Department of Chemical Engineering
Texas A&M University
College Station, Texas, USA.

email: emredemirel@tamu.edu

AIChE Fall Annual Meeting, Minneapolis, MN
October 30, 2017

Building Block Superstructure

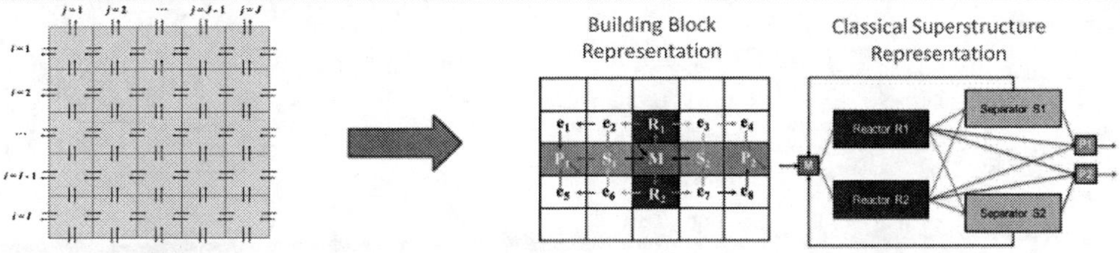

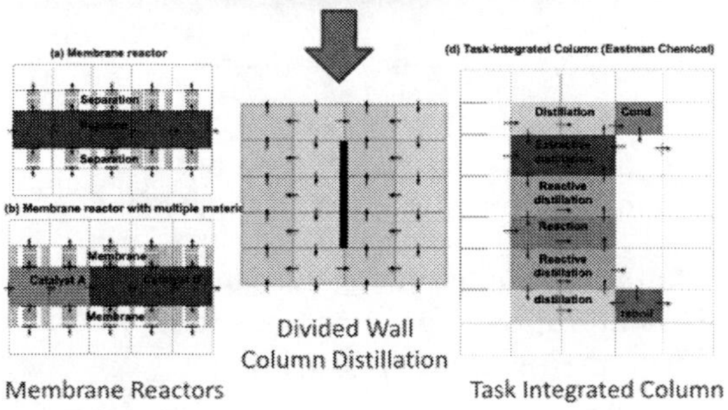

Trade-offs between process design and energy demand: Simultaneous Heat Integration

Research Objective:
Provide a unified superstructure for simultaneous process synthesis and integration with automated flowsheet generation.

2

[1] Demirel, et al. 2017, CACE.

Simultaneous Process Synthesis and Heat Integration Model

Flow representation

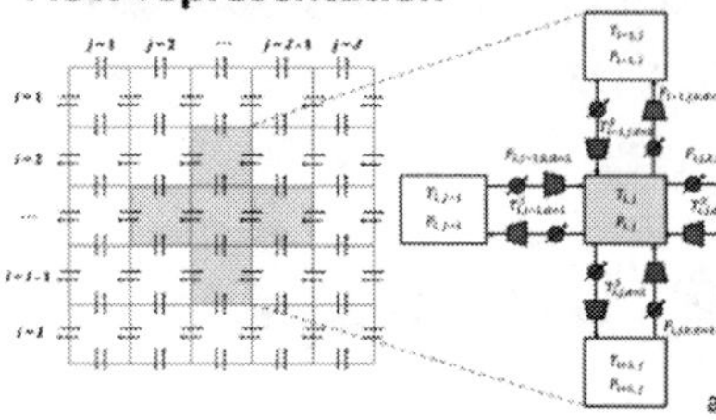

Stream identities (hot/cold) are not known *a priori* (unclassified streams).

Available models for minimum utility targeting with unclassified streams:

- Dynamic temperature intervals (Kong et al., 2017, CACE)
  - Pinch Location Method (Quirante et al., 2018, CACE)

Mixed-Integer Nonlinear Optimization (MINLP) Problem

Block Superstructure Model Key Constraints

Material Balance around each block:

$$F_{i,j-1,k} - F_{i,j,k} + R_{i-1,j,k} - R_{i,j,k} + G_{i,j,k} + \sum_{f\in FS} M_{i,j,k,f} - \sum_{p\in PS} P_{i,j,k,p} = 0$$

Energy Balance around each block:

$$EF_{i,j-1} - EF_{i,j} + ER_{i-1,j} - ER_{i,j} - EP_{i,j} + EM_{i,j} + Q_{i,j} + W_{i,j} = 0$$

Unit Assignments (Binary Variables):

$$z_{i,j}^{unF} + z_{i,j}^{srF} = 1 \qquad z_{i,j}^{srF} = \sum_s z_{i,j,s}^{sF} \qquad \sum_c z_{i,j,c}^{rxn} = 1$$

Corresponding min utility consumption ↑ ↓ Inlet & Outlet Temperature Stream Flow rates

Classification of streams:

$$z_w^{hot} + z_w^{cold} = 1$$

Overall Heat Balance:

$$Q^c = \sum_w F_w (T_w^+ - T_w^-) + Q^h$$

Q^C for pinch candidates:

$$Q^c = \sum_w F_w \max(0, T^p - TS_w^{out}) - \max(0, T^p - TS_w^{in})$$

Heat Integration Model Key Constraints

3

Automatic Flowsheet Generation with Simultaneous Heat Integration

Reference Superstructure (Kong et al., 2017, CACE)

Hot Utility: 14 MW
Cold Utility: 0 MW

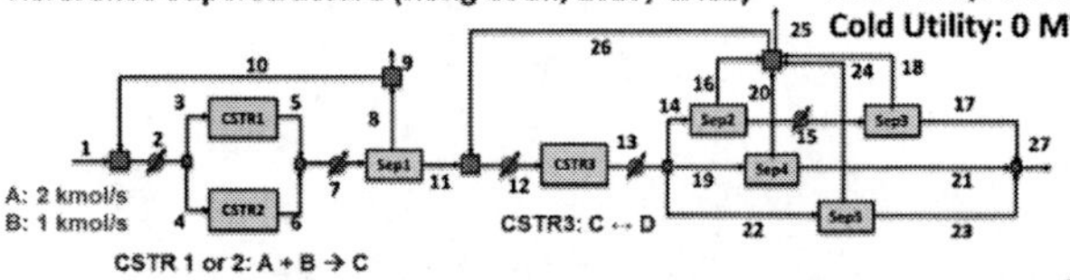

Maximize Profit = Revenue – Capital cost – Utility Cost

Reference solution: $14.50 MM/y

Solution of the problem with 4x3 Block Superstructure

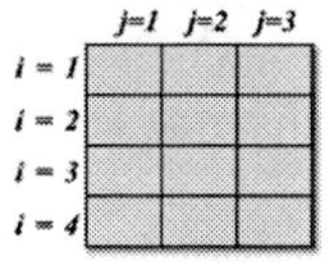

Final Profit: $14.95 MM/y

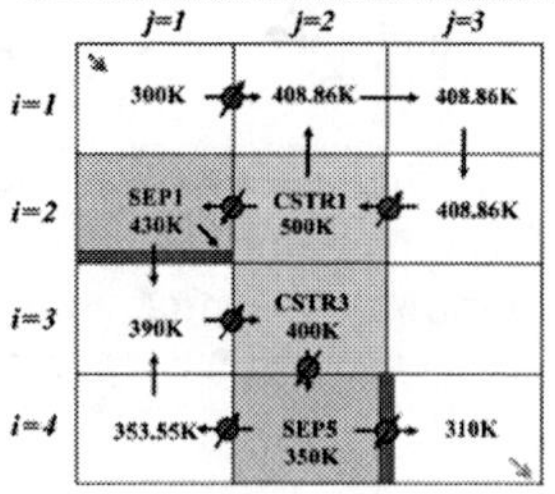

Final Flowsheet:

Hot Utility: 12.7 MW
Cold Utility: 0 MW

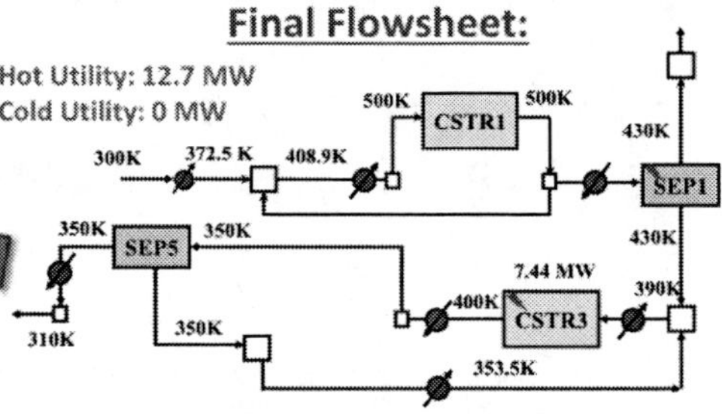

$450,000/y improvement:

- Change in the position of recycle stream
- Identification of product stream cooler

Automated flowsheet generation and identification of position and identity of the process streams.

4

The University of Kansas
Chemical & Petroleum Engineering

MOSAIC: Parallel Computing, Multi-objective Optimization Applications

Bridgette Befort and Kyle V. Camarda
Chemical and Petroleum Engineering, The University of Kansas, Lawrence, KS
Email: Bridgette.Befort@ku.edu

The University of Kansas
Chemical & Petroleum Engineering

About MOSAIC

- Free, web-based, open-source modeling, simulation, and optimization environment
- Tools
  - Modeling
  - Simulation
    - Visualization—Flowsheeting
  - Optimization
- LaTeX-Style Entry
- Solves and creates code in many languages
  - Import/Export
- CAPE-Open Compliant

- Modeling and Simulation of Chemical Processes, steady state or dynamic
- High Performance Computing
- Design and Build Algorithms
- System Produces Code to Export
  - GAMS
  - Matlab
  - C++
  - Python

MOSAIC

The University of Kansas
Chemical & Petroleum Engineering

Complex Optimization Analysis

- Chemical Engineering Systems
  - Large and Complex Processes
  - Need to simultaneously improve parameters
- Multi-objective Optimization
  - Method in which two or more objective functions are optimized to Pareto optimality
  - No objective can be improved without weakening the other objectives
- High Performance Computing—Parallel techniques which allow complex problems to be solve quickly and efficiently
  - Algorithms—Stochastic, Parallel Tabu Search

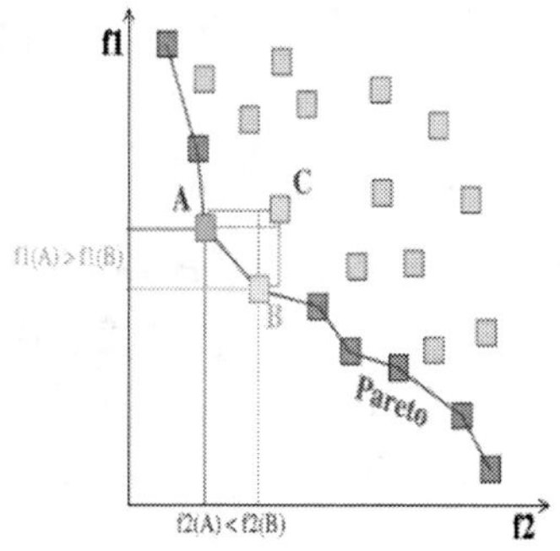

Pareto Optimality Frontier for minimizing *f1* and *f2*. Squares define feasible solutions. (Figure obtained from commons.wikimedia.org/wiki/File:Front_pareto.svg)

The University of Kansas
Chemical & Petroleum Engineering

The use of MOSAIC and high performance computing techniques provide the opportunity to perform complex optimization analyses of large chemical engineering systems for the purpose of simultaneously improving multiple plant parameters.

- Construct and Solve Problems using chemical plant design specifications and custom algorithms built in MOSAIC
- Example: Williams-Otto Process Objectives
  - Profit
  - Environmental Performance
  - Safety

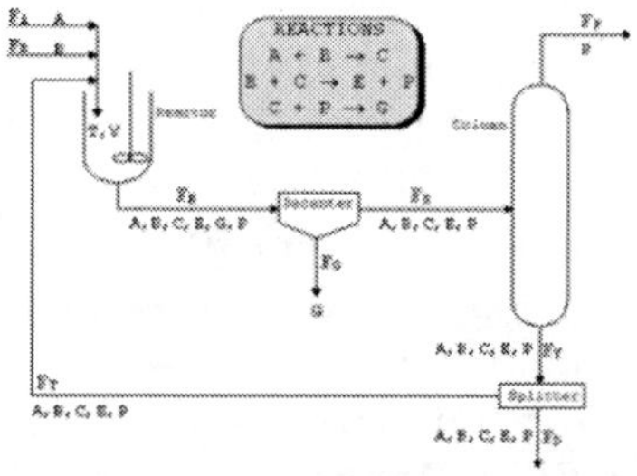

Williams-Otto Process (Figure obtained from che.boun.edu.tr)

493663 A Multi-Stake Holder Approach for the Optimal Planning and Integration of the Supply Chain of Fuels Involving CO_2 Capture

Abstract:

The global energy demand currently is satisfied by fossil fuels, which can cause environmental problems such as the global warming. Several alternatives to reduce carbon emissions during energy production have been proposed, for instance the solar energy use, biofuels production and carbon capture technologies. One the main reasons to use biomass is that this raw material can capture most of the CO_2 emitted by biofuels. Additionally, the carbon emissions caused by the fuels production can be captured by forest biomass, which can be seen as a carbon capturing industry. Therefore, a system capable to integrate the fuels based on biomass and fossil resources as well as a carbon capture system through forest plantations is an attractive alternative. Nevertheless, there are several involved players such as biomass producers, society, government and businessmen with diverse priorities. Thus, this paper proposes a mathematical formulation for the optimal planning and integration of a fuel production system considering carbon capturing under a multi-stakeholder scheme. The carbon capture is proposed to be accomplished via forest plantations, which can receive a monetary benefit from the pollutant industries. The multi-stakeholder approach contemplates benefits and affectations in each one of the entities of the involved supply chains, and determining how the interactions between the different stakeholders take place. The present methodology takes into account profits and number of new jobs obtained by biorefineries, refineries and forest plantations, as well as the net emissions caused by the system. Also, the methodology contemplates features such as the project life time, bioresources availability, amounts and type of products to be produced, involved plantsâ€™ location and capacity. The mathematical model was applied to a case study from Mexico considering a set of locations for biorefineries, forest plantations, refineries, distribution centers, biomass suppliers and well oils. Furthermore, diverse processing technologies and types for bioresources were considered. On the other hand, the multi-stakeholder approach considered 20 individual stakeholders, which are associated to diverse weight factors generated systematically in order to define priorities. Results are presented through several points in a 3D figure with net emissions, generated jobs and economic objective function as axes. Each point represents a supply chain configuration with a dissatisfaction value in comparison with the reported and utopic solutions. This way, this approach allows proposing a solution with the lowest dissatisfaction ratio taken into account diverse interests. Finally, the model consists of 6,407 continuous variables, 2,161 binary variables and 5,431 constraints. It was coded in the GAMS software. This model was solved using the solver CPLEX with a computer with a processor Intel Core i7 at 2.90 GHz with 16 GB of RAM. The average CPU time for each solution of the mathematical approach was around 0.156 s.

Optimal Control Structure Design for Cyber-Physical Systems (Poster: 716)

Temitayo Bankole[1], Paolo Pezzini[2], Nor Farida[3], Kenneth Bryden[2], David Tucker[3], D. Bhattarcharrya[1]

[1]Department of Chemical and Biomedical Engineering
West Virginia University

[2]Ames Laboratory
Ames, USA

[3]National Energy Technology Laboratory
US Department of Energy

tsbankole@mix.wvu.edu
Oct. 2017

Introduction & Problem Statement

- Why HyPer?
  - Hybrid Gas turbine -fuel cell system promise high efficiency
  - Waste heat from Fuel cell is fed to turbine.

Cyber -Physical

- Numerical model simulates the fuel cell.

Focus:

- Find optimal controlled variables.
- Flexible
- highly Unstable
- Numerous candidates

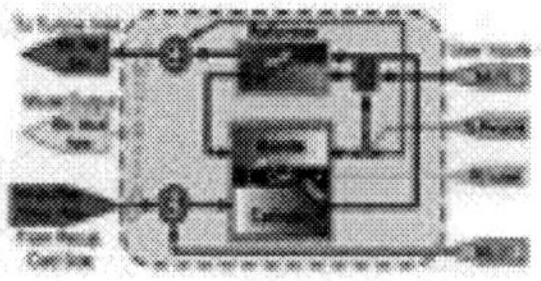

Controlled Variable Selection

- APriori

$$y = G_p u + G_d d + W_n e$$
$$\left\|(G_p)_i\right\|_\infty \geq 1 \; \forall \, i$$
$$\left\|(G_p)_i\right\|_\infty \geq \left\|(G_d)_i\right\|_\infty \; \forall \, i$$
$$\tau\left(u_i, y_j\right) \leq \xi_j$$

$$J\left(\frac{\$}{h}\right) = \sum_i \alpha_i \dot{W}_i - \sum_j \beta_j \dot{Q}_j$$

Prescreen poor candidates

Optimization:

$$\min_{P_n}\left(L(P_n), J_c(P_n)\right)$$

$$J_c(P_n) = \underline{\sigma}^{-1}\left(\hat{G}(P_n)\right)$$

$$L = \frac{1}{6n_u}\left\|\left(G_p J_{uu}^{\frac{1}{2}}\right)^{-1}\left(G_p J_{uu}^{-1} J_{ud} - G_d\right) W_d \, W_n\right\|_F^2$$

Posteriori Analysis:

Remove correlated variables

Check performance

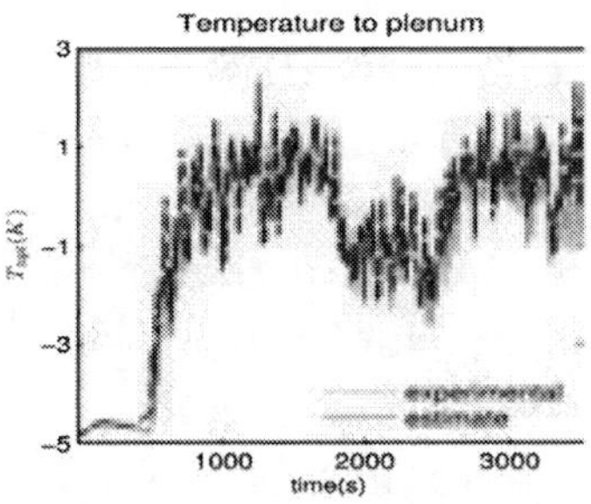

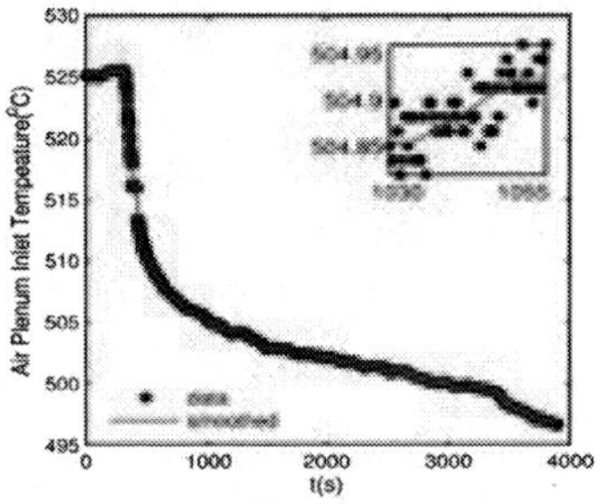

Results & Conclusions

- Branch and Bound
  - Node consists of F_n and C_n
  - Prune branches
  - Time consuming - 3hrs 20 minutes

  Multiagent Optimization framework
  - Metaheuristic methods
  - Communication protocol
  - Agents
    - Genetic Algorithm
    - Simulated annealing
    - Ant Colony optimization
  - < 5 minutes

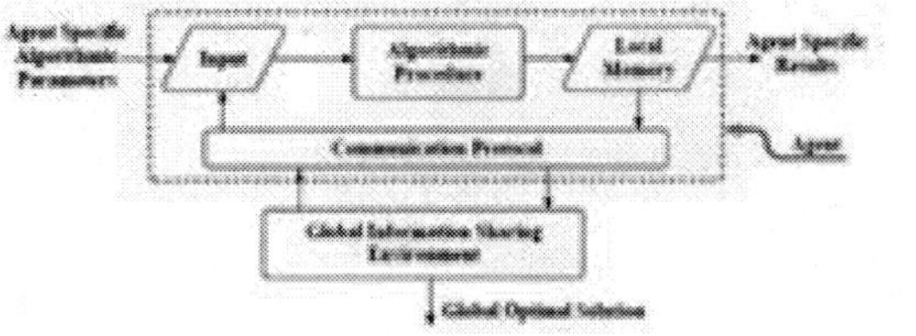

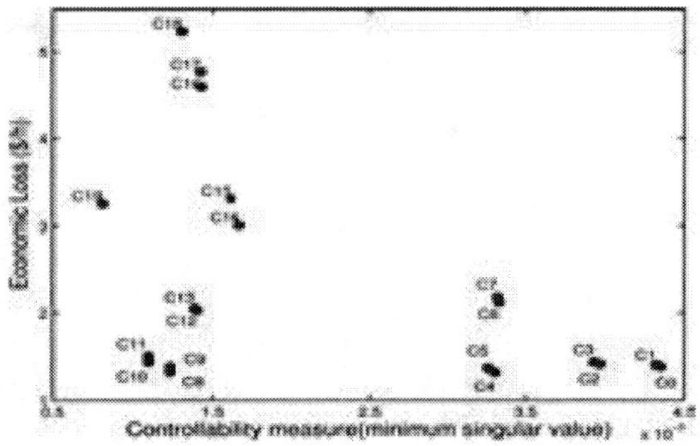

- Acknowledgement

The authors gratefully acknowledge support from NETL DOE through grant no. DE-FE0012451 titled AOI 1: Development of Integrated Biomimetic Framework with Intelligent Monitoring, Cognition and Decision Capabilities for Control of Advanced Energy Plants

- Disclaimer

499419 A Study for Integration of Procurement Planning and Short-Term Scheduling in Petroleum Refineries

Abstract:

Mathematical programming has been implemented into decision making since 1950s for crude planning to optimize both production and crude purchasing plan. Long-term planning and short-term scheduling constitute two important decision elements for petroleum refineries. Planning is responsible for creating profitable plans exploiting the constantly changing market opportunities, while scheduling is required to execute feasibly and optimally as much as possible according to obtained plans. It has been studied and practiced for a long time, especially in the last two decades driven by increasingly intensive global competitions, more volatile feedstock and product markets, as well as stricter environmental regulations. Especially, the availability of new crudes will lead to the evaluation of refinery assets to leverage refinery procurements in order to maximize refinery profits.

Unfortunately, due to the differences between planning and scheduling, such as different time scopes and process constraints, discrepancies always exist, which initiate inherent gap between generated plans and schedules. Therefore, the communications between planning and scheduling need to be enhanced in order to validate the generated plans and make any necessary adjustments for the plans to be feasible under real operation settings. Existing publications are either neglecting the ever-changing crude supplies or product demands, variable shipment sizes and demurrage costs [1] or focused on an aggregated planning problem without considering realistic operating practices and connection decisions between units [2]. And the aforementioned works disregard the continuous operation of crude distillation units. The reason for these sacrifices lies in the extreme difficulties when solving a full space problem with integrated formulation of planning and scheduling.

In this paper, an iterative framework integrating refineries crude procurement planning and detailed scheduling is proposed. The general objective of the integrated model is to maximize crude refining profits while satisfy all crude transfer and operation constrains. The long-term planning is decomposed into sub-periods representing different scheduling horizons. Hierarchically, the planning problem serves as a master sub-problem while the detailed scheduling is a salve sub-problem. The crude purchasing and refining plans from master problem are provided as inputs to the slave problem. Under the circumstances that any planned operations are impossible with detailed processing constraints, a feedback loop from the slave problem to the master problem will be added in the form of extra constraints. The goal of such feedback is to find the true optimal high-level decisions. In addition, the master problem can provide an increasing lower bound while the sub-problem can provide an upper bound. Thus, iterative methods can lead to optimal solutions if solved until the gap is closed [3]. Also a rolling forward time horizon framework will be created in such a way that once a portion of the time horizon has been scheduled, no additional modifications are allowed for the given period of time horizon. Thus, the number of iterations required to schedule the entire time horizon will be minimized. The finalized refining profits will be compromised due to consistent refinement from detailed scheduling problem. A detailed case study will be conducted to validate the efficacy of the proposed integration methodology framework.

References

1. Oddsdottir, T.A., M. Grunow, and R. Akkerman, *Procurement planning in oil refining industries considering blending operations.* Computers & Chemical Engineering, 2013. **58**: p. 1-13.

2. Zhang, J., Y. Wen, and Q. Xu, *Simultaneous Optimization of Crude Oil Blending and Purchase Planning with Delivery Uncertainty Consideration.* Industrial & Engineering Chemistry Research, 2012. **51**(25): p. 8453-8464.

3. Maravelias, C.T. and C. Sung, *Integration of production planning and scheduling: Overview, challenges and opportunities.* Computers & Chemical Engineering, 2009. **33**(12): p. 1919-1930.

495892 Hybrid Modeling of Bioreactor Systems Using First Principles and Deep Neural Networks with Constraints of Validity Domain for Optimization of Feeding Strategy

Abstract:

Bioreactors in industry often grow genetically engineered cells and their operations often switch between different modes. Hence, identification of a reliable model can offer the potential for improving the economic performance and operational sustainability of the bioprocess. However, finding a proper model structure and identifying its parameters is an onerous task for practical applications because there is still a lack of fundamental understanding of the biological mechanisms governing the process. In addition, the existing growth kinetic models are mainly designed for wild-type cells and have a limited range of validity only to describe particular dynamics such as the exponential growth.

A framework combining first principles and data-based models to describe unknown kinetics has been studied as a remedy to the problems in the field of bioprocess optimization. In general, hybrid modeling approaches use first principles models, e.g., mass balances, as the main structure of the model, and data-based models to estimate kinetic parameters in net generation terms [1] [2]. With the component of first principles modeling, this hybrid model can be potentially exploited beyond the region of identification data and is amenable to physical interpretations. With ongoing advances in sensor technology, massive amounts of operational data are becoming available for industrial processes. The data sets often include different types of measurements such as optical density, light intensity, concentrations, etc. under different operating conditions from various strains of a cell. These characteristics make it difficult to analyze the bioreactor system using a simple nonlinear regression model like feed-forward neural networks.

This study proposes a hybrid modeling approach that combines a first principles model with deep neural networks (DNN) that can utilize the complex data and describe dynamic behaviors of a bioreactor system. The proposed approach first compresses the measurements using principal component analysis (PCA) and then feed the compressed data into the DNN model. The projected variables can reduce the dimension of input, remove the feature extraction steps, and make DNN learn faster. It is also important to identify valid domains for the model prediction if the hybrid model is to be used for other tasks such as optimization. The proposed approach includes criteria for checking â €œvalid domainâ€ of the hybrid model to assess whether the prediction is from undue extrapolation or not. The criteria are represented as convex hull and confidence interval [3] so that they can be readily integrated into real-time optimization techniques such as model predictive control. As an illustration of industrial operations, feed-rate optimization of a fed-batch bioreactor will be presented and show the efficacy of the proposed scheme.

[1] von Stosch, Moritz, et al. "Hybrid semi-parametric modeling in process systems engineering: Past, present and future."Â *Computers & Chemical Engineering*Â 60 (2014): 86-101.

[2] Oliveira, R. "Combining first principles modelling and artificial neural networks: a general framework." *Computers & Chemical Engineering* 28.5 (2004): 755-766.

[3] Kahrs, O., and W. Marquardt. "The validity domain of hybrid models and its application in process optimization." *Chemical Engineering and Processing: Process Intensification* 46.11 (2007): 1054-1066.

Dynamic Modeling and Control of a Natural Gas Combined Cycle Power Plant with CO_2 Capture

Yifan Wang, Debangsu Bhattacharyya, Richard Turton
Chemical and Biomedical Engineering Department
West Virginia University

yw0001@mix.wvu.edu
Debangsu.Bhattacharyya@mail.wvu.edu
Richard.Turton@mail.wvu.edu

WestVirginiaUniversity.

Motivation

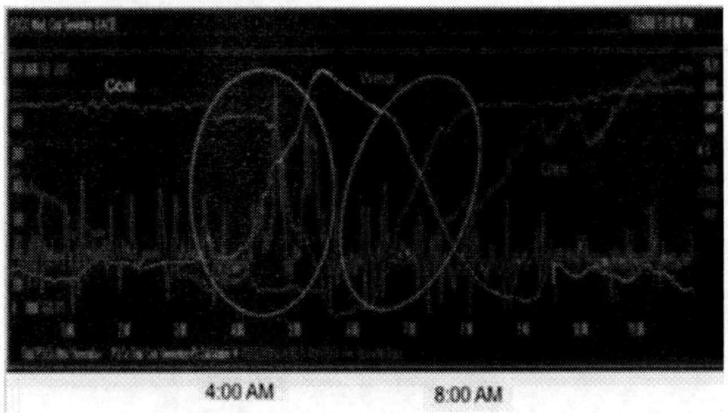

Power Plant Cycling Operation

- Decreased efficiency
- Increased emissions
- Additional cost and maintenance

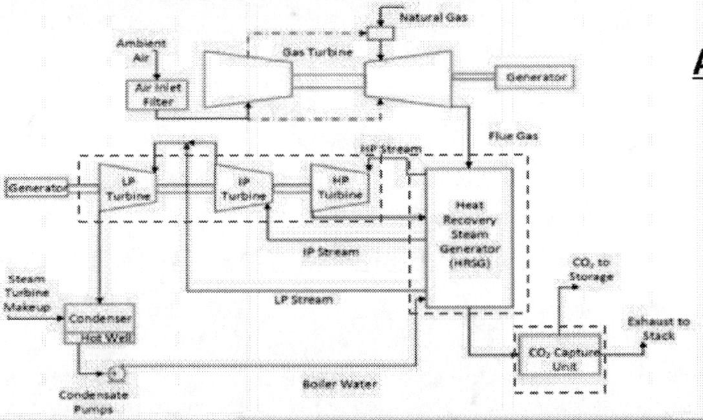

Approach

- NGCC plant with rigorous equipment level models
- CO_2 capture process by MDEA/PZ

Model Development

HRSG Design

- Heat section arrangement
- Geometric design

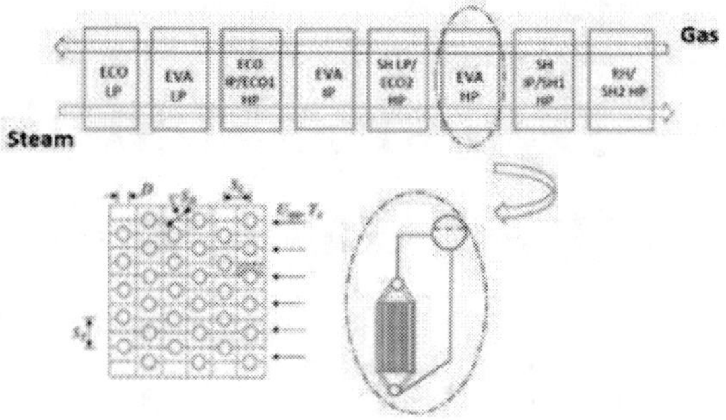

Steam Turbine Model in APD

- Stage by stage model
- Moisture effect

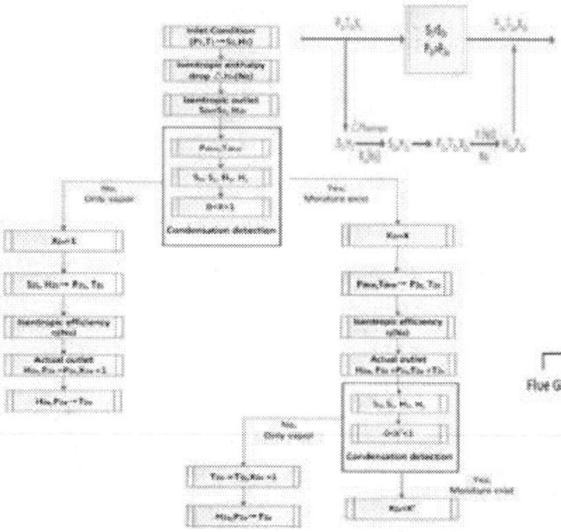

CO_2 Capture Process by MDEA/PZ

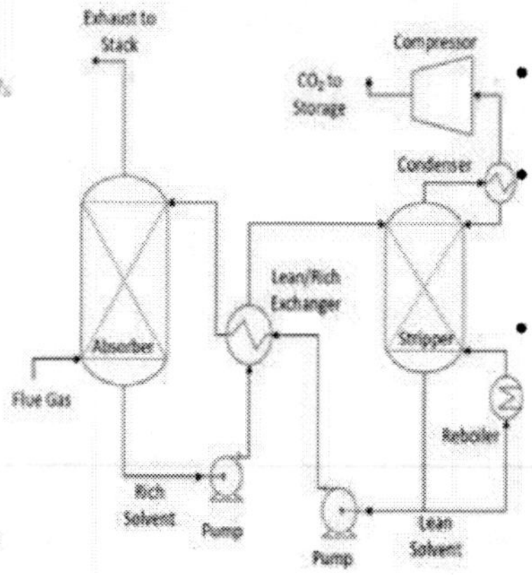

- Rate-based model in steady state
- Equilibrium model with stage efficiency in dynamics
- Steam consumption for lean solvent regeneration

Dynamic Simulation and Control

Evaporator and Condenser

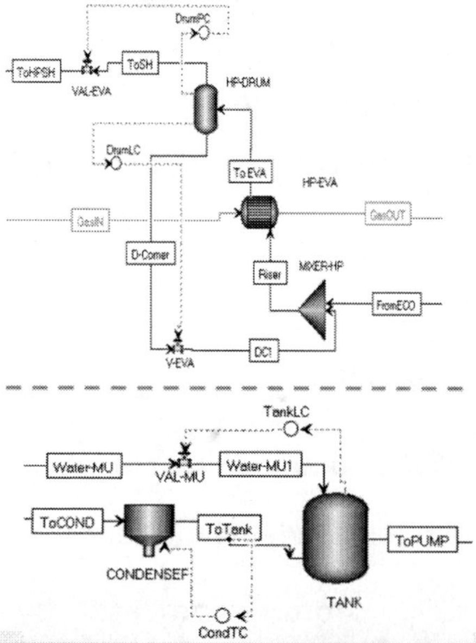

CO_2 Capture Process

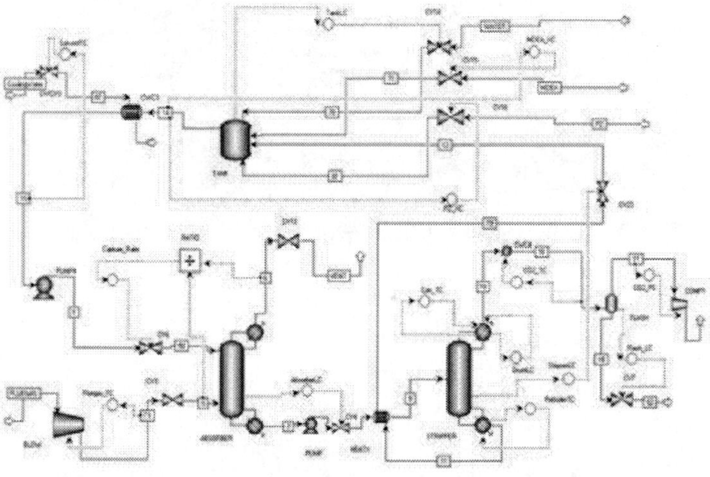

For more details

Poster 718, Exhibit Hall B, 15:15- 16:45

West Virginia University

499656 Exposure Reconstruction of Multiple Chemicals from Human Biomonitoring Data Using Markov Chain and Differential Evolution Monte Carlo

Abstract:

Humans are exposed to a wide range of chemicals that can be found in multiple environmental and microenvironmental media such as dust, soil, food, air, daily health care products and through multiple routes namely through the skin, inhalation and orally. These chemicals are emitted from human activities and they can have a significant contribution to short and long term exposure events causing adverse effects to human. However, the external exposure and the intake can be evaluated and estimated using Physiologically Based Biokinetic models assimilating biomonitoring data. This approach is known as reverse dosimetry or exposure reconstruction and its major and defining benefit is the retrospective interpretation of biomonitoring data allowing the estimation of cumulative and aggregate exposure.

The current study aims at the estimation of external and target tissue exposure to 17 different chemicals, including both rapidly (BPA, DEHP, triclosan) and non-rapidly (PCBs, BDEs, HCB, DDT, metals) metabolized compounds, starting from human biomonitoring (HBM) data and focusing on perinatal and childhood exposure.

The simulations were carried out using the INTEGRA platform, a software that provides realistic exposure scenarios coupled with a generic physiologic based bio-kinetic (PBBK) model and numerical "reverse engineering" techniques for exposure reconstruction. The applied exposure reconstruction algorithm is based on the Markov chain Monte Carlo and Differential Evolution Monte Carlo techniques. The process starts from ancillary exposure-related data that are fed into the exposure model taking into account multiple exposure routes. The results are evaluated against the biomonitoring data distributions, aiming at the reduction of uncertainty in back-calculating doses, by minimizing the error between the predicted and the actual biomonitored data. Parameterization of the model to cover a large chemical space is feasible using quantitative structure–activity relationship (QSAR) models provide by INTEGRA. Ancillary exposure parameters were obtained from the INTEGRA database.

The methodological framework was based on reverse dosimetry modelling and creation of exposure probability distributions. The distributions were consistent with collected human biomonitoring (HBM) data. HBM data were obtained from cohort and biomonitoring studies from several Mediterranean countries, namely Spain, Italy, Croatia and Slovenia. Then, the reconstructed exposure estimates were used to feed the PBBK model that was executed in forward-mode using two different dietary scenarios to derive internal doses of chemicals in target tissues.

The first exposure scenario consisted of one exposure event for newborn until the 4th year of his/her life. The main assumption for the newborns until the first six months of age was that they were fed exclusively with breast milk. Then, it was assumed that from the 6th until the 18th month daily diet consisted of 6 different meals and from the age of 18 months to 4 years it consisted of differentiated meals every 3 hours. It was also assumed that the contribution of each meal to the daily intake dose is the same. This assumption was based on the fact that the modelled chemicals have long half-life time, and consequently they accumulate for years in the human body.

The second exposure scenario consisted of one exposure event lasting 30 years assuming a daily food consumption based on the actual dietary schedule of the generic European adult population. In this case the main assumption was that the generic population consumes 3 daily meals. The exposure scenario of the simulation was assumed to start at the age of 15 years.

The examined chemical substances were PCB and PBDE congeners (PCB28, PCB52, PCB99, PCB101, PCB153, PCB180, BDE28, BDE47, BDE153, BDE154), HBC, pp'DDT, mercury, arsenic, BPA, DEHP and triclosan.

The results of the simulations showed that the predicted intake dose is commensurate with intake estimates found in the literature for both short- and long-term exposure scenarios in the European population. For the cases of BPA, DEHP and triclosan external intake estimates (e.g. for BPA 0.4 μg/kg_bw/d) were significantly lower than the respective tolerable daily intake (TDI). The estimated internal dose and the respective concentration in breast milk of BPA, DEHP and triclosan was very low because of their rapid metabolism. Similar were the results on exposure of neonates and infants.

Regarding non-rapidly metabolized chemicals, the BDE-47 simulation for Spain population showed that in steady state, at the age of 30, in utero concentration in the fetus of a pregnant mother had a median (min – max) value of 0.2 (0.1 - 0.8) ug/L. In Spain for the PCBs the results showed that in steady state and at the age of 4 years, PCB-153 concentration in the liver and in the brain are similar with a median value respectively of 1.3 (0.4 – 4.2) ug/L and 1.5 (0.6 – 3.8) ug/L while the concentration levels in gastrointestinal tract were one order of magnitude lower (median 0.1 (0.01 – 0.7) ug/L). Additionally, the simulation results for PCB28 and PCB101 showed brain concentrations with median value 1.9(0.9 – 3.0) and 0.9(0.0 – 4.0) ug/L. Liver concentration for PCB52 and PCB180 was 0.5(0.8 – 3.5) and 2.1(0 – 14) ug/L respectively. Moreover, PBBK simulations for pp'-DDT were carried out for the Menorca and Valencia population. Results revealed that in steady state condition and at the age of 30 years, the concentration of pp'-DDT in brain and in utero are about 1.5 - 2 times higher for the population of Menorca than in Valencia. Results showed that in Menorca the concentration in the brain and in utero was 3.5(0.2 – 12.1) and 2.5(0 – 2.2) ug/L. Last but not least, the HCB concentration in the adult brain of Valencia population had median value 2.1(0.6 – 7.7) ug/L.

Examining the concentration of arsenic in Slovenian and Italian populations in skin tissue were 5.5 (1.0 – 29.7) and 18.0(6.8 - 47.3) ug/L respectively. Additionally, arsenic concentration in the lung tissue for the Slovenian and Italian population had median values of 1.7(0.4 – 7.1) and 4.1(1.1 – 15.2) ug/L. Moreover, the concentration of mercury in Slovenian and Croatian populations in the brain was 46.5(12.1 – 174.8) and 9.2(2.8 – 29.0) ug/L respectively. Also, for the above mentioned populations the gastrointestinal concentration was calculated 53.8(9.6 – 58.4) and 134.9(56.0 – 321.1) ug/L.

From the methodological point of view, using an integrated exposure framework for linking mechanistically external and internal exposure, provides a comprehensive overview on how realistic exposure scenarios are translated into internal dose to humans, accounting for the age-dependent and route specific bioavailability differences. To this aim a generic PBTK modelling framework that captures lifetime internal exposure is a valuable tool with many applications in chemical risk assessment, by exploiting the continuously growing wealth of in vitro testing and biomonitoring data. It is worth mentioning that the advantage of using an approach based on PBBK modeling is that it allows the estimation of internal doses of xenobiotics that exceed levels associated with biological pathway alterations and, eventually, induction of biological perturbations that may lead to health risk. This represents a more appropriate metric to be linked with adverse health outcomes providing a more mechanistic and biology-based approach to human health risk assessment.

Assessing exposure at multiple scales across the source-to-dose continuum, needs to take into account the actual complexity of the environmental and biological/physiological processes that are critical to the proper description of the phenomena involved. This results in targeted interventions and consequently more cost efficient risk management. In addition, a comprehensive integrated exposure framework estimating tissue dosimetry for the various relevant exposure scenarios, could be of great use in exploiting the in vitro HTS (High-Throughput Screening) results rapidly produced by ToxCast21, advancing thus both exposure science and toxicology towards serving the needs of risk assessment in the 21st century. It is remarked that through PBBK models it can be effectively predicted and estimated the uptake dose and the deposition in tissues although the uncertainty that exists to the design and develop of the model.

Overall, our comprehensive modelling framework supports the association of a variety of environmental, exposure and biomonitoring data, as well as the incorporation of recent advances of in vitro toxicology using high-throughput systems in the risk assessment process enhancing thus significantly the artillery of environmental health science and chemical safety assessors and managers.

501385 Combined Dynamic Simulation and Scheduling Optimization of Shutdown Procedures of Ethylene Plants

Abstract:

Shutdown and startup procedures are important operations for maintenance of olefin plants. Both procedures involve complicated facility operations, take long time and may produce huge amount of flare emissions. Optimize shutdown procedures could shorten duration time, reduce flare emissions and decrease cost by saving intermediate materials.

In this presentation, a new shutdown methodology for olefin plants is proposed. The key issue is to recover remaining liquid and vapor materials in facilities during the shutdown procedure. To achieve that, CGC (Charge Gas Compressors) must be kept running during recovering, and recovered hydrocarbon is sent to other plants as feed. Dynamic simulations of the whole procedure are used to determine possibility and to provide necessary operation data.

Moreover, the shutdown procedure can be considered as a scheduling problem. Thus, an MIP (Mixed-Integer Programming) model is developed to get an optimized shutdown schedule for facilities. By optimizing shutdown arrangements for facilities, the overall shutdown time could be reduced by 25%. Combining with the new shutdown methodology, over 90% of hydrocarbon could be saved, and flare emissions could also be reduced by 90%.

500191 Single and Multi-Objective Optimizations Using Parallelized Process Simulators

Abstract:

Equation-oriented (EO) optimization methods are difficult to apply to rigorous thermodynamic methods, discontinuous variables, or systems where the number of equations is not constant. Despite these challenges, EO methods are still quite commonly used for process optimization, due in part to a perception that EO methods are faster than process simulation-based methods.
We have found that by integrating parallel processing techniques into modern process simulation and optimization, we can achieve rapid results comparable to the speed of more traditional EO modeling methods. We can attain these speeds even while using first-principle models, without risking validity of the model through too many simplifying assumptions. In this presentation, we will consider several process examples to explore the advantages of using simulators for optimization projects.

499146 Term Elimination and Optimal Experiments for Model Reduction

Abstract:

Models of complex systems, particularly thermofluid models of chemical processes, tend to be computationally and mathematically complex, consisting of many differential algebraic equations. Use of such models for applications beyond simple simulation, including optimization and controls development, can be infeasible due to their complexity. Model reduction methods are applied to develop reduced order models that adequately reflect the input-output behavior of higher-fidelity models, with significantly less computational cost. Currently practiced model reduction techniques are frequently based on projections [1], regressions, or surrogate modeling [2] [3]. The abstracted reduced models produced with these methods are effective for approximating outputs, but do not generally retain the physically meaningful parameters of the original model or system.

This work presents a novel method for the creation and evaluation of reduced order models, using a high-fidelity model as a basis. The proposed method consists of two key algorithmic processes. The first is a method for generating reduced models from an original model by heuristically directed term elimination. This process selectively removes or replaces parameters, variables, functions, and equations form the original model, creating a new candidate reduced model with each manipulation. The second key process is the selection of the most suitable reduced models from the candidate set by the evaluation of optimal experiments using a T-optimal design criterion. The candidate set of reduced models for the optimal experiments are those that are found to be suitable representations of the original for a nominal experiment. Similar formulations using different design criteria have been previously explored for use in the selection of kinetic models [4] [5] and fault detection and isolation [6].

Eqs. (1) and (2) enumerate the optimal experiment design problem, wherein are the outputs, state variables, equations, initial conditions, and parameters of the original model while correspond to the reduced model. The input trajectory is selected to maximize the weighted error in the outputs of the reduced model as compared to the original, integrated over the duration of the experiment . The parameter values of the reduced model, , are adjusted to minimize error and better approximate the original model. The resulting optimal design vector characterizes the experiment which will demonstrate the worst-case performance of the parameterized candidate reduced model at representing the original model within the state space of interest. If the observed maximum error from the optimal experiment is within a specified threshold, the reduced model can be considered as suitable.

(1)

(2)

The proposed method is demonstrated by application to some classic chemical process models, including homogenous and spatially discretized reactor models. It is observed that the method can autonomously recreate and select reduced models which match those identified in existing literature. It is also shown that the optimal experiment design approach can more reliably identify unsuitable reduced models as compared to experiments with nominal inputs or those with inputs at minimum or maximum values. Application to larger models of varying structure is performed in order to quantify the computational cost advantages achievable by the selected reduced models.

Acknowledgment

This work was sponsored by the UTC Institute for Advanced Systems Engineering (UTC-IASE) of the University of Connecticut and the United Technologies Corporation. Any opinions expressed herein are those of the authors and do not represent those of the sponsor.

References

[1] B. Moore, "Principal component analysis in linear systems: Controllability, observability, and model reduction," *IEEE Transactions on Automatic Control,* vol. 26, no. 1, pp. 17-32, 1981.

[2] M. Schmidt and H. Lipson, "Distilling free-form natural laws from experimental data," *Science,* vol. 324, no. 5923, pp. 81-85, 2009.

[3] A. Cozad, N. V. Sahinidis and D. C. Miller, "Learning surrogate models for simulation-based optimization," *AIChE Journal,* vol. 60, no. 6, pp. 2211-2227, 2014.

[4] L. Han, Z. Zhou and G. M. Bollas, "Model-based analysis of chemical-looping combustion experiments. Part I: Structural identifiability of kinetic models for NiO reduction," *AIChE Journal,* vol. 62, no. 7, pp. 2419-2431, 2016.

[5] L. Han, Z. Zhou and G. M. Bollas, "Model-based analysis of chemical-looping combustion experiments. Part II: Optimal design of CH4-NiO reduction experiments," *AIChE Journal,* vol. 62, no. 7, pp. 2432-2446, 2016.

[6] K. A. Palmer, W. T. Hale, K. D. Such, B. R. Shea and G. M. Bollas, "Optimal design of tests for heat exchanger fouling identification," *Applied Thermal Engineering,* vol. 95, pp. 382-393, 2016.

494518 Scalable Modeling and Solution of Stochastic Multiobjective Optimization Problems

Abstract:

We present a scalable computing framework for the solution stochastic multiobjective optimization problems. The proposed framework uses a nested conditional value-at-risk (nCVaR) metric to find compromise solutions among conflicting random objectives. We prove that the associated nCVaR minimization problem can be cast as a standard stochastic programming problem with expected value (linking) constraints.We also show that the solutions satisfy stochastic dominance of second order. Finally, we demonstrate that these problems can be implemented in a modular and compact manner using PLASMO (a Julia-based structured modeling framework) and can be solved efficiently using PIPS-NLP (a parallel nonlinear solver). We apply the framework to a combined heat and power design study in which we seek to find compromise solutions that trade-off cost, water, and emissions in the face of uncertainty in electricity and water demands.

501324 Optimal Refinery Crude Scheduling with Considerations of Crude Mixing Along with the Pipeline Transportation

Abstract:

Front-end crude scheduling is a crucial issue for petroleum industries as the crude is transported through multi-stage facilities before entering crude distillation units (CDU), which normally include berth loading at port sites, various tanks for storage and blending, pipeline transportation, as well as CDU processing. Since the crude supply and consumption are subject to the more volatile market nowadays, the optimal crude scheduling becomes more increasingly important for the probability of refiners. The pipeline system is a vital transportation mode for petroleum industries due to its flexible operation, economic cost, high efficiency and low environmental impact.Â Distinguished from other transportation modes, like trucks and ships, the pipeline system is featured with continuous operations, thus itâ€™s very suitable and economically beneficial for the transportation in large quantity and long distance. Therefore, it has been widely used for the transportation of crude oil, LNG, petroleum products and their derivatives.

The pipeline system is rarely considered in studies of crude scheduling. [1,2] Recently, some works are published incorporating the pipeline system into crude scheduling, which makes models more realistic.[3] However, these studies did not taken into account the crude mixing interface, the mixing at interface sections of two different types crudes. Usually, the crude mixing interface would generate a segment with combined properties of two mixed crudes. For crude refining, sulfur and heavy metal in crude oil are treated as key components, whose levels should be rigorously controlled before feeding CDUs. These key component concentrations in the mixed oil segment may cause off-spec oil properties inside charging tanks if their impacts are neglected, and thus jeopardize the CDU processing. For the safe and optimal operations of CDUs and the whole process, the crude mixing segment should be precisely identified together with the key component concentrations before the blending in the charging tanks and processing in the CDUs. Based on our literature survey, none published work has systematically addressed the oil mixing issues along with the pipeline transportation for refinery front-end crude scheduling.

In this study, a novel refinery crude scheduling model has been developed with the consideration of crude mixing along with the pipeline transportation, where the continuous-time representation is utilized to present the whole front-end crude movement scope. A detailed scheduling covering crude unloading from vessel, crude transportation through berths, storage tanks, pipeline system and charging tanks, and finally crude charging to CDUs is determined toward the maximal profit. In this work, CDUs with different requirements on key component concentrations are modeled and subject to manufacturing limits. Meanwhile, industrial operation constraints, such as multiple crude oils, crude blending and brine settling time, are taken into account to make the developed model more realistic and applicable in realities.Â In addition, some technical improvements on model, like the duration of tank holdup for inventory cost and the tighten bounds for transportation flowrate are also programmed to make model solution more practical. The efficacy and performance of the proposed model has been tested and demonstrated through various case studies.

References

[1] Lee, H., Pinto, J., Grossmann, I., & Park, S. (1996). Mixed-integer linear programming model for refinery short-term scheduling of crude oil unloading with inventory management. Industrial and Engineering Chemistry Research, 35(5), 1630â€"1641.

[2] Chen, X., Grossmann, I., Zheng, L., (2012)A comparative study of continuous-time models for scheduling of crude oil operations in inland refineries. Computer and Chemical Engineering, 44, 141-167

[3] Zhang, S., Xu, Q., (2015) Refinery continuous-time crude scheduling with consideration of long-distance pipeline transportation. Computer and Chemical Engineering, 75, 74-94.

499611 Optimal Planning Under Uncertainty for a Supply Chain Focused on Residual Biomass Conversion Using Geographic Information Systems and Mathematical Programming

Abstract:

Biomass is a renewable resource with attractive characteristics for manufacturing many types of products such as biofuels and generation of energy. One of the main reasons to use residual bioresources for biofuels production is that biomass can capture a significant part of emissions during its growing. In this regard, various studies have addressed diverse important features such as development of technologies for biomass treatment and processing, schemes to determine location of biorefineries and harvesting sites, models to evaluate the economic and environmental impacts, approaches capable of considering the associated dynamic in biomass and products storage and supply, between others. However, the most of the works have omitted other important issues such as location of the potential supply chain's nodes considering geographic constraints as well as the uncertainty associated with the residual biomass availability and price over time, mainly because this may increase drastically the number of options, variables and constraints to be taken into account. Therefore, this paper presents an approach for the optimal planning of a supply chain focused on biofuels production from agricultural residues, which involves a methodology to locate potential sites for biomass extraction considering environmental, social and geographic limitations using a tool based on geographic information systems, and a formal model based on mathematical programming. It is important to mention that planning problem for any supply chain consists on determining raw materials, products, processing technologies, processed and produced amounts, facilities' location, among other issues to obtain diverse benefits. On the one hand, mathematical model is a Mixed Integer Linear Programming Problem and consists on mass balances for all considered biomass types, constraints for biomass availability and product demand, cost equations and functions to evaluate environmental and social impacts. On the other hand, the geographic constraints are the protected areas exclusion, terrain's slope, wetland zones' location, distance to water bodies, distance to human communities, frequency of weather phenomena like torrential rain and hurricanes, and distance to transportation and electrical infrastructure. Additionally, the proposed approach includes the variation in the biomass availability and price over time through a geoprocessing model by generating several scenarios based on historical information. The method was tested for a nationwide case study in Mexico to determine potential locations for biomass suppliers, processing plants, consumption regions and processing technologies to obtain biofuels from agricultural residues. The case study contemplates the location and production of several agricultural residues such corn stover, sugar cane bagasse, sorghum straw, residue from agave and wheat straw to produce diverse products such as Fisher Tropsch liquids, ammonia, formaldehyde, methanol, among others via different gasification alternatives. Results show the variation of potential locations for biomass suppliers to be used in the proposed biomass processing system, while the availability of bioresources changes. Furthermore, results illustrate the influence of the type of raw material in the location of these suppliers. The geoprocessing model was useful to identify potential locations to be taken into account in the subsequent supply chain planning problem decreasing considerably the number of options to choose accounting social, environmental and geographic issues, for instance for the case of corn stover the considered initial municipalities were 2337, nevertheless the geoprocessing model can reduce them to around 200 municipalities with high potential to install process facilities and biomass suppliers to be included in the mathematical model for optimal planning.

An Optimization Approach to Ordinary-Fractional Multi-Compartmental Models with Applications to Pharmacokinetics and Optimal Drug Usage

Vicente Rico-Ramirez

vicente@iqcelaya.itc.mx

Instituto Tecnologico de Celaya, Mexico

17AIChE

Fractional Calculus and Compartmental Models in Pharmacokinetics

- **Fractional calculus introduces derivatives and integrals of non-integer order. Non-local operators of Fractional Calculus are useful to model the behavior of state variables presenting memory effects.**

$$\frac{d^\alpha f}{dx^\alpha}$$

$$_aD_x^\alpha f(x) = \frac{d^n}{dx^n}\, _aI_x^{n-\alpha} f(x) = \frac{1}{\Gamma(n-\alpha)}\left(\frac{d^n}{dx^n}\right)\int_a^x (x-t)^{n-\alpha-1} f(t)d$$

Left Riemann-Lioville Fractional Derivative (RLFD)

- ***This work is concerned with dynamic systems including both fractional as well as ordinary differential equations. Recent applications have been reported for Compartmental Modeling in Pharmacokinetics***

Dynamic flow of chemicals within the body: Drug absorption, distribution and elimination .

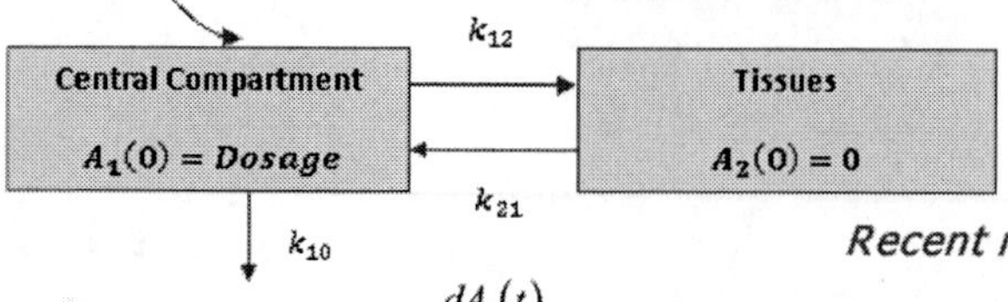

Conventional models involve ordinary dynamics

$$\frac{dA_2(t)}{dt} = k_{12}A_1(t) - k_{21}A_2(t)$$

$$\frac{dA_1(t)}{dt} = -k_{12}A_1(t) + k_{21}A_2(t) - k_{10}A_1(t) + u(t)$$

Recent reports suggest fractional dynamics

$$\frac{dA_1(t)}{dt} = -(k_{12}+k_{10})A_1(t) + k_{21}\, {}_0^CD_t^{1-\alpha} A_2(t) + u(t) \qquad \frac{dA_2(t)}{dt} = k_{12}A_1(t) - k_{21}\, {}_0^CD_t^{1-\alpha} A_2(t)$$

17AIChE 17AIChE

Modeling and Optimization of Fractional-Ordinary Dynamic Systems

- *This paper focuses on compartmental models in Pharmacokinetics, but our approach can be applied to the modeling and Optimization of general fractional-ordinary dynamic systems.*

Four Steps in Our Approach

1 Reformulation of Compartmental Dynamics as a Set of Fractional-Ordinary Differential Equations (SFODE)

Define artificial variables and use the formula for successive fractional derivation:

$$ {}_0^C D_t^{1-\alpha} A_2(t) = B \qquad \frac{dA_1(t)}{dt} = -(k_{12}+k_{10})A_1(t) + k_{21}\,B(t) + u(t) \qquad {}_0^C D_t^{\alpha} B = k_{12}A_1(t) - k_{21}\,B(t) $$

2 Development and Implementation of a Predictor-Corrector Integration Technique for the SFODE

The technique for fractional problems could be used for the ordinary case, but unnecessary significant computational effort would be involved in the integration

$$ \frac{dy_k}{dt} = G_k(\mathbf{y},\mathbf{z},t) \qquad \mathbf{y}(0) = \mathbf{y}_0 $$

$$ {}_0^C D_t^{\alpha_l} z_l = F_l(\mathbf{y},\mathbf{z},t) \qquad \mathbf{z}(0) = \mathbf{z}_0 $$

Predictor steps for OFDE

$$ y_k^{\,P}(t_{n+1}) = y_k(t_n) + hG_k(t_n,\mathbf{y}(t_n),\mathbf{z}(t_n)) $$

$$ z_l^{\,P}(t_{n+1}) = z_{l,0} + \frac{1}{\Gamma(\alpha_l)}\sum_{j=0}^{n} b_{j,n+1,l}F_l(t_j,\mathbf{y}(t_j),\mathbf{z}(t_j)) $$

17AIChE

17AIChE

Modeling and Optimization of Fractional-Ordinary Dynamic Systems

3 Non-linear fitting techniques + ordinary-fractional integration + experimental data to attain kinetics parameters and fractional orders

Parameters for the Fractional Model of Amiodarone	
α=0.5870	k_{12}= 2.9533 day^{-1}
k_{10}=1.4913 day^{-1}	k_{21}=0.4854 day$^{-\alpha}$

4 Dynamic Optimization: Combined numerical/theoretical strategy for Fractional-Ordinary Optimal Control Problems

Generalized optimality conditions for fractional-ordinary optimal control problems were derived

The resulting two-point boundary value problem involving a SOFDE (from optimality conditions) was solved by using a modified version of the gradient method

End conditions for adjoint variables following fractional dynamics were also derived

$$ \min \quad J = \int_0^{t_f} \left[A_2(t) - \delta\right]^2 dt $$

$$ {}_0^C D_t^{1-\alpha} A_2(t) = B $$

$$ \frac{dA_1(t)}{dt} = -(k_{12}+k_{10})A_1(t) + k_{21}\,B(t) + u(t) $$

$$ {}_0^C D_t^{\alpha} B = k_{12}A_1(t) - k_{21}\,B(t) $$

$$ A_1(t) \le \theta $$

$$ A_2(t) \le \theta $$

$$ 0 \le u \le u_{\max} $$

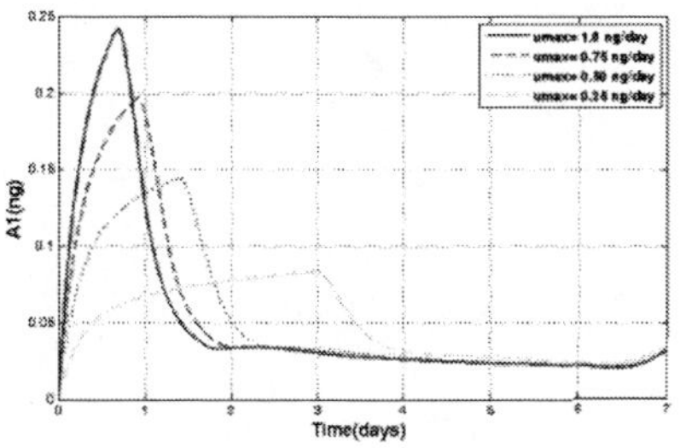

17AIChE

493111 GPU Parameter Tuning for Dense Linear Least Squares Problems

Abstract:

Linear least squares problems (LLSPs) routinely arise in many types of scientific and engineering problems. As the size of datasets used to generate linear models increases, there exists a need to efficiently solve large dense linear least squares problems. One of the fastest ways to solve large LLSPs is to utilize the parallel processing power of graphics processing units (GPUs) and GPU parallel LLSP solvers such as cuSolverDN [5] and communication avoiding QR [3]. There has also been work to develop hybrid algorithms that take advantage of the processing power of both CPUs and GPUs through the MAGMA project [7]. However, these parallel algorithms are typically designed and optimized for one GPU architecture and may be unusable or suboptimal on another GPU. To design algorithms that are portable between devices, most algorithms are designed with tunable parameters that can be adjusted to avoid having to rewrite most of an algorithm to remain optimal on any GPU architecture. Traditionally, these parameters have been chosen through extensive experimentation or through heuristics that come from the knowledge of solver designers.

The primary contribution of this work is to propose systematic methods for tuning GPU or hybrid CPU/GPU algorithms through the use of derivative-free optimization (DFO) [6] and simulation optimization (SO) [1]. A second contribution of our work is to provide a comparison of thirty-one DFO and four SO solvers in the context of tuning GPU algorithms.

To determine a baseline of performance, we evaluated the performance of a few of the most well-known dense linear algebra libraries: LAPACK [2], a multicore solver PLASMA [4], a GPU only algorithm cuSolverDN [5], and a hybrid implementation MAGMA [7]. We evaluated each of these solvers over a wide range of different sized square and tall and skinny matrices. Tall and skinny matrices commonly arise when solving LLSPs, and are typically challenging to obtain high performance on because of the matrix structure. For square matrices, the solver MAGMA was able to outperform all of the other solvers. However, for tall and skinny matrices MAGMA was not able to perform as well as the other solvers.

Our computational results show that the best DFO solver is able to speed up the performance of MAGMA by 1.67x compared to default MAGMA parameters. After tuning MAGMA through the proposed approach, MAGMA was able to outperform all other solvers for tall and skinny matrices.

References cited

[1] S. Amaran, N. V. Sahinidis, B. Sharda, and S. J. Bury. Simulation optimization: A review of algorithms and applications. Annals of Operations Research, 240:351–380, 2016.

[2] E. Anderson, Z. Bai, C. Bischof, L. S. Blackford, J. Demmel, J. Dongarra, J. Du Croz, S. Hammarling, A. Greenbaum, A. McKenney, and D. Sorensen. LAPACK users' guide (third ed.). Society for Industrial and Applied Mathematics, Philadelphia, PA, USA, 1999.

[3] M. Anderson, G. Ballard, J. Demmel, and K. Keutzer. Communication-avoiding QR decomposition for GPUs. In Parallel & Distributed Processing Symposium (IPDPS), 2011 IEEE International, pages 48–58, 2011.

[4] B. Hadri, H. Ltaief, E. Agullo, and J. Dongarra. Tile QR factorization with parallel panel processing for multicore architectures. In Parallel & Distributed Processing (IPDPS), 2010 IEEE International Symposium on, pages 1–10, 2010.

[5] NVIDIA Corporation. cuSolver, Current as of 28 December, 2016. http://docs.nvidia.com/cuda/cusolver/#axzz4SZ3ssQJO.

[6] L. M. Rios and N. V. Sahinidis. Derivative-free optimization: A review of algorithms and comparison of software implementations. Journal of Global Optimization, 56:1247–1293, 2013.

[7] S. Tomov, J. Dongarra, and M. Baboulin. Towards dense linear algebra for hybrid GPU accelerated manycore systems. Parallel Computing, 36:232–240, 2010.

494680 Phase Change Transient Model for Predicting Frost Formation in Ambient Air Vaporizer

Abstract:

Natural gas is continuously getting attention as an eco-friendly and sustainable energy source mainly composed of methane. Natural gas is liquefied into liquefied natural gas (LNG) in order to facilitate transportation and then regasified to natural gas (NG) to be supplied to the user at the terminal. LNG regasifiers used in terminals include ORV, SCV, AAV, etc. ORV uses seawater as a heat source and SCV burns fuel to provide heat. Among them, Ambient Air Vaporizer (AAV) is LNG regasification technology using air in the atmosphere as a heat source. The ORV increases the temperature of seawater and affects the marine ecosystem, and SCV burns the fuel, which adversely affects the atmospheric environment. On the other hand, AAV uses the atmospheric air directly, so it is eco-friendly and offers reasonable price and high operating efficiency. However, in the process of transferring thermal energy from air to LNG, the phenomenon of condensation of water vapor contained in the air into frost on the AAV surface occurs due to cryogenic temperature of LNG. This frost acts as a thermal resistance that interferes with the heat transfer between the air and the LNG, thereby lowering the heat transfer efficiency between the LNG and the air. As time goes on, the frost accumulates more and more. As a result, after a certain period of time, the heat transfer efficiency of the AAV becomes very low and the target NG output temperature is not reached. Thus, the AAV will alternate between operation and defrosting. Therefore, for the design of AAV and its application to LNG regasification of AAV, it is essential to understand how frost is formed in AAV. However, research on AAV and studies on frost growth in AAV are lacking. Therefore, this study presents a model for frost growth using CFD, and applies this model to AAV to analyze the effect of frost growth on AAV performance on AAV.
The proposed model is an Eulerian-Eulerian multiphase flow model of ice and moist air simulating frost formation and growth using CFD. At below freezing point, the water vapor contained in the moist air is phase-changed to Ice, and the generated ice accumulates without flowing. That is, mass transfer with phase change from air to frost occurs over time, changing the thickness and density of the frost and affecting the flow of moist air. Therefore, this model can observe the thickness, density, and shape of the frost with time, and also the effects of frost on the flow of air and its effect on heat transfer.

495741 Simulation of the Effect of Contact Area Loss in All-Solid-State Batteries

Abstract:

Maintaining the physical contact between the solid electrolyte and the electrode is important to improve the performance of all-solid-state batteries. Imperfect contact can be formed during cell fabrication and will be worsened due to cycling, resulting in degradation of the battery performance. In this paper, the effect of imperfect contact area was incorporated into a 1ÂD Newman battery model by assuming the current and Li concentration will be localized at the contacted area. Constant current discharging processes at different rates and contact areas was simulated for a film-type Li|LiPON|LiCoO2 allÂsolidÂstate battery. The capacity drop was correlated with the contact area loss. It was found at lower cutoff voltage, the correlation is almost linear with a slope of 1; while at higher cut off voltage, the dropping rate is slower. To establish the relationship of the applied pressure and the contact area, Perssonâ€™s contact mechanics theory was applied, as it uses self-affined surfaces to simplify the multi-length scale contacts in all-solid-state batteries. The contact area and pressure was computed for both film-type and bulk-type all-solid-state Li-ion batteries. The model is then used to suggest how much pressures should be applied to recover the capacity drop due to contact area loss.

496443 Multi-Phase Mathematical Modeling of Compressed CO_2 Expansion through a Coanda Nozzle

Abstract:

A Coanda nozzle can be used to produce a high velocity mixture of air, gaseous CO_2, and dry ice particles from a supply of liquid CO_2. Such a process is effective (for instance) for residue-free rapid cooling or precision cleaning. A multiphase thermodynamic and computational fluid dynamics (CFD) analysis of this flow is presented for the purpose of optimizing the process parameters, which include the temperature of the liquid CO_2 supply, the flow rate, and the pressure and nozzle and air configuration.

Temperature, pressure, velocity and mass fraction profile distributions have been modeled on the spray-nozzle system. Flow characteristics have been quantified for dry ice particle sizes of $1.0\text{x}10^{-3}$m, $1.0\text{x}10^{-4}$m and $1.0\text{x}10^{-5}$m. Maximum velocity and residence time as a function of inlet temperature have been simulated and quantified. The analysis suggests that as inlet CO_2 temperature increases, the maximum velocity increases and the particle residence time decreases. Decreasing dry ice particle size decreases the residence time in the nozzle channel. Overall, this study provides a basis for optimizing the temperature and velocity profiles, and thus the cooling or cleaning capacity of such Coanda devices.

498349 Stochastic Modeling of CTB-GM1 Binding Mechanisms

Abstract:

Cholera toxin (CTx) is a toxin protein, which can lead to lethal cholera. CTx is an AB_5 protein that is consisted of an enzymatic A-subunit and five identical cholera toxin B-subunits (CTB) [1]. CTB binds with gangliosides such as GM1 on host cell membrane, which facilitates the endocytosis of CTx and the development of cholera. Despite extensive experimental and theoretical work on understanding the CTB structure and function, there is a lack of systematic framework that relies on fundamental understanding of the binding mechanisms of CTB at the microscopic level and utilizes such information to elucidate the detailed mechanism of CTB-GM1 interactions, which is the prerequisite for CTx cell entry. It is worth noting that a pentavalent CTB can simultaneously binds to multiple copies of GM1 in cell membranes. Also, the positive binding cooperativity between each binding subunit can further enhance the overall binding avidity. Since microscopic binding events highly dependent on the current surface configuration, the kinetic Monte Carlo methodology is applied to simulate the complex interactions between CTB and GM1 microscopically.

The goal of this research focuses on modeling and simulation of CTB-GM1 interactions. In this work, GM1 is modeled as an entity constrained to a 2 dimensional host cell membrane, whereas CTB is modeled as an entity capable of binding and detaching from the membrane surface. Three types of microscopic events are considered: 1) molecular attachment, 2) detachment, and 3) migration events on host cell membranes. The implementation of the kinetic Monte Carlo methodology requires knowledge of association, dissociation and migration reaction constants. Previous work computed a range of values to these parameters until satisfactory agreement between the calculated and the experimental binding rates was achieved [2, 3]. Due to the overwhelming likelihood of migration events, migration has been decoupled from the event selection process in order to improve computational efficiency [4]. For molecular attachment, it is assumed that each lattice site is available for attachment, and thus the attachment rate is equal over the lattice. Conversely, for molecular detachment and migration events, the rates are dependent on the local environment. Each local environment comprises up to 6 nearest neighbors. This allows us to classify our local environment into seven classes (zero to six nearest neighbors) in order to increase the computational efficiency when calculating the rates associated with executing a kinetic Monte Carlo event. We will also apply the kinetic Monte Carlo model to examine other host cell receptors such as GM2, which exhibits a distinct CTB binding kinetics. The specific details of how the GM1 concentration affects the CTB-GM1 interactions as well as steric hindrance between the CTB molecules on the membrane surface will be presented.

Reference

1. de Haan, L.; Hirst , T.R. Cholera toxin: A paradigm for multi-functional engagement of cellular mechanisms. *Molecular Membrane Biology* **2004**, *21*, 77-92.
2. Lauer, S.; Goldstein, B.; Nolan, R.L.; Nolan, J.P. Analysis of cholera toxin â€“ ganglioside interactions by flow cytometry. *Biochemistry* **2002**, *41*, 1742-1751.
3. Lin, H.; Kitova, E.N.; Klassen, J.S. Measuring positive cooperativity using the direct ESI-MS Assay. Cholera toxin B subunit homopentamer binding to GM1 pentasaccharide. *J. Am. Soc. Mass. Spectrom.* **2014**, *25*, 104-110.
4. Crose, M.; Kwon, J.S.; Nayhouse, M.; Ni, D.; Christofides, P.D. Multiscale modeling and operation of PECVD of thin film solar cells. *Chem. Eng. Sci.* **2015**, *136*, 50-61.Â

500043 Numerical Investigation of a Water Droplet on Vibrating Surface with and without Inclination

Abstract:

Modeling of droplet-surface interaction remains a challenge in fluids research, because of its complex, multiscale nature that involves both microscale interactions and macroscale flows. To understand its underlying mechanisms is important for the application design. The interactions amongst liquid/gas/solid molecules dominate the contact line dynamics, and determines the stationary and dynamic contact angles. The interactions between liquid/solid molecules also governs friction on the surface. We propose a novel numerical method to handle the droplet on a superhydrophobic surface, and validate our model with experiments on a 3mm water droplet sitting on a vibrating surface. A variable surface tension field is used to model the superhydrophobic behavior, and a term based on the gradient of surface tension is formulated to compensate the surface tension force. Different cases have be investigated for validating our methods and understanding of the underlying physics. Although the vibration-induced wetting transition was investigated in some recent studies, the vibration mechanism of droplet shedding has not yet been fully understood. This research quantitatively considers the effect of vibration on droplet shedding under various vibration resonance conditions, providing a possible way to effectively shed droplet off surfaces in condensation applications.

501527 Novel Optimization-Based Adaptive Sparse-Grid Methods for Numerical Integration

Abstract:

The Smolyak algorithm (also known as Sparse-Grid, Hyperbolic Cross-Points or Boolean method) was derived in 1963 by Russian mathematician Smolyak, and was based on the fact that spaces of n-dimensional continuous linear functionals contain tensor products of spaces of continuous linear functionals on the one-dimensional space [1]. This development was followed by a plethora of advances in theory, methods and applications, due to the realization that high-dimensional tensor product problems can be solved efficiently using accurate n-dimensional approximations which can be generated from univariate approximations [2]. Sparse-grid methods have been used for approximation, integration and optimization in mathematics, physics, economics and engineering design [3-5]. Two of their main advantages are (a) sparse-grid tensor products do not grow exponentially with dimension n (curse-of-dimensionality), and (b) convergence rates and error-bound analysis are available as a function of smoothness of the function within a bounded region.

In this work, we aim to show the power of this extremely mathematically-rich theory, and extend the capabilities of existing methods by combining existing theory with theory from deterministic optimization. Specifically, we have developed a computational tool for multidimensional integration of functions, using the Smolyak algorithm with a variety of univariate quadrature rules, such as Clenshaw-Curtis, Gauss-Patterson and Fejer rules. These rules are based on roots or extrema of orthogonal polynomials and lead to non-equidistant sparse tensor grids, which perform exceptionally well with fewer number of sample points, when compared to equidistant full-tensor grid points. We have incorporated a newly derived formula for the efficient calculation of the weights for the calculation of multidimensional integrals. This derivation circumvents a main hurdle in the Smolyak algorithm, namely the problem of redundant weights which significantly increases the computational cost of the method [3]. This development allows us to push the limits of high dimensional integration, which we will show through benchmark cubature problems. Furthermore, we have incorporated an approach for the further reduction of the total number of points required for accurate multidimensional integral calculation, using ideas from adaptive-sparse grids [6-7] and optimization. Overall, this talk aims to introduce significant theoretical and methodological advances, which have a wide range of applications in the field of chemical engineering.

1. Smolyak, S. A. (1963). Quadrature and Interpolation Formulas for Tensor Products of Certain Classes of Functions. Soviet Math. Dokl., 4, 240-243

2. Novak, E., Ritter, K. (1996). High dimensional integration of smooth functions over cubes, Numer. Math. 75, 79-97

3. Judd, K.L., Maliar, L., Maliar, S., Valero, R. (2014). Smolyak method for solving dynamic economic models: Lagrange interpolation, anisotropic grid and adaptive domain. J of Economic Dynamics & Control, 44, 92-123

4. Bohn, B., Garcke, J., Griebel, M. (2016). A Sparse-Grid Based Method for Generative Dimensionality Reduction of High-Dimensional Data. J of Comp Physics, 309, 1-17

5. Malin, B., Krueger, D., Kubler, F. (2011). Solving the multi-country real business cycle model using a Smolyak-collocation method. J of Economic Dynamics & Control, 35, 229-239

6. Gerstner, T., Griebel, M. (2003) Dimension-Adaptive Tensor-Product Quadrature. Computing, 71, 65-87

7. Bungartz, H.J., Dirnstorfer, S. (2003). Multivariate Quadrature on Adaptive Sparse Grids. Computing, 71, 89-114

500800 Comparison of Machine Learning Approaches for Process Model Development from Big Data

Abstract:

Over the last decade, the development of advanced sensing technologies coupled with powerful computing tools and modern automation infrastructure has tremendously increased the amount and variety of data collected in laboratory and industrial settings. For example, basic process control systems (BPCS), which are designed to keep the processing variables within their normal operating ranges, contain databases with longitudinal information for all process variables along with all recorded alarms. Massive amounts of data, which describe physical and chemical properties of process variables, are recorded. To extract useful and relevant information from these vast amounts of data, data analytics and machine learning techniques are gradually penetrating the process and pharmaceutical industries. They are used to improve the understanding of the process, to develop surrogate process models, for fault-diagnosis, and to learn about the health and stability of the overall process (e.g., [1-4]).

Surrogate process models, which are also known as metamodels or emulators, in this context, are approximate models that are constructed to statistically relate a set of input process variables to a set of output process variables. There are a number of machine learning techniques that can be used to construct surrogate models such as Extreme Learning Machines (ELMs) [5], Artificial Neural Networks (ANNs) [6], and Automated Learning of Algebraic Models using Optimization (ALAMO) [7], but little work has been done to systematically compare their ability to learn the response of complicated models with different characteristics, such as those generated by chemical and pharmaceutical processes.

This study compares eight surrogate-model construction approaches using computational experiments. The construction approaches considered include: ANNs, ALAMO, Radial Basis Networks (RBNs) [8], ELMs, Gaussian Progress Regression (GPR) [9], Random Forests (RFs) [10], Support Vector Regression (SVR) [11], and Multivariate Adaptive Regression Splines (MARS) [12]. Each approach is used to construct surrogate models for predicting the outputs of thirty-four test functions, which can be found in the Virtual Library of Simulation Experiments (https://www.sfu.ca/~ssurjano/optimization.html), and that have with various shapes and numbers of inputs. The input-output data that is employed for training the surrogate models is generated using Latin Hypercube, Sobol and Halton sampling methods.

The performance of the surrogate models for each challenge function were compared using maximum absolute error (MAE) and root mean squared error (RMSE). The results revealed that, at large sample sizes, the sampling method applied to generate the training data set did not have a statistically significant impact on the performance measures. However, when the results were examined in groups constructed based on the number of inputs and shape of the test functions, the surrogate-models constructed using ANN, ALAMO and ELM yielded smaller MAE and RMSE than the other surrogate-model construction approaches. It is also worth noting that the models constructed using ALAMO had consistently simpler functional forms that ANN and ELM models.

1. Machin, M., L. Liesum, and A. Peinado, Implementation of modeling approaches in the QbD framework: examples from the Novartis experience. Eur Pharm Rev, 2011. 16: p. 39-42.

2. Kirdar, A.O., et al., Application of near-infrared (NIR) spectroscopy for screening of raw materials used in the cell culture medium for the production of a recombinant therapeutic protein. Biotechnology Progress, 2010. 26(2): p. 527-531.

3. Kirdar, A.O., et al., Application of Multivariate Analysis toward Biotech Processes: Case Study of a Cell-Culture Unit Operation. Biotechnology Progress, 2007. 23(1): p. 61-67.

4. Kirdar, A.O., K.D. Green, and A.S. Rathore, Application of Multivariate Data Analysis for Identification and Successful Resolution of a Root Cause for a Bioprocessing Application. Biotechnology Progress, 2008. 24(3): p. 720-726.

5. Huang, G.-B., D.H. Wang, and Y. Lan, Extreme learning machines: a survey. International Journal of Machine Learning and Cybernetics, 2011. 2(2): p. 107-122.

6. Haykin, S., Neural networks and Learning Machines. 3 ed. 2009: Prentice Hall.

7. Cozad, A., N.V. Sahinidis, and D.C. Miller, Learning surrogate models for simulation-based optimization. AIChE Journal, 2014. 60(6): p. 2211-2227.

8. Park, J. and I.W. Sandberg, Universal approximation using radial-basis-function networks. Neural computation, 1991. 3(2): p. 246-257.

9. Rasmussen, C.E., Gaussian processes for machine learning. 2006.

10. Breiman, L., Random forests. Machine learning, 2001. 45(1): p. 5-32.

11. Basak, D., S. Pal, and D.C. Patranabis, Support vector regression. Neural Information Processing-Letters and Reviews, 2007. 11(10): p. 203-224.

12. Friedman, J.H., Multivariate adaptive regression splines. The annals of statistics, 1991: p. 1-67.

498619 Application of Machine Learning Algorithms for the Selection of Jet Fuel from Hydrocarbon Blends

Abstract:

Application of Machine Learning Algorithms for the Selection of Jet Fuel from Hydrocarbon Blends

Rajib Mukherjee[1], Noof Abdalla [2], Nasr Mohamed [2], Marwan El Wash[2] , Nimir O. Elbashir[2], Mahmoud El-Halwagi[1]
1. Gas and Fuels Research Center, Texas A&M Engineering Experiment Station, College Station, TX 77843, USA

2. Department of Chemical Engineering, Texas A&M University, Qatar, Education City, Doha

Abstract

Blending is an essential part in transportation fuel preparation. In this paper application of machine learning algorithms will be presented for accurate prediction of the properties synthetic fuel composed of blend of hydrocarbons obtained from gas to liquid (GTL) technology. Statistical measures have been used to ensure that the blended fuel meets American Standard for Testing Materials (ASTM) D1655 standard. The analysis is performed with GTL fuel building blocks: *normal* paraffins, *iso* paraffins as well as *cyclo* paraffins and aromatics. Four different properties of the blend are monitored including density, freezing point, flash point, and heat of combustion for the comparison of synthetic fuel with aviation grade fuel standard. Limited experimental data on estimated property as obtained from different composition across a broad range of different components of the synthetic blends is used for our analysis. Multivariate regression with partial least squares (PLS) regression method for linear as well as support vector machine (SVM) for nonlinear properties have been used. A multivariate statistical tool has also been developed using Principal Component Analysis (PCA) to investigate any new composition for its quality. This machine learning algorithms based blend selection can identify optimal composition for any number of components. This is followed by blend property characterization using already developed regression models to confirm the qualification of the selected blends as aviation grade fuel. Experimental verifications are also conducted on the selected blends for model validation.

498553 Determination of the Optimal Number of Eigenfunctions in Proper Orthogonal Decomposition Based on Machine Learning

Abstract:

Determination of the optimal number of eigenfunctions in Proper Orthogonal Decomposition based on Machine Learning

Harwinder Singh Sidhu[1],[2], Abhinav Narasingam[1],[2], Prashanth Siddhamshetty[1],[2],

Joseph Sang-Il Kwon[1],[2]

[1] Artie McFerrin Department of Chemical Engineering, Texas A&M University, College Station, TX 77845 USA

[2] Texas A&M Energy Institute, Texas A&M University, College Station, TX 77845 USA

Spatiotemporal data, whether captured through remote sensors or large-scale simulations has always been 'Big'. However, recent advances in next generation smart manufacturing have dramatically intensified data generation through networked information-based technologies throughout the chemical industry and other manufacturing enterprises, making spatiotemporal data even bigger. Therefore, we need an algorithm that is able to extract the underlying process dynamics from large datasheets to make decisions for safe and efficient operations.

Proper Orthogonal Decomposition (POD) [1] is a widely used model-reduction technique which extracts dominant spatial patterns from spatiotemporal data obtained via experiments or large-scale simulations of distributed parameter systems (DPSs). In POD, the essential information of a large-scale complex system, consisting of m variables, is extracted in '*d*' retained eigenfunctions. The main idea of this technique is based on a simple observation that very often the underlying signal of high-dimensional data belongs to a low-dimensional space that is spanned by '*d*' eigenfunctions. Thus, *d* is much smaller than *m*. Nevertheless, the selection of these optimal number of eigenfunctions remains a key challenge. Although there are several guidelines [2] available in literature for selecting these eigenfunctions, they may not necessarily be optimal. For example, one may retain the eigenfunctions whose eigenvalues are larger than a prescribed value. Such a selection may discard certain low-energy states which can have a large influence on the accuracy of a developed low-dimensional system [3].

Motivated by the penalized regression methods for simultaneous variable selection and coefficient estimation, like elastic net, we have formulated the following regularization problem to determine the optimal number of eigenfunctions, '*d*', that minimizes the energy functional consisting of two terms:

(1)

The first term is the square of the distance from a snapshot x_j to the *d* dimensional subspace *V* spanned by the orthogonal eigenfunctions. This term represents the quality of the subspace to capture the information embedded in the original spatiotemporal data. $\beta(d)$, satisfying $\beta(0) = 0$, acts as a penalizer that measures the complexity of the reduced model. The regularization parameter, $a=0$, quantifies the relative trade-off between the complexity and the error of the reduced model. Techniques such as cross validation can be employed to select an appropriate *a*. Now, for a given *a*, a larger *d* will minimize the first term but will increase the model complexity and vice-versa. Therefore, the overall energy functional is a convex function, and the minima will give us the optimal number of eigenfunctions.

We implement the proposed methodology for order reduction of a hydraulic fracturing process described by multiple nonlinear parabolic PDEs with time-dependent spatial domains [4]. First, a representative ensemble of solutions is constructed by solving a high-order discretization of the PDEs. Then, the proposed method is applied to derive an optimal set of empirical eigenfunctions, which are subsequently used as basis functions within a model reduction framework, like Galerkin's method, to derive low-order ODE systems that accurately describe the dominant dynamics of the hydraulic fracturing process. These ODE systems are then used to compute approximate solutions (i.e., spatiotemporal profiles) to the original system.

References:

[1] Berkooz G., Holmes P., Lumley J. The proper orthogonal decomposition in the analysis of turbulent flows. *Ann. Rev. Fluid. Mech.* 1993;25:539-575.

[2] Jolliffe IT. Principal Component Analysis. New York: Springer; 2002.

[3] Noack, B.R., Schlegel, M., Ahlborn, B., Mutschke, B., Morzynski, M., Comte, P., Tadmor, G.

A finite-time thermodynamics formalism for unsteady flows. *J. Non-Equilib. Thermodyn.* 2008;33:103-148.

[4] Yang, S., Siddhamshetty, P., Kwon, J.S. Optimal pumping schedule design to achieve a uniform proppant concentration level in hydraulic fracturing. *Compt. Chem. Eng.*, 2017;101:138-147.

500245 A Scalable Statistical Machine Learning Method: Application for Fault Detection and Fault Propagation Pattern Inference in the Tennessee Eastman Process

Abstract:

Central elements of statistical machine learning algorithms (SMLAs) are joint probability distributions (JPDs). The performance and convergence of SMLAs greatly depend on how JPDs are estimated from historical data. Because many existing JPD estimation methods and their resulting SMLAs suffer from the curse of dimensionality phenomenon, their scalability and applicability for big data analytics are seriously questioned. A general method of JPD estimation is needed to develop the next generation of SMLAs that can cope with the complex, high-volume, high-dimensional nature of big data. Recently we introduced an efficient method of estimating JPDs of continuous random variables with arbitrary relationships [1, 2]. As the backbone of the method is a monotonizing transformation that â€˜rolls outâ€™ (monotonizes) the relationships, the method was named the rolling pin (RP) method. This method allows for estimating JPDs without any knowledge of the actual causal structure of the attributes. The method offers many advantages over its well-known counterparts such as the original parametric copula method, moment-based density estimation, and nonparametric techniques of joint probability estimation. An RP-estimated JPD can be incorporated effectively in a variety of big-data statistical learning tasks such as classification, regression, clustering, pattern recognition, dimensionality reduction, and probabilistic modeling and inference. In probabilistic inference, it is computationally more efficient than Bayesian networks.

In this paper, we present a study on the scalability of the JPD estimation method. In particular, we apply the method to the large-scale Tennessee Eastman (TE) process [3, 4]. This process has a total of 91 variables (12 manipulated, 38 state, and 41 measured variables). It has five unit operations (a two-phase reactor, a condenser, a flash separator, a recycle compressor, and a product stripper). We show that the RP method is easily scalable, is computationally efficient and flexible, and allows for reliably estimating JPSs of large-scale highly nonlinear processes such as the TE process. Also, it is demonstrated that the RP method provides a computationally efficient and flexible framework for performing probabilistic inference in highly nonlinear systems with non-monotonic variable interdependencies. The advantages of this inference framework over Bayesian networks are presented.

References

[1] Mohseni Ahooyi, T., Arbogast, J.E., and Soroush, M. (2015b). An Efficient Copula-Based Method of Identifying Regression Models of Non-Monotonic Relationships. *Chem. Eng. Sci.*, 136(2), 106.

[2] Mohseni Ahooyi, T., Arbogast, J.E., and Soroush, M. (2015a). Applications of the Rolling Pin Method: 1. an Efficient Alternative to Bayesian Network Modeling and Inference. *Ind. & Eng. Chem. Research*, 54(16), 4316.

[3] Downs, J. J., Vogel, E. F. (1993). A Plant-Wide Industrial Process Control Problem. *Comput. & Chem. Eng.*, 17(3),

245.

[4] Yu, H., Khan, F., Garaniya, V. (2015). A Probabilistic Multivariate Method for Fault Diagnosis of Industrial Processes. *Chem. Eng. Research & Design*, 104, 306.

500771 Decision Support Platform for the Automation of Processing Recipes and Process Scheduling Management Supported By Knowledge Management

Abstract:

Nowadays industries seek smart solutions for deal with information and data management. The effective and timely use of enterprise wide optimization models requires robust and reliable data acquisition systems to extract model relevant parameters and to drive the proposed enterprise wide coordination strategies. This work aims to create a knowledge-based platform for systematic and standardized management of the general, site and master recipes within the process industry. Besides, the system deals with data extraction from process documentation, facilitating the management of data exchange at scheduling level with different process levels. The platform allows for the creation of a master recipe ready for the production planning and scheduling, as well as for the process management. As a result, the recipe management functionalities are supported by a traceable and reliable system. Thus, different information and data required for the scheduling task is delivered to the specific analytical program as required. Finally, the framework could easily react to production system changes or analytical modeling changes, with flexible data and information structures.

An ontological model based on ANSI/ISA-88 and ANSI/ISA-95 standards is the basis for the development of the proposed algorithms in the recipe management functions, since it allows information sharing, and semantic searches within available information are possible. Additionally, the recipes and procedural elements derived from the different recipe management functions are instantiated in the aforementioned ontological model. The algorithms have been programmed in Jython and embedded with the semantic model within a software platform.

On the one hand, recipe management activity consists of five main functions. Firstly, The definition of recipe procedural elements function, consist of defining process stages, operations and actions. Then, resulting recipe building blocks are necessary for the creation of the general and site recipes, and for the definition of the master recipe procedural elements (unit procedures, operations and phases). Even more, one of the most critical issues consists of mapping of the process stages, process operations and process actions defined in the site recipes into unit procedures, operations and phases contained in the master recipe.

On the other hand, four control activities are considered in this work, such as, recipe management activity creates, master recipes, production orders processing function from ANSI/ISA-95 standard providing additional information for the production planning and scheduling, process management providing the status of the process cell and the batch progress, as well as, information to the unit supervision for batch execution, and finally, production information management reporting all available information collected from the other control functions.

In addition, the detail of the master recipe has been specified in the semantic model in order to identify the alternative scheduling problem features related to mass balances, batch assignment, sequencing and timing constraints. The main advantage of this framework consists of using a common representation approach, namely the ontological model, which can be used to retrieve the required information for optimizing any scheduling problem.

501551 Machine Learning Based Uncertainty Quantification of Erosion in Pipeline Transportation

Abstract:

The transport of solids in multiphase flows is common practice in energy industries because of the unavoidable extraction of solids from oil and gas bearing reservoirs. For safe and efficient operation, reliable estimates of the erosion that occurred inside the pipelines is required. However, the phenomenon that leads to erosion, especially in multiphase flow systems, is very complex and depends on many factors. Consequently, the developed erosion-rate prediction models cannot exactly capture this complicated process. In some cases, there are considerable model uncertainties in the erosion-rate predictions.

The uncertainty in model predictions can stem from three sources in general: (1) uncertainty in experimental measurements of input conditions, (2) model form uncertainty (i.e., incomplete presentation of the actual system due to lack of knowledge or imprecise experimental observations), and (3) model parameter uncertainty. The experimental data uncertainty consists of measurement errors due to both instrumental and human errors. The reliability of the models largely depends on the ability of model form to capture the details of erosion process in enough granularity. The uncertainty in the model parameters results from an inability to accurately quantify the parameters of a model (Shrestha, 2009).

In previous studies, we introduced a systematic framework to quantify erosion-rate prediction uncertainty for operating conditions where experimental data are not available (Dai and Cremaschi, 2016). The framework incorporates the impacts of model form and parameter uncertainties to estimate prediction uncertainties, and combines data clustering and Gaussian Process Modeling (GPM) (Rasmussen and Williams, 2006) approaches with Monte Carlo Simulation. The results of these previous studies revealed that our systematic framework reduced estimated prediction uncertainties by 30% when compared to applying GPM alone. Despite this reduction, the estimated prediction uncertainties remain relatively large, because the training data - erosion-model prediction discrepancy â€" span over six orders of magnitude. In our previous studies, the training data was normalized to [0.1, 1] to reduce its large range. Although this reduced the range of the training data, it also masked some data features, especially in regions around origin, i.e., where erosion-model prediction discrepancies were small. There were significant inconsistencies in GPM predictions in these regions, and, GPM failed to estimate the correct sign of the model-discrepancy resulting in contradicting conclusions on whether the model under or over-predicted the erosion rate. The differentiation of model over- or under- predictions has significant engineering consequences for erosion applications. Higher orders of over-predictions may result in unnecessary overdesign, increase capital costs, and limit the production efficiency. On the other hand, under-estimated erosion rates may cause facility integrity issues, and, may result in considerable downtime for unexpected repairs.

This talk will discuss a novel approach developed for overcoming these challenges. The approach distributes the data into a finer grid, yet still resolves the challenges associated with large data ranges, and hence, it enhances

differentiability of data when constructing GPM. The developed approach combines GPM classification and log transformation, and introduces them to our previously developed systematic framework for model-uncertainty quantification. The training data that is used to construct GPM (erosion-model prediction discrepancy) is the difference between the erosion-model predictions and the experimental measurements for a large set of experimental and field conditions. In the developed approach, GPM-based classification (Rasmussen and Nickisch, 2010) is applied to estimate the sign of the discrepancy, while GPM regression (Shi and Choi, 2011) trained using log-transformed absolute values of the discrepancy data is employed to estimate the magnitude of the discrepancy. The approach, using our systematic framework, is employed to estimate prediction uncertainty of a well-known erosion model (Mazumder et al., 2005), which is routinely used by upstream oil and gas industry. Our analysis indicates that using the log-transformed data as inputs improves the uncertainty quantification of erosion-rate modeling, and resulted in an additional 20% reduction in prediction uncertainties when compared to the previous approach.

Acknowledgements

This work is supported by the Chevron Energy Technology Company. Discussions and comments from the Haijing Gao and Janakiram Hariprasad of Chevron and Brenton McLaury, Siamack Shirazi of E/CRC at the University of Tulsa were highly acknowledged.

Reference

Dai, W. and Cremaschi, S., 2016, A Data-Mining Framework for Uncertainty Analysis in Pipeline Erosion Modeling, 2016 AIChE Annual Meeting, 13-18, November 2016, San Francisco, CA, USA.

Mazumder, Q., Shirazi, S.A., McLaury, B., Rybicki, E., and Shadley, J., 2005, Development and Validation of a Mechanistic Model to Predict Solid Particle Erosion in Multiphase Flow, International Journal of Wear.

Rasmussen, C.E. and Nickisch, H., 2010, Gaussian processes for machine learning (GPML) toolbox. J Mach Learn Res 11:3011â€"3015.

Rasmussen, C.E. and Williams, C.K. I., 2006, Gaussian Processes for Machine Learning, The MIT Press.

Shi, J. and Choi, T., 2011, Gaussian Process Regression Analysis for Functional Data, U.K., London:Chapman & Hall.

Shrestha, D.L., 2009, Uncertainty Analysis in Rainfall-Runoff Modelling: Application of Machine Learning Techniques, PhD. Dissertation, UNESCO-IHE, the Netherlands.

496152 Generating Ontology-Based Process Models Automatically

Abstract:

The objective of the project is to develop a modelling tool that generates code for model-based computational engineering applications automatically. The models are to be centralised and derived from a common ontology that captures the behaviour of a class of systems. We have in our development focused mainly on controlled multi-scale physical-chemical-biological processes. In contrast to many other applications, we synthesise our process models directly from our ontology.

The construction of the ontology is a bootstrap from a very limited set of meta terms, consecutively implementing a hierarchical set of objects that define a superstructure of the class of processes being captured by the ontology. The result is that we can present a new approach to multi -scale multi-disciplinary representation and generation of process models.

The ontology extends with adding additional classes, attributes and relations to capture the full complexity of multi-scale, multidisciplinary process models. The relations captures the behaviour of the process models and is described by equations. The basic ideas of a network representation of controlled physical-chemical-biological models using ontologies were described in [1].

Having defined the above- mentioned ontology, the process of building the models is initiated by selecting attributes defined in the ontology and combine the attributes to form a set of minimal models purely based on the information captured in the ontology. We refer to these minimal model as primary models. The primary models are then to be used as building blocks of an interactive graph editor that constructs a multi-network representation where each network define time and length scales.
The primary models are connected into a structure rendering the modelled plant's behaviour. This structure describes the topology of the model and is represented by a directed graph. The nodes of the directed graph represent capacities of states of the model while the arcs represent interactions affecting the state of a capacity. An example of such interactions for physical models is the transport of mass and energy. The output from the interactive graph editor is incidence matrices representing the topology of the different states of the model, a list of nodes and a list of arcs with descriptions of which building blocks that were selected to represent each node and arc. The modelling concepts behind the directed graph were described in [2].

The directed graph with the building blocks provides a structure to the model including the states or interactions of states for each building block. The ontology provides the relations between the states of the building blocks in the form of equations. An equation is a recipe how to produce a value of a variable based on other variables and mathematical operators. The variable and equation relationship is implemented as a tree of other variables and mathematical operators. The modelling tool traverses the variable/equation tree to determine, using some defined rules, which equations and variables that have to be included in the final model. This selection have to be manually assisted by the user in some cases, for example, if there exist alternative equations to calculate the same property. The automatically generated executable code is based on templates for operators to the make the equations into expressions written in the syntax of the targeted language. The sequence in which the equations are calculated is determined based on variable classes identified in the ontology. The variables are implemented as vectors and matrices with the dimensionality defined by the directed graph representing the topology. Initial conditions for the states of nodes and values of constants have to be provided by the user.

The resulting executable code can be included in a simulator environment that again can be used for design or optimisation purposes. The definition of networks allows for multi-scale modelling. Each of the networks will be calculated at its scales, and a separate communication network handles the communication between different networks. This network separation facilitates for distributed computing. The modelling tool is, at is current state, able to produce executable Python code for a dynamic simulator using a simulator template and the executable implementation of the equations.

This contribution will focus on how to combine the available information from the ontology, the directed graph and the operator templates to produce executable code automatically.

1. Heinz A. Preisig, Arne Tobias Elve, Ontology construction for multi-network models, Computer Aided Chemical Engineering, Elsevier, 2016, Volume 38, Pages 1087-1092, ISSN 1570-7946, ISBN 9780444634283, http://dx.doi.org/10.1016/B978-0-444-63428-3.50186-7
2. Arne Tobias Elve & Heinz A Preisig, (2017), Graph-based Mathematical Modelling - Concepts, in â€˜FOCAPO/CPC 2017â€™

495928 Combination of Stochastic NMPC/DRTO and Dynamic (re-)Evaluation of the Description of Model Uncertainty in the Optimization of Batch Campaigns

Abstract:

Robust dynamic optimization/adaptive control has been proposed to overcome a significant drawback of conventional dynamic optimization (DRTO) and optimal control (NMPC), namely, the requirement for accurate process models. One of the existing types of robust NMPC/DRTO is stochastic NMPC/DRTO (Rossi et al., 2016). This approach relies on stochastic optimization techniques and approximates the uncertainty in the process model using an appropriate set of uncertain parameters (SUP) and their joint probability distribution function (PDFUP). Although this type of strategy is applicable to a large variety of systems, we will consider the case of batch processes.

It is well-known that the performance of stochastic NMPC/DRTO primarily depends on the selection of the SUP and on the estimate of the PDFUP. The latter can be updated periodically based on the measurements acquired in successful batch cycles, using, for instance, Monte-Carlo approaches or Bayesian frameworks (Mockus et al., 2015). However, the PDFUP mainly affects the performance of the stochastic NMPC/DRTO algorithm and not its robustness. The selection of the SUP is the more crucial and challenging aspect because it may strongly affect both the robustness and the performance of the stochastic NMPC/DRTO method. Moreover, in order to meet real time computational constraints, the SUP can only include a limited number N_p ($N_p = 4, 5$) of uncertain parameters.

Typically, the SUP is identified statically using sensitivity analysis (Saltelli et al., 2005) or based on empirical knowledge of the batch process. However, the use of a fixed SUP is often sub-optimal because the importance of any given uncertain parameter can be expected to vary over time, even over a single batch cycle. This is because the relevance of the uncertain parameters may depend on both their stochastic properties (PDFUP) and the operating conditions of the batch process. While the PDFUP does not significantly change over time in a single batch cycle, the same does not apply to the operating conditions of the process, which are continuously modified to compensate for process disturbances and to improve process performance.

Therefore, we identify the SUP in real-time, using a three-stage, hierarchical procedure (Figure 1). First, we employ conventional methods based on sensitivity analysis and/or empirical knowledge of the batch process to construct a set of N_p^{tot} parameters called SUP^{tot} ($N_p^{tot} = 10, 15$) (Phase I). This set includes those parameters, which have the most impact on the model predictions over a wide range of operating conditions. Secondly, after every batch cycle, we employ the algorithm called PDFE&U (Probability Distribution Function Estimation & Update), presented at the 2016 AIChE Annual Meeting, to re-estimate in parallel the joint probability distributions of all the subsets of N_p uncertain parameters (SUPs) included in SUP^{tot} (Phase II). In this phase, we also define proper indicators of both the statistical dispersion of all the uncertain parameters and their correlation. Note that the application of parallel computing allows us to carry out this phase in the idle time between two consecutive batch cycles. Finally, after every control action

(specifically, in the idle time between two consecutive actions), we apply the novel algorithm called RTSASUP (Real-Time Selection of the Active Set of Uncertain Parameters) to select the optimal SUP (SUP^{opt}) (Phase III). This newly estimated set of uncertain parameters is used, in the stochastic NMPC/DRTO engine, to compute the control action after the next one (in other words, the SUP^{opt} identified at the end of control action n is used to compute control action $n+2$).

Figure SEQ Figure * ARABIC 1: Architecture of the proposed strategy for the online identification of the optimal uncertainty set.

Note that RTSASUP chooses SUP^{opt} as the set of N_p uncertain parameters that have the most influence on the predictions of the process model, show the largest statistical dispersion and exhibit the smallest correlation among them. In order to do this, the algorithm measures the sensitivity of the process model to the uncertain parameters by solving a series of decoupled small-scale dynamic optimization problems in parallel, and then combines this latter information with the indicators of the statistical dispersion and the correlation of the uncertain parameters, estimated in Phase II. The use of parallel computing allows RTSASUP to be efficient enough to run in the idle time between two consecutive control actions (approximately a minute or two of clock time).

We demonstrate the proposed approach to the online identification of the optimal SUP (SUP^{opt}) using a batch version of the Tennessee Eastman challenge. The case study confirms that the online re-estimation of the SUP carried out via RTSASUP benefits both the performance and the robustness insured by stochastic NMPC/DRTO frameworks. Such efficiency improvements appear to be more pronounced in the presence of process perturbations, i.e. when the operating conditions of the process experience larger variations in single batch cycles.

References:

L. Mockus, J.J. Peterson, J.M. Lainez, G.V. Reklaitis, 2015, Batch-to-batch variation: a key component for modeling chemical manufacturing processes, Organic Process Research & Development, 19, 908-914.

F. Rossi, G. Reklaitis, F. Manenti, G. Buzzi-Ferraris, 2016, Multi-scenario robust online optimization and control of fed-batch systems via dynamic model-based scenario selection, AIChE Journal, 62, 3264-3284.

A. Saltelli, M. Ratto, S. Tarantola, F. Campolongo, 2005, Sensitivity analysis for chemical models, Chemical Reviews, 105, 2811-2828.

496627 Autocovariance-Based Model Mismatch Diagnosis for MPC with State Estimation

Abstract:

Model predictive control (MPC) is widely used in the process industries for complex, multivariate control problems [1],[2]. MPC offers significant advantages over conventional multi-loop PID control schemes; however, as a model-based method, its performance depends greatly on the quality of the process model. If the model used in the MPC does not provide an accurate representation of the plant, control performance will suffer. Recognizing this, practitioners invest significant time and resources in the model building phase of an MPC implementation, performing, e.g., extensive open loop step testing to the plant dynamics.

Nevertheless, chemical process systems are subject to a variety of physical phenomena (corrosion, fouling, catalyst deactivation) and other changes (to equipment, process operation, etc.) that cause the plant dynamics to drift in time, such that the response at a given time after the system identification effort will be different from the model used in the MPC. Since frequent model updating iscostly, re-identification efforts are generally only pursued once controller performance has degraded severely.

A method for detecting plant-model mismatch before this stage would thus be highly desirable, and MPC controller performance monitoring (CPM) has become an area of vigorous research activity in recent years. Many proposed approaches are extensions of previously established CPM methods to MPC systems. Notably, data-driven methods like Principal Component Analysis and Partial Least Squares regression have been proposed [3],[4]. More recently, approaches based on input-output partial correlation analysis [5] and statistical analysis of higher moments of the output and error distribution [6] have been introduced to improve specificity; they attempt to trace the source of the mismatch to a subset of variables. In our previous work, we provided further specificity, proposing an output autocovariance-based method for both locating and quantifying plant-model mismatch from routine operation data, i.e. without requiring disruption to the process [7],[8],[9].

In this work, we extend the autocovariance-based approach to the important class of systems consisting of state-space MPC controllers employing Kalman filtering for state estimation. Since, in contrast to previous frameworks, these systems incorporate both state and output noise models, we represent the closed loop outputs as Vector Moving Average processes from which the theoretical autocovariance function can be derived [10]. Then, the sample estimator for the autocovariance of process outputs obtained from routine operating data is used to formulate a minimization problem, whose result is an asymptotically consistent estimator of the true plant-model mismatch. We present case studies demonstrating the application of our approach for MPC systems with state estimation.

References

[1] S. J. Qin, "Statistical process monitoring: Basics and beyond," Journal of Chemometrics, 17:480–502, 2003.

[2] M. L. Darby and M. Nikolaou, "MPC: Current practice and challenges," Control Engineering Practice, 20(4):328–342, 2012.

[3] J. Yu and S. J. Qin, "Statistical MIMO controller performance monitoring. Part I: Data-driven covariance benchmark," Journal of Process Control, 18:277–296, 2008.

[4] A. AlGhazzawi and B. Lennox, "Monitoring a complex refining process using multivariate statistics," Control Engineering Practice, 16:294–307, 2008.

[5] Abhijit S. Badwe, Ravindra D. Gudi, Rohit S. Patwardhan, Sirish L. Shah, and Sachin C. Patwardhan, "Detection of model-plant mismatch in MPC applications," Journal of Process Control, 19(8):1305–1313, 2009.

[6] V. Botelho, J.O. Trierweiler, M. Farenzena, R. Duraiski, "Perspectives and challenges in performance assessment of model predictive control," Canadian Journal of Chemical Engineering, 94: 1225–1241, 2016.

[7] S. Wang, J. M. Simkoff, M. Baldea, I. Castillo, D. Stanley, L. Chiang, and R. Bindlish, "Autocovariance-based plant-model mismatch estimation for linear model predictive control," Systems and Control Letters, vol. DOI:10.1016/j.sysconle.2017.03.002, 2017.

[8] S. Wang, J. M. Simkoff, M. Baldea, I. Castillo, D. Stanley, L. Chiang, and R. Bindlish, "Autocovariance-based MPC model mismatch estimation for systems with measurable disturbances," Journal of Process Control, vol. DOI:10.1016/j.jprocont.2017.03.002, 2017.

[9] J. M. Simkoff, S. Wang, M. Baldea, I. Castillo, D. Stanley, L. Chiang, and R. Bindlish, "Plant-model mismatch diagnosis and estimation from closed-loop data for state-space MPC," AIChE Journal, 2017 (*submitted*).

[10] G. Box, G. Jenkings, and G. Reinsel, Time Series Analysis: forecasting and control, Prentice-Hall, Inc., 3rd edition, 1994.

498049 A Tractable Method for Closed-Loop Active Fault Diagnosis of Stochastic Linear Systems

Abstract:

Reliable fault detection and isolation in increasingly complex engineering systems is critical for ensuring stability and reliability of system operation in the face of anomalies such as sensor and actuator malfunctions, system failures, and drifts in system parameters. Various sources of system uncertainty, including incomplete knowledge of system faults, exogenous disturbances, and measurement noise, increase the challenges of the diagnosis problem. Methods for fault diagnosis can be broadly categorized as being data- or model-based [1]. The latter category generally seeks to detect the system fault status based on measurements obtained during operation, commonly referred to as passive fault diagnosis. The reliability of passive fault diagnosis can, however, be adversely impacted by system uncertainties and the presence of feedback controllers that can mask the impact of faults on system outputs. This observation motivates active fault diagnosis (AFD), which involves designing an input sequence that, when applied to the system, facilitates 'correct' (with high probability) fault diagnosis in the presence of uncertainty [2].

When the system uncertainty is deterministic and bounded, AFD methods generally aim to design input sequences that are robust to worst-case realizations of the uncertainty. Most of these methods, primarily developed for linear time-invariant systems, separate all reachable sets of the fault models to guarantee diagnosis [3, 4, 5]. With probabilistic descriptions of system uncertainty, the AFD problem is more naturally addressed in a stochastic setting. Probabilistic AFD formulations utilize distributional information on the uncertainties, potentially reducing the conservatism commonly observed in worst-case deterministic AFD. An early approach to open-loop (offline) AFD for linear time-variant systems with additive stochastic disturbances and measurement noise is developed in [6], which relies on minimizing an upper bound on the probability of misclassification of fault models that is defined in terms of the Bhattacharyya coefficient [7].

In this talk, we present a computationally efficient method for closed-loop AFD of multiple faults in stochastic linear systems with uncertain initial conditions. The proposed AFD method relies on computing an optimal open-loop input sequence that is applied to the system in a receding-horizon fashion by updating the input design problem online based on the system measurements. A key challenge in this approach is ensuring the online solution time is significantly shorter than the sampling time. This challenge is addressed by defining the finite-horizon AFD problem as maximizing the pairwise Bhattacharyya distances (a statistical distance measure) between the predicted output probability distributions, subject to system constraints represented by a convex polytope. The main advantage of the proposed closed-loop AFD method is that the global solution can be determined efficiently through exhaustive enumeration of a small number of vertices. This approach contrasts with formulations that directly minimize the nonconvex error bound, proposed in [6], that require solving a nonconvex problem either to some local optimum or to its global optimum using algorithms such as [8] with (possibly) excessive computational requirements. Furthermore, the computational complexity of the proposed method is independent of the number of fault models and the number of states in each model. We demonstrate the performance of the closed-loop AFD method on a benchmark stochastic fault diagnosis problem with five models.

References

[1] V. Venkatasubramanian, R. Rengaswamy, K. Yin, and S. N. Kavuri, "A review of process fault detection and diagnosis. Part I: Quantitative model-based methods," *Computers & Chemical Engineering*, vol. 27, no. 3, pp. 293–311, 2003.

[2] M. Blanke, M. Kinnaert, J. Lunze, and M. Staroswiecki, *Diagnosis and Fault-Tolerant Control*, second edition. Springer, 2006.

[3] S. L. Campbell and R. Nikoukhah, *Auxiliary Signal Design for Failure Detection*. Princeton University Press, 2004.

[4] M. Sampath, S. Lafortune, and D. Teneketzis, "Active diagnosis of discrete-event systems," *IEEE Transactions on Automatic Control*, vol. 43, no. 7, pp. 908–929, 1998.

[5] A. E. Ashari, R. Nikoukhah, and S. L. Campbell, "Auxiliary signal design for robust active fault detection of linear discrete-time systems," *Automatica*, vol. 47, no. 9, pp. 1887–1895, 2011.

[6] L. Blackmore and B. Williams, "Finite horizon control design for optimal discrimination between several models," in Proceedings of the *IEEE Conference on Decision and Control*, San Diego, CA, pp. 1147–1152, 2006.

[7] T. Kailath, "The divergence and Bhattacharyya distance measures in signal selection," *IEEE Transactions on Communication Technology*, vol. 15, no. 1, pp. 52–60, 1967.

[8] M. Tawarmalani and N. V. Sahinidis, "A polyhedral branch-and-cut approach to global optimization," *Mathematical Programming*, vol. 103, no. 2, pp. 225–249, 2005.

502512 Set-Membership Nonlinear Regression Approach to Parameter Estimation

Abstract:

Most commonly, regression problems in nonlinear parameter estimation are posed as optimization problems, where selected parameter values are tuned so that the model predictions match the available measurements as close as possible, for instance in the maximum-likelihood sense. Incorporation of measurement uncertainty into the regression exercise has spawned several different schools of thought. Statistical approaches can be broadly classified as *Frequentist* or *Bayesian*. The former seek to determine confidence regions around the regressed parameter values, considered as the `true' parameter values [1]. They are capable of dealing with large-scale models, but confidence analysis based on model linearization can be misleading, especially for highly nonlinear models or when the available measurements are scarce. Bayesian approaches meanwhile construct a probability distribution of the parameters based on the available observations, from which credibility regions can be inferred [2]. These approaches are appealing, but they rely on a so-called prior for the parameter estimates, and existing algorithms such as Markov-chain Monte-Carlo (MCMC) can be time consuming for problems having upwards of 5-10 parameters. An alternative to these statistical approaches is *set-membership* estimation, which seeks to determine the set of all possible parameter values that are consistent with bounded-error measurements [3]. Such problems have been shown to be tractable using complete-search methods with ca. 10 parameters, e.g., based on interval analysis or higher-order inclusion techniques [4,5]. Issues may arise however, as model mismatch or outlying measurements can lead to situations where the resulting parameter set is empty.

We propose a new approach to nonlinear regression, called *set-membership nonlinear regression* (SMNLR). Given a residual function $\mathcal{R}(\mathbf{p}|\mathbf{e})$, where stands for the (unknown) model parameters and the measurement errors, the problem addressed involves characterizing the set *P* given by

(1)

where $P_0 \subset \mathbb{R}^n$ is a (non-empty) prior set for the model parameters; and a given measurement-error set (e.g., bounds on the measurement errors, level-sets of a joint probability distribution, *etc*). We study the properties of SMNLR, with a view to relating problem (1) with existing parameter estimation approaches. In the context of bounded-error measurement estimation, SMNLR presents the advantage that the set *P* is always non-empty, regardless of the model mismatch or measurement error. In the context of statistical inference, *P* may be interpreted as parameter estimability sets for a given confidence level-set *E* on the measurement errors [6]. Moreover, we derive conditions under which the SMNLR regions are asymptotically equivalent to the classical Wald confidence regions.

Notwithstanding, SMNLR is a particular arduous problem as it involves finding *all* global minimizers to *all* parameter estimation problems with $\mathbf{e} \in E$. We describe how this problem can be tackled using complete-search methods (branch-and-prune), and we discuss several simplifications for reducing the computational burden. Comparisons are made with the classical set-membership estimation approach as well as classical confidence analysis methods for selected parameter estimation problems.

Bibliography

[1] S.P. Asprey, S. Macchietto, 2000. Statistical tools for optimal dynamic model building. *Computers & Chemical Engineering* **24**:1261-1267.
[2] D. Alspach, H. Sorenson, 1972. Nonlinear Bayesian estimation using Gaussian sum approximations. *IEEE Transactions on Automatic Control* **17**(4):439-448.
[3] E. Walter, 1990. Special Issue on Parameter Identification with Error Bounds. *Mathematics & Computers in Simulation* **32**:447-607.
[4] L. Jaulin and E. Walter, 1993. Set inversion via interval analysis for nonlinear bounded-error estimation. *Automatica* **29**(4):1053-1064.
[5] B. Chachuat, B. Houska, R. Paulen R, N.D. Peric, J. Rajyaguru, M.E. Villanueva, 2015. Set-theoretic approaches in analysis, estimation and control of nonlinear systems. *IFAC-PapersOnLine* **48**(8):981-995.
[6] K.A. McLean and K.B. McAuley, 2012. Mathematical modelling of chemical processes - obtaining the best model predictions and parameter estimates using identifiability and estimability procedures. *The Canadian Journal of Chemical Engineering* **90**(2):351-366.

490302 Model Predictive Control Under Model Structural Uncertainty

Abstract:

Model predictive control is a widely used control strategy in the process control industry. Model predictive control, as the name suggests uses a model to predict the future behaviour of the process and compute optimal input moves in a receding horizon fashion. Models that are used in the controller may be developed based on first principles or from experimental data or a combination of both. However, models are inherently uncertain due to model simplification or lack of knowledge. In order to handle the uncertainty, many different forms of MPC have been developed in the past, such as robust MPC, stochastic MPC, feedback min-max MPC and adaptive MPC. Almost all the works in the literature on MPC under uncertainty, focuses on model parameters and external disturbances as the primary source of uncertainty. However, more commonly, plant-model mismatch occurs due to model simplification and/or lack of knowledge. Alternatively, step response models may be identified using experimental data with the model order chosen rather arbitrarily. This leads to model structural uncertainty, where the underlying model structure itself is uncertain.

Structural uncertainty however, is often disregarded in model based control design and has received little attention in the past. This is due to the fact that, unlike model parametric uncertainty, it is more challenging to handle structural uncertainty. A robust MPC framework considering structural uncertainty for linear systems has been presented in [3] and very recently, an explicit dual control framework for handling structural uncertainty was presented in [4]. However, in this work, we consider the scenario-based MPC framework presented in [1].

In this work, we show that scenario-based MPC is an efficient tool to handle model structural uncertainty. Scenario-based MPC or feedback min-max MPC is based on the evolution of the uncertainty represented by a scenario tree. It was first introduced as a means to handle uncertainty in the model parameters and future disturbances [1],[2]. Traditionally in this approach, the uncertain parameters, either belonging to a compact set or described by a probability distribution function, are sampled to form a discrete set of scenarios. The scenario tree is then used to predict a cone of state trajectories to represent the propagation of uncertainty. An optimization problem is then solved to compute optimal control trajectories. The notion of feedback is introduced by letting the computed optimal input vary according to the different scenarios. The closed-loop optimization nature of the scenario-based MPC has led to reduced conservativeness compared to robust MPC methods [2]. Similarly, we apply the same principle to handle model structure uncertainty. In the case of model structure uncertainty, the cone of predicted state trajectories is based on different model structures, instead of a single model structure with sampled parameters. The optimization problem is then solved over the entire scenario tree, hence taking into account the different model structures. Each model can be given different weights to prioritize between the different models.

Previous work considering scenario-based MPC either assumes uncertainty in the form of constant model parameters as in [2] or future external disturbances as in [1]. In this work, we show that scenario-based MPC is not just restricted to these forms of uncertainty, but is also an efficient tool to handle model structural uncertainty, hence opening up a new class of uncertainty that can be handled in MPC problems.

An offshore oil and gas production optimization problem is used as a case study to demonstrate the proposed method. Oil well models are often subject to structural uncertainty due to lack of knowledge or model simplifications. Consider the reservoir inflow performance model that describes the flow rate of fluids from the reservoir that enters the well. Due to the lack of measurements in the wellbore, it is challenging to develop empirical model based on data and thus, one usually relies on physical models. Different physical models exist that describe the inflow performance. For example, the linear PI model [5], Vogel inflow model [6], Fektovich inflow model [7] and Jones inflow model [8] are often used in the oil and gas industry. Such model structure uncertainty is often ignored, which may lead to suboptimal performance or even constraint violations. Scenario-based MPC can be used in such cases to account for this uncertainty.

References

[1] Scokaert, P., Mayne, D., 1998. Min-max feedback model predictive control for constrained linear systems. IEEE Transactions on Automatic control 43 (8), 1136–1142.

[2] Lucia, S., Finkler, T., Engell, S., 2013. Multi-stage nonlinear model predictive control applied to a semi-batch polymerization reactor under uncertainty. Journal of Process Control 23 (9), 1306–1319.

[3] Feng L, Wang JL, Poh EK. Improved robust model predictive control with structured uncertainty. Journal of Process Control. 2007 Sep 30;17(8):683-8.

[4] Heirung TA, Mesbah A. Perspectives On Stochastic Predictive Control With Autonomous Model Adaptation For Model Structure Uncertainty. FOCAPO-CPC, 2017.

[5] Evinger, H.H. and Muskat, M. 1942. Calculation of Theoretical Productivity Factor. Trans., AIME 146: 126.

[6] Vogel, J.V. 1968. Inflow Performance Relationships for Solution-Gas Drive Wells. J Pet Technol 20 (1): 83–92. SPE 1476-PA.

[7] Fetkovich, M.J.: "The Isochronal Testing of Oil Wells," paper SPE 4529 presented at the 1973 SPE Annual Meeting, Las Vegas, Nevada

[8] Jones, L.G., Blount, E.M., and Glaze, O.H. 1976. Use of Short Term Multiple Rate Flow Tests To Predict Performance of Wells Having Turbulence. Presented at the SPE Annual Fall Technical Conference and Exhibition, New Orleans, Louisiana, 3-6 October 1976. SPE-6133-MS

490403 Passivity-Based Observer and Application in Reaction Heat Estimation

Abstract:

Title: Passivity-Based High Gain Observer with Application to Reaction Heat Estimation

This work presents a passivity-based approach to estimate rapidly varying parameters in partially modelled dynamical systems. The proposed observer can be applied to a wide range of practical problems. The main assumption is that the measurements are not "too noisy" and that the interconnections between the system that is modeled and the un-modelled system has the same dimension as the observation vector. Noise is a problem since the observer uses derivatives. A trade-off occurs between the speed of parameter tracking and the noise level, as been stated in [1] where a sliding-mode differentiator is designed for noisy measurements. The observer is applied to estimate reaction heat in an unsteady state polymerization reactor. Some industrial results will be discussed.

Estimation of time-varying parameter is important for process control, process monitoring and fault detection. Luenberger theory [2] provides the foundation for observer design in linear systems. Kazantzis and Kravaris [3] generalized this theory to nonlinear systems using Lyapunov's auxiliary theorem. One important assumption for these observers, the Kalman filter and the closely related moving horizon estimation schemes, is that the system dynamics are modeled. In many practical applications, it may be difficult to model the system dynamics completely.

In this work, we present the design of a passivity-based observer that can be applied to some classes of systems with un-modeled internal dynamics. To illustrate, consider the following problem:

(1)

(2)

(3)

Equation (1) represents the un-modelled dynamics, equation (2) represents the model and equation (3) represents the measurement. The objective is to estimate the time-varying interconnection parameter . The important point to note that is that the measurement vector includes the measurement of x as well as the estimate of the time derivative of . The bounded signal represents the uncertainty in the derivative estimate. The problem we consider is how to infer the interconnecting parameter, through the measurement vector y and model equations (2).

The proposed approach treats the parameter estimation problem as a control problem. In this approach, we construct and error between the observations and the output produced from the observer equation (2). The system is passive and fast tracking can be achieved by high gain feedback. A Lyapunov function argument in conjunction with the time derivative estimation due to Levant [1] provides theoretical justification for the application. The theory shows that there will be a trade-off between the accuracy of the tracking estimate and the noise level and that the error converges to zero in the limit of no noise. It is quite straightforward to generalize the theory to problems where the dynamics of system (1) has high order and the dynamics of system (2) are nonlinear as long as the input-output passivity condition is maintained.

The practical problem is motivated by an industrial application to polymerization reactor control. In this application, the monomer concentration is inferred by estimating the heat of reaction through a nonlinear energy balance written on the form of equation (2). The temperature and its derivative are estimated using a Savitzky-Golay filter. Simulation results and results from industrial trials show that the estimator converges quickly in semi-batch polymerization.

[1] Levant, Arie. Robust exact differentiation via sliding mode technique. Automatica 34(3): 379-384 (1998).

[2] Luenberger, David. An introduction to observers. IEEE Transactions on automatic control 16(6): 596-602 (1971).

[3] Kazantzis, Nikolaos, and Costas Kravaris. Nonlinear observer design using Lyapunov's auxiliary theorem. Systems & Control Letters 34(5): 241-247 (1998).

491978 Parameter Estimation and Process Optimization of Heterogeneous Batch Reactors Under Uncertainty

Abstract:

Heterogeneous solid-liquid batch reactors are widely applied to produce fine chemicals in metallurgical, agrochemical and pharmaceutical processes. Compared with fundamental homogenous reactions, solid-liquid reactions are usually coupled with mass transfer, such as liquid phase diffusion and solid particle dissolution. However, a detailed reaction mechanism is typically inaccessible in a real industrial plant due to limited prior knowledge and difficulty of measurements. Understanding reaction mechanisms and constructing acceptable models based on obtainable information are desired to facilitate process optimization and control. In this study, we first introduce a model identification and parameter estimation procedure and implement it on our batch reactors with limited real plant/experimental data. Optimal control is then implemented on the estimated model to reduce total batch time while satisfying operation constraints and final concentration requirements.

Two particular mechanisms -- solid surface reaction models and dissolution controlled models are considered in the targeted solid-liquid reactions. A uniform kinetic model is developed with a model indicating factor to differentiate mechanisms. Particle morphology is considered similar to that described in [1] by introducing a shape factor. In addition, several other parameters are also included, such as activation energies, diffusion coefficients and heat transfer coefficients. A detailed procedure for parameter estimation [2] and model identification [3] is implemented, including pre-selection of estimable parameter subset, simultaneous estimation by using measured data, estimation quality evaluation, model simplification and model discrimination. In our case, only reactor temperature profiles and highly limited concentrations are measured. Small confidence intervals and good data fitting are obtained by the identified model.

The motivation behind the model identification effort is to develop an optimal control solution to minimize batch time and increase process capacity. The end of batch is defined by satisfying final conversion requirements. Feeding of liquid reactive components and reactor/jacket inlet temperatures are treated as decision variables. Optimal control is obtained by solving the optimality conditions of the Hamiltonian together with the corresponding adjoint system. A dynamic model is derived with multiple stages describing different batch operation steps. Orthogonal collocation on finite elements is applied to discretize the problem, resulting in a large scale nonlinear programming problem formulated and solved in AMPL/IPOPT. Finally, to address potential parametric uncertainties, disturbances and model mismatch, a robust optimization strategy is applied by adding back-off terms to ensure that constraints are satisfied in the worst-case scenario.

References

[1] Salmi, Tapio, Henrik Grénman, Johan Wärnå, and Dmitry Yu Murzin. "New modelling approach to liquid–solid reaction kinetics: From ideal particles to real particles." *Chemical Engineering Research and Design* 91.10 (2013): 1876-1889.

[2] Lin, Weijie, Lorenz T. Biegler, and Annette M. Jacobson. "Modeling and optimization of a seeded suspension polymerization process." *Chemical Engineering Science* 65.15 (2010): 4350-4362.

[3] Stewart, W. E., Y. Shon, and G. E. P. Box. "Discrimination and goodness of fit of multiresponse mechanistic models." *AIChE journal* 44.6 (1998): 1404-1412.

[4] Diehl, Moritz, Hans Georg Bock, and Ekaterina Kostina. "An approximation technique for robust nonlinear optimization." *Mathematical Programming* 107.1 (2006): 213-230.

493774 Rapid and Accurate Fault Detection and Diagnosis for Uncertain Nonlinear Systems Using Advanced Set-Based State Estimation Techniques

Abstract:

In applications such as wind energy, industrial robotics, and chemical processing, increases in complexity and automation have made component malfunctions and other abnormal events (i.e., faults) an ever-present threat to safety and reliability. Thus, fault detection and diagnosis (FDD) algorithms have become an essential feature of modern control systems, leading to significant decreases in downtime, maintenance costs, and catastrophic failures. However, while well-established statistical methods are effective in many cases, they often fail to make the critical distinction between faults and normal process disturbances. An attractive alternative is to exploit detailed process models that, at least in principle, can be used to characterize the outputs consistent with normal operation, providing a rigorous basis for FDD. Methods that furnish a guaranteed enclosure of these outputs (e.g., using set-based state estimators) are particularly attractive because they eliminate the possibility of costly false alarms, and can even guarantee FDD for certain faults using active inputs (i.e., excitations). However, such methods are currently impractical for systems with strong nonlinearites or large uncertainties. For such systems, existing set-based estimation techniques often produce enclosures that are far too conservative to be useful for FDD, or avoid this only at excessive computational cost. Thus, there is a critical need for advanced algorithms that can rapidly detect and diagnose faults for realistic nonlinear systems, and do so rigorously in the presence of disturbances, measurement noise, and large model uncertainties.

In this contribution, we will present recent advances in set-based state estimation for uncertain nonlinear systems, and demonstrate their application to provide fast and accurate fault detection and diagnosis for such systems. Specifically, these advances are based on a powerful new method for enclosing the reachable sets of nonlinear systems recently developed by the authors. This algorithm achieves high efficiency (necessary for online implementation) by using simple interval enclosures, but overcomes the high conservatism typically associated with interval methods using a technique called the addition of states and invariants (ASI). In ASI, additional state variables are defined as certain linear or nonlinear combinations of the original states, and the original dynamics are augmented with additional differential or difference equations for these new states. By construction, the states of this augmented system satisfy a set of linear or nonlinear equality constraints at all times, called solution invariants. The key idea is then to compute an enclosure of the solutions of the augmented system at each time step using a cheap (and often conservative) interval method, and subsequently refine this enclosure using the solution invariants. Our recent results demonstrate that, for very many systems of practical interest, there exist choices of new state variables that lead to dramatically tighter enclosures of the reachable set through ASI, with only modest additional computational cost relative to standard interval methods. This contribution will discuss a preliminary extension of this technique to set-based state estimation rather than reachability analysis (the former incorporates process measurements), and the application of the resulting estimators to achieve significantly faster and more accurate fault detection and diagnosis than is achievable with existing set-based methods. The proposed approach will be demonstrated using two CSTR and batch reactor case studies (and potentially others), and compared with existing state-of-the-art set-based FDD algorithms based on more

complex zonotopic and polytopic set enclosures.

493871 Basbl: Branch-and-Sandwich Bilevel Solver. Implementation and Computational Study Using Basblib Test Set

Abstract:

Mathematical programming and optimization are key activities in process systems engineering, with applications in synthesis, control, design, operations, and planning [1]. Many chemical engineering problems are bilevel optimization problems [1, 2], that is, hierarchical problems with one constraint in the main (outer) optimization problem that is an optimization problem. Examples of bilevel optimization problems are design or parameter estimation problems in which phase equilibrium must be satisfied as a constraints, or problems in which the optimum economic design is sought while minimizing energy consumption. The numerous applications of bilevel programming problems provide a strong incentive for developing efficient solvers for this large class of problems.

In this talk, we discuss BASBL, an implementation of the deterministic global optimization algorithm Branch-and-Sandwich for mixed-integer nonconvex/nonlinear bilevel problems, within the open-source MINOTAUR framework [7]. The solver stems from the original Branch-and-Sandwich algorithm [3, 4] and modifications proposed in our recent works [5, 6]. We also introduce BASBLib, an extensive online library of bilevel benchmark problems collected from the literature. The library is designed to enable contributions from the bilevel optimization community. We use the problems from BASBLib, including problems derived from practical applications, to study the performance of BASBL using different algorithmic options, including a variety of bounding schemes, branching, and node selection strategies.

References

1. Biegler, L.T., Grossmann, I.E.: Retrospective on optimization. Comput. Chem. Eng. 28, 1169–1192 (2004).
2. Floudas, C.A., Gounaris, C.E. A review of recent advances in global optimization. J Glob Optim, 45, 3–28 (2009)
3. Kleniati, P.-M., Adjiman, C.S.: Branch-and-Sandwich: a deterministic global optimization algorithm for optimistic bilevel programming problems. Part I: Theoretical development. J Glob Optim, 60(3), 425–458 (2014).
4. Kleniati, P.-M., Adjiman, C.S.: Branch-and-Sandwich: a deterministic global optimization algorithm for optimistic bilevel programming problems. Part II: Convergence analysis and numerical results. J Glob Optim, 60(3), 459–481 (2014).
5. Paulavičius, R. and Adjiman, C.S.: BASBL: Branch-And-Sandwich BiLevel solver. I. Algorithmic improvements and extensions, (2017). Submitted.
6. Paulavičius, R., Kleniati, P.-M., and Adjiman, C.S. BASBL: Branch-And-Sandwich BiLevel solver. II. Implementation and computational study with the BASBLib test set, (2017). Submitted.
7. Mahajan, A., Leyffer, S., Munson, T., Linderoth, J., Luedtke, J.: MINOTAUR: Toolkit for Mixed Integer

Nonlinear Optimization Problems, http://wiki.mcs.anl.gov/minotaur/index.php/Main_Page.

497217 Robust Optimization for Problems with Endogenous Uncertainty

Abstract:

Mitigation of the risk that stems from uncertainty in model parameters is important to take into account during decision-making via rigorous mathematical optimization. To this end, Robust Optimization (RO) has emerged as a systematic approach to handle such risk, by seeking to identify solutions that remain feasible under any realization of the parameters from within an uncertainty set [1]. This set, which is chosen by the modeler a-priori, constitutes a comprehensive collection of all parameter realizations that the modeler wishes to insure against, and it can be linked to a confidence interval for probabilistic constraint satisfaction [2].

Traditionally, the uncertainty set in RO is a constant set, being unaffected by the optimizer's decisions. In many cases, however, this is not realistic, as one's decisions may affect the level and type of risk one is actually exposed to. These cases are referred to as cases of endogenous uncertainty [3], and they have been investigated thoroughly in the Stochastic Programming literature [4–7]. In contrast, how to best handle endogenous uncertainty has been a more recent endeavor in the field of RO [8].

In this talk, we review the state-of-the-art in RO for problems with endogenous uncertainty, and we present novel decision-dependent uncertainty sets to model these settings [9] as well as a computational framework to optimize in light of such sets. We also present extensive computational results from a number of benchmark problems, including multi-stage problems that involve decision-making with recourse [10-11].

References:

[1] A. Ben-Tal, L. El Ghaoui, and A. Nemirovski, *Robust optimization*. Princeton University Press, 2009.

[2] Y. A. Guzman, L. R. Matthews, and C. A. Floudas, "New A Priori and A Posteriori Probabilistic Bounds for Robust Counterpart Optimization: I. Unknown Probability Distributions," *Comp. Chem. Eng.*, vol. 84, pp. 568–598, 2016.

[3] T. W. Jonsbråten, R. J. B. Wets, and D. L. Woodruff, "A class of stochastic programs with decision dependent random elements," *Ann. Oper. Res.*, vol. 82, pp. 83–106, 1998.

[4] V. Gupta and I. E. Grossmann, "A new decomposition algorithm for multistage stochastic programs with endogenous uncertainties," *Comput. Chem. Eng.*, vol. 62, pp. 62–79, 2014.

[5] V. Gupta and I. E. Grossmann, "Solution strategies for multistage stochastic programming with endogenous uncertainties," *Comput. Chem. Eng.*, vol. 35, no. 11, pp. 2235–2247, 2011.

[6] S. Solak, J. P. B. Clarke, E. L. Johnson, and E. R. Barnes, "Optimization of R&D project portfolios under endogenous uncertainty," *Eur. J. Oper. Res.*, vol. 207, no. 1, pp. 420–433, 2010.

[7] R. M. Apap and I. E. Grossmann, "Models and computational strategies for multistage stochastic programming under endogenous and exogenous uncertainties," *Manuscr. Prep.*, 2015.

[8] M. Poss, "Robust combinatorial optimization with variable cost uncertainty," *Eur. J. Oper. Res.*, vol. 237, no. 3, pp. 836–845, 2014.

[9] N. H. Lappas and C. E. Gounaris, "The Use of Decision-dependent Uncertainty Sets in Robust Optimization," *Proc. FOCAPO-2017*, paper F71, pp. 1–6, 2017.

[10] D. Bertsimas and A. Georghiou, "Design of Near Optimal Decision Rules in Multistage Adaptive Mixed-Integer Optimization," *Oper. Res.*, vol. 63, no. 3, pp. 610–627, 2015.

[11] N. H. Lappas and C. E. Gounaris, "Multi-stage Adjustable Robust Optimization for Process Scheduling Under Uncertainty," *AIChE J.*, vol. 62, no. 5, pp. 1646–1667, 2016.

488878 Heuristics with Performance Guarantees for the Minimum Number of Matches in Heat Recovery Networks

Abstract:

Optimal heat exchanger network synthesis improves heat recovery in chemical processes [1, 2, 3, 4]. Heat exchanger network synthesis is a nonconvex mixed-integer nonlinear optimization problem that may be solved simultaneously, i.e. to generate an optimal network without decomposition [5, 6, 7, 8]. An alternative, sequential method, decomposes the synthesis problem into: (i) minimizing utility cost, (ii) minimizing the number of matches, (iii) deriving many possible superstructures, (iv) minimizing the investment cost, and (v) configuring the final network [9]. The sequential method cannot guarantee global optimality of the original problem, but it reduces the computational burden.

We investigate the *minimum number of matches* subproblem of the sequential method. This subproblem is an NP–hard mixed-integer linear program (MILP) exhibiting: combinatorial explosion in the possible hot and cold stream configurations and symmetric mathematical structure creating unnecessarily large search trees [10, 11, 12]. Because the minimum number of matches subproblem still poses computational difficulty for moderately-sized networks [13], we develop provably efficient approximation algorithms with guaranteed solution quality and run-time bounds.

Previous work achieves a 2-approximation ratio for a single temperature interval and a non-constant approximation ratio for multiple temperature intervals scaling with the number of hot and cold streams [14]. This presentation develops new approximation algorithms with improved performance guarantees and proves their tightness. In the sequential method, many possible stream configurations are required to evaluate the minimum overall cost. So a complementary contribution of this presentation is developing (provably) efficient algorithms producing multiple solutions. The proposed algorithms and novel MILP formulations are validated experimentally.

We close the presentation by reflecting on the tremendous contribution of Professor Christodoulos A. Floudas to both the algorithmic methodology and practical application of process systems engineering. We show that Professor Floudas' legacy, which is integral to this current work, also offers a fruitful path to the future of multiscale systems engineering.

1. C.A. Floudas and I.E. Grossmann. Synthesis of flexible heat exchanger networks with uncertain flowrates and temperatures. *Comput Chem Eng*, 11(4):319–336, **1987**.
2. J. A. Elia, R. C. Baliban, and C. A. Floudas. Toward novel hybrid biomass, coal, and natural gas processes for satisfying current transportation fuel demands, 2: Simultaneous heat and power integration. *Ind Eng Chem Res*, 49(16):7371–7388, **2010**.
3. R. C. Baliban, J. A. Elia, R. Misener, and C. A. Floudas. Global optimization of a MINLP process synthesis model for thermochemical based conversion of hybrid coal, biomass, and natural gas to liquid fuels. *Comput Chem Eng*, 42:64–86, **2012**.
4. M. Escobar and J. O. Trierweiler. Optimal heat exchanger network synthesis: A case study comparison. *Appl Therm Eng*, 51(1-2):801–826, **2013**.
5. T. F. Yee and I. E. Grossmann. Simultaneous optimization models for heat integration – II. Heat exchanger network synthesis. *Comput Chem Eng*, 14(10):1165–1184, **1990**.
6. A. R. Ciric and C. A. Floudas. Heat exchanger network synthesis without decomposition. *Comput Chem Eng*, 15(6):385–396, **1991**.
7. K. P. Papalexandri and E. N. Pistikopoulos. Synthesis and retrofit design of operable heat exchanger networks. 1. Flexibility and structural controllability aspects. *Ind Eng Chem Res*, 33(7):1718–1737, **1994**.
8. M. Mistry and R. Misener. Optimising heat exchanger network synthesis using convexity properties of the logarithmic mean temperature difference. *Comput Chem Eng*, 94:1–17, 2016.
9. C. A. Floudas, A. R. Ciric, and I. E. Grossmann. Automatic synthesis of optimum heat exchanger network configurations. *AIChE J*, 32(2):276–290, **1986**.
10. K. C. Furman and N. V. Sahinidis. Computational complexity of heat exchanger network synthesis. *Comput Chem Eng*, 25:1371 – 1390, **2001**.
11. C. A. Floudas. *Nonlinear and Mixed-Integer Optimization: Fundamentals and Applications Topics in Chemical Engineering*. Topics in Chemical Engineering. Oxford University Press, New York, **1995**.
12. G. Kouyialis and R. Misener. Detecting symmetry in designing heat exchanger networks. In C. Maravelias, J. Wassick, E. Ydstie, and L. Megan, editors, *Proceedings of Foundations of Computer Aided Process Operations / Chemical Process Control*, **2017**.
13. Y. Chen, I. E. Grossmann, and D. C. Miller. Computational strategies for large-scale MILP transshipment models for heat exchanger network synthesis. *Comput Chem Eng*, 82:68–83, **2015**.
14. K. C. Furman and N. V. Sahinidis. Approximation Algorithms for the Minimum Number of Matches Problem in Heat Exchanger Network Synthesis. *Ind Eng Chem Res*, 43(14):3554– 3565, **2004**.

495015 Recent Contributions By Floudas Lab at the Interface of Chemical Engineering and Computational Biology

Abstract:

Professor Floudas' Lab applied optimization to address problems at multiple scales. A theme of the lab focused on applications in the computational biology domain [1]. In this talk, we will focus on key recent contributions in the area of modeling, simulating, and designing proteins with Post-translational modifications (PTMs) and non-canonical amino acids (NCAAs) leveraging mixed-integer and nonlinear optimization techniques. Additionally, the talk will highlight contributions made in the protein structure prediction domain.

Statistics of PTMs contained in the Swiss-Prot database were curated to identify those most frequently occurring [2]. We developed Forcefield_PTM [3], a set of AMBER charge and torsion forcefield parameters consistent with ff03 for 32 common PTMs. The parameterization methodology was extended to create charge parameters for 147 NCAAs in Forcefield_NCAA[4]. The forcefields were validated against experimental hydration free energies through thermodynamic integration calculations in both the TIP3P and TIP4P-Ew water models finding comparable correlations and errors to ff03 for natural amino acids [5]. The forcefields were integrated into a new protein design method that combines integer linear optimization and molecular simulations, and they were used to help design several new analogs of Compstatin [6]. Several analogs exhibited enhanced activity and/or solubility compared to the natural amino acid-containing starting compound. Due to their potency and solubility, these peptides are promising candidates for therapeutic development in complement-mediated diseases.

The lab contributed immensely to the protein folding problem. Recently, the lab participated in a "Coopetition", WeFold [7], collaborating with 13 labs worldwide strategically combining methods and resources to develop innovative hybrid pipelines to test in double-blind experiments. Many webtools have been created to disseminate Floudas Lab's research to a broader audience [8, 9]. The methods developed by Chris' intellectual progeny are actively being used to discover new inhibitors and to predict and refine protein structures.

References

1. Khoury, G.A., *Computational methods & forcefields for protein design, structure prediction, & refinement with natural & modified amino acids.* 2015, Princeton University.

2. Khoury, G.A., R.C. Baliban, and C.A. Floudas, *Proteome-wide post-translational modification statistics: frequency analysis and curation of the swiss-prot database.* Scientific reports, 2011. **1**: p. 90.

3. Khoury, G.A., et al., *Forcefield_PTM: Ab Initio Charge and AMBER Forcefield Parameters for Frequently Occurring Post-Translational Modifications.* Journal of Chemical Theory and Computation, 2013. **9**(12): p. 5653–5674.

4. Khoury, G.A., et al., *Forcefield_NCAA: ab initio charge parameters to aid in the discovery and design of therapeutic proteins and peptides with unnatural amino acids and their application to complement inhibitors of the compstatin family.* ACS synthetic biology, 2014. **3**(12): p. 855-869.

5. Khoury, G.A., N. Bhatia, and C.A. Floudas, *Hydration free energies calculated using the AMBER ff03 charge model for natural and unnatural amino acids and multiple water models.* Computers & Chemical Engineering, 2014. **71**: p. 745-752.

6. Gorham Jr, R.D., et al., *New compstatin peptides containing N-terminal extensions and non-natural amino acids exhibit potent complement inhibition and improved solubility characteristics.* Journal of medicinal chemistry, 2015. **58**(2): p. 814.

7. Khoury, G.A., et al., *WeFold: A Coopetition for Protein Structure Prediction.* Proteins: Structure, Function, Bioinformatics, 2014(10.1002/prot.24538): p. In Press.

8. Khoury, G.A., et al., *Princeton_TIGRESS: Protein geometry refinement using simulations and support vector machines.* Proteins: Structure, Function, and Bioinformatics, 2014. **82**(5): p. 794-814.

9. Smadbeck, J., et al., *Protein WISDOM: A Workbench for In silico De novo Design of BioMolecules.* Journal of Visualized Experiments, 2013(77): p. e50476.

497746 Advances in Modeling, Synthesis, and Global Optimization of Hybrid Energy Systems

Abstract:

Significant challenges against a sustainable energy future include energy availability, economics, and the generation of lower carbon energy [1]. To address these challenges, multi-scale modeling and optimization approaches are necessary toward the production of transportation fuels and petrochemicals from alternative feedstocks (e.g., coal, biomass, natural gas, and municipal solid waste - MSW) [1].

At the reactor design level, the effect of key variables are captured through first principles or data analysis while keeping model complexity compatible with large-scale process optimization [2,3]. A first-principles based dynamic optimization framework for the steam cracking of hydrocarbons is highlighted, followed by novel data-driven approaches for microchannel steam reforming and MSW gasification where fundamental models describing these phenomena are too complex. These resulting models are incorporated within a process synthesis superstructure designed to compare novel and competing refinery alternatives for the production of liquid fuels and chemicals from carbonaceous feedstocks [4-7].

A comprehensive process synthesis superstructure is developed to convert alternative feedstocks (e.g., coal, biomass, natural gas, MSW) into liquid transportation fuels, olefins, and aromatics. This superstructure enables the simultaneous evaluation of many competing process alternatives (>1E9) within a single large-scale (>20K variables and constraints) nonconvex mixed-integer nonlinear model (MINLP). A tailored branch-and-bound global optimization algorithm is developed to determine the optimal refinery topology at the highest profit or lowest cost [4-7]. Several case studies are illustrated to demonstrate the capability of the global optimization algorithm and to quantify the economic viability of the optimal topologies. The process synthesis approach is further extended to handle uncertainties and synthesize nationwide energy supply chains.

References:

[1]: Floudas, C. A.; Niziolek, A. M.; Onel, O.; Matthews, L. R. Multi-scale systems engineering for energy and the environment: Challenges and opportunities. AIChE Journal 2016, 62 (3), 602-623.

[2]: Onel, O.; Niziolek, A. M.; Hasan, M.; Floudas, C. A. Municipal solid waste to liquid transportation fuelsâ€"Part I: Mathematical modeling of a municipal solid waste gasifier. Computers & Chemical Engineering 2014, 71, 636-647.

[3]: Onel, O.; Niziolek, A. M.; Butcher, H.; Wilhite, B. A.; Floudas, C. A. Multi-scale approaches for gas-to-liquids process intensification: CFD modeling, process synthesis, and global optimization. Computers & Chemical Engineering, 2017, DOI: 10.1016/j.compchemeng.2017.01.016

[4]: Onel, O.; Niziolek, A. M.; Elia, J. A.; Baliban, R. C.; Floudas, C. A. Biomass and natural gas to liquid

transportation fuels and olefins (BGTL+C2_C4): Process Synthesis and Global Optimization. Industrial & Engineering Chemistry Research 2014, 54, 359-385.

[5]: Niziolek, A. M.; Onel, O.; Elia, J. A.; Baliban R. C.; Floudas, C. A. Coproduction of Liquid Transportation Fuels and C6_C8 Aromatics from Biomass and Natural Gas. AIChE Journal 2015, 61, 831-856.

[6]: Onel, O.; Niziolek, A. M; Floudas, C. A Optimal Production of Light Olefins from Natural Gas via the Methanol Intermediate. Industrial & Engineering Chemistry Research 2016, 55 (11), 3043-3063.

[7] Niziolek, A. M.; Onel, O.; Floudas, C. A. Production of Benzene, Toluene, and Xylenes from Natural Gas via Methanol: Process Synthesis and Global Optimization. AIChE Journal 2016, 62 (5), 1531 â€" 1556.

505376 On a Mathematical Modelling and Optimization Approach for the Systematic Synthesis and Development of Integrated Value Chains for Biorefineries

Abstract:

The biorefinery concept addresses the holistic use of biorenewables and is gaining increasing attention with applications that include the production of fuels and chemicals from different resources. The design of product portfolios remains a critical decision that depends on the nature of resources, the processing technologies, the infrastructures installed and the markets available. Energy efficiency and integration across production paths is critical in selecting product portfolios. However, the synthesis of product portfolios accounts for a combinatorial problem of significant complexity as the number of products and feedstocks are large, several alternative technologies are available to choose, and the ways to integrate the processing paths explode in combinatorial manner.

The work highlights a combined use of graph theory and cascade analysis (thermodynamics) to first conceptualize and subsequently formulate the problem using mixed-integer programming technology. Directed graphs are introduced to systematically translate biorefinery value chains into processing networks where thermodynamics is used to define energy containers of integration. The latter is addressed at the process level but, primarily, through patterns of process-to-process integration. The mathematical approach accounts for all available combinations and could accommodate preferences for portfolio profiles (e.g. mix of fuels and chemicals) or particular types of products (e.g. neutraceuticals, polymers). The mathematical approach considers integration for the entire plant each time delivering the product mix to match the minimum (plant) energy or the most profitable production. The research steps are illustrated with small problems but results are presented from three large real-life biorefineries where the approach has been applied to guide developments.

501381 Variable and Term Selection of Approximations for Data-Driven Optimization

Abstract:

The motivation of this work is ultimately enabling design and optimization of processes/systems using high-dimensional input-output data streams, through the use of surrogate or approximating functions. In other words, we aim to optimize problem $\{min\ f(x);\ s.t.\ g(x)\leq 0;\ x\ \ [x\_lo,x\_up\ ];\ x\ \ R\}$, for which any of the real objective function ($f(x)$) and/or the constraints ($g(x)$) cannot be algebraically expressed and purely depend on data. Due to the lack of the real functions (f,g), problem $\{min\ f'(x);\ s.t.\ g'(x)\leq 0;\ x\ \ [x\_lo,x\_up\ ];\ x\ \ R\}$ is formulated, which uses approximate functions, or surrogate functions $f'(x)$ and $g'(x)$. Surrogate functions have been proven to be very useful as the intermediate step towards the search of optimal points. At the same time, their accuracy and complexity is critical towards the performance of data-driven optimization methods [1,2].

In this work, several challenges of surrogate-based optimization will be studied through a comprehensive comparison of the performance of various types of approximation methods for problems with increasing dimensionality and in the absence or presence of noise. First, most existing data-driven optimization methods [3] (also known as black/grey-box optimization or derivative-free optimization) require the *a-priori* knowledge of the important degrees of freedom, which may not be true in real systems or industrial applications. Second, as the dimensionality of data-driven systems increases, the curse-of-dimensionality currently limits the performance and applicability of surrogate-based optimization methods. Lastly, oftentimes input-output data contains noise that may significantly affect surrogate model fitting and prediction accuracy.

The above challenges will be studied in this work via the simulation of synthetic data sets from a large set of benchmark problems for optimization, ranging from problems with tens to hundreds of variables with and without noise. First, we investigate the simultaneous variable selection and surrogate term selection by using support vector regression (SVR) [4,5]. Support vector machines (SVM) have been successfully used for high dimensional feature selection, dimensionality reduction and function estimation through the SVR extension [6]. SVM-based methods rely on a ε-insensitive loss function that has the ability to reduce the risk of overfitting as well as reduce the complexity of the surrogate function via regularization. Here, we use the effects of ε-sensitivity and regularization to develop surrogate functions that are specifically targeted for optimization. More specifically we develop (a) functions that are more accurate in regions where minimum/maximum values are observed, and (b) functions with a low number of terms and reduced complexity. The performance of the proposed SVR approach is compared to several popular methods for surrogate model identification for optimization, such as kriging or Gaussian process modeling, quadratic approximations, polynomial approximations, and radial basis functions.

References:

1. Forrester, A.I.J., A. Sóbester, and A.J. Keane, *Engineering Design via Surrogate Modelling - A Practical Guide.* 2008: John Wiley & Sons.

2. Cozad, A., N.V. Sahinidis, and D.C. Miller, *Learning surrogate models for simulation-based optimization.*AIChE Journal, 2014. 60(6): p. 2211-2227.

3. Boukouvala, F., R. Misener, and C.A. Floudas, *Global optimization advances in Mixed-Integer Nonlinear Programming, MINLP, and Constrained Derivative-Free Optimization, CDFO.*European Journal of Operational Research, 2016. 252(3): p. 701-727.

4. Vapnik, V., Golowich, S. E., and Smola, A. J. Support vector method for function approximation, regression estimation and signal processing. In Advances in Neural Information Processing Systems, 1997. 9:p. 281-287.

5. H. Drucker, C. J. C. Burges, L. Kaufman, A. J. Smola, and V. Vapnik. Support vector regression machines. In Advances in Neural Information Processing Systems, 1997. 9:p. 155-161.

6. V. N. Vapnik. The Nature of Statistical Learning Theory. Springer-Verlag New York, New York, NY, USA, 1995.

500082 Data-Driven Approximation of Feasible Region, Constrained Design of Experiments, and Optimization

Abstract:

We present a data-driven approximation of the feasible region of constrained black-box problems. In the design of many engineering systems, one often needs to choose parameters such that the performance is maximized while satisfying physical, operational, and regulatory constraints. The performance objective and constraints may be obtained only through expensive function evaluation or process plant data. The derivatives in these cases are not always available or expensive to obtain. In such cases, it is often better to use derivative-free method for optimization. Although there is abundance of works/algorithms developed for bound-constrained or unconstrained black-box problems, there is limited literature that addresses constrained problems [1-6]. Recently, Zhang et al. [7] approximated the feasible region by polytopes and formulated it as a mixed-integer linear program. In this work, we approximate the feasible region (convex or nonconvex, continuous or disjoint) as the region described by the convex hull of feasible samples and their neighboring points, while subtracting the infeasible samples and their neighboring points from the convex hull. We represent each sample and its neighboring points using a Euclidean ball with the sample as its center and radius proportional to the minimum constraint satisfaction (if the sample is feasible) or violation (if the sample is infeasible). A linear programming problem is solved to choose the largest ball that avoids any intersection between feasible and infeasible samples. The procedure for the approximation of a feasible region proceeds iteratively. At each iteration, new samples are obtained via a computational design of experiments (DoE). The DoE is posed as a packing problem, which is a quadratically constrained program (QCP) and is solved to optimality using a global solver. This procedure is repeated in an iterative manner until the entire domain is mapped by set of convex-hulls and Euclidian balls. Once the feasible region is approximated, optimization is performed by developing surrogate model for the objective function. The algorithm is applied to several test problems and compared to existing solvers to test the effectiveness of the algorithm.

References:

[1] Liuzzi, G.; Lucidi, S. A derivative-free algorithm for inequality constrained nonlinear programming via smoothing of an penalty function. *SIAM Journal on Optimization*, 20(1):1-29, 2009.

[2] Bertsekas, D.P. Constrained Optimization and Lagrangian Multiplier Methods. *Academic Press, New York*, 1982.

[3] Xue, D.; Sun, W. On convergence analysis of a derivative-free trust region algorithm for constrained optimization with separable structure. *Science China Mathematics*, 57(6):1287-1302, 2014.

[4] Eason, J.P.; Biegler, L.T. A trust-region filter method for glass box/black box optimization. *AICHE Journal*, 62(9):3124-3136, 2016.

[5] Sampaio, Ph.R; Toint, Ph.L. Numerical experience with a derivative-free trust-funnel method for nonlinear

optimization problems with general nonlinear constraints. *Optimization methods and Software*, 31(3):511-534, 2016.

[6] Augustin, F.; Marzouk, Y.M. NOWPAC: A provable convergent derivative-free nonlinear optimizer with path-augmented constraints. Technical report, Massachusetts Institute of Technology, 2014.

[7] Zhang, Q.; Grossmann, I. E.; Sundaramoorthy, A.; Pinto, J. M. Data-driven construction of convex region Surrogate models. *Optimization and Engineering*, 17(2):289-332, 2015.

500691 Data-Driven Model Building of Zeolite Adsorption Processes with Uncertainty Quantification and Propagation to Dynamic Simulations of CO_2 Adsorption

Abstract:

The prediction and propagation of uncertainty in complex mathematical model simulations is a research subject attracting increasing attention in recent years [1,2]. Computer models, while a great tool to predict the behavior of a variety of specific applications, are full of unknown parameters whose values are inherently uncertain. Additionally, the model itself may not be a perfect description of the modeled process, resulting in unreliable predictions even with the best-fitting parameter values [3]. Therefore, advanced calibration tools are required to incorporate uncertainty into parameter estimation to enhance the performance of the model. Bayesian calibration is an invaluable calibration technique that can be used to address model uncertainty. It consists of making inferences about the unknown parameters and model form uncertainty based on experimental data, which results in posterior distributions of the parameters. By using models calibrated in this manner, numerical simulations produce predictive distributions of key process variables, enabling decision makers to appreciate the incertitude and spend less time and resources during scale-up by designing fewer and more effective incremental steps [4].

Post-combustion CO_2 capture is a technology for which the interest in up-scaling for commercial applications has gained significant momentum of late [5]. In this study, a rigorous complete dynamic mathematical model of a fixed bed adsorber for CO_2 adsorption onto NaX zeolite (also known as zeolite 13X) is developed and implemented in gPROMS®. The small-scale adsorption model that governs the dynamic phase equilibrium between the gas and the solid phase is constructed from adsorption isotherm and calorimetry data using a model form based on the Langmuir isotherm equipped with Gaussian process stochastic functions. The Bayesian approach to model construction and calibration quantifies uncertainty in the adsorption model. The physical adsorption model is then implemented into the process-scale fixed bed adsorber model, thus propagating uncertainty to the dynamic, non-isothermal fixed bed adsorber model.

To save computational time, two simplified mathematical models of the fixed bed adsorber are formulated. From a deterministic perspective, the simplified models are shown to be in close agreement with the complete model. In order to evaluate whether the simplified models can be used as substitutes of the complete model for uncertainty propagation, an ANOVA analysis is conducted at a significance level of 0.05, followed by linear contrast analysis to test specific hypothesis related to the output distributions of the three models.

A number of case studies are performed and for each case, the uncertainty is quantified in output variables of interest – namely, CO_2 breakthrough times and concentration and temperature profiles. The case studies show that the uncertainty inherent to the physical adsorption model scales up differently depending on the operating conditions. Some operating conditions give rise to increased process-scale uncertainty, while others to decreased uncertainty. An additional case study demonstrates that accounting for model form discrepancy at the small scale (in contrast with accounting for parameter uncertainty only) significantly reduces the uncertainty in the process-scale variables.

References

1. Bhat K. S., Mebane D. S., Storlie C. B., and P. Mahapatra, "Upscaling Uncertainty with Dynamic Discrepancy for a Multi-scale Carbon Capture System". *Journal of the American Statistical Association*, 2017.
2. Kalyanaraman J., Kawajiri Y., Lively R. P., and M. J. Realff. "Uncertainty Quantification via Bayesian Inference Using Sequential Monte Carlo Methods for CO2 Adsorption Process". *AIChE Journal*, 2016, 62 (9) pp. 3352-3368.
3. Kennedy M. C., and A. O'Hagan, "Bayesian calibration of computer models", *Journal of the Royal Statistical Society Series B*, 2001, 63 (3) pp. 425-464.
4. Miller D. C., Syamlal M., Mebane D. S., Storlie C., Bhattacharyya D., Sahinidis N. V., Agarwal D., Tong C., Zitney S. E., Sarkar A., Sun X., Sundaresan S., Ryan E., Engel D., and C. Dale (), "Carbon Capture Simulation Initiative: A Case Study in Multiscale Modeling and New Challenges", *Annual Review of Chemical and Biomolecular Engineering*, 2014, 5 pp. 301-23.
5. Miller D. C., Agarwal D. A., and C. Tong, "CCSI and the role of advanced computing in accelerating the commercial deployment of carbon capture systems", *SciDAC Conference*, June 2011.

493742 Data-Driven Stochastic Robust Optimization: General Modeling Framework and Efficient Computational Algorithm for Handling Labeled Multi-Class Uncertainty Data

Abstract:

Optimization under uncertainty has attracted tremendous attention from both academia and industry [1]. A plethora of mathematical programming techniques, such as stochastic programming [2], and robust optimization [3], have been proposed. These techniques have their respective strengths and weaknesses, which leads to different scopes of applications [4-7]. Stochastic programming is an approach that focuses on the expected performance of a solution by leveraging probability distributions. However, it becomes computationally intractable as the number of scenarios increases [8]. Robust optimization provides an alternative paradigm that does not require an accurate knowledge of probability distributions, and offers an attractive benefit of computational tractability. Nevertheless, the robust optimization solution usually suffers from the conservatism issue. The state-of-the-art approaches leverage the synergy of different optimization methods to inherit their corresponding advantages and complement respective drawbacks. However, these approaches fall short of leveraging uncertainty data to benefit decision making processes. Thanks to massive amounts of data and great advancements in machine learning, data-driven optimization emerges as a promising paradigm [9, 10].

Uncertainty data in large data sets are often collected from various conditions, which are encoded by class labels. By leveraging the label information, we propose a novel data-driven stochastic robust optimization (DDSRO) framework that systematically and automatically handles the labeled multi-class uncertainty data. To account for the uncertainty data with different labels, we develop a modeling framework that leverages the synergy of stochastic programming and robust optimization. Stochastic programming is nested in the outer problem to optimize the expected objective over categorical classes. Operational decisions need to hedge against the worst case to ensure the robustness of operations. Besides, estimating a joint probability distribution of a high-dimensional uncertainty is much more challenging than estimating a categorical distribution. Therefore, robust optimization, which does not require an accurate knowledge of probability distribution, is nested in the inner problem. A bag of Dirichlet process mixture models is employed to construct uncertainty sets from the labeled multi-class uncertainty data [11, 12]. This data-driven nonparametric uncertainty set could adjust its complexity to that of data automatically, thus truthfully capturing the nature of uncertainty within the same class. The state-of-the-art machine learning technique is seamlessly integrated with the stochastic robust optimization framework in the form of a multi-level mixed-integer linear program (MILP). We further develop a tailored column-and-constraint generation (C&CG) algorithm [13] to solve the DDSRO problem efficiently. It is worth noting that the proposed DDSRO framework is an extension of data-driven adaptive nested robust optimization (DDANRO) [9], which was recently proposed to handle uncertainty data homogeneously without considering the label information. DDSRO greatly enhances the capabilities of DDANRO, and expands its scope of applications to more complex data structures. To demonstrate the advantages of the proposed framework and the efficiency of the computational algorithm, a motivating numerical example and two case studies on strategic planning of process networks [14] under supply and demand uncertainties are presented.

References

[1] I. E. Grossmann, R. M. Apap, B. A. Calfa, P. García-Herreros, and Q. Zhang, "Recent advances in mathematical programming techniques for the optimization of process systems under uncertainty," *Computers & Chemical Engineering,* vol. 91, pp. 3-14, 2016.

[2] J. R. Birge and F. Louveaux, *Introduction to stochastic programming*: Springer Science & Business Media, 2011.

[3] A. Ben-Tal, L. E. Ghaoui, and A. Nemirovski, *Robust Optimization*: Princeton University Press, 2009.

[4] H. Shi and F. You, "A computational framework and solution algorithms for two-stage adaptive robust scheduling of batch manufacturing processes under uncertainty," *AIChE Journal,* vol. 62, pp. 687-703, 2016.

[5] J. Gong, D. J. Garcia, and F. You, "Unraveling optimal biomass processing routes from bioconversion product and process networks under uncertainty: An adaptive robust optimization approach," *ACS Sustainable Chemistry & Engineering,* vol. 4, pp. 3160-3173, 2016.

[6] D. Yue and F. You, "Optimal supply chain design and operations under multi-scale uncertainties: Nested stochastic robust optimization modeling framework and solution algorithm," *AIChE Journal,* vol. 62, pp. 3041-3055, 2016.

[7] J. Gong and F. You, "Optimal processing network design under uncertainty for producing fuels and value-added bioproducts from microalgae: Two-stage adaptive robust mixed integer fractional programming model and computationally efficient solution algorithm," *AIChE Journal,* vol. 63, pp. 582-600, 2017.

[8] B. H. Gebreslassie, Y. Yao, and F. You, "Design under uncertainty of hydrocarbon biorefinery supply chains: Multiobjective stochastic programming models, decomposition algorithm, and a Comparison between CVaR and downside risk," *AIChE Journal,* vol. 58, pp. 2155-2179, 2012.

[9] C. Ning and F. You, "Data-driven adaptive nested robust optimization: General modeling framework and efficient computational algorithm for decision making under uncertainty," *AIChE Journal,* 2017. DOI: 10.1002/aic.15717

[10] D. Bertsimas, V. Gupta, and N. Kallus, "Data-driven robust optimization," *Mathematical Programming,* 2017. DOI:10.1007/s10107-017-1125-8

[11] T. Campbell and J. P. How, "Bayesian nonparametric set construction for robust optimization," in *2015 American Control Conference (ACC)*, 2015, pp. 4216-4221.

[12] D. M. Blei and M. I. Jordan, "Variational inference for Dirichlet process mixtures," *Bayesian analysis,* vol. 1, pp. 121-143, 2006.

[13] B. Zeng and L. Zhao, "Solving two-stage robust optimization problems using a column-and-constraint generation method," *Operations Research Letters,* vol. 41, pp. 457-461, 2013.

[14] F. You and I. E. Grossmann, "Stochastic inventory management for tactical process planning under uncertainties: MINLP models and algorithms," *AIChE Journal,* vol. 57, pp. 1250-1277, 2011.

501708 Dynamic Canonical Correlation Analysis for the Extraction and Diagnosis of Plant-Wide Oscillations

Abstract:

In this work, we present a newly developed dynamic data modeling algorithm, that is, DiCCA (dynamic-inner canonical component analysis) to extract a set of latent variables that have maximized auto-correlations. This DiCCA algorithm is different from prior versions of dynamic extensions of CCA as it derives explicit inner dynamic models with the objective to maximize the canonical correlations. As a consequence, the extracted dynamic components are most predictable from their past values.

Since oscillating signals are one of the most predictable types of signals, DiCCA is effective to extract oscillating latent variables among a set of process variables. The Tennessee Eastman oscillating dataset has been studied for root cause analysis. However, in most literature, only low frequency oscillations are examined. Higher frequency oscillations are either undiscovered or ignored. In fact, due to feedback and cascade controls, oscillations at different frequencies often interact with each other and result in a compounding effect. Therefore, a comprehensive analysis of oscillations at all frequencies is necessary for understanding the process and diagnosing the root cause.

In the Tennessee Eastman oscillating dataset, two different oscillations are discovered after DiCCA modeling. One is the well known low frequency oscillation with a period of 340 samples, the other one is a high frequency oscillation with a period of 18 samples. This high frequency oscillation has not been discovered previously.

Pairwise spectral granger causality analysis is performed on the variables with high frequency oscillations for the root cause analysis, and PC2.OP and PC2.PV are identified as the root cause. By applying the curve fitting method, it is further confirmed that valve stiction, rather than outside disturbance or aggressive tuning, causes the high frequency oscillations. The same analysis is conducted on the variables with low frequency oscillations as well, where LC2.OP and LC2.PV are identified as the root cause, and valve stiction is diagnosed as the physical reason. This corroborates with the conclusions in the paper of Thornhill et al. (2003), where LC2 control valve is confirmed as the root cause of the low frequency oscillations after an online test. This consistency demonstrates the effectiveness of DiCCA based analysis.

In addition, we analyze a set of data collected after LC2 valve was fixed. After DiCCA modeling, it is found that there is an oscillation with a period of 133 samples. Although the PC2 control valve was not fixed, the high frequency oscillations became less pronounced in this fixed dataset. One possible explanation for this observation is that oscillations at different frequencies interact with each other due to feedback and cascade controls. Therefore, fixing one control valve can change the other oscillation patterns.

In this work, DiCCA is demonstrated as an effective tool to analyze oscillating data. The newly discovered high frequency oscillations are studied and the root cause is identified. This analysis can be used as a standard way to study plant-wide oscillating datasets in the future.

498592 Design of a Supervisory Control System for Inter-Networked Facilities Operation

Abstract:

Operation of highly networked systems with interdependent influent streams can result in suboptimal network performance, where individual system operability and optimization are the primary concern of plant operators and engineers. The authors have designed a supervisory control system for operation of county-wide facilities transport through utilization of machine learning and connectionist system modeling for the prediction of unsteady state influent behavior. Black box neural network models were generated from available multi-year data collected by individual plant SCADA systems, and a supervisory control system was designed to manage both influent routing for inter-networked facilities as well as individual plant-based operation.

Time-series artificial neural network models were generated for each of the seven advanced wastewater treatment facilities in Hillsborough County, Florida, where each facility receives influent flow from a series of centralized pumping stations that are engineered to divert flow via opening and closing of flow gates from one facility to another during instances of abnormally high influent flow rates. Due to current gate operation, there is necessarily a large amount of flow variance at any one instance, which leads to significant operational challenges at the plant level to manage and treat that supply in real time. Sudden changes in influent—coupled with long residence times for the wastewater treatment process—can be detrimental to meeting EPA regulations, since the primary control strategy involves manipulating dissolved oxygen concentration for the nitrification-denitrification reaction via mechanical aeration (a very slow process). It is often necessary to know several hours in advance of any significant changes to plant operation in order to affect a change that will keep the plant effluent in compliance or prevent expensive fines due to excess wastage to retention ponds.

As a result of the pumping stations and gate operation, influent data reported by the SCADA system is generally fraught with noise; influent flow rates vary by ±2 million gallons per day (MGD) within a 5-minute span. In order to create any meaningful models from available data, it is necessary to pre-process the data via profile smoothing, data cleaning, feature extraction, and feature generation. Then, a closed-loop time-series neural net can be trained from a subset of existing data to predict future events—often up to a day in advance. In combination with margins of error calculations at each time step, the predictive capability of the neural network can account for non-standard operational behavior (e.g., holidays) and anomalous activity (e.g., major sporting events or illegal industrial dumping). Machine learning aspects of the project allow neural networks to be retrained over time to account for gradual changes in plant performance and shifts in population behavior.

Finally, information gained from neural net models was abstracted to design an energy efficient supervisory control system for operators at the individual plant level and engineers in charge of flow diversion for the entire facilities network. The information was conveyed to operators in the form of SCADA-interfaced GUI with real-time prediction and rules-based control strategies for engineers and operators for optimizing network-level and plant-level performance.

494526 A Graph-Based Modeling and Optimization Framework for Complex Systems

Abstract:

This talk presents PLASMO (Platform for Scalable Modeling and Optimization), a Julia-based [1] and open-source modeling platform that facilitates the creation of complex optimization applications. A distinctive feature of PLASMO is that it uses a graph abstraction in which optimization models live over nodes that are connected through edges though coupling constraints. This abstraction enables the creation of hierarchical graph models in which a node can be a graph model itself. PLASMO also uses a virtual graph abstraction in which optimization models interact with one another on-the-fly by exchanging information. Under this virtual graph abstraction, nodes are tasks and edges are communication links.

PLASMO enables modularization of complex models, because the graph topology is independent of the node models. PLASMO also facilitates the construction of coupled heterogenous networks (e.g., supply chains and natural gas and electric networks), compartmentalizes automatic differentiation and model processing tasks, manipulates graphs to perform diverse model analysis tasks (e.g., network partitioning and aggregation) and manipulates the graph so that the model can be solved solved with standard solvers (e.g., Gurobi,Ipopt) or with structured solvers (e.g., PIPS-NLP [2] and DSP [3]). Furthermore, the virtual graph abstraction facilitates the creation of algorithms such as model predictive control [4], stochastic dual dynamic programming [5], and Lagrangian relaxation [6].

This presentation will focus on the modeling aspects of PLASMO and discusses how to use the platform to model and solve coupled infrastructure networks and multi-stage stochastic programming models. The talk will conclude with a motivating example on the centralized and decentralized control of a regional power grid and gas network [7]. We also show how to create and solve multi-stage stochastic formulations for battery models.

[1] Bezanson, J., Edelman, A., Karpinski, S. & Shah, V. B. Julia: A fresh approach to numerical computing. arXiv 1–37 (2015).

[2] N. Chiang, C. G. Petra, and V. M. Zavala. Structured nonconvex optimization of large-scale energy systems using PIPS-NLP. In Proc. of the 18th Power Systems Computation Conference (PSCC), Wroclaw, Poland, 2014.

[3] K. Kim and V. M. Zavala. Algorithmic innovations and software for the dual decomposition method applied to stochastic mixed-integer programs. Optimization Online, 2015

[4] J. B. Rawllings and D. Mayne. *Model predictive control*: theory and design. Nob Hill Publishing, 2009.

[5] Shapiro, A. Analysis of stochastic dual dynamic programming method. Eur. J. Oper. Res. 209, 63–72 (2011).

[6] Ruszczyński, Andrzej P. *Nonlinear optimization.* Vol. 13. Princeton university press, 2006.

[7] Jalving, J., Abhyankar, S., Kim, K., Hereld, M., Zavala, V. M. A Graph-Based Computational Framework for Simulation and Optimization of Coupled Infrastructure Networks. To Appear in *IET Generation, Transmission & Distribution,* 2016.

498515 Modeling LPS-Induced TNF-α Production in Macrophages

Abstract:

Sepsis results in about 200,000 deaths per year in United States alone, primarily due to the excessive immune response to the infecting pathogen [1]. Immune cells such as macrophages detect the presence of pathogen-derived molecules through pattern-recognition receptors (PRR). Toll-like receptor 4 (TLR4) is one such PPR that binds bacterial lipopolysaccharide (LPS), which is the major component of the cell wall in gram-negative bacteria, and initiates multiple intracellular signaling pathways to remove the pathogen [2]. Among different signaling pathways initiated by LPS binding to TLR4, the pathway leading to activation of the transcription factor NFκB is critical because NFκB mediates the production of several pro-inflammatory cytokines to eliminate the pathogens or PPR. One such cytokine is TNFα that stimulates pro-inflammatory signaling and cytokine production in other cells and leads to sustained inflammation. Since chronic inflammation can itself be detrimental to the organism, it is important to elucidate the molecular mechanisms underlying TNFα production and its regulation in the macrophage response to LPS.

Previous studies have constructed mathematical models to predict LPS-induced NFκB signaling pathway to study dynamic interactions among different components in the LPS-mediated NFκB signaling pathway [3, 4]. In this work, we updated the deterministic model developed by Caldwell et al. [3] with the *in vitro* measurements so that the model could simulate the NFκB dynamics accurately under the wide range of LPS concentrations. Data on the levels of IκBα, which binds and sequesters NFκB activation, and TNFα production were obtained in RAW264.7 macrophages in response to different concentration of LPS using intracellular staining and flow cytometry. A part of this data set was used to estimate parameters in the model while the remaining was used for validation. Since the number of data points was far less than that of parameters, the parameter estimation problem is ill-posed; that is, we could not estimate all parameters in the model. In order to acquire a robust solution from an ill-conditioning problem, the parameter selection method was employed [5]. Specifically, the parameter selection method, which involves the sensitivity analysis and the Gram-Schmidt orthogonalization, chose and estimated only some of the model parameters. In this study, six parameters, which were closely related to the transcription of IκBα and TNFα mRNA, were selected for the estimation. After estimating the selected parameters, the resultant deterministic model was able to qualitatively and quantitatively predict the dynamics of TNF and IκBα in macrophages upon LPS stimulation. Furthermore, the updated deterministic model can simulate cell-to-cell variability by differences in parameter values. First of all, a distribution matching methodology [6] is employed to estimate distributions of key parameter values with flow cytometry data. Next, a set of values is picked from the distribution corresponding to different parameters and assigned to the model to simulate the response of different cells in the population. As a result, the heterogeneous responses within cell population can be modeled by the collections of distinct responses from individual cells. Consequently, the final model will be capable of simulating both single-cell and population-average response to LPS.

Reference

1. Deng, M.; Scott, M.J.; Loughran, P.; Gibson, G.; Sodhi, C.; Watkins, S.; Hackam, D.; Billiar, T. R. Lipopolysaccharide clearance, bacterial clearance, and systemic inflammatory responses are regulated by cell type-specific functions of TLR4 during sepsis. *J. Immunol.* **2013**, *190 (10)*, 5152-5160.
2. Kawasaki, T.; Kawai, T. Toll-like receptor signaling pathway. *Front. Immunol.* **2014**, *5*, 461.
3. Caldwell, A.B.; Cheng, Z.; Vargas, J.D.; Birnbaum, H.A.; Hoffmann, A. Network dynamics determine the autocrine and paracrine signaling functions of TNF. *Genes & Dev.* **2014**, *28*, 2120-2133.
4. Kellogg, R.A.; Tian C.; Lipniacki, T.; Quake, S.R.; Tay, S. Digital signaling decouples activation probability and population heterogeneity. *eLife.* **2015**, *4*, e08931.
5. Kravaris, C.; Hahn, J.; Chu, Y. Advances and selected recent devlopments in state and parameter estimation. *Comp. & Chem. Eng.* **2013**, *51*,111-123.
6. Hasenauer, J.; Waldherr, S.; Doszczak, M.; Radde, N.; Scheurich, P.; Allgower, F. Identification of models of heterogeneous cell populations from population snapshot data. *BMC Bioinformatics* **2011**, *12(1)*, 125.

500906 Reaction Network Structure and Flux Analysis for Thin-Film Deposition Processes

Abstract:

Reaction network structure and flux analysis for thin-film deposition processes.

Hossein Salami, Raymond A. Adomaitis

Chemical Vapor Deposition (CVD) and Atomic Layer Deposition (ALD) are vapor-based processes for production of thin solid films in which heterogeneous and homogenous reactions with widely ranging time scales are involved. Proposing an accurate reaction mechanism consisting of adsorption/desorption and reactions within various phases that results in a well-posed species balance is among the first steps in studying these processes. A valid reaction network (RN) combined with accurate kinetic data is vital for process modeling and optimization and forms the backbone of further engineering analysis.

In this contribution, we will introduce an analysis approach based on RN graphs and linear algebra methods which provides a mechanistic framework for answering four critical questions regarding the validity of a proposed reaction RN prior to the often time-consuming step of calculating individual reaction rates and any attempt to solve the system of differential equations produced by species balances:

1. Can the equilibrium and finite-rate reaction time scales be decoupled and independently measured?

2. What is the physical origin of time-independent modes in the system's dynamic behavior and how they can be used to validate the proposed RN?

3. Does the proposed RN result in a unique film stoichiometry that is independent of reaction rates?

4. Does the RN pose a self-limiting mechanism essential to a true ALD process?

The first question can be answered by finding a transformation which essentially separates the effect of different reaction rates and defines new variables in the form of reaction variants and invariants that are linear combination of original concentrations [1]. However, this approach does not provide answers to the remaining questions and additional analysis is required. The approach we will describe is based on the species-reaction graphs introduced by Craciun and Feinebrg [2] and provides a set of rules which can be used to answer above questions using the graph constructed from the RN under study without requiring any knowledge regarding individual reaction rates.

Furthermore, we demonstrate the application of our RN graphs in material flux analysis and in studying the sensitivity of the RN to the process parameters such as the reactor temperature and precursor partial pressure. We also will assess its potential for extension to other areas in chemical engineering research, such as metabolic engineering or

polymerization reactions.

In this talk, we use the RN associated with the titania ALD process based on titanium isopropoxide ($Ti(iPrO)4$ or TTIP) / water precursors as a representative case to show the applicability of the proposed approach. This process has been shown to deviate from an ideal self-limiting ALD at temperatures greater than $250\ ^{o}C$ where the rate of hydrolysis reactions begins to decrease while pyrolysis reactions are activated [3,4]. The relative contribution of ALD versus CVD growth mode to $TiO2$ film deposition will be assessed using our flux analysis method and the results will be presented in the context of the RN graphs.

Figure 1. The sub-graph associated with the initial adsorption of TTIP molecule on a hydroxylated surface in titania ALD process. The blue color highlights the path which can be used to derive the formula for the time-independent mode associated with conservation of titanium atoms in the reactor. (g) and (b) refer to bulk and gas phases, others are surface species.

[1] D. Rodrigues, S. Srinivasan, J. Billeter, D. Bonvin, Comput. Chem. Eng., **2015**, 73, 23.

[2] G. Craciun, M. Feinberg, SIAM J. Appl. Math., **2006**, 66(4), 1321.

[3] A. Rahtu, M. Ritala, Chem. Vap. Deposition, **2002**, 8, 21.

[4] M. Reinke, Y. Kuzminykh, P. Ho_mann, J. Phys. Chem., **2016**, 120(8), 4337.

493579 Hierarchical Modeling and Control of ERK Signaling

Abstract:

Cell signaling is the process by which extracellular information is transmitted into the cell. Cell signaling pathways are interconnected networks of signaling proteins that communicate through complex molecular mechanisms. As a case in point, ERK (extracellular signal-regulated kinases) signaling plays an important role in numerous cellular processes such as proliferation, apoptosis, DNA synthesis and differentiation. Signaling faults due to mutations and failures in the regulatory mechanisms are known to result in different types of cancers. A complex network of cellular interactions and feedback loops are usually arranged in a hierarchical fashion in space and time to transform the information content of extracellular signals to the DNA in the nucleus of the cell which in turn leads to several cellular functions (see Fig.1).

Activation of EGFR (Epidermal Growth Factor Receptor) by binding of its specific ligands, namely epidermal growth factor and transforming growth factor a (TGF a), starts ERK signaling pathway, as shown in Fig.1.Upon ligand binding, two subunits of EGFR are dimerized, leading to increase in the enzymatic activity of its cytoplasmic tyrosine kinase domain. Adaptor protein Grb2 binds to the phosphorylated RTK (Receptor tyrosine kinase) followed by recruitment of SOS forming Grb2-SOS complex Membrane-bound protein Ras, which is a small GTP binding protein, interacts with Grb2-SOS complex, and it is transformed to its active conformation by exchanging GDP for GTP. Active Ras acts as an important switch which starts the sequential phosphorylation of MAPK pathway that consists of the Raf/MEK/ERK signaling cascade.

Activated ERK, when translocated to the nucleus, phosphorylates nuclear transcription factors (e.g. Myc), that govern cellular responses. Phosphorylated transcription factors stimulate transcription of genes which are responsible for encoding different proteins, including ones required for cell cycle progression (e.g. Cyclin D)

We decompose and analyze the ERK signaling pathway in two interacting functional subsystems with distinct characteristics: First, EGFR-Ras-MAPK subsystem, and second nuclear reactions triggered by activated ERK, including transcriptional processes. These subsystems operate at different time scales (e.g. the nuclear reactions are much slower than EGFR signaling throughout cytoplasm) and they interact with each other via several inhibiting and activating feedback loops. For instance, autocrine loop with positive feedback regulates amplitude and duration of MAPK activation, by incorporating a ligand-releasing protease (i.e. TACE). We have modeled the complete ERK signaling pathway shown in Fig.1 using mass-action kinetics and conservation laws. Numerous models exist in the literature for different parts of the pathway. Our model is more comprehensive as it combines all the significant parts and treats the system as a whole. It is well-established that each specific cellular response is determined by the duration, magnitude, frequency and localization of ERK signaling.

The developed dynamic nonlinear model exhibits very rich dynamics that is needed to tune the steady-state and dynamic characteristics of ERK signaling. Simulations show that ERK signaling can be modulated by a nested set of negative and positive feedback loops to deliver the desired cellular action. Bistability (i.e. ability to switch between two stable steady-states separated by an unstable steady-state) is the hallmark of cellular responses. Using bifurcation analysis, we establish conditions for the existence of bistability in different parts of the network and illustrate how the desired cellular function can be regulated by modulating the bistability characteristics and on-off switching behavior of ERK signaling.

Using this model, we are able to explain the biological phenomena and the clinical observations reported in the literature and postulate possible new mechanisms to improve understanding of an important cellular signaling network.

Avraham, R., & Yarden, Y. (2011). Feedback regulation of EGFR signaling: decision making by early and delayed loops. *Nature Reviews. Molecular Cell Biology*, *12*(2), 104–17. https://doi.org/10.1038/nrm3048.

Dhillon, A., Hagan, S., Rath, O., & Kolch, W. (2007). MAP kinase signalling pathways in cancer. *Oncogene*, *26*, 3279–3290. https://doi.org/10.1038/sj.onc.1210421

Figure 1. The hierarchy of ERK signaling. Dashed lines represent the feedback signals with arrow ends for activation and blunt ends for inhibition. DUSP: Dual-specificity phosphatase; TACE: (TNF)-alpha converting enzyme, also called ADAM17.

496045 Parameter Estimation and Sensitivity Analysis in Beer Fermentation Modelling and Dynamic Optimisation

Abstract:

Growing economic pressure experienced by the UK alcohol industry as a result of a global surge in beer product supply has created an extremely competitive environment for breweries, many of whom are seeking to reap the benefits of model-based process intensification and optimisation (Rodman and Gerogiorgis, 2016). While beer production is a multistage process, improving fermentation efficiency has the strongest potential for return on both capital and time investment on the entire process, as the batch time often exceeds 120 hours and is the throughput bottleneck. Improvements to beer production have been historically obtained by incrementally altering proven recipes via trial and error, in an effort to achieve a more desirable product, as a result of the limited understanding this complex chemical system. Nowadays it is essential to employ accurate models of the fermentation process in order to achieve reliable dynamic simulations and subsequent process optimisation: trial and error is no longer a competitive strategy, as empirical investigative campaigns would imply a prohibitively high cost, not only in terms of equipment use and personnel time for sampling and analysis, but also due to lost beer production as result of the necessary process downtime and off-spec production of undesirable (hence non-marketable) output.

The complexity of the fermentation biochemical system (Vanderhaegen et al., 2006) renders comprehensive dynamic model formulation cumbersome and the required parameter estimation prohibitive. Numerous previous publications (Engasser, 1981; Gee and Ramirez, 1988; de Andrés-Toro 1998; Trelea et al., 2001) have postulated reduced order models, which are concise enough to allow for parameter estimation on the basis of experimental campaign data, but also suitably descriptive by considering the most essential subset of important biochemical reactions via selective aggregation. Subsequent process optimisation studies have been successfully performed on the basis of these foregoing models, primarily addressing batch time minimisation or ethanol yield maximisation (de Andrés-Toro, 1998; Carrillo-Ureta, 2001), or both (Rodman and Gerogiorgis, 2016). A necessary assumption of process optimisation studies is that all parameter datasets published in the original fermentation modelling papers are accurate and consistent.

The present paper has relied on a comprehensive compilation of original time-dependent concentration data for beer fermentation campaigns (Gee, 1990; de Andrés-Toro, 1996) in order to perform an parameter estimation study (Dochain, 2003) for the respective reduced-order dynamic models. The Simulink environment has been used within MATLAB®, enabling the implementation of a nonlinear least squares method on the basis of the trust-region reflective algorithm for systematic parameter vector estimation. Remarkable discrepancies have been identified between the originally published parameter vectors and the corresponding ones determined herein, with the latter (our own results) universally improving dynamic model fidelity - and remarkably so, for certain critical parameters. Furthermore, we have performed systematic sensitivity analyses in order to assess and elucidate the relative significance of parametric discrepancy on the validity of dynamic simulation and optimisation results; the evolution of certain concentration observables is of particular interest, inasmuch as the latter govern process performance (: ethanol) and flavour (: diacetyls, ethyl acetate).

This comprehensive sensitivity analysis illustrates how state trajectory model predictions are affected by individual model parameters (DiStefano, 2013). Our new parameter set is subsequently used for further dynamic simulations towards attainable fermentation envelope visualisation. Multi-objective dynamic optimisation (time minimisation and ethanol maximisation) has also been performed with the new parameter vector we have determined, to investigate how and to what extent optimal operation is affected by parametric uncertainty, and thereby assess the validity of prior optimisation studies.

LITERATURE REFERENCES

Carrillo-Ureta, G., P. Roberts, and V. Becerra, 2001. Genetic algorithms for optimal control of beer fermentation. Proc. IEEE Int. Symp. Intell. Control, 391-396.

de Andrés-Toro, B., 1996. Modelling, Simulation, and Optimal Control of and Industrial Beer Fermentation Process (in Spanish). PhD Thesis, University Complutense of Madrid, Spain.

de Andrés-Toro, B., et al., 1998. A kinetic model for beer production under industrial operational conditions. Math. Comput. Simulat. 48(1): 65-74.

DiStefano, J, 2013. Dynamic Systems Biology Modelling and Simulation, Academic Press.

Dochain, D., 2003. State and parameter estimation in chemical and biochemical processes: a tutorial, J. Proc. Control 13(8): 801-818.

Engasser, J. M., Marc, I., Moll & M. Duteurte, B. Kinetic modelling of beer fermentation. Proceedings of the 18th Congress of the European Brewery Convention, 1981 Copenhagen. 579-86.

Gee, D. A., Ramirez, W. F. 1988. Optimal temperature control for batch beer fermentation. Biotechnol. Bioengineering, 31, 224-234.

Gee, D. A., 1990. Modelling, optimal control, state estimation, and parameter identification as applied to a batch fermentation process, PhD Thesis, The University of Colorado, Boulder, USA.

Rodman, A. D., Gerogiorgis, D.I., 2016. Multi-objective process optimisation of beer fermentation via dynamic simulation, Food Bioprod. Proc. 100(A): 255-274.

Trelea, I. C., Titica, M., Landaud, S., Latrille, E., Corrieu, G. & Cheruy, A. 2001. Predictive modelling of brewing fermentation: From knowledge-based to black-box models. Math. Comput. Simulat.Â 56, 405-424.

Vanderhaegen, B., et al., 2006. The chemistry of beer aging â€" a critical review. Food Chem. 95(3): 357-381.

Xiao, J., Zhou, Z., Zhang, G., 2003. Ant colony system algorithm for the optimization of beer fermentation control. J. Zhejiang Univ. 5(12): 1597-1603.

497407 Modeling the Influence of the HPA Axis and the Circadian Clock on the Regulation of the Cell Cycle

Abstract:

The circadian clock is a cell-autonomous, time-keeping system that orchestrates the activity of critical biological mechanisms enabling organisms to anticipate and adapt to fluctuating demands in response to a variety of environmental changes in a timely manner (1). The central circadian clock resides within the hypothalamic suprachiasmatic nucleus (SCN) and can be entrained by a number of environmental cues, including, the light-dark cycle and temperature. The SCN â€œgatesâ€ the activity of peripheral clocks in the liver, muscle, kidneys etc in a tissue-specific manner by relaying temporal information to the peripheral clocks via systemic signaling mechanisms such as the sympathetic nervous system as well as humoral factors (2).

Recently, there has been an increasing appreciation of the importance of the circadian clock in the temporal regulation of the cell cycle (3). Previous research has shown that the cell cycle is coupled to the circadian clock, with circadian rhythmicity in cell proliferation being observed in a number of organisms, including cyanobacteria, zebrafish, rodents and even humans, suggesting a strongly evolutionarily conserved link between the two systems (4). For instance, the DNA-replicative, S- and mitotic phases of the cell cycle tend to occur during the night, which is generally thought to enable the protection of these critical processes from harmful UV radiation during the day (5). In addition to the cell-autonomous influence of the circadian clock on the cell cycle, systemic signals such as growth factors, nutrients and hormones are also known to regulate the cell cycle. Often these systemic signals are also under the control of the circadian clock. One such critical mediator, cortisol, the primary stress response hormone, has both inhibitory and stimulatory influences on cell proliferation and the progression of the cell cycle (6). In addition, cortisol exhibits pronounced circadian rhythmicity and plays an important role in relaying temporal information from the central clock in the SCN to the peripheral circadian clocks.

It has been proposed that dysregulated entrainment of the cell cycle by the circadian clock might be involved in the development of cancer (7). Furthermore, evidence suggests that cell cycle desynchrony in a population of proliferating cells leads to an impaired would healing response. Along these lines, glucocorticoid administration has been shown to synchronize the cell cycle resulting in more controlled proliferation in ex vivo disease models of asthma (8). In the present work, we develop a mathematical model, to study the influence of both the circadian clock as well as endogenous cortisol rhythms on the regulation of the cell cycle. Systems biology approaches are particularly suited to study such problems, given the complexity of feedback between the three dynamical systems. Our current model accounts for the dual influences of these two systems on the cell cycle by coupling a semi-mechanistic model for light entrainment on peripheral clock genes through the HPA axis previously developed by our group (9), to a skeletal model of the cyclin network developed by Gerard and Goldbeter (10). Furthermore, we model a population of cells in order to study the influence of cortisol and the circadian clock on the synchronization of the cell cycle. Current model results correctly predict a circadian dependence of the cell cycle, with most cells entering the DNA-replicative S phase towards the end of the light phase and the beginning of the dark phase. Moreover, we find a time-of-day dependent variation in the extent of cell cycle synchronization, with cells being most synchronized in the G1 phase in the early parts of the light phase. Importantly, the model predicts that the extent of synchronization in the population is dependent on the circadian dynamics of cortisol, with increasing cell cycle synchrony during the rising phase of cortisol in comparison to the falling phase. These results emphasize the ability of cortisol and the circadian clock to regulate cell cycle synchronization in a time-of-day dependent manner and provide a framework to assess how deviations from homeostatic cortisol rhythms might perturb the cell cycle.

1. Albrecht U. Timing to Perfection: The Biology of Central and Peripheral Circadian Clocks. Neuron. 2012;74(2):246-60. doi: 10.1016/j.neuron.2012.04.006.

2. Dibner C, Schibler U, Albrecht U. The mammalian circadian timing system: organization and coordination of central and peripheral clocks. Annu Rev Physiol. 2010;72:517-49. doi: 10.1146/annurev-physiol-021909-135821. PubMed PMID: 20148687.

3. Feillet C, van der Horst GT, Levi F, Rand DA, Delaunay F. Coupling between the Circadian Clock and Cell Cycle Oscillators: Implication for Healthy Cells and Malignant Growth. Front Neurol. 2015;6:96. doi: 10.3389/fneur.2015.00096. PubMed PMID: 26029155; PubMed Central PMCID: PMCPMC4426821.

4. Morf J, Schibler U. Circadian cell-cycle progression: Cracking open the gate. Cell. 2010;140(4):458-9. doi: 10.1016/j.cell.2010.02.002. PubMed PMID: 20178739.

5. Masri S, Cervantes M, Sassone-Corsi P. The circadian clock and cell cycle: interconnected biological circuits. Curr Opin Cell Biol. 2013;25(6):730-4. doi: 10.1016/j.ceb.2013.07.013. PubMed PMID: 23969329; PubMed Central PMCID: PMCPMC4573394.

6. Dickmeis T, Foulkes NS. Glucocorticoids and circadian clock control of cell proliferation: at the interface between three dynamic systems. Mol Cell Endocrinol. 2011;331(1):11-22. doi: 10.1016/j.mce.2010.09.001. PubMed PMID: 20833224.

7. Gerard C, Goldbeter A. Temporal self-organization of the cyclin/Cdk network driving the mammalian cell cycle. Proc Natl Acad Sci U S A. 2009;106(51):21643-8. doi: 10.1073/pnas.0903827106. PubMed PMID: 20007375; PubMed Central PMCID: PMCPMC2799800.

8. Freishtat RJ, Watson AM, Benton AS, Iqbal SF, Pillai DK, Rose MC, et al. Asthmatic airway epithelium is intrinsically inflammatory and mitotically dyssynchronous. Am J Respir Cell Mol Biol. 2011;44(6):863-9. doi: 10.1165/rcmb.2010-0029OC. PubMed PMID: 20705942; PubMed Central PMCID: PMCPMC3135846.

9. Mavroudis PD, Corbett SA, Calvano SE, Androulakis IP. Mathematical modeling of light-mediated HPA axis activity and downstream implications on the entrainment of peripheral clock genes. Physiological Genomics. 2014;46(20):766-78. doi: 10.1152/physiolgenomics.00026.2014.

10. Gerard C, Goldbeter A. A skeleton model for the network of cyclin-dependent kinases driving the mammalian cell cycle. Interface Focus. 2011;1(1):24-35. doi: 10.1098/rsfs.2010.0008. PubMed PMID: 22419972; PubMed Central PMCID: PMCPMC3262247.

488641 Superstructure Optimization of Bio-Refineries Using Metabolic-Network Models

Abstract:

Major environmental concerns such as global warming and climate change are attributed to increased atmospheric carbon dioxide due to excessive use of fossil fuels in the past century. Thus, attention has shifted to clean energy resources and renewable fuels. Biomass-derived fuels have proven effective in reducing greenhouse gas emissions [1]. Algal biofuels are among the most promising renewable energy resources due to unique properties of algal species. The resilience of algae to grow in harsh environments is highly desirable. Moreover, algae can grow in wastewater or seawater, reducing the dependence on freshwater recourses. Green algae are highly suitable for second-generation biofuels because their cultivation does not rely on terrestrial and food feedstocks [2]. Since algae are CO_2-consuming photosynthetic organisms, algal biorefineries are usually designed and constructed within carbon sequestration networks (CSNs). CSNs comprise several processing sections, including carbon capture, cultivation, harvesting, extraction, upgrading, and remnant treatment. Technologies used in the cultivation section mostly control the performance of CSNs. However, current cultivation technologies still do not have desirable productivities. Nevertheless, previous studies on the optimization of carbon-sequestration plants showed that a competitive price-per-gallon for biofuel can be achieved if environmental considerations such as global warming and CO2-mitigation are accounted for [3]. However, optimization models were based on approximate empirical correlations for the cultivation section.

Recent advances in genomics have led to the development of metabolic-network-reconstruction techniques, which have proven effective in studies of microbial communities. Using these techniques, quantitative models have been constructed to describe the behaviour of algal and microbial communities under various environmental conditions, such as light intensity and substrate concentrations [4]. In this study, we undertake a superstructure optimization of an algal CSN, including five major processing sections. This model leads to a mixed-integer nonlinear programming problem where the net present value is maximized with respect to design parameters and mass flowrates of the network streams. The genome-scale metabolic network of *C. reinhardtii* is used to model algae growth in the cultivation section. Closed photobioreactors are less susceptible to contamination, offering better control and higher growth rates then open ponds; however, compared to closed reactors, open ponds are easy to construct, inexpensive, and require minimal maintenance, making them desirable choices for large-scale production [3]. Therefore, we adopt open ponds as the technology of choice for the cultivation section. Here, open ponds are modelled as CSTRs, where the consumption rate of CO_2 and production rate of biomass are directly obtained from the metabolic-network model. This furnishes a reliable basis for the coupled optimization of design parameters and flowrates, which, in turn, leads to more realistic economic assessments under various operating conditions, such as light intensity and substrate concentrations.

Â [1] Williams, P. J. L. B., & Laurens, L. M. L. (2010). Microalgae as biodiesel & biomass feedstocks: Review & analysis of the biochemistry, energetics & economics. *Energy & Environmental Science*, *3*(5), 554.

[2] Yue, D., You, F., & Snyder, S. W. (2014). Biomass-to-bioenergy and biofuel supply chain optimization: Overview, key issues and challenges. *Computers & Chemical Engineering*, *66*, 36â€"56.

[3] Gebreslassie, B. H., Waymire, R., & You, F. (2013). Sustainable design and synthesis of algae-based biorefinery for simultaneous hydrocarbon biofuel production and carbon sequestration. *AIChE Journal*, *59*(5), 1599â€"1621.

[4] Chang, R. L., Ghamsari, L., Manichaikul, A., Hom, E. F., Balaji, S., Fu, W.,Â Papin, J. A. (2011). Metabolic network reconstruction of Chlamydomonas offers insight into light-driven algal metabolism. *Mol Syst Biol*, *7*(518), 518.

The Role of Community Structures in Network Control

A Case for the Evolution of Modular Networks in Biology

Presentation on 2017 AIChE Annual Meeting

October 31, 2017

Wentao Tang, Prodromos Daoutidis

Department of Chemical Engineering and Materials Science

UNIVERSITY OF MINNESOTA

1

Networks

- Networks – nodes & edges
- Network science – macroscopic, statistical features

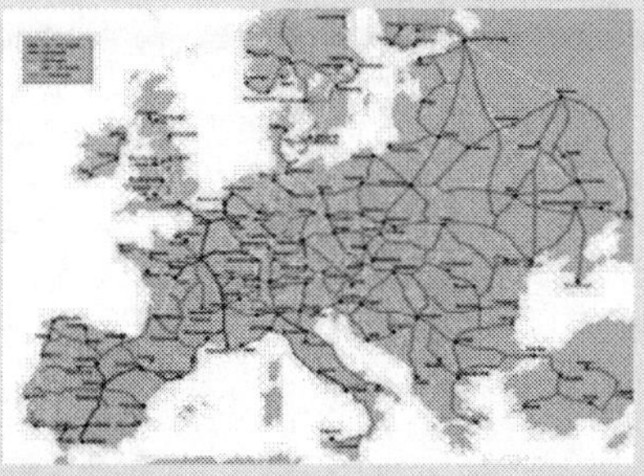

High-speed railway system of Europe

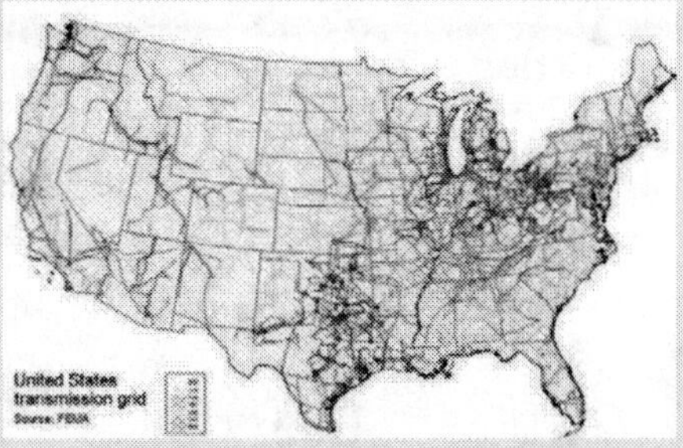

Electric grid of the U.S.

Newman, M. E. J. (2010). *Networks: An Introduction.* Cambridge University Press.
Barabási, A. L. (2016). *Network science.* Oxford University Press.

2

Community structures

- Community structures in biological and social networks
- Community detection: modularity maximization[2-4]

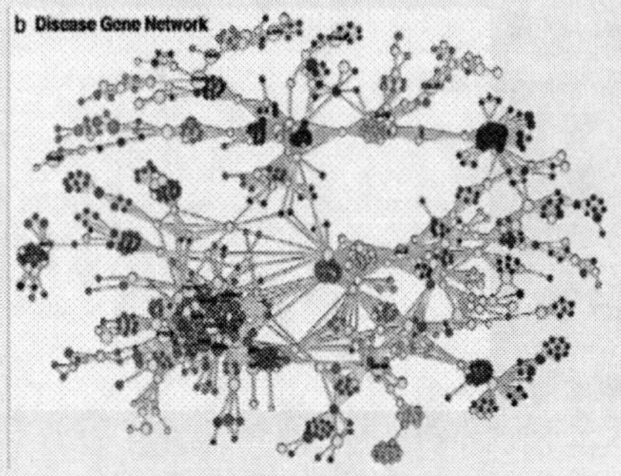

Genes relevant to diseases[1]

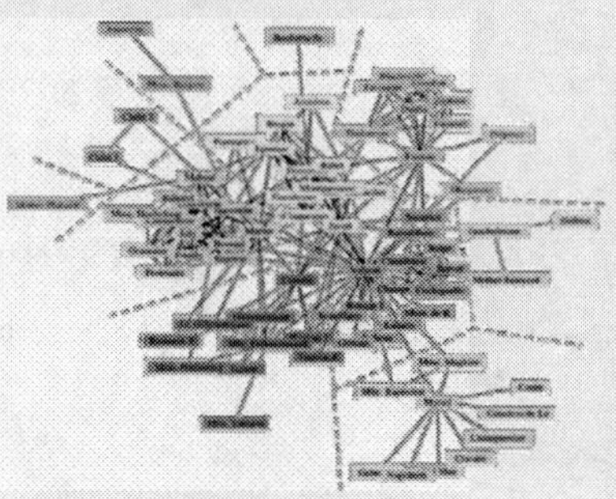

Relationship between characters in Les Misérables[2]

[1] Goh, K. I., *et al.* (2007). *PNAS* 104(21), 8685-8690.
[2] Newman, M. E. J., & Girvan, M. (2004). *Phys. Rev. E* 69(2), 026113.
[3] Newman, M. E. J. (2006). *PNAS* 103(23), 8577-8582.
[4] Fortunato, S. & Hric, D. (2016). *Phys. Rep. 659*, 1-44.

3

Why are networks modular?

- Research in systems biology – Adaptability
- Logical electric circuits = Biological networks
- Rewiring / Modification = Mutation
- Circuits giving prescribed signal = Survivors of natural selection
- Modular networks: easier to adapt

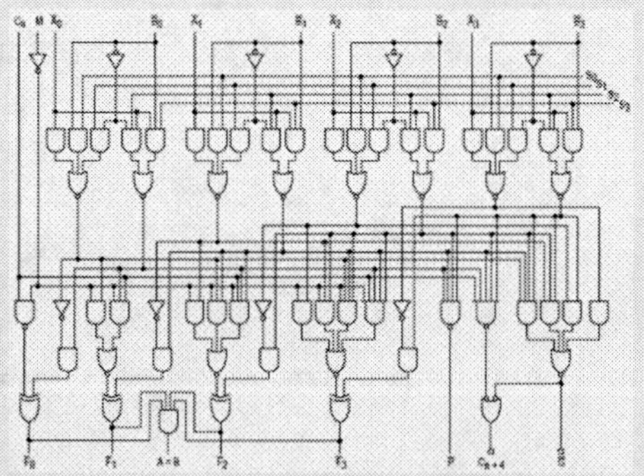

Logical circuits

Kashtan N., & Alon, U. (2005). *PNAS* 102(39), 13773-13779.
Espinosa-Soto, C., & Wagner, A. (2010). *PLoS Comput. Biol.* 6(3), e1000719.
Clune, J., Mouret, B., & Lipson, H. (2013). *Proc. R. Soc. B* 280(1755), 20122863.

4

Perspective of Controllability

- Networks should evolve into more controllable structures
- Hypothesis: Modularity contributes to controllability

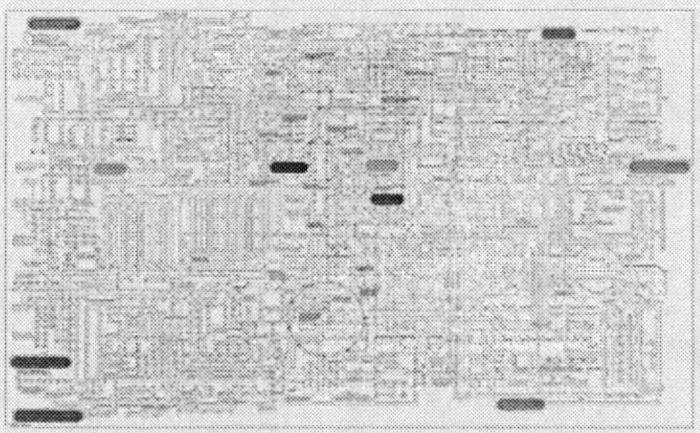

Human metabolic map

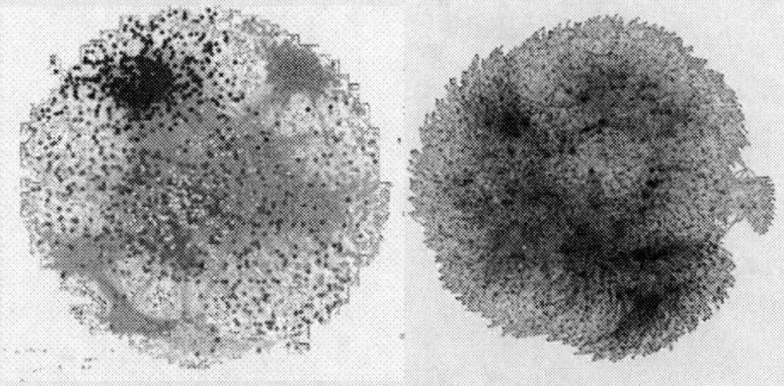

Protein interaction and gene regulatory network

5

Network Generation

- Creating networks with different modularity values (Q)
- Artificial synthesis: simulated annealing (100 nodes, 300 edges)
- Higher modularity = Clear community structures
- Laplacian dynamics

$$\dot{x}_i(t) = -\sum_{j=1}^{n} a_{ij}(x_i(t) - x_j(t)) + u_i(t) + d_i(t)$$

disturbance

difference with neighbors *control input*

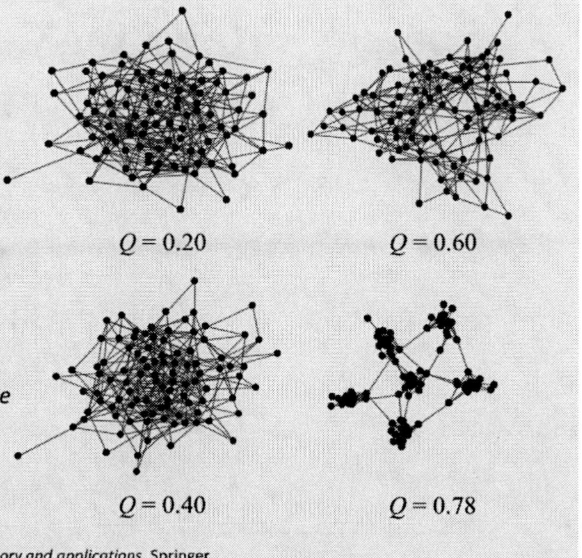

$Q = 0.20$ $Q = 0.60$ $Q = 0.40$ $Q = 0.78$

van Laarhoven, P. J. M., & Aarts, E. H. L. (1987). *Simulated annealing: Theory and applications*. Springer.

6

Sparsity-Promoting Control

Lin, F., Fardad. M., & Jovanović, M. R. (2013). *IEEE Trans. Automat. Control* 58(9), 2426-2431.

$$\min_F \quad J(F) \quad + \gamma \quad \mathrm{card}(F)$$

Feedback gain | Performance: Effect of disturbances | Sparsity: Number of nonzero elements in F

- Tradeoff: control performance versus feedback sparsity
- Parameter γ promotes sparsity

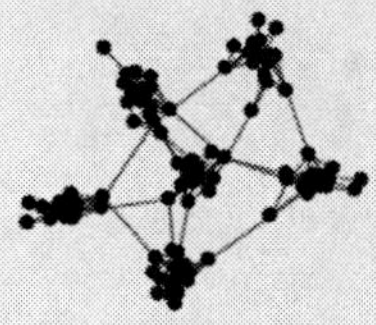

$\gamma = 10^{-6}$ $\gamma = 10^{-4}$ $\gamma = 10^{-2}$

- Sparsity is promoted by compromising inter-community feedbacks rather than intra-community ones.

7

Sparsity-Promoting Control

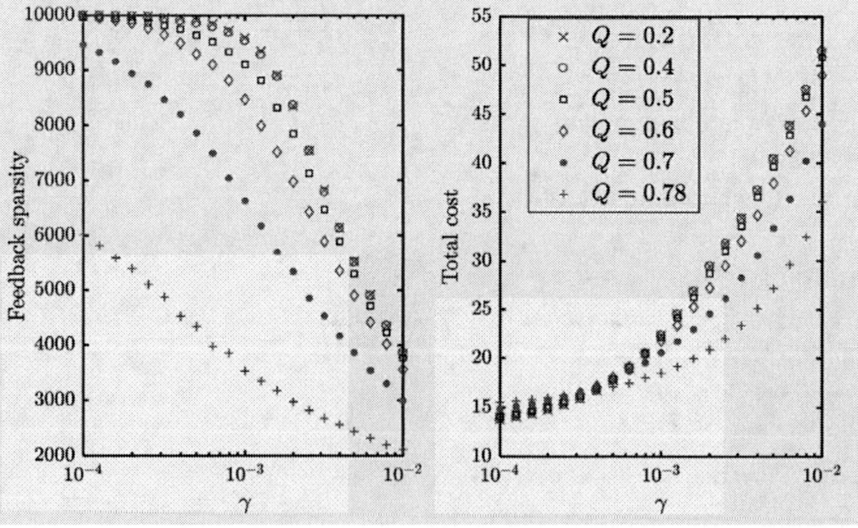

- Sparsity increases with modularity.
- Total cost is reduced by modularity when sparsity cost is significant.

8

Simulated Evolution

- Method: Simulated annealing aiming at decreasing total control cost

> • The correlation of modularity with the cost of control explains the emergence of modularity.

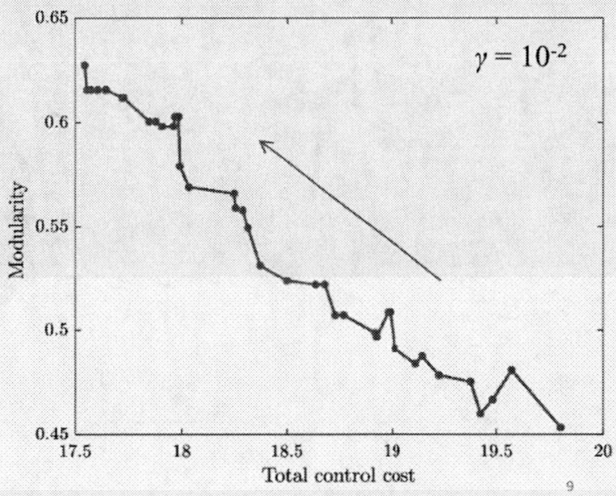

9

Structured (Decentralized) Optimal Control

Lin, F., Fardad. M., & Jovanović, M. R. (2011). *IEEE Trans. Automat. Control* 56(12), 2923-2929.

Control performance under structure constraint

$\mathcal{S} =$

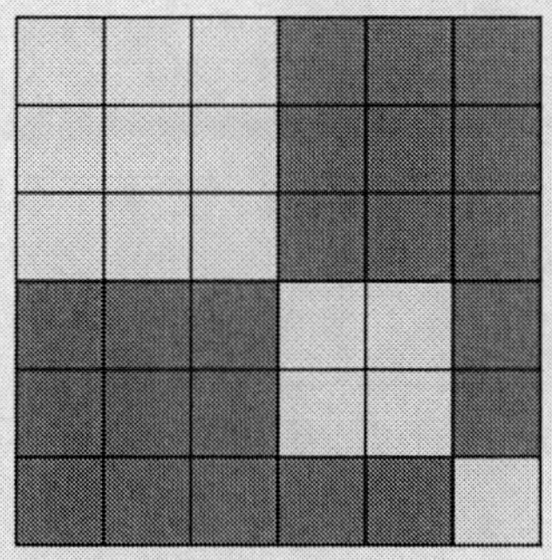

Block diagonal = decentralized control architecture

$$\min_{F} \quad J(F)$$
$$F \circ \mathcal{S} = 0$$

- Similarity of any decentralized structure $\mathcal{S}$ to community structure $\mathcal{S}_0$

$$\text{Similarity} = \text{card}(\mathcal{S}_0 \circ \mathcal{S})/\text{card}(\mathcal{S}_0)$$

- Control performance deterioration of $\mathcal{S}$

$$p = (J - J_0)/J_0$$

10

Structured (Decentralized) Optimal Control

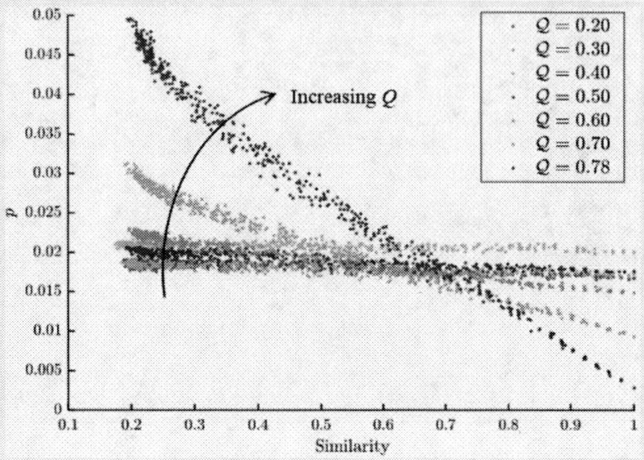

- For non-modular networks, performance is insensitive to the control architecture.
- For modular networks, performance correlates strongly with the architecture.
- For modular networks, communities give the optimal decentralized control.

11

Conclusion

Community structures in networks act as the "core" of feedback control.

→ Sparse controllability by intra-community feedback.
→ Advantage over non-modular networks.
→ Performance-preserving decentralized control.

Evolution of modularity in biological networks

12

Discussions

- Understanding other network features

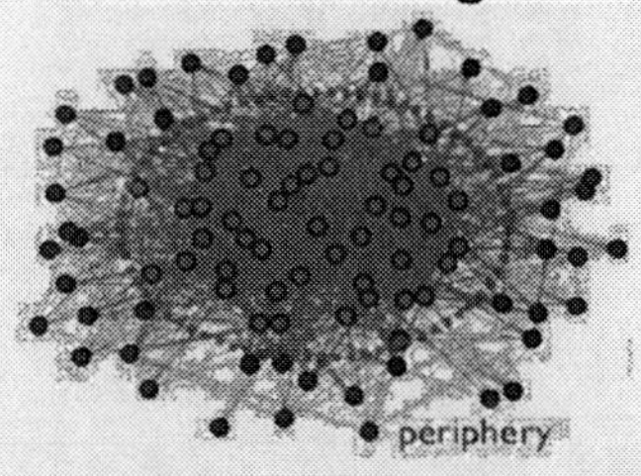

Core-periphery structure[1]

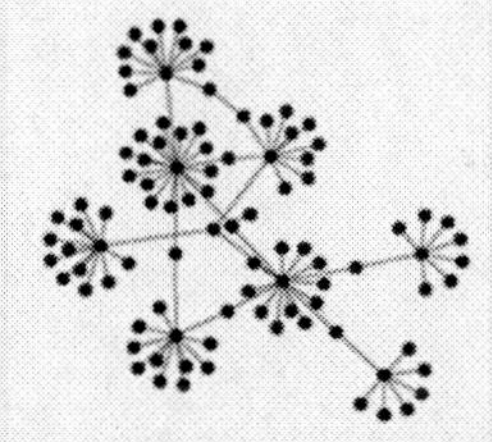

Assortativity/disassortativity[2]

[1] Holme, P. (2005). *Phys. Rev. E* 72(4), 046111.
[2] Newman, M. E. J. (2002). *Phys. Rev. Lett.* 89(20), 208711.
[3] Daoutidis, P., Tang, W., & Jogwar, S. S. (2017). *Comput. Chem. Eng.*, in press.

- Exploiting network modularity for the decomposition of process control[3]

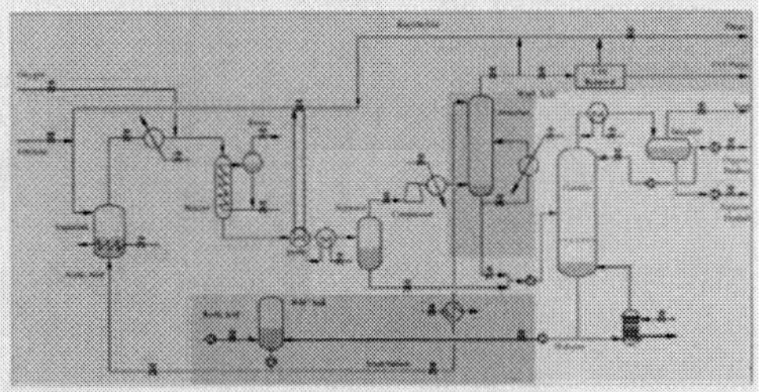

13

Acknowledgment

- Financial support from NSF-CBET

14

THANK YOU

15

491318 K-Ath - Towards a Multi-Tissue Kinetic Model of *Arabidopsis thaliana*

Abstract:

k-ath - Towards a Multi-tissue Kinetic Model of *Arabidopsis thaliana*

Wheaton Schroeder and Rajib Saha

The University of Nebraska – Lincoln, Lincoln, NE

Arabidopsis thaliana has been a model system for modern plant science for the past three decades due to its small genome, short lifecycle, and ease of genetic manipulation. Its prominent role in omics and plant science research makes it an ideal choice for the creation and integration of emerging omics-based tools, including metabolic modeling of metabolism. Such computational modeling of metabolism is now an indispensable tool to drive the processes of understanding, discovering, and redesigning of any biological systems. Although Flux Balance Analysis (FBA) is the primary tool used for this purpose, it has significant limitations due to the lack of reaction kinetics, chemical species concentration, and metabolic regulation. These are critical to the understanding and subsequent design of engineering interventions directed to overproduction of specific bioproducts or improvement of plant performance, particularly in the presence of feedback regulation. Several metabolic models for *A. thaliana* already exist, and most regulatory interactions are well studied and documented. In addition, omics data is widely available for *A. taliana* as a model plant system including genomics, fluxomics, transcriptomics, and metabolomics. Hence, combining and advancing this knowledge in a kinetic model framework will allow us to study the interactions among biosynthesis pathways and individual tissue-types, thus proposing the *in silico* design of useful engineering interventions for *A. thaliana*. Moreover, this research will showcase the value of a plant-scale kinetic model for answering these questions in other plant species.

In this work, we are developing a plant-scale core metabolic network including such pathways as starch and sucrose metabolism; glycolysis and gluconeogenesis; citrate cycle; all amino acid biosynthesis pathways; pyruvate metabolism; the pentose phosphate pathway; photosynthesis and photorespiration pathways; and various transport and uptake pathways. At present, we have completed the draft development of three tissue-specific metabolic models: leaf (282 genes, 109 metabolites, and 191 reactions), seed (274 genes, 153 metabolites, and 222 reactions), and stem (12 genes, 54 metabolites, and 91 reactions). Moving forward, we will develop a root metabolic model, combine these four models in a plant-scale metabolic model, and using fluxomic, metabolomics, and transcriptomic data we will develop kinetic equation form for each reaction. Note that these metabolic models are restricted to core primary metabolism mainly due to the high computational requirements of the kinetic model framework. By decomposing each reaction to mass action kinetics, the rate constants for each mass action step will be determined with respect to fluxomic data, allowing us to solve for the kinetic constants of each kinetic equation in the system. Future work for this project includes sub-dividing existing tissue models, such as stem into xylem and phloem, and adding additional tissue types. Finally, this kinetic modeling framework of a model plant is transferable to other plant systems (provided the omics datasets are available for that species) to provide *in silico* design of useful engineering interventions.

500591 Technical Economic Analysis of an Intensified Integrated Gasification Combined Cycle Plant Design Featuring Membrane and Adsorptive Reactors

Abstract:

Currently, coal-based power plants are the largest source of electricity both globally and in the United States, and despite there being an abundant supply of low-priced natural gas, the Energy Information Administration (EIA) projects that by 2040, coal-based power plants will still provide 254.1 GWe of the generation capacity of the United States, and will account for approximately 36% of the total electricity generation capacity globally1,2. For this reason, coal-based power plants have begun considering the adoption of carbon capture technology in order to reduce the quantity of CO2 emissions, since each year in the United States 6,870 million metric tons of greenhouse gases (GHG) are emitted into the atmosphere, and CO2 represents over 80% of these GHG emissions3.

Because of the environmental concerns these emissions evoke, several carbon capture technologies have been proposed and integrated into these coal power plants with the goal to reduce the CO2 emissions by at least 90%. The main process utilized for carbon separation and capture employs absorption with amine-type solvents, whereby CO2 is separated through the use of a Dual-Stage Selexol unit4.

In this work, the design of an Integrated Gasification Combined Cycle (IGCC) plant featuring membrane and adsorptive reactors is first developed and then analyzed from an economic viewpoint. The design features membrane and adsorptive reactors that have over 90% carbon capture capability without the use of a Dual-Stage Selexol unit. The resultant intensified design is able to produce power more efficiently than traditional IGCC plants with Carbon-Capture Storage (CCS) technology. The UNISIM software (Honeywell trademark) is used to create an intensified flowsheet for the proposed power production plant design, and heat integration studies are presented. A technical economic analysis (TEA) is carried out on the proposed flowsheet to determine economic feasibility of the novel design. The economics of the intensified designs will be compared to a traditional IGCC plant with CCS to identify the extent of process intensification over the baseline design.

1. Annual Energy Outlook 2013, Department of Energy, U.S. Energy Information Administration, 2013.
2. Annual Energy Outlook 2014, Department of Energy, U.S. Energy Information Administration, 2013.
3. Inventory of U.S. greenhouse gas emissions and sinks: 1990-2014. Washington, DC: U.S. Environmental Protection Agency, n.d. Print.
4. Department of Energy/National Energy Technology Laboratory. Cost and Performance Baseline for Fossil Energy Plants Volume 1b: Bituminous Coal (IGCC) to Electricity Revision 2b â€“ Year Dollar Update. Document Number: DOE/NETL- 2015/1727.

501648 An Electrochemical Method to Remove Aqueous Sulfide from Swine Manure

Abstract:

Sulfide is synthesized by sulfur reducing bacteria during the anaerobic degradation of swine manure. The year-long pit storage of swine manure allows the accumulation of sulfide, which eventually results in a high risk of hydrogen sulfide (H 2 S) releasing to the air during the manure pumping. H 2 S is an extremely hazardous and toxic gas; and a number of casualties have been reported due to H 2 S poisoning in the past several years. In order to minimize the sulfide level in swine manure and create a safe working environment, an effective way is presented in this study. By applying a certain voltage onto low carbon steel electrodes partially immersed in the swine manure storage, a 90% or above sulfide removal can be achieved after 1 to 2 weeks treatment. In our experiment, the optimal voltage was investigated to be 0.7 V, which allowed the majority removal of sulfide, and meanwhile a guarantee of less electrode material consumption. The effects of potential parameters, including manure solid content, electrode surface area, distance between electrodes, treatment temperature, biogas composition in the reactor headspace, and sulfide recovery after shutting down the power, were investigated. This electrochemical treatment has shown a significant aqueous sulfide removal from swine manure. Optimization based on mentioned parameters can further improve the treatment feasibility in reality.

499078 Pilot-Scale Testing of Electrochemical Removal of Hydrogen Sulfide in Deep-Pit Swine Manure Storage

Abstract:

The presence of hydrogen sulfide (H_2S) is a widespread issue in organic waste storage, and this toxic air contaminant causes serious concerns over environment, health and safety. One example is the deep-pit manure storage, a common practice for swine facilities in upper-Midwestern states. It occasionally experiences human or animal fatal overexposure to manure gas that contains H_2S. In this study, we developed an electrochemical-based oxidation method to remove or greatly reduce sulfide generation in the stored liquid manure. The treatment system needed some equipment installation and retrofit of the current storage pit with little maintenance requirement, low electricity consumption, and low chemicals usage. Lab-scale experiments were conducted in simulated swine manure storage, and systematically analyzed sulfide removal kinetics at varying total solids contents, initial sulfide concentrations, pH, applied voltages, electrode surface areas, electrode distances, and temperatures. An on-farm pilot-scale study was then performed to further confirm the treatment performance in close-to-real environment, and provide necessary data for techno-economic assessment. An electrochemical treatment device was designed and installed in two manure pump-out accesses in a testing barn that held about 200 growing-finishing pigs. Each pit access was divided to two compartments, one for control and the other for treatment installation. Perforated sheets of two electrode materials of low carbon steel (LCS) and stainless steel Type AISI 304 (SS) were installed in accesses and tested in 97 daysâ€™ operation. From the monitored H_2S concentration in pit headspace during the treatment, it can be seen that the LCS treatment decreased a substantial amount of H_2S concentration, achieving removal efficiency of 56%-78% depending on the applied voltage of between 0.6 and 0.7 V. SS treatment removed 13%-38% of H_2S. LCS treatment required a lower voltage application but achieved better removal performance, and is more economically feasible compared to SS installation according to techno-economic assessment. However, electrode material cost accounted for 76%-87% of the net present cost of the treatment assuming 20 yrsâ€™ life-time of the treatment device. Cheaper material, lower surface area, or more effective material will have to be optimized in order to decrease the treatment cost.

495035 Economic and Environmental Impact Analyses of Integrated Dehydrogenation and Hydroformylation in Gas-Expanded Liquid Media

Abstract:

Economic and Environmental Impact Analyses of Integrated Dehydrogenation and Hydroformylation in Gas-Expanded Liquid Media

Dupeng Liu,[1,2] Raghunath V Chaudhari[1,2] and Bala Subramaniam[1,2]

[1]Department of Chemical & Petroleum Engineering, [2]Center for Environmentally Beneficial Catalysis, University of Kansas

Lawrence, KS 66047

Increased shale gas production in the U. S. has also enhanced the availability of C_2-C_5 alkanes (the so-called natural gas liquids) as collateral products, and opportunities for their utilization in novel ways. For example, propylene production capacity via propane dehydrogenation (PDH) is slated to grow rapidly over the next several years.[1] However, for a typical PDH process, the per pass conversion is only about 50%. The separation of propane and propylene mixtures is highly energy and capital intensive. Previously, we had reported the beneficial effects of propane expanded liquids (with propylene/propane mixtures as feed) for propylene hydroformylation with Rhodium-based catalysts. At sufficiently high pressures (>10 bar), the pressurized propane/propylene mixture forms a gas-expanded liquid phase that promotes regio-selective hydroformylation of propylene with homogenous Rh complex. In addition, compared with conventional propylene hydroformylation process to produce butyraldehyde using polymer grade propylene as feedstock, the feed in the propane-expanded liquid (PXL) process is 50% cheaper.[3]

In this talk, economic and environmental analyses of the integrated dehydrogenation and hydroformylation in PXL process are presented and benchmarked against simulated conventional hydroformylation processes for which data were obtained from the Ruhrchemie/Rh™ne-Poulenc and Dow processes.[4-5] Preliminary simulation results indicate that capital investment and production cost associated with the PXL process is lower than the conventional processes (by up to 40% and 25%, respectively) due to the elimination of the C_3 separation section. Life cycle analysis (LCA) was also performed based on plant scale simulation of conventional and PXL processes. These analyses have provided guidance in catalyst design and in choosing materials and operating conditions that favor process economics while lessening environmental footprint.

1. Nawaz Z. Light alkane dehydrogenation to light olefin technologies: a comprehensive review[J]. Reviews in Chemical Engineering, 2015, 31(5): 413-436.

2.Nawaz Z. Dynamic modeling of catofin¨ fixed-bed iso-butane dehydrogenation reactor for operational optimization[J]. International Journal of Chemical Reactor Engineering, 2016, 14(1): 491-515.

3. Liu, D., Chaudhari, R.V., Subramaniam, B. Kinetic Modeling of Propene Hydroformylation in Propane-Expanded Liquid with Rh-Based Complexes, 2016 AIChE annual meeting San Francisco, CA.

4. Wattimena F, Heijmen H J. U.S. Patent 4,521,630[P]. 1985-6-4.

5. Devon T J, Phillips G W, Puckette T A, et al. U.S. Patent 5,332,846[P]. 1994-7-26.

492359 Bio-Energy with Carbon Capture and Storage (BECCS): Are Inefficient Power Plants a Better Option?

Abstract:

Negative emissions technologies (NETs) in general, and bio-energy with carbon capture and storage (BECCS) in particular, have been identified as key technologies in the context of limiting climate change to no more than 1.5 – 2°C (Fuss *et al.*, 2014). By actively removing CO_2 from the atmosphere, NETs provide a service, and thus investors would expect an appropriate risk-adjusted rate of return, varying as a function of the quantity of public money involved. In addition to providing negative emissions, BECCS facilities also generate firm low carbon power. However, as the electricity sector is integrating an increasing share of intermittent renewable energy (iRE), such as wind and solar, the dispatch load factor of such BECCS facilities may be small relative to their capacity. This has the potential to significantly underutilise these assets for their primary purpose of removing CO_2 from the atmosphere. In this study, we present a techno-economic-environmental evaluation of BECCS plants with a range of operating efficiencies, considering their full- and part-load operation relative to a national-scale annual CO_2 removal target, electricity price and negative emissions credit. We find that in all cases, a lower capital cost, lower efficiency BECCS plant is superior to a higher cost, higher efficiency facility from both environmental and economic perspectives. We show that it may be preferable to operate the BECCS facility in base-load fashion, constantly removing CO_2 from the atmosphere and dispatching electricity in a demand-following manner. We show that the use of this "spare capacity" to produce hydrogen for, *e.g.*, injection to a natural gas system for the provision of low carbon heating, can add to the overall environmental and economic benefit of such a system. The sensitivity of the model to biomass carbon footprint is also presented. The only point where our conclusions are altered is where the CO_2 emissions associated with the biomass supply chain are sufficiently large so as to eliminate BECCS CO_2 removal (Mac Dowell & Fajardy, 2017).

References

Fuss, S., Canadell, J. G., Peters, G. P., Tavoni, M., Andrew, R. M., Ciais, P., Jackson, R. B., Jones, C. D., Kraxner, F., Nakicenovic, N., Le Quere, C., Raupach, M. R., Sharifi, A., Smith, P. & Yamagata, Y. (2014). Betting on negative emissions. Nature Climate Change, 4 (10), 850–853. https://doi.org/10.1038/nclimate2392

Mac Dowell, N., & Fajardy, M. (2017). Inefficient power generation as an optimal route to negative emissions via BECCS. Environmental Research Letters, at press: https://doi.org/10.1088/1748-9326/aa67a5

APPLICATION OF SOME PREDICTIVE THERMODYNAMIC MODELS FOR INDUSTRIALLY IMPORTANT SYSTEMS

Yizu Zhu
Mohammad Shafiei

AIChE Annual Meeting, Minneapolis, MN

October 31, 2017

CHEMISTRY THAT MATTERS™

THERMODYNAMICS CRITICAL TO INTEGRATED PROCESS ENGINEERING & ECONOMICS

Physical Properties play very important roles in sustainable process engineering and in improving companies' bottom line.

Critical phase equilibrium data are unavailable for compounds under many circumstances:

- Design new products
- Screen conceptual design processes
- Scaleup and commercialize new technologies
- Trouble-shoot & optimize manufacturing plants
- Support techno-economics

Reliable "Predictive Equation Of State" models can be very helpful in tackling the above challenges.

No. 1

IMPACT OF PHYSICAL PROPERTIES ON SYSTEM ECONOMICS

A Close Boiling System

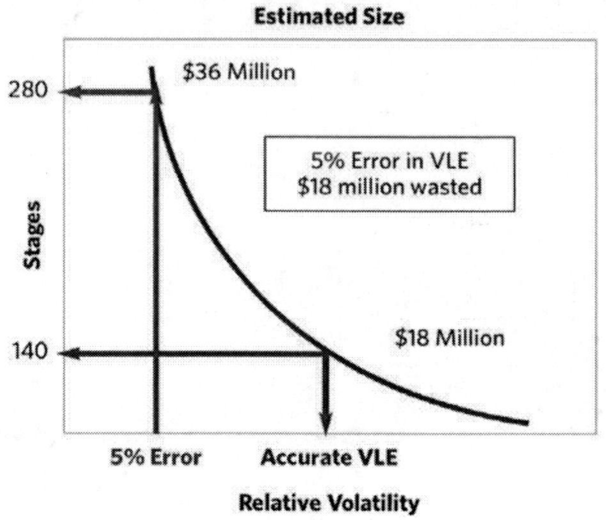

Dhole, et al, Aspen Tech White Paper, 2012

No. 2

IMPACT OF PHYSICAL PROPERTIES ON SYSTEM ECONOMICS

Ammonia Reaction System using Aspen Plus V8.8

	FEED	PRODUCT		
		"Best available" model with parameters regressed from Data	Using A+ PSRK Prediction	Using Ideal Gas Model
Temperature C	450	450	450	450
Pressure bar	90	150	150	150
Vapor Frac	1	1	1	1
Mole Flow kmol/hr	4	3.272	3.263	3.3
Mass Flow kg/hr	34.06	34.06	34.06	34.06
Volume Flow cum/hr	2.67	1.36	1.36	1.32
Enthalpy Gcal/hr	0.012	0.003	0.003	0.003
Mass Flow kg/hr				
NH3		12.39	12.55	11.92
H2	6.05	3.85	3.82	3.93
N2	28.01	17.82	17.69	18.21
Mass Frac				
NH3		0.364	0.368	0.350
H2	0.178	0.113	0.112	0.115
N2	0.822	0.523	0.519	0.535
Difference%		0.00%	1.28%	-3.82%

The productivity impact is ~17.5$MM/year for a 1,000KTA Ammonia plant (Feb 2017 retail price).

No. 3

THERMODYNAMIC CHALLENGES AND PREDICTIVE MODELS

- Industry presents challenges for representing
  - Chemically non-ideal systems
  - Involving supercritical components
  - In multiphase regions at extreme temperature and pressures
  - Specialty chemicals and other systems of rare compounds with no VLE-LLE data

- Equation Of State approach is good for high pressure, but until recently, is not-applicable to highly non-ideal chemical systems.

- New Mixing Rules for EOS emerged in the 1990s to represent the interactions in liquid phases of complex mixtures:
  - Predictive SRK (Holderbaum and Gmehling, 1991)
  - Modified Huron Vidal (MHV2) (Dahl, Fredenslund, Rasmussen, 1991)
  - Wong-Sandler (1992)
  - Etc.

No. 4

PENG - ROBINSON EOS

$$P = \frac{RT}{V_m - b} - \frac{a}{V_m(V_m + b) + b(V_m - b)}$$

Mixing Rule: $a = \sum_i \sum_j x_i x_j (a_i a_j)^{0.5}(1 - k_{ij}), \quad b = \sum_i x_i b_i$

$$a_i = \alpha_i 0.45724 \frac{R^2 T_{ci}^2}{P_{ci}}, \quad b_i = 0.07780 \frac{RT_{ci}}{P_{ci}}$$

$$\alpha_i(T) = [1 + m_i(1 - T_{ri}^{0.5})]^2, \quad m_i = 0.37464 + 1.54226\,\omega_i - 0.26992\,\omega_i^2$$

Binary interaction parameter k_{ij} needs to be regressed from phase equilibrium data

No. 5

APPLICATION OF PREDICTIVE EOS MODELS FOR HIGH PRESSURE AND HIGHLY NON-IDEAL SOLUTION SYSTEMS

SRK: $P = RT/(v-b) - a/v(v+b)$

PSRK's Mixing Rule:

$$b = \sum x_i b_i$$

$$a = b[g^E/A_1 + \sum x_i a_i/b_i + (RT/A_1)\sum x_i \ln(b/b_i)]$$

g^E : Excess Gibbs Energy from UNIFAC Group Contribution Activity Coefficient Model

A_1 : constant

(Holderbaum and Gmehling, 1991; Dahl and Michelson, 1990)

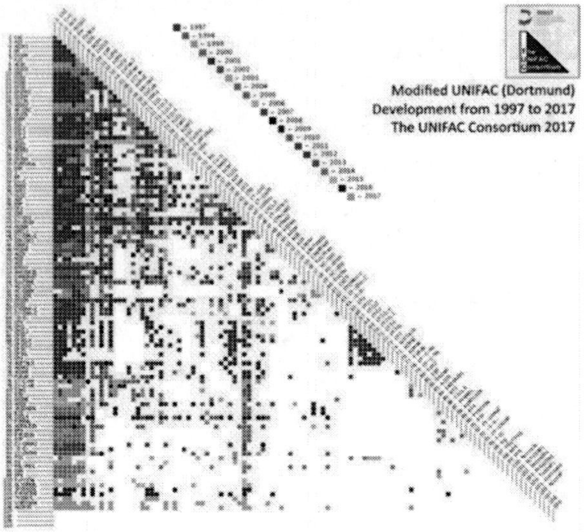

No. 6

KEY FEATURES OF THE NEW MIXING RULES

- A composition-dependent term F(x)
- F(x) is related to Excess Gibbs Energy or Helmholtz Energy
- Allows a single Equation Of State model to represent the phase behavior of:
  - Low Density Fluid (usually by EOS)
  - High Density Fluid (usually by Activity Coefficient Models)
  - Polar Fluid including highly non-ideal LLE systems

No. 7

سابك
sabic

PREDICTION FOR MEOH-H2S PHASE BEHAVIOR

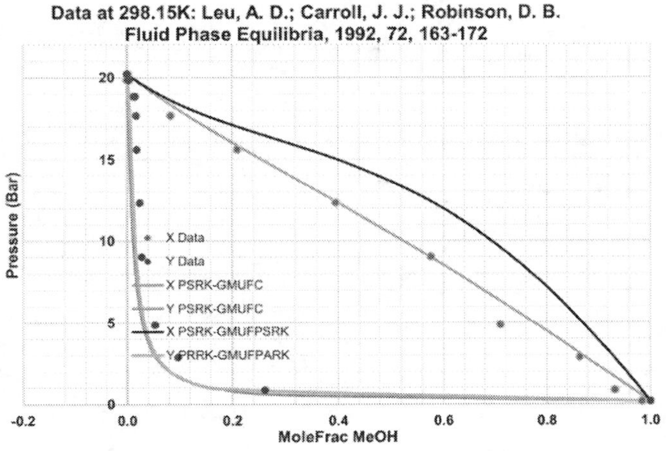

سابك
sabic

METHANE-N-BUTANE P-XY AT HIGH PRESSURE

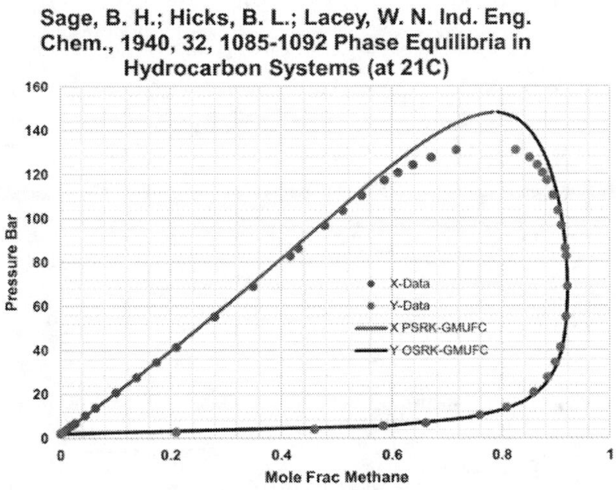

EtAc–HEXANE SYSTEM T-XY PHASE DIAGRAM

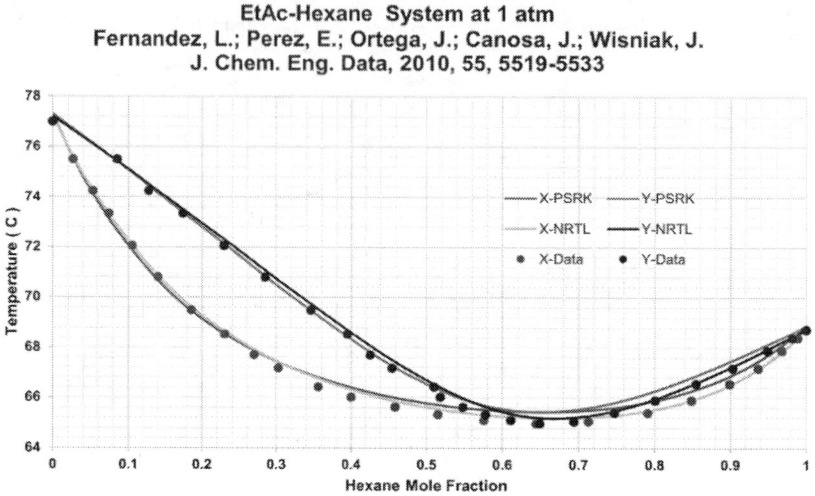

PSRK PREDICTION FOR HIGH P, STRONGLY NON-IDEAL SYSTEMS

MEOH-CO2 SYSTEM

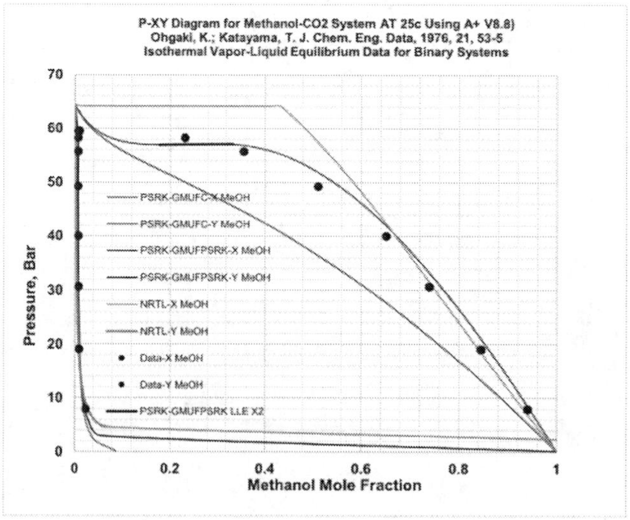

CO2-CYCLOHEXANE SYSTEM AT HIGH PRESSURE

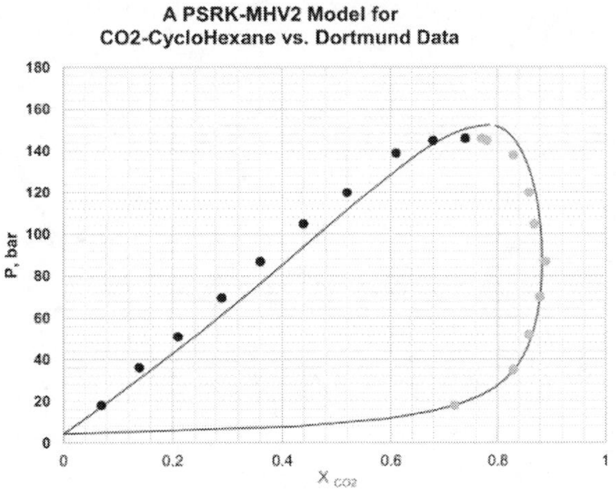

No. 12

NH3-WATER SYSTEM AT TYPICAL PLANT OPERATION CONDITION

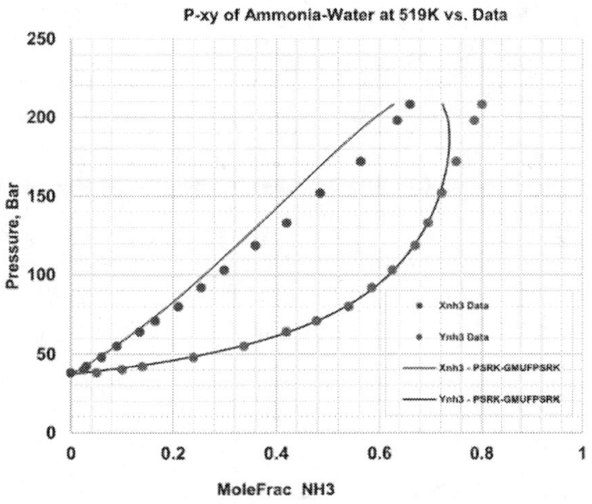

No. 13

N-BUOH – WATER HETEROGENEOUS AZEOTROPE PREDICTION

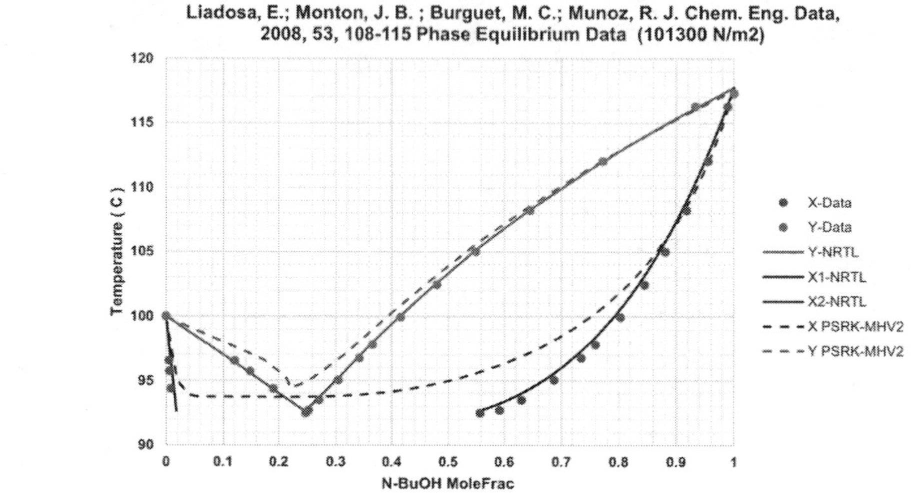

No. 14

SEPARATION IN AN ETHYLENE PLANT WITH PSRK

	Tower Reflux	Tower Bottom Liquid Rate	Tower Top Temperature
	t/h	t/h	C
Plant Data	12.00	52.30	-98.50
RKS	12.31	51.00	-100.10
PSRK	12.99	52.40	-99.50

No. 15

TEMPERATURE IMPACT ON HYDROCARBON CONCENTRATION IN RESIN USING POLYSRK FOR PRODUCT QUALITY IMPROVEMENT AND CAPACITY INCREASE

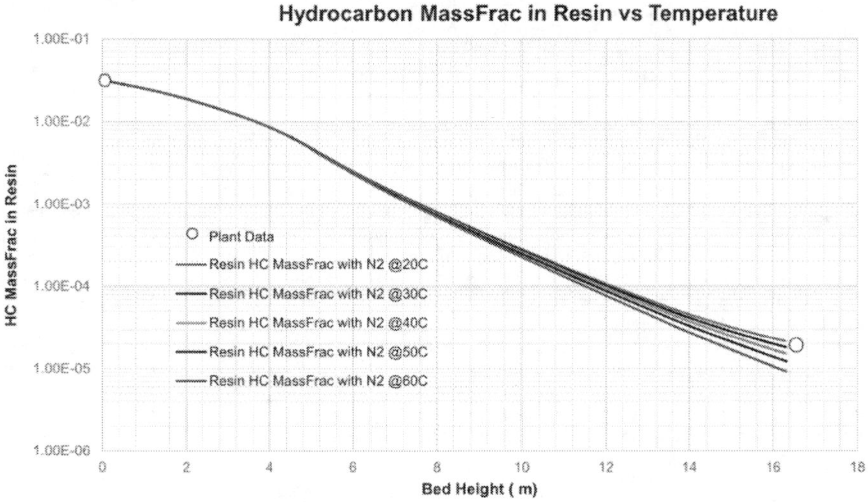

No. 16

CONCLUSIONS

- Physical properties have critical impact to the economics of process design and manufacturing operations.
- Predictive Equation of State models provide a powerful tool in representing highly non-ideal systems under high pressure where data are unavailable and other types of thermodynamic models are unable to handle.
- Predictive EOS models can be used with Excess Gibbs Free Energy of various liquid activity coefficient models available in simulation software such as Aspen Plus where binary interaction parameters are available (such as NRTL, WILSON, UNIQUAC, etc).
- This provides a powerful tool to handle a variety of complex systems where data are sparsely available:
  - Conceptual design of new products and processes
  - R&D
  - Scale-up and commercialization of innovative technologies
  - Manufacturing retrofitting and performance optimization

No. 17

DISCLAIMER

DISCLAIMER: THE MATERIALS, PRODUCTS AND SERVICES OF SAUDI BASIC INDUSTRIES CORPORATION (SABIC) OR ITS SUBSIDIARIES OR AFFILIATES ("SELLER") ARE SOLD SUBJECT TO SELLER'S STANDARD CONDITIONS OF SALE, WHICH ARE AVAILABLE UPON REQUEST. INFORMATION AND RECOMMENDATIONS CONTAINED IN THIS DOCUMENT ARE GIVEN IN GOOD FAITH. HOWEVER, SELLER MAKES NO EXPRESS OR IMPLIED REPRESENTATION, WARRANTY OR GUARANTEE (i) THAT ANY RESULTS DESCRIBED IN THIS DOCUMENT WILL BE OBTAINED UNDER END-USE CONDITIONS, OR (ii) AS TO THE EFFECTIVENESS OR SAFETY OF ANY DESIGN OR APPLICATION INCORPORATING SELLER'S MATERIALS, PRODUCTS, SERVICES OR RECOMMENDATIONS. UNLESS OTHERWISE PROVIDED IN SELLER'S STANDARD CONDITIONS OF SALE, SELLER SHALL NOT BE RESPONSIBLE FOR ANY LOSS RESULTING FROM ANY USE OF ITS MATERIALS, PRODUCTS, SERVICES OR RECOMMENDATIONS DESCRIBED IN THIS DOCUMENT. Each user is responsible for making its own determination as to the suitability of Seller's materials, products, services or recommendations for the user's particular use through appropriate end-use and other testing and analysis. Nothing in any document or oral statement shall be deemed to alter or waive any provision of Seller's Standard Conditions of Sale or this Disclaimer, unless it is specifically agreed to in a writing signed by Seller. Statements by Seller concerning a possible use of any material, product, service or design do not, are not intended to, and should not be construed to grant any license under any patent or other intellectual property right of Seller or as a recommendation for the use of any material, product, service or design in a manner that infringes any patent or other intellectual property right.

No. 18

THANK YOU AND Q&A

No. 19

504433 Multiscale Optimization Strategies for the Integrated Design, Planning and Scheduling of Process Systems

Abstract:

Chris Floudas devoted the last decade of his life in the multiscale optimization of energy systems. Multiscale optimization is a very challenging problem that requires novel representations and effective solution strategies. We explore in this presentation several strategies for the integration of design, planning and scheduling in multiscale process systems.

In this presentation, we first give a general overview of multilevel decision making in the chemical process industry, in which strategic, tactical and operational decisions must be integrated. We discuss a network structure that helps to identify both the material and information flows that interconnect models associated at each level to facilitate their integration. We first start with integration and planning in batch multiproduct plants with sequence dependent changeovers. We show how the incorporation of traveling salesman constraints in an MILP planning model can capture the main effects of changeover at the detailed scheduling level. We next address the long term planning of electric power systems in which the challenge lies in the spatial and temporal multiscale nature of the problem. We describe strategies based on spatial aggregation and time sampling that allow effective long term strategic planning through a multiperiod MILP model for the selection of coal, natural gas, nuclear, wind and solar technologies for power generation over a very long term horizon. Finally, we address the multiscale production routing problem, which considers the coordination of production, inventory, distribution, and routing decisions in multicommodity supply chains with complex production facilities. We describe an iterative MILP model involving two different time grids, one at the finer level for the detailed mode-based production scheduling model, and one for a restricted vehicle routing subproblem that is considered in each time period of the coarse time grid. For each of the three cases, we report computational results that illustrate the effectiveness of the proposed multiscale optimization strategies.

Memorial Paper to Remember Chris Floudas

Supercritical CO_2 in Production of Biodiesel from Algae: Multiscale Processing

Warren D. Seider
Cory S. Silva
Geetanjali Yadav
University of Pennsylvania

Lindsay Soh
Lafayette College

Julie Zimmerman
Yale University

Oct. 31, 2017

Chris arrived at Princeton (~1993)
we talked about multiphase equilibrium -
A thermodynamics challenge - not just $Q = UA\Delta T_{LM}$ in heat-exchanger networks

const. T and P:

$$\min_{n} G = \sum_{\ell=1}^{P} \sum_{i=1}^{C} n_{i\ell} \bar{G}_{i\ell}$$

$$\text{ST: } n_{if} = \sum_{\ell=1}^{P} n_{i\ell} \qquad \forall\, i$$

$$n_{i\ell} \geq 0 \qquad \forall\, i, \ell$$

$$\text{Vapor - } \bar{G}_{il} = \frac{\partial G}{\partial n_{il}} = \mu_{il} = G_{il} + RT \ln \frac{\phi_{il} P y_{il}}{f_{il}}$$

$$\text{Liquid - } \bar{G}_{il} = \quad = G_{il} + RT \ln \frac{\gamma_{il} P_i^s x_{il}}{f_{il}}$$

or using Eq. of State

$$= G_{il} + RT \ln \frac{\phi_{il} P x_{il}}{f_{il}}$$

Challenge – to minimize G and properly locate the tangent plane

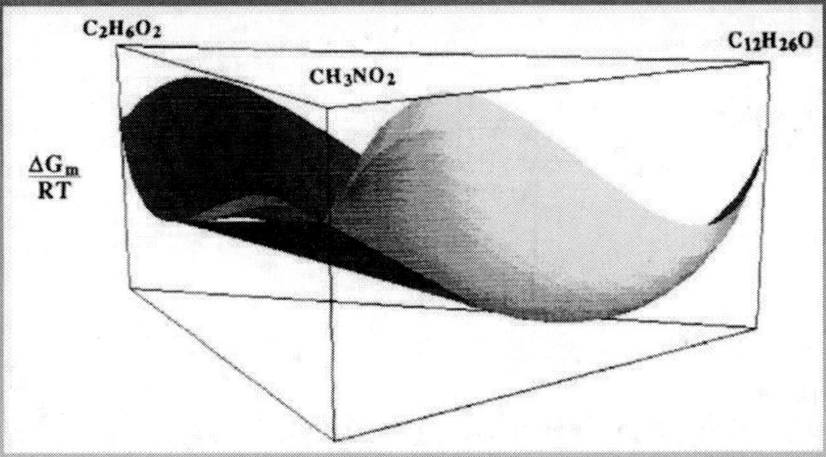

Often do not locate the global minimum in G – obtain incorrect number of phases

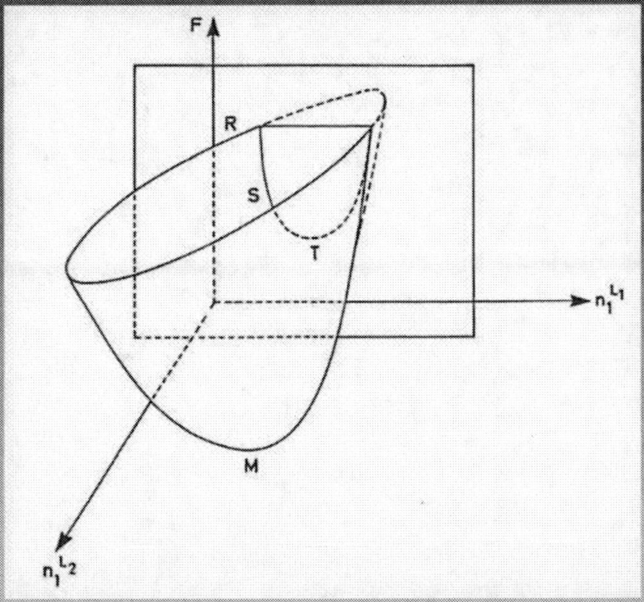

Need phase stability analysis -

Michelsen (1982) - Must check the tangent-plane distance function

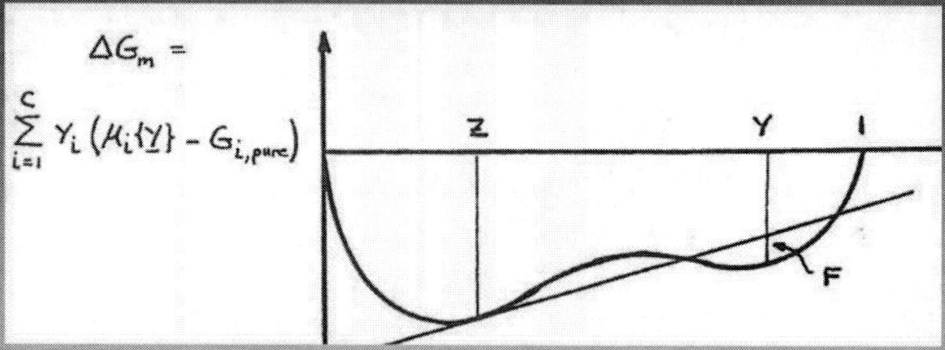

For phase stability:

$$F\{\underline{Y}\} = \sum_i Y_i \left(\mu_i\{\underline{Y}\} - \mu_i\{\underline{Z}\}\right) \geq 0 \qquad \forall Y$$

A global optimization expert, Floudas advised Conor McDonald:

For the UNIFAC and ASOG group contribution methods, they introduced a branch-and-bound global optimization method to:

1. Minimize the Gibbs free energy, and
2. Minimize the tangent-plane distance function

Implemented in their GLOPEQ package

Subsequently, their algorithm was implemented for the NRTL and UNIQUAC equations

More recently, Floudas advised Rick Baliban on methods involving the synthesis, global optimization, and topology analysis of processes for the conversion of biomass and natural gas to liquid transportation fuels

They produced synthesis gas either:

1. Directly using Fischer-Tropsch synthesis, or
2. Indirectly using catalytic conversion of methanol using ZSM-5

and carried out process synthesis including simultaneous heat, power, and water integration – with constraints to produce 50% less CO_2 emissions than a petroleum-based refinery

Supercritical CO2 in Production of Biodiesel from Algae 7

Our energy venture – the NAABB project to produce biofuel from algae; i.e., biodiesel from algae

Using supercritical CO_2 to accelerate the rate of transesterification of triolein with methanol, and CO_2 microbubbles to lyse the cell walls –

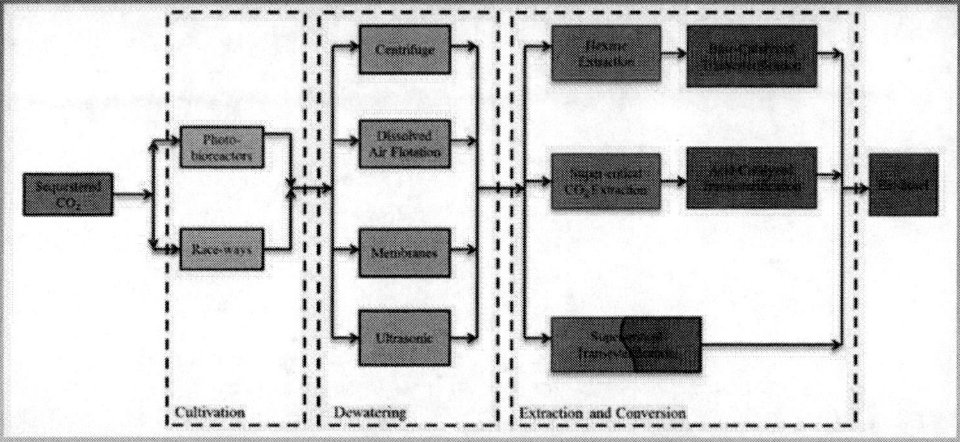

Supercritical CO2 in Production of Biodiesel from Algae 8

Accelerate the transesterification reactions

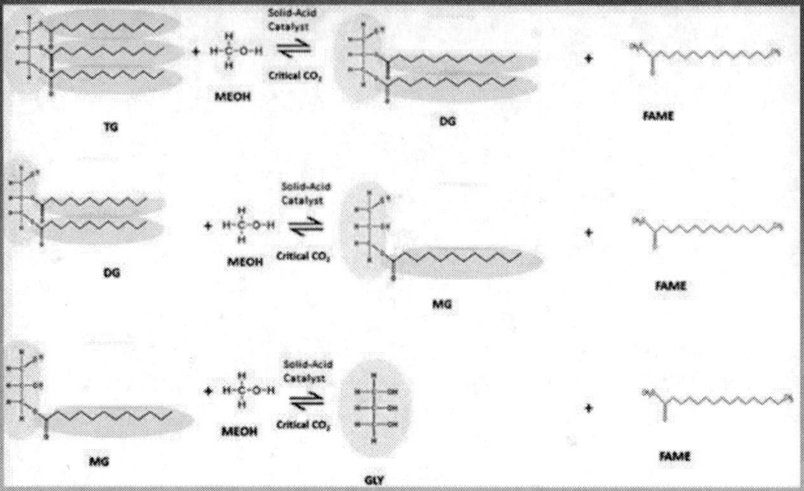

and optimize a model that requires multiphase equilibrium calculations using the SAFT equation-of-state as the reactions proceed.

Prior Work –

High pressure CO_2 (9.65 MPa) was experimentally found to be effective and selective in converting algae oil triglycerides (Soh et al., 2015)

Microbubbles look promising for lysing the cell walls (Origin Oil, 2011; and W. Zimmerman and Yadav).

CO_2 was an effective co-solvent in transesterification reactions –

yielding high conversion at relatively low T

Soh et al. (2014, 2015) carried out experiments for the trans-esterification of triolein to methyl-oleate. Only total concentration data were measured for triolein, diolein, monolein, and methyloleate species.

In his work (Silva et al., 2015), calculations included H_2O because algae cells are grown in aqueous raceways.

He worked with a relatively simple system:

Eight species assumed to be present:

Triolein, diolein, monolein, water, methanol, methyl-oleate, glycerol, and CO_2

Objective – to model multiphase equilibria properly using the SAFT EOS:

PC-SAFT – in ASPEN PLUS
SAFT-γ Mie – in gPROMS

Note in prior work, crude models, like the Redlich-Kwong EOS in ASPEN PLUS, were used –

but are not suitable in the critical region

Up to 3 phases considered –

vapor, polar liquid, non-polar liquid

must estimate phase concentrations correctly to obtain accurate conversions and reaction rates

Supercritical CO2 in Production of Biodiesel from Algae 11

Using PC-SAFT in ASPEN PLUS and gSAFT in gPROMS

Ternary LLE Mixtures – typical of effluent streams from biodiesel reactors – data from an ASPEN PLUS database

Water, Methanol, Methyl-oleate – 318.2 K, 1 atm

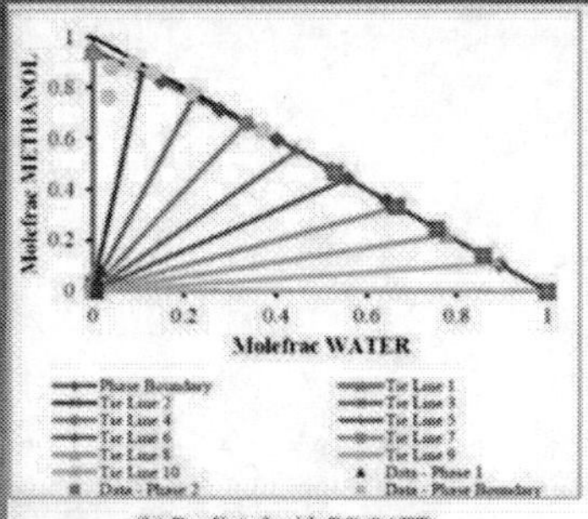

(b) Predicted with PC-SAFT

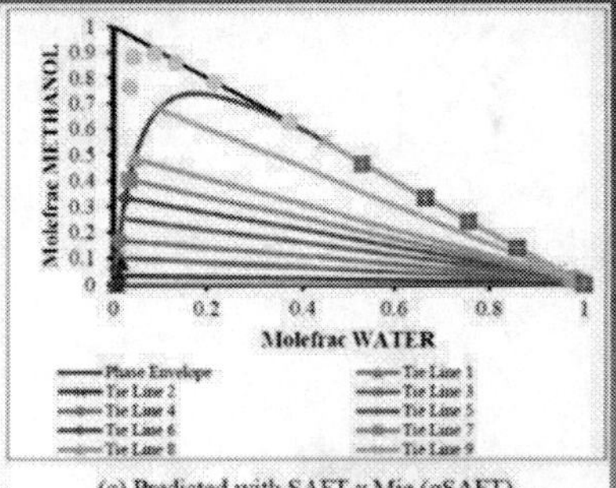

(c) Predicted with SAFT-γ Mie (gSAFT)

Good agreement of both SAFT EOS – but some deviations at high methanol and methyl-oleate conc. – gSAFT does best for phase envelope

RK-ASPEN – significant deviations at high water conc. – doesn't accurately represent strongly associating mixtures

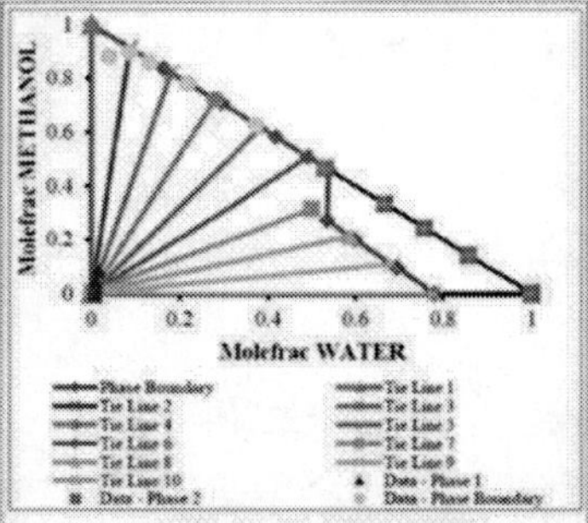

(a) Predicted with RK-ASPEN

Supercritical CO2 in Production of Biodiesel from Algae 12

Glycerol, Methanol, Methyloleate – 333 K, 1 atm

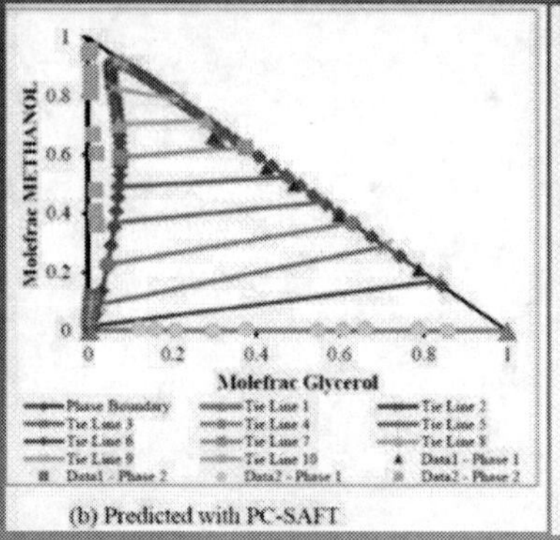

(b) Predicted with PC-SAFT

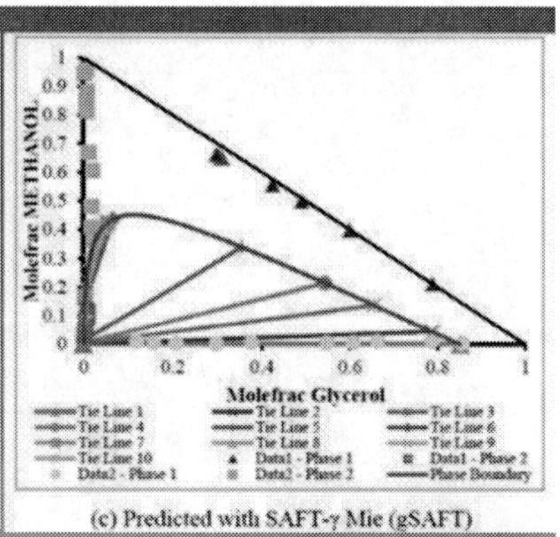

(c) Predicted with SAFT-γ Mie (gSAFT)

Good agreement with PC-SAFT – but not gSAFT

Again, poor agreement with RK-ASPEN

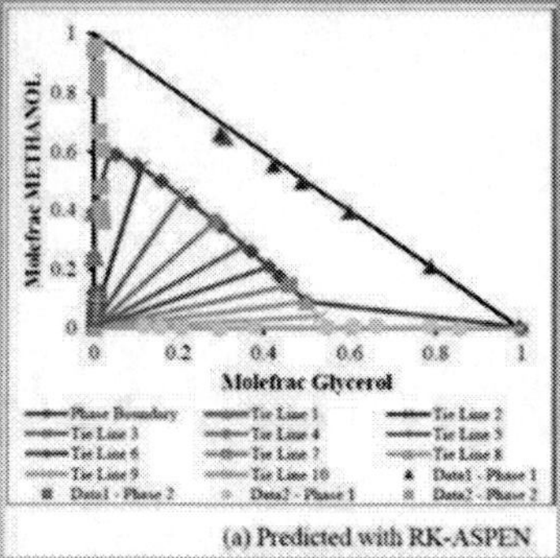

(a) Predicted with RK-ASPEN

Supercritical CO2 in Production of Biodiesel from Algae 13

Ternary Inlet Mixtures – typical for a biodiesel reactor using CO_2 to increase conversion rates

Triolein, Methanol, CO_2 – 353 K (
– large excess of methanol to speed-up lab reactor)

2014) –
orange circles – from supercritical to LLE
grey crosses – from LLE to VLLE

Vapor, Liquid, VLE, LLE, VLLE entries computed using *Gibbs flash method (RGIBBS)* in ASPEN PLUS

Open spaces – no convergence

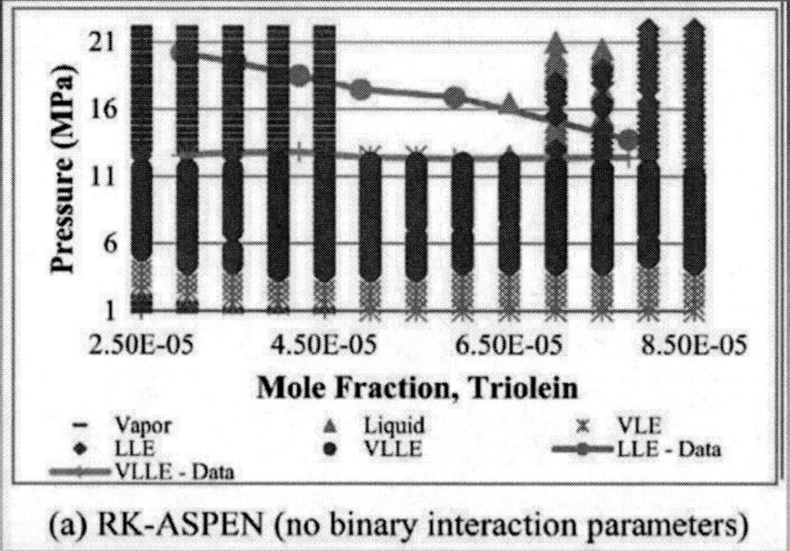

(a) RK-ASPEN (no binary interaction parameters)

Supercritical CO2 in Production of Biodiesel from Algae 14

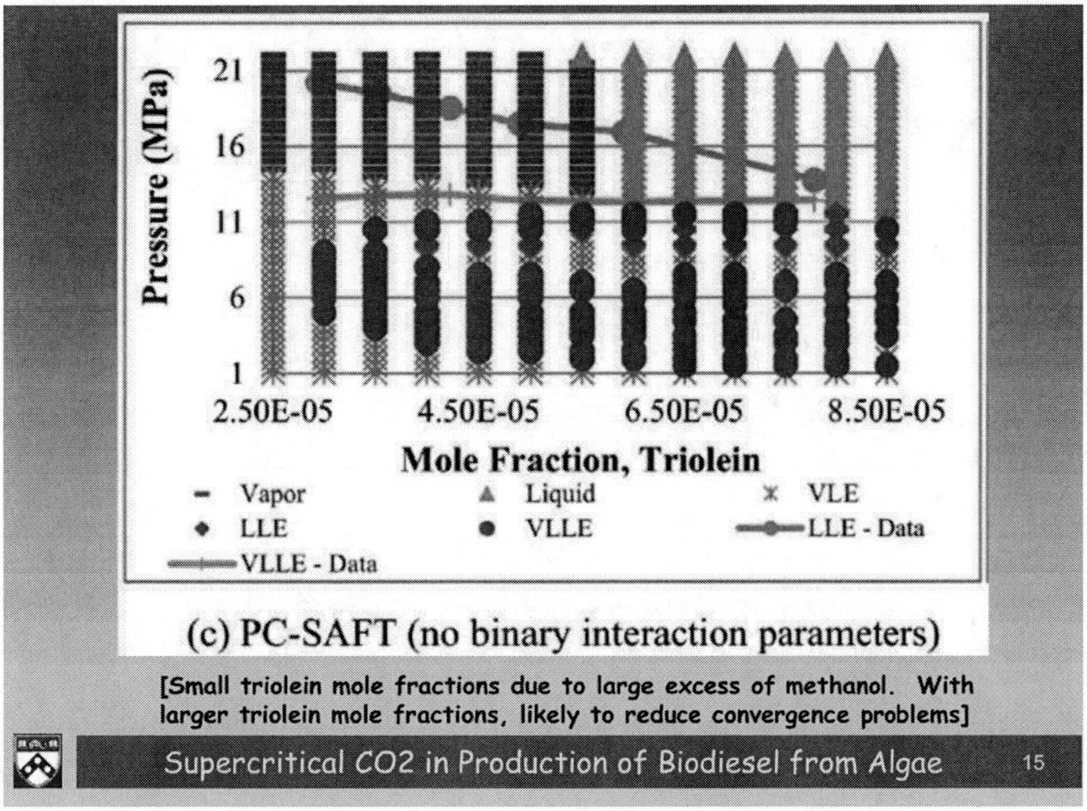
Pressure (MPa)
21
16
11
6
1
2.50E-05
4.50E-05
6.50E-05
8.50E-05
Mole Fraction, Triolein
Vapor
Liquid
VLE
LLE
VLLE
LLE - Data
VLLE - Data
(c) PC-SAFT (no binary interaction parameters)
[Small triolein mole fractions due to large excess of methanol. With larger triolein mole fractions, likely to reduce convergence problems]
Supercritical CO2 in Production of Biodiesel from Algae
15

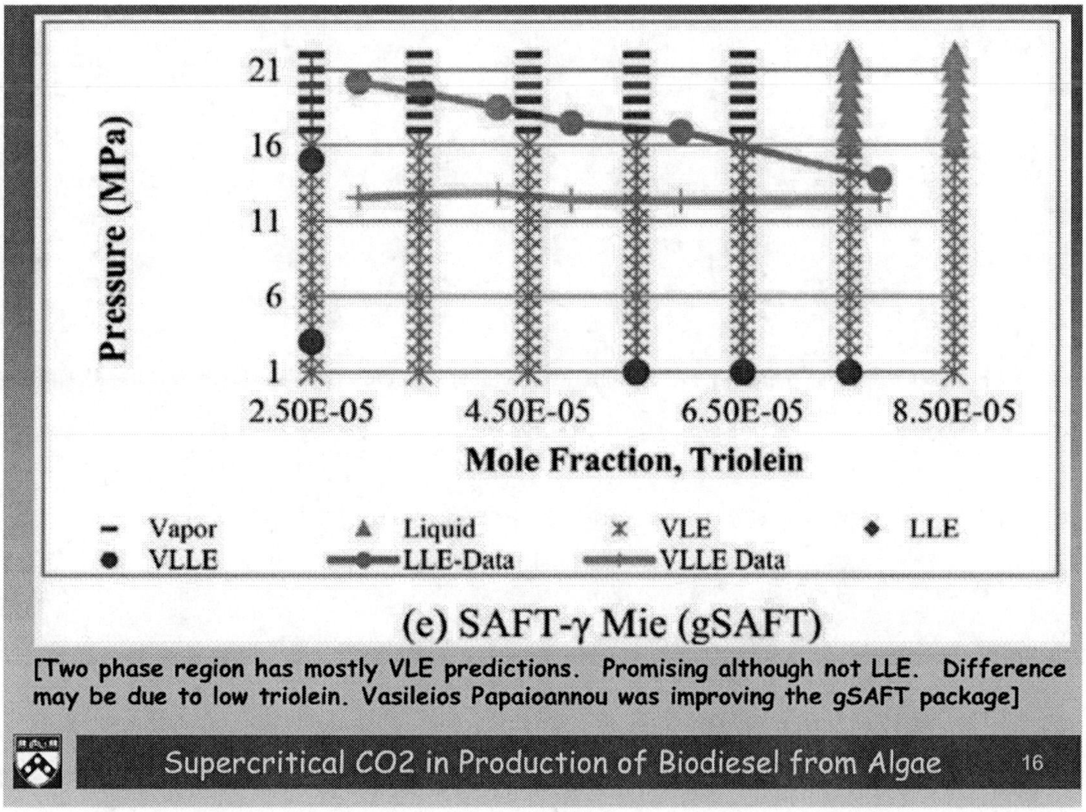
Pressure (MPa)
21
16
11
6
1
2.50E-05
4.50E-05
6.50E-05
8.50E-05
Mole Fraction, Triolein
Vapor
Liquid
VLE
LLE
VLLE
LLE-Data
VLLE Data
(e) SAFT-γ Mie (gSAFT)
[Two phase region has mostly VLE predictions. Promising although not LLE. Difference may be due to low triolein. Vasileios Papaioannou was improving the gSAFT package]
Supercritical CO2 in Production of Biodiesel from Algae
16

Tangent-plane-distance regions of stability and instability (in two polygons) - computed using FORTRAN program by Prof. Romain Privat

Triolein, Methanol, CO_2 at 353.15 K - phase equilibria computed using

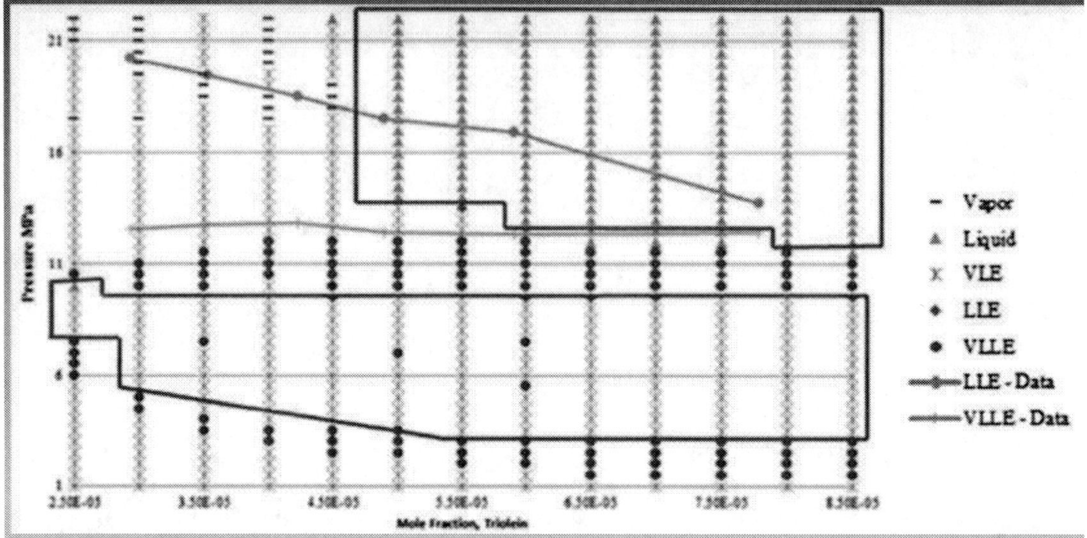

Currently - we're planning to:

recompute phase equilibria using PC-SAFT and gSAFT - 2-years later. Seek to help improve the results.

take experimental data for binary pairs involving diolein and monolein to obtain binary interaction coefficients - seek to improve accuracy of LLE and VLLE calculations

take expt. data at higher ratios of triolein/methanol - seek triolein phases that dissolve significant methanol concentrations

Examine microbubble power requirements and effectiveness, understanding

impact of bubble generation rate; e.g., ~20,000 bubbles/sec

local pressure increases as bubble radius decreases

bubble radius pulsation with sonic waves

cavitation accompanying bubble collapses

Will improve calculations for batch transesterification reaction involving triolein and methanol with CO_2 (Silva et al., 2016) using 3-reaction kinetic mechanism

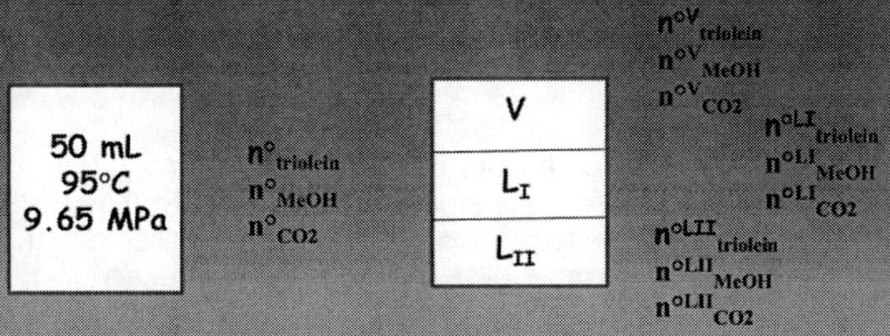

First, minimize G to obtain phase equilibria; i.e., the moles of each species in each phase and phase volumes

For each phase (V, L_I, L_{II}), compute the intrinsic rate of reaction substituted in the mass balances:

$$\frac{dC_{ij}}{dt} = \nu_{R_{ij}}\, r_j$$

where i is the species counter in reaction (R_j), j = 1, 2, 3.

For each phase (V, L_I, L_{II}), integrate over the next time-step, h:

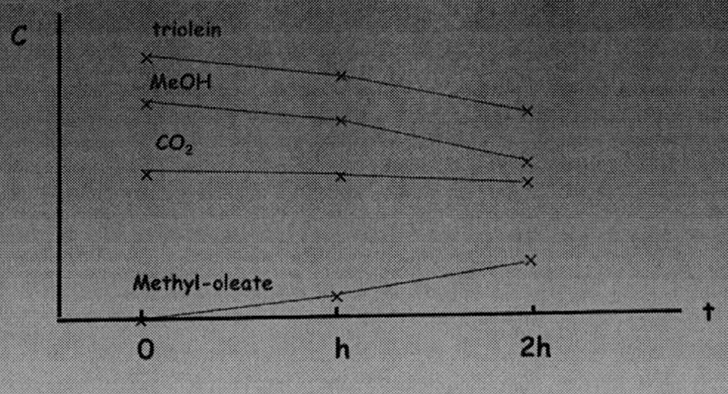

At each sampling time, calculate multiphase equilibrium to determine phase distribution. Compare with experimental data.

501635 A Computational ODE Model for the Evaluation of Complement System Activation, Function, and Regulation in Homeostasis and Disease

Abstract:

The complement system is an ancient form of innate immunity that eliminates invading pathogens and maintains host homeostasis. It is a very effective defense mechanism that is constantly on the "look out" but once activated unleashes a local inflammatory response capable of rapidly removing the stimulus. However, due to the potency of the complement system, regulators present in fluid phase and bound to membranes are also constantly on the "look out" for improper complement activation on host cells. Once the delicate balance between activation and regulation is compromised, a cascade of host cell damage follows leading to autoimmune diseases. Thus, understanding the dynamics involved in the biochemical reactions of the complement system is crucial. In light of this, we have a developed a comprehensive quantitative model of the complement system that incorporates all pathways (alternative, properdin, classical, lectin, and common) and their respective membrane bound and fluid phase components. Our model demonstrates the action of the complement system against invading pathogens, and the coordination of complement activators and regulators that shields host cells from improper complement attack. The insights obtained from this model may aid in the development of patient-specific diagnostic tool for pathological conditions that involve complement deficiencies, polymorphisms, deletions or mutations. Furthermore, the model makes it possible to determine points of intervention with drugs, and to evaluate the efficacies of single drugs or combinations of drugs targeting different complement proteins, protein fragments, or protein complexes. We will present a unified model that consists of a system of over 500 ordinary differential equations, which functions as a basis to derive autoimmune disease-specific models. We will also present basic mechanistic insights for selected complement-mediated diseases, and the effects of drugs under development in our lab for these diseases.

505328 Optimal Distribution of Byproduct Gases, Steam and Power in an Iron and Steel Plant

Abstract:

Byproduct gases, steam and electricity play an important role in providing energy for production units in the iron and steel industry. Optimal distribution of byproduct gases, steam and electricity in an iron and steel plant can significantly decrease energy cost and reduce CO_2 emissions. However, such optimal distribution is not trivial because it involves many production units, steam turbines, combined heat and power units, and waste heat and energy recovery units, and several realistic operational features such as byproduct gas mixing, gasholder level control, minimum heating and energy requirement, and maximum allowable burner switches, resulting in a large complex combinatorial problem. In this presentation, we develop a novel multi-period mathematical model for optimal distribution of byproduct gases, steam, and power in an iron and steel plant. The consuming rates of byproduct gases are variable. Different byproduct gases are allowed to be mixed in order to satisfy minimum heating and energy requirement of production units. Electricity purchase and sale decisions with each having different price are correctly modelled with new binary variables. The burner switching operation is correctly modelled with fewer binary variables allowing turning on and off at any time. Several important practical features such as fuel selection, gasholder level control, ramp rate variation, piecewise constant generation rates of byproduct gases, and piecewise constant demand profiles of byproduct gases, steam and electricity are also incorporated. The computational results demonstrate that the optimal operating cost is obtained within 2 CPU seconds for an industrial example using the proposed model, which is reduced by 6.25% compared to that from actual operation.

505372 Global Optimization of Metabolic Reaction Networks

Abstract:

Modeling and analysis of metabolic networks have been central to research in systems biology, biotechnology, and medical sciences. The core methods in these investigations involve optimization methods which formulate a metabolic objective for the function of the metabolic networks (subject to constraints that are defined by the stoichiometry and the data from transcriptomics, proteomics, metabolomics, and physiology), and they analyze their properties under the optimal operation of these objectives. We will present a class of metabolic objectives and their formulations as global optimization problems. We will discuss how the solution of these problems can provide insights into the function of the metabolic networks and can guide the design of metabolic engineering and the identification of drug targets.

1

Virtual Metrology as a Big Data solution to Semiconductor Manufacturing

Kerul Suthar
Devarshi Shah
Q. Peter He

Department of Chemical Engineering
Auburn University

2

Outline

- Motivation
- Virtual Metrology (VM)
- Existing VM approaches
- Statistics pattern analysis (SPA)
- Industrial case study
- Conclusions

3

Motivation-Big data challenges

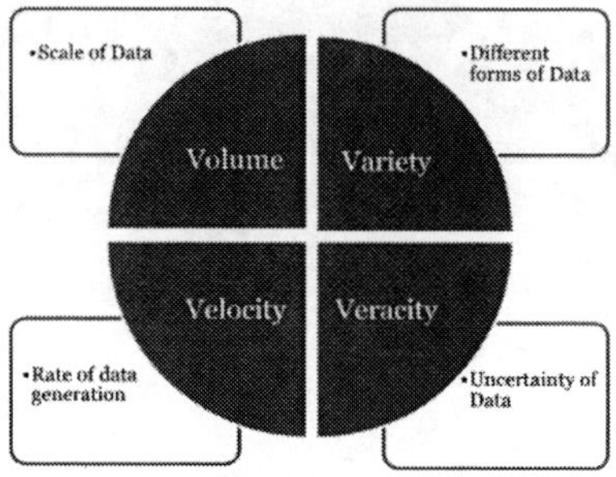

- A major focus for the next round of the transformation of advanced manufacturing
- Big data evolution: An opportunity for managing significantly larger amounts of data and acting on it with analytics for improved diagnostics and prognostics.

Moyne, J. and Iskandar, J., 2017, *Processes*, 5(3), p.39.

4

Application to Semiconductor industry

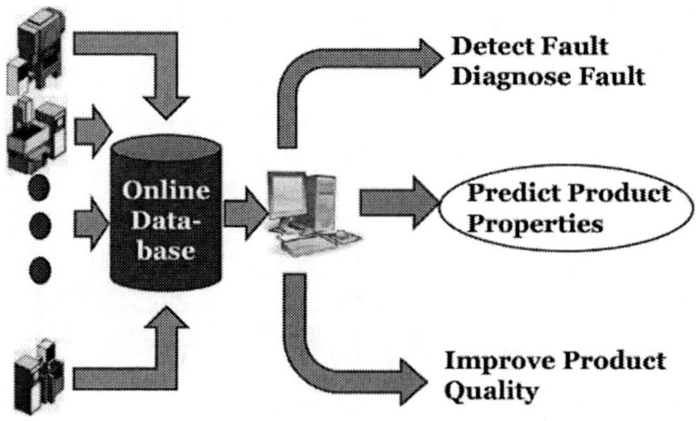

- Big data generated from thousands of devices, at the volume that are being produced , simply cannot be analyzed through traditional analytical tools.
- Become proactive instead of being reactive, regardless of the volume.

Metrology

- Introduction to Metrology
- The need for wafer to wafer control instead of lot to lot process control
- Leads to an increase in cost and production cycle time
- Solution?

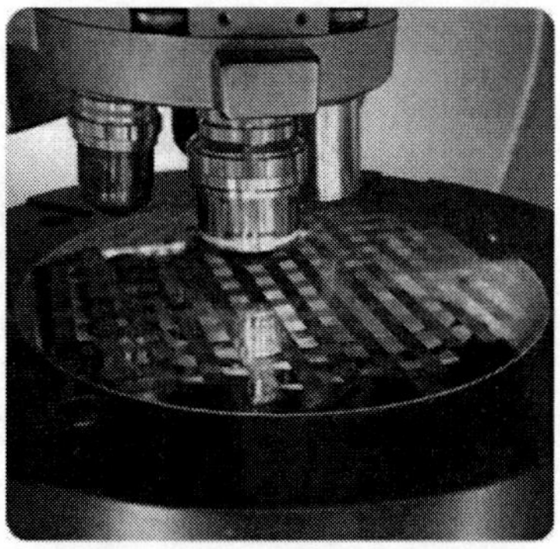

Optical 3d Microscope – www.bruker.com/products/surface-and-dimensional-analysis/3d-optical-microscopes

Outline

- ~~Motivation~~
- Virtual Metrology (VM)
- Existing VM approaches
- Statistics pattern analysis (SPA)
- Industrial case study
- Conclusions

Solution: Virtual Metrology

- Predict the product properties based on the relationship between previous metrology data of quality variable and process data, without physical measurement.
- Objective: Achieve total quality management & run-to-run control.
- Identification of most predictive features of a target value.

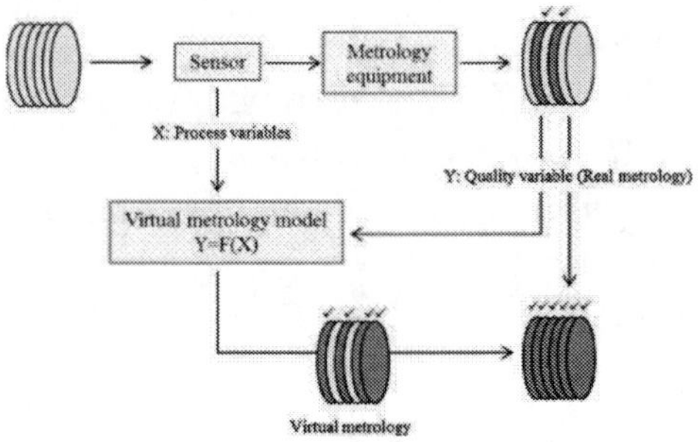

Park, C & Kim, SB 2016, *Journal of Process Control*, vol 42

Advantages?

- Reduced cost
- Reduced production time as there are no delays due to Metrology measurement
- Predictive maintenance
- Detection of faults faster

Outline

- ~~Motivation~~
- ~~Virtual Metrology (VM)~~
- Existing VM approaches
- Statistics pattern analysis (SPA)
- Industrial case study
- Conclusions

Existing VM approaches

- Model driven & data driven
- Why data driven?
  - Easy implementation online
  - Enabled by data driven modelling such as multivariate statistical methods
- Key focus on correlation between process variables and response variable
- Data driven methods:
  1. Multiple linear regression (MLR)
  2. Principal component regression (PCR)
  3. Partial Least Squares (PLS)
  4. Time series analysis (TSA)
  5. Kalman filter (KF)
  6. **Statistics pattern analysis (SPA)**

What are the challenges?

- Preprocessing:
  1. Data unfolding
  2. Trajectory alignment
  3. Trajectory mean shift
- Require human intervention and extensive offline data preprocessing
- Time consuming
- Automation not possible in real time application without delay.

http://www.simplycommunicate.co.za/wp-content/uploads/2017/06/How-to-Handle-Incident-Investigation.jpg

Outline

- ~~Motivation~~
- ~~Virtual Metrology (VM)~~
- ~~Existing VM approaches~~
- Statistics pattern analysis (SPA)
- Industrial case study
- Conclusions

Statistics pattern analysis

- How is it different ?
  - Extract correlation between batch statistics and metrology measurements
- Batch statistics
  - Mean
  - Standard deviation
  - Skewness
  - Kurtosis
  - Correlation
  - Cross correlation
- Advantages?
  - Eliminates pre-processing steps
  - Superior prediction performance
- Semiconductor process data - Severe Non-Normality in data
  - Batch statistics (In SPA) satisfy gaussian distribution much better in comparison to process variables

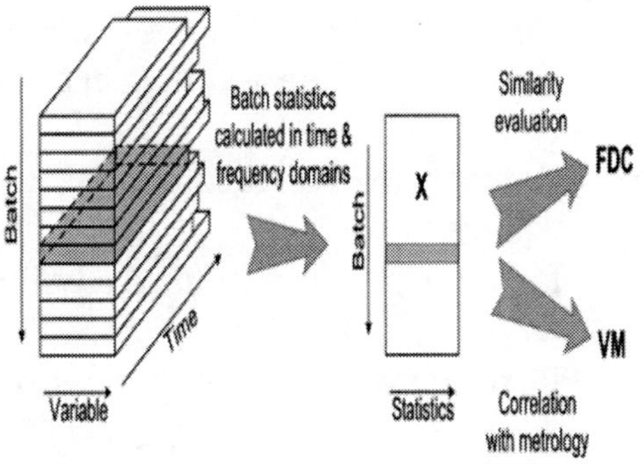

He, Q. Peter, et al. *American Control Conference (ACC), 2012*. IEEE, 2012.

He, Q.P. and Wang, J., 2011, *AIChE journal*, *57*(1), pp.107-121.
He, Q. Peter, et al. *American Control Conference (ACC), 2012*. IEEE, 2012.

Outline

- ~~Motivation~~
- ~~Virtual Metrology (VM)~~
- ~~Existing VM approaches~~
- ~~Statistics pattern analysis (SPA)~~
- Industrial case study
- Conclusions

Case study: Plasma etch process

- Objective: Predict sheet resistance using Optical emission spectroscopy (OES) data
- Dataset: 18 OES signals for 1120 wafers
- Data for 784 wafers used a training data
- Rest 336 wafers – testing data

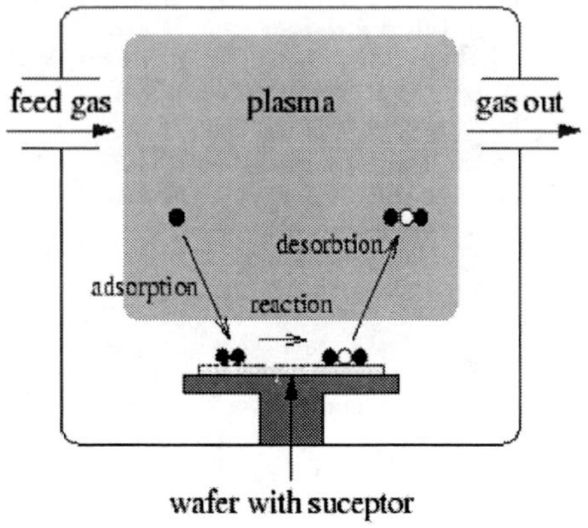

http://www.gs68.de/tutorials/plasma/img52.png

Performance evaluation

- Performance metric
  - R^2
  - Mean absolute percentage error (MAPE)

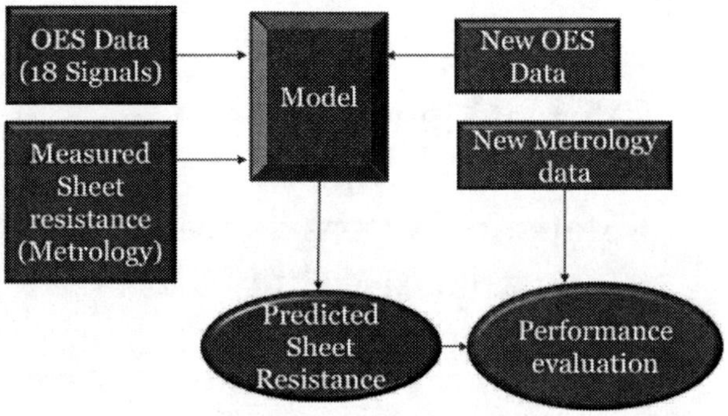

17

Prediction methods

- Statistics pattern analysis flow chart

18

Predictions: Static methods

Static methods		
Method	R^2	MAPE
PLS	0.6287	6.6578
PCR	0.6154	6.6737
MLR	0.4045	7.526
TSA	-0.8861	13.7194
★ SPA	0.7311	5.4850

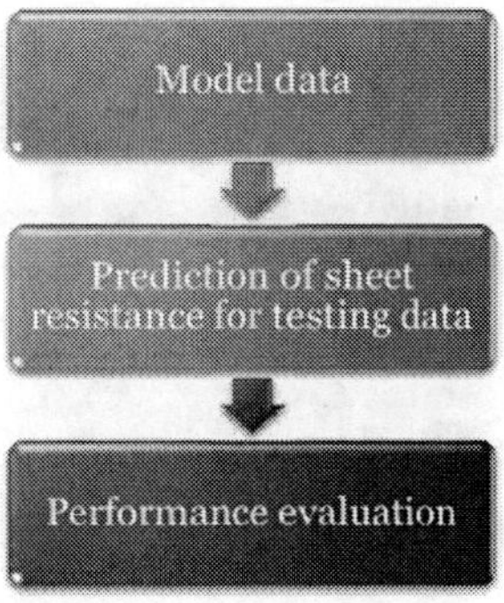

Predicted vs Measured Sheet resistance

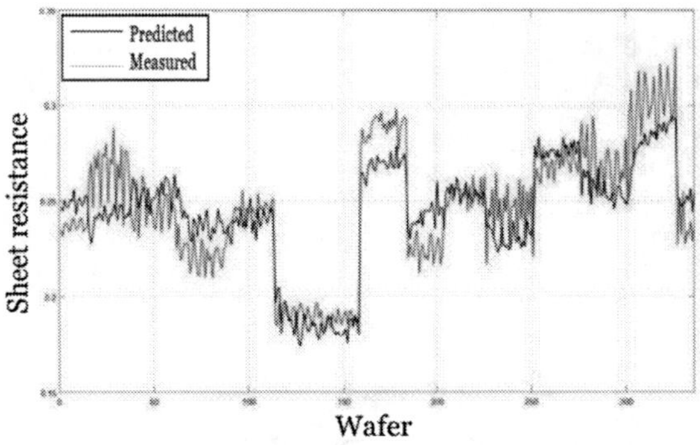

SPA

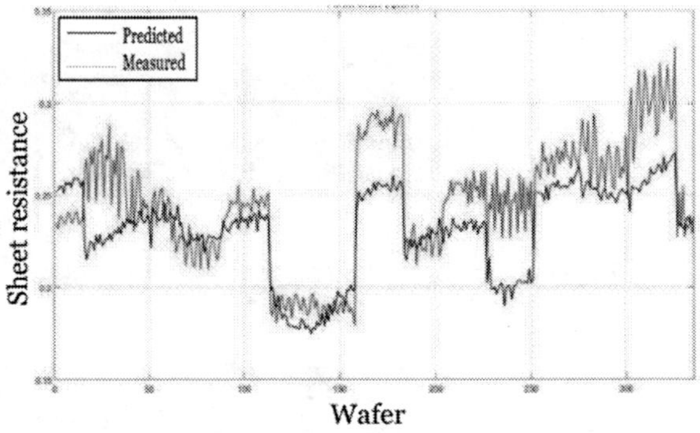

PLS

Predictions: Lot to lot methods

Recursive (Lot to Lot) methods		
Method Name	R^2	MAPE
RPLS	0.6137	6.4770
KF	-5.2447	24.98
★ Recursive SPA	0.6955	5.9632

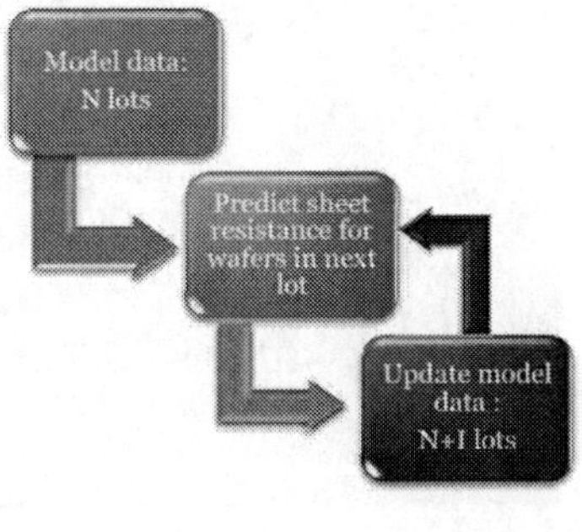

Predicted vs Measured sheet resistance

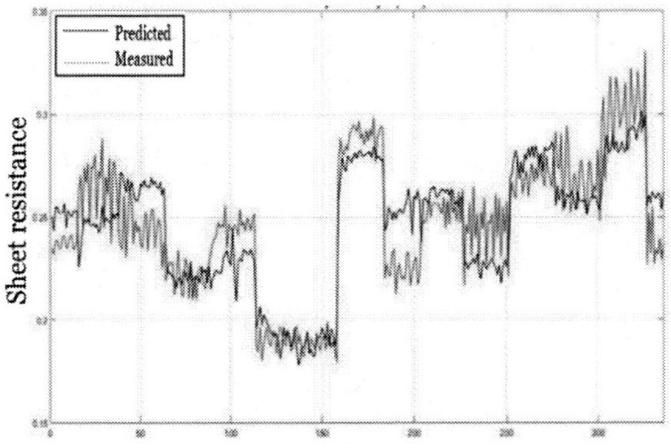

SPA

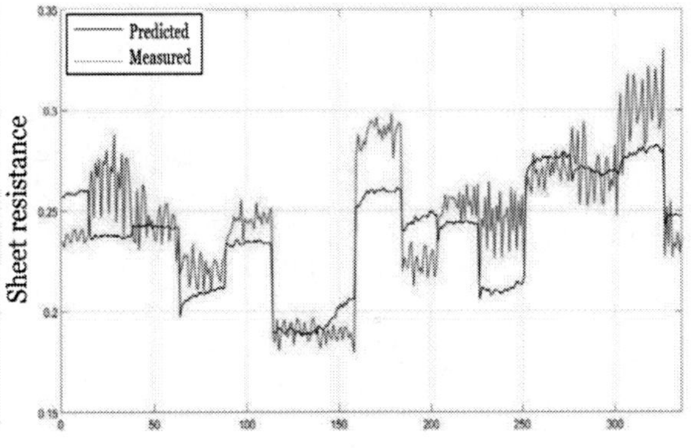

RPLS

Predictions: Wafer to Wafer methods

Recursive (Wafer to Wafer) methods		
Method Name	R^2	MAPE
KF	0.7242	5.0685
RPLS	0.699	5.6714
★ Recursive SPA	0.8002	4.5976

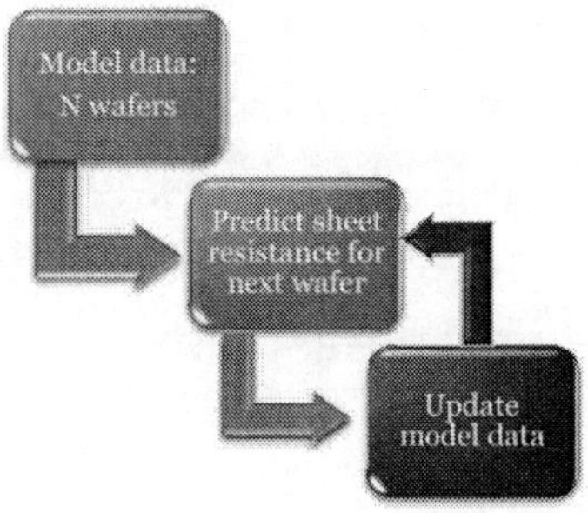

Predicted vs measured sheet resistance

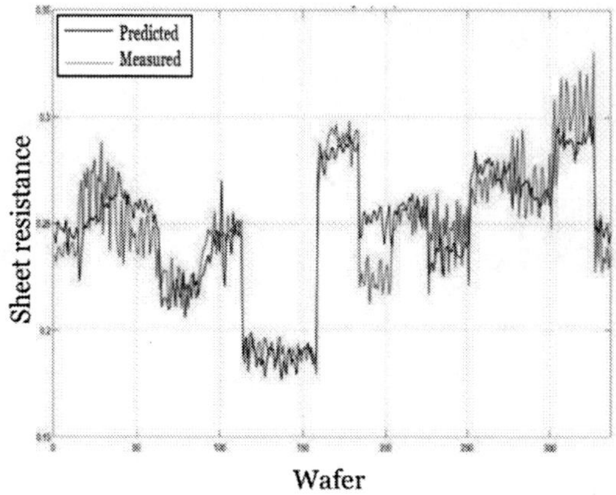

SPA

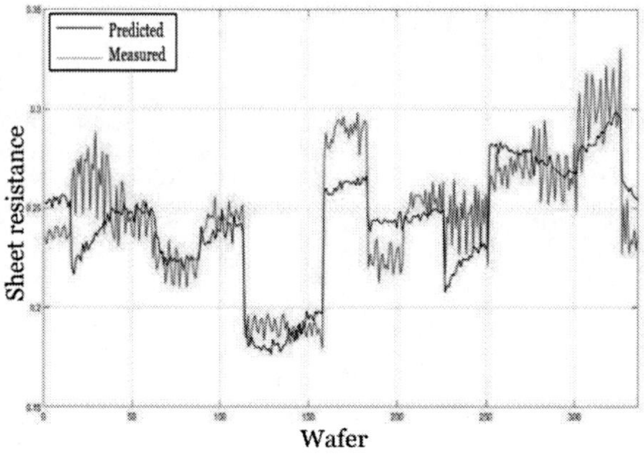

RPLS

Outline

- ~~Motivation~~
- ~~Virtual Metrology (VM)~~
- ~~Existing VM approaches~~
- ~~Statistics pattern analysis (SPA)~~
- ~~Industrial Case Study~~
- Conclusions

Conclusions

- Metrology incapable for real time wafer to wafer control on account of delay.
- Challenges related to handling of big data for semiconductor industry are addressed by Virtual Metrology methods.
- Virtual Metrology proven to be efficient technique for execution without delay.
- Statistics pattern analysis proven to be an efficient technique as it not only eliminates data preprocessing steps but also shows superior predictive performance over other data driven approaches.

Acknowledgments

- Dr. Jin Wang (Auburn University)
- Other group members
- NSF 126481 and 1547163

AUBURN UNIVERSITY

Questions?

500510 Enabling Discovery and Integration of Process Models and Data Using Ontology in the Domain of Biorefining

Abstract:

Process System Engineering (PSE) has been recognised as a method to design and operate chemical processes [1]. PSE has played an essential role to significantly improve our understanding of the processes and increasing efficiency of those processes by enabling the use of modelling, simulation, optimisation, planning and control. Due to the challenges in developing efficient and sustainable biorefining processes, the PSE tools have provided the vital links with notions of process systems engineering and biorenewables to address a variety of products and services [2].

With increasing number of specialising tools and methods [3], the ability to search different type of models and data involved in biorefining processes and concomitant integration of these models and data remains a challenge. The best way to retain these valuable models and data that often resides only within individuals, is to store in such a way to make their future retrieval and reuse as easy as possible. The InterCAPEmodel ontology represents knowledge of biorefining models and data in a systematic approach, where identifying, capturing, retrieving, sharing and effectively reusing these models and data are possible [4]. The ontological approach ensures the full utilisation of modelling knowledge that can be coupled with the potential of individual expertise to identify the best suited model for the integration of CAPE (computer aided process engineering) models and data in the domain of biorefining.

In this paper, an ontological approach towards the discovery of biorefining models and data for the purpose of integration is proposed. The InterCAPEmodel ontology is introduced as a foundation to support the discovery of the process models and data as well as their integration. The ontology provides a common vocabulary to eliminate any heterogeneity for effective communication among users. The modelling knowledge, both implicit knowledge that reflects modeller's intuition and their experience in modelling, as well as explicit knowledge that identifies type, characteristics, and associated properties of the models and data are captured in the form of taxonomy and relations that clarify the semantic stronger, more coherent, and more broadly applied. In addition, the ontology supports the registration process where common reference is established for integration of the models and data in the domain of biorefining. The ontology guides user to navigate through the taxonomy and assists in registering a model or data (i.e. process, energy, economic, environmental models or data) as an instance, which can be searched, retrieved and reused by the community. The "input-output matching" [5] is employed as a method of model integration, which is a process of instance matching. The properties that have the most meaningful association in integration are established as a common reference characterising inputs and outputs of the model and/or data. This method of integration permits partial matching, which is more flexible in discovery of models and data, which further increases the reusability of the models and data. To highlight the main functionality of the ontology approach, the design of InterCAPEmodel ontology, as well as the semantic methods used to measure relevance are validated by a number of case studies using models and data related to biorefining process.

[1] M. Alvarado-Morales, N. Al-Haque, K. Gernaey, J. Woodley, R. Gani, 2008, CAPE methods and tools for systematic analysis of new chemical product design and development, Computer Aided Chemical Engineering, on CD. vol. 25, pp. 997-1002.

[2] R. Gani, I. E. Grossman, 2007, Process Systems Engineering and CAPE – What Next?, 17th ESCAPE Proceeding Book, Bucharest, Romania, vol. 24, pp. 1-5.

[3] M. Matzopoulos, 2011. Dynamic Process Modeling: Combining Models and Experimental Data to Solve Industrial Problems, Process Systems Engineering: Dynamic Process Modeling.

[4] L. Koo, N. Trokana, F. Cecelja, 2017. A Semantic Framework for Enabling Model Integration for Biorefining. Computers & Chemical Engineering, 38, pp.733-738.

[5] N. Trokanas, F. Cecelja, T. Raafat, 2014. Semantic input/output matching for waste processing in industrial symbiosis. Computers & Chemical Engineering, 66, pp.259–268.

489175 Multivariable Adaptive Subspace Identification of Blood Glucose Concentration Dynamics for People with Type 1 Diabetes Mellitus

Abstract:

The Artificial Pancreas (AP) system is one of the efficient therapeutic methods for people with type 1 diabetes mellitus (T1DM). This automated system is designed to regulate the blood glucose concentration by using a continuous glucose monitor (CGM), a control system, and an insulin pump [1]. The ultimate goal of an AP system is to simplify the treatment for people with T1DM and reduce the risks associated with hypoglycemia or hyperglycemia. AP research has produced successful results in clinical and outpatient studies, but there are still many limitations that must be addressed. These limitations include exclusive use of glucose information and lack of accurate personalized models to describe the glucose-insulin dynamics in the body over time for use in AP control laws.

One of the challenges to achieve a fully automated and reliable AP is the lack of an accurate model to represent the dynamic changes in the patient's physiology under various conditions such as exercise and meals. The individualization of the insulin-glucose model is needed because each person has a unique glucose-insulin dynamic behavior. Also, the glucose-insulin dynamics and insulin sensitivity of a person can change from day to day. Moreover, daily activities such as eating and doing physical activities cause large variations in glucose concentration excursions that may not be described accurately by a generalized model. Hence, models with generalized parameters cannot reflect the dynamic behavior of the specific patient in different situations and adaptive personalized glucose-insulin models are necessary for the AP systems.

Subspace model identification (SMI) methods are effective data-driven techniques for identifying state-space models from multi-input and multi-output (MIMO) measurements of a dynamic system. These methods build structured block Hankel matrices using input and output data to retrieve certain subspaces that are related to the system matrices. In this work, a novel closed-loop/open-loop recursive SMI algorithm based on the optimized version of the Predictor-Based Subspace Identification (PBSID) method, the so-called PBSIDopt method is used to identify the glucose–insulin model [2, 3]. Since the techniques for data-based modeling utilize input and output data to identify a model, for a number of cases, although the working system is inherently stable, some samples with high noise levels or unknown disturbances may cause the identified model to be unstable. So, different steps have been suggested to guarantee the stability of the identified system.

The proposed method is able to provide a stable time-varying individualized state space model with glucose concentration values as output (from a CGM sensor) based on estimation of plasma insulin concentration and meal absorption rate (from an Unscented Kalman filter based on Hovorka's glucose-insulin dynamic model) and biometric variables (Metabolic Equivalent of Task values reported by BodyMedia SenseWear sports armband) as inputs [4, 5, 6]. Adaptive identification (every 5 minutes) allows the model to be valid over various daily life conditions without providing meal and exercise information.

Model identification and validation are based on clinical data from closed-loop experiments. The models are evaluated by means of various performances indices: Variance accounted for (VAF), Root mean square error (RMSE), Normalized root mean square error (NRMSE), Mean square error (MSE) and Normalized mean square error (NMSE). The proposed method provides a stable personalized time-varying state space model over time. The approach proposed in this work has shown a strong potential to identify a consistent glucose–insulin model in real time for use in an AP system.

References

[1] Cinar, A.; Turksoy, K.; Hajizadeh, I. Multivariable artificial pancreas method and system. 2016; US Patent App. 15/171, 355.

[2] Houtzager, I.; van Wingerden, J.-W.; Verhaegen, M. Fast-array recursive closed-loop subspace model identification. IFAC Proceedings Volumes, 2009, 42 (10), 96-101.

[3] Hajizadeh, I.; Rashid, M.; Turksoy, K.; Samadi, S.; Feng, J.; Sevil, S.; Frantz, N.; Lazaro, C.; Maloney, Z.; Littlejohn, E.; Cinar, A. Multivariable Recursive Subspace Identification with Application to Artificial Pancreas Systems. IFAC-Papers Online. 2017, Accepted.

[4] Hajizadeh, I.; Turksoy, K.; Cengiz, E.; Cinar, A. The extended and unscented Kalman filtering methods for real-time plasma insulin concentration estimation in an artificial pancreas control system for patients with type 1 diabetes. Annual Meeting AIChE. 2016.

[5] Hajizadeh, I.; Turksoy, K.; Cengiz, E.; Cinar, A. Real-time estimation of plasma insulin concentration using continuous subcutaneous glucose measurements in people with type 1 diabetes. Proceedings of the American Control Conference (ACC). Seattle, WA, 2017.

[6] Hovorka, R.; Canonico, V.; Chassin, L. J.; Haueter, U.; Massi-Benedetti, M.; Federici, M. O.; Pieber, T. R.; Schaller, H. C.; Schaupp, L.; Vering, T.; Wilinska, M. E. Nonlinear model predictive control of glucose concentration in subjects with type 1 diabetes. Physiol. Meas. 2004, 25, 905.

493352 Model Reduction and Approximation for Simultaneous Design, Control, and Scheduling

Abstract:

The simultaneous consideration of process design, process control and process operation/scheduling, has started to receive attention in the open literation[1,2] as it is well accepted that it may oer distinct opportunities for cost reduction, process intensication, and production imporvement. However, it is a challneging problem due to its multiscale nature, the different type of decisions that have to be considered at the different time scales, and the underlying model complexities that arise by connecting the three components (design, control, and scheduling)[2]. Model reduction and approximation is considered to be a key enabling technology to make the problem of simultaneous design, control, and scheduling more tractable and amenable to efficient solution strategies[3]. In this work, we study the impact of different model reduction strategies on the design-dependent and control-aware scheduling/operational strategies, based on our recently introduced PAROC framework[4]. In particular, we employ model reduction and approximation technologies such as piecewise linearization[5, 6] for control and a time scale bridging model[7, 8] for scheduling, based on which feasibility and optimality of the obtained simultaneous solutions is assesed in the presence of uncertainty; a multiproduct isothermal continuously stirred tank reactor (CSTR) is used to illustrate the key features of the proposed methodology.

References

[1] Diangelakis, N. A., Model-based multi-parametric programming strategies towards the integration of design, control and operational optimization (2017) Diss. Imperial College London.

[2] Patil, B.P., Maia, E., Ricardez-Sandoval, L.A. Integration of scheduling, design, and control of multiproduct chemical processes under uncertainty (2015) AIChE Journal, 61 (8), pp. 2456-2470.

[3] Nie, Y., Biegler, L.T., Villa, C.M., Wassick, J.M. Discrete time formulation for the integration of scheduling and dynamic optimization (2015) Industrial and Engineering Chemistry Research, 54 (16), pp. 4303-4315.

[4] Pistikopoulos, E. N., Diangelakis, N. A., Oberdieck R., Papathanasiou M. M., Nascu I.,Sun M., 2015. PAROC - An integrated framework and software platform for the optimisation and advanced model-based control of process systems, Chemical Engineering Science, Vol. 136, 115-138.

[5] Wang, Y., Yang, P. Application of a control strategy based on PWA model in CSTR system (2016) Huagong Xuebao/CIESC Journal, 67 (3), pp. 865-870.

[6] Rivotti, P., Lambert, R.S.C., Pistikopoulos, E.N. Combined model approximation techniques and multiparametric programming for explicit nonlinear model predictive control (2012) Computers and Chemical Engineering, 42, pp. 277-287.

[7] Subramanian, K., Maravelias, C.T., Rawlings, J.B. A state-space model for chemical production scheduling (2012) Computers and Chemical Engineering, 47, pp.97-110.

[8] Diangelakis, N. A., Burnak, B., Pistikopoulos, E. N. A multi-parametric programming approach for the simultaneous process scheduling and control - Application to a domestic cogeneration unit. (2017) Foundations of Computer Aided Process Operations / Chemical Process Control, In Press.

494187 Handling Delayed and Irregular Measurements inÂ Batch Subspace Model Identification Framework

Abstract:

Batch processes form an integral part of the manufacturing industry, either to directly produce specialty chemicals, or as lead up to continuous operation (in the form of process startups). Traditional approach for control of batch processes has been to use open loop, recipe based policy. In this approach, a predefined input trajectory is applied to each batch [1]. The approach assumes that the desired product quality can be obtained by repeating historically successful input profiles. Although these approaches are easy to implement and do not require a model for the process, they are incapable of rejecting disturbances that affect the process. This has motivated the use of feedback control strategies.

A classical closed loop control approach in batch process is trajectory tracking. In this approach, a predefined set-point trajectory for a measured process variable, often obtained from past successful batches, is tracked. PI controllers are often used for implementing trajectory tracking control. When implementing trajectory tracking approaches (with PI or otherwise) even with perfect tracking, the desired quality may not be obtained, as the relationship between the measured/tracked variable and final quality may change significantly with changes in the process conditions.

These challenges are addressed by control strategies that are cognizant of the causal relationship not only between the manipulated and the (online) measured variables, but also between the manipulated and the final quality variable. A popular model based control approach, model predictive control (MPC), has increasingly been studied for the control of batch processes [2-5]. The reasons for popularity of these control schemes are twofold: first, the feedback controller can counter the model uncertainties that are associated with model simplifications and measurement errors, the other being their ability to handle the ever present input/output constraints. In implementing these formulations, where possible, good first principles models are preferred due to the ability to predict process behavior beyond the data set used to estimate the model parameters. However, first principles models are difficult to develop as it is often impractical to estimate the parameters through experiments, particularly because of the associated cost. Further, these first principles models, for instance, in case of distributed parameter systems such as batch particulate processes, pose computational challenges for direct use in predictive control formulation.

These issues have motivated the use of simpler, often linear, models derived from past batches database. A variety of approaches for development of data-driven models have been proposed. One excellent approach is partial least squares (PLS), which models the process in a projected latent space [6]. These models are essentially time-varying linear models, linearized around mean past trajectories, and therefore require the batches to be of same length, or to recognize an appropriate alignment variable. To account for these limitations, a multi-model approach was proposed in [5]. These models were based on the 'current measurements' of the process instead of the 'time'. These developments were followed by contributions in the area of integration of these data-driven models with the advanced control formulations [4-5]. More recently a subspace identification based batch control approach was proposed in [7] where a LTI state-space model of the batch process is estimated. Subspace identification methods, in general, are a class of model identification methods that are non-iterative in nature and models a process as a state-space linear time invariant system, identifying the system matrices along with the order of the system. These methods utilize efficient matrix factorization methods such as singular value decomposition (SVD) and QR factorization for their implementation. This is in contrast to classical system identification techniques, which are iterative optimization based algorithms that minimize the prediction errors (PEM) using efficient methods such as maximum likelihood estimation (MLE) and expectation maximization (EM) algorithm. These methods offer well established theoretical properties and have been widely studied [8-11]. However, most of the existing classical and subspace identification approaches focus on continuous operation. The subspace identification approach in [7] was different from conventional subspace identification approach for continuous process in the way Hankel matrices are constructed to utilize data from multiple batches. This formulation will henceforth be referred to as batch subspace identification.

The existing formulations of batch subspace identification method have, however, dealt generally with all process variables being measured at a fixed rate [7,12,13]. Recently, [14] discussed the formulation for a multi rate process, to handle the case where some of the measurements are available at different rate than others. However, these results do not explicitly consider delayed measurements in the batch subspace identification formulation. Further, the formulation of batch subspace identification needs to be generalized for the case of measurements arriving at a unsteady rate during the batch, that is, the rate at which variables are measured is not same during the entire batch duration.

Motivated by these considerations, this work presents a subspace identification based state-space modeling and control approach for batch process with delayed and infrequent measurements. The key idea is to adapt the recently developed batch subspace identification approach to explicitly handle delayed and irregular measurements. The ideas proposed in this work will be illustrated through a particulate batch process.

[1] D. Bonvin, B. Srinivasan, D. Hunkeler, Control and optimization of batch processes, IEEE Control Systems Magazine 26 (6) (2006) 34-45.

[2] D. Shi, N. H. El-Farra, M. Li, P. Mhaskar, P. D. Christofides, Predictive control of particle size distribution in particulate processes, Chemical Engineering Science 61 (1) (2006) 268-281.

[3] D. Bonvin, Optimal operation of batch reactors- a personal view, Journal of Process Control 8 (5-6) (1998) 355-368.

[4] S. Aumi, P. Mhaskar, Integrating Data-Based Modeling and Nonlinear Control Tools For Batch Process Control, AIChE Journal 58 (2012) 2105-2119.

[5] S. Aumi, B. Corbett, P. Mhaskar, and T. Clarke-Pringle. Data-based modeling and control of nylon-6, 6 batch polymerization. Control Systems Technology, IEEE Transactions on, 21(1):94-106, Jan 2013.

[6] J. Flores-Cerrillo, J. F. MacGregor, Control of particle size distributions in emulsion semibatch polymerization using mid-course correction policies, Industrial and Engineering Chemistry Research 41 (7) (2002) 1805-1814.

[7] B. Corbett, P. Mhaskar, Subspace identification for data-driven modelling and quality control of batch processes, AIChE Journal 62 (2016) 1581-1601.

[8] L. Ljung, System Identication: Theory for the User, Pearson Education, 1998.

[9] H. Raghavan, A. K. Tangirala, R. Bhushan Gopaluni, S. L. Shah, Identication of chemical processes with irregular output sampling, Control Engineering Practice 14 (5) (2006) 467-480.

[10] X. Yang, W. Xiong, B. Huang, H. Gao, Identication of Linear Parameter Varying Systems with Missing Output Data Using Generalized Expectation-Maximization Algorithm, IFAC Proceedings Volumes 47 (3) (2014) 9364-9369.

[11] O. Wu, K. Hariprasad, B. Huang, J. F. Forbes, Identication of linear dynamic errors-in-variables systems with a dynamic uncertain input using the EM algorithm, in: 2016 IEEE 55th Conference on Decision and Control (CDC), 2016, pp. 1229-1234.

[12] Abhinav Garg, P. Mhaskar, Subspace Identification Based Modeling and Control of Batch Particulate Processes, Industrial and Engineering Chemistry Research (2017), Submitted.

[13] Abhinav Garg, Brandon Corbett, Prashant Mhaskar, Gangshi Hu, Jesus Flores-Cerrillo, High Fidelity Model Development and Subspace Identification of a Hydrogen Plant Startup Dynamics, Computers and Chemical Engineering (2017), Submitted.

[14] Rashid, M. M., Mhaskar, P., & Swartz, C. L. (2017). Handling multiâ€rate and missing data in variable duration economic model predictive control of batch processes, AIChE Journal, in press.

495031 Design of Experiments for Identification of Model That Is Suitable for Use in Control System

Abstract:

Advanced process control critically relies on a multivariable model of the controlled process. Identifying such a model from input-output data requires an estimate for the dimension of the model order (dimension of state space) and estimation of the related parameters. The accuracy of a model order estimate depends strongly on the kind of input-output data available. It has long been demonstrated (P. Misra & Nikolaou, 2003) that standard identification experiments with PRBS inputs may not generate data suitable for estimation of system order. In fact, for ill-conditioned systems it was demonstrated that such experiments may yield data that may produce model order estimates ranging from completely wrong (unsuitable for use in controller design) to indeterminate. Experiments using *appropriately proportioned rotated* pseudo random binary sequence (PRBS) inputs were demonstrated to produce accurate model order estimates. However, such inputs depend on the very process to be identified. Therefore, some sort of an adaptive procedure is needed, through which an initial experiment is conducted and model identified, which in turn is used to design and conduct the next experiment from which an updated model is identified, and so on.

In this presentation we develop such a procedure. In addition, we generalize the preceding ideas and put them in a more rigorous framework that allows further development (S. Misra & Nikolaou, 2017).

We test the proposed procedure on two case studies namely a high-purity distillation column, and a fluidized catalytic cracking (FCC) reactor-regenerator system. For the latter, the true order of the system, equal to 15, is captured accurately with the proposed procedure, whereas standard approaches completely fail to produce any reasonable number.

References:

Misra, P., & Nikolaou, M. (2003). Input design for model order determination in subspace identification. Aiche Journal, 49, 2124-2132.

Misra, S., & Nikolaou, M. (2017). Adaptive design of experiments for model order estimation in subspace identification. Computers and Chemical Engineering, 100, 119-138.

495766 How Does a Control Room Operator Identify the Process? Insights Using a Cognitive Engineering Approach

Abstract:

Industrial accidents in the past such as those in BP Texas City Refinery (2005), Three Mile Island (1979) and BayerCrop Science (2008) that have caused loss of money and lives and environmental damage have served as unremitting impetus for process plants to enhance safety. As a result, industries have developed and equipped themselves with several technological advancements such as distributed and redundant sensing, sophisticated control algorithms and automation techniques, advanced alarm management systems and support and guidance systems. These technologies collect, process and present information regarding the process state at all times to plant operators, typically stationed in a distributed control system (DCS) room. Operators, the consumer of the above information have the responsibility of operating the process under normal conditions so as to maximize profits as well as handling abnormal situations to prevent accidents.

In carrying out these tasks, operators use their *knowledge* of process dynamics (including nominal operating values of variables, interrelationship between variables) along with fault diagnosing *capability*, which are consolidated in what is called the operator's mental model of the process. Operators typically acquire their mental model through operator training systems (OTS) and/or on-the-job experience. An OTS consists of a simulated environment of the process wherein trainee operators are exposed to typical scenarios that might occur in the real plant and constitutes a major component of training programs. These systems are required to be able to accurately simulate process dynamics in a range of operating conditions in real time and research efforts have identified avenues for improving OTSs from a computational point of view (Patle et al., (2014)). For example, an expert guidance based intelligent training system was proposed in (Shin and Venkatasubramanian, (1996)) that enhances diagnosing skill of operators, and an assessment technique of operators during training was proposed based on the deviations of operator actions from predefined sequences of actions (Lee et al., 2000). These techniques emphasize on improving simulation capabilities of an OTS, enhancing diagnosing skills of operators and assessing the operator's ability to follow standard guidelines.

The key element of a training program, i.e., the operator and the ultimate aim of a training program, i.e. imparting an *adequate* mental model of the process to the operator is not addressed explicitly in the above techniques. However, it is important to take into account the cognitive abilities of trainees in designing tasks so as to ensure a successful training program. With advancement of technology that have led to techniques such as model predictive control (MPC), the roles of operators also change (Kluge et al., (2014)) which further increases the importance of accounting operator's abilities. Understanding the cognitive behavior of operators has been emphasized as being critical to enhancing their skills and abilities (Bullemer and Nimmo, (1994)). Previous work has also identified the importance of cognitive processes of operators during an OTS task in evaluating their ability (Sharma et al., (2016)).

In this work, we propose a framework that accounts for the operator's abilities in designing training tasks

and that is focused on imparting an adequate mental model through an online method of knowledge and capability assessment. We use eye tracking as a tool to peek into the cognitive processes of the operator during a training task and assess their mental models as they evolve during/across tasks. We follow a systems engineering approach to operator training that comprises three components – (1) Design suitable training tasks, (2) measure the operator' response and analyze the same with respect to their actions to examine their mental models and (3) update their mental model during the task which can then be used to identify gaps for designing suitable tasks. This translates to an adaptive training program that accounts for the operator's abilities during the task in an online manner and design tasks so as to ensure an adequate model imparted to the operator. The proposed framework is partly used with human subjects to assess their mental models during a simulated training exercise.

An ethanol production plant is simulated and training tasks are designed in three stages – (1) manipulation of valves to observe the effect on process variables, (2) tracking a process variable to a desired set point, and (3) tracking multiple process variables at the same time. The first stage is aimed at allowing the operator to identify the relation between manipulated valves and process variables while the latter two stages require the operator to identify appropriate valve(s) to achieve certain objective(s). Eye tracking data is collected during the tasks and the operator's gaze data is analyzed to obtain association graphs at different stages of a task, representative of the operator's mental model of the process (or the part of the process being targeted by a task). Association graphs are directed graphs derived from eye gaze data where nodes represent the variables the operator has gazed, and edges represent an association between two nodes. The strength of each edge is set to be proportional to the gaze transition probability between the two nodes, representing the extent to which an association is examined by the operator. Evolution of these mental models can be studied during a task, and across multiple tasks. Knowledge and capability can be assessed through their performance in unguided tracking tasks. This monitoring and assessment technique can be performed online and gaps in operator's mental model identified at the end of each task so as to adaptively design subsequent training tasks that can ensure an adequate model imparted to the operator.

References

Patle, D.S., Ahmad, Z. and Rangaiah, G.P., 2014. Operator training simulators in the chemical industry: review, issues, and future directions. Reviews in Chemical Engineering, 30(2), pp.199-216.

Shin, D. and Venkatasubramanian, V., 1996. Intelligent tutoring system framework for operator training for diagnostic problem solving. Computers & chemical engineering, 20, pp.S1365-S1370.

Lee, S., Jeong, I. and Il, M., 2000. Development of evaluation algorithms for operator training system. Computers & Chemical Engineering, 24(2), pp.1517-1522.

Kluge, A., Nazir, S. and Manca, D., 2014. Advanced applications in process control and training needs of field and control room operators. IIE Transactions on Occupational Ergonomics and Human Factors, 2(3-4), pp.121-136.

Bullemer, P.T. and Nimmo, I., 1994, October. Understanding and supporting abnormal situation management in industrial process control environments: a new approach to training. In Systems, Man, and Cybernetics, 1994. Humans, Information and Technology., 1994 IEEE International Conference on (Vol. 1, pp. 391-396). IEEE.

Sharma, C., Bhavsar, P., Srinivasan, B. and Srinivasan, R., 2016. Eye gaze movement studies of control room operators: A novel approach to improve process safety. Computers & Chemical Engineering, 85, pp.43-57.

500477 Predictive Control of Lake Levels and Estimation of Stream Flows for the Adaptive Management Complex Natural Watersheds. Part 1. Estimation

Abstract:

This paper presents an overview of potential role of predictive control and estimation in the adaptive management of lake levels and river flows in a complex watershed. The study area comprises the Rainy River watershed straddling the Canada/United States border between Minnesota and Ontario. This 70,225 sq. km watershed is managed by the International Joint Commission as a component of the Lake of the Woods basin. The Rainy River basin includes wilderness areas and national parks on both sides of the borders, ultimately provides drinking water for 750,000 people, and is a source of electricity generation, recreation, and tourism. This watershed is a part of the Canadian Shield - a large landmass in North America that comprises 20% of the world's fresh water.

Outflow from the two largest lake systems within the watershed, Rainy Lake and the Namakan Reservoir, are regulated by dams operated by commercial entities. Under terms of a treaty between the Canada and USA encompassing all transboundary waters, when feasible dam operators are required to manage lake levels to remain with 'rule curves' that establish desired upper and lower bounds on lake levels as a function of calendar date. From time to time the rule curves are reviewed and updated, the most recent changes occurring in 1970 and a 2000. Rule curves incorporate complex tradeoffs among stakeholders in the region with profound consequences for the region's ecological well-being. A review of the rule curves is currently underway for the purpose of proposing changes to take effect within the next few years.

This project was initiated following a series of high water events following the rule curve revision of 2000, including major flooding events in 2002 and 2014. The initial work used historical data for the period from 1948 to 2015 to establish that qualitative change in lake level control has occurred in recent decades.

The project now encompasses three main efforts:

- Development of a tracking filter to estimate net inflows to the major reservoirs in the watershed using available lake level gauges and limited stream flow measurements.
- Development of multivariable predictive control strategy for the integrated control of the two major dams in the watershed. Constraints on dam operation include management of emergency high water and low water events, hydrological conveyance constraints upstream of the dams, ecological and property considerations, and constraints on river bank erosion downstream of the dams.
- Development of a revised rule curve order for the adaptive management of the water levels and flows that is feasible, can accommodate more variable seasonal flows resulting from the effects of climate change, and that accommodate the biological requirements of key sentinel species representative of the regional ecology.

For the purpose of estimating the net inflow to Rainy Lake, a novel Kalman tracking filter has been developed that uses available level gauges and incomplete stream flow data to estimate inflows. The serendipitous occurrence of redundant lake measurements in the historical database provided a statistical model for the level measurement errors. The error model was used to tune the tracking filter to produce maximum likelihood estimates of inflows to Rainy Lake. The inflow estimates show a clear change in inflows between the periods 1970-2000 and 2000-2015

The statistical significance of this change was verified using an empirical stage-frequency diagram for the flow on upper Rainy River. The KS Kolmogorov–Smirnov) statistic verifies the statistical significance of this change. Local precipitation data for the same periods are not substantially different by the same statistic. This provides strong empirical evidence for the proposition that the change in seasonal flows caused by the 2000 rule curve change induced high water events on Rainy Lake due to the unique discharge characteristics of upper Rainy River.

500675 Predictive Control of Lake Levels and Estimation of Stream Flows for the Adaptive Management Complex Natural Watersheds. Part 2. Economic Model Predictive Control

Abstract:

The purpose of this presentation is to demonstrate the utility of advanced, multivariable control to better regulate lake levels and stream flows in the complex system of lakes and control points comprising the the Rainy - Lake of the Woods watershed. Previous work demonstrated the construction and tuning of Kalman filters for estimating unmeasured inflows to Rainy and Namakan lakes, and demonstrated the benefit of coordinating control of the dams at International Falls and Kettle Falls for the purpose of regulating lake levels within existing rule curves.

This study extends that work by implementing advanced control strategies to control lake levels subject to current rule curves. Model predictive control is a strategy that uses available measurements of Rainy Lake and Namakan Lake levels, a tracking filter to measure ungaged inlet flows, a model of the two reservoir system, and a prediction horizon to estimate an optimal control trajectory. In this instance the control variables are the daily configurations of the dams located at International Falls, Kettle Falls, and the Seine River system. Real-time feedback control is achieved by daily updates of the optimal control trajectory in response to new measurements.

Economic Model Predictive Control (EMPC) is an essential aspect of the proposed control system. The objective incoporates a complex set of water level requirements for key sentinel species including Wild Rice, Walleye and Northern Pike, Loon, and Muskrat.

The primary goal of adaptive management is regulate lake levels and river flows in response to measured events rather than to fixed dates on the calendar. The driving forces behind changes in the hydrology of watersheds in North America is large climate change. In the Rainy River basin, the change in climate has led to earlier snow melt and ice out of the regionâ€™s lakes, and more intense precipitation events in late Spring and early Summer.

MULTI-OBJECTIVE OPTIMIZATION OF CRUDE DISTILLATION UNIT, VACUUM DISTILLATION UNIT AND HYDROCRACKING REACTOR IN DIESEL PRODUCTION PROCESS TO MAXIMIZE PROFITABILITY AND REDUCE GREENHOUSE GASES USING NSPSO AND NSGAII BIOINSPIRED TECHNIQUES

PhD Stu. Johana Orjuela
PhD Stu. Camilo Monroy

20/11/2017

AICHE ANNUAL MEETING - 2017

2

CONTENT

- PROBLEM STATEMENT
- OBJECTIVES
- METHODOLOGY
- RESULTS
- CONCLUSIONS

Chemical and Environmental Engineering Department
Universidad Nacional de Colombia

Environmental Engineering
Universidad de Cundinamarca

3

PROBLEM STATEMENT

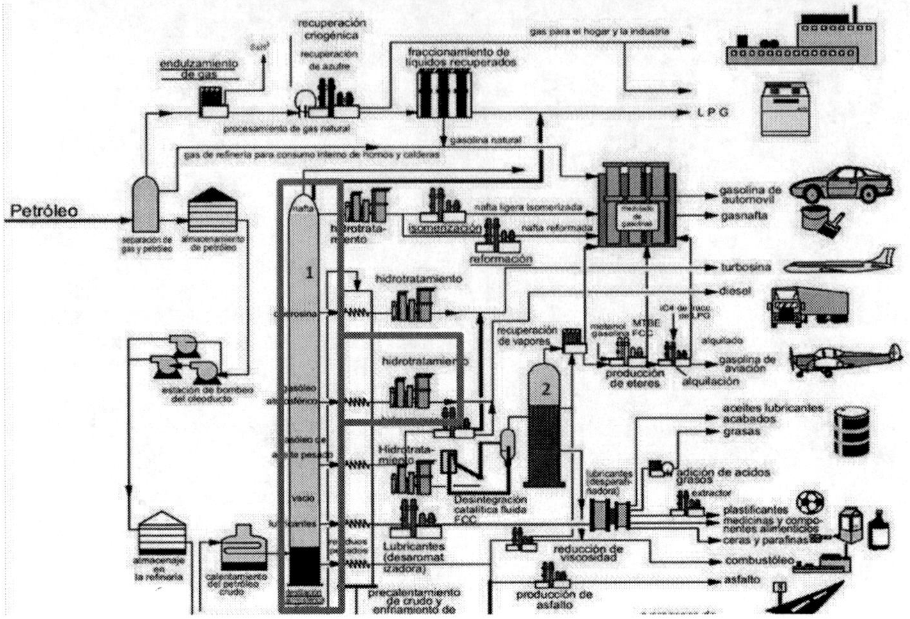

Chemical and Environmental Engineering Department
Universidad Nacional de Colombia

Environmental Engineering
Universidad de Cundinamarca

4

PROBLEM STATEMENT

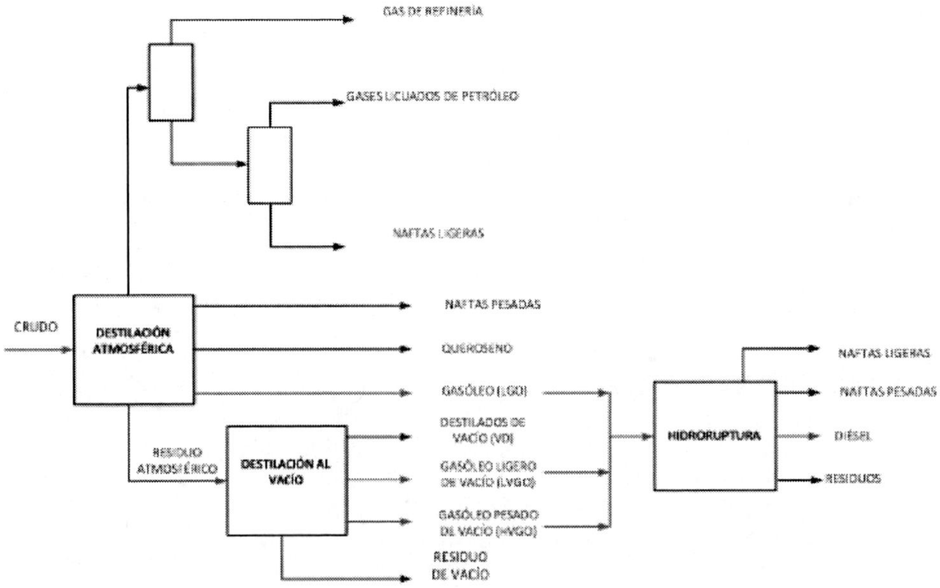

Chemical and Environmental Engineering Department
Universidad Nacional de Colombia

Environmental Engineering
Universidad de Cundinamarca

5

PROBLEM STATEMENT

OPTIMIZATION

- Reduce greenhouse gas emissions
- Decrease energy consumption
- Increase the Diesel produced
- Reduce operational costs
- Increase profitability

Chemical and Environmental Engineering Department
Universidad Nacional de Colombia

Environmental Engineering
Universidad de Cundinamarca

6

General Objective

Evaluate through meta-heuristic multiobjective optimization, the operational and processing intervals, in order to minimize the emission of greenhouse gases and maximize the profitability of the process.

Chemical and Environmental Engineering Department
Universidad Nacional de Colombia

Environmental Engineering
Universidad de Cundinamarca

7

METHODOLOGY

Aspen-HYSYS ⟺ MATLAB®

Modeling and Process Simulation

- Southeast Asian refinery conditions (validation of the simulation)
- Sensitivity analysis
- Evaluate the objective functions

Programming Bioinspired Algorithms

- Extract the results of Hysys
- Obtaining optimal pareto fronts

Chemical and Environmental Engineering Department
Universidad Nacional de Colombia

Environmental Engineering
Universidad de Cundinamarca

8

METHODOLOGY

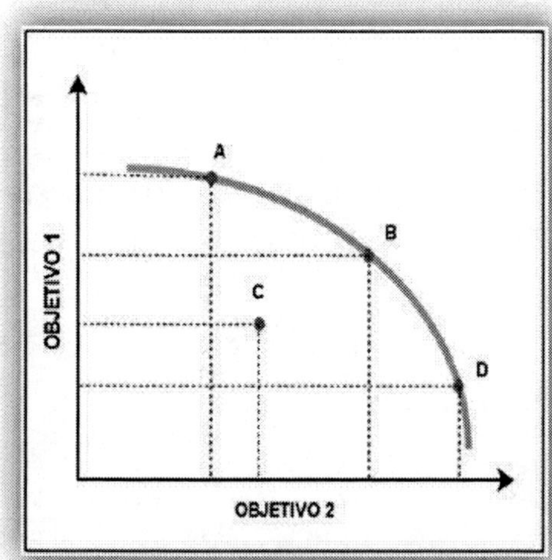

NSGA-II **m-NSPSO**

Chemical and Environmental Engineering Department
Universidad Nacional de Colombia

Environmental Engineering
Universidad de Cundinamarca

9

RESULTS

- SENSITIVITY ANALYSIS
- APPROACH TO THE OPTIMIZATION PROBLEM
- SELECTION OF THE OPTIMIZATION ALGORITHM
- OPTIMIZATION OF THE PROCESS

Chemical and Environmental Engineering Department
Universidad Nacional de Colombia

Environmental Engineering
Universidad de Cundinamarca

10

SIMULATION

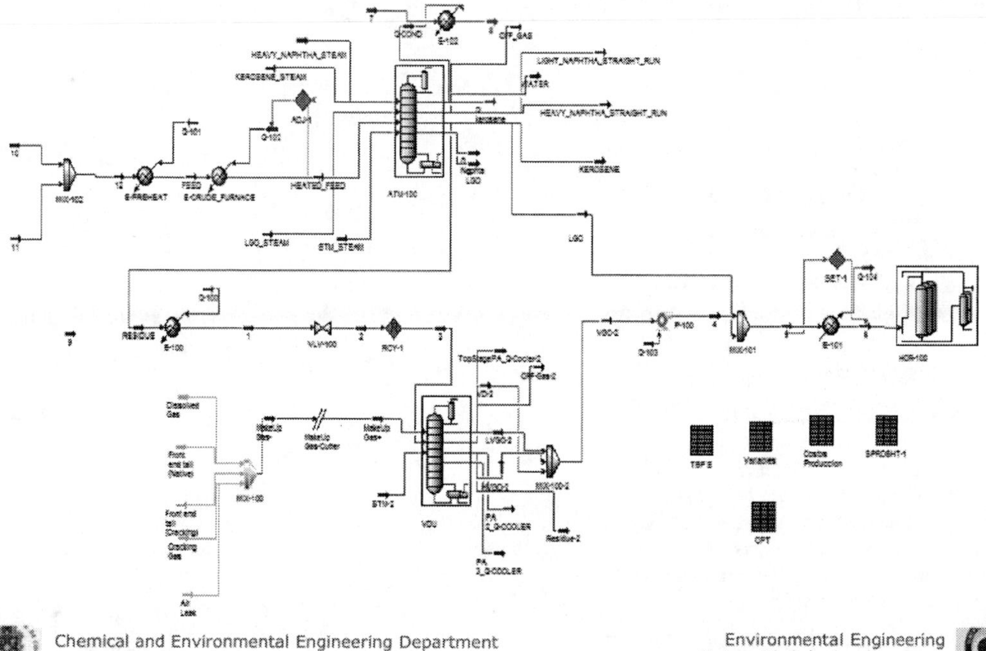

Chemical and Environmental Engineering Department
Universidad Nacional de Colombia

Environmental Engineering
Universidad de Cundinamarca

11

SENSITIVITY ANALYSIS

- 20 manipulated variables (independent variables)
- 30 output variables

$$\% \, Variation = \left[\frac{\max value - \min \text{value}}{\max value} * 100\right]$$

Flow (kg/h)	Greenhouse effect gases					
	Feed HC	kerosene	LGO	LVGO	HVGO	VGO
Top gas	30,46	55,34	59,22	44,37	29,87	40,56
Light Naphta	68,64	72,73	72,23	70,54	48,01	59,50
Heavy Naphta	36,93	76,48	64,51	60,84	41,95	38,03
Kerosene	81,13	92,47	93,08	81,71	38,81	51,32
LGO	80,98	59,41	89,53	90,16	86,63	86,96

12

OPTIMIZATION PROBLEM

Profit function: The profit function is calculated as follows

INAMDAR et al 2014

$$\phi = P_1 - P_2 - P_3 - \text{Energy Cost}$$

where

ϕ = Profit (Rs./day)

P_1 = Selling price of the products (Rs./day)

$= \sum_i (\text{Prod}_i \times \text{Price}_i); \; i = 1, 2, \ldots, 7,$

P_2 = Cost of raw materials (Rs./day)

$= (F \times \text{Cost}_f)$

P_3 = Operating cost (Rs./day)

$= \sum (RF \times RC + PA \times PC + Q \times UC)$

RF: Reflux flow rate (MT/day)
RC: Cost of pumping reflux (Rs./MT)
PA: Total pump around flow rate (MT/day)
PC: Cost of pumping PA flows (Rs./MT)
Q: Quantity of utility, e.g., steam rate (MT/day)
UC: Utility cost (Rs./MT)

13 OPTIMIZATION PROBLEM

- Max Profitability
- Min Greenhouse Gases.

- Subject to:

- Top gas flow [10000 - 40000 kg / h].
- Light naphtha flow [30000 - 150000 kg / h].
- Heavy naphtha flow [10000 to 50000 kg / h].
- Kerosene flow [10000 to 80000 kg / h].
- LGO flow [50000 to 200000 kg / h].
- Crude mix percentage [20% to 80%].
- Restriction in the quality of solvents

Chemical and Environmental Engineering Department
Universidad Nacional de Colombia

Environmental Engineering
Universidad de Cundinamarca

14 OPTIMIZATION PROBLEM

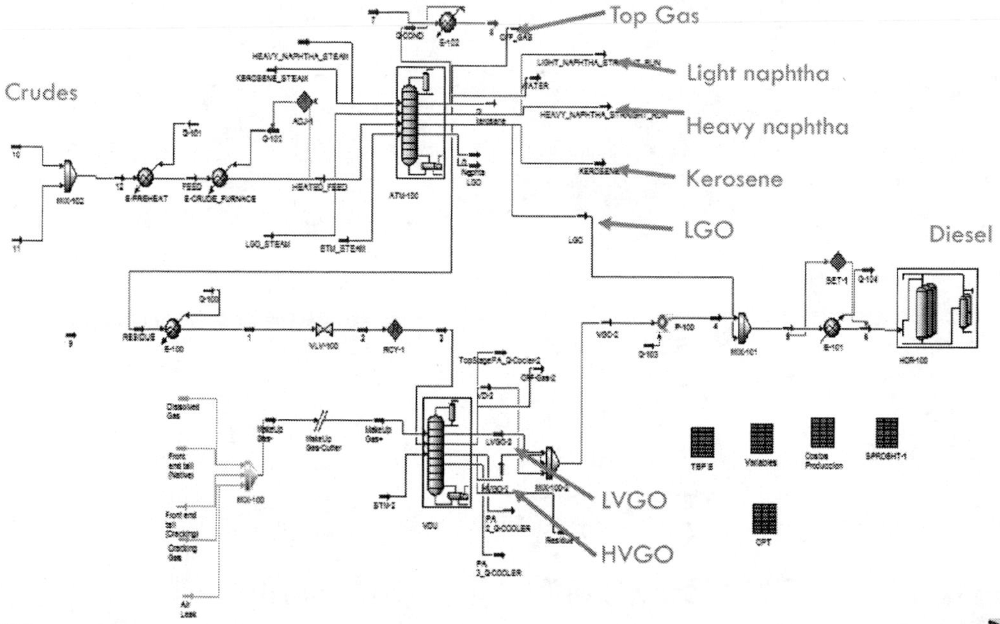

Chemical and Environmental Engineering Department
Universidad Nacional de Colombia

Environmental Engineering
Universidad de Cundinamarca

15

OPTIMIZATION PROBLEM

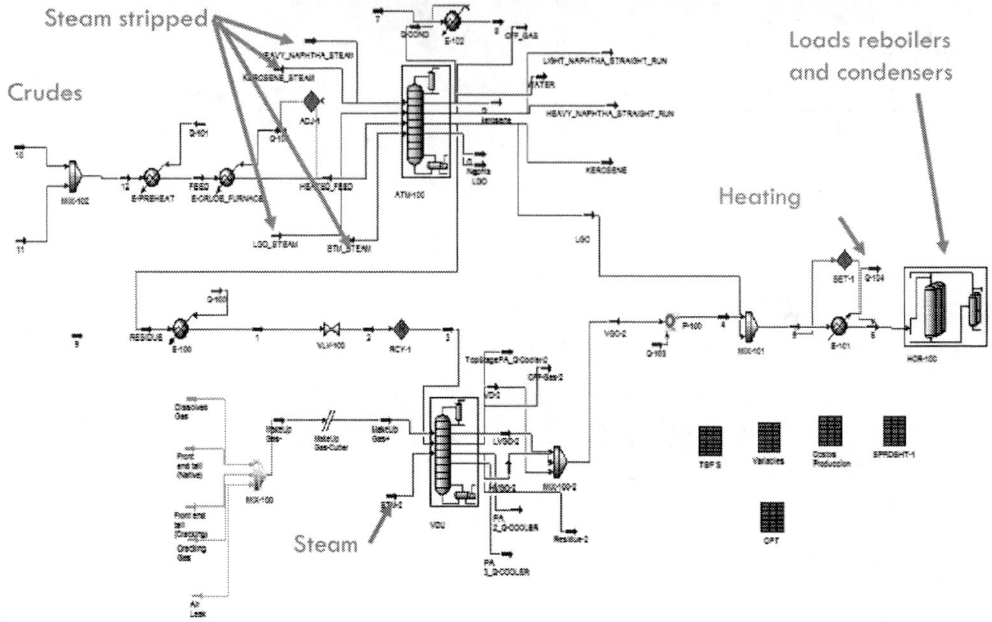

Chemical and Environmental Engineering Department
Universidad Nacional de Colombia

Environmental Engineering
Universidad de Cundinamarca

16

Mono-objective Test Functions

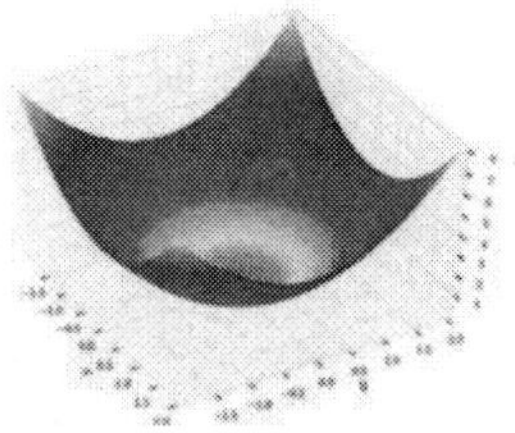

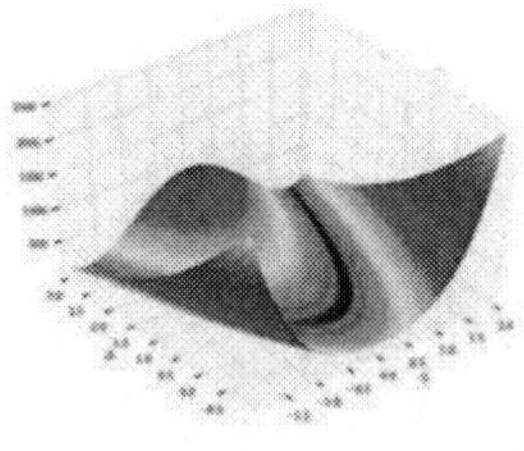

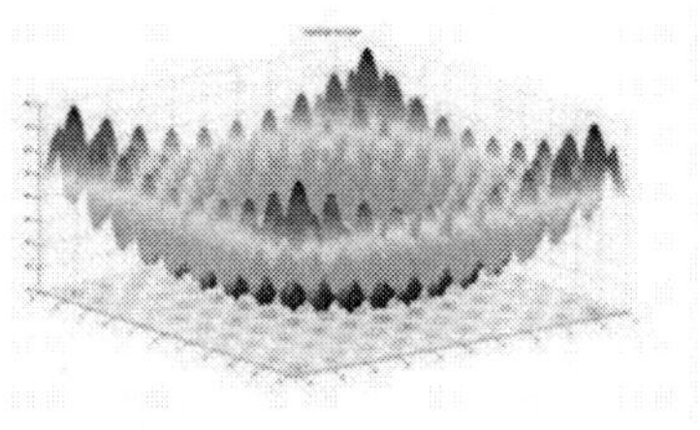

- Cuadratic Rosenbrock Rastrigin

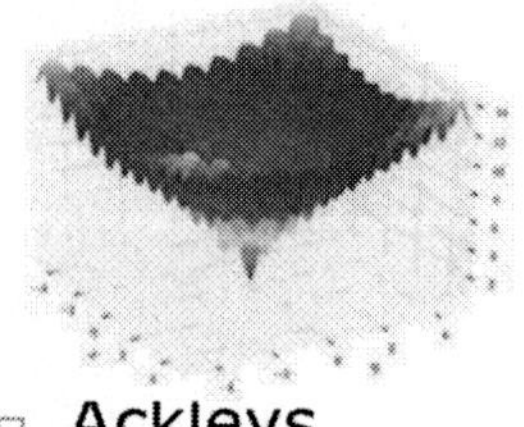

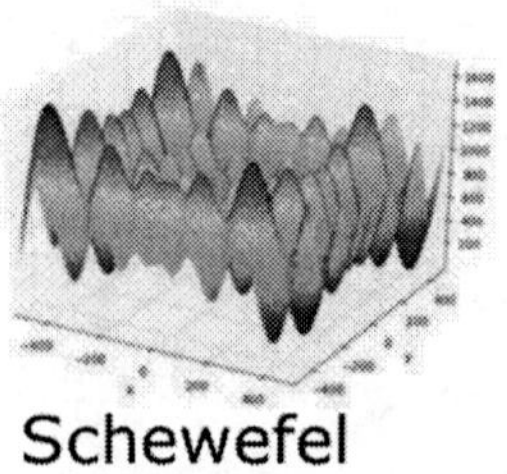

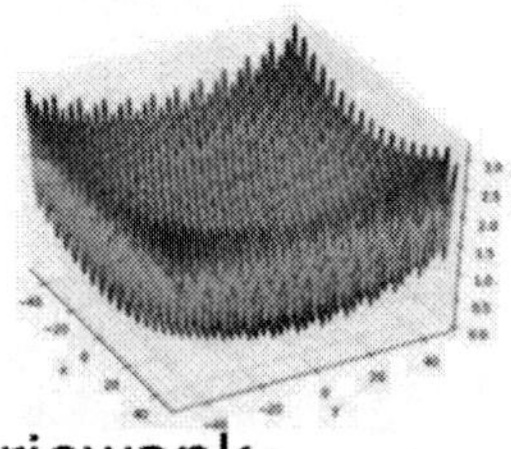

- Ackleys Schewefel Griewank

Chemical and Environmental Engineering Department
Universidad Nacional de Colombia

Environmental Engineering
Universidad de Cundinamarca

17 Mono-objective Test Functions

- KUR

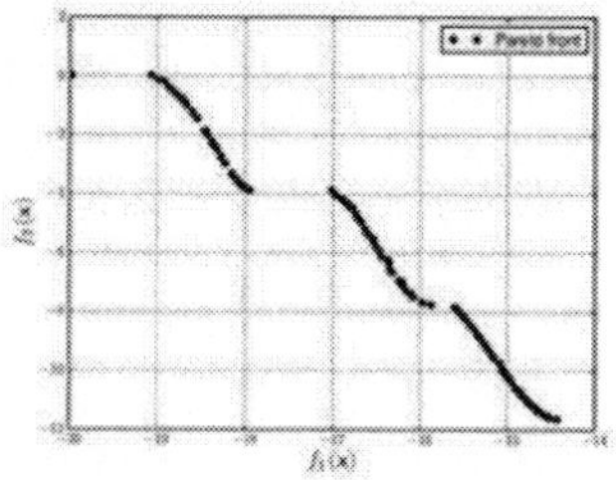

ZDT3

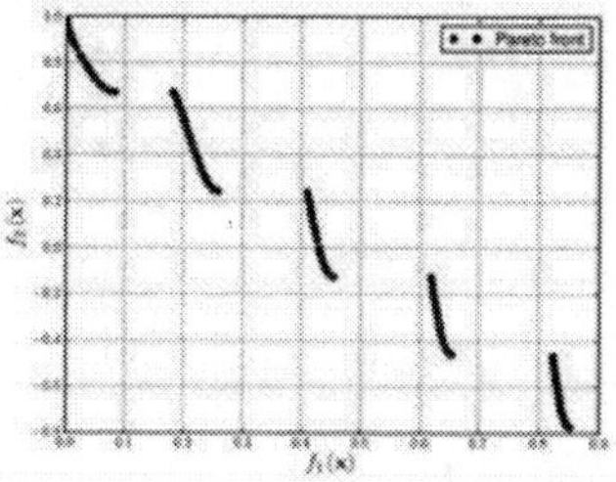

- ZDT4

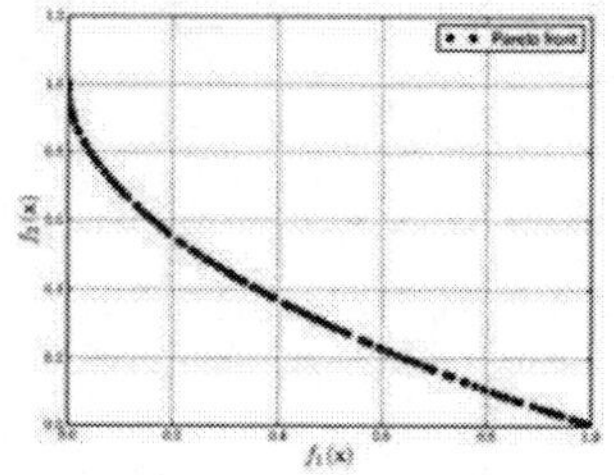

ZDT6

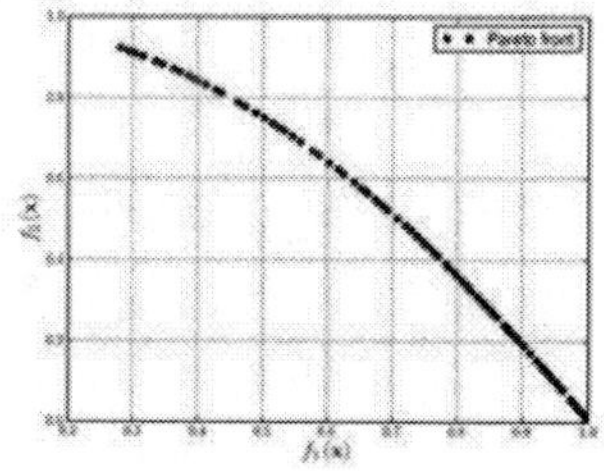

Chemical and Environmental Engineering Department
Universidad Nacional de Colombia

Environmental Engineering
Universidad de Cundinamarca

18 Algorithm selection

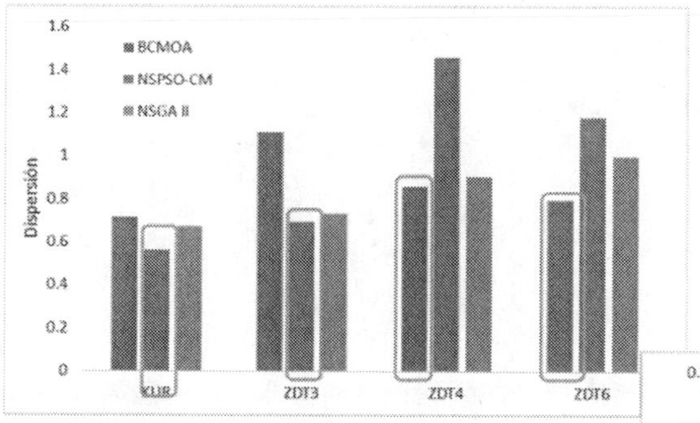

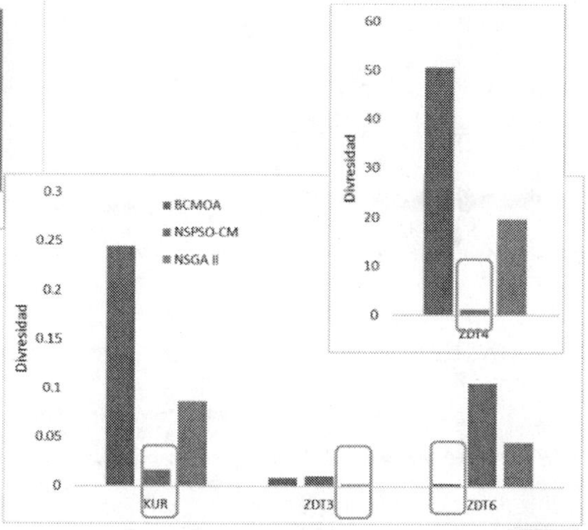

Chemical and Environmental Engineering Department
Universidad Nacional de Colombia

Environmental Engineering
Universidad de Cundinamarca

19

Algorithm selection

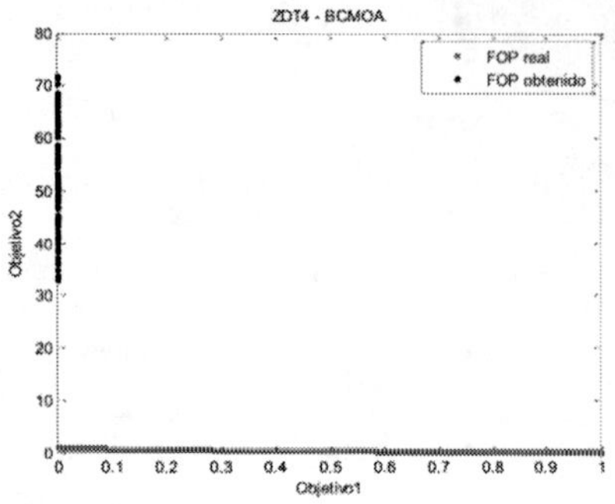

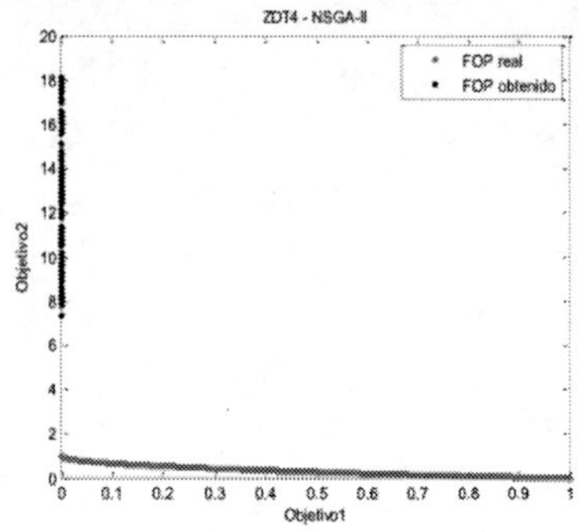

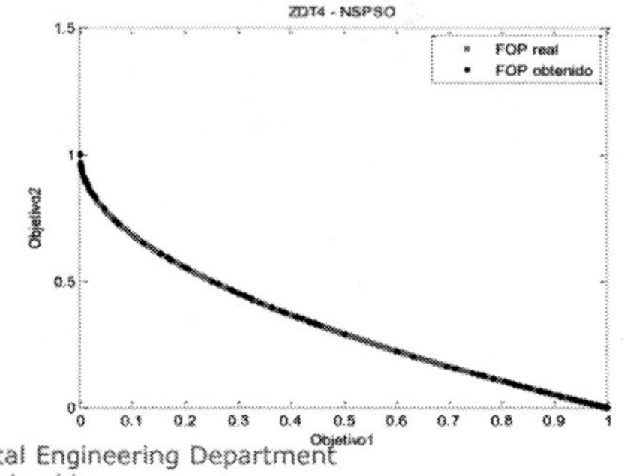

Chemical and Environmental Engineering Department
Universidad Nacional de Colombia

Environmental Engineering
Universidad de Cundinamarca

20

Results Optimization

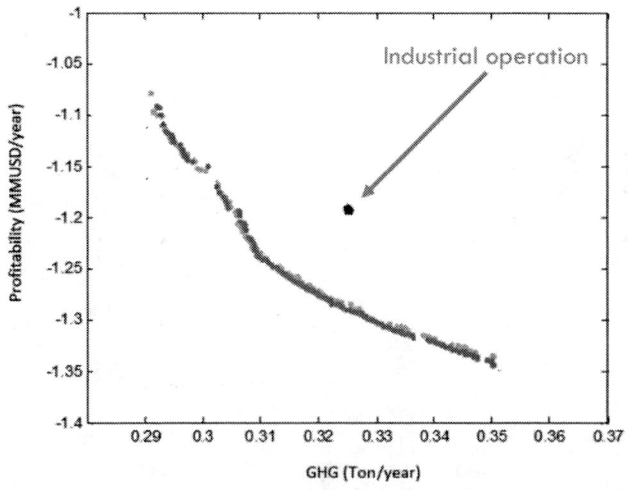

Figura 2. FOP obtained with the NSGA-II algorithm (red), NSPSO (green) and current point operation (black)

Chemical and Environmental Engineering Department
Universidad Nacional de Colombia

Environmental Engineering
Universidad de Cundinamarca

21

Resultados Optimización

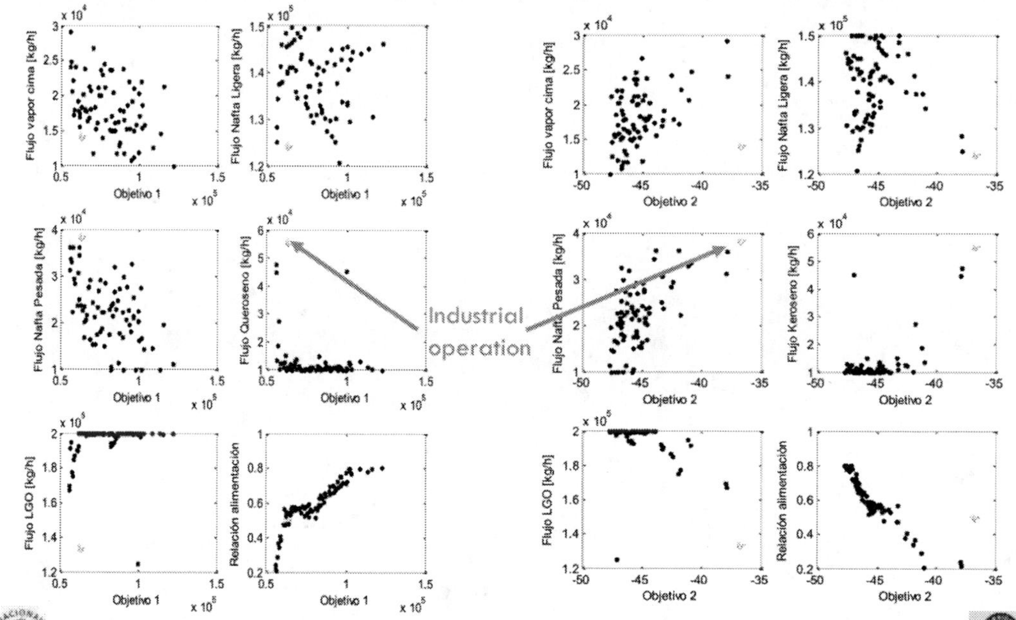

Chemical and Environmental Engineering Department
Universidad Nacional de Colombia

Environmental Engineering
Universidad de Cundinamarca

22

Conclusions

- It was found that the mNSPSO algorithm obtained the best performance
- We obtain a non-dominated set of optimal solutions that allow the maximization of profitability and the maximization of the emission of greenhouse gases.
- Two points of operation stand out from this set of optimal solutions: The first maximizes the quality of the products: peak steam flow (23900 kg / h), light naphtha flow (125130 kg / h), heavy naphtha flow (36089 kg) / h), kerosene flow (47475 kg / h), LGO flow (166960 kg / h) and percentage of crude mixture (22%). The second maximizes the flow produced by diesel: peak steam flow (10000 kg / h), light naphtha flow (146100 kg / h), heavy naphtha flow (10000 kg / h), kerosene flow (10000 kg / h) , flow of LGO (199730 kg / h) and percentage of crude mixture (80%). Both solutions improve the current ones in the refinery.

Chemical and Environmental Engineering Department
Universidad Nacional de Colombia

Environmental Engineering
Universidad de Cundinamarca

23

Chemical and Environmental Engineering Department
Universidad Nacional de Colombia

Environmental Engineering
Universidad de Cundinamarca

Development of a Computational Fluid Dynamic modelling concept for the RecoDust-Process

Franz Edler [a], Wofgang Reiter [a], Bernhard Geier [a], Johannes Rieger [a], Christoph Spijker [b], Harald Raupenstrauch [b]

[a] K1-MET GmbH, Linz, Austria

[b] Chair of Thermal Processing Technology, Montanuniversitaet Leoben, Leoben, Austria

Corresponding author: Franz Edler, Franz-Josef-Strasse 18, 8700 Leoben, Austria

E-mail: franz.edler@k1-met.com

Keywords

RecoDust, Flashreactor, CFD modelling, industrial burner, dust treatment, steel industry, atmosphere particle kinetic model, advanced Flamelet Equilibrium Hybrid model with discrete ordinate model, steel mill dust recycling, zinc reduction

Abstract

The **RecoDust process** offers new possibilities for minimum waste strategies in the iron and steel industry. The idea was realized in the Flash-Reactor pilot plant, whereas the **RecoDust process** represents an innovative pyrometallurgical treatment for steel mill residuals such as dusts with low zinc contents (~ 5 wt-% to 15 wt-%). In the framework of this project, the goal is to further develop the Flash-Reactor regarding the burner geometry, the dust feed charging system and the reactor design. This will be realized by means of Computational Fluid Dynamics (CFD) case studies whereas detailed simulation models are implemented. For a scale-up of the current pilot plant a mathematical CFD model is set-up.

RecoDust-Process

For the steel mill dusts with varying zinc contents a recycling process is limited or impossible without additional treatment. The **Waelz-Kiln** as state of the art treatment is only economical with zinc contents over 20 %. The dust from the blast furnace route possesses zinc contents between 5 wt-% and 15 wt-%, so they have to be recycled externally or landfilled. Aim of the innovative **RecoDust process** is the production of two fractions: A zinc rich oxide fraction and an iron rich fraction. The purity of the products has to be high enough to ensure that both fractions can be used as secondary resources in the zinc or iron industry. The process uses a reducing atmosphere provided by an oxygen/natural gas burner with temperatures around 1700 °C. Within these conditions, the fine dust melts in a reaction vessel and the zinc oxide is reduced and vaporized.

The zinc gas leaves the reaction chamber with the exhaust gas and oxidizes to zinc oxide dust in a post combustion tube. In a final step, the exhaust gas is cooled and subsequently zinc oxide is separated from the flue gas in a bag filter system. The non-volatile components gathered on the bottom of the reaction vessel are discontinuously tapped as a slag. The benefit of the **RecoDust process** is the absence of foregoing treatment steps for the dusts.

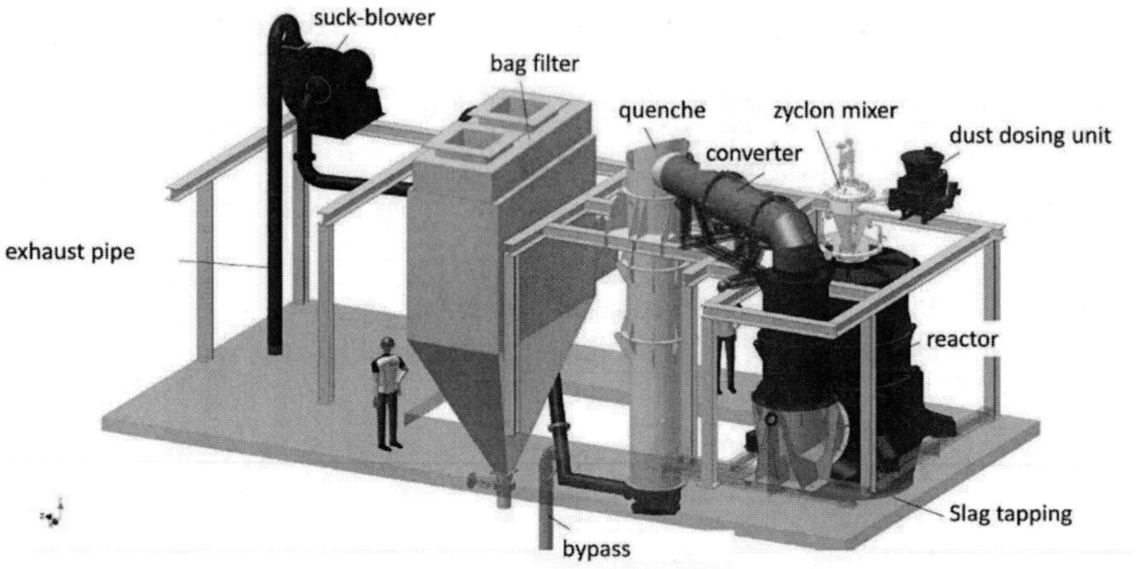

Figure 1: 3D-sketch of the Flash-Reactor pilot plant.

However, to ensure a fast melting procedure, a free floating and dry dust is limited to a grain size less than 1 mm. Unlike most other established processes there are no additives charged into the process, and the reducing conditions are only controlled by the oxygen flow (air excess ratio) in the mixing cyclone and the burner lance. The Flash-Reactor pilot plant as shown in Figure 1 installed in the technical center at the Chair of the Thermal Processing Technology at Montanuniversitaet Leoben has a dust mass throughput up to 300 kg/h.

Simulation concept

The direct-fired dust reactor is in the focus of the development with CFD. A sub-stoichiometric methane / oxygen flame with an air excess ratio < 1 is used in the burner. The main process parameters are particle size and composition, gas composition, particle and gas temperature and burner and reactor geometry. The simulation concept has two blocks, Block A and Block B.

Block A includes fast models to support the campaign planning and evaluating the results on the pilot plant. With the **"Atmosphere Particle Kinetic Model"**, it is possible to check the reactions in range of the calculation settings [1] as shown in Figure 2.

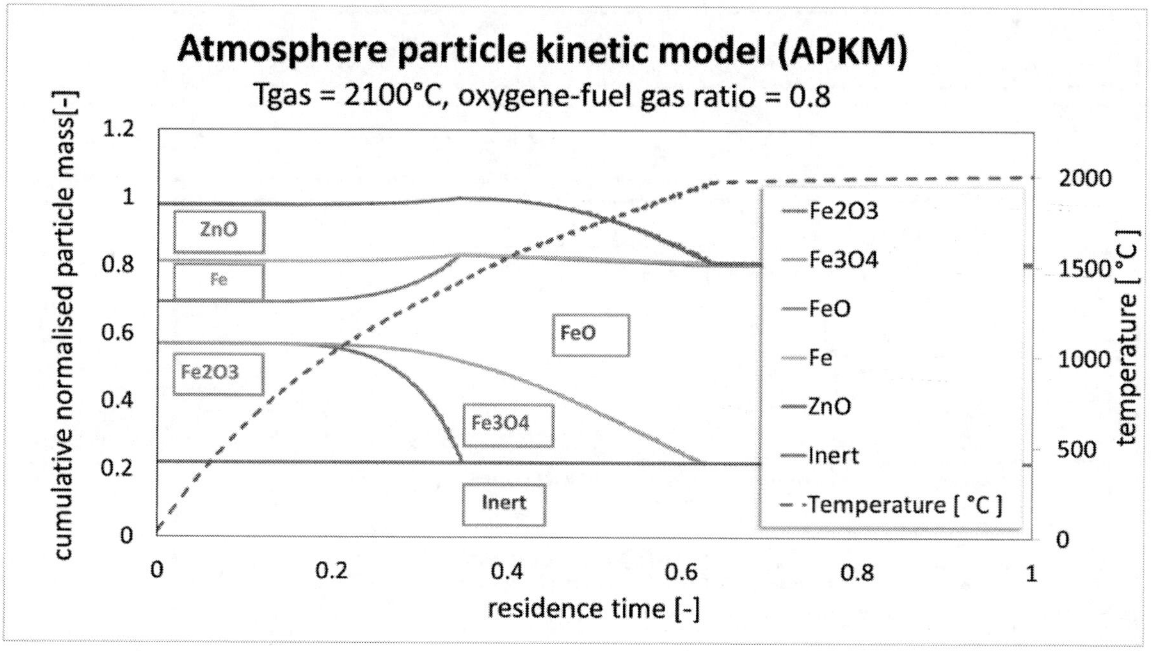

Figure 2: Results of the atmosphere particle kinetic model. Cumulate mass and particle temperature of the composition in dependence on the particle residence time.

The first test case was calculated by using the computational fluid dynamic program ANSYS Fluent v15. The fluid transport is solved in Eulerian framework and the dispersed phase is modeled by using Lagrangian approach. Turbulence model used in this work is realizable-k-ε model, which produces the most stable results. The flue gas composition is calculated with a reduced 17 species reaction mechanism [2] designed to model natural gas combustion. User defined function has been used for particle reactions and heat and mass transfer between fluid and dispersed phase. The mathematical description is based on an assumption of a non-porous dust particle with different compositions and size distributions.

In **Block B** of the simulation concept, the CFD program OpenFoam v2.4 [3] is used. For considering the chemical reaction kinetics, a Flamelet solver based on LibOpenSMOKE [4] [5] is implemented. An advanced Flamelet Equilibrium Hybrid model with discrete ordinate model (DOM) for the radiation [6] and transient thermo parcels, with higher stability of the calculation system are integrated. The flue gas composition is calculated with the above-mentioned reduced species mechanism [2]. The fluid transport is solved in an Eulerian framework and the dispersed phase with a Lagrangian approach. The realizable-k-ε model was used to describe turbulence phenomena [4].

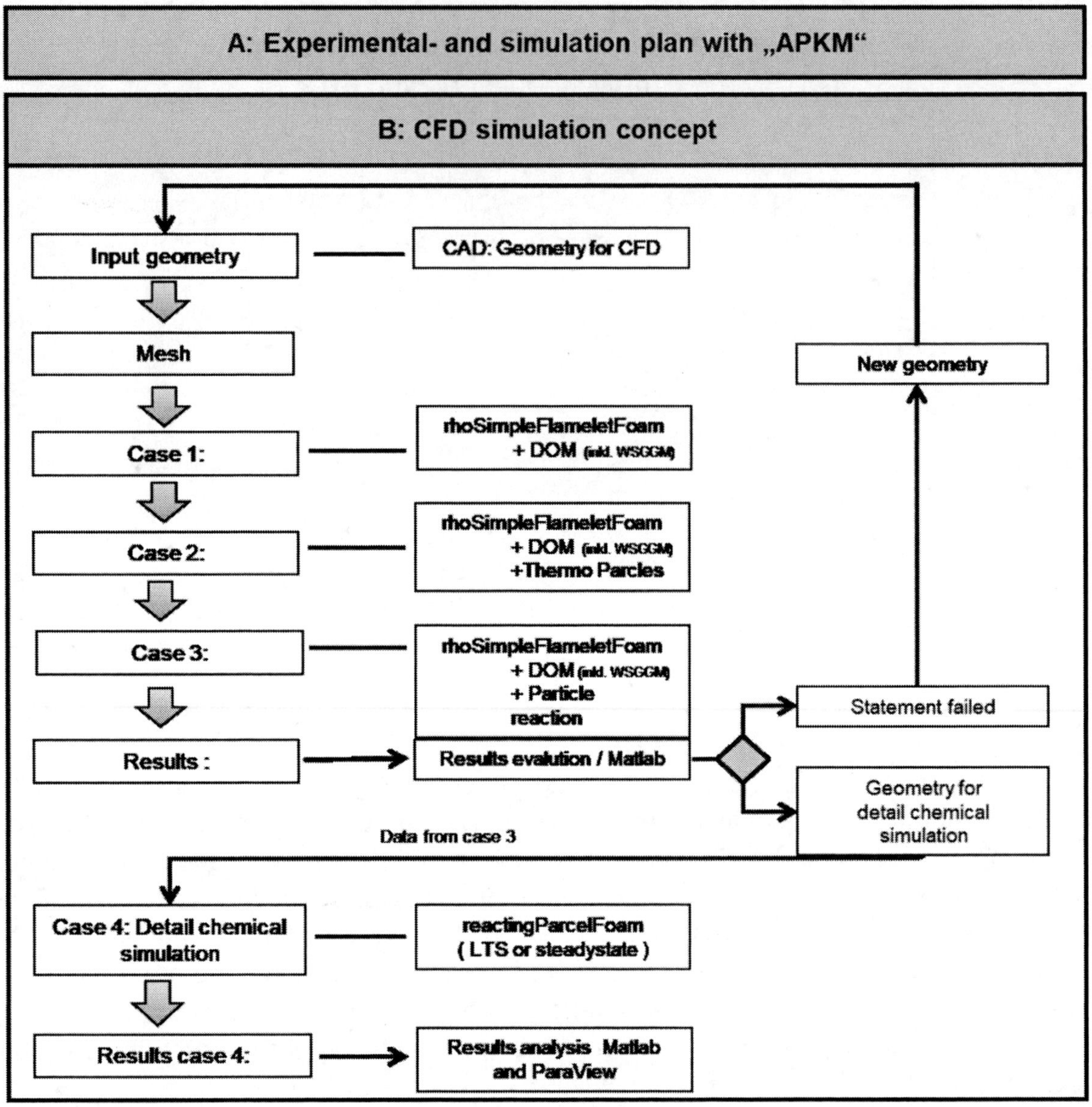

Figure 3: Scheme of Simulation concept.

Figure 3, shows a schematic view of the simulation concept, which describes the order of calculation results from Case 1 to Case 3 to analyze the influence of geometry on process efficiency. With the Flamelet Equilibrium Hybrid model including transient thermo parcel it is possible to check new burner designs and reactor designs in case of gas flow and particle flow through the Flash-Reactor. The velocity distribution, temperature distribution and the particle transport out of the reactor (= carry over) are important system parameters to evaluate the geometry design. In case that the assumed reactor design and process parameters meet the expectations in terms of process efficiency, next step of the simulation concept can be started (= Case 4). This case includes the calculation of the particle kinetic coupled with the fluid phase. The solution from the Flamelet Case 3 transferred to an OpenFOAM Partially Stirred Reactor model with transient parcel tracking. After the detailed chemical calculation, it is possible to evaluate the particle and the slag distribution from the wall film, the bottom slag or the particle carry over into the exhaust pipe, respectively. These results enable a comparison of calculated with experimental data.

Results and discussion

It is mandatory for the scale up of the pilot plant to understand the conditions inside the reactor. In Figure 4, the particle distribution colored by particle diameter in the reactor and in the pipe to the converter. The particles follow the velocity streamlines depending on the particle size and density. In Figure 4, the particles colored by diameter within the fluid volume are plotted.

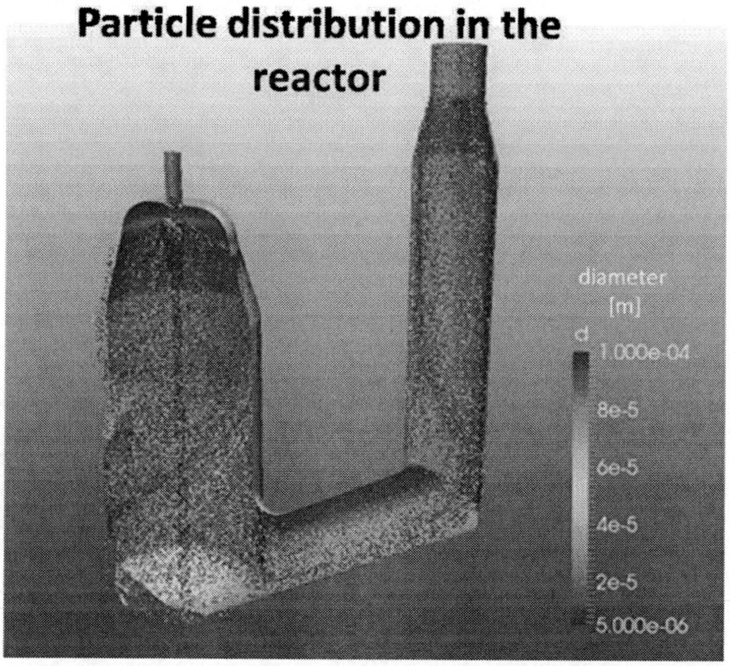

Figure 4: Particle distribution in the reactor

In the Figure 5, explicit chosen calculated results of the three-dimensional CFD case studies are shown. As discussed above, the fluid condition and the flame shape are important for the process. The experimental data quantifies the particle carry over out of the reactor is being influenced by the velocity distribution inside the reactor. The shape of the flame and high gradients from the temperature distribution, as well as the residence time are important for the reduction of the zinc oxide. As shown in Figure 5 below on the right, the velocity streamlines confirm the inhomogeneous conditions in the reactor. This means that due to the burner geometry and the resulting mixing behaviour inside the reactor, an intensive contact between solids and gas is realized leading to a high zinc conversion rate. In Figure 5 on the top right above, the input of zinc oxide for different particle classes in the wall film is shown. The particle size is in the range between 1 um (class 1) and 1000 um (class 10).

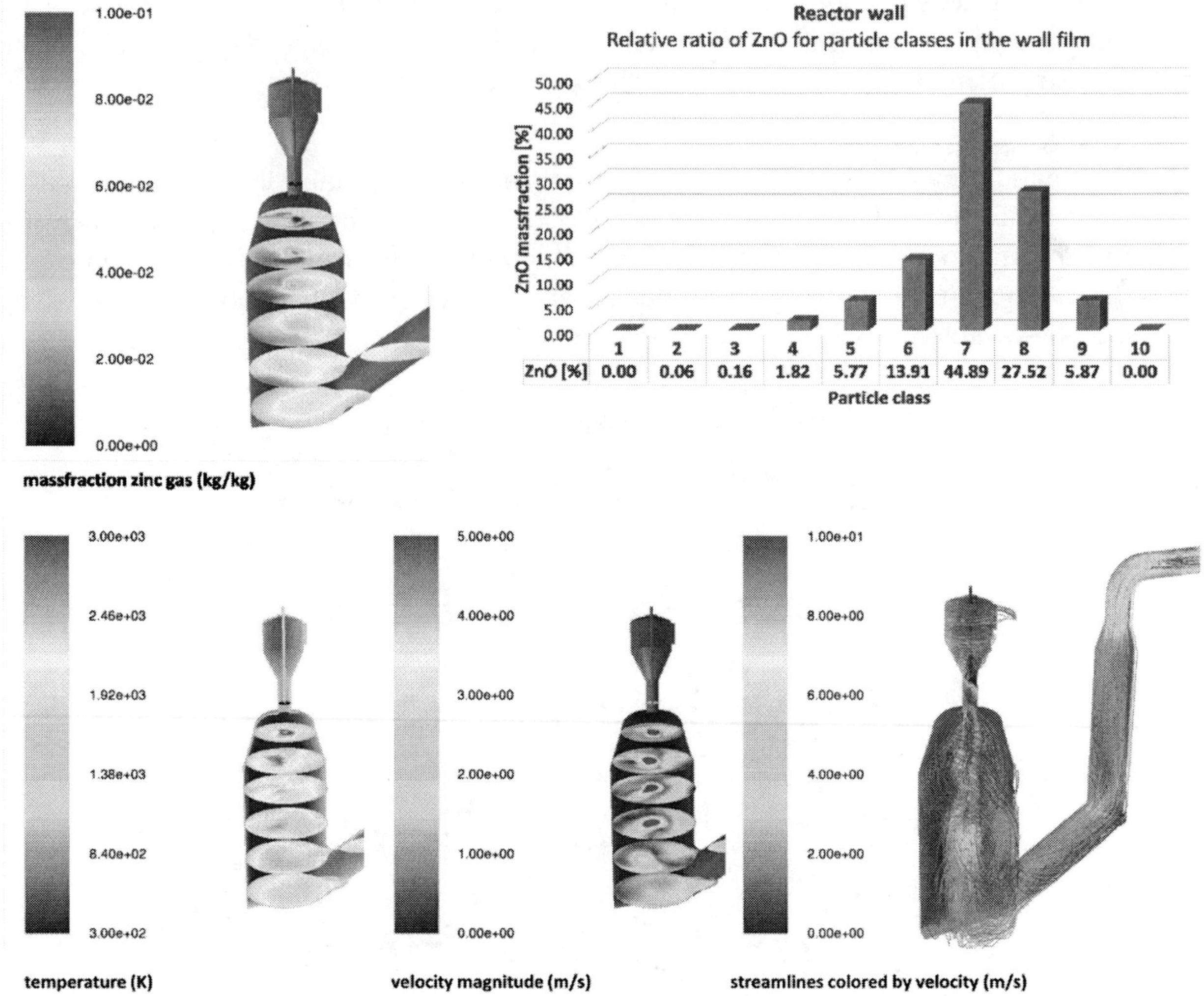

Figure 5: Results from the CFD calculation and Data evaluation with Matlab. Figure on the top left above: Mass fraction of zinc (g) in different cross sections of the reactor. Figure on the top right above: Data evaluation with MATLAB regarding the relative ratio of ZnO contained in each particle class at the reactor wall film. Figure below on the left: Temperature distribution in different cross section of the reactor. Figure below in the middle: Velocity magnitude in different cross sections of the reactor. Figure below on the right: Streamlines colored by velocity magnitude.

Outlook

The implementation of simple Flamelet models for case 1 to case 3 is completed. For future work, the particle reactions in the OpenFOAM "partial stirred reactor model solver" will be implemented. Finally, all Simulations are able to run on OpenFOAM and a routine for result analysis could be programmed.

Acknowledgements

The authors gratefully acknowledge the funding support of K1-MET GmbH, metallurgical competence center. The research program of the competence center K1-MET is supported by COMET (Competence Center for Excellent Technologies), the Austrian program for competence centers. COMET is funded by the Federal Ministry for Transport, Innovation and Technology, the Federal Ministry for Science, Research and Economy, the province of Upper Austria, Tyrol, and Styria and the Styrian Business Promotion Agency.

References

[1] F. Edler, "Development of an atmosphere particle kinetic model for particle reactions in a combustion Flash-Reactor using CFD- methods.," in *Energy Procedia*, 2017.

[2] "Cantera," [Online]. Available: http://www.cantera.org. [Accessed 11 04 2017].

[3] S. Thynell, "Discrete-ordinates method in radiative heat transfer," *International Journal of Engineering Science Volume 36,* pp. Issues 12-14, pp. 1651-1675, 1998.

[4] C. Spijker, "Unsteady Laminar Flamelet Modellierung zur Beschreibung von Mündungsmischbrennern," Montan Universitaet Leoben, 2010.

[5] "OpenFOAM," [Online]. Available: http://www.openfoam.com/. [Accessed 11 04 2017].

[6] T. Holzmann, "www.holzmann-cfd.de," [Online]. Available: http://www.holzmann-cfd.de/index.php/de/flamelet-model/14-development. [Accessed 11 04 2017].

[7] A. Couci, A. Frassoldati , T. Faravelli und E. Ranzi, „libOpenSMOKE, User-Guide," 2012.

499467 Dynamic Simulation and Optimization for Gas Expanded Liquid Phase of Ethylene Oxide Production with Free Carbon Dioxide Emission

Abstract:

Ethylene oxide production is considered as one of the processes that produce carbon dioxide as byproduct. In response to the global warming issue, the following work illustrates alternative synthesis to produce ethylene oxide with free carbon dioxide emission. The new procedure was found by The Center for Environmentally Beneficial Catalysis (CEBC), University of Kansan and improved by Laboratory of Integrated Systems Engineering (LISE), Lamar University. The ethylene oxide is produced by reacting ethylene and hydrogen peroxide which is the reaction oxidant over methyltrioxorhenium catalyst (MTO) in a continuous stirred reactor at 104 âµ'F and 725.18 psia.

At LISE, the new steady state design has been built on Aspen Plus version 8.8, focusing on solving some of the problems that have been mentioned by CEBE and improving the distillation and purification process. Therefore, the design is consisted of two continuous stirred reactors, a flash drum, a trays column, and two packed bed columns.Â In response to the waste of unreacted hydrogen peroxide, a better design layout set has been developed to recycle and regenerate that hydrogen peroxide back to the main epoxidation reactor. Fortunately, these adjustments lead to optimize the alternative ethylene oxide synthesis, it decreases the production cost to 52 ¢/lb. of ethylene oxide, and the payback period for new ethylene oxide production plant is 3.48 year which is evaluated by Aspen Process Economic Analyzer.

For further studies on the process safety and process operation stability, a dynamic well controlled model is built in Aspen Dynamic version 8.8. The safety of the process has been assigned as the first priority, and the process capability is the second priority in designing the control scheme and control parameters tuning for the dynamic model. First, the conversion in the main epoxidation reactor is controlled by the reactor temperature to specified value, so the catalyst stays as active as possible. Second, with the hydrogen peroxide decomposer, the decomposition reaction is controlled by the reactor temperature to ensure that all the hydrogen peroxide is decomposed before entering the separation and purification step. As well as, the reactor vapor outlet composition is controlled by the reactor pressure to keep that out of the flammability zone for ethylene oxide, ethylene, and methanol (which is reaction solvent). Third, with the hydrogen peroxide recycle stream, the decomposed oxygen is regenerated to hydrogen peroxide; therefore, the main control objective is to make sure all oxygen is converted to hydrogen peroxide before they reenter the epoxidation reactor. Fourth, with the distillation columns, the main objective for the first tower is to recycle most of the unreacted ethylene back to the epoxidation reactor. On the second column, the objective is to get the water in the bottom stream to be pure as much as 99.995% mol. On the last column, the objective is to achieve the desired market purity for ethylene oxide. Finally, by applying all the previous objectives and specifications to the model, the process operation stability could respond on any disturbance that is generated by upstream set point changes up to 15% of its steady state values.

Also, we have gone further, and we rebuilt the control system in the DeltaV DCS system to control the simulated process through OPC Client server, so we can have a more realistic and reliable model to the real plant operation.

Right now, we are working on alarm rationalization study using DeltaV system, and it will be done by the conference time. Beside that work, we will make a model predictive control model which will be generated by Aspen DMC Plus and DeltaV that will be used to do further study on dynamic optimization and real time optimization.

500617 Inter-Individual Variability in Physiological Response to Lipid Infusions Predicts Considerable Heterogeneity in Outcomes for Lipid Resuscitation: A Physiologically Based Pharmacokinetic-Pharmacodynamic Study in a Virtual Population

Abstract:

Local anesthetic systemic toxicity (LAST) is a life-threatening event in which the introduction of anesthetic to the bloodstream induces cardiac arrest. Anesthetic-induced cardiac arrest is refractory to standard life support measures and is thus often fatal. A bid to understand the etiology of hypersensitivity to local anesthetics led to the discovery of a potential antidote for LAST: intravenous lipid emulsions. These suspensions, composed primarily of vegetable oil, egg phospholipids, and water, are freely available in clinical settings. Their off-label use as antidotes to anesthetic toxicity first yielded a successful clinical resuscitation in 2006. Two years later, a lipid suspension was used to revive a patient suffering from >70 minutes of cardiac arrest that was unresponsive to all other resuscitation efforts. Intriguingly, the offending toxin in this case was not an anesthetic, but a combination of 2 other pharmaceutical compounds. An ensuing series of small animal studies and anecdotal case reports hints at the possibility of lipid emulsions being broad-spectrum antidotes to drug toxicity.

Unfortunately, our understanding of when and how lipid interventions prove beneficial is still limited to clinical anecdotes and small animal experiments. Both sources of information are plagued by heterogeneous outcomes and limited exploration of the multivariate parameter space characterizing cardiac life support. Hence, even the underlying mechanism of therapeutic efficacy has long been the subject of speculation and heuristic reasoning with limited possibility for experimental validation. What we now know about the multifunctional mechanism of action has been obtained through our groupâ€™s integrative studies incorporating mechanistic mathematical modeling.

In our most recent work, we sought to further demonstrate how mathematical analysis can be used to gain insight into the likely success of lipid-mediated resuscitation. Importantly, we embraced the inherent variability between pre-clinical subjects and established how estimation of relevant physiological parameters on an individual basis can inform treatment protocols by permitting the analysis of outcome heterogeneity over simulated virtual populations. Using data previously collected in a study involving rats (N=7), we employed uncertainty quantification approaches to identify a subset of 2 key parameters that could be estimated for each subject. One parameter was determined to be correlated with the magnitude of cardiac response to lipid infusion, while the other played a role in determining the smoothness of the response over time. To account for the dynamic nature of these underlying physiological processes, we adjusted our existing pharmacokinetic-pharmacodynamic model to allow time variation of the 2 selected parameters; these were permitted to vary according to four linear splines of equal length. Model comparison tests were employed to ascertain that the choice of 4 splines was most appropriate for our data set.

From our individualized estimates, we constructed population distributions for each parameter. We then sampled from these distributions to create a virtual population (N=10,000) undergoing emulsion therapy after a simulated overdose of the anesthetic bupivacaine. The virtual population exhibited a wide range of expected outcomes, with only a subset of the population expected to recover baseline cardiac function upon treatment. We thereby demonstrate the limitations of using mean responses from small preclinical studies to heuristically posit guidelines for lipid-mediated cardiac resuscitation.

496390 Effect of Circadian Disruption on Hepatic Gluconeogenesis

Abstract:

The circadian machinery is biological time-keeping mechanisms which orchestrate the biological changes that follow a 24-hour cycle. In mammals, the superchiasmatic nuclei (SCN) in the hypothalamus takes the light/dark cycle as the entraining signal and acts as the central pace maker that synchronizes the clock genes in the peripheral organs, such as muscles, liver, heart, and kidneys. The rhythmicity of hormonal and metabolic signals play an important role in immune [1] and metabolic functions [2] by maintaining and controlling the phase relations among the various clocks. The robustness of circadian rhythms is associated with a homeostatic state of the host while disruption in the circadian rhythms is observed in a diseased state of the host [3].

Aside from the light/dark cycle, food intake is a strong *zeitgeber*, a cue given by the environment on host to reset the internal body clock. Feeding rhythms are a more potent *zeitgeber* compared to the light/dark cycle eventually synchronizing the peripheral clocks away from the SCN, especially in organs governing the energy homeostasis of the host such as the liver. Therefore, an interconnection between the circadian clocks and metabolism is evident, and it is supported by a multitude of studies. At a clinical level, several studies correlate the disruption of circadian rhythms to increased rate of metabolic syndrome. Shift-work and sleep deprivation dampen the circadian rhythms, and are linked with obesity and high level of triglycerides [4]. At a molecular level, Clock mutant mice showed decreased metabolic rate [5] while liver-specific *Bmal1* knockout disrupted glucose homeostasis in mice [6]. These studies suggest that a bidirectional relationship exists between the circadian clocks and metabolism, probably as parts of a network of complex signaling cascades that intricately regulate one another.

Moreover, one can realize from these observations that a broader environmental or behavioral cues (such as feeding patterns) can influence specific, low-level targets (such as liver enzymes). To explore this hypothesis, we would like to demonstrate by a mathematical model how the feeding/fasting cycle influences a specific metabolic function in relation to the circadian rhythms entrained to the light/dark cycle. As a first step toward answering this question, we have previously demonstrated via mathematical modeling how the rhythms of the peripheral clock genes (PCGs) in the liver can be completely entrained to the feeding rhythms, synchronizing away from the light/dark cycle. We further developed this model to explore the effects of light-feeding phase relations on hepatic metabolism. We model the transcription of two genes, *G6pc* and *Pck1*, as a surrogate for glucose metabolism in the human liver.

In our model, PPARγ coactivator-1a (PGC1-a) is a key molecule that incorporates the effects of light and feeding to the transcription of the gluconeogenic genes. Transcription of PGC-1a is activated by forkhead box protein O1 (FOXO1) binding upstream of *Pck1* and *G6pc* [7]. When in the nucleus, PGC-1a is deacetylated by SIRT1, which results in an increased activity [8]. The glucocorticoid receptor is also required along with HNF-4a for full activation of *Pck1* promoter [9]. PGC-1 has shown to be induced synergistically in primary liver cultures by cAMP and glucocorticoids. Additionally, *Pck1* promotor requires co-activation of the glucocorticoid receptor and the liver-enriched transcription factor HNF-4a by PGC-1 for full transcriptional activation. HNF-4a is necessary for the action of PGC1, as hepatocytes from mice lacking HNF-4a loses the ability of PGC-1 to activate key genes of gluconeogenesis such as *Pck1* and *G6Pc* [10]. The role of HNF-4a and PGC-1a were lumped for simplicity in our model. FOXO1 also plays an important role in gluconeogenesis. PGC-1a binds and co-activates FOXO1 in a manner inhibited by Akt-mediated phosphorylation [11]. FOXO1 is known to be required for robust activation of gluconeogenic gene expression in hepatic cells and in mouse liver. Since FOXO1 binds to a DNA motif upstream of the gluconeogenic genes, we used it to model the signal transmission from PGC-1a to transcription of gluconeogenic genes [11]. Finally, PGC-1a also activates the transcription of Bmal1 (Equation 7) [12].

Our model builds upon earlier works [13] of a semi-mechanical model for light entrainment on the peripheral clock genes through the HPA axis. The main driver for oscillations in the HPA axis are the negative feedback between glucocorticoid and corticotropin-releasing hormone (CRH)/adrenocorticotropic hormone (ACTH) using a modified Goodwin oscillator. The glucocorticoid released from the central compartment regulates the peripheral clock genes (*Per/Cry*, *Bmal1*, and *Clock*). The feeding/fasting cycle influences the periphery by modifying the redox reactions between NAD^+ and NADH [14]. The feeding-entrained NAD^+ activates human sirtuin 1 (SIRT1), a class II histone deacetylase. SIRT1 then acts in coordination with CLOCK/BMAL1 complex, interacting with the PCGs in several ways. It facilitates the transcription of NAMPT, which is the rate-limiting enzyme for the NAD^+ salvage cycle, regulated by the CLOCK/BMAL1 complex. NAD^+, in turn, activates SIRT1, as well as entraining glucocorticoids through a transit compartment that represents neural input through the ventromedial arcuate nucleus [15]. SIRT1 also facilitates the degradation of PER/CRY proteins in the nucleus, exerting more influence on the peripheral clock genes. Finally, the transcription of *G6pc/Pck1* (gluconeogenic gene) is controlled by binding of glucocorticoid-receptor complex to the glucocorticoid response element on the promotor region, as well as FOXO1 binding also to the promotor region.

Current model successfully predicts some observations from experimental time-restricted feeding studies. Some basic predictions include that restricting feeding during the active phase results in phase shift of total gluconeogenic enzyme amount and the increase of amplitude in PEPCK total enzyme amount [16]. Another interesting prediction is that when the system is subjected to constant lighting, the oscillation in G6Pase activity becomes less robust [17]. Furthermore, overexpression of CRY protein results in decreased transcription of gluconeogenic genes [18]. Disruption of circadian clocks by the means of Clock gene knockout resulted in over-expression of *G6pc* and *Pck1*, while the nominal expression level could be partially recovered by altering the light-feeding phase relations. To analyze this complex network, sensitivity analysis of the model was performed in three parts: 1) local sensitivity analysis of all parameters; 2) selection of important elementary effects by Morris method; and 3) global uncertainty analysis employing the high dimensional model representations (HDMR). Interpreting and integrating the results from all three methods revealed that transcription and activation of PGC-1a and FOXO1 along with *Per/Cry* transcription had the greatest influence over the expression level of gluconeogenic genes, demonstrating the close relationship between circadian rhythms and hepatic metabolic activities.

References

1. Silver, A.C., et al., *The circadian clock controls toll-like receptor 9-mediated innate and adaptive immunity.* Immunity, 2012. **36**(2): p. 251-61.

2. Feillet, C.A., U. Albrecht, and E. Challet, *"Feeding time" for the brain: a matter of clocks.* J Physiol Paris, 2006. **100**(5-6): p. 252-60.

3. Fang, M., et al., *Enhancement of NAD(+)-dependent SIRT1 deacetylase activity by methylselenocysteine resets the circadian clock in carcinogen-treated mammary epithelial cells.* Oncotarget, 2015. **6**(40): p. 42879-91.

4. Spiegel, K., et al., *Effects of poor and short sleep on glucose metabolism and obesity risk.* Nat Rev Endocrinol, 2009. **5**(5): p. 253-61.

5. Turek, F.W., et al., *Obesity and metabolic syndrome in circadian Clock mutant mice.* Science, 2005. **308**(5724): p. 1043-5.

6. Lamia, K.A., K.F. Storch, and C.J. Weitz, *Physiological significance of a peripheral tissue circadian clock.*

Proc Natl Acad Sci U S A, 2008. **105**(39): p. 15172-7.

7. Daitoku, H., et al., *Regulation of PGC-1 promoter activity by protein kinase B and the forkhead transcription factor FKHR.* Diabetes, 2003. **52**(3): p. 642-9.

8. Schwer, B. and E. Verdin, *Conserved metabolic regulatory functions of sirtuins.* Cell Metab, 2008. **7**(2): p. 104-12.

9. Yoon, J.C., et al., *Control of hepatic gluconeogenesis through the transcriptional coactivator PGC-1.* Nature, 2001. **413**(6852): p. 131-8.

10. Rhee, J., et al., *Regulation of hepatic fasting response by PPARgamma coactivator-1alpha (PGC-1): requirement for hepatocyte nuclear factor 4alpha in gluconeogenesis.* Proc Natl Acad Sci U S A, 2003. **100**(7): p. 4012-7.

11. Puigserver, P., et al., *Insulin-regulated hepatic gluconeogenesis through FOXO1-PGC-1alpha interaction.* Nature, 2003. **423**(6939): p. 550-5.

12. Green, C.B., J.S. Takahashi, and J. Bass, *The meter of metabolism.* Cell, 2008. **134**(5): p. 728-42.

13. Mavroudis, P.D., et al., *Mathematical modeling of light-mediated HPA axis activity and downstream implications on the entrainment of peripheral clock genes.* Physiol Genomics, 2014. **46**(20): p. 766-78.

14. Diaz-Munoz, M., et al., *Anticipatory changes in liver metabolism and entrainment of insulin, glucagon, and corticosterone in food-restricted rats.* Am J Physiol Regul Integr Comp Physiol, 2000. **279**(6): p. R2048-56.

15. Yi, C.X., et al., *Ventromedial arcuate nucleus communicates peripheral metabolic information to the suprachiasmatic nucleus.* Endocrinology, 2006. **147**(1): p. 283-94.

16. Perez-Mendoza, M., J.B. Rivera-Zavala, and M. Diaz-Munoz, *Daytime restricted feeding modifies the daily variations of liver gluconeogenesis: adaptations in biochemical and endocrine regulators.* Chronobiol Int, 2014. **31**(7): p. 815-28.

17. Ashman, P.U. and J.R. Seed, *Biochemical studies in the vole, Microtus montanus. II. The effects of a Trypanosoma brucei gambiense infection on the diurnal variation of hepatic glucose-6-phosphatase and liver glycogen.* Comp Biochem Physiol B, 1973. **45**(2): p. 379-92.

18. Lamia, K.A., et al., *Cryptochromes mediate rhythmic repression of the glucocorticoid receptor.* Nature, 2011. **480**(7378): p. 552-6.

498747 A PK/PD Model of ACE Inhibition in Kidney Cells for Treatment of Diabetic Tissue Damage

Abstract:

Diabetes is the primary cause of chronic kidney disease (CKD). High glucose conditions have been known to initiate and exacerbate the pathophysiology of CKD particularly through damage to podocytes. Podocytes are terminally differentiated renal cells that play a key role in maintaining the structure and function of the kidneys. Podocytes express a local renin-angiotensin system (RAS) that is altered in diabetes. Studies have shown that Angiotensin II (ANG II), a blood pressure regulating hormone, produced by the RAS is upregulated in diabetic conditions and triggers podocyte injury and apoptosis. Angiotensin converting enzymes (ACE) inhibitor drugs inhibit the production of ANG II. However, these drugs are not well characterized for use in CKD. The progression of CKD could be slowed by controlling the ANG II levels to prevent irreversible podocyte loss. Due to sparse experimental evidence for glucose-dose-dependency of ANG II, especially for podocytes, we propose and parameterize a mathematical model to describe the glucose-stimulated RAS in podocytes that produces ANG II. The model can connect drug concentration to the inhibition of the ANG II in RAS in podocytes for the ACE inhibitor drugs in the cases of normal and impaired renal function.

We define a model for the pharmacokinetic RAS network that triggers the synthesis of ANG II and the pharmacodynamic physiological response to ACE inhibitors in a time and dose-dependent manner for normal and impaired kidney. The model takes glucose dynamics and drug dosage as the input. The glucose-sensitive podocyte RAS pathway is represented by a system of ordinary differential equations to track ANG II and other RAS peptides. Glucose-dependence is added through different enzyme activity parameters. A one-compartment model with first-order absorption from the site of administration and first-order elimination is used here for the oral administration of ACE inhibitor drugs. Previous experimental studies are used to estimate the unknown parameters and kinetic constants for the model. The parameters are analyzed by local and global sensitivity analyses along with other validation cases. MATLAB is used to solve the system of ordinary differential equations, fit parameters to data, and to develop a graphical user interface (GUI).

The model fits a mathematical function to experimental data for glucose dynamics and uses it to predict the corresponding change in ANG II concentrations with the pharmacokinetic RAS network model. The pharmacodynamic part of the model shows that ANG II concentration is lowered by the drugs. The model is used to explore the effects of different drug dosage on the levels of the diacid form of the drug and ANG II in diabetic or normal patients with normal or impaired kidney function over time.

We have developed a physiologically relevant PK/PD model for ACE inhibition parameterized for local RAS in podocytes for drugs in normal and impaired renal conditions. The model can use experimental data for glucose dynamics from normal and diabetic subjects and predict the dosage for each drug in normal or impaired kidney to control the ANG II levels. The model is packaged as a MATLAB app to facilitate reuse for research and educational purposes. The result of this work will enable further studies on the impacts of ACE inhibition in chronic kidney disease or other diseases.

495632 Dynamic Analysis of the DNA Sensing Pathway Predicts Host Immune Response

Abstract:

Introduction

To effectively respond to a pathogen, the host intracellular immunity must dynamically assess and change with the environment. Dysregulation of the immune response can increase disease pathology and impair viral clearance. This can lead to chronic, systemic infections which—left untreated—can be fatal [1]. It is therefore imperative to develop a predictive model describing how pathogens invade the host and abrogate the immune response.

Significant research has gone into identifying key host proteins that perpetuate the signaling cascades responsible for innate immunity. The recently identified DNA sensing molecule cGAS recognizes viral DNA and invokes a type I interferon response. This leads to paracrine signaling of the JAK-STAT pathway and subsequent transcription of interferon stimulated genes (ISGs). ISGs effect various elements of the virus life cycle, many of which have specific mechanisms that are not yet known [2]. While this is an excellent description, it fails to accurately characterize the dynamics of the viral DNA response. Understanding these dynamics can help explain observations that are not addressed by static, qualitative descriptions. For example, it has been well established that viruses with similar genotypes exhibit distinct kinetics which alter the course of infection [3]. Modeling will elucidate the discrepancies in experimental observation like this and motivate new hypotheses.

Methods

To characterize the DNA sensing pathway's behavior in response to viral DNA, ordinary differential equations (ODEs) are employed to simulate how host proteins coordinate and activate ISGs. The model is comprised of 16 differential equations that represent the forward and reverse rates for each species (proteins and mRNAs) in the pathway. Typical formulations for biochemical reactions are used to define these rates including mass action, Michaelis Menten, and Hill kinetics. The latter is useful for modeling highly cooperative binding events such as the interactions between cGAS, viral DNA, ATP, and GTP to form the signaling molecule cGAMP. The model contains 24 rate constants, some of which have estimates from literature while others are unknown.

Traditional parameter estimation techniques that rely on gradient descent or maximum likelihood estimation which can be insufficient for large nonlinear ODE systems [4]. To compensate, a Bayesian approach is implemented to survey the parameter landscape and return probabilistic distributions as estimates. Adaptive Markov chain Monte Carlo (MCMC) simulation provides a computationally sufficient method to estimate the posterior distributions for all the unknown parameters. Uniform priors were used to avoid biasing parameter estimation [5]. To perform simulations for MCMC, the R package Flexible Modelling Environment (FME) was utilized [6]. Data for parameter fitting comes from time-course gene expression experiments. A transcriptionally deficient variant of Herpes Simplex Virus (HSV) was used to infect human fibroblasts. This type of experiment is informative because it controls for the negative regulation from viral proteins and allows for the observation of the unabated DNA sensing pathway.

In addition to parameter fitting, global and local sensitivity analyses were performed to determine the robustness of the model. Relatively large sensitivities, while also revealing stability properties, can be effective for motivating new experimental hypotheses. A parameter sensitive in the model is a likely candidate for viral manipulation.

Results

Model simulations capture the relative timing and shape of the dynamics seen in experiment and literature [7]. Typical upregulation of interferon beta occurs within four to six hours given an input of viral DNA and follows a saturation curve. Precursor protein complexes exhibit low accumulation due to the high input and output rates. It is possible that this is an accurate description of the biology, but also could be an indication that model reduction is needed. The production of ISGs show a delayed upregulation starting at 24 hours and downregulation after 48 hours. The eventual decline is controlled by the negative regulators (SOCS1, SOC2, USP18) that overtake the interferon signal and shut down the pathway. In many instances biologically, chronic inflammation occurs from aberrant ISG production that was not controlled by the negative regulators [8]. The model predicts this behavior with continual DNA stimulation.

We show how perturbations in highly sensitive kinetic parameters abrogate immune responses making their associated proteins likely targets for viral inhibition. Feedback mechanisms are also excellent targets due to their ability to amplify signals and stabilize systems.

Future work aims to expand this model to incorporate the transcription of interferon stimulated genes (ISGs) that are positively autoregulated, as well as ISGs that are not directly activated by the interferon response. These subsets of ISGs can be determined by different knockout experiments along the DNA sensing pathway. This analysis is vital for modeling to determine the timeframes of individual ISG activation.

Additionally, the model can be implemented to simulate different types of viral infections and autoimmune diseases caused by the detection of self-DNA. Understanding the different contexts in which this signaling pathway is stimulated can provide additional information into the structure of the pathway and varying kinetics observed.

Bibliography

[1] D. Kobasa, S. M. Jones, K. Shinya, J. C. Kash, J. Copps, H. Ebihara, Y. Hatta, J. Hyun Kim, P. Halfmann, M. Hatta, F. Feldmann, J. B. Alimonti, L. Fernando, Y. Li, M. G. Katze, H. Feldmann, and Y. Kawaoka, “Aberrant innate immune response in lethal infection of macaques with the 1918 influenza virus,” *Nature*, vol. 445, no. 7125, pp. 319–323, Jan. 2007.

[2] J. W. Schoggins, “Interferon-stimulated genes and their antiviral effector functions,” *Curr. Opin. Virol.*, vol. 1, no. 6, pp. 519–525, 2011.

[3] J. E. Medeiros-Filho, “Differences in viral kinetics between genotypes 1 and 3 of hepatitis C virus and between cirrhotic and non-cirrhotic patients during antiviral therapy,” *World J. Gastroenterol.*, vol. 12, no. 45, p. 7271, 2006.

[4] M. Ashyraliyev, Y. Fomekong-Nanfack, J. A. Kaandorp, and J. G. Blom, “Systems biology: parameter estimation for biochemical models,” *FEBS J.*, vol. 276, no. 4, pp. 886–902, Feb. 2009.

[5] K. S. Brown and J. P. Sethna, “Statistical mechanical approaches to models with many poorly known parameters,” *Phys. Rev. E*, vol. 68, no. 2, p. 21904, Aug. 2003.

[6] K. Soetaert and T. Petzoldt, “Inverse Modelling, Sensitivity and Monte Carlo Analysis in *R* Using Package

FME," *J. Stat. Softw.*, vol. 33, no. 3, pp. 1–28, 2010.

[7] M. Fribourg, B. Hartmann, M. Schmolke, N. Marjanovic, R. A. Albrecht, A. García-Sastre, S. C. Sealfon, C. Jayaprakash, and F. Hayot, "Model of influenza A virus infection: Dynamics of viral antagonism and innate immune response," *J. Theor. Biol.*, vol. 351, pp. 47–57, 2014.

[8] G. Trinchieri, "Type I interferon: friend or foe?," *J. Exp. Med.*, vol. 207, no. 10, 2010.

500733 Computational Modeling of Tuberculosis Granuloma Activation

Abstract:

Tuberculosis (TB) is one of the most common infectious diseases and deadliest diseases worldwide. It is estimated that one-third of the world's population is infected with TB, and 1.5 million TB-related deaths were reported in 2014. TB is spread by aerosol droplets containing Mycobacterium tuberculosis (Mtb). The Mtb bacteria enter through the respiratory system and are attacked by the immune system in the lungs. The bacteria are clustered and contained by aveolar macrophages into cellular aggregates called granulomas. These granulomas can hold the bacteria dormant for long periods of time in a condition called latent TB. The bacteria can be activated when the granulomas are compromised by other immune response events in a host, such as cancer, HIV, or aging. MMP-1 dysregulation has been recently implicated in TB activation through experimental studies, but the mechanism is not well understood. Animal and human studies currently cannot probe the dynamics of activation, so a computational approach is proposed to fill this gap. The overall objective of the study is to predict TB cavity formation (a hallmark of activation) in response to the dynamics of MMP-1 dysregulation.

A previously developed mathematical model of the immune response to Mtb in the lungs [1] is used as the basis for this work. This previous model can track the levels of T cells, bacteria, cytokines and chemokines and can simulate the possible outcomes of the infection: clearance, latency, and reactivation. The infection outcome of interest in this work is latency. The previous model is a system of 16 ordinary differential equations (ODEs). To model the effect of MMP-1 dysregulation, 2 additional ODEs for the concentration of MMP-1 and collagen were developed, and a term was added to a preexisting ODE to represent the leakage of extracellular bacteria. The preexisting model and new ODEs were incorporated into a Python class, and an ode solver included in the Scipy package was used to solve the system of ODEs. A global sensitivity analysis using SALibâ€™s implementation of Saltelli sampling and Sobol analysis was done on the model.

With the new equations to the model, a long term progression towards a leaking case is possible. Using a set of parameters that generates a latent case in the previous model, the new model at first starts to tend towards the latent case. After the bacterial population establishes itself, production of MMP-1 is increased. This drives the degradation of collagen, which allows bacteria to escape the granuloma. The case ultimately tends towards a steady state with a constant rate of bacterial leakage.

The model proposed here is intended to provide a possible mechanism for the link between MMP-1 dysregulation and TB activation. The new equations added into the model serve to add a leaking case that represents the formation of a necrotic cavity, which is characteristic of activation after a period of latency. Further work includes validating individual model parameters, sensitivity analysis, and model results under other conditions.

References: [1] Sud, Bigbee, Flynn & Kirschner. 2006. *The Journal of Immunology*, 176 (7), 4296â€“4314.

Acknowledgments: This work is supported in part by the Oklahoma Center for Respiratory Infectious Diseases,

supported by the National Institute of General Medical Sciences of the National Institutes of Health under Award Number P20GM103648.

498646 Investigating Cholera Toxin Binding Mechanism with Gangliosides Via Kinetic Modeling and Experimental Measurements

Abstract:

Cholera toxin (CTx), which is secreted by *Vibrio cholera,* is a toxin protein that is the main culprit of cholera. CTX is an AB_5 toxin, containing an enzymatic A-subunit and five identical cholera toxin B-subunits (CTB) [1]. CTB binds the gangliosides such as GM1 on host cell membranes leading to the endocytosis of CTx. In cytoplasm, the toxic A-subunit adversely increases the intracellular cyclic AMP level in the host cell [2]. The elevated cyclic AMP concentration results in severe fluid loss, which can be life-threatening if persists. In order to better cope with CTx invasion, the detailed mechanism of CTB-GM1 interactions, which is the prerequisite for CTx cell entry, needs to be elucidated. It is worth noting that a pentavalent CTB can simultaneously binds to multiple copies of GM1 in host cell membranes. The positive cooperativity between each binding subunit can further enhance the overall binding avidity [1, 3]. The conventional ligand-receptor analysis (monovalent binding model) is not sufficient to describe the complex multivalent CTx binding mechanisms. In this study, both *in silico* and *in vitro* approaches are employed to investigate complex interactions between CTB and GM1 in a cell-membrane-mimicking environment.

Lauer et al. developed a five-step kinetic model for CTB binding to GM1 on lipid bilayer [3]. However, the model did not account for the aforementioned positive cooperativity among CTB subunits. Here, we integrated the five-step model with other binding mechanism proposed by Turnbull et al. [1]. The new kinetic model incorporated association/dissociation constants with cooperativity factors. Based on the number of bound GM1 per CTB and the local configuration, seven CTB-GM1 complexes were considered in our simulation. The accuracy of kinetic model were verified by a novel protein binding assay. We used a quantitative nanocube-based biosensor to collect CTB-GM1 binding kinetics in a cell membrane mimicking environment [5]. The high-throughput and easy-to-use features of nanocube sensor allow collect massive data at various experimental conditions. In addition, we applied the same methodology to examine other host cell receptors such as GM2 and fucosyl-GM1, which exhibit distinct CTB binding kinetics. In summary, we integrated theoretical modeling and experimental measurements to explore cholera pathogenesis. Our work offers a systematic tool for the biomedical community to reveal the pathogenesis of other diseases.

Reference

1. Turnbull, W.B.; Precious, B.L.; Homans, S.W. Disseting the cholera toxin-ganglioside GM1 interaction by isothermal titration calorimetry. *J. Am. Chem. Soc.* **2004**, *126*, 1047-1054.
2. de Haan, L.; Hirst , T.R. Cholera toxin: A paradigm for multi-functional engagement of cellular mechanisms. *Molecular Membrane Biology* **2004**, *21*, 77-92.
3. Lauer, S.; Goldstein, B.; Nolan, R.L.; Nolan, J.P. Analysis of cholera toxin â€" ganglioside interactions by flow cytometry. *Biochemistry* **2002**, *41*, 1742-1751.
4. Lin, H.; Kitova, E.N.; Klassen, J.S. Measuring positive cooperativity using the direct ESI-MS Assay. Cholera toxin B subunit homopentamer binding to GM1 pentasaccharide. *J. Am. Soc. Mass. Spectrom.* **2014**, *25*, 104-110.
5. Worstell, N. C.; Krishnan, P.; Weatherston, J.D.; Wu, H. Binding cooperativity matters: A GM1-like ganglioside cholera toxin B subunit binding study using a nanocube-based lipid bilayer array. *Plos One* **2016**, *11 (4)*, e0153265.

498782 Biomarker Identification in Autism Spectrum Disorder: Common Pitfalls and Emerging Strategies

Abstract:

Biomarkers are measurements obtained from a patient that are predictive of some phenomenon, such as a disease [1] or environmental exposure [2]. Biomarkers link complex observations to a quantitative measurement (e.g., loss of consciousness in diabetes with hypoglycemia) and can be used for a variety of purposes, such as assigning diagnostic status, stratifying patients, or measuring clinical outcomes [3]. Consequently, biomarkers are integral to the understanding and treating disease. However, biomarkers are frequently assessed with inappropriate analytics methods [4] and restricted to single measurements [5], which contributes to the underwhelming numbers of proposed biomarkers that lead to clinical translation [5]. These issues with current biomarker development/reporting are further discussed below.

Distributions of biomarkers in the diseased and healthy cohorts are typically described by measures of central tendency and some measure of the dispersion or range. Furthermore, the most commonly reported difference in the biomarker distributions for diseased and healthy cohorts is a difference in the measure of central tendency [3]. However, this "statistical significance" is usually misinterpreted as being synonymous with "clinical significance". Instead, diagnostic biomarkers should be assessed on the separability of the diseased and healthy distributions (e.g. sensitivity/specificity, receiver-operating-characteristic curve, etc.) in order to achieve clinically-relevant biomarkers.

Potential biomarkers usually arise from discussions of or mathematical modeling of a biological network and how this network connects to the phenomenon of interest. Traditionally, only single measurements or metrics are used to construct a potential biomarker and the remaining portions of the underlying network is ignored [3]. While this approach can work well for simpler diseases, more complex diseases lacking a completely understood disease mechanism (e.g., neurological diseases) can benefit from incorporating multiple measurements into the potential biomarker without specific relationships between multiple measurements imposed *a priori* [3], [6].

These issues with current biomarker reporting/development are highlighted through case studies of biomarker identification from Autism Spectrum Disorder (ASD). It is shown that effective biomarkers can be developed from multiple measurements and that p-Values from Student's t-tests can lead to incorrect conclusions regarding biomarker quality. In comparison to this, multivariate statistical analysis and modeling techniques can result in the identification of biomarkers that can be used for disease classification and severity prediction even when univariate approaches fail [6].

[1] T. J. Lyons and A. Basu, "Biomarkers in Diabetes: Hemoglobin A1c, Vascular and Tissue Markers," *Transl. Res.*, vol. 159, no. 4, pp. 303–312, Apr. 2012.

[2] N. L. Benowitz, "Cotinine as a biomarker of environmental tobacco smoke exposure," *Epidemiol. Rev.*, vol. 18, no. 2, pp. 188–204, 1996.

[3] J. C. McPartland, "Considerations in biomarker development for neurodevelopmental disorders," *Curr. Opin. Neurol.*, vol. 29, no. 2, pp. 118–122, Apr. 2016.

[4] A. D. Barker, C. C. Compton, and G. Poste, "The National Biomarker Development Alliance accelerating the translation of biomarkers to the clinic," *Biomark. Med.*, vol. 8, no. 6, pp. 873–876, Jul. 2014.

[5] G. Poste, "Bring on the biomarkers," *Nature*, vol. 469, no. 7329, pp. 156–157, Jan. 2011.

[6] D. P. Howsmon, U. Kruger, S. Melnyk, S. J. James, and J. Hahn, "Classification and adaptive behavior prediction of children with autism spectrum disorder based upon multivariate data analysis of markers of oxidative stress and DNA methylation," *PLOS Comput. Biol.*, vol. 13, no. 3, p. e1005385, Mar. 2017.

491442 Stability Analysis of Stochastic SchlöGl Model

Abstract:

The behavior of chemical reaction networks is influenced by the number of molecules and the size of the system. When the number of molecules is asymptotically reaching large values, the system is in the macroscopic (thermodynamic) limit. In such cases, a deterministic modeling formalism is more suitable to describe the system. Away from it, at the microscopic limit, a stochastic approach is more suitable since the system's size is small and thermal molecular fluctuations become significant [1].

We have previously discussed the lack of knowledge of the behavior of networks in the mesoscopic area, where the system is sizable yet still under the influence of thermal noise [2]. This lack is largely due to numerical and computational difficulties in stochastically modeling mesoscopic system [2]. We have also provided insight into how these issues can be resolved with the help of ZI-closure scheme method [3]. This is a numerical method to solve chemical master equations by maximizing the entropy of the reacting system.

We employ the Schlögl model reaction network, as a general example of bistable systems. A wide variety of system sizes is studied with ZI-closure. In the cases that the system has one stable attractor, the stochastic system recovers the deterministic behavior for large system sizes. However, the stochastic mesoscopic behavior only partially matches the deterministic solution in the region of kinetic parameter values that result in two stable deterministic attractors. In such cases, the stochastic system recovers only one of the deterministic attractors in the mesoscopic limit.

Stability analysis of the stochastic system provides a better insight into the system's behavior. Eigenvalues of the stochastic system are calculated through the Jacobian matrix of the linearized system around the steady state [4]. The correlation between the eigenvalues of the stochastic system at the mesoscopic limit and the ones of the deterministic formalism reveals the underlying stability dynamics of the system. The stochastic formalism can distinguish between attractors and thus retrieves the more stable one. In contrast, the deterministic formalism finds all the stable attractors indistinguishably.

1. Constantino PH, Vlysidis M, Smadbeck P, Kaznessis YN, "Modeling stochasticity in biochemical reaction networks", J Phys D: Appl Phys 2016 Feb 1; 49(9): 093001 10.1088/0022-3727/49/9/093001
2. Vlysidis M, Kaznessis YN "The Mesoscopic Behavior of Stochastic Schlögl Model", 2016 AIChE Annual Meeting, San Francisco, CA
3. Smadbeck P, Kaznessis YN, "A Closure Scheme for Chemical Master Equations", Proc Natl Acad Sci U S A. 2013 Aug 27; 110(35): 14261-5 10.1073/pnas.1306481110
4. Smadbeck P, Kaznessis YN, "On a theory of stability for nonlinear stochastic chemical reaction networks", J Chem Phys 2015 May 8; 142: 184101 10.1063/1.4919834

497810 A Surrogate-Based Method for Constrained Optimization with Black-Box Noisy Simulations

Abstract:

A Surrogate-based Method for Constrained Optimization with Black-box Noisy Simulations

Zilong Wang, Marianthi Ierapetritou

Department of Chemical and Biochemical Engineering, Rutgers University, Piscataway, NJ, USA

Over the past decade, simulation and modeling techniques have been extensively investigated and implemented by the pharmaceutical manufacturing industry in order to enhance the process knowledge and improve the process efficiency [1]. Together with experimental work, the development of process models is considered as a key component to facilitate the modernization of pharmaceutical manufacturing [2], which leads to reduction of the total cost, improvement of product quality, and increase of the overall process agility, flexibility, and robustness [3].

However, the increased complexity in process models also increases the difficulty of utilizing such models for optimization of pharmaceutical processes. A first issue that needs to be considered is the increasing computational complexity. This arises from the growing amount of calculations that are needed to account for the effects of different material properties and process conditions on the final product qualities. Additionally, in some cases the relationship between input factors and output variables is not explicitly expressed, and we may not have access to the closed-form expressions for the objective function and constraints [4]. Furthermore, since the variability is playing an important role in the pharmaceutical product quality, it can be modeled in the simulation by incorporating a noise term, which is introduced at the beginning of an integrated process (e.g. feeders) and propagates along the whole process. The variability in pharmaceutical processes is commonly heteroscedastic in nature [5], and the relationship between variability and input factors is usually unknown after it is propagated in the simulation. To address those issues, an efficient approach is needed to solve the constrained optimization problem based on black-box noisy simulations for pharmaceutical manufacturing processes.

In this work, we propose a surrogate-based strategy to solve black-box constrained optimization problems. Two surrogate models are constructed separately for the objective function and the constraints. Stochastic Kriging [6] is used as the surrogate to account for the heteroscedastic noise that is inherent to the black-box simulations we encounter. We utilize an efficient adaptive sampling method to sequentially improve the two surrogate models and direct the search to a feasible optimum. This adaptive sampling is performed in two stages. In the first stage, it attempts to locate the feasible region boundaries that are defined by the black-box constraints. Then, in the second stage, the focus is on searching for the feasible optimum subject to the feasibility characterized in the first stage. The algorithm is suitable for black-box noisy simulations, and its performance is demonstrated with a case study on continuous pharmaceutical manufacturing process.

References

[1] Ierapetritou M, Muzzio F, Reklaitis G. Perspectives on the continuous manufacturing of powder-based pharmaceutical processes[J]. AIChE Journal, 2016, 62(6): 1846-1862.

[2] Lee S L, Oâ€™Connor T F, Yang X, et al. Modernizing pharmaceutical manufacturing: from batch to continuous production[J]. Journal of Pharmaceutical Innovation, 2015, 10(3): 191-199.

[3] Oâ€™Connor T F, Lawrence X Y, Lee S L. Emerging technology: A key enabler for modernizing pharmaceutical manufacturing and advancing product quality[J]. International journal of pharmaceutics, 2016, 509(1): 492-498.

[4] Wang Z, Escotet-Espinoza M S, Ierapetritou M. Process analysis and optimization of continuous pharmaceutical manufacturing using flowsheet models[J]. Computers & Chemical Engineering, 2017. (in press)

[5] Engisch W E, Muzzio F J. Method for characterization of loss-in-weight feeder equipment[J]. Powder technology, 2012, 228: 395-403.

[6] Ankenman B, Nelson B L, Staum J. Stochastic kriging for simulation metamodeling[J]. Operations research, 2010, 58(2): 371-382.

493981 Efficient Solution of Mixed-Integer Multistage Stochastic Programs for the Optimal Design of Smart Manufacturing Systems Using "Smooth-in-Expectation" Decision Rules

Abstract:

The ability to operate manufacturing plants in a flexible manner has long been motivated by the need to accommodate long-term market fluctuations and equipment failures. However, two contemporary trends have greatly heightened the significance of flexibility. First, increasingly volatile markets and the growing utilization of renewable resources (both energy and chemicals) are introducing unprecedented levels of variability and uncertainty into process operations. Second, advances in automation and process intensification are making it possible to envision systems that are much more responsive and adaptable than ever before. Together, these factors are creating an environment in which flexible operating policies have the potential to achieve dramatic increases in efficiency and reliability in many critical applications, including distributed power generation, (bio)refining, and chemical manufacturing in multi-product plants. Indeed, this potential has placed flexibility astride information technology as a central theme of the smart manufacturing paradigm.

However, the design and optimization of flexible systems poses tremendous challenges because it necessarily integrates design, scheduling, and control problems that are conventionally solved independently. Specifically, insofar as the quality of a design is determined by its ability to adapt, long-term design decisions are strongly dependent on operational details that occur on much shorter time-scales. Thus, design decisions must be optimized simultaneously with a potentially huge number of decisions representing short-term scheduling and control actions over long time horizons, and in multiple future scenarios. In general, this leads to large-scale nonlinear multistage stochastic programs with mixed-integer recourse decisions.

Although the effort to optimally integrate design and operations has a long history, the solution of such multistage problems remains a serious outstanding challenge. Broadly, tractable formulations still require significant limitations on the modeling of either short time-scale effects (e.g., coarse time discretization, static or linearized models), uncertainty (e.g., few scenarios, two-stage rather than multi-stage decisions), or permissible operational strategies (e.g., heuristic scheduling policies, fixed control laws). Critically, these approximations degrade optimization models in exactly the aspects that are most essential for assessing the efficiency and reliability of a smart manufacturing system, and for properly weighing these factors against its cost. Thus, there is a need for new theory and algorithms that can effectively solve multistage stochastic programs with mixed-integer recourse.

In this contribution, we will present recent progress towards a new approach for efficiently solving such programs using "smooth-in-expectation" decision rules. In this approach, recourse decisions are no longer formulated as explicit optimization variables, but are rather determined (suboptimally) through decisions

rules (i.e., functions of the random variables and system state). Although these rules have fixed functional forms, they are parameterized by additional 'first-stage' decisions in order to impart some flexibility. This type of approximation was originally developed for linear stochastic programs with continuous recourse variables, and is now commonly used to solve very large-scale linear problems with many stages, which are intractable by other means. However, in the case of integer recourse variables, which are critical for smart manufacturing systems (e.g., scheduling and unit commitment decisions), the use of decision rules has been much less successful. Specifically, this is because decision rules that make integer decisions are inherently discontinuous. Thus, replacing integer recourse decisions with decision rules results in a highly discontinuous optimization model, which hardly simplifies matters.

To overcome this limitation, a key insight in our approach is that the expected value of the objective function can be a smooth function of the decision variables despite the presence of discontinuous decision rules in the model. This can occur because the integral of a discontinuous function need not be discontinuous, and requires that (i) the random variables are continuously distributed, and (ii) the decision rules satisfy a set of 'non-degeneracy' conditions that we have characterized theoretically. We call a set of mixed-integer decision rules that satisfy these conditions "smooth-in-expectation." Applying smooth-in-expectation mixed-integer decision rules to a multistage stochastic program results in a single-stage program with a continuously differentiable objective function. In particular, the overwhelming discrete character of the original problem (assuming mixed-integer recourse) is dramatically reduced, leaving only first-stage integer decisions from the original formulation. On the other hand, the objective of this reformulation is an expected value that typically cannot be evaluated precisely, but rather must be estimated by Monte Carlo methods. In our approach, this problem is solved using a custom stochastic gradient decent algorithm.

In this talk, we will discuss the construction of general-purpose smooth-in-expectation decision rules and the efficient solution of the resulting smooth stochastic program. Case studies in chemical manufacturing and distributed power generation will be used to demonstrate the advantages of our approach relative to existing nonlinear multistage stochastic programming formulations.

501783 A Generalized Knapsack-Problem Based Decomposition Heuristic to Solve Large-Scale Multistage Stochastic Programs

Abstract:

Heuristic algorithms are one approach to address the computation complexity of multistage stochastic programs (MSSPs). Increases in the number of uncertain parameters or improving accuracy by adding addition possible realizations for each uncertain parameter causes exponential growth in problem size. Solutions to problems considered in literature are generally limited to cases considering between two and 15 uncertain parameters. When considering real-world size problems there can be hundreds of uncertain parameters. As such, several authors have explored the use of heuristics to generate good solutions for multistage stochastic programs. Colvin and Maravelias (2009) explored a rolling-horizon heuristic approach. The approach divides the planning horizon of the pharmaceutical clinical trial planning problem into a finite number of subsets and generates MSSPs for each subset. Gupta and Grossmann (2014) implemented a Lagrangean decomposition scheme using a scenario grouping strategy. The scenario grouping strategy allowed partial decomposition of the full space model. Scenarios are grouped based on differences in one endogenous parameter. The approach was applied to the oil field planning problem. More recently, Apap and Grossmann (2016) considered a sequential scenario decomposition strategy to solve MSSPs with endogenous and exogenous uncertainty. The strategy solved a series of endogenous problems to approximate the solution of the original problem. The authors tested the strategy using the oil field planning problem and the process synthesis problem.

In this presentation, we describe a generalized knapsack-problem based decomposition heuristic to solve large-scale multistage stochastic programs. The algorithm builds upon our previous work (Christian and Cremaschi, 2015), and it is expanded to accommodate multistage stochastic programs with both endogenous and exogenous uncertain with non-trivial recourse actions. The heuristic algorithm strategically solves a series of knapsack problems along the planning horizon of multistage stochastic programs. The items for the knapsack problems are constructed using the decision variables of the MSSP. The item values are estimated using the expected gains associated with the value of the decision variable. Constraints from the original MSSP are modified and persist as weight constraints on the knapsack problems. The heuristic starts by generating and solving a knapsack problem at the first time period. Based on the items selected as the solution of this first knapsack problem, uncertainty is realized, and new knapsack problems corresponding to each of these realizations are generated and solved. The algorithm terminated when the end of the planning horizon is reached.

We apply this generalized knapsack-problem based decomposition heuristic to solve three problems: (1) pharmaceutical clinical trial planning problem (Colvin and Maravelias, 2008), (2) artificial lift infrastructure planning problem (Zeng and Cremaschi, 2017), and (3) new technology investment planning problem, which seeks to determine the optimal investment strategy for new technology develop given uncertainty in the performance and success of the technology. The results show that, similar to the clinical trial planning problems solved in Christian and Cremaschi (2015), the generalized heuristic provides tight feasible solutions for the MSSPs of the new technology investment planning problems, and the optimum solutions

for the MSSPs of the artificial lift planning problems. For all problems considered, the tight feasible solutions are generated several orders of faster than solving the original MSSPs.

References

Apap, R. M., & Grossmann, I. E. (2016). Models and Computational Strategies for Multistage Stochastic Programming under Endogenous and Exogenous Uncertainties. *Computers & Chemical Engineering*. https://doi.org/http://dx.doi.org/10.1016/j.compchemeng.2016.11.011

Christian, B., & Cremaschi, S. (2015). Heuristic solution approaches to the pharmaceutical R&D pipeline management problem. *Computers and Chemical Engineering*, *74*, 34–47.

Colvin, M., & Maravelias, C. T. (2008). A stochastic programming approach for clinical trial planning in new drug development. *Computers & Chemical Engineering*, *32*(11), 2626–2642. https://doi.org/http://dx.doi.org/10.1016/j.compchemeng.2007.11.010

Colvin, M., & Maravelias, C. T. (2009). Scheduling of testing tasks and resource planning in new product development using stochastic programming. *Computers & Chemical Engineering*, *33*, 964–976. https://doi.org/10.1016/j.compchemeng.2008.09.010

Gupta, V., & Grossmann, I. E. (2014). Multistage stochastic programming approach for offshore oilfield infrastructure planning under production sharing agreements and endogenous uncertainties. *Journal of Petroleum Science and Engineering*, *124*, 180–197. https://doi.org/10.1016/j.petrol.2014.10.006

Zeng Z, & Cremaschi S. Artificial lift infrastructure planning for shale gas producing horizontal wells. *Foundations of Computer Aided Process Operations / Chemical Process Control*, 2017

492750 Mixed-Integer Nonlinear Programming Models for Optimal Design of Reliable Chemical Plants

Abstract:

Mixed-integer nonlinear programming models for optimal design of reliable chemical plants

Yixin Ye[1], Ignacio E. Grossmann[1], Jose M. Pinto[2], Sivaraman Ramaswamy[2]

[1]Department of Chemical Engineering, Carnegie Mellon University, Pittsburgh, PA 15213

[2]Business and Supply Chain Optimization R&D, Praxair, Danbury, CT 06810

Abstract

Plant availability has always been a critical consideration for the design and operation of chemical processes, for it represents the expected fraction of normal operating time, which impacts directly the ability of meeting customer demands. The failure of a key unit is likely to cause the shut-down of the entire plant. An effective way to reduce the impact of equipment breakdown on the entire plant is to install back-up units in the design phase. Current state-of-the-art design reliability (regarding number and selection of back-up units) for these plants or other chemical processes is to use simulation tools to assess different design alternatives. Currently, discrete event simulation tools are used to evaluate reliability/availability of selected alternatives in terms of equipment redundancy, additional inventory and preventive maintenance to simulate the behavior of every asset in a plant using historical maintenance data and statistical models (Sharda and Bury, 2008). However, this approach does not systematically consider all the alternatives as it would be the case in an optimization approach.

Research has been done to quantitatively analyze the reliability of chemical plants (Thomaidis and Pistikopoulos, 1995). Goel et al. (2003b) consider both design and planning of production and maintenance in an MILP model with variable reliability parameters and fixed system configuration. Aguilar et al. (2008) address reliability issue in utility plant design and operation and report the optimization result of an MILP model for a specific example considering two failure situations. Terrazas-Moreno et al. (2010) formulate an MILP model using Markov chains to optimize the selection of alternative plants and the design of intermediate storage for an integrated production site.

However, there are virtually no general mixed-integer programming models proposed for the optimal design of reliable chemical processes. In response to this gap, this work develops optimization models for reliable design of chemical plants, where the major trade-off considered is between plant availability and the investment cost of back-up equipment. Mixed-integer models are formulated to select redundant equipment for serial systems in terms of availability, in order to maximize system availability (i.e. probability that the system performs without failures), and hence sales revenue, while minimizing the total cost of the system.

First, we consider a basic scenario where all the stages need only one unit to work properly. A set of potential units for each stage are given with certain availabilities (i.e. the probability of finding the unit available), operating priorities, which means that a unit can only become active when all installed units that have higher priorities have failed, and cost rates. Based on the parameters provided above, the relationship between the availability of stage and its design decision is established. On the basis of the basic scenario, an additional standby strategy is applied. Here, two units each with half capacity can work at the same time in one stage, such that when one of them fails, there is still half of the original output left. System availability is also redefined to capture the new situations of having partial capacity.

Two non-convex MINLP models for maximizing system net profit are formulated regarding the two situations, respectively. In addition, another non-convex -constrained MINLP model is formulated for the first scenario by maximizing availability and minimizing the total cost. It is shown that this model can be reformulated as a convex MINLP in order to generate the corresponding Pareto-optimal solutions.

The proposed MINLP models are applied to several case studies that include sensitivity analysis of the availabilities predicted for each unit. Furthermore, the proposed models for availability evaluation are incorporated in the synthesis of methanol and hydro-alkylation processes that involve optimizing their flowsheet superstructures.

Reference

Aguilar, O., Kim, J.-K., Perry, S., and Smith, R. (2008). Availability and reliability considerations in the design and optimization of flexible utility systems. Chemical Engineering Science, 63(14):3569-3584.

Goel, H. D., Grievink, J., and Weijnen, M. P. (2003b). Integrated optimal reliable design, production, and maintenance planning for multipurpose process plants. Computers & chemical engineering, 27(11):1543-1555.

Sharda, B. and Bury, S. J. (2008). A discrete event simulation model for reliability modeling of a chemical plant. In Proceedings of the 40th Conference on Winter Simulation, pages 1736-1740. Winter Simulation Conference.

Terrazas-Moreno, S., Grossmann, I. E., Wassick, J. M., and Bury, S. J. (2010). Optimal design of reliable integrated chemical production sites. Computers & Chemical Engineering, 34(12):1919-1936.

Thomaidis, T. V. and Pistikopoulos, E. (1995). Optimal design of flexible and reliable process systems. IEEE transactions on reliability, 44(2):243-250.

492757 Network Design with Uncertain Edge Failures: Two-Stage Robust Optimization for Single-Commodity Networks

Abstract:

The design of networks has applications in a wide variety of fields, and the tools of process systems engineering are well suited for this task. For example, mixed-integer linear optimization techniques have been shown to be a suitable framework for generating networks with specific requirements regarding their properties [1,2]. In other cases, such as in single-commodity networks, a general graph of nodes is known *a priori*, and a final network of edges must be built between these nodes such that the demands at each location are met [3]. For this latter problem, uncertainty may arise when constructed edges randomly fail after being built. Generally known as *survivable* or *resilient* network design [4,5], optimization that takes into account the possibility of edge failures can prevent disastrous logistical issues which can arise through purely deterministic designs.

The deterministic single-commodity network design problem involves the minimization of the total cost of the network, including the investment cost and operational costs. The capacities of each edge of the network, as well as the actual flows along each edge, must be determined and their associated costs calculated. When edges can fail, the uncertain problem naturally separates into two stages. The initial capacities of each edge must be chosen in the first stage, while the flows may only be determined in the second stage, after one has observed which edges have failed. We formalize this problem as a two-stage robust optimization problem [6,7,8] using an uncertainty set of binary random variables that indicate whether an edge has failed. The two-stage robust optimization formulation is solved using a suitably modified column and constraint generation algorithm [9] in order to determine the minimum cost network that remains feasible for a given number of possible simultaneous edge failures.

The two-stage robust optimization formulation and column and constraint generation algorithm for the network design problem under demand uncertainty is presented and used to solve various problems adapted for our robust optimization context from the Survivable Network Design Library [10]. A preprocessing algorithm is presented which analyzes the initial graph and simplifies it when possible to reduce problem complexity. Results are obtained for various levels of robustness; that is, optimal network topologies are determined for different allowable extents of possible edge failures. The required computational effort for optimization with our modified column and constraint generation algorithm is also discussed.

[1] Gounaris, C.E.; Rajendran, K.; Kevrekidis, I.G.; Floudas, C. A. Generation of networks with prescribed degree-dependent clustering. Optimization Letters, 2011, 5 (3), 435-451.

[2] Gounaris, C.E.; Rajendran, K; Kevrekidis, I.G.; Floudas, C.A. Designing Networks: A Mixed-Integer Linear Optimization Approach. Networks, 2016, 68 (6), 283-301.

[3] Cacchiani, V.; Jünger, M.; Liers, F.; Lodi, A.; Schmidt, D. R. Single-commodity robust network design with finite and Hose demand sets. Math. Program. Ser. B, 2016, 157, 297-342.

[4] Tomaszewski, A.; Pióro, M.; Żotkiewicz, M. On the complexity of resilient network design. Networks, 2010, 55, 108–118.

[5] Orlowski, S; Pióro, M. Complexity of column generation in network design with path-based survivability mechanisms. Networks, 2012, 59, 132–147.

[6] Ben-Tal A.; El Ghaoui L.; Nemirovski A. Robust optimization. Princeton University Press, 2009.

[7] Lappas, N.H.; Gounaris, C.E. Multi-Stage Adjustable Robust Optimization for Process Scheduling Under Uncertainty. AIChE Journal, 2016, 62 (5), 1646-1667.

[8] Gounaris, C.E. Advances in Robust Optimization and Opportunities for Process Operations. In: Proceedings of the Foundations of Computer-Aided Process Operations/Chemical Process Control (FOCAPO 2017/CPC IX), C.T. Maravelias, L. Megan, J. Wassick and E.B. Ydstie (Eds.), 2017, Paper ID IF110.

[9] Zeng, B.; Zhao, L.; Solving two-stage robust optimization problems using a column-and-constraint generation method. Operations Research Letters, 2013, 41, 457-461.

[10] Orlowski S.; Wessäly R.; Pióro M.; Tomaszewski A. SNDlib 1.0—survivable network design library. Networks, 2010, 55(3), 276-286.

500726 Robust Process Intensification and Optimization: Application to Carbon Capture Systems

Abstract:

The Intergovernmental Panel on Climate Change (IPCC) estimates that the world must limit global temperature increases to less than 2°C above preindustrial levels to avoid irreversible damage to the ecosystem. At current CO_2 emission rates, the IPCC also estimates global temperatures would reach this limit in just under 20 years. The largest point sources of concentrated CO_2 exist in the fossil power industry, motivating a rapid switch to zero carbon energy via CO_2 capture on fossil based power, nuclear and low carbon renewable deployments. While nuclear and renewable power generation is increasing, fossil based power generation will continue to play a significant role especially in developing countries and so CO_2 capture solutions must be provided. However, the cost and performance of state of the art CO_2 capture technology are insufficient to support broad scale deployment. Development of CO_2 capture technology for fossil power plant emissions must be extremely aggressive to be cost effectively deployed, especially at the rates required to limit global warming to less than 2°C. Developers must entertain novel chemistries and systems to significantly improve performance, however doing so often results in complex systems with novel components that tend to increase cost. Moreover, novel technology and complex systems often exhibit great uncertainty due to low maturity levels.

Even with rigorous deterministic optimization, uncertainty in system performance and cost remains a barrier to timely industrial adoption of otherwise high performance designs. Furthermore, in novel systems full probabilistic characterization of uncertainty is at best case limited, reducing the justification to employ stochastic optimization approaches in these cases. In this work we apply a recently proposed optimal control formulation for intensification of a pre-combustion CO_2 capture system with the aim of consolidating thermodynamic operations to reduce overall costs. Assuming novel system design efforts for which no useful uncertainty characterization is available, we extend recently developed nonlinear robust optimization techniques to develop steady state intensified designs that operate feasibly, safely and satisfy all performance criteria under parametric uncertainty. A central objective of robust optimization is to generate designs conservative enough to ensure feasible operation amidst all uncertainty realizations. However, solutions to current nonlinear robust formulations often rely on approximations that result in a degree of over-conservatism that may offset any cost reduction provided by process intensification efforts. We propose flexible statistically-derived â€œback-offâ€ constraints that ensure feasibility yet eliminate the over-conservatism of most current approaches. Moreover, we propose to leverage the aforementioned method to directly and simultaneously mitigate the worst possible impact of user-specified uncertainty on the objective function.

499518 Robust Optimization Using Polyhedral Norm and General Asymmetric Uncertainty Set

Abstract:

Robust optimization technique has been applied to handle the uncertainty inherent in many decision making problems. Uncertainty set construction is a key step for applying robust optimization since it affects the conservativeness and computational tractability of the deterministic robust counterpart problem. The choice of uncertainty set is naturally problem-dependent. However, general guidelines for constructing uncertainty set include that it should not lead to overly conservative solution, and computationally tractable deterministic robust counterpart formulations are preferred.

In the past, various types of uncertainty set ranging from simple boxes (Soyster, 1973) to more advanced conic-representable sets have been studied. Ben-Tal and Nemirovski (2000) introduced ellipsoidal type of uncertainty set for robust linear optimization. Bertsimas et al. (2004) introduced norm induced uncertainty set for robust linear optimization. This can lead to various symmetric set under different type of norms. Li et al. (2011) made a comparative study of various symmetric set induced robust optimization models. Yuan et al. (2016) studied robust linear optimization under correlated uncertainty, and demonstrated the advantage of introducing correlation information into the uncertainty set construction. A major motivation behind those different methods is that tractable linear robust counterpart optimization problems are obtained, which brings significant computational advantage.

In the existing studies, uncertainty set has been generally defined as the symmetric type. However, they may not be appropriate or be too conservative if the actual distribution is asymmetric. To capture the asymmetric distribution, Chen et al. (2007) introduced deviation measures to capture distributional asymmetry, however they assumed that the primitive uncertain parameters are independent. In this work, we investigate the robust linear optimization under uncertainty set induced from various type of polyhedral norms, including l-1 norm, l-infinity norm, D-norm, CVaR norm, deltoidal norm and general polyhedral norm. Furthermore, we introduced a novel method for asymmetric uncertainty set construction based on the distributional information or sampling data of uncertain parameters. The proposed method can be applied to general uncertainty distributions, where correlation between primitive uncertain parameters can be easily incorporated into the polyhedral norm induced uncertainty set. We also derived robust formulations under various polyhedral norms, which all lead to linear optimization problems. For all of those norm induced robust counterpart formulations, model complexity is compared to provide practical guidelines for uncertainty set construction. The asymmetric set induced robust optimization model is compared with the classical symmetric set induced robust optimization model. Numerical example and process design example are studied. Results demonstrate that using asymmetric uncertainty sets leads to less conservative robust solution.

Reference:

[1] Soyster, A. L. (1973). Technical note: convex programming with set-inclusive constraints and applications to inexact linear programming. Operations research, 21(5), 1154-1157.

[2] Ben-Tal, A., & Nemirovski, A. (2000). Robust solutions of linear programming problems contaminated with uncertain data. Mathematical programming, 88(3), 411-424.

[3] Bertsimas, D., & Sim, M. (2004). The price of robustness. Operations research, 52(1), 35-53.

[4] Bertsimas, D., Pachamanova, D., & Sim, M. (2004). Robust linear optimization under general norms. Operations Research Letters, 32(6), 510-516.

[5] Li, Z., Ding, R., & Floudas, C. A. (2011). A comparative theoretical and computational study on robust counterpart optimization: I. Robust linear optimization and robust mixed integer linear optimization. Industrial & engineering chemistry research, 50(18), 10567-10603.

[6] Yuan, Y., Li, Z., & Huang, B. (2016). Robust optimization under correlated uncertainty: Formulations and computational study. Computers & Chemical Engineering, 85, 58-71.

[7] Chen, X., Sim, M., & Sun, P. (2007). A robust optimization perspective on stochastic programming. Operations Research, 55(6), 1058-1071.

502544 Modeling and Reaction Kinetics of Reactors for Polyesters Production

Abstract:

Scale-up of polymerization processes for the production of co-polyesters is complicated by the wide range of raw material units that can be used. For this reason, the knowledge of kinetics, mass transfer and thermodynamic characteristics of the reaction medium are important for the design of new polymers.

In this paper, an approach for the determination of reaction kinetics at the laboratory scale will be presented. The results will then be used in the reactor models for the scale-up of new polymer recipes to the commercial scale reactor.

A comparison of the modeling and experimental data at different scales will be finally presented.

490805 Modeling and Simulation of Nonwovens Manufacturing

Abstract:

It is of interest to increase the efficiency of fiber web materials used in products ranging from filters to thermal and acoustic insulators. Particular areas of optimization include throughput increase, final product quality and waste reduction. Both analytical modeling and Computational Fluids Dynamics (CDF) simulations are used as tools to optimize nonwovens manufacturing. Analytical modeling is based on a transport phenomena approach using the momentum, continuity and heat transfer equations. Multiphase CFD simulations using open source software and internally developed codes are used to simulate processing conditions and conduct optimization while accounting for turbulence and polymer rheology. Material properties and process conditions are henceforth tailored in the quest of optimum webs.

SOME STUDIES ON THE INHERENT RESILIENCE FOR A GAS SWEETENING UNIT

Sirshendu Guha, Engineers India Limited, New Delhi, India

Abstract

Mitchell and Mannan (2007) evaluated resilience of simple process systems like steam pipe, water pipe, water pump and heat exchanger etc., and considered physical failures of these systems. However, they did not provide any mathematical background and their approach was too simplistic. Hence, it could not be directly applied to complex process systems such as absorber, regenerator and distillation columns. Linh et al. (2012) reported resilience analysis considering physical failure of the process systems to improve the safety of process plants.

In contrast to physical failure of process systems, quantification of system resilience has been carried out considering performance failure of process systems under this work. The inherent resilience for process systems can be conceptualized from the perspectives of material resilience.

Correlations have been developed to assess resilience properties of constituent process systems pertaining to a Gas Sweetening Unit (GSU) as a case study. Fig.1 shows the schematic flow diagram of the Gas Sweetening Unit (GSU).

A steady state condition has been considered and system stress and system strain equations have been developed to quantify the system resilience.

This study has enabled developing a new method for identification of thermodynamically efficient operating capacity regime for optimization of energy usage in process systems. Optimum energy usage for a process system is achieved when it is operated at the most thermodynamically efficient capacity regime and this is evaluated by estimation of newly conceived Thermodynamic Coefficient of Performance (THCOP) data at various capacities.

This work provides general formulae for quantification of system resilience properties of one absorber column, one regenerator column, one shell and tube heat exchanger and associated one process as well as one utility pipeline system pertaining to a GSU (Fig.1).

It is found that the GSU (Fig.1) operating capacity range of 80%-100% is the most thermodynamically efficient operating capacity regime. The minimum operating capacity for the GSU is considered as 1% whereas the maximum operating capacity is 150%. The design capacity (i.e., 100%) of the GSU is 27814 kg/hr. The GSU is designed to sweeten sour natural gas containing 1.28 wt% of H_2S to sweet natural gas containing less than or equal to 1 ppmw H_2S.

It is found that absorber column and regenerator column systems under study possesses inherent resilience of around 5% and 15% respectively with regard to variation in upstream feed sour gas flow rate beyond 100% design flow rate, i.e., 27814 kg/h.

It is also found that the lean-rich exchanger system under study possesses inherent resilience of around 10% with regard to variation in upstream feed sour gas flow rate beyond 100% design flow rate.

It is noticed that the inlet feed sour gas pipeline system (process pipeline) indicates drastic fall in modulus of resilience value (*Ur*) and great increase in modulus of elasticity (*Esys*) value above feed gas mass flow rate of 100% indicating attainment of allowable pressure drop and limiting or yielding stress.

It is also found that the utility pipeline system (which carries superheated steam) shows a plateau at very high steam flow rate (57600 kg/hr-64800 kg/hr) indicating attainment of allowable pressure drop and limiting or yielding stress region for the system in case of flow variation.

Results also indicate that similar to a material, all the process systems under study (i.e., absorber, regenerator, lean-rich exchanger, process pipeline and utility pipeline) of Gas Sweetening Unit(GSU, Fig.1) demonstrate inverse relationship of modulus of resilience (*Ur*) with modulus of elasticity (*Esys*) in all applicable operating variable deviation regimes.

Computer simulation using a process simulator SIMULATION SCIENCES INC, Pro/II has been utilized for this study.

Finally, one example is given regarding design procedure in relation to incorporation of 50% over capacity factor or inherent resiliency in the absorber column by augmentation of number of column trays.

Introduction

Quantification of "resilience" in the domain of process systems is still elusive. Fiksel (2006) understood "system resilience as the capacity of a system to tolerate disturbances while retaining its structure and function" (Fiksel, 2006).Material resilience is the extent to which energy may be stored in a material by elastic deformation (Mitchell and Mannan, 2007). Quantification of resilience for a material can be easily done by evaluating the area under the stress-strain curve from zero stress value to the yield stress value (Mitchell and Mannan, 2007).

Likewise a material, resilience of a process system may also be defined as the quantity of energy, the system will be able to store before attaining the "point of instability" (Mitchell and Mannan, 2007).System instability may be understood as inability of the system to perform efficiently at its targeted performance levels.

Flexibility and resiliency of process systems were first described by Morari (1983). Guha and Das (2014) reported that in an insulated pipe segment system which carries superheated steam as process fluid, magnitudes of inherent system resiliencies are 927.8 kJ/m^3-s, 43 kJ/m^3-s and 31.5 kJ/m^3-s for variation of mass flow rate, inlet pressure and inlet temperature respectively.

In this work, the resilience magnitudes of process systems have been quantified by evaluation of the area under the system stress -system strain curve from the system stress value corresponding to a minimum operating condition to the system stress value corresponding to a maximum allowable operating condition. System stress corresponding to a maximum allowable operating condition will be limited by the design condition up to which the system will be able to maintain its targeted performance levels.

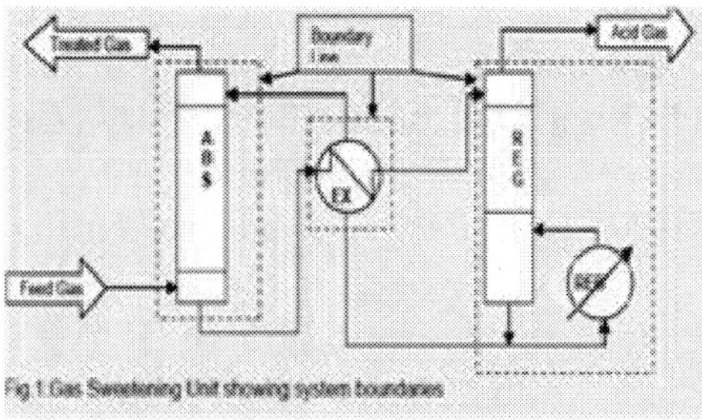

Fig 1 Gas Sweetening Unit showing system boundaries

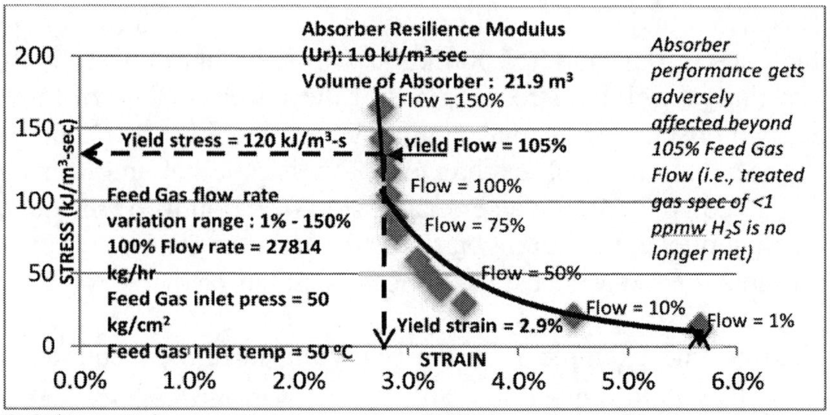

Fig.2: Stress-strain curve of absorber system of GSU for variation of feed gas mass flow rate

Research Objectives

In this work, following research objectives have been envisaged.

(a) Review of mathematical equations that can be used to assess resilience properties of a material.
(b) Development of mathematical equations for quantification of resilience properties for complex process systems pertaining to a Gas Sweetening Unit (GSU, Fig.1).
(c) Determination of thermodynamically efficient operating capacity regime for a GSU.

Methodology

(a) Review of mathematical equations that can be used to assess resilience properties of a material.

The amount of work which is required to deform unit volume of a material to its elastic limit was considered as the resilience of the material (Mitchell and Mannan, 2007). For a linearly elastic material, the modulus of resilience (U_R) can be represented as (Mitchell and Mannan, 2007).

$$U_R = \Omega^2_{yield} / 2E \qquad (1)$$

In (Eq.1), Ω_{yield} is yield stress of the material and E is modulus of elasticity of the material.

Development of (Eq.1) is based on the assumptions that elongation of the material has been carried out in an ideal reversible way maintaining equilibrium at all times and the cross-sectional area of the material does not alter during the process of elongation.

Since actual elongation process can never be carried out following an ideal reversible path and the cross-sectional area of the material changes during elongation process, one can construe that the area under the curve of *true material stress* and *true material strain* from zero stress to the yield stress; should be used for determination of the modulus of resilience (U_R) of the material.

If we consider a rod of length '*L*' and 'A' cross section which is elongated, with '*X*' representing the extent of elongation, we can state that for infinitesimal elongation '*dX*', the infinitesimal work done '*dW*' on the rod (-ve work as per convention) can be represented as,

$$dW = - \Omega.A.\ dX = -\Omega\ dV \quad (2)$$

In the above equation, 'Ω' is material stress; 'V' is material volume.

From (Eq.2), one can define *true material stress* 'Ωt' in terms of irreversible or actual work $W(i)$ & change in material volume (ΔV) as ,

$$\Omega t = -dW/dV = dU/dV = - W(i)/\Delta V\ (dU = -dW \text{ when } dQm = 0) \quad (3)$$

$$Or\ \Omega t = -Actual\ Work/Volume\ Change \quad (3a)$$

In (Eq.3), 'Qm' is heat transferred into the material; 'U' is internal energy of the material.

Using 1st and 2nd laws of thermodynamics and experimental results (Raghavan, 1990) one can obtain following relation for lost work pertaining to a material during the process of elongation (*Ta* being environment temperature, *S* being entropy of the material, F being force applied on the material).

$$F = -Ta\ (dS/dX)_{at\ constant\ Ta} \quad (3b)$$

$$Or\ -Ta\ dS = FdX = \Omega t\ .A.dX \quad (3c)$$

$$Or\ d(X/L) = -TadS/\ \Omega t.A.L \quad (3d)$$

$$Or\ d\varepsilon t = -TadS/\ \Omega t.V = -TadS\ (dV/V)/\ dU \quad (4)$$

True material strain is 'εt' which can be defined in terms of lost work $W(L)$, fractional volume change (C) & actual work $W(i)$ as,

$$\varepsilon t = C\ (-TadS\ )/\ dU = C\ W(L)/-W(i) \quad (5)$$

$$\text{Or } \varepsilon t = [Vol.Change\ Fraction]\ x\ [Lost\ Work]/[-Actual\ Work] \quad (5a)$$

In (Eq.5), $C = dV/V$, which is a dimensionless quantity, and is not equal to dL/L as cross-sectional area changes during an elongation process. The entropy change is due to change in configuration of the material during the process of elongation. For non-linearly elastic materials, one can use values of true stress, computed from (Eq.3) and values of true strain εt, computed from (Eq.5) for graphical evaluation of material modulus of resilience 'U_R'. Using (Eq.3) and (Eq.5), "U_R" for a linearly elastic material which obeys the yield stress-strain relation:

$\Omega_{yield} = E\ \varepsilon_{yield}$, can be expressed as,

$$U_R = [-C\ Ta/2\ dS/dV]_{yield} = [\ Ta\ (S1-S2)/2V1]_{yield} = \Omega^2_{yield}/\ 2E \quad (6)$$

In (Eq.6), S1, S2 are initial and final entropies of the material. V1 is initial material volume.

(b) Development of mathematical equations for quantification of resilience properties for complex process systems pertaining to a Gas Sweetening Unit (GSU, Fig.1)

System Stress: In line with (Eq.3) of a material, the stress equation of a system for unsteady flow condition can be derived by using the first law of thermodynamics as (Guha and Das, 2015),

$$S_{su} = 1/V_{sys}[d(MU)_{sys}/dt] = [\sum(mh)_i + \sum(mu^2/2)_i + \sum(mgz)_i + \sum q_j - \sum w_l]/V_{sys} \quad (7)$$

Similarly, for steady flow condition, the system stress equation can be written as (Guha and Das, 2015),

$$S_{ss} = [\sum(mh)_{in} + \sum(mu^2/2)_{in} + \sum(mgz)_{in} + (\sum q_j)_{in} - (\sum w_l)_{in}]/ V_{sys} \quad (8)$$

System Strain: Likewise a material (Eq.5), system strain can also be viewed as response of the system to application of a load in terms of change in its original entropy and is necessarily dimensionless. One can define system strain or response as ratio of "Waste work" to "$Exergy_{in}$" (input exergy) and the general equation for system strain or response can be represented as (Guha and Das, 2015),

$$S_{sn} = T_o\beta/Exergy_{in} = Waste\ Work/Exergy_{in} \quad (9)$$

where $T_o\beta = \sum[m(h-T_os]_i + \sum(1-T_o/Tj)q_j - \sum w_l$ = Waste Work

Based on the system stress and system strain definitions given above, correlations are developed for the quantification of system resilience modulus (Ur) for constituent complex process systems (absorber, lean-rich exchanger & regenerator) of a Gas Sweetening Unit (Fig.1, Fig.2):

System Stress Equations:

$$S_{s(abs)} = [\sum(mh)_{in}]/V_{(abs)} = 1/ V_{(abs)}[L_{sl}h_{sl} + V_{fg}h_{fg}] \quad (10)$$

$$S_{s(LRRxch)} = [(mh)_{in}]/V_{(LRExch)} = 1/ V_{(LRExch)}[L_{sr}h_{sr} + L_{sl}\ h_{sl}]_{at\ inlet} \quad (11)$$

$$S_{s(reg)} = [(mh)_{in} + \sum q_j]/V_{(reg)} = 1/ V_{(reg)}[L_{sr}h_{sr} + Q_{reb}] \quad (12)$$

System Strain Equations:

$$S_{sn(abs)} = [L_{sl}b_{sl} + V_{fg}b_{fg}] - [V_{tg}b_{tg} + L_{sr}b_{sr}] / L_{sl}[(h_{sl} - h^0_{sl}) - T_0(s_{sl} - s^0_{sl})] + V_{fg}[(h_{fg} - h^0_{fg}) - T_0(s_{fg} - s^0_{fg})] \quad (13)$$

$$S_{sn(LRExch)} = T_0\ [L_{sr}\ (s_{srout}-s_{srin}) + L_{sl}(s_{slout}-s_{slin})] / L_{sr}[(h_{sr} - h^0_{sr}) - T_0(s_{sr} - s^0_{sr})] + L_{sl}[(h_{sl} - h^0_{sl}) - T_0(s_{sl} - s^0_{sl})] \quad (14)$$

$$S_{sn(reg)} = [L_{sr}b_{sr}] - [\ V_{ag}b_{ag} + L_{sl}b_{sl}] + (1-T_0/T_{reb})Q_{reb} /\ L_{sr}[(h_{sr} - h^0_{sr}) - T_0(s_{sr} - s^0_{sr})] + (1-T_0/T_{reb})Q_{reb} \quad (15)$$

System resilience modulus (Ur) can be quantified by evaluation of the area under the system stress - system strain curve from the system stress value corresponding to a minimum operating condition to the system stress value corresponding to a maximum allowable operating condition. Soft computing simulation technique, using a process simulator needs to be utilized for generation of system stress-strain data for varying operating conditions.

Process Simulation

Soft computing process simulation technique has been used utilizing a high accuracy SIMULATION SCIENCES INC process simulator, Pro/II for this study. The process simulator has been used for steady state simulation of the Gas Sweetening Unit (GSU) that has been designed to sweeten natural gas which contains 1.28 wt% H_2S. Gas sweetening is accomplished by usage of Di-Ethanol amine as solvent. The unit is designed for a treated gas H_2S specification of < 1 ppmw. The absorber and regenerator columns of the GSU have been simulated considering equilibrium stages. The chosen thermodynamic model which has been considered for the entire simulation input file is "special amine package".

The physical and thermodynamic property data of all process streams (external and internal of absorber column, regenerator column and external of lean-rich exchanger) generated in the process simulator output files are utilized in the quantitative correlations which are developed and used in order to estimate variation of stress and strain with varying operating conditions for the absorber column, regenerator column and lean rich exchanger systems.

Results & Analysis

Fig.2 indicates that the absorber column system under study possesses inherent resilience of around 5% with regard to variation in upstream feed sour gas flow rate beyond 100% design flow rate, i.e., 27814 kg/h. Hence, this absorber column may be designed for 95% feed gas capacity even if the column is actually operated for 100% feed gas capacity. Conversely, if the column is designed for 100% feed gas capacity, it may be actually operated at 105% feed gas capacity. The absorber column system is a tray column of 16.5m height (TL-TL) and 1.3m (1300 mm) internal diameter corresponding to 100% feed gas capacity.

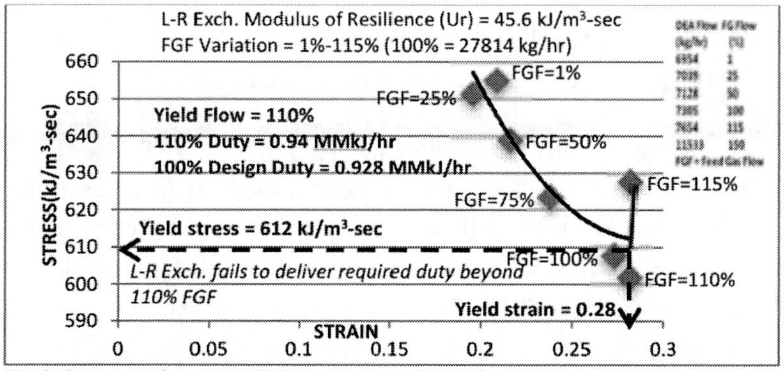

Fig.3: Stress-strain curve of LR Exchanger of GSU for variation of feed gas flow rate.

Fig.3 shows that the lean-rich exchanger system under study possesses inherent resilience of around 10% with regard to variation in upstream feed sour gas flow rate beyond 100% design flow rate. Similar to the absorber column, the lean-rich exchanger may be designed for 90% feed gas capacity although it is actually operated for 100% feed gas capacity. In other words, the lean-rich exchanger may be operated for 110% feed gas capacity even if it is designed for 100% feed gas capacity. In case of 100% feed gas capacity, the designed lean-rich exchanger is a TEMA AES type (1- shell pass and 2-tube passes) heat exchanger with design duty (for 100% feed gas capacity) of 0.928×10^6 kJ/hr corresponding to design U*A (overall heat transfer coefficient being U and heat transfer surface area being A) of 4830 kJ/hr-°C with MTD (mean temperature difference) of 192.2 °C.

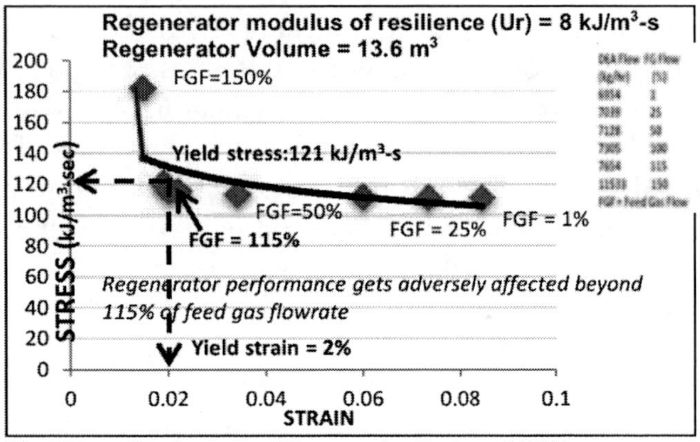

Fig.4 Stress-Strain curve of Regenerator System of GSU for variation of feed gas flow rate

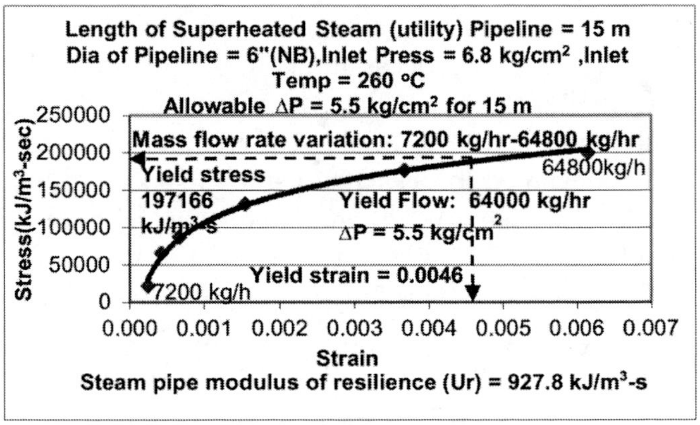

Fig.5 Stress-strain curve of superheated steam pipeline system for mass flow rate variation

Fig.4 indicates that the regenerator column system under study possesses inherent resilience of around 15% with regard to variation in upstream feed sour gas flow rate beyond 100% design flow rate, i.e., 27814 kg/h. The regenerator column may be designed for 85% feed gas capacity in spite of the same being actually operated for 100% feed gas capacity. Alternatively, the regenerator column may be operated for 115% feed gas capacity although it is designed for 100% feed gas capacity. The regenerator column system is a tray column of 21.5m height (TL-TL) and 0.9m (900 mm) internal diameter corresponding to 100% feed gas capacity.
It may be noted that since the limiting inherent resilience is 5% pertaining to the absorber column, overall, for the entire GSU, minimum inherent resilience or flexibility of 5% can be allowed even if the unit is designed for 100% design capacity (i.e., 27814 kg/hr).

Fig.5 shows one utility pipeline system (which carries superheated steam) stress versus strain for steam mass flow rate variation of 7200 kg/hr to 64800 kg/hr. The curve shows a plateau at very high steam flow rate region of 57600 kg/hr to 64800 kg/hr indicating attainment of practically allowable maximum pressure drop of 5.5 kg/cm^2 pertaining to 15 m pipeline length at yield or limiting stress of 197166 kJ/m^{3}s corresponding to yield or limiting flow of 64000 kg/hr for the utility pipeline system in case of flow variation. The physical significance of attainment of the yield or limiting flow of 64000 kg/hr is that the 6" (NB) pipeline system will not be able to carry superheated steam of more than 64000 kg/hr as long as inlet pressure of this steam is fixed at 6.8 kg/cm^2.

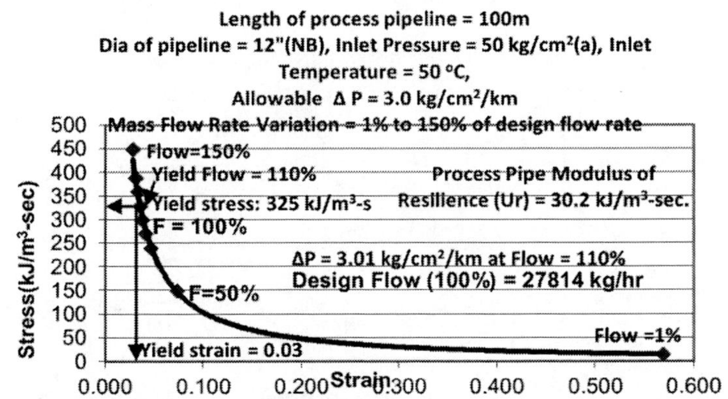

Ur (kJ/m³-s)	*Esys (kJ/m³-s)*	*Flow range (%)*
19.41	272	1-50
7.4	4175	50-100
2.27	**13382**	**100-130**

Fig.6 Stress versus strain curve for inlet sour feed gas pipeline system for mass flow rate variation

Fig.6 shows the inlet sour feed gas pipeline system (process pipeline) stress versus strain for deviation in mass flow rate of feed sour gas from 1% of normal throughput to 150% of the normal throughput (100% normal throughput = 27814 kg/hr). The curve indicates drastic fall in modulus of resilience value (*Ur*) and great increase in modulus of elasticity (*Esys*) value at mass flow rate variation range of 100% -130% of feed gas indicating attainment of allowable pressure drop of 3.0 $kg/cm^2/km$ at yield or limiting stress of ~ 325 kJ/m^3s. Allowable pressure drop of 3.0 $kg/cm^2/km$ is a defined limiting parameter based on which the pipeline size of 12" (NB) has been computed pertaining to 100% mass flow rate (27814 kg/hr) of feed gas. The curve also demonstrates inverse relationship of modulus of resilience (*Ur*) with modulus of elasticity (*Esys*) for the pipeline system of sour feed gas.

(c) Determination of thermodynamically efficient operating capacity regime for a GSU.

Optimum energy usage for a process system is achieved when it is operated at the best thermodynamically efficient capacity regime which also ensures operation in minimum waste work capacity range (Guha, 2015). Best thermodynamically efficient capacity regime can be evaluated by estimation of Thermodynamic Coefficient of Performance (THCOP) data at various operating capacities (Guha, 2015).

Thermodynamic Coefficient of Performance (THCOP) is defined as (Guha, 2015),

$$THCOP = [1 + \beta/\sum (m_i s_i)_{in}] \quad (16)$$

Where, β = rate of entropy creation in a system and $(m_i s_i)_{in}$ = rate of entropy inflow with i^{th} stream.

If THCOP = 1.0, the system is best thermodynamically efficient. That is rate of irreversible entropy creation (β) = 0. Waste work for a separation process can be evaluated as (Guha, 2015),

$$(T_0\beta) = \text{Waste work} = \text{Total work} - \text{Minimum work} \quad (17)$$

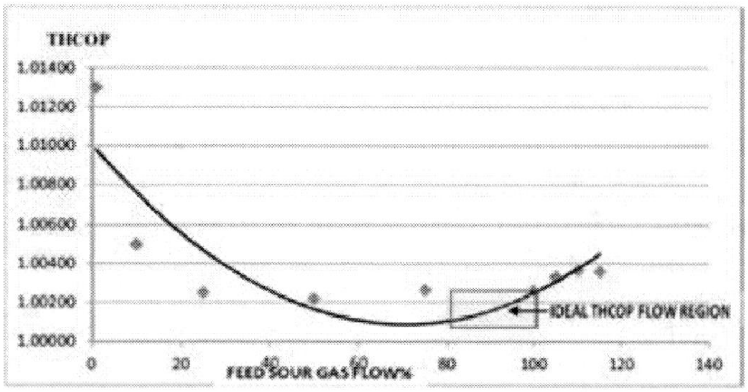

Fig.7 THCOP variation with Feed Sour Gas Flow%

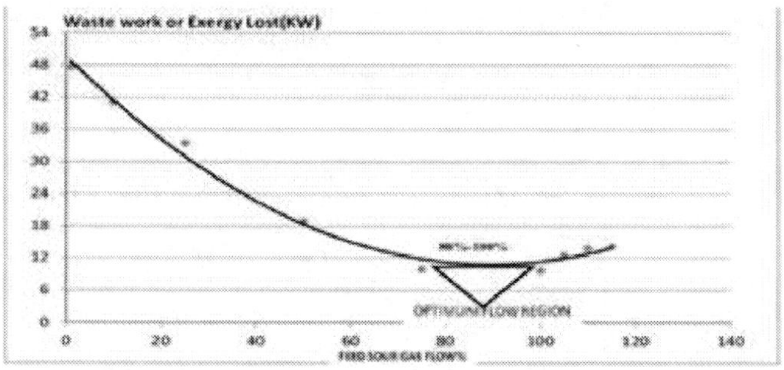

Fig.8 Waste work variation with Feed Sour Gas Flow%

In order to evaluate best thermodynamically efficient operating capacity regime of a GSU (Fig.1), THCOP (Fig.7) as well as Waste work (Fig.8) variation against variation in Feed sour gas flow rate pertaining to the regenerator column is to be determined as in a GSU (Fig.1), energy usage takes place in the regenerator column system only (steam is used in regenerator re-boiler).

Fig.7 and Fig.8 clearly indicate that the GSU operating capacity range of 80%-100% is the best thermodynamically efficient operating capacity regime. Moreover, since optimum or ideal flow region has been evaluated as 80%-100% (Fig.7, Fig.8) for the regenerator column and in turn for the entire GSU, the GSU turndown capacity should not be less than 80% of its design capacity. In other words, operation of the GSU below 80% of its design capacity will incur higher percentage of wastage of valuable work or energy in spite of spending lesser total work or energy at these capacities.

Variation of Modulus of Resilience (Ur) with Modulus of Elasticity (Esys)														
Absorber			Regenerator			Lean-Rich Exch.			Steam Pipeline			Sour feed gas Pipeline		
Ur	Esys	Flow range*	Ur	Esys	Flow range*	Ur	Esys	Flow range*	Ur	Esys x 10^6	Flow range#	Ur	Esys	Flow range*
0.739	1738	1-50	2.6	44.6	1-25	4.8	2137	1-50	28.49	154.8	7200-28800	19.41	272	1-50
0.184	22852	50-100	2.9	42.5	25-50	35.5	548	50-100	430	29.9	28800-57600	7.4	4175	50-100
0.077	73844	100-150	1.5	195	50-100	5.3	112391	100-115	469	9.3	57600-64800	2.27	13382	100-130
			0.5	1760	100-115									
			0.5	14520	115-150									

*Ur in kJ/m³-s, Esys in kJ/m³-s, *Flow range in %,#Flow range in kg/hr*

Application to Process Design

Design procedure in relation to incorporation of 50% over capacity factor or inherent resiliency in the absorber column by augmentation of number of column trays can be found in literature (Guha and Das, 2015).

Conclusions

Inherent resilience for process systems have been conceptualized from the perspectives of material resilience for development of mathematical equations which are used to assess resilience properties of constituent process systems pertaining to a GSU (Fig.1).

The study has enabled developing a new method for identification of thermodynamically efficient operating capacity regime for optimization of energy usage in a GSU. GSU operating capacity range of

80% to 100% has been determined as the optimum operating capacity regime with regard to energy usage for its operation. The frame work of application of resilience analysis into process equipment design with regard to identification of thermodynamically efficient operating capacity regime for the purpose of energy usage optimization is applicable to any process system as well as process unit. Furthermore, usage of this concept may provide roadmaps for minimizing CAPEX (capital expenditure) at the cost of little enhancement in OPEX (operating expenditure) for any process plant in CPI.

Notation

A = surface area of L-R exchanger,
b_{sl} = availability function of lean solvent,
b_{tg} = availability function of treated gas,
b_{ag} = availability function of acid gas,
g_i = stream i acceleration due to gravity,
h_{sr} = enthalpy of rich solvent,
h^o_{fg} = enthalpy of feed gas at datum cond.,
h_i = stream i enthalpy,
h^o_{sl} =enthalpy of lean solvent at datum cond.
L_{sr} = rich solvent flow rate,
M_{sys} = system mass,
m_i = stream i mass flow rate,
q_{jin} = heat input rate,
S_{su} = system stress (unsteady state),
S_{sn} = system strain (steady state),
$S_{sn(reg)}$ = regenerator strain,
s_i = stream i entropy,
$S_{s(LR\ Exch)}$ = Lean-Rich Exch. stress,
s_{sl} = entropy of lean solvent,
s_{fg} = entropy of feed gas,
s_{srout} = entropy of out flowing rich solvent,
s_{slout} = entropy of out flowing lean solvent,
s_{sr} = entropy of rich solvent,
t = time,
T_o = environment/datum temperature,
U = over-all heat transfer co-eff. of L-R Ex.
u_{in} = input stream velocity,
Ur = system modulus of resilience,
V_{fg} =feed gas flow rate,
V_{reg} = regenerator volume,
V_{sys} = system volume,
V_{ag} = acid gas flow rate,
wl = work per unit time for part of system,
z_{in} = input stream elevation,

b = (h-T_os) = availability function.
b_{fg} = availability function of feed gas.
b_{sr} = availability function of rich solvent.
$Exergy_{in}$ = exergy input.
g_{in} = input stream acceleration due to gravity.
h_{fg} = feed gas enthalpy.
h_{in} = enthalpy of input stream.
h_{sl} = enthalpy of lean solvent.
h^o_{sr} = enthalpy of rich solvent at datum cond.
L_{sl} = lean solvent flow rate.
m_{in} = input stream mass flow rate.
Q_{reb} = reboiler duty.
q_j = heat exch. rate between system & external reservoir.
S_{ss} = system stress (steady state).
$S_{s(reg)}$ = regenerator stress.
$S_{sn(abs)}$ = absorber strain.
$S_{s(abs)}$ = absorber stress.
$S_{sn(LR\ Exch)}$ = Lean-Rich Exch. strain.
s^o_{sl} = entropy of lean solvent at datum conditions.
s^o_{fg} = entropy of feed gas at datum conditions.
s_{srin} = entropy of in flowing rich solvent.
s_{slin} = entropy of in flowing lean solvent.
s^o_{sr} = entropy of rich solvent at datum conditions.
T_j = environment/external reservoir temperature.
T_{reb} = reboiler temperature.
u_i = stream i velocity.
U_{sys} = system internal energy.
Esys = system modulus of elasticity.
$V_{(LR\ Exch)}$ = Lean-Rich Exch. volume.
$V_{(abs)}$ = absorber volume.
V_{tg} = treated gas flow rate.
wl_{in} = work input.
z_i = stream i elevation.
β = rate of entropy creation.

References

1. Fiksel, J., "Sustainability and Resilience: Towards a systems approach", ***Sustainability: Science, Practice & Policy, 2, 14-21,(2006).***
2. Guha, S., and Das, S.K., "Resilience Analysis of a Pipe Segment System Carrying Superheated Steam", ***Advanced Materials Research, 917, pp. 232-243, (2014).***
3. Guha, S. and Das, S.K., "Thermodynamic Availability Analysis: Usage in Quantification of Resilience Modulus, Modulus of Elasticity and Yield Stress for an Absorber Column System", ***IJCSIT, Vol. 6(4), (2015),3630-3649.***
4. Guha, S., "Simulation Studies for Analysis of Inherent Resilience Properties of an Amine Regeneration Column Used in a Gas Sweetening Unit", ***Procedia Computer Science, 62 (2015), 593-602.***
5. Linh T.T. Dinh, Hans Pasman, Xiaodan Gao, M. Sam Mannan (2012), "Resilience engineering of industrial processes: Principles and contributing factors", ***Journal of Loss Prevention in the Process Industries, 25,233-241.***
6. Mitchell S. M., M. Sam Mannan., Ph.D. Thesis, "Resilient Engineered Systems: The Development of an Inherent System Property", ***Texas A&M University, Texas (2007).***
7. Morari, M., "Flexibility and resiliency of process systems", ***Comp. Chem. Engg., 7, 423-437,*** **(1983).**
8. V. Raghavan (1990), "Materials Science and Engineering", ***Prentice Hall, 3rd Edition.***

491609 Tubular Multi-Tube Cooled Reactor As a Fractional Heat Exchanger â€“ Modeling Reactive Heat Transfer

Abstract:

Many industrially important vapor phase reactions are equilibrium controlled and highly exothermal in nature which leads to high reactor exit temperature. Very often product gas requires cooling in a tubular heat exchanger before it is sent to the next unit operation. Residual reaction could occur under reduced temperature within the heat exchanger and end product profiles and final gas conditions are determined by the relative rates of heat transfer and reaction in such a unit. If the gas final cooled product compositions and conditions are important to the next unit operation, proper prediction, monitoring and control of the unit through rigorous process modeling is critical.

If the inlet gas to heat exchanger temperature is extremely high, then heat transfer by radiation should be considered since the effect of this heat transfer mode cannot be neglected when inlet gas temperature is very high. Exchanger outlet gas temperature can be controlled by properly design the residence time of gas inside the exchanger. Any reactions after the cooling require certain gas compositions and temperature which should be properly handled together with the gas cooler. Optimization of the combined unit operations will result in potentially reduced capital investment and operational costs, along with obtaining more desired final product compositions and conditions.

If the purpose of the cooling is purification of the gas stream, reactions occurring in either vapor or condensed liquid phase should be evaluated carefully to identify the extent of reaction and its impact on the separation performance. When a phase change exists, both phase and chemical equilibrium should be considered and mass transfer across the interface should be included along with heat transfer. Optionally chemical reaction kinetics and fluid flow of the condensed liquid should be considered.

This talk will present rigorous modeling of reactive heat transfer which may be useful to process intensifications.

494706 Thermal Stability of Ethylene Polymerization in Fluidized-Bed Reactor

Abstract:

Ethylene polymerization in fluidized bed reactor is a simple process and is easier for production separation. Thus fluidized bed reactor is widely used in ethylene and propene polymerization. In order to improve the production rate of polyethylene in fluidized bed reactor, condensed mode operation and super-condensed mode operation are developed. With inert condensed compound such as i-pentane or n-hexane or comonomer such as n-hexene, the recycle stream can become a gas-liquid mixture under the reasonable temperature, thus more heat of reaction can be removed by the recycle stream and the production rate will be dramatically increased.

Ethylene polymerization is a rapid and highly exothermic process. As the melting temperature of polyethylene near the reaction temperature, thermal instability of fluidized bed may cause the oscillation of reactor temperature, and even induce polyethylene melting and plant stop. Therefore, thermal stability and dynamical behavior of fluidized bed not only concern design and scale up of fluidized bed reactor, but also concerns the real time monitoring, handling of abnormal conditions and safety production.

Several methods have been employed to analyze the stability and dynamic behaviors, such as mathematics deduction method, nonlinear analysis method, process simulation method, etc. The mathematics deduction method is based on strict mathematical derivation, however, it cannot be used to solve relatively high dimensional systems. Nonlinear analysis method is used to solve relatively high dimensional systems. Meanwhile, the process simulation method is applied to solve large complex chemical process but this method still exist some disadvantages in stability analysis. In this work, a simple model of fluidized bed reactor is proposed to get the thermal stability criterion. Then effects of structural parameters of fluidized bed and operation parameters on thermal stability of ethylene polymerization process are investigated based on a detailed mathematical model of fluidized bed.

In the simple model of fluidized bed reactor, concentration of reactants and reactor pressure are assumed to be constant. Therefore, the simple model is a first order ordinary differential equation group with respect to catalyst concentration and reactor temperature. With mathematics deduction, the coefficient matrix of and thermal stability criterion are obtained. The calculation results show that space-time yield of polyethylene on the Hopf bifurcation points is proportion to catalyst activation energy, reactor height and square of reaction, and is in inverse proportion to superficial velocity and density of recycle gas.

Based on the two phase model, a multi zone model is proposed in order to describe ethylene polymerization in fluidized bed in condensed mode and super-condensed mode. In the multi zone model, fluidized bed is composed of emulsion phase and bubble phase, and the emulsion phase is divided into gas-liquid-solid zone and gas- solid zone. The evaporation of liquid occurs in the gas-liquid-solid zone with a certain thickness liquid film on the surface of solid particles. Furthermore, the liquid evaporation rate obeys the equilibrium evaporation principles. Then the structural parameters and operation parameters of fluidized bed reactor are chosen to trace solution curves and

bifurcation points referred to the catalyst feed rate.

The calculation results of multi zone model show that there are Hopf bifurcation points and limit points in the solution curves, which divide the solution curve into three parts-stable part with low temperature, unstable part and stable part with high temperature. Moreover, the steady state multiplicity of fluidized bed reactor occurs and the temperature range of stable part with low temperature shrinks with increase of activation energy of catalyst and condensed compounds in the recycle stream. In addition, thermal stability criterion based on a simple model fit well with the prediction of multi zone model.

492382 From Tool to Discipline- Developing a Model-Centric Process Engineering Culture

Abstract:

Process simulators and first-principles models are widely used in the commodity chemicals industry to drive process engineering and design activities. The applicability of these tools in the specialty chemicals and polymer industry is challenging due to inadequate fundamental thermodynamic and kinetic data for chemical systems of interest, complexity of chemistries practiced, utilization of batch and semi-batch operations and "non-traditional" unit operations and equipment. A successful application of a model-based approach can result in more robust scale-up efforts for new process development and provide fundamental insight to guide process optimization efforts for existing plant assets.

A model-centric approach aims to put first-principles models at the center of all process engineering activity. The layered approach involves integrating different sets of expertise such as applied thermodynamics and phase equilibria, experimental design, reaction engineering and computational chemistry with the goal of developing a core template for process development efforts and asset templates for plant support activities. The objective is to develop a fundamental understanding of existing assets and technologies practiced and to "de-risk" new process development. This presentation will touch upon the experience of developing such a program at SABIC Specialties.

497158 Christos Georgakis: His Work and Legacy over 40 Years in Research and Education

Abstract:

From his undergraduate diploma research on fuels to his present interests in statistical modeling from data, monitoring, diagnosis and control, Christos Georgakis academic work has spanned a multitude of subjects including among others; reaction engineering, optimization and control of batch processes, interaction of process design and control, plant-wide control, and nonlinear control of chemical reactors. In all these endeavors he has contributed pivotal ideas that have led to patented control systems for industrial practice, novel educational programs, and forward looking conferences in process systems engineering. In this presentation I will try to give an overarching view of his 40-year activity and account for his many contributions and impact to our profession.

494521 Optimal Control Laws for Batch and Semi-Batch Reactors Using the Concept of Extents

Abstract:

Optimal Control Laws for Batch and Semi-batch Reactors Using the Concept of Extents

D. Rodrigues, J. Billeter and D. Bonvin

Laboratoire d'Automatique, Ecole Polytechnique Fédérale de Lausanne

EPFL – Station 9, 1015 Lausanne, Switzerland

dominique.bonvin@epfl.ch

In the context of dynamic optimization of batch and semi-batch reactors, this contribution presents a method that uses the concept of extents to generate solutions that satisfy the necessary conditions of optimality given by Pontryagin's maximum principle. The method is divided in two parts. In the first part, the reactor model is written in terms of decoupled extents, and adjoint-free optimal control laws are generated for all possible types of arcs that may occur in the optimal solution. In the second part, the correct sequence of arcs is determined, and, for each sequence, the optimal switching times and initial conditions are computed numerically.

The optimal control problems (OCPs) are formulated in Mayer form, with n_u piecewise-continuous inputs $\mathbf{u}(t)$, n_x states $\mathbf{x}(t)$ described by the differential equations $(t) = \mathbf{f}(\mathbf{x}(t),\mathbf{u}(t))$, $\mathbf{x}(t_0) = \mathbf{x}_0$, the cost function $\varphi(t_f,\mathbf{x}(t_f))$, the n_t terminal constraints $\boldsymbol{\psi}(t_f,\mathbf{x}(t_f)) \leq \mathbf{0}$, the n_g mixed path constraints $\mathbf{g}(\mathbf{x}(t),\mathbf{u}(t)) \leq \mathbf{0}$ and the n_h first-order pure-state constraints $\mathbf{h}(\mathbf{x}(t)) \leq \mathbf{0}$. Note that this class of OCPs is not restrictive in most dynamic optimization problems dealing with reactors.

The optimal input trajectories are composed of a (typically finite) number of arcs. For each arc and for each input, the optimal input is determined by either an active path constraint or a condition that expresses a physical compromise that depends exclusively on the dynamics of the system [1].

Let the input u_j be one element of $\mathbf{u}$. The goal is to find an expression that relates the optimal input u_j, or one of its time derivatives, to the states, the inputs or the time derivatives of the inputs, thus resulting in an *adjoint-free* optimal control law. For each arc, one of the following cases is possible:

1. The optimal input u_j is determined by the active path constraint $g_k(\mathbf{x},\mathbf{u}) = 0$.

2. The optimal input u_j is determined by the active path constraint $h_k(\mathbf{x}) \leq 0$, and u_j is obtained such that $\partial h_k/\partial \mathbf{x}(\mathbf{x})\ \mathbf{f}(\mathbf{x},\mathbf{u}) = 0$.

3. Otherwise, the optimal input u_j is determined by the condition $\det(M_u) = 0$, where

$$M_u := [\partial \mathbf{f}^u / \partial u_j(\mathbf{x},\mathbf{u}) \;\; \Delta_j \partial \mathbf{f}^u / \partial u_j(\mathbf{x},\mathbf{u}) \;\; \Delta_j^{\rho -1} \partial \mathbf{f}^u / \partial u_j(\mathbf{x},\mathbf{u})],$$

and the operators $\Delta_j, \ldots, \Delta_j^{\rho -1}$ are defined as

$$\Delta_j \mathbf{v} := \partial \mathbf{v}/\partial \mathbf{x}\, \mathbf{f}(\mathbf{x},\mathbf{u}) - \partial \mathbf{f}^u / \partial \mathbf{x}^u\, (\mathbf{x},\mathbf{u})\, \mathbf{v} + \mathbf{\Sigma}_{n\geq 0}\, \partial \mathbf{v}/\partial \mathbf{u}^{(n)}\, \mathbf{u}^{(n+1)},$$

$$\Delta_j^{l} \mathbf{v} := \Delta_j (\Delta_j^{l-1} \mathbf{v}), \text{ if } l = 2,\ldots, \rho_j - 1,$$

for any vector field $\mathbf{v}$ of dimension ρ_j, with $\mathbf{x}^u$ being the ρ_j-dimensional vector of states that can be influenced by manipulating u_j, and $\mathbf{f}^u(\mathbf{x},\mathbf{u})$ such that $^u = \mathbf{f}^u(\mathbf{x},\mathbf{u})$.

However, the input u_j and its time derivatives may not appear explicitly in the function $\det(M_u)$. Hence, as a general approach to find the optimal input u_j when it is not determined by an active path constraint, the function $\det(M_u)$ is subject to time differentiation until u_j or one of its time derivatives appears in $d^r(\det(M_u))/dt^r$, for some r_j. Let $u_j^{(\xi)}$ be the highest-order time derivative of u_j that appears in $d^r(\det(M_u))/dt^r$. Then, the optimal input $u_j^{(\xi)}$ is obtained such that $d^r(\det(M_u))/dt^r = 0$.

The mass and heat balances for batch and semi-batch reactors can be written using the concept of extents [2]. Let us consider a homogeneous batch or semi-batch reactor with R independent reactions and p independent inlets ($p = 0$ for batch reactors), where $\mathbf{u}_{in}(t)$ is the p-dimensional vector of inlet flowrates, and $q_{ex}(t)$ is the exchanged heat power. The numbers of moles $\mathbf{n}(t)$ and the heat $Q(t)$ can be expressed as a linear combination of extents, according to

$$\mathbf{n}(t) = \mathbf{N}^{\mathrm{T}}\, \mathbf{x}_r(t) + \mathbf{W}_{in}\, \mathbf{x}_{in}(t) + \mathbf{n}_0,$$

$$Q(t) = (-\Delta\mathbf{H})^{\mathrm{T}}\, \mathbf{x}_r(t) + {}_{in}^{\mathrm{T}}\, \mathbf{x}_{in}(t) + x_{ex}(t) + Q_0,$$

where $\mathbf{n}_0$ is the S-dimensional vector of initial numbers of moles, Q_0 is the initial heat, $\mathbf{N}$ is the $R\times S$ stoichiometric matrix, $\Delta\mathbf{H}$ is the R-dimensional vector of heats of reaction, $\mathbf{W}_{in}$ is the $S\times p$ inlet-composition matrix, $_{in}$ is the p-dimensional vector of inlet specific enthalpies, $\mathbf{x}_r(t)$ is the R-dimensional vector of extents of reaction, $\mathbf{x}_{in}(t)$ is the p-dimensional vector of extents of inlet, and $x_{ex}(t)$ is the extent of heat exchange.

The state vector of dimension $n_x := R + p + 1$ is

$$\mathbf{x}(t) := [\mathbf{x}_r(t)^{\mathrm{T}}\; \mathbf{x}_{in}(t)^{\mathrm{T}}\; x_{ex}(t)]^{\mathrm{T}},$$

while the input vector of dimension $n_u := p + 1$ is

$$\mathbf{u}(t) := [\mathbf{u}_{in}(t)^{\mathrm{T}}\; q_{ex}(t)]^{\mathrm{T}}.$$

The dynamic equations can be written compactly as

$$(t) = \mathbf{f}(\mathbf{x}(t),\mathbf{u}(t)),\; \mathbf{x}(t_0) = \mathbf{0}_{R+p+1},$$

by defining

$\mathbf{f}(\mathbf{x}(t),\mathbf{u}(t)) := [\mathbf{r}_V(\mathbf{x}(t))^{\mathrm{T}}\ \mathbf{u}(t)^{\mathrm{T}}]^{\mathrm{T}}$,

where $\mathbf{r}_V(\mathbf{x}(t))$ is the R-dimensional vector of reaction rates. In batch and semi-batch reactors, with $_j =$ $f_j(\mathbf{x},\mathbf{u}) := u_j$, it is possible to define the following vectors of dimension $\rho_j := R+1$:

$\mathbf{x}^u \ := [\mathbf{x}_r^{\mathrm{T}}\ x_j]^{\mathrm{T}}$,

$\mathbf{f}^u\ (\mathbf{x},\mathbf{u}) := [\mathbf{r}_V(\mathbf{x})^{\mathrm{T}}\ u_j]^{\mathrm{T}}$.

Note that, since this system is input-affine, $\det(\mathrm{M}_u\ )$ and its time derivatives are polynomial functions of u_j and its time derivatives, thus resulting in a finite number of solutions, and typically a single solution that satisfies the condition in Case 3.

Let us define the state vector $_j$ as the complement of the state x_j (all states $\mathbf{x}$ except x_j), and the vector $_j(\mathbf{x},\mathbf{u})$ as the corresponding complement of $f_j(\mathbf{x},\mathbf{u})$. Then, one can prove that, when the optimal input u_j is not determined by an active path constraint:

1. For reactors with a single independent reaction, the input u_j is determined by

$d(\det(\mathrm{M}_u\ ))/dt = \partial(\partial\mathbf{r}_V/\partial x_j(\mathbf{x}))/\partial x_j\ u_j + \partial(\partial\mathbf{r}_V/\partial x_j(\mathbf{x}))/\partial_j\ _j(\mathbf{x},\mathbf{u})$,

since u_j and its time derivatives do not appear in

$\det(\mathrm{M}_u\ ) = \partial\mathbf{r}_V/\partial x_j(\mathbf{x})$.

2. For reactors with 2 independent reactions, the input u_j is determined by

$\det(\mathrm{M}_u\ ) = \det([\partial\mathbf{r}_V/\partial x_j(\mathbf{x})\ \partial(\partial\mathbf{r}_V/\partial x_j(\mathbf{x}))/\partial x_j])\ u_j$

$+ \det([\partial\mathbf{r}_V/\partial x_j(\mathbf{x})\ \text{-}\partial\mathbf{r}_V/\partial\mathbf{x}_r(\mathbf{x})\ \partial\mathbf{r}_V/\partial x_j(\mathbf{x}) + \partial(\partial\mathbf{r}_V/\partial x_j(\mathbf{x}))/\partial_j\ _j(\mathbf{x},\mathbf{u})])$

One can use symbolic computation software to evaluate the function $\det(\mathrm{M}_u\ )$ and its time derivatives and obtain the optimal input u_j or one of its time derivatives when it is not determined by an active path constraint.

The advantage of the proposed approach is that it reduces the set of possible arcs to a finite number of possibilities. This, in turn, results in a finite number of arc sequences if one assumes an upper bound on the number of arcs present in the optimal solution. Hence, instead of solving the original infinite-dimensional problem, one can simply perform numerical optimization for each arc sequence, using the switching times between arcs and the initial conditions as decision variables.

Note that the dynamic state equations can be integrated forward in time, since it is possible to evaluate the corresponding inputs without knowledge of the adjoint variables. Once the forward integration is complete, one can integrate backward in time to obtain the corresponding adjoint variables, which enables the computation of the gradients with respect to the switching times and initial conditions of the arcs.

The proposed approach will be illustrated to maximize the final quantity of product in an acetoacetylation reaction, subject to constraints on the final concentration of reactants and by-products [3].

[1] B. Srinivasan, S. Palanki, and D. Bonvin. Dynamic optimization of batch processes: I. Characterization of the nominal solution, *Comput. Chem. Eng.*, 27:1-26, 2003.

[2] D. Rodrigues, S. Srinivasan, J. Billeter, and D. Bonvin. Variant and invariant states for chemical reaction

systems, *Comput. Chem. Eng.*, 73:23-33, 2015.

[3] B. Chachuat, B. Srinivasan, and D. Bonvin. Adaptation strategies for real-time optimization, *Comput. Chem. Eng.*, 33:1557-1567, 2009.

500576 Extensive, Intensive and Invariant Variable Control Systems

Abstract:

In the current paper we review the basics of the Legendre-Fenchel dual and how to use this concept to motivate a variety of methods proposed for process control and control of networks during the past six decades. Many of these methods have faded into the background, some have found industrial application and still in practical use and some provided basis for fruitful research that continues to this day. The objective of this paper is to seek some communality among these diverse method and provide foundations for further development and discussion.

The main idea behind the use of extensive variables and invariants for process control go back to the mass an energy balance control systems developed and applied by Buckley and Shinskey in the 1960ies. The use of process invariants (symmetries) was proposed by Fjeld et al. (1974) and the Extensive Variable Control method was proposed by Georgakis (1986) for the synthesis of multivariable and (or) nonlinear control structures. The Generic Model Control (GMC) method was proposed by Lee and Sullivan (1988), partial control by Shinnar et al. (1996) and Inventory control by Farschman et al. (1998). The latter was based on the observation that material and enenrgy balances could be viewed as special cases of passive systems. Tyreus proposed Dominant Variable Control and more recently Bonvin (2005) developed reaction and flow invariants and Adomaitis et al (2016) applied invariants to modeling and control of CVD reactors. All these papers posed the question of how to control complex chemical processes and networks of processes using only a small subset of all the variables used to model such systems. They also aim to address question such as which variables to measure and control and in this sense the methods are more focused on control strategies rather than which algorithm to use.

The main idea that emerges in all these applications is that chemical processes are governed by conservation principles (symmetries) that will hold at every length and time scale. These symmetries are in their simplest form expressed as conservation of mass, energy and mol-numbers for example. This property can also be expressed in terms of duality and the optimization of specifc potentials and ths provides a more general way of expressing dynamics and control. The main problems that emerge are the questions of how to use these principles for control and optimization, how to present convincing arguments to demonstrate that variables that are not controlled, the so-called zero dynamics, are stable and finally how to satisfy tracking objectives. These problems are by no means completely addressed at present. Some open research problems will be presented.

489181 Customer Feedback Control: Design and Implementation on a Nanocomposite Manufacturing Process

Abstract:

Actual feedback from customers (in the form of qualitative/quantitative data) is the most directly relevant indicator of manufactured product performance in end-use—whether or not the product actually meets customer requirements as indicated by the manufacturing process specifications. Consequently, it is necessary to use such customer feedback data directly and explicitly within a control scheme in order to take rational corrective action when the product fails to perform in end-use as specified. However, to date, such a control scheme does not exist, as a result of a variety of challenges. In this presentation, we address these challenges and discuss the design and implementation of such a customer feedback regulator, which utilizes customer feedback information (z) (fundamentally binary, since the product either performs acceptably, or not) to determine appropriate target set-points for product properties (w). Specifically, we introduce techniques for modeling the relationship between measured product properties (continuous independent variables, w) and the customer feedback data (binary dependent variable, z); we also introduced concepts of "target/design probability of acceptance" and "achievable probability of acceptance"; and then develop theoretical results for determining optimal adjustments to the product property set-points needed to achieve the target probability of acceptance, given property measurements and customer feedback data. The performance of the control scheme is illustrated *in simulation* via an example involving the manufacture of an industrial polymer nanocomposite for a packaging film application. The simulation results show how, in response to an unmeasured disturbance that compromised the product performance in end use, the customer feedback regulator sequentially adapted the operating conditions and successfully increased the probability of acceptance to the desired level.

496582 Process Operability Analysis of High Dimensional Systems

Abstract:

Conventionally the plant and control designs are done sequentially, but a preferred approach would be to address control issues early in the process design. Operability analysis is one of the bridges that link process design and control. The Operability index framework introduced by Vinson and Georgakis[1] laid a good foundation for this purpose. This methodology can be used to design new processes, and compare competing designs. It can also be used in the debottlenecking and assessment studies of existing processes. In this framework, the designer evaluates the input to output and output to input maps through elaborate simulation or design calculations.

Subramanian and Georgakis[2] applied OI methodology to primarily two dimensional nonlinear systems using one dimensional boundary marching algorithms with heuristics to map the interior domains. In the present work, high dimensional continuation method is used for computing the entire solution manifold of algebraic systems. This approach systematically computes the entire region of interest. The power of this method is demonstrated by revisiting some of the processes studied in [2]. An open source software library called Multifario [3] is used for exploring the high dimensional input, output and/or disturbance spaces. It computes the solution manifold as a set of overlapping neighbourhoods of polytopes. The computation proceeds by establishing the boundary points and computing an additional polytope that is in the area of interest. This procedure is repeated until the entire solution space of interest is obtained.

The ideal Continuous Stirred Tank Reactor (CSTR) system is considered with a single first order reaction of type taking place. The system parameters are taken from [2]. The coolant flow-rate and the volume of the reactor are taken as inputs and, the exit concentration and the temperature of the reactor are considered as outputs. The disturbance variables are the feed concentration and the feed temperature. The high dimensional computational procedure was validated by reproducing the input to output and output to input maps that were reported in [2]. Interestingly, the overall Desired Input Space (DIS) that involved two inputs, two outputs and two disturbance variables was computed in one full calculation run and it matched well with the previously reported result that was approximated as a union of finite set of maps. Later, the method was applied to a CSTR with series reaction system. Here the new method identified the output maps from a given available input space along with the multiple steady-states.. The results from the new approach matched well with the previously reported results.

Having studied square systems that have the same number of inputs and outputs, a non-square system where one of the output variables is specified to be interval controlled rather than set-point controlled. (e.g. reactor temperature). The additional degree of freedom made available with such specifications, can be exploited to lower the input requirements. Competing design (input) options can be compared with a design objective function, for example that is least away (measured by norm) from the nominal operating point or that requires minimum reactor volume, etc. From the results computed from Multifario for the first order system, we select a group of points that approximate the minimal DIS through heuristics. The results reasonably compare with the theoretical calculation. We are pursuing to fine tume the

approach to improve the accuracy.

The effects of key parameters in the software that control the accuracy and speed of computations were studied and will be reported. It has to be pointed out that these calculations are computationally intensive as higher dimensional manifolds of interest are explored fully. At present, we have been able to apply this approach for systems of containing upto eight variables within a few hours of computational time.

References :

[1] D.R. Vinson, C. Georgakis, J. Process Control, **10,** 185, 2000.

[2] Subramanian, S. and Georgakis, C., Chemical Engineering Science, 56, 5111â€"5130, 2001.

[3] M.E. Henderson, Int. J. Bifurcat Chaos, **12**, 451, 2002.

Study of Moment-Based MPC formulations and their Connection to Classical Control

R. Zhang*, M. B. Saltik*, L. Özkan*
* Eindhoven University of Technology, 5600 MB Eindhoven, The Netherlands

Model predictive control (MPC) is a well-established modern control technology used in diverse applications to provide (sub)optimal operating conditions while incorporating safety and performance constraints. In MPC, the control action at the current time instant is obtained by solving a finite horizon optimal control problem according to the forecasts of the future process behavior. Hence the quality/validity of predictions generated from process models determine the performance of these controllers ([1]). Although models based on first principles are expected to provide better predictions for a wider range of operating conditions ([2]), the model predictions should also incorporate the effects of uncertain deviations to overcome the adverse effects. To this end, robust model predictive control techniques are developed in order to reduce the effect of uncertainty ([3,4]) in dynamical processes.

Several distinct robust model predictive control techniques have established over the last decades, which can be classified according to the modeling and treatment of uncertainty realizations. A very well known technique of establishing robustness is the worst-case optimization techniques, e.g., ([5]). In this technique, a min-max problem is constructed which suppresses the effect of all possible uncertainty realizations over the prediction horizon. Recently, the so-called chance constrained, e.g., ([6]), and the scenario based optimization ([7]) paradigms have been introduced. In these approaches the MPC specifications or the constraints are allowed to be violated, but one controls the chance of specification violations. However, all of these mentioned approaches are yet to be adopted into the industrial applications. The essential reason for this gap between industry and academia is the huge deterioration of the closed-loop performance and the required computational power to solve the robust optimization problem. In order to surpass these problems, relatively recently, moment based MPC formulations are introduced ([8,9]) which consider the linear combinations of the (centralized) moments of the uncertain (stochastic) predictions within the MPC problem. In this approach, the ability to discard rare-and-severe realizations is adjusted by tuning parameters weighting the moments of the cost function. Hence, these parameters balance the suppression of uncertain effects versus the on average the closed-loop behavior.

In this contribution, we discuss the effects of tuning parameters in moment based MPC problems on the closed-loop time and frequency domain characteristics with the help of simulations. We present the effects resulting from Mean-MPC, for the plant-model mismatch case, and the Mean-Variance

MPC, for the additive perturbations case, while comparing them with the nominal MPC response. Since the effect of moments of cost function leads to an MPC problem with stage-wise changing state and input weights, one can also construct the closed-loop transfer function of the unconstrained MPC controller, thus compare the effects the moment parameters in the frequency domain. More specifically, we first demonstrate that the MV-MPC introduces a virtual (low pass) filter into the prediction model. This leads to introduction of cross terms of states (over the prediction stages) with state (equivalently output) weighting matrix accumulating the effects of uncertainty. This re-adjusts the spread (variance) versus on average (mean) performance of the MPC controller. Moreover, by adjusting the weight parameter, we show improvements in the bandwidth or the phase response of the closed-loop system. Lastly, we demonstrate the backing-off effect the mean-MPC for plant-model mismatch case. In this case the resulting MPC controller reduces the aggressiveness of the controller more and more as the trust in model predictions reduces. This effect is adjusted via additional terms accumulating in the input actions' weighting matrix. We demonstrate the proposed moment based MPC techniques on a quadruple tank system.

Acknowledgement:

This work has been done within the project 'Improved Process Operation via Rigorous Simulation Models (IMPROVISE)' in the Institute for Sustainable Process Technology (ISPT).

References:

[1] – Qin, S.J. and Badgwell, T.A., 2003. A survey of industrial model predictive control technology. *Control Engineering Practice*, *11*(7), pp.733-764.

[2] – Özkan, L, Kothare M.V., Georgakis,C., 2003. Control of a solution copolymerization reactor using multi-model predictive control. *Chemical Engineering Science, 58*(7), pp.1201-1221.

[3] – Goodwin, G.C., Kong, H., Mirzaeva, G. and Seron, M.M., 2014. Robust model predictive control: reflections and opportunities. *Journal of Control and Decision*, *1*(2), pp.115-148.

[4] – Mayne, D.Q., 2014. Model predictive control: Recent developments and future promise. *Automatica*, *50*(12), pp.2967-2986.

[5] – Kothare, M.V., Balakrishnan, V. and Morari, M., 1996. Robust constrained model predictive control using linear matrix inequalities. *Automatica*, *32*(10), pp.1361-1379.

[6] – Schwarm, A.T. and Nikolaou, M., 1999. Chance-constrained model predictive control. *AIChE Journal*, *45*(8), pp.1743-1752.

[7] – Bernardini D, and Bemporad A. Scenario-based model predictive control of stochastic constrained linear systems. In the 28th Chinese Control Conference. CDC/CCC 2009. IEEE.

[8] – Saltik M.B. et al., On the moment based robust MPC formulations, AICHE Annual Meeting, San Francisco, USA, 2016.

[9] – Saltik M.B. et al., Moment Based Model Predictive Control for Systems with Additive Uncertainty, Accepted for presentation in American Control Conference 2017.

494143 New Directions on Process Operability: Bilevel and Parallel Programming Approaches for Process Intensification and Modularity

Abstract:

Process operability emerged in the last decades as a powerful tool for the design and control of chemical processes, including a variety of academic and industrial-scale applications from companies such as Air Products and Chemicals (APCI) and DuPont (Lima et al., 2010; Georgakis et al., 2003). In particular, process operability was proposed as a method for verifying the ability of a process design, defined by the available input set (AIS), to reach an achievable output set (AOS) that considers production targets (Vinson and Georgakis, 2000). Recent efforts in process operability have been focused on the calculation of the desired input set (DIS) for process design and intensification of natural gas utilization applications described by nonlinear models. Among the applications addressed were a catalytic membrane reactor (MR) for the direct methane aromatization (DMA) conversion to hydrogen and benzene, as well as a natural gas combined cycle (NGCC) system for power generation (Carrasco and Lima, 2017a, b, c). These past results indicate that operability can be a tool for enabling process intensification of nonlinear chemical and energy processes, facilitating the realization of the concept of modular manufacturing. However, there is still a gap in terms of problem dimensionality that nonlinear operability methods can handle.

In this presentation, a historical perspective on process operability will be given including how the past concepts have evolved to address the current challenges and potential future directions. In particular, the incorporation of bilevel and parallel programming approaches into the classical process operability concepts will be discussed. Bilevel programming enables the formulation of a hierarchical structure for combining design and intensification objectives in one step. Additionally, parallel computing is proposed as the key to tackle the challenge of computational time limitation. Such approaches provide a systematic framework for process intensification of chemical and energy systems towards modularity. Modular systems have the potential to transform the US economy, by utilizing local stranded gas without the need of transporting it from the exploration areas to the utilization plants. However, the design and intensification of modular energy systems is a challenging task as these processes are represented by complex and nonlinear models, with large numbers of differential and algebraic equations that demand high computational effort for their solution. The proposed operability approach addresses this dimensionality gap by employing parallel programming tools.

Preliminary results on the implementation of the proposed method show a reduction in computational time up to 2 orders of magnitude, when compared to the original results without parallelization. For parallel computing purposes, the MATLAB Distributed Computing Server at WVU with 63 workers (cores) is employed. In terms of intensification, the results for a 4-D DMA-MR system show that different reactor designs with similar levels of performance may have significantly different process footprints (up to 90%) in terms of reactor volume. Also, a modular NGCC power plant design with a dramatic reduction in size, from 400 to 0.5 [MW], is produced by specifying conditions of the gas and steam turbine cycles, while still keeping the high net plant efficiency of 55 [%]. The current research results open new avenues for operability future research, such as: i) formulation of an integrated framework for the design, intensification and control of modular energy systems; ii) incorporation of dynamic operability ideas (Georgakis et al., 2003) for intensification and drug administration; iii) development of new operability indicators; and iv) employment of stochastic tools to consider design and control under uncertainties.

References

Carrasco J. C. and Lima F. V. "Novel operability-based approach for process design and intensification: Application to a membrane reactor for DMA". AIChE J. 2017a; 63(3): 975-983.

Carrasco J. C. and Lima F. V. "An optimization-based operability framework for process design and intensification of modular gas utilization systems". Accepted for publication in Comput. Chem. Eng. 2017b; DOI: CACE-5645.

Carrasco J. C. and Lima F. V. "Operability-based approach for process design, intensification, and control: application to high-dimensional and nonlinear membrane reactors". In proceedings of the FOCAPO/CPC 2017c.

Georgakis C., Uzturk D., Subramanian S., and Vinson D. R. "On the operability of continuous processes". Control Eng. Prac. 2003; 11(8): 859–869.

Lima F. V., Georgakis C., Smith J. F., Schnelle P. D., and Vinson D. R. "Operability-based determination of feasible control constraints for several high-dimensional nonsquare industrial processes". AIChE J. 2010; 56(5): 1249-1261.

Vinson D. R. and Georgakis C. "A new measure of process output controllability". J. Proc. Cont. 2000; 10(2-3): 185-194.

501505 A Method to Reduce DimensionalityÂ of Powder Flow Characterization

Abstract:

Background: Material flow properties often have substantial impact on the performance of the manufacturing processes. For solids materials, the flow behavior is a multi-dimensional characteristic. Currently, a practical way is to measure powder properties using several standardized lab scale tests so that the flow behavior of a material can be described using a large group of indices. However, this approach is often time-consuming, and some of the indices obtained may be redundant. **Purpose:** The purpose of this study is to examine the intrinsic relations between different flow indices in a multivariate approach, and to develop a methodology that identifies an optimal set of measurements depending on the availability of time, materials, and instrument. **Method:** The proposed methodology includes techniques to characterize material flow properties and multivariate analysis. Pharmaceutical materials with varying flow properties were firstly characterized. The flow properties of each material was represented by multiple flow indices. To reduce the dimensionality of the database, the mathematical correlations between flow indices within each test were firstly examined. Computational iterations of principal component analysis was then performed. Depending on availability of instruments, the amount of materials, and time needed for each test, each flow index can be associated with an assigned weight. The criteria of selecting the optimal set was based on correlation coefficient of the Weighted Euclidean distance. **Results:** Results suggested that by assigning different weights to each flow index depending on the availability of time, materials, and instrument, different optimal sets of characterization measurements can be identified. The optimal set significantly reduced the cost of measurements. In addition, compared to the original dataset, the optimal set was able to capture maximal information of the material flow behavior. The optimal set of measurements can also be used prior to establishing the predictive correlations between material flow properties and process performance. **Conclusion:** The work presented here has shown an efficient approach to reduce dimensionality of material flow testing. This approach is especially powerful when the amount of time, instrument, or a material is limited, or when latent variables representing material flow properties are needed to develop process control strategy.

Using Mechanistic Modeling of Chromatography to Increase Process Understanding

Tim Fattor, Steve Hunt*, Jonathan Rocher*, and Bob Todd**
**KBI Biopharma*

Although the past decade has seen a rise in the use of mechanistic modeling of chromatography unit operations to increase process understanding, the biopharmaceutical industry's preferred approach to characterization of these processes is still the resource-intensive and experimentally-based design of experiments (DoE). Although this traditional approach does increase process understanding to a limited degree, resource pressures invariably limit the size (number of experiments) and scope (number of parameters investigated) of these studies and therefore result in a diminished level of process understanding. Due to inherent limitations, DoE based process characterization studies often do not adequately address important sources of process variability, including the impact of: feed stream variability, difficult to control parameters (e.g. column bed height, resin-lot), and variability in buffer preparation. Further, DoE studies are not well suited to identify the edges-of-failure for a process. As mechanistic modeling of chromatography unit operations is uniquely suited to address all of these limitations, we have modeled an intermediate cation exchange purification step using a targeted set of calibration experiments. Using this model, we demonstrate the increased process understanding that can be achieved through mechanistic modeling of chromatography unit operations.

OUTLINE

1. INTRODUCTION
   - Continuous processes: set up, monitoring and control
   - Multivariate monitoring
2. METHODOLOGY
   - Construction of the multivariate model
   - Informatics set up
   - Monitoring
3. RESULTS
   - R&D site: real time monitoring and retrospective analysis
   - Manufacturing site: real time monitoring
4. IMPLEMENTATION
5. FUTURE PERSPECTIVES
6. CONCLUSIONS

1. INTRODUCTION

Continuous processing in pharmaceutical industry

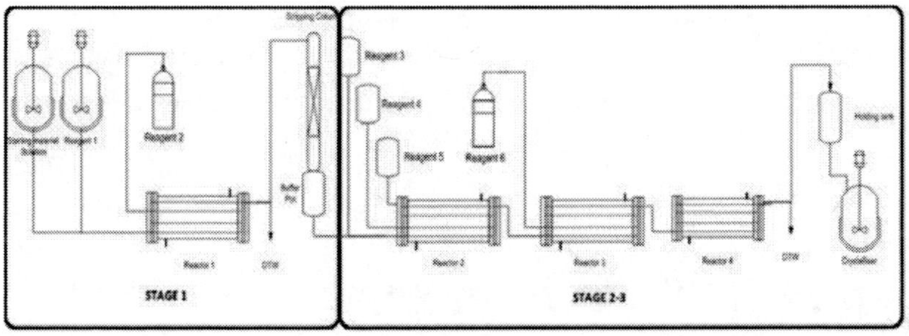

- Smaller footprint
- No inventory
- Flexible batch size
- Lower solvent and energy consumption
- Consistent quality

- Sensitive to solid formation
- Little regulatory precedents

Multivariate monitoring of a continuous manufacturing process for API: enhancing the power of real time data 3

1. INTRODUCTION

Set up of continuous API synthesis: process and monitoring

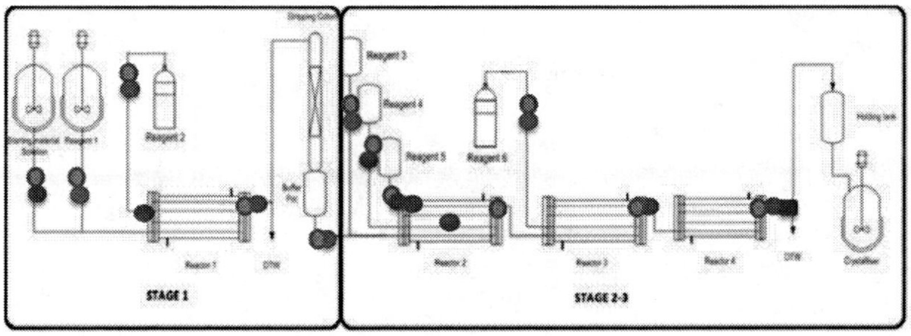

- Conductivity sensors
- Pressure sensors (each pump, reagent and reactor)
- Flow rate sensor (each pump and reagent)
- Pump speed sensor (each pump: not shown)
- Temperature sensors across reactors and heat transfer fluid (not shown)
- Online HPLC

R&D: 30 sensors
Manufacturing: 29 sensors

R&D: 40 sensors
Manufacturing: 52 sensors

Multivariate monitoring of a continuous manufacturing process for API: enhancing the power of real time data 4

1. INTRODUCTION

Set up of continuous synthesis: monitoring and control

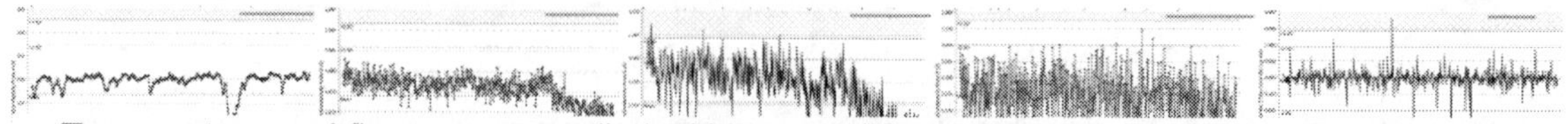

- Temperature and flow rate controlled within predefined ranges in order to control product quality (feedback loops) → actively monitored: excursion triggers alarm and divert to waste
- All other sensors → passively monitored: investigated for RCA, process change etc.

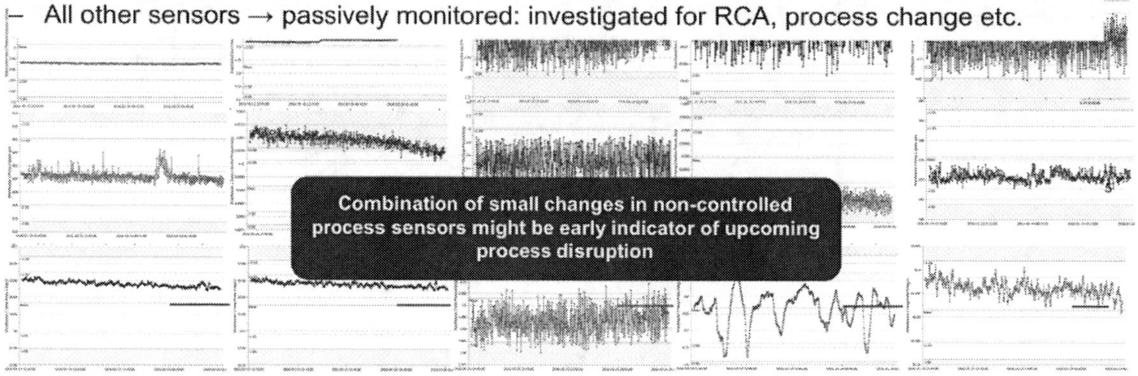

1. INTRODUCTION

Multivariate monitoring

- MSPC model can summarize variability of multiple process sensors readings over time
- Summarizes main trends and correlation patterns in a new trend and calculates a limit based on the 95% confidence interval of normal distributed data.
  - Simplified visualization of process performance
  - Understand natural versus impactful process variation
  - Enhanced process control resulting in more consistent quality
  - Early drift detection can trigger correction actions and prevent forced shut down
  - Facilitates root cause analysis

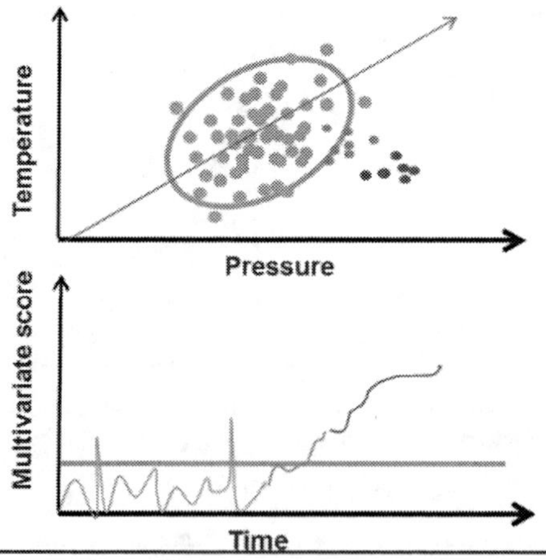

2. METHODOLOGY

1. Construction of the multivariate model

- Data selection
  - State-of-control measurements only (↑sensitivity)
  - Fixed process settings (↑sensitivity)
  - Sufficient process length to ensure representative for future processes (↑robustness), e.g. include full feed bottle cycle

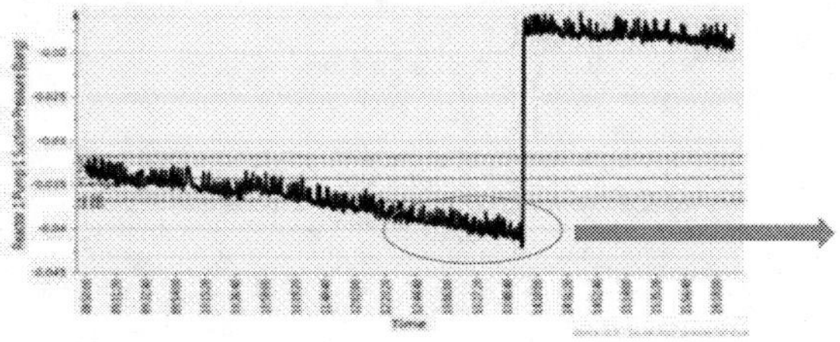

2. METHODOLOGY

1. Construction of multivariate model

- Pre-processing
  - Process sensors different scales and ranges → scale to unit variance (divide by SD)
  - Controlled sensors: observed standard deviation <<< accepted standard deviation → down weight
- Model selection
  - Cross validation prone to over fit model due to collinearity of data (i.e. consecutive time points not independent) → check validity of selected model components (e.g. >5%)

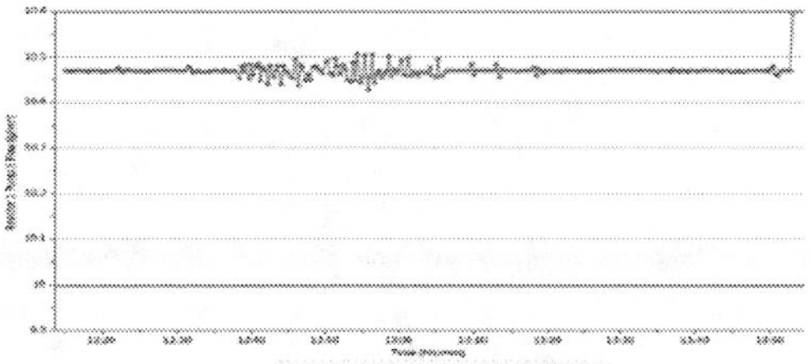

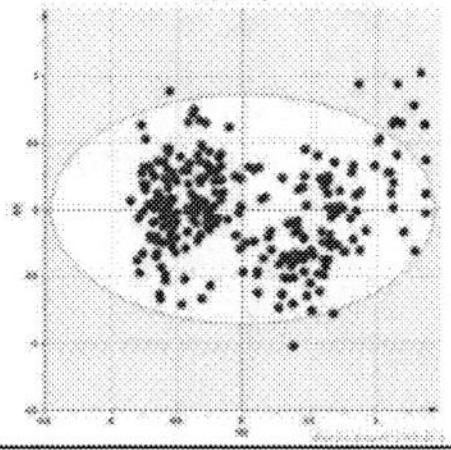

2. METHODOLOGY

Informatics set up

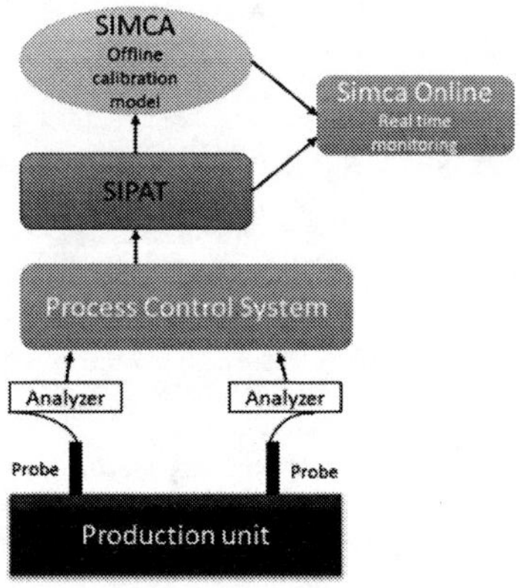

- Sensor readings from probe every second
- Sensor readings collected by SIPAT every 20 seconds
  - Statistical analysis confirmed no information lost by decreased frequency
- Historical data extracted via SIPAT and imported to Simca to build model defining operational space off line
- Multivariate model imported to Simca Online and set up sensor and alarm configuration
- Monitoring in real time started via Simca Online

2. METHODOLOGY

3. Monitoring

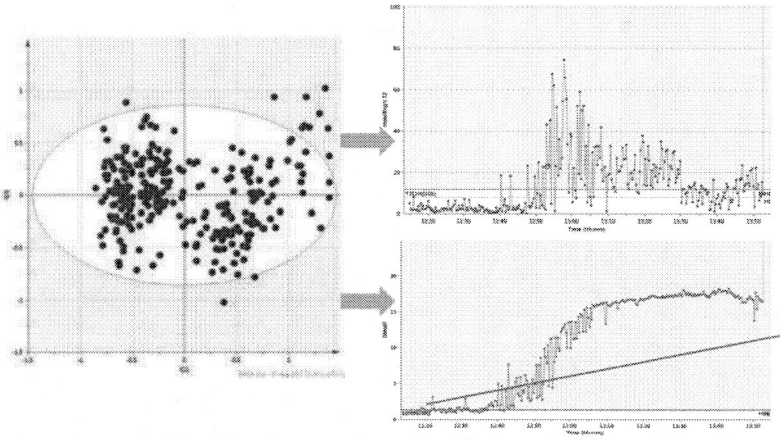

By default multivariate alarm flagged if:

- Three consecutive measurements exceed 99% statistical limit
- 10 out of 100 consecutive measurements exceed 99% statistical limit

Needs optimizing to reduce false positive alarms, whilst maintaining sensitivity for relevant process changes

3.RESULTS

1. R&D site monitoring: evaluation during development campaign

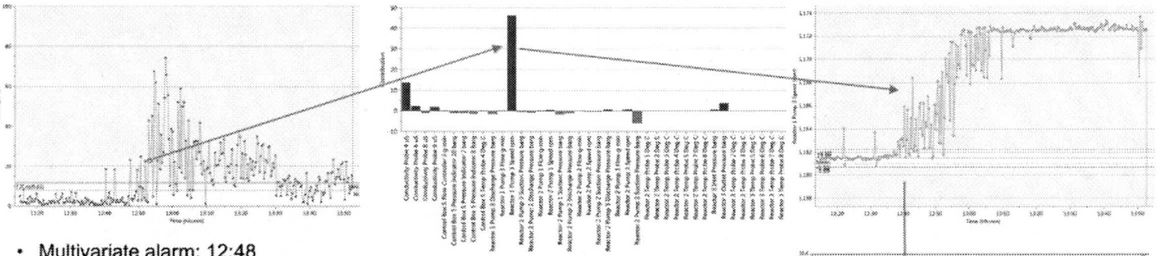

- Multivariate alarm: 12:48
- Root cause multivariate alarm: 15 rpm change in pump speed
- Univariate alarm: 13:52
- Action: pump switch at 13:52
- Increase in pump speed detected one hour before the univariate flow rate alarm → pump switch based on multivariate alarm could prevent flow rate excursion and divert to waste

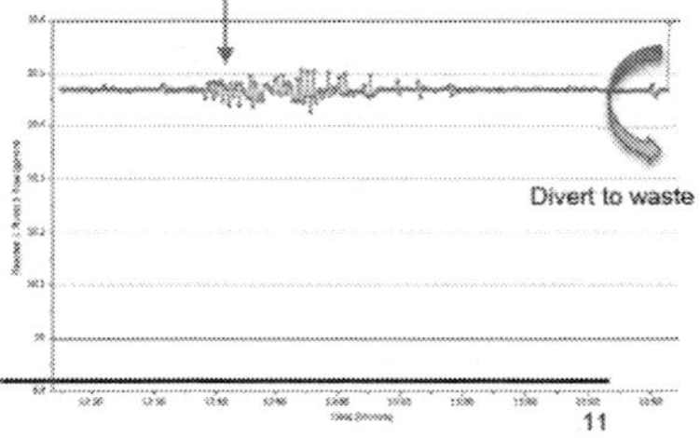

Multivariate monitoring of a continuous manufacturing process for API: enhancing the power of real time data

11

3. RESULTS

1. R&D site monitoring: evaluation during development campaign

Observed process failure	Multivariate detection	Process change	Univariate Alarm	Action
Stage 2 feed pump failure	12:48	15 rpm increase pump speed	13:52	Switch pump
Reagent 2 flow failure	13:18	0.6 barg increase upstream p	15:28	Increase pressure on reagent + divert to waste
7 more similar flow failures for reagent 2 flow				
Reagent 3 pump failure	02:19	12 rpm increase in pump speed	06:38	None (naturally restored)

Multivariate monitoring of a continuous manufacturing process for API: enhancing the power of real time data

12

3. RESULTS

1. R&D site: retrospective RCA for recurrent reagent stream blockage

Largest process changes before blockage:

- Controller differential pressure
- R3 downstream pressure
- Column pressure

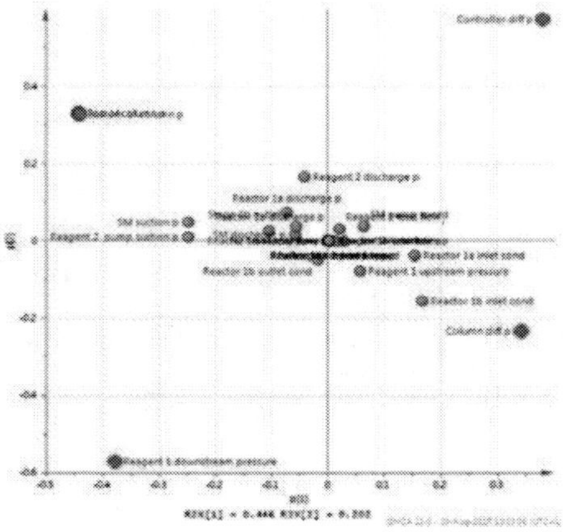

- MVA confirmed root cause blockage
  - Originates from reagent stream
  - Does not start in the reactor

Wealth of real time process data facilitates retrospective root cause analysis

3. RESULTS

2. Manufacturing site monitoring: evaluation during development campaign

Observed process failure	Multivariate detection	Process change	Univariate Alarm	Action
Reagent 2 flow failure	14:53	0.35 barg increase p after flow controller	14:05 (24 h later)	Divert to waste
2 more reagent 2 flow failures as described above				
Reagent 3 pump failure	12:16	Spikes (2 g/min increase) flow rate	12:36	Switch pump
Reagent 1 pump failure	06:06	0.015 bar increase supply filter differential p	06:23	Switch pump
Reagent 6 depletion	23:14	3 % increase flow control output	23:27	Divert to waste and top up reagent 6

4. IMPLEMENTATION

Plan: routine implementation for predictive maintenance by operators

- Procedural flow and decision tree
- List with specific actions in case of detected drift, e.g. switch pump in case of 5 consecutive pump pressure excursions

Aim: detect early onset process failures enabling actions to maintain consistent process performance and avoid divert to waste

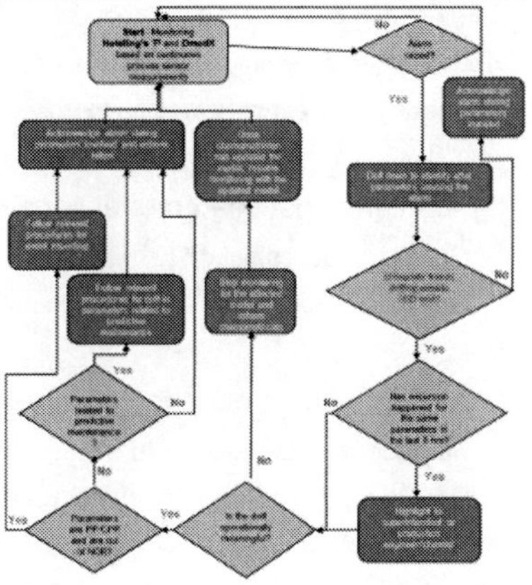

5. FUTURE PERSPECTIVES

Learning

- Increased understanding model sensitivity/robustness and improve implementation strategy
- Combine process sensor data with quality data (e.g. online HPLC): enable detailed quality information at high frequency
- Combine system models and multivariate process monitoring to investigate opportunities for process control

Implementation

- Implement for all high volume processes (API and drug product) containing multiple continuous synthesis stages or unit operations as part of the control strategy (quality assurance and batch release)
- Develop model validation and maintenance plan compliant with regulatory requirements and accommodating high flexibility (e.g. frequent model updates to enable set point changes within the design space)

Advanced Control

- Real-time control of product quality using feedback and feedforward loops

6. CONCLUSIONS

- Multivariate monitoring enabled early detection onset process failure
  - Prolonged time window from detection till process excursion
  - Enables preventative actions to ensure process performance and avoid divert to waste
- Right balance between model sensitivity and model robustness (minimized false alarms) is critical:
  - Optimize data selection to define operational space
  - Optimize monitoring diagnostics limits and alarm settings based on manufacturing experience
- Routine implementation for predictive maintenance by operators
  - Standardized procedures (prevent subjective decisions, actions changing product quality, e.g. set point change)
  - Rigorous yet agile plan for model validation and maintenance accommodating frequent updates
- Future: embed more quality aspects or achieve real time optimization and control

6. ACKNOWLEGEMENTS

- Malcolm Berry: project lead
- Hannah Robinson, Peter Clements for engineering advice
- Peter Shapland for chemistry advice
- Peter Hamilton and Christian Airiau: platforms support and vision
- Ian Barylski: SIPAT and SOL support

500198 Global System Analysis of Interconnected Flow Sheet Models for Drug Product Manufacturing to Performance

Abstract:

Flowsheet models developed using gPROMS platform presents a holistic modeling and simulation approach to design drug product manufacturing processes. Often, these models involve estimation of several parameters using experimental data and measurements. Global Systems Analysis (GSA) is a powerful tool which can be used to identify the cardinal factors of the different parts of the process and their effect on critical quality attributes. In this presentation, we will share our recent experience using GSA through two case studies. First case study will focus on the development of API dry milling model and application of GSA in identifying inputs which are most influential for the model predictions. In the second case study, a comprehensive flowsheet that combines API crystallization, downstream solids operations and drug performance is thoroughly analysed using GSA.

499041 Closed Loop Dynamics of Ribbon Density in a Dry Granulation Process

Abstract:

End to end continuous processing is currently a key focus area in pharmaceutical manufacturing. The increase in research and development is evident from recent FDA approvals of drug products manufactured using continuous downstream processes [1]. Dry granulation is one of the unit operations added for improving material handling and blend uniformity in the continuous tableting line.

Dry granulation is an inherently continuous process. The particle size distribution and bulk density of the granules are critical quality attributes of the process, which depend primarily on the ribbon relative density. The ribbon density is hence a crucial intermediate quality attribute of the unit operation. It is thereby valuable to monitor and control the ribbon density for ensuring desired granule qualities.

Assurance of product quality in real-time requires satisfactory process models with a robust sensor network and control architecture. A dynamic, simplified process model along with real-time monitoring capability is important to the implementation of supervisory control of the process to achieve Quality by Control.

The Johanson model [2] describes the fundamental solid mechanics of roller compaction as a standalone steady state process. FEM analyses have suggested modifications to the model for improved prediction of ribbon density [3]. A process scale model, however, needs to consider the production rate and requires integration of previous unit operations that feed into the roller compactor, in addition to the solid mechanics of the roller compactor.

The process model by Reynolds et al [4] integrates the screw conveyor with the roller compaction system, describing the effect of screw speed on the ribbon density and production rate at steady state. A dynamic process model is, however, essential for the implementation of a control system, as proposed by Hsu et al [5]. The model did not include the flow rates through the system and lacked experimental validation.

This work focuses on a modified process model incorporating the solid mechanics, process dynamics and throughput. The model combines the dynamics of material flow from Hsu et al (2010a) and Reynolds et al (2010) and solid mechanics suggested by Liu and Wassgren (2016). Demonstration of real-time ribbon density monitoring using NIR and microwave sensors [6,7] enable the validation of such a dynamic model. Additionally, the model evaluates the integrated system of a hopper and screw conveyor feeding into the roller compactor.

Blends comprising of acetaminophen and microcrystalline cellulose at inlet flow rates of 8-12 kg/h and varying roll pressures are used for real-time monitoring of ribbon density. The solid mechanics material parameters are estimated using ribbons. The process variables and ribbon density are recorded and monitored using DeltaV DCS.

This work progresses to demonstrate the implementation of ribbon density as proposed by Hsu et al. The integrated

process model and real time ribbon density monitoring enable the development of control strategies for the desired ribbon density and eventually the granule properties.

References:

1. https://blogs.fda.gov/fdavoice/index.php/2016/04/continuous-manufacturing-has-a-strong-impact-on-drug-quality/
2. Johanson JR. A Rolling Theory for Granular Solids. *J Appl Mech*. 1965;32(4):842-848
3. Liu Y, Wassgren C. Modifications to Johansonâ€™s roll compaction model for improved relative density predictions. *Powder Technol*. 2016; 297:294-302.
4. Reynolds G, Ingale R, Roberts R, Kothari S, Gururajan B. Practical application of roller compaction process modeling. *Comput Chem Eng*. 2010;34(7):1049-1057
5. Hsu S-H, Reklaitis G V., Venkatasubramanian V. Modeling and Control of Roller Compaction for Pharmaceutical Manufacturing. Part I & II: Process Dynamics and Control Framework. *J Pharm Innov*. 2010;5(1-2):14-36
6. Gupta A, Austin J, Davis S, Harris M, Reklaitis G. A Novel Microwave Sensor for Real-Time Online Monitoring of Roll Compacts of Pharmaceutical Powders Online-A Comparative Case Study with NIR. *J Pharm Sci*. 2015;104(5):1787-1794.
7. McAuliffe MAP, Oâ€™Mahony GE, Blackshields CA, et al. The Use of PAT and Off-line Methods for Monitoring of Roller Compacted Ribbon and Granule Properties with a View to Continuous Processing. *Org Process Res Dev*. 2015;19(1):158-166.Â

493080 Integrated Control and Data Management System for Continuous Pharmaceutical Manufacturing Process

Abstract:

Continuous manufacturing (CM) is evolving as a preferred platform for pharmaceutical products involving solid dosages forms. One advantages of CM among others is that the product quality can be controlled in real time and thereby its open the possibilities of achieving Quality by Control (QbC) and real time release (RTR) paradigm. However, there are different levels of challenges associated with design and implementation of an efficient control system into pharmaceutical manufacturing plant. Identification of right control strategies such as feedforward (FF), feedback (FB), or combined FF/FB[1] as well as right control algorithms such as PID, MPC, or hybrid MPC-PID is often a tough decision that need to be made prior to implementation of integrated control system into the manufacturing plant[2]. Selection of a right control strategy and algorithm is mandatory task both for efficient manufacturing as well as regulatory point of view. Therefore, the different control options need to be evaluated and the best integrated control system needs to be implemented into the pilot-plant. The integrated control system is then need to be coupled with material traceability framework for real time identification of batches so that the product can be recall in case if needed. The continuous manufacturing is a data rich process and therefore a systematic data management framework is also desired to be implemented to collect, store and manage all the relevant data in real time.

In this work, different control strategies for direct compaction continuous tablet manufacturing process has been developed and implemented into our pilot-plant facility using industry standard control platform. The critical control variables are drug concertation, powder level, main compression force, tablet weight and hardness. Both, feedback (FB) as well as feedforward (FF) control strategies have been implemented into the pilot-plant. Feedback (FB) control strategy has been compared with a combined feedforward/feedback (FF/FB) control strategy. A novel method for real time monitoring of powder density has been developed and integrated with feedforward control loop. The feedforward control loop takes the powder bulk density signal as an input and mitigate the effect of variations in powder bulk density on CQAâ€™s proactively and thereby reduces the rejections significantly in compare to only feedback control strategy. Three control algorithms (PID, MPC, hybrid MPC-PID) have been also implemented and compared for feedback loops. The flexible control system is designed in such a way that any control scheme can be easily selected. The control system is also modular in nature that can be easily transferred to any other control platform and manufacturing plant. The continuous tablet manufacturing pilot-plant runs under each control scheme and their performance are evaluated and thereby a best control system has been identified. A systematic methodology for material traceability has been also developed and coupled with integrated control system. The methodology starts with material and process specifications followed by scanning the powder drums and transmitting the material and time information to the control platform. The control platform then utilizes the information stored in the database, generated through residence time distribution model and information collected through real time monitoring to track the material during continuous manufacturing. All the relevant data generated during continuous manufacturing has been systematically collected, stored and organized in a data hub (OSI PI). The PAT data has been collected through PAT

data management tool (synTQ) and then transferred to the PI system while the other operational data has been collected in control platform (DeltaV) and send to PI system via OPC communication. Finally, the data has been transferred to the cloud system for enterprise wise efficient sharing and applications.

The objective of this presentation is two-fold: first to evaluate the performance of integrated control system of continuous pharmaceutical tablet manufacturing process and then to highlight the material traceability and data management framework.

References

1. Singh, R., Muzzio, F., Ierapetritou, M., Ramachandran, R. (2015). A combined feed-forward/feed-back control system for a QbD based continuous tablet manufacturing process. PROCESSES Journal, 3, 339-356.
2. Haas, N. T., Ierapetritou, M., Singh, R. (2017). Advanced model predictive feedforward/feedback control of a tablet press. Journal of Pharmaceutical Innovation, 1-14, DOI 10.1007/s12247-017-9276-y.

497291 *In silico* Process Characterization of Drug Product Filling Operations: Predicting Quality Attributes from First Principles

Abstract:

Process characterization of drug product filling operations is typically a time- and labor-intensive process requiring substantial manufacturing line down-time due to the large number of process inputs and lack of mechanistic understanding of the fluid dynamics and filler systemsâ€™ sensitivities to drug product properties (e.g. viscosity, density, surface tension), or variations in site, equipment type, settings, and ambient conditions.

As an alternative to empirical process characterization, we have developed a computational model which predicts process quality attributes â€“ for instance flow rate, fill volume/weight, cavitation, droplet formation, and product concentration â€“ from first principles of product dynamic properties, fluid dynamics, structural mechanics and control theory. Formulation of the model couples finite element and finite volume solutions to structural deformation and fluid dynamics in pinch valves, and leverages analytical solutions to ordinary differential and algebraic equations accounting for orifice and pipe flow as well as equipment control laws.

We present an application of the model to a Time/Pressure filling operation, including formulation, deployment as a software application, and development of a workflow for in-silico process characterization utilizing design space response studies, local sensitivity studies, and uncertainty analysis.

490467 Easy Advanced Global Optimization (EAGO): An Open-Source Platform for Robust and Global Optimization in Julia

Abstract:

The use of generalized McCormick relaxations in global optimization problems presents significant advantages in process systems engineering applications.. These include the generation of tight relaxations and quadratically-convergent iterative methods which help to avoid clustering. Additionally, problems may even be solved with a reduction in dimensionality, in contrast to the state of the art auxiliary-variable-methods (AVM) [1-4].

However, the usage of these techniques has been limited due to: the need to utilize a non-smooth solver, complexity in formulating relaxations of implicit functions, and the necessity of object-oriented programming approaches [1,2]. These object-oriented approaches have previously required a trade-off between accessibility and speed. Favoring performance, current implementations are available in C++ [1,2,6-8] which makes accessibility an issue as higher-level programming languages, such as MATLAB® [5], have come to predominate end-user applications.

The first limitation has recently been eliminated by a novel method for constructing differentiable relaxations [6]. Implicit function theory has recently been advanced within a generalized McCormick relaxation framework for addressing global optimization problems in reduced space [7,8]. In this paper, we address the third major limitation by introducing an easy advanced global optimization (EAGO) package for Julia [9].

As a Julia implementation, EAGO offers a user-friendly front-end while maintaining speed comparable to the C language. This software package makes use of modern techniques in McCormick-type relaxations and validated interval arithmetic to solve globally nonlinear optimization problems with a guaranteed certificate of optimality. The package contains three main components:

- C1) Routines which make use of differentiable and implicit McCormick-based relaxation theory to solve optimization problems efficiently in reduced space.
- C2) Implementations of meta-algorithms for the solution of nonconvex semi-infinite programs (SIP) are also made available in this package [7,10]. The routine can be utilized with user-specified solvers such as BARON [3-4] in additional to the provided global optimization library.
- C3) Also included in the software package are Julia implementations of extended-interval arithmetic, McCormick-relaxations (both standard and differentiable), reverse-interval propagation via computational graphs, and a branch & bound library.

This paper will focus on demonstrating the capabilities of EAGO. Use-cases for SIP meta-algorithms will be provided to demonstrate the ease with which robust optimization problems may be formulated and efficiently solved. Additional examples will be given illustrating the application of EAGO solvers to large systems that commonly arise in model-based system engineering applications. Benchmark results will be provided and compared to state-of-the-art optimization software (BARON [3-4], ANTIGONE [12]) based on select samples from the COCONUT library [11].

Industrially-relevant process systems engineering examples will be presented illustrating both ease-of-use and performance aspects of EAGO.

[1] Scott, J.K., Stuber, M.D., and Barton, P.I. Generalized McCormick Relaxations. *J Global Optim*, 51:569-606, 2011.

[2] Mitsos, A., Chachuat, B., and Barton, P.I. McCormick-Based Relaxations of Algorithms. SIAM J. Optim. 20(2): 573-601.

[3] Tawarmalani, M. and N. V. Sahinidis, A polyhedral branch-and-cut approach to global optimization, Mathematical Programming, 103(2), 225-249, 2005.

[4] Sahinidis, N. V., BARON 14.4.0: Global Optimization of Mixed-Integer Nonlinear Programs, User's manual, 2014.

[5] MATLAB version 9.0.0. Natick, Massachusetts: The MathWorks Inc., 2016

[6] Khan, K.A., Watson, H.A., and Barton, P.I. Differentiable McCormick Relaxations. *J Global Optim*, 67(4):687-729, 2017.

[7] Stuber, M.D., and Barton, P.I. Semi-Infinite Optimization with Implicit Functions. *Ind. Eng. Chem. Res.*, 54:307-317, 2015.

[8] Stuber, M.D., Scott, J.K., and P.I. Barton. Convex and Concave Relaxations of Implicit Functions. *Optimization Methods and Software*. 30(3), 424-460, 2014.

[9] Bezanson, J. et al. Julia: A fresh approach to numerical computing. *SIAM Review*. 59, 65-98, 2017.

[10] Mitsos, A. Global Optimization of Semi-Infinite Programs via Restriction of the Right-Hand Side. *Optimization*. 60:10-1,1291-1308.

[11] O. Shcherbina, et al. Benchmarking Global Optimization and Constraint Satisfaction Codes, In *Global Optimization and Constraint Satisfaction*, Bliek, Ch., Jermann, Ch. and Neumaier, A. Springer, Berlin 2003, 212-222.

[12] Misener, R. and Floudas, C.A. ANTIGONE: Algorithms for coNTinuous / Integer Global Optimization of Nonlinear Equations. *J Global Optim*, 59:503-526, 2014.

497914 Proto: Platform for Robust Optimization

Abstract:

In recent years, robust optimization has grown tremendously as a tool for incorporating parameter uncertainty into optimization models [1,2]. Robust optimization centers on the use of uncertainty sets, which define the uncertain parameter realizations which must be considered in the model. These uncertainty sets are deterministically imposed on the model, resulting in robust counterparts which replace the uncertain constraints in the original formulation. Depending on the probability distributions of the uncertain parameters and the uncertainty sets used, a wide variety of probabilistic bounds have been derived which provide guarantees on the probability of constraint violation [3-6]. These tools for handling uncertainty can be extremely valuable in answering important questions in process systems engineering [7].

In order to ease the implementation of robust optimization and the application of probabilistic bounds, a Platform for Robust OpTimizatiOn (PROTO) has been developed that performs these computations for a user based on information regarding the constraints, uncertainty sets, parameter distributions, and optimal solutions. PROTO exists as a webtool and has three main functionalities. Before solving an optimization problem, a user may provide information regarding the type of uncertainty set used and uncertain parameter distributions, if known. PROTO will then automatically determine which *a priori* bounds are applicable and apply them to determine the appropriate size of the uncertainty set. After solving an optimization problem, a user may provide information regarding the optimal solution and uncertain parameter distributions, and the best applicable *a posteriori* bound will be determined and used to calculate the probability of constraint violation. In either case, a user can also specify the bounds which they would like to utilize. Finally, if an optimization model with specified uncertain parameters is provided in GAMS format, the robust counterpart model will be formulated and a high quality robust solution will be generated at desired *a posteriori* probabilities using an iterative algorithm [8].

PROTO will provide probabilistic bounds for box, ellipsoidal, and polyhedral type uncertainty sets and uncertain parameters with unknown, normal, uniform, triangular, and raised cosine distributions. As some of these distributions are bounded while others are not, a natural question arose in the creation of PROTO regarding how to handle constraints with both bounded and unbounded uncertain parameters. Thus, new uncertainty sets and robust counterparts were derived for generalized uncertainty sets, which combined the interval behavior of traditional uncertainty sets for bounded parameters with unbounded uncertainty set geometries for unbounded parameters [9]. These new uncertainty sets and robust counterparts will be presented along with practical examples of the usage of PROTO which will demonstrate its potential impact for users with parameter uncertainty in their optimization problems.

[1] Li, Z.; Ding, R.; Floudas, C.A. A Comparative Theoretical and Computational Study on Robust Counterpart Optimization: I. Robust Linear Optimization and Robust Mixed Integer Linear Optimization. Industrial & Engineering Chemistry Research 2011, 50, 10567-10603.

[2] Ben-Tal A.; El Ghaoui L.; Nemirovski A. Robust optimization. Princeton University Press, 2009.

[3] Li, Z.; Tang, Q.; Floudas, C.A. A Comparative Theoretical and Computational Study on Robust Counterpart Optimization: II. Probabilistic Guarantees on Constraint Satisfaction. Industrial & Engineering Chemistry Research 2012, 51 (19), 6769-6788.

[4] Guzman, Y.A.; Matthews, L.R.; Floudas, C.A. New a priori and a posteriori probabilistic bounds for robust counterpart optimization: I. Unknown probability distributions. Computers & Chemical Engineering 2016, 84, 568-598.

[5] Guzman, Y.A.; Matthews, L.R.; Floudas, C.A. New a priori and a posteriori probabilistic bounds for robust counterpart optimization: II. A priori bounds for known symmetric and asymmetric probability distributions. Computers & Chemical Engineering 2017, 101, 279-311.

[6] Guzman, Y.A.; Matthews, L.R.; Floudas, C.A. New a priori and a posteriori probabilistic bounds for robust counterpart optimization: III. Exact and near-exact *a posteriori* expressions for known probability distributions. Computers & Chemical Engineering 2017, 103, 116-143.

[7] Floudas, C.A.; Niziolek, A.M.; Onel, O.; Matthews, L.R. Multi-scale systems engineering for energy and the environment: Challenges and opportunities. AIChE Journal 2016, 62 (3), 602-623.

[8] Li, Z.; Floudas, C.A. A Comparative Theoretical and Computational Study on Robust Counterpart Optimization: III. Improving the Quality of Robust Solutions. Industrial & Engineering Chemistry Research 2014, 53 (33), 13112-13124.

[9] Matthews, L.R.; Guzman, Y.A.; Floudas, C.A. Generalized Robust Counterparts for Uncertain Constraints with Bounded and Unbounded Uncertain Parameters. 2017, In Preparation.

497516 Recent Developments in Pyomo

Abstract:

Computational tools for modeling mathematical programs are widely used within both academia and industry. Available commercial and open-source modeling packages support generic modeling by separating modeling constructs from instance data through concepts like sets, parameters, and parameterized constraints. However, their model representations are limited to constructs that directly correspond to established solver inputs. In general, this implies that mathematical programs must be expressed as either linear or nonlinear algebraic models; that is, a list of variables, an algebraic objective expression, and a list of algebraic constraints.

In contrast, the Pyomo environment is designed to be an open Python-based environment for developing and exploring complex, optimization-based analytical strategies. This includes support for extending the modeling environment to include non-algebraic modeling constructs, developing automated model transformation routines, and automating problem decomposition and interrogation.

Pyomo development this year has been driven by the Institute for the Design of Advanced Energy Systems (IDAES), a DOE/Fossil Energy-funded University/National Laboratory partnership developing advanced modeling, simulation, optimization, and design capabilities for innovative advanced energy systems through the use of process systems engineering tools and approaches. In this presentation we highlight some of the recent developments in the Pyomo project, including improved performance, improved support for building domain-specific modeling environments on Pyomo, new model simulation and initialization strategies, and improved support for problem decomposition.

496333 Semantic Repository for Biorefining Model Integration

Abstract:

The increasing availability of biorefining models and the heterogeneity characterising them necessitates efficient acquisition, discovery and integration tools to enable their reuse. This has been addressed by the CAPE-OPEN standard which significantly facilitates model reusability and interoperability (Braunschweig et al., 2004). Still, efficiency and flexibility of model discovery remain a challenge. This is primarily because the model interface semantics are implicit, understandable by humans, but difficult to allow for automation controlled by machines. In order to enhance model discovery and subsequently model integration, we propose a semantic repository biorefining models benefiting from a semantic annotation and implemented as a semantic web service.

The semantic annotation in the form of a full description of models is accomplished by the biorefining domain ontology developed in Web Ontology Language (Koo et. al 2017), hence allowing for a precise description of model functionality, their inputs, outputs, precondition and effects (IOPE). Models are stored as instances of ontology top level classes *Model By Functionality*, *Model By Input*, *Model By Output*, *Model By Platform,* or their respective subclasses.

Model integration, including model discovery is enabled by introducing an established the Web Ontology Language Semantic Mark-Up for Services (OWL-S) framework (Martin et. al 2004). In consequence, the models are recorded as web services and described by OWL-S which uses three interconnected sub-ontologies (Profile.owl, Process.owl, Service.owl), each linked to a model instance in the biorefining domain ontology.

The proposed model repository operates by following five stages:

i) *Model Registration* for registering a model and hence providing all IOPE-s for that model; the registered model becomes an instance of the biorefining ontology;

ii) *Model Publishing* for sharing all registered model;

iii) *Model Discovery* for discovering other models and concomitantly integrating it with other models; model discovery is enabled by an I/O semantic matchmaking with I inputs a requesting model. Candidate models are ranked by the level of semantic similarity with a full explanation enabling the user an informed selection. Increased flexibility and a wider range of options are achieved by the introduction of partial matching, which, again, are fully described for their relevance;

iv) *Model Selection,* the stage where the user chooses the most appropriate candidate model; offered for integrating it with requesting model;

v) *Model composition* triggers the synchronization algorithm for execution and data exchange between the selected models.

The functionality of the proposed semantic repository was tested using several biorefining related scenarios, i.e. modelling of bioethanol production and modelling several bioprocesses.

References

Braunschweig, B., Fraga, E., Guessoum, Z., Marquardt, W., Nadjemi, O., Paen, D., Pinol l, D., Roux, P. Sama, S., Serra, M., Stalker, I.,Yang, A., 2004. CAPE web services: the COGents way, Comput. Aided Chem. Eng., 18 (C) , pp. 1021â€"1026

Koo, L., Trokanas, N., and Cecelja, F. , 2017. A semantic framework for enabling model integration for biorefining. Computers & Chemical Engineering. http://doi.org/10.1016/j.compchemeng.2017.02.004

Martin, D., Burstein, M., Lassila, O., Paolucci, M., Payne, T. and McIlraith, S., 2004. Describing Web Services using OWL-S and WSDL. http://www.daml.org/services/owl-s/1.1/owl-s-wsdl.html

(448e) Programmable process structures, generated from a network and from functional meta-prototypes

Monika Varga and Bela Csukas
Research Group on Process Network Engineering, Kaposvar University, Hungary

Position

This paper is about a non-conventional process modeling approach, implemented temporarily in declarative, logical programming. Considering the usual trade-off principle of Information Technology, it is well structured, but not effective, currently. It is an in-house used experimental tool, tried for dynamic simulation based analysis of multi-scale, hybrid processes in a broad range of applications, from cellular signaling based biosystems to ecological, technological, agri-food and environmental process systems.

1 Background and motivation

Process Systems Engineering (PSE) originally evolved from the field of chemical engineering about fifty years ago. At millennium, Grossmann and Westerberg (2000) highlighted the research challenges, as well as formulated a broader definition for PSE. Chemical engineers often play an interfacing role between the field-specific chemical & related engineering, and the method-specific applied mathematical and computer sciences (Klatt and Marquardt, 2009). Stephanopoulos and Reklaitis (2011) gave a comprehensive and detailed overview about the foundation from the roots until the recent drivers of development. By now, having exceeded the "chemical supply chain", the ***problem domain extended to directions of both nano- and tera-scales***. Regarding nano-scale, for example in Systems Biology PSE contributed significantly to the analysis, understanding and design of biological systems (e.g. Stephanopoulos and Schmitt, 2001). Regarding tera-scale, the utilization of PSE methods for solution of Water-Food-Energy Nexus related problems (e.g. Garcia and You, 2016) can be mentioned.

There is an up-to-date challenge to ***build bridge*** (or rather to find cooperative solutions) ***between*** the basically ***functional*** approach of algebraic or differential equations ***and*** the ***structural approach*** of networks and nets. In terms of functional modelling, we can start from the notion of process systems in the sense of Kalman's State Space Model (Kalman et al, 1969). Accordingly, this class of systems can be declared by the time dependent input, state and output vectors, connected by the output and transition functions. Similarly we can interpret the subsequent states of an abstract automaton for the discrete process systems in the discrete time. The respective PSE models are usually described by a set of ODE's, PDE's or DAE's. Considering structural features, the topological characteristics of the networks offer useful insights into the structure of the studied system (e.g. Strogatz, 2001). Dynamic characteristics might be considered by combining network analysis with the linear state space models. (e.g. Liu et al, 2011), however, linearity seems to be an arbitrary assumption for complex, multiscale, non-linear, partly event driven processes. From another aspect, the essential structural features of process models can be represented by nets. While networks can be interpreted as relations within a set of elements, nets can be defined as relations between two sets (Petri, 1962 and 1966). In the past decades, extensive efforts have been made for the further development e.g.in the form of higher order and Quantitative Petri Nets (Chen and Hofestädt, 2003).

The hybrid, multiscale models may be more complex, than any specific mathematical apparatus. The presented ***Programmable Structures are motivated*** by the need that engineering analysis, design and control of the complex multidisciplinary process systems require easily modifiable, extensible and connectable methodologies, with common representation of functional and structural features.

2 Methodology of Direct Computer Mapping based Programmable Structures

In ***Direct Computer Mapping*** (DCM) of process models (Csukas, 1998; Csukas et al, 2011) the natural building blocks of the elementary states, actions and connections are mapped onto a unified set of building elements determining an executable computer code, directly, without any specified mathematical apparatus. This principle was successfully applied for various processes from technological process systems (e.g. Csukas et al, 1999; Temesvari et al, 2004) up to the multi-sectoral agri-food processes (e.g. Varga et al, 2012).

Based on these previous and on the ***recently developed multidisciplinary applications*** it is obvious, that the structure of processes is more sophisticated, than the networks in sense of network science, on the one hand. However, the case-specific local functionalities of the elementary building blocks of these structures are less complicated, than the systems of integral and partial differential equations, combined with discrete events, on the other. As an intermediate solution, we developed ***Programmable Structure*** (Varga et al. 2016a; Varga et al, 2016b; Varga et al, 2017; Varga and Csukas, 2017) that can be generated from a network description and from two functional meta-prototypes. The meta-prototypes and the derived process structures are prepared for the semantically distinguished, but syntactically uniform representation of the "*model specific conservation law based, additive*" measures, and of the "*over-writable*" signals.

In this approach the usual network or net based formalization of process structures is extended to a special structure of unified state and transition elements, containing dedicated input and output slots for additive measures and signals, as well as parameter slots for parameters. The actual structure is defined by standardized connecting elements, corresponding to increases and decreases of (extensive) measures, as well as to reading and overwriting of the intensive properties and other qualitative or quantitative signals.

In various applications many state and transition elements can usually be modelled with the same local programs. Accordingly, we distinguish the actual elements and the program defining prototypes. The prototype elements contain symbolic input, parameter and output variables, as well as local programs, using these variables and limited number of global data. During the simulation, the actual elements start with the initial conditions and parameters, and the output values are recalculated stepwise with the knowledge of input and parameter data, according to the associated local program prototype. The distinguished input and output slots for extensive/intensive data and for signals support the combined execution of the balance-based and rule-based functionalities.

The data flow between the elements is determined by the connections. Actually, the receiving side of the connections select the topology-determined output message(s) from the output slots of sending element, according to the reading operator. Next, it delivers the message(s) to the input slot of receiving element according to the sending side and the respective writing operator.

Both the actual elements, as well as the receiving and sending sides of the connections may be specified by a spatial identifier, determining the compartment and/or the level and/or the scale of the given entity. Also both the elements and the connections may have timing that defines temporal constraints for the execution, supporting the combined event-driven and time-driven execution of models.

Summarizing the methodology, the ***programmable structure*** of process models are generated from the description of the (optionally multilevel or multiscale) process network and from two functional meta-prototypes. The case-specific, actual functional prototypes, containing the local models may be copied and edited from the general meta-prototypes. The actual state and transition elements are parameterised and initialised concerning their case-specific prototypes. The actual model elements are executed by their associated prototypes. This execution and the connection-based communication between the state and transition elements of the ***programmed structure*** are solved by a general purpose kernel program.

In the proposed model representation the detailed description of possibility (design) space, as well as the measurements and the evaluating functions of the optionally multiple objectives may be embedded in the state and transition elements. Accordingly, in identification and optimization tasks the robust simulator may easily be combined with external meta-heuristic optimizers.

3 Experimental implementation of Programmable Structures

The code of the recent implementation is written in declarative, logical Prolog language that is a limited subset of first order predicate calculus. Prolog makes the easy description and the unification based execution of the local program prototypes possible. This feature, combined with the applied, restricted and standardized way of model description, supports the application of a general kernel for the quite different actual models. In addition, this declarative language offers a formal description *eo ipso*, which fosters also the possible future implementations.

The automatic generation of the programmable structures starts from the GraphML based graphical implementation of state and transition meta-prototypes, as well as from the textual or graphical description of the process network. Generation means multiplying the meta-prototype elements, supplemented with the connections between them. (It is to be noted, that the expert user can also start with the GraphML editor, from scratch.) The generated graphical models can be edited by various (actually yEd) GraphML editors. In this step the expert user makes copies from the meta-prototypes, next declares the case-specific programs. This is followed by initialization and parameterization of the actual elements. Next a Prolog program interprets the GraphML description into the facts and clauses, declaring the dynamic model. Finally the general purpose Prolog kernel program executes the various modes of dynamic simulation. This means a cyclically repeated processing of state elements, state → transition connections, transition elements and transition → state connections.

4 Applications

Programmable Structures were successfully applied for quite different complex, multidisciplinary processes from the low-scale cellular biosystems (Varga et al, 2017) through ecological (Varga and Csukas, 2017) and technological process systems (Csukas et al, 2013) up to the Recirculating Aquaculture Systems (Varga et al, 2016), and environmental process networks

(Varga et al, 2016). The advantage of this approach is that it gives flexibility for both structural and functional modification and extension of the process model.

The methodology supports the unified generation of reusable process models. The generated models are reconfigurable and extensible, as well as the models can be incremented from any "final" state. The robustness of dynamic simulation is based on the built-in control mechanisms, initiated by the automatic checking of feasibility bounds for the underlying "model specific conservation law based" measures.

The simulation models can be transformed into State Space Models, where the measurements (i.e. the input data) of the control come from the output variables of the model, while the interventions (i.e. the output data) of the control usually modify the parameters of the model.

The robust simulation fosters the dynamic model based (sub)optimization with various, externally implemented meta-heuristic methods (like Genetic Algorithm). The basic principle, in accordance with the engineering way of thinking, is that instead of searching for an exact global optimum with a simplified model, it might be better to generate multi-objective suboptimal solutions from a more detailed model.

The unified approach helps the model based comparison of the sustainability and resilience specific features of natural and human built processes. The keynote elements are the "model specific conservation law based" balances, the inherently local and rationally global process architectures, as well as the cooperative evaluation feedback between the functionally connected subsystems. These can be applied in the model based organization of sustainable process networks, consciously.

5 Summary

Based on the multidisciplinary applications, the ***characteristic features of Programmable Structures*** can be summarized, as follows (Varga et al, 2017; Varga and Csukas, 2017):

- According to the architecture of process systems, there are two main classes of elements, determining the states (measures, simple or complex signals) and the transitions (i.e. transformations, transportations, rules, etc.).
- Both classes of elements are determined by input, parameter and output slots (connectors) for functionally dedicated sets of data triplets, characterized by the identifying names, by the numerical or symbolic values and by the additional instructions.
- In both classes there are meta-prototypes, prototypes and actual elements. Meta-prototypes determine the structure and the character of the input, output and parameter slots (connectors). The prototypes contain a local program, i.e. a locally executable functionality that determines output variables with the knowledge of the input and parameter ones. The actual elements are "empty" multiplications of the prototypes, where only the parameter and the state input slots are filled with data triplets, containing actual numerical or symbolic values.
- There are predefined types (reading, writing, modifying, increasing, decreasing, etc.) of state > transition and transition > state connecting elements, which are the structure determining entities.
- The input and output slots, as well as the respective connections may be associated semantically (for model specific conservation law based extensive and intensive properties, as well as for over-writable signals).

- State and transition elements can be associated with spatial coordinates, determining the compartment, level or scale of the given entity. The appropriate connections span the respective multi-compartment, multi-level and/or multi-scale structure.
- Various kinds of temporal behaviour (e.g. discrete times, time intervals, etc.) can be associated with both the elements and the connections.
- Description of the possibility (design) space and the components of the evaluating actions may be embedded in the elements, while special configuring and evaluating elements may be added, which may communicate with the optional external optimizer.
- The detailed editable model can automatically be generated from the simple description of the (optionally multi-compartment, multi-level or multi-scale) network structure and from the respective programmable meta-prototypes or prototypes. Having edited the model, another interpreter generates the dynamic database of the actual model.
- The elements and connections of the model database can be executed cyclically by a general purpose main program with automatically controlled time steps. The execution supports keeping the model within the feasibility space of the model specific conservation law based stoichiometric measures.

The ***declarative model implementation*** makes possible the application of list structures and recursive unification at any level of the description. Accordingly, the general description of the model is prepared for optional number of list members, without the previous declaration of their length. Consequently we can extend the model with new components in form of d(Name, DataList, Dimension) triplets in any place, at any time. This makes possible the qualitative or quantitative tracking and tracing for various added-on components.

The ***major advantages*** of the developed methodology come from the inherent and plausible coupling between the structural and functional knowledge, as well as from the extensibility and connectivity of the models.

In line with the well-known trade-off principle between being well-structured and effective, the ***major disadvantage*** of the methodology is less effectiveness, compared with the case-specifically developed and implemented procedural methods and tools. These disadvantages may be compensated by the forthcoming micro-granularly parallel hardware / software implementations later on, because the well structured representation of local program prototypes supports these implementations.

6 Planned further development

Recently, there are Big Data with limited predictive abilities, on the one hand, and large models with lack of appropriate parameters, on the other. We are to collaborate in building "tandem models" based on mutual, cooperative evaluation feedback between 'a priori' modelling and data mining. Hence the cognitive simulation model could be continuously identified and validated by the accumulating data, while data based reasoning could be extended stepwise by the capabilities of the continuously evolving dynamic simulator.

We plan collaboration to apply the same approach for the implementation of stochastic and fuzzy models.

In other collaborations we ought to search for the more effective software implementations of Programmable Structures, using possible new programming languages and environments, as well as model-friendly NoSQL databases.

According to some preliminary tests, this non-conventional modelling approach might be familiar with some methods of micro-granular parallelism, e.g. in sense of software/hardware co-design, so we seek for collaboration also in this field.

Acknowledgement
This work was supported by EFOP-3.6.2-16-2017-00018 program.

References

(N.B. The major references, outlining Programmable Structures are highlighted with bold letters.)

Chen, M., Hofestädt, R., 2003. Quantitative Petri net model of gene regulated metabolic networks in the cell. In Silico Biol. 3(0029), 347–365.

Csukas, B. 1998. Simulation by Direct Mapping of the Structural Models onto Executable Programs. In: AIChE Annual Meeteng 1998, Miami, Paper 239/9.

Csukas, B., Balogh, S., Kovats, S., Aranyi, A., Kocsis, Z., Bartha, L. 1999. Process design by controlled simulation of the executable structural models. Comp.Chem.Engng. 23(Suppl), 569-572.

Csukas, B., Varga, M., Balogh, S. 2011a. Direct computer mapping of executable multiscale hybrid process architectures. In: Proceedings of Summer Simulation Multiconference'2011, Den Haag, The Netherlands, 2011.06.26-2011.06.29. p. 87–95. (ISBN:1-56555-345-4).

Csukas, B., Varga, M., Miskolczi, N., Balogh, S., Angyal, A., Bartha, L., 2013 b. Simplified dynamic simulation model of plastic waste pyrolysis in laboratory and pilot scale tubular reactor. Fuel Process. Technol. 106, 186–200. doi:10.1016/j.fuproc.2012.07.024

Garcia, D.J., You, F., 2016. The Water-Energy-Food Nexus and Process Systems Engineering: A New Focus. Comput. Chem. Eng. doi:10.1016/j.compchemeng.2016.03.003

Grossmann, I.E., Westerberg, A.W., 2000. Research challenges in process systems engineering. AIChE J. 46, 1700–1703. doi:10.1002/aic.690460902

Kalman, R.E., Falb, P.L., Arbib, M.A. 1969. Topics in mathematical system theory. McGraw Hill Book Co.

Klatt, K.U., Marquardt, W., 2009. Perspectives for process systems engineering-Personal views from academia and industry. Comput. Chem. Eng. 33, 536–550. doi:10.1016/j.compchemeng.2008.09.002

Liu, Y.Y., Slotine, J.J., Barabasi, A.L., 2011. Controllability of complex networks. Nature. 473, 167-173.

Petri, C.A. 1966. Communication with automata. Supplement1 to Techn.Report RADC-TR-65-377, Vol. 1 Griffiss AFB, New York. Original in German (Kommunikation mit Automaten, Univ. of Bonn, 1962)

Petri, C.A. 1977. General Net Theory. In: Proc. of 1976 Joint IBM/Univ. of Newcastle upon Tyne Seminar on Computing System Design, Newcastle upon Tyne, Computing Laboratory (1977) 131–169. http://homepages.cs.ncl.ac.uk/brian.randell/GeneralNetTheory.pdf

Stephanopoulos, G., Reklaitis, G. V., 2011. Process systems engineering: From Solvay to modern bio- and nanotechnology.. A history of development, successes and prospects for the future. Chem. Eng. Sci. 66, 4272–4306. doi:10.1016/j.ces.2011.05.049

Stephanopoulos, G., Schmitt, W.A., 2001. Systems Biology: an emerging theme in biological research. Comput. Aided Chem. Eng. 9, 55–68. doi:10.1016/S1570-7946(01)80006-7

Strogatz, S.H., 2001. Exploring complex networks. Nature. 410, 268–276.

Temesvári, K., Aranyi, A., Csukas, B., Balogh, S. 2004b. Simulated Moving Bed Separation of a Two Components Steroid Mixture. Chromatographia, 60: 189-199.

Varga M, Csukás B, Balogh S. 2012. Transparent Agrifood Interoperability, Based on a Simplified Dynamic Simulation Model. In: Mildorf T, Charvat K (eds.) ICT for Agriculture, Rural Development and Environment: Where we are? Where we will go?. Prague: Czech Centre for Science and Society, 2012. pp. 155-174. (ISBN:978-80-905151-0-9)

Varga, M., Balogh, S., Csukas, B., 2016b. An extensible, generic environmental process modelling framework with an example for a watershed of a shallow lake. Environ. Model. Softw. 75, 243-262. doi:10.1016/j.envsoft.2015.10.022

Varga, M., Balogh, S., Wei, Y., Li, D., Csukas, B., 2016a. Dynamic simulation based method for the reduction of complexity in design and control of Recirculating Aquaculture Systems. Inf. Process. Agric. 3, 146–156. doi:10.1016/j.inpa.2016.06.001

Varga, M., Csukas, B. 2017. Generation of extensible ecosystem models from a network structure and from locally executable programs. Ecological Modelling, *accepted manuscript*

Varga, M., Prokop, A., Csukas, B. 2017. Biosystem models, generated from a complex rule/reaction/influence network and from two functionality prototypes. BioSystems, 152, 24-43. http://dx.doi.org/10.1016/j.biosystems.2016.12.005

495297 Pyosyn - a Python Tool for General Process Synthesis

Abstract:

Pyosyn - a Python tool for general process synthesis

Qi Chen[1], Anthony Burgard[2], John Eslick[2], Andrew Lee[2],
John Siirola[3], David Miller[2], Ignacio Grossmann[1]

[1]Department of Chemical Engineering, Carnegie Mellon University, 5000 Forbes Avenue, Pittsburgh, PA 15213, USA

[2]National Energy Technology Laboratory, 1268 Wallace Road, South Park Township, PA 15129, USA

3Center for Computing Research, Sandia National Laboratories, P.O. Box 5800, Albuquerque, NM 87185, USA

Abstract

Process synthesis allows industrial practitioners to identify optimal configurations among process flowsheet alternatives and enables researchers to guide development of new technologies towards areas with the greatest promise. One of the main approaches to synthesis is superstructure optimization, which facilitates the representation, modeling, and solution of diverse synthesis problems as mathematical programs [1]. However, few computational tools are available to support mathematical programming-based conceptual design. The most prominent software implementations for conceptual design include PROSYN [2], ICAS [3], and MIPSYN [4]. PROSYN and ICAS predominantly utilize heuristic rules driven by thermodynamic insights and general rules-of-thumb. MIPSYN is one of the few mathematical-programming-oriented implementations. However, these tools have had limited penetration in industrial practice. As a result, mathematical programming-based conceptual design has seen little adoption in industry [5]. In order to encourage industrial adoption of optimization-based synthesis methods, an easy-to-use general purpose software platform for synthesis is needed.

In this paper, we present implementation progress for Pyosyn, a high-level Python-based tool for process synthesis. Process synthesis problems involve both nonlinear, nonconvex equations and discrete design decisions. A general purpose solution must therefore effectively manage the complexities associated with modeling and solving these problems. Pyosyn supports superstructure optimization by providing a toolset for defining unit models, linking them to form a superstructure, generating an algebraic representation, and applying one of several optimization algorithms to solve the synthesis problem. The Pyosyn toolset leverages the Pyomo optimization modeling library [6] in Python, which allows for the programmatic modification and manipulation of modeling objects such as variables and constraints. This enables development of generally applicable automatic model transformations to reformulate, relax, and/or constrict the optimization problem, all in one language, which can reduce the number and complexity of programmatic interfaces. It also enables the implementation of meta-solvers that iteratively apply a set of transformations and sub-solvers to find a solution, such as logic-based outer approximation (LOA) [7, 8].

LOA decomposes the synthesis problem into a mixed-integer linear program (MILP) master problem which selects new flowsheet configurations and a series of reduced space nonlinear program (NLP) sub-problems to optimize selected flowsheets. Outer approximation with equality-relaxation (OA/ER) cuts [9] generated at the solution of each sub-problem augment the quality of the MILP master problem linear approximation with each iteration. Solution of the NLP sub-problems in reduced space—only the variables and constraints for active flowsheet units are included—decreases the dimensionality and thus complexity of sub-problems. Use of reduced space NLPs also avoids issues from zero-flow singularities that arise when attempting to solve equations for inactive flowsheet units. Pyosyn includes a built-in implementation of LOA.

Pyosyn also features the ability to formulate and solve synthesis problems in a high-level, intuitive manner. Unit models in Pyosyn group together variables, parameters, and constraints that represent a particular processing task or equipment in the superstructure. Unit models can contain other unit models, allowing for hierarchical description of the superstructure. In conjunction with modeling using Generalized Disjunctive Programming (GDP) [7], this preserves the logical structure of the synthesis problem for reformulations or transformations. The unit models are also independently solvable, allowing for granular testing and facilitating model initialization procedures during the solution process.

Use of Pyosyn is demonstrated in detail through two case studies: the synthesis of a chemical looping process and a desulfurization process [10]. We explain the implementation of LOA as a series of model transformations and show that Pyosyn facilitates joint modeling efforts by both expert and non-expert optimization users.

References

[1] Chen Q, Grossmann IE. 2017. Recent developments and challenges in optimization-based process synthesis, *Annu. Rev. Chem. and Biomol. Eng.* 8:12.1-12.35. doi:10.1146/annurev-chembioeng-080615-033546

[2] Schembecker G, Simmrock KH, Wolff A. 1994. Synthesis of chemical process flowsheets by means of cooperating knowledge integrating systems. Presented at 4th Eur. Symp., *Comput. Aided Process Eng.*, Inst. Chem. Eng. Symp. Ser. 133:333–41. Dublin, Ireland.

[3] Gani R, Hytoft G, Jaksland C, Jensen AK. 1997. An integrated computer aided system for integrated design of chemical processes. *Comput. Chem. Eng.* 21(10):1135–46. doi:10.1016/S0098-1354(96)00324-9

[4] Kravanja Z, Grossmann IE. 1996. A computational approach for the modeling/decomposition strategy in the MINLP optimization of process flowsheets with implicit models. *Ind. Eng. Chem. Res.* 35(6):2065–70. doi: 10.1021/ie950424f

[5] Harmsen GJ. 2004. Industrial best practices of conceptual process design. *Chem. Eng. Process. Process Intensif.* 43(5):677–81. doi:10.1016/j.cep.2003.02.003

[6] Hart WE, Laird C, Watson J-P, Woodruff DL. 2012. *Pyomo - Optimization Modeling in Python*, Springer US, Boston, MA. doi:10.1007/978-1-4614-3226-5.

[7] Turkay M, Grossmann IE. 1996. Logic-based MINLP algorithms for the optimal synthesis of process networks, *Comput. Chem. Eng.* 20: 959-978. doi:10.1016/0098-1354(95)00219-7.

[8] Duran MA, Grossmann IE. 1986. An outer-approximation algorithm for a class of mixed-integer nonlinear programs, *Math. Program.* 36: 307. doi:10.1007/BF02592064.

[9] Viswanathan J, Grossmann IE. 1990. A combined penalty function and outer-approximation method for MINLP optimization. *Comput. Chem. Eng.* 14(7):769–82. doi:10.1016/0098-1354(90)87085-4

[10] Qian Z, Chen Q, Grossmann IE. 2017. Optimal synthesis of rotating packed bed reactor, *Comput. Chem. Eng.* To appear. doi:10.1016/j.compchemeng.2017.02.026

489832 Super-O: A Tool for Processing Network Synthesis Using Superstructure Optimization

Abstract:

A business and engineering framework for processing network synthesis was proposed by Quaglia et al. (2012) and extended by Bertran et al. (2017). The framework is based on a superstructure optimization approach, including four key elements: (i) a superstructure representation named Process Step-Interval Network (PSIN); (ii) a generic process model; (iii) a solver from an optimization environment; and (iv) a database for data management.

A software implementation of the framework, named Super-O, guides the user through the steps for formulating and solving synthesis problems of different processing networks by integrating the necessary in-house and commercial tools, consisting of models, databases and commercial solvers. The use of Super-O allows for the reduction of the time needed for the formulation and solution of network optimization problems.

A generic process model is used for each processing alternative in the superstructure and represents the set of equality constraints in the optimization problem. The generic process model is based on a series of processing tasks, namely mixing, reaction, waste removal, separation and utility consumption. The purpose of the databases is to provide a common platform for different users to store, search and retrieve data for the formulation and solution of biorefinery synthesis problems. Each database is built on a specifically designed data structure that consists of three main data sections, namely a basic data section, a section for data related to the material, and a section containing process data. This data structure has been used to set up databases in different application fields, for example, a biorefinery database that contains 10 types of biomass, over 100 processing alternatives and 9 products.

A number of case studies have been solved using Super-O, these span over various problem types (feedstock selection, product selection, processing route selection, plant allocation, etc.) and industries (chemicals and bio-chemicals, biorefineries, CO2 utilization, wastewater treatment, etc.).

References:

- Quaglia, A., Sarup, B., Sin, G., Gani, R., 2012. Integrated business and engineering framework for synthesis and design of enterprise-wide processing networks. Comput. Chem. Eng. 38, 213â€"223.

- Bertran, M.-O., Frauzem, R., Sanchez-Arcilla, A.-S., Zhang, L., Woodley, J.M., Gani, R., 2017. A generic methodology for processing route synthesis and design based on superstructure optimization. Comput. Chem. Eng.Â DOI: 10.1016/j.compchemeng.2017.01.030

490561 Robust Optimization with Decision-Dependent Uncertainty Sets

Abstract:

Robust Optimization (RO) is a systematic approach for mitigating risk from parameter uncertainty in optimization models. Its main distinctive property is that it enforces the problem constraints for any realization of the uncertain parameters within the prescribed uncertainty set, which is typically defined as a constant set [1]. In many cases, however, uncertainty can be affected by the decision maker's strategy (endogenous uncertainty) [2] . While this class of uncertainty has been investigated thoroughly in the Stochastic Programming (SP) literature (e.g. [3]–[6]), little has been done in the context of RO [7]. Motivated by this fact, we introduce broadly applicable decision-dependent polyhedral uncertainty sets, which allow us to capture functional changes in correlations induced by given decisions, as well as to eradicate conservatism e ects from parameters that become irrelevant in view of the optimal decisions.

In this context, we investigate the numerical performance of applying the standard, duality-based reformulation approach to solve RO problems with decision-dependent sets. We also present a novel algorithmic solution approach based on the Kelley's cutting plane method [8] within a tailored branch-and-bound framework, which can improve tractability. Furthermore, we show the capability of our proposed methodology to incorporate for the first time near-optimal recourse actions, such as piece-wise linear or piece-wise constant decision rules [9], in multi-stage stage setups with endogenous uncertainty. This feature is of great importance since it has been shown [10] that deferring a subset of the decisions for later ("wait-and-see") can lead to more profitable—yet equally robust—solutions compared to traditional RO, where all of the decisions are considered as "here-and-now".

The modeling capabilities afforded to us by using these new decision-dependent sets is showcased via a number of case studies, for which we quantify the additional benefits that can be gained by introducing recourse actions in a non-anticipative fashion to multi-stage problems involving endogenous uncertainty.

References:

[1] A. Ben-Tal, L. El Ghaoui, and A. Nemirovski, *Robust optimization.* Princeton University Press, 2009.

[2] T. W. Jonsbråten, R. J. B. Wets, and D. L. Woodruff, "A class of stochastic programs with decision dependent random elements," *Ann. Oper. Res.*, vol. 82, pp. 83–106, 1998.

[3] V. Gupta and I. E. Grossmann, "A new decomposition algorithm for multistage stochastic programs with endogenous uncertainties," *Comput. Chem. Eng.*, vol. 62, pp. 62–79, 2014.

[4] V. Gupta and I. E. Grossmann, "Solution strategies for multistage stochastic programming with endogenous uncertainties," *Comput. Chem. Eng.*, vol. 35, no. 11, pp. 2235–2247, 2011.

[5] S. Solak, J. P. B. Clarke, E. L. Johnson, and E. R. Barnes, "Optimization of R&D project portfolios under endogenous uncertainty," *Eur. J. Oper. Res.*, vol. 207, no. 1, pp. 420–433, 2010.

[6] R. M. Apap and I. E. Grossmann, "Models and computational strategies for multistage stochastic programming under endogenous and exogenous uncertainties," *Manuscr. Prep.*, 2015.

[7] M. Poss, "Robust combinatorial optimization with variable cost uncertainty," *Eur. J. Oper. Res.*, vol. 237, no. 3, pp. 836–845, 2014.

[8] J. E. Kelley, Jr., "The Cutting-Plane Method for Solving Convex Programs," *J. Soc. Ind. Appl. Math.*, vol. 8, no. 4, pp. 703–712, 1960.

[9] D. Bertsimas and A. Georghiou, "Design of Near Optimal Decision Rules in Multistage Adaptive Mixed-Integer Optimization," *Oper. Res.*, vol. 63, no. 3, pp. 610–627, 2015.

[10] A. Ben-Tal, A. Goryashko, E. Guslitzer, and A. Nemirovski, "Adjustable robust solutions of uncertain linear programs," *Math. Program.*, vol. 99, no. 2, pp. 351–376, 2004.

497995 Optimization Under Uncertainty Using Surrogate Models for Confidence Evaluation

Abstract:

This work presents a novel approach to solving optimization under uncertainty problems. These problems are important because many real-life engineering processes have significant inherent uncertainty. The uncertainty in these problems make it difficult to optimize using common optimization methods. The proposed method determines the range for which a possible solution may lay in by estimating a confidence value to that range. The confidence value is determined by testing the percentage of surrogate quadratic models that would have solutions inside the range of interest. Uncertainty in the objective function is accommodated using randomly selected quadratic models and soft constraint penalties. The size of the range of interest and the overall objective function are minimized to yield a range for which the solution to the objective function lies with a specified confidence value. The paper outlines case studies conducted using the novel optimization method as well as a more relevant example involving the optimization of a closed-loop dynamic diabetic patient model system.

499450 Multistage Adaptive Conditional Value at Risk Optimization Using Piecewise Linear Decision Rule

Abstract:

Many practical decision making problems in process operations can be represented as multi-stage adaptive optimization problem, where some decisions can be adjusted based on the information revealed at the time of decision making. Adaptive/adjustable optimization has received an increased level of research interest since the seminal paper of Ben-Tal et al. [1]. The framework includes non-adjustable variables interpreted as "here and now" decisions and adjustable variables interpreted as "wait and see" variables. While the fully-adjustable counterpart problem is generally intractable, applying some restrictions on the form of future-stage decisions can reduce the problem complexity. Among those restrictions, linear decision rule based approximation is widely applied to adjustable variables to obtain tractable approximation, where the future-stage decisions are restricted to be affine function of the uncertain parameters. Furthermore, to improve the solution optimality of linear decision rule based model while maintaining the problem tractable, various improved decision rule approximation methods were studied. For example, linear decision rule approximation can be improved by extending the affine dependency to the auxiliary variables that are used in describing the uncertainty set [2].

In the literature, decision rule based approximation has received lots of study for stochastic programming (SP) and robust optimization (RO). Each method portrays a specific meaning of the cost (objective function): stochastic programing finds the optimal expected cost, while robust optimization computes the worst-case cost. In this work, we investigate the multistage adaptive optimization with conditional value at risk (CVaR) objective. CVaR determines the average cost conditional on being on the bad tail of the distribution (expected cost that exceeds the corresponding value at risk). It provides a trade-off between stochastic programming and robust optimization, and quantifies the level of risk with the optimal solution. It also represents a mean to relate the decreased cost with the increased level of risk.

We propose a method for handling conditional value at risk objective in multistage adaptive/adjustable optimization under uncertainty. Furthermore, we addressed approximation quality improvement through piecewise linear decision rule. The idea of lifted uncertainty set [3] is implemented in the proposed CVaR optimization formulation. The proposed CVaR optimization method is further compared with stochastic programming and robust optimization formulations. A computational study is performed using two classical problems in operation research: transportation and news vendor problems. It is observed that the optimal solution to SP is not accompanied with risk level and the RO solution leads to the most conservative results. On the other hand, for CVaR objective based formulation, it provides a trade-off between the SP and RO. The numerical investigation illustrates the possibility of using CVaR as a tool to select risky versus neutral decision policies. Studies on larger scale problems are also performed to highlight the flexibility of using CVaR method, and the nature of cost improvements it brings.

References:

[1] Ben-Tal, A., Goryashko, A., Guslitzer, E., Nemirovski, A.: Adjustable robust solutions of uncertain linear programs. Mathematical Programming. 2004, 99(2), 351–376.

[2] Chen, X., Zhang, Y.: Uncertain linear programs: Extended affinely adjustable robust counterparts. Operations Research. 2009, 57(6), 1469–1482.

[3] Georghiou A, Wiesemann W, Kuhn D. Generalized decision rule approximations for stochastic programming via liftings. Mathematical Programming. 2015, 152(1-2), 301-38.

489376 Derivation of Generalized Affine Decision Rules for Mixed Integer Linear, Quadratic and Nonlinear Adjustable Robust Optimization Problems By Multi-Parametric Programming

Abstract:

Classical robust optimization (RO) is an approach for incorporating uncertainty in optimization problems, and traditionally assumes that all decisions must be made before the realization of uncertainty (referred to as "here-and-now" decisions), a strategy which may be overly conservative (1). A more realistic approach is adjustable robust optimization (ARO) which involves recourse decisions (i.e. reactive actions after the realization of the uncertainty, "wait-and-see") as functions of the uncertainty, typically posed in a two-stage stochastic setting (2). Solving the general ARO problems is very difficult, therefore ways to reduce the computational effort have been proposed, with the most popular being the affine decision rules, where "wait-and-see" decisions are approximated as affine adjustments of the uncertainty (2, 3); an approximation which is exact and optimal only for the case of linear ARO problems.

In this work we propose a novel method for the derivation of generalized affine decision rules for linear/quadratic/nonlinear and mixed-integer ARO problems through multi-parametric programming. The ARO problem is treated as a multi-level programming problem (4) and it is then solved using M-POP, which is a novel algorithm for the exact and global solution of multi-level mixed-integer linear or quadratic programming problems (5, 6). The main idea behind the proposed approach is to solve the lower optimization level of the ARO problem parametrically, by considering "here-and-now" variables and uncertainties as parameters. This will result in a set of affine decision rules for the "wait-and-see" variables as a function of "here-and-now" variables and uncertainties for their entire feasible space. A set of illustrative numerical examples are provided to demonstrate the potential of the proposed novel approach.

References:

(1) Ben-Tal, A., Ghaoui, L., Nemirovski, A. Robust optimization (2009) Princeton University Press, Princeton and Oxford.

(2) Ben-Tal, A., Goryashko, A., Guslitzer, E., Nemirovski, A. Adjustable robust solutions of uncertain linear programs (2004) Mathematical Programming, 99 (2), pp. 351-376.

(3) Kuhn, D., Wiesemann, W., Georghiou, A. Primal and dual linear decision rules in stochastic and robust optimization (2011) Mathematical Programming, 130 (1), pp. 177-209.

(4) Ning, C., You, F. Data-driven adaptive nested robust optimization: General modeling framework and efficient computational algorithm for decision making under uncertainty (2017) AIChE Journal; In Press (onlinelibrary.wiley.com/doi/10.1002/aic.15717/pdf)

(5) Faisca, N.P., Dua, V., Rustem, B., Saraiva, P.M., Pistikopoulos, E.N. Parametric global optimisation for bilevel programming (2007) Journal of Global Optimization, 38 (4), pp. 609-623.

(6) Avraamidou, S., Diangelakis, N. A., Pistikopoulos, E. N. Mixed Integer Bilevel Optimization through Multi-parametric Programming (2017) Foundations of Computer Aided Process Operations / Chemical Process Control; In Press. (http://folk.ntnu.no/skoge/prost/proceedings/focapo-cpc-2017/FOCAPO-CPC%202017%20Contributed%20Papers/73_FOCAPO_Contributed.pdf)

499478 A Predictor-Corrector Algorithm for Projection-Based Derivative-Free Optimization

Abstract:

With the advent of high-fidelity, complex process modeling and simulation in many areas including computational fluid dynamics-based reactor design [1], Nonlinear Algebraic and Partial Differential Equations (NAPDE)-based multiscale optimization [2] and image processing [3], data-driven methods are gaining more importance than ever. Most often, these optimization problems are formulated as black-box problems where the objective function and/or constraints cannot be expressed analytically as explicit functions of decision variables. This implies that the values of the objective and/or constraints become available only after performing a full-scale simulation and, therefore, derivative information of the objective and/or constraints is expensive. To this end, derivative free optimization (DFO) [4] has emerged as a promising approach to optimize black-box problems [5]. DFO techniques is being increasingly applied in process systems engineering for superstructure optimization [6], process optimization [7-9] and flowsheet optimization [10]. Existing model-based trust region approaches generally take less number of evaluations for smooth problems but they often converge to local minima.

In this work, we present an algorithm to address bound-constrained and multi-dimensional black-box problems based on projection onto a one-dimensional space. Specifically, we consider a problem of minimizing $G(t)$, which can be interpreted as the projection of the original objective function $f(x_i)$, $i = 1, \ldots, n$ taken in the t-space such that t is linear combination of x_i. Additionally, we also provide useful properties of the univariate function $G(t)$ and show that the minimum of $G(t)$ also corresponds to the minimum of $f(\mathbf{x})$. This enables us to solve the original optimization problem in two steps. The first step involves identifying $G(t)$ and is referred as the inner loop [11]. In this work, the inner loop is applied at discrete values of $t(t_1, t_2, \ldots, t_k)$. Once the solution at the previous $t(t_{k-1})$ is known, sensitivity theorem [12] is utilized to obtain close solution for the inner loop at next $t(t_k)$. The second step is optimizing the univariate function $G(t)$ to find the solution of the original problem. We also propose an alternate projection by normalizing t. We compare the performance of the two projections with existing DFO solvers. The step length, α, where $\alpha = t_k - t_{k-1}$, is another critical parameter in determining the number of evaluations taken by the algorithm. We have considered and compared two strategies related to the step length. The first strategy uses a fixed or uniform step length. The second strategy uses an adaptive approach where the value of α is revised as the algorithm proceeds, based on the number of evaluations taken in the previous inner loop.

We have tested the algorithm to solve several high-dimensional test problems. For instance, we are able to find the global solution of a 50-dimensional test problem (emfl_eps) in 40,747 evaluations. We also obtain the solutions of explin2, expquad and qrtquad (each of which are 120-dimensional problems) using 85521, 62760 and 89484 function evaluations, respectively. The algorithm is applied to a large set of test problems taken from [5] and compared to existing model-based DFO solvers. The proposed approach performs superior relative to ORBIT [13] and SNOBFIT [14].

References:

[1] Zimmermann, S.; Taghipour, F. CFD Modeling of the Hydrodynamics and Reaction Kinetics of FCC Fluidized-Bed Reactors. *Industrial and Engineering Chemistry Research*, 44 (26):9818–9827, 2005.

[2] Hasan, M. M. F.; First, E. L..; and Floudas, C. A. Cost-effective CO2 capture based on in silico screening of zeolites and process optimization. *Physical Chemistry Chemical Physics*, 15(40):17601-17618, 2013.

[3] Oeuvray, R.; Bierlaire, M. A new derivative free algorithm for the medical image registration problem. *International Journal of Modelling and Simulation*,27(2):115-124, 2007.

[4] Conn, A. R.; Scheinberg, K.; and Vicente, L. N. Global convergence of general derivative-free trust-region algorithms to first and second order critical points. *SIAM Journal on Optimization*, 20(1):387-415, 2009.

[5] Rios, L. M.; Sahinidis, N.V. Derivative-free optimization: a review of algorithms and comparison of software implementations. *Journal of Global Optimization*, 56(3):1247–1293, 2013.

[6] Henao, C. A.; Maravelias, C. T. Surrogate-based superstructure optimization framework. *AIChE Journal*, 57(5):1216-1232, 2011.

[7] Nuchitprasittichai, A.; Cremaschi, S. Optimization of CO2 capture process with aqueous amines using response surface methodology. *Computers & Chemical Engineering*, 35(8):1521-1531, 2011.

[8] Agarwal, A.; Biegler L.T. A trust-region framework for constrained optimization using reduced order modeling. *Optimization and Engineering*, 14(1):3-35, 2013.

[9] Boukouvala, F.; Hasan, M. M. F.; Floudas, C. A. Global optimization of general constrained grey-box models: new method and its application to constrained PDEs for pressure swing adsorption. *Journal of Global Optimization*, 67(1):3-42, 2017.

[10] Caballero, J. A.; Grossmann, I. E. An algorithm for the use of surrogate models in modular flowsheet optimization. *AIChE Journal*, 54(10):2633-2650, 2008.

[11] Bajaj, I.; Hasan, M.M.F. A Novel Derivative-Free Optimization Method based on Single Dimension Projection. In the *Proceedings of FOCAPO/CPC*, 2017.

[12] Fiacco, A.V. Introduction to sensitivity and stability analysis in nonlinear programming. *Academic Press, New York*, 1984.

[13] Wild, S.M.; Regis, R.G.; Shoemaker, C.A. ORBIT: Optimization by radial basis function interpolation in trust-regions. *SIAM Journal on Scientific Computing*, 30(6):3197-3219, 2008.

[14] Huyer, W.; Neumaier, A. SNOBFIT – Stable Noisy Optimization by Branch and Fit. *ACM Transactions on Mathematical Software*, 35(2):1-25, 2008.

498113 Optimization of Black-Box Problems Using Smolyak Grids and Polynomial Approximations

Abstract:

A new black-box surrogate-based optimization method is proposed, which aims to converge to the global optimum of box-constrained problems using only input-output data. Optimization methods that are based on the exchange of designed data-streams between the optimization solver and the data-generating system (simulation or experiment) have gained significant interest recently [1-3]. The number of applications of data-dependent optimization methods is already large, spanning a wide range of areas such as mechanical and structural design, complex flowsheet optimization and process synthesis, geosciences, parameter estimation of large ODE systems, and many more. Several challenges still exist in terms of the computational tractability of the methods as the problem size increases, especially since the cost of the simulation/experiment limits the amount of data that can be exchanged. Moreover, due to the lack of algebraic forms and derivatives, convergence to optimality reliably and in a finite number of iterations is hard to achieve for a wide range of problems. As a result there is a high interest for black-box optimization methods that locate better solutions by globally searching the feasible space and converging to optimal solutions with improved convergence rates and a limited number of samples.

The proposed work tests the performance of a new surrogate-based optimization algorithm that is based on Smolyak-grid sampling and polynomial interpolation, which have been predominantly used for numerical integration [4]. The algorithm starts with a global search of the multidimensional space, using a Smolyak grid [5] which is constructed using Chebyshev extrema in the one-dimensional space. The collected samples are used to fit polynomial interpolants, which are used as surrogates towards the search for the optimum. The developed algorithm adaptively refines the grid by collecting new points in promising regions, and iteratively refines the search space around the incumbent sample until the search domain reaches a minimum size and convergence has been attained. The algorithm is tested on a large set of benchmark problems and its performance is compared to a recent algorithm for global optimization of grey-box problems using quadratic, kriging and radial basis functions [6]. It is shown that the proposed algorithm has a consistently reliable performance for the majority of test problems, which does not depend on the initialization of the sampling set. It will be shown that this reliability in performance is attributed to the use of Chebyshev-based sparse grids, which have been shown to approximate complex functions with low error bounds.

This work is dedicated to the authors' mentor and postdoctoral advisor, Christodoulos Floudas, who introduced the authors to this area.

References:

1. Conn, A.R., K. Scheinberg, and L.N. Vicente, *Introduction to derivative-free optimization*. 2009, Philadelphia: SIAM.

2. Boukouvala, F., R. Misener, and C.A. Floudas, *Global optimization advances in Mixed-Integer Nonlinear Programming, MINLP, and Constrained Derivative-Free Optimization, CDFO*. European Journal of Operational Research, 2016. **252**(3): p. 701-727

3. Rios, L.M. and N.V. Sahinidis, *Derivative-free optimization: a review of algorithms and comparison of software implementations*. Journal of Global Optimization, 2013. **56**(3): p. 1247-1293

4. Gerstner, T. and M. Griebel, *Numerical integration using sparse grids*. Numerical Algorithms, 1998. **18**(3-4): p. 209-232.

5. Smolyak, S.A. *Quadrature and interpolation formulas for tensor products of certain classes of functions*. in *Dokl. Akad. Nauk SSSR*. 1963.

6. Boukouvala, F. and C.A. Floudas, *ARGONAUT: AlgoRithms for Global Optimization of coNstrAined grey-box compUTational problems*. Optimization Letters, 2016: p. 1-19.

495996 On the Impact of Solution Representations for Stochastic Optimisation of Control Trajectories of Industrial Fermentation Processes

Abstract:

Despite their widespread use, limitations may arise when using deterministic dynamic optimisation methods in optimal control problems. Required CPU time may be significant and solution trajectories may display extremely strong sensitivity to initialisation. Stochastic optimisation methods have versatility advantages which can be explored towards overcoming such limitations when observed (Xiao, Zhou and Zhang, 2003). Nevertheless, systematic guidelines for formulating a control trajectory representation in order to ensure compatibility with stochastic optimisation algorithms have not been hitherto developed. The present paper considers an industrial control vector optimisation problem, namely the temperature manipulation during beer fermentation (Rodman and Gerogiorgis, 2016), and applies a multi-objective plant propagation algorithm in order to investigate the influence of problem formulation and solution representation on algorithmic efficiency.

Fermentation is a key step in the production of all beer products, and a possible production bottleneck due to its significant duration (often exceeding 1 week). Therein lies a strong incentive for process optimisation and batch duration reduction; nevertheless, a clear trade-off is evident beween attainable ethanol concentration and required batch time (Carrillo-Ureta et al., 2001). Biochemical system operation and fermentation progress is sensitive to yeast cellular activity, which in turn depends on fermenter operating conditions. Consequently, it is desirable to compute and determine temperature manipulations protocols (control profiles) for improved plant performance. The complexity of the chemical system (>600 species: Vanderhaegen et al., 2006) renders comprehensive parameter estimation infeasible and dynamic optimisation (even for reduced models) quite cumbersome. A validated reduced-order dynamic model of the fermentation process (de Andrés-Toro, 1998) has been previously analysed (Rodman and Gerogiorgis, 2016) and confirmed to be adequately descriptive to capture the key chemical pathways and facilitate a stochastic optimisation study.

The Plant Propagation Algorithm employed is The Strawberry Algorithm (Salhi and Fraga, 2011), which emulates the behaviour of strawberry plants by encompassing the two key aspects of effective global optimisation algorithms: solution exploration and intensification. In nature, plants which are in a favourable environment will reproduce with greater frequency, within their local vicinity. Conversely, plants which are not as well situated will reproduce through longer distances. This inspires the Strawberry evolutionary algorithm: each member of the population (an individual solution) is evaluated (objective function computed) and a fitness function is used to rate the performance of each (Fraga and Amusat, 2016). A fitness-based selection procedure is used to identify those solutions for propagation. More fit solutions generate new solutions in the local neighbourhood; less fit solutions generate fewer solutions but further away.

The present beer fermentation control problem is formulated so that a vector of decision variables corresponds to a member of the population at each generation. The decision variable vector must be suitably formulated to allow for an

expressive as well as inexpensive portrayal of the (temperature manipulation) profile for beer fermentation control: accordingly, a range of different solution representation strategies have postulated, implemented and evaluated. We hereby demonstrate that the specific formulation of the decision vector has strong influence on stochastic algorithm performance: the probability of returning favourable solutions after a fixed number of generations as well as the required number of generations to consistently achieve convergence vary significantly, as evidenced by the required CPU time as well as the quality of computed solutions. Additional benefits of select solution representations include inherent suitability for industrial application of the corresponding temperature manipulation profiles. These new stochastic optimisation results allow us to elucidate and present heuristics for effective implementation of stochastic algorithms, particularly in achieving optimal biochemical process performance under reduced-order dynamic models.

LITERATURE REFERENCES

Carrillo-Ureta, G., P. Roberts, and V. Becerra, 2001. Genetic algorithms for optimal control of beer fermentation. Proc. IEEE Int. Symp. Intell. Control, 391-396.

de AndrÃ©s-Toro, B., et al., 1998. A kinetic model for beer production under industrial operational conditions. Math. Comput. Simulat. 48(1): 65-74.

Vanderhaegen, B., et al., 2006. The chemistry of beer aging â€" a critical review. Food Chem. 95(3): 357-381.

Xiao, J., Zhou, Z., Zhang, G., 2003. Ant colony system algorithm for the optimization of beer fermentation control. J. Zhejiang Univ. 5(12): 1597-1603.

Rodman, A. D., Gerogiorgis, D.I., 2016. Multi-objective process optimisation of beer fermentation via dynamic simulation, Food Bioprod. Proc. 100(A): 255-274.

Salhi, A., Fraga, E. S., 2011, Nature-inspired optimisation approaches and the new Plant Propagation Algorithm, The International Conference on Numerical Analysis and Optimization (IceMATH 2011), Yogyakarta, Indonesia, K2-1-8.

Fraga, E., and Amusat, O. 2016. Understanding the impact of constraints: A rank-based fitness function for evolutionary methods. Advances in Stochastic and Deterministic Global Optimization. Springer, 243-254.

497580 Seamless Particle-Based Modeling of Blood Clotting

Abstract:

We propose a new multiscale framework that seamlessly integrate four key components in the process of thrombosis namely, blood rheology, cell mechanics, coagulation kinetics and transport of species, and platelet adhesive dynamics. We use transport dissipative particle dynamics (tDPD) which is an extended form of original DPD as the base solver to model both blood flow and the reactive transport of chemical species in the coagulation cascade. Further, we use a coarse-grained representation of blood cell's membrane that accounts for its mechanics; both red blood cells and platelets are resolved at sub-cellular resolution, and stochastic bond formation/dissociation are included to account for platelet adhesive dynamics at the site of injury. Our results show good agreement with in vivo experiments. The numerical framework allows us to perform systematic analysis on different mechanisms of blood clotting. In addition, this new multiscale particle-based methodology can open new directions in addressing different biological processes from sub-cellular to macroscopic scales.

493245 Incorporating Computational Fluid Dynamics into the Study of Cell Damage Due to Hydrodynamic Stress in Bioreactors and Pumps

Abstract:

With the ever growing health care market, the demand for cost-effective systems for the cultivation of various types of animal cells continues to grow. Recently, the sub-field of "cell therapy" has demonstrated significant growth which presents unique demands on the market: instead of the cells making products, the cells are the product. Since the product is the cell, which is hygienically and nutritionally more sensitive than the other products in the health care market, greater regulatory restrictions as well as specific cellular functions are needed which create greater demands on bioprocessing. For this study, the hydrodynamic stress imposed upon the growing cells in bioreactors and pumps will be investigated, focusing on the interaction between rotating impeller and the Chinese Hamster Ovary (CHO) cells. A commercial blood pump, which supports hemodynamic stabilization, was connected to a standard 3L bioreactor to run CHO cells at 1100, 1600, 2200 and 3000 RPM for 10 days, while the LDH level, viable cell density and cell size were measured to quantify the cell damage. Computational fluid dynamics (CFD), which employs finite element method to solve numerical flow dynamics problems, was used to simulate the experimental condition for comparison with the experimental data. Energy dissipation rate (EDR) was calculated using post-CFD translation software inside the pump to quantify and visualize the cell damage. Furthermore, a 3-D printer will be incorporated to test the feasibility of 3-d printing the blood pump and the rotor. While the change in blood pump's RPM appeared to have no effect on the LDH level and cell size, an increase in RPM from 2200 to 3000 resulted in a significant drop in viable cell density. Therefore, the simulation data of 2200 and 3000 RPM was further investigated to compare the EDR values between the two rates. Based on the histogram analysis, the two rate showed similar energy dissipation rate distribution except for the region between $10^4 - 10^5$ W/m^3, which is in agreement with the reported literature values. Use of post-CFD translation software allows users to visualize the user-defined variable (EDR), and the visual analysis pointed our attention to the escape neck area of the blood pump, where high EDR values were reported for both 2200 and 3000 RPM. After analyzing the results, we have come to a conclusion that different geometrical design of escape neck area might lead to less damaging blood pump. Therefore, the geometry of the pump was recreated in the 3 dimensional computer-aided design to be 3-D printed, and the product that displayed the same physical characteristic was made, showing the feasibility of incorporating 3d printing into the further study. There are several advantages for incorporating CFD and 3-D printing into the study of cell damage due to hydrodynamic stress. It is relatively straight forward to design/modify current pump designs, simulate them with CFD, potentially modify the design, and then actually produce a pump with a 3-D printer. This easy, iterative process will allow researchers in the field of health care to develop medical device involving flow with limiting resources. Furthermore, accurate description of the stress interaction between the impeller blades and the growing cell will benefit significant number of bioprocesses involving shear-sensitive cellular materials. The bio-industry will be able to design their product or process with less damage on cells, which will lead to more efficient process and benefit both the health care industry and the customers.

487856 A New Patient-Specific Targeted Pulmonary Drug Delivery Method to Treat Lung Cancer Using E-Cigarette Technology

Abstract:

A New Patient-Specific Targeted Pulmonary Drug Delivery Method to Treat Lung Cancer using E-Cigarette Technology

Keywords: Targeted Pulmonary Drug Delivery; Computational Fluid-Particle Dynamics (CFPD); Precise Treatment Plan

1. Background

Lung cancer strikes 225,000 people every year and accounts for $12 billion in health care costs in the United States. Inhalation of therapeutic drug aerosols is now becoming a novel way to administer micro/nano- particles or vapors with anti-cancer agents to treat lung cancer as well as other lung and systemic diseases. However, due to certain design deficiencies, existing pulmonary drug delivery devices still have poor efficiencies for delivering drugs to the lung tumor sites. Major portions of the aggressive medicine deposit on healthy tissues, which causes severe side effects and induces extra health care expenses. Therefore, it is an urgent need to understand the drug particle dynamics better and develop a revolutionary patient-specific pulmonary drug delivery method and device to significantly improve therapeutic outcomes by improving drug delivery efficacy to target lung tumors while minimizing drug deposition on healthy tissues. Improved drug delivery will reduce unexpected side-effects and overall health care costs. Electronic cigarettes (ECs) show the possibility to be employed as the ready-made prototype of a next-generation device with built-in design advantages for targeted pulmonary drug delivery. Compared with existing drug inhalers, ECs have the following built-in advantages: (1) low inspiratory flow rate to trigger a 100% dose delivery (2.25 L/min), (2) flexible mouthpiece configuration design, and (3) wide choices of drug formulations. The above advantages provide physicians full control of the "air-drug stream" in human respiratory tracts to direct the drugs to the specific sites. As part of the preclinical study, it is important to evaluate the feasibility and effectiveness of the new methodology. Compared to experimental investigations, an accurate and realistic computer simulation model, i.e., a Computational Fluid-Particle Dynamics (CFPD) model governed by natural laws of physics would significantly contribute to reducing the research time and cost, and visualize drug-aerosol transport and deposition patterns with high resolution to advance the fundamental understanding of the underlying physics.

2. Research Objective

Motivated by the in-house preliminary studies and the potential benefits of EC, the primary research objective of this study is to investigate the validity and generality of the new inhalation therapy using EC technologies, and identify the optimal operational conditions for targeted drug delivery to treat lung cancers on a patient-specific level. Integrating the advantages of EC product design and CFPD modeling, a computer-aided targeted drug methodology and protocol is created to precisely control the drug trajectories in the human respiratory tract. Ideally, an individual's CT-scan of his/her lung morphology and diseased site are known *a priori,* and the optimal drug targeting operation can be predicted with the CFPD simulations of drug particle transport and deposition for each individual.

3. Numerical Setup

ANSYS CFX and Fluent 18.0 were employed for the CFPD simulations. A Virtual Human (VH v1.0) was created to restore the realistic drug inhalation scenario. Three human upper airways were selected for the inter-subject variability study to validate the general feasibility of the targeted drug delivery method. Each human upper airway configuration contains an oral cavity, pharynx, larynx, and the same tracheobronchial tree. Inlet centers are all located at (x,y,z)=(0,0,0), and the gravity is along the positive x-direction. Realistic physical properties of pulmonary drug particles treating lung cancers were employed. Specifically, the particle diameter d_p is from 1 to 5 micron, and the particle density ρ_p is 1000 kg/m^3. The inhalation flow rate Q_{in} varies from 2 to 10 L/min. The Euler-Lagrange scheme enhanced by in-house C programs was employed to simulate airflow and drug particle transport and deposition.

4. Results and Discussion

An experimental validated CFPD model was used to simulate drug particle transport and deposition in the three subject-specific human upper airway models with different geometric and operational parameters. These parameters include nozzle diameter, drug releasing velocity, drug releasing angle, EC-like inhaler position, and angle to oral cavity centerline, and breathing patterns. The general feasibility of the deeper lung tumor targeting methodology was validated. The optimal geometric and operational parameters for the new inhaler design were also determined to achieve the most effective targeting performance. A "drug particle release map" was generated for each case to correlate specific targeting site with a specific drug release position by a "backtracking" strategy. Specifically, the drug particle release maps were generated by backtracking the drug trajectories to link their release positions with their deposition locations using the high-resolution data provided by CFPD simulations. It can be observed from the parametric analyses that lower inhalation flow rate is preferred to guarantee the more regulated laminar flow regimes, which is easier for doctors to direct the air-drug streams to a specific site with minimum disturbance by the airflow fluctuation. Furthermore, the high morphological complexity of the human respiratory tract will generate more airflow fluctuation, thereby reducing the drug targeting effectiveness.

5. Summary

This study validated the general feasibility of the "controlled air-drug stream" method to target lung tumors by confirming the significant improvement of drug delivery efficiency at specific sites of lesions. Furthermore, the simulation process paves the way of a new protocol to seek personalized lung cancer treatment on a patient-specific level aided by CFPD simulations. Improved drug delivery will reduce unexpected side-effects and overall health care costs. The proposed research will provide non-invasive high-resolution data for patient-specific drug transport and deposition, a critical step towards translational work needed to maximize the efficiency of animal studies, to develop animal-to-human transfer functions, and to test development in clinical trials. This work is expected to support wider applications for creating new targeting strategies using EC product designs as the next-generation drug delivery device in combating a broad range of lung and systemic diseases.

6. Acknowledgements

The authors are grateful for the financial support by NIH (No. P20GM103648), and the use of ANSYS Software (ANSYS Inc., Canonsburg, PA) as part of the OSU-ANSYS Academic Partnership.

493718 A Computational Fluid Dynamics Analysis of the Carotid Artery

Abstract:

In the human body, the carotid artery provides the brainâ€™s main blood supply carrying oxygenated blood from the heart to the head. As the carotid ascends the neck, it splits into two main branches referred to as the carotid bifurcation. This bifurcation contains a bulge, the carotid sinus, which causes a region of flow reversal to develop during the cardiac cycle. While this phenomena has been medically observed, it is not yet fully understood from a fluid dynamics perspective. The goal of this project was to characterize the velocity profiles and pressure gradients in the carotid artery to better understand the biological significance of the carotid sinus. Computational Fluid Dynamics (CFD - ANSYS Fluent) was used as the link between medicine and engineering, allowing us to quickly and easily visualize blood flow in this region under a variety of important physiological conditions. By creating a biologically accurate model of the carotid and applying relevant boundary conditions, blood flow and pressure contours were observed and applied to medically relevant issues. This work served to help better characterize blood flow in the carotid artery. The simulations run in this experiment illustrated the same velocity profiles seen in previous studies, but in much more detail. The observed pressure contours also provided a cause for the flow reversal, illustrating an adverse pressure gradient within the carotid sinus. Baro- and chemo- receptors in the carotid sinus were also simulated to try and find biological significance for flow reversal. Further refinement is still needed, but these simulations show promise for future work. Applications of this research to atherosclerosis were then studied. Since this is a medically relevant phenomenon that is not fully characterized, this project could serve as a starting point for a much larger study. Future work includes obtaining actual patient scans to use their specific measurements to model actual patient arteries. The carotid bifurcation is very vulnerable to plaque build up, and as the carotid leads to the brain, these plaques breaking off are a major cause of blood clots and stroke. Overall, CFD allowed for a medical phenomenon to be visualized and better characterized with great promise for future clinical significance.

499100 CFD-Based Validation of Drug Metabolism from Clinical Data in 3D Configuration

Abstract:

Successful evaluation of efficiency and safety of new medicines is a time consuming and expensive process. The incorporation of *in silico* methods in preclinical stages of drug development has increased significantly over the recent years as a result of low cost and high capacity data producing abilities. Additionally, recent trends suggest a sway toward a “learn and confirm” process, rather than the traditional multi-phased progression, that could lead to increased success. However, there is currently no systematic validation approach that bridges the gap between experimental data, computational simulation, and patient results by examining metabolic profiles. With the extensive amount of experimental data available, it is vital to find a way to connect small-scale, experimental data to large-scale, clinical data. In this regard, we evaluated the metabolism of the popular drug, acetaminophen (APAP), using kinetic data from literature in a computational simulation. We evaluated the metabolism of APAP to produce three major metabolites well studied in literature. First we converted available pharmacokinetic data on sulfation, glucourinidation and CYP pathway based on CYP 3A4/5, 2D6, 2A6, 2E1 and 1A2 isoforms to per cell basis and to the whole liver. Then simulations were performed using commercially available computational fluid dynamics (CFD) software (Comsol Multiphysics 5.2a) in a smaller 3D porous space that could be cultured in standard 24-well and 96-well plates. We considered contributions from CYP isoforms in four different configurations: i) individual, ii) accounting for activity, iii) accounting for abundance, and iv) grouped average. Using the results we compare the output to clinical data. We also tested the effect of dosage and cell number that are typically varied in cell culture to the output metabolic profiles. These evaluations show that one could test the values of pharmacokinetic data, CYP isoform information, and compare to clinical data. Activity-based isoform analysis largely agrees with clinical data. Also, upon validation these profiles can be utilized in the development of synthetic liver constructs.

On the Numerical Validity of Commonly-Employed Scaffold Mimics for Shear Stress and Flow Field Calculations in Tissue Engineering Models

Olufemi Kadri, Cortes Williams III, Robert L. Shambaugh, V.I. Sikavitsas and R.S.Voronov

NJIT
New Jersey's Science & Technology University
THE EDGE IN KNOWLEDGE

Tissue Engineering (TE)

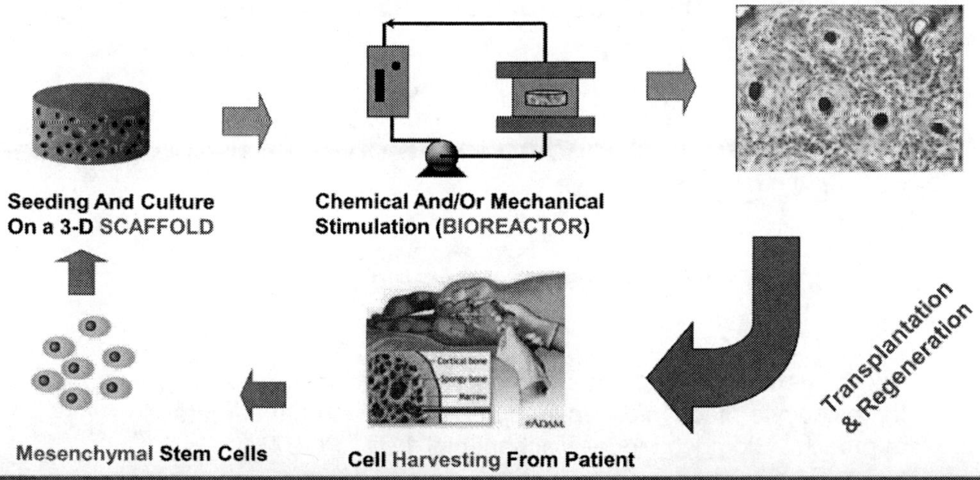

NJIT
New Jersey's Science & Technology University
THE EDGE IN KNOWLEDGE

Fluid shear plays a prominent role

- **Mechanical Stimulation of Cells by Fluid Shear:**

✓ Perfusion of culture media.

✓ Direct mechanical compression.

NJIT New Jersey's Science & Technology University THE EDGE IN KNOWLEDGE

Target stress range

- Physiologic stress experienced by cells in canaliculi, **1 - 30 dynes/cm^2**

- Stress corresponding to the onset of tissue formation, **1 - 22 dynes/cm^2**

- Not all stress levels are desirable.

Verbruggen, S. W., T. J. Vaughan and L. M. McNamara (2014). "Fluid flow in the osteocyte mechanical environment: a fluid-structure interaction approach." Biomech Model Mechanobiol **13**(1): 85-97.

NJIT New Jersey's Science & Technology University THE EDGE IN KNOWLEDGE

Computational methodology

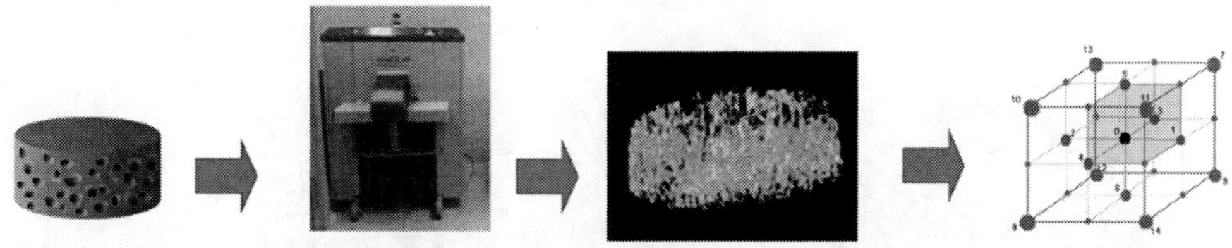

Voronov, R. (2010). Fluid Shear Stress and Nutrient Transport effects via Latiice-Boltzmann and Lagrangian Scalar Tracking Simulations of cell culture media perfusion through artificial bone engineering constructs imaged with microCT. PhD Doctoral Dissertation, University of Oklahoma.

This study compares two modeling approaches

Differ by:

- Simulation domain size
- Boundary conditions

Approach 1

Models Representative Volume Element(RVE) within scaffold

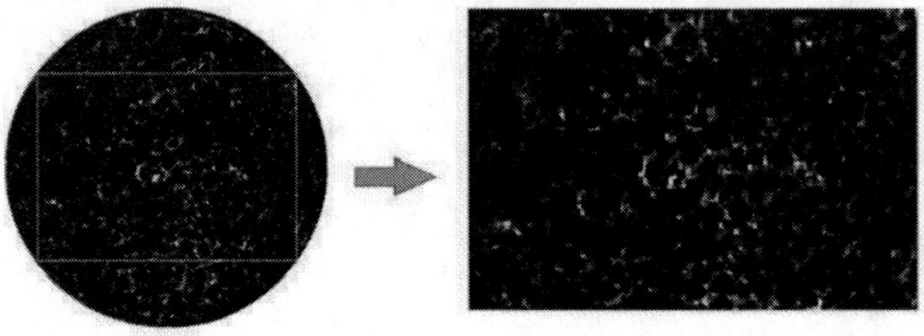

Periodic boundary conditions are applied in all directions to approximate the full scaffold domain.

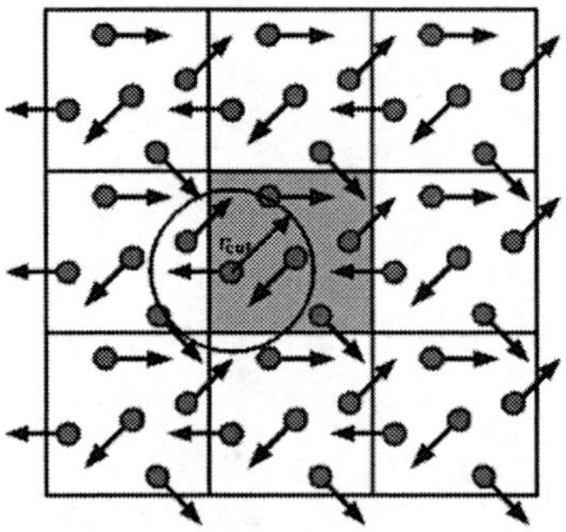

Common due to low computation requirements

Image from http://www.compsoc.man.ac.uk/~lucky/Democritus/Theory/pbc-mi.html

Approach 2

- Simulation domain is scaffold placed inside a pipe.
- Uniform flow at inlet and zero pressure at outlet.

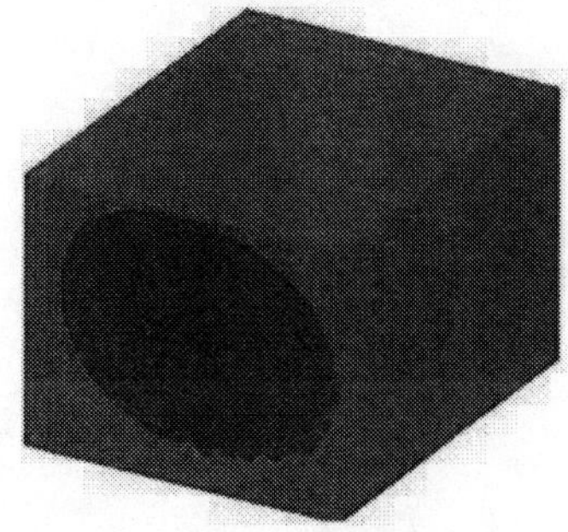

Accurate but computationally expensive!

Scaffold architecture is complex!

FIBER SALT LEACHED

FLOW FIELD CALCULATION: Lattice Boltzmann method

$$f_i(\vec{x}+\vec{e}_i\Delta t, t+\Delta t) - f_i(\vec{x},t) = \Omega_i(\vec{x},t) \pm ff$$

$f_i(\vec{x},t)$ **Particle distribution function**

$\vec{e}_i$ **Directional velocity vector**

Ω_i **BGK Collision operator**

ff **Forcing factor**

$$\rho = \sum_{i=0}^{n} f_i$$

$$\rho \vec{U}_i = \sum_{i=0}^{n} f_i \vec{e}_i$$

NJIT New Jersey's Science & Technology University THE EDGE IN KNOWLEDGE

Simulation details

- Ten scaffolds total:

Six salt leached and Four fibers

- Flow rate of 0.15mL/min

- Shear stress is estimated:

$$\sigma = \upsilon \frac{1}{2}\left(\nabla \vec{U} + \nabla \vec{U}^T\right)$$

NJIT New Jersey's Science & Technology University THE EDGE IN KNOWLEDGE

Results

Shear Stress Statistics(dynes/cm^2)

Scaffold type	Porosity	Average	Maximum
Salt leached	0.8719		
Full		6.30E-2	1.601
RVE		5.17E-2	5.19E-1
Fiber	0.9472		
Full		1.22E-1	3.4201
RVE		5.78E-2	7.8E-1

Underpredicted average stresses by factors of:

- 1.22 for salt leached
- 2.1 for fiber

*Results shown for one of fiber and salt leached

NJIT New Jersey's Science & Technology University THE EDGE IN KNOWLEDGE

Similar stress patterns for salt leached!

Stress maps: Full domain vs RVE

NJIT New Jersey's Science & Technology University THE EDGE IN KNOWLEDGE

Stress patterns are similar but greater underestimation for fiber

Sectional view

SALT LEACHED

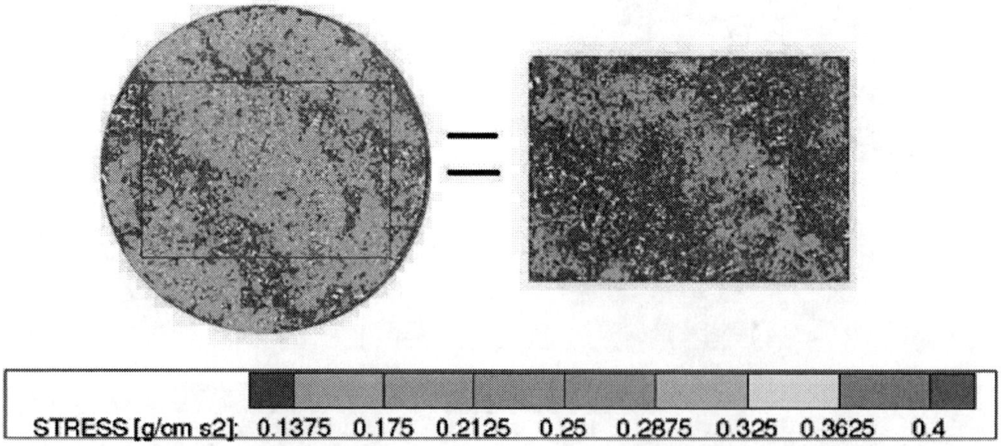

FULL DOMAIN ACTUAL RVE

NJIT
New Jersey's Science & Technology University
THE EDGE IN KNOWLEDGE

FIBER

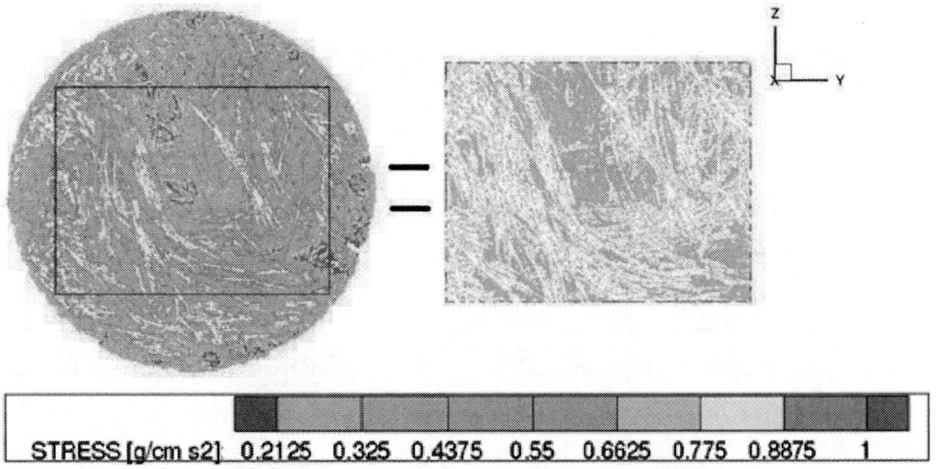

FULL DOMAIN ACTUAL RVE

- Similar stress distributions for salt leached
- Less similar for fiber

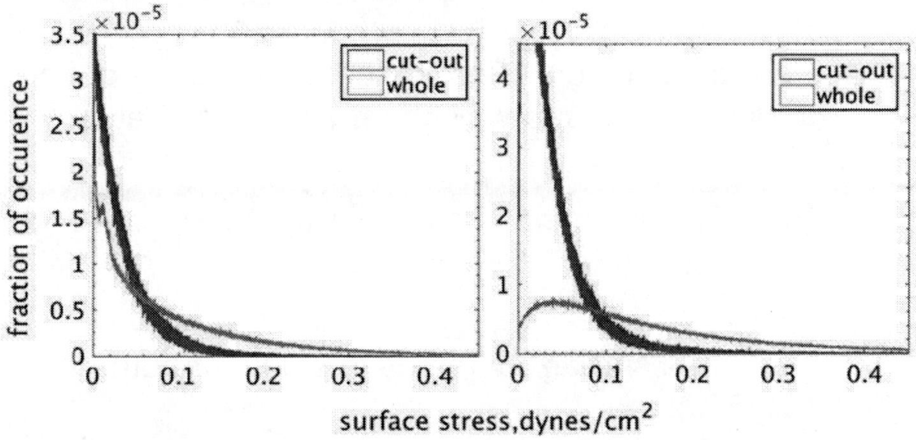

Salt Leached Fiber

Conclusions

- RVE approximations underestimate wall shear stresses.
- The approximations yielded better results for salt leached scaffolds.
- Although stress patterns are similar, RVE approximations are not able to completely characterize heterogeneity in the scaffolds.

NJIT New Jersey's Science & Technology University THE EDGE IN KNOWLEDGE

FUTURE WORK

- Perform study on different fiber scaffold cut-out sizes to obtain improved wall shear stress estimates.
- Run simulations on scaffolds with grown cells and neo-tissue.
- Perform a similar comparative study for nutrient transport.

NJIT New Jersey's Science & Technology University THE EDGE IN KNOWLEDGE

Acknowledgments

- Dr. Roman Voronov
- University of Oklahoma Supercomputing Center for Education & Research (OSCER).
- Texas Advanced Computing Center (TACC).

NJIT New Jersey's Science & Technology University THE EDGE IN KNOWLEDGE

THANK YOU!

NJIT New Jersey's Science & Technology University THE EDGE IN KNOWLEDGE

498228 A Computational Multiphase Flow Model to Predict the Transport and Deposition of Inhaled Flu Virus-Laden Droplets in Human Respiratory Tracts for Early Infection Diagnosis

Abstract:

Keywords: Influenza Virus; Computational Fluid-Particle Dynamics (CFPD); Oxidative Stress; Noninvasive Early Diagnosis

Introduction

Influenza A virus (IAV) is highly transmissible as one of the leading causes of pulmonary infections such as pneumonia. 5 to 20% of the population in the United States contract influenza every year. Three circulating subtypes, i.e., type A/H1N1, type A/H3N2, and type B, can infect humans and cause massive global epidemics. IAVs pass among individuals through the air in droplets when someone sneezes or coughs. The airborne transmission of infectious substances expelled by the respiration system of an infected patient is commonly known to be the key contagion mechanism. Specifically, in the elderly (>65 years), in infants (0-4 years), and in people with chronic diseases, influenza is associated with especially high mortality. Influenza virus replicates in the epithelial cells throughout the respiratory tree, with the virus being recoverable from both the upper and lower respiratory tract of people naturally or experimentally infected. The epithelium of the trachea, bronchi and the pulmonary alveoli were primary sites of influenza viral replication. Moreover, lower airway infections can result in flooding of the alveolar compartment, development of acute respiratory distress syndrome and death from respiratory failure. Current approaches to diagnosis and surveillance rely heavily on clinical case definitions and a variety of laboratory assays, which are too slow. Therefore, early diagnosis of IAV infection is ***significant*** to improve treatment outcomes and decrease the use of antibiotics, especially for children. It can also aid the surveillance for the assessment of the current contamination situation and control the pandemic disease. The ***long-term goal*** is to develop a noninvasive and accurate early diagnostic methodology which will be able to detect the infection before the onset of symptomatic disease. The method should also be sensitive and specific, which can be used to identify the virus type, a primary infected region in the respiratory tracts, and the infection severity. Oxidative stress plays a key pathophysiological role. Influenza Infection may result in increased oxidative stress and induced increased excretion of VOC biomarkers in the breath such as alkanes and methylated alkanes. Enlightened by this fact, the ***research objective*** of this study is to virtually design and test the noninvasive diagnostic framework using a new multiscale computational model.
The ***central hypothesis*** is that the exhaled VOC pattern variations can reflect the details of virus infection status in human pulmonary tracts. To virtually test the feasibility of the proposed diagnosis, the multiscale numerical model should be capable of capturing the transport and deposition dynamics of the inhaled virus-laden droplets, and the after-deposition dynamics (or within-host dynamics) of the virus to track the change of VOC biomarkers in human respiratory tracts. However, it is still a lack of effective tools to predict the human uptakes of the flu virus due to different exposure conditions, Significant knowledge gaps still exist regarding the quantitative relationships between exhaled VOC biomarkers and the influenza infections. Therefore, in this study, a computational multiphase flow model

based on a Euler-Euler scheme is developed to predict the transport and deposition of inhaled flu virus-laden droplets in human respiratory tracts. The amount of pathogens inhaled and the regional deposition in upper airways and deeper airways are quantitatively predicted. Additionally, a compartment model is also developed for the within-host dynamics of the virus in epithelial cells.

The Multiscale Computational Model

A Computational Fluid-Particle Dynamics (CFPD) model is developed based on the Euler-Euler scheme to simulate the transport and deposition of flu virus-laden droplets transport in human respiratory systems. Realistic transient breathing conditions (Gupta et al., 2010) and indoor ventilation conditions will be applied (Ge et al., 2013). Physiochemical properties of virus-laden droplets were retrieved from realistic cough and sneeze aerosols (Gupta et al., 2010; Kwon et al., 2012; Thatiparti et al., 2017). Simulations will be performed in subject-specific human respiratory configurations from mouth to G6, which are reconstructed from CT/MRI data. Specifically, droplet diameter ranges from 0.4-30 μm, and droplet density is 1.388 kg/m^3. Numerical solutions with appropriate boundary conditions were achieved by using a user-enhanced, commercial finite-volume based program, i.e., ANSYS Fluent and CFX 18.0 (ANSYS Inc., Canonsburg, PA). The shear stress transport (SST) transition model was employed to solve the laminar-to-turbulence airflow fields.

To bring the simulation process to the health endpoint, we will also establish a compartment model to describe the after-deposition dynamics in epithelial cells, i.e., within-host dynamics (Pawelek et al., 2012). It will include tracking the number changes of uninfected epithelial cells, infected epithelial cells, refractory cells, free virus, and IFN. Major mechanisms include infection, virus production, epithelial cell apoptosis, IFN-induced antiviral effect, as well as the concentrations of VOC biomarkers excreted.

Proposed Results Mainstream and secondary flow patterns and their influence on virus-laden transport will be visualized and discussed. Parametric analyses will be performed for the impacts of morphological variations and breathe patterns on virus local deposition in human respiratory tracts. The relationship between direct virus deposition and lower airway infection will be investigated based on the quantitative deposition data. The time courses of the flu virus and VOC biomarkers excreted due to the infection will also be tracked using the compartment model. Furthermore, inter-subject variability study will be performed numerically to evaluate the generality of the proposed diagnostic methodology.

Summary

Using the experimentally validated multiscale computational model, the lung aerosol dynamics simulation with realistic exposure conditions (coughing and indoor ventilation) provides us accurate and high-resolution data of virus transport and deposition in the human respiratory tract, which will be informative for the determination of the relationship between exposure, intakes and uptakes of the virus. The quantitative data pave the way for multiscale modeling include the within-host dynamics of the virus. The future multiscale model will provide a noninvasive and time-saving numerical tool guide the noninvasive diagnosis of the early detection of different influenza infection by providing information for virus transport, deposition, production, and elimination, as well as the VOC biomarker generation dynamics. It will also significantly contribute to: (1) optimize the indoor ventilation to control the transmission of the airborne infectious diseases mechanically; and (2) Evaluate the effectiveness of vaccines to cure pulmonary diseases.

Acknowledgements

The authors are grateful for the financial support by NIH (No. P20GM103648), and the use of ANSYS Software (ANSYS Inc., Canonsburg, PA) as part of the OSU-ANSYS Academic Partnership.

References

Ge, Q., Li, X., Inthavong, K., & Tu, J. (2013). Numerical study of the effects of human body heat on particle transport and inhalation in indoor environment. *Building and Environment*, *59*, 1-9.

Gupta, J. K., Lin, C. H., & Chen, Q. (2010). Characterizing exhaled airflow from breathing and talking. *Indoor air*, *20*(1), 31-39.

Kwon, S. B., Park, J., Jang, J., Cho, Y., Park, D. S., Kim, C., ... & Jang, A. (2012). Study on the initial velocity distribution of exhaled air from coughing and speaking. *Chemosphere*, *87*(11), 1260-1264.

Pawelek, K. A., Huynh, G. T., Quinlivan, M., Cullinane, A., Rong, L., & Perelson, A. S. (2012). Modeling within-host dynamics of influenza virus infection including immune responses. *PLoS Comput Biol*, *8*(6), e1002588.

Thatiparti, D. S., Ghia, U., & Mead, K. R. (2017). Computational fluid dynamics study on the influence of an alternate ventilation configuration on the possible flow path of infectious cough aerosols in a mock airborne infection isolation room. *Science and Technology for the Built Environment*, *23*(2), 355-366.

501881 Optimal Mechano-Electric Treatment of Ventricular Fibrillation

Abstract:

Ventricular fibrillation (VF) [1] is a life-threatening cardiac arrhythmia (irregular heartbeat) resulting in uncoordinated contraction of the ventricular myocardium and causing a failure of the heart to pump blood to the rest of the body, this leads to sudden cardiac death, which is one of the most common causes of death in the industrialized world. In its turn, contraction of the heart causes cardiac tissue deformation which feeds back on the wave propagation and affects electrophysiological properties via mechano-electric feedback (MEF) [2]. Electrical-excitation and contraction of the heart can be linked by an electromechanical model of cardiac tissue.

Strong electrical shocks that are given to the heart to reset the heartbeat back to its normal beat have remained the most reliable approach to terminate VF, this is known as defibrillation. This electrical therapy can damage tissue and leave scars which are considered as major sources of substrates for cardiac arrhythmias. These electric-based realization algorithms have not considered mechanical properties of cardiac tissue, despite the fact that mechanical deformation is shown to influence electrical activity of the heart tissue. Many studies have shown the importance of the MEF and, for instance, mechanical impact on chest, in the area directly over the heart, can either cause Commotio Cordis when chest receives a blow, or terminate cardiac arrhythmia such as VF using the human fist on the chest during precordial thump. We [5] recently presented a novel mechanical perturbation algorithm to control cardiac alternans [3], which is a perturbation in the heart rhythm, in relevantly sized cardiac tissues using an electromechanical model that couples the mathematical models of cardiac mechanics and electrophysiology.

Controlling the complex spatio-temporal dynamics underlying life-threatening cardiac arrhythmias is extremely difficult [4]. Computational simulation is increasingly recognized as an important tool for the investigation and termination of ventricular arrhythmias, including fibrillation [4]. In this work a strongly realistic electromechanical model where the excitable behavior is modeled using a bidomain description of cardiac tissue. The bidomain model consists of a system of coupled parabolic and elliptic PDEs for two potentials in the cardiac muscle, coupled with a nonlinear system of ODEs describing the ionic currents flowing across the cardiac membrane. The Ten Tusscher [5] model is used to represent the electrophysiological properties, and the mechanical properties of the passive myocardium are described using the exponential Guccione passive elasticity model [6] which is the most cited model in the literature. The active tension that couples the electrophysiological model with the cardiac mechanics model is generated using the Niederer-Hunter-Smith [7] model which is advanced model. In this work, we will explore the possibility of Cardiac Defibrillation by combining the electrical shock and mechanical stimuli delivered to the heart in order to minimize defibrillation energy and avoid adverse side effects associated with high voltage shocks. This is the first attempt to combine electro-physiologically relevant cardiac models of electrical wave propagation and contractility of cardiac tissue in a synergistic effort to terminate VF.

[1]- Zevitz, M.E. Ventricular fibrillation [online]. eMedicine. http://www.emedicine.com/med/topic2363.htm (2004).

[2] M.J. Lab, Mechanoelectric feedback (transduction) in heart: concepts and implications, Cardiovasc.Res, vol.32, p.

3-14, 1996.

[3] - Azzam Hazim, Youssef Belhamadia, Stevan Dubljevic, Control of cardiac alternans in an electromechanical model of cardiac tissue, Computers in Biology and Medicine 63 (2015) 108-117.

[4]- Alfio Quarteroni. Modeling the Heart and the Circulatory System. Springer, New York, 2015

[5] Ten Tusscher KH, Panfilov AV. Alternans and spiral breakup in a human ventricular tissue model. Am J Physiol Heart Circ Physiol 291: H1088-H1100, 2006.

[6] J. M. Guccione, A. D. McCulloch, L. K. Waldman, Passive material properties of intact ventricular myocardium determined from a cylindrical model, J. Biomech. Eng. 113 (1) (1991) 42-55.

[7] S. A. Niederer, P. J. Hunter, and N. P. Smith, A quantitative analysis of cardiac myocyte relaxation: a simulation study, J Biophysics, vol. 90, p. 1697-722, 2006.

489748 Distributed Model Predictive Control of Complex Plants: A Systematic Study of Decomposition Effects

Abstract:

The decomposition of a large scale complex plant into constituent subsystems is an essential step for the synthesis of distributed model predictive control (DMPC) architectures, which directly affects the closed-loop performance and computational effort [1-3]. While there have been some initial attempts to address the decomposition problem for specific classes of systems [4, 5], developing a systematic framework for optimal decomposition into subsystems with desirable interaction characteristics is still an open problem.

From a network theory perspective, this problem can be posed as identifying weakly-connected subsystems whereby the variables of each subsystem are strongly connected. Such a point of view has been adopted to generate hierarchies of system decompositions with different degrees of decentralization based on input-output connectivity, following an agglomerative or divisive clustering procedure [6]. An alternative method based on the maximization of the modularity of the system equation graph was recently proposed to decompose the network into "communities" that include inputs, states, and outputs [7]. The latter approach minimizes the interactions between the resulting communities and appears well suited for distributed control.

In this work, we provide a systematic evaluation of the impact of system decomposition on DMPC designs for a benchmark integrated plant. An iterative DMPC architecture is applied to a reactor-separator system containing two continuous stirred tank reactors and a vapor-liquid separator. The performance and computational effort of implementing the iterative DMPC for different decompositions identified through community detection methods [6, 7] or by intuition [3, 8], are analyzed and compared to those obtained by a centralized model predictive control (CMPC) design. State/output regulation and tracking case studies are considered. In each case, it is documented that the optimal decomposition obtained by the community detection method, which minimizes the interactions between the subsystems, enables closed-loop performance close to that of the CMPC, while reducing the computation effort significantly.

[1] P.D. Christofides, R. Scattolini, D. Munoz de la Pena, and J. Liu. Distributed model predictive control: A tutorial review and future research directions. Computers & Chemical Engineering, 51:21–41, 2013.

[2] B.T. Stewart, S.J. Wright, and J.B. Rawlings. Cooperative distributed model predictive control for nonlinear systems. Journal of Process Control, 21:698–704, 2011.

[3] P.D. Christofides, J. Liu, D. Munoz de la Pena. Networked and Distributed Predictive Control. Springer-Verlag, London, 2011.

[4] W. Al-Gherwi, H. Budman, and A. Elkamel. Selection of control structure for distributed model predictive control in the presence of model errors. Journal of Process Control, 20:270–284, 2010.

[5] N. Motee and B. Sayyar-Rodsari. Optimal partitioning in distributed model predictive control. In Proceedings of the American Control Conference, 5300–5305, Denver, CO, 2003.

[6] S. Heo and P. Daoutidis. Control-relevant decomposition of process networks via optimization-based hierarchical clustering. AIChE Journal, 62(9):3177–3188, 2016.

[7] S.S. Jogwar and P. Daoutidis. Community-based synthesis of distributed control architectures for integrated networks. Chemical Engineering Science, Submitted, 2016.

[8] M.J. Tippett, J. Bao. Distributed model predictive control based on dissipativity. AIChE J., 59(3):787–804, 2013.

500103 Dynamic Real-Time Optimization of Distributed MPC Systems Using Rigorous Closed-Loop Prediction

Abstract:

Dynamic real-time optimization (DRTO) has been developed as a supervisory layer within the process control architecture that computes set-point trajectories for an underlying regulatory control system based on economic optimization and a dynamic process model. Unlike traditional steady-state RTO systems, it does not require the plant to be at steady state prior to execution, thus making it well-suited for systems with recycle, with slow dynamics and which undergo frequent grade changes. Tosukhowong et al. propose a DRTO formulation to capture slow dynamics of a system using a linear model [1], while Kadam and Marquardt develop a DRTO formulation that estimates uncertain parameters and performs fast updates based on a sensitivity technique [2]. Most DRTO systems are based on an open-loop model of the process that does not consider the dynamic effects of the underlying regulatory controller. However, DRTO formulations based on the closed-loop prediction of the plant under constrained MPC have recently been developed [3, 4].

Large process plants are generally not controlled using a single, monolithic MPC system, with a distributed MPC architecture favored to take the modularity of a large process into consideration and provide additional flexibility [5]. Thorough reviews on the architecture and formulations of distributed MPCs have been conducted in [6, 7, 8]. Camponogara et al. propose a coordination scheme that allows communication among MPC subsystems which is allowed once for each local problem and multiple times before implementation of the control actions [9]. Liu et al. introduce distributed Lyapunov-based MPCs that adopt a sequential and iterative approach with one or bi-directional strategies, under the assumption of known state feedback for all controllers [10]. Stewart et al. present a cooperative scheme for distributed MPCs that ensures the closed-loop stability of the system upon termination of iterations for all subsystems [11].

In this work, a closed-loop DRTO formulation is developed for the economic coordination of distributed MPC systems by assigning computed set-point trajectories based on the prediction of future interactions between the plant dynamics and distributed MPCs. This closed-loop formulation naturally results in a multilevel optimization problem whose primary optimization problem computes the reference trajectories for each MPC subsystem based on a full plant dynamic model under an economic objective, and whose MPC optimization sub-problems compute optimal control actions to be applied to the full plant dynamic model based on partial models each of which contains model equations relevant to its associated subsystem. Such a multilevel optimization problem can be transformed into a single-level optimization problem by applying the Karush-Kuhn-Tucker (KKT) optimality conditions to transform the MPC optimization sub-problems into sets of algebraic constraints embedded in the primary optimization problem, resulting in a single mathematical program with complementarity constraints (MPCC).

Relevant linear and nonlinear case studies are conducted to explore two modes of operation of the DRTO formulation: target tracking and economic optimization. Results show that for economic optimization, due to the flexibility of set-point trajectories and ability to foresee future interactions, the closed-loop DRTO formulation has the capacity to coordinate distributed MPCs to achieve similar economic performance when compared to the centralized counterpart, and for target tracking, each output is able to achieve the desired target value rapidly compared to both centralized and decentralized MPC cases.

Due to the large scale and level of detail modeled for each subunit of a process, approximation techniques for the closed-loop DRTO formulation would be useful in order to efficiently coordinate multiple distributed MPCs at various locations. Strategies of this type are under development, with economic and computational performance comparisons planned.

References

[1] T.Tosukhowong, J. Lee, J. Lee and J. Lu, "An introduction to a dynamic plant-wide optimization strategy for an integrated plant," *Comp. Chem. Eng.* , vol. 29, no. 1, pp. 199-208, 2004.

[2] J. Kadam and W. Marquardt, "Sensitivity-based solution updates in closed-loop dynamic optimization," in *Proceedings of the 7th International Conference on the Dynamics and Control of Process Systems - DYCOPS 7, Boston, Massachusetts, USA*, Boston, 2004.

[3] M. Z. Jamaludin and C. L. E. Swartz, "Approximation of Closed-loop Prediction for Dynamic Real-time Optimization Calculations," *Computers and Chemical Engineering,* vol. 103, pp. 23-38, 2017.

[4] M. Z. Jamaludin and C. L. E. Swartz, "Dynamic Real-time Optimization with Closed-loop Prediction," *AIChE Journal,* Accepted Author Manuscript. doi:10.1002/aic.15752, 2017.

[5] G. Pannocchia, "Distributed Model Predictive Control"," in *Encyclopedia of Systems and Control*, London, Springer London, 2013, pp. 1-9.

[6] R. Scattolini, "Architectures for distributed and hierarchical Model Predictive Control â€“ A review," *J. Process Control,* vol. 19, no. 5, pp. 723-731, 2009.

[7] P. Christofides, R. Scattolini, D. MuÃ±oz de la PeÃ±a and J. Liu, "Distributed model predictive control: A tutorial review and future research directions," *Comp. Chem. Eng.,* vol. 51, pp. 21-41, 2013.

[8] J. Rawlings and B. Stewart, "Coordinating multiple optimization-based controllers: New opportunities and challenges," *J. Process Control,* vol. 18, no. 9, pp. 839-845, 2008.

[9] E. Camponogara, D. Jia, B. Krogh and S. Talukdar, "Distributed model predictive control," *IEEE Control Systems,* vol. 22, no. 1, pp. 44-52, 2002.

[10] J. Liu, X. Chen, D. Muñoz de la Peña and P. Christofides, "Sequential and iterative architectures for distributed model predictive control of nonlinear process systems," *AIChE J.,* vol. 56, no. 8, pp. 2137-2149, 2010.

[11] B. Stewart, A. Venkat, J. Rawlings, S. J. Wright and G. Pannocchia, "Cooperative distributed model predictive control," *Syst. Control Lett.,* vol. 59, no. 8, pp. 460-469, 2010.

494837 Graph-Theoretic Approach for the Synthesis of Distributed Control Architecture

Abstract:

Designs with multiple material recycle and energy integration loops are quite common in modern chemical plants. Such designs promise favourable economics, however, their operation and control is quite challenging due to strong interactions between process units. For such networks, traditional decentralized controllers have limited effectiveness due severe control loop interactions. On the other hand, the design and implementation of a fully centralized controller is impractical due to large system size. To this end, distributed control has been exploited as an efficient control framework [1]. Most of the work in this area has addressed issues like feasibility, stability and optimality of the control solution for a heuristically selected architecture.

In this talk, we present a systematic framework for the synthesis of distributed control architecture using principles of graph theory. To this end, the control problem of an integrated network is abstracted as an equation graph with the various input, output and state variables as nodes and their (static or dynamic) interdependence as edges. In this setting, the distributed architecture synthesis problem is equivalent to decomposing the equation graph into â €˜communitiesâ€™ with strong connectivity between nodes belonging to each community and weak inter-community edges. Initially, it is considered that all the edges are equally important. The distributed structure obtained in this case thus results in minimum inter-controller interactions. Subsequently, strength of interaction between a variable pair is used as an edge weight. The distributed architecture obtained in this case results in optimal input-output pairing.

The direct decomposition of the equation graph allows for deriving state-space descriptions for the individual sub-controllers. The framework uses minimum system information and is highly scalable due to advances in community detection algorithms. The effectiveness of the proposed framework is illustrated with the help of a case study on an integrated network.

References:

[1] Liu, J., Munoz de la Pena, D., Christofides, P. D. Distributed model predictive control of nonlinear process systems. AIChE J. 55 (5), 1171-1184.

494271 Event-Based Networked Control of Distributed Process Systems with Sensor-Controller Communication Failures

Abstract:

The design and analysis of event-based control systems have emerged in recent years as the focus of significant attention in both the academic and industrial circles in networked control research. Interest in event-based control approaches stems, in part, from their promising capabilities to achieve the desired levels of control system performance with reduced utilization of the resources of the sensor- controller communication medium. Maintaining a balance between control performance, on the one hand, and extent of network utilization, on the other, is an important objective in the design of any networked control system, particularly in the context of applications involving large-scale sensor and actuator networks, where reducing network utilization can lead to savings in the energy expenditure of battery-powered wireless devices, thereby prolonging their service life and enabling their deployment in a wide range of technological applications [1].

A key idea in event-triggered model-based control is to compute the control action based on the predictions of an approximate model of the plant, and to suspend or restore sensor-controller communication in response to events that tie the model estimation error to a certain stability threshold. Sensor-controller communication is established only to update the model state when the prescribed stability threshold is breached; otherwise communication is suspended and network resources are saved. Compared with time-triggered communication strategies (e.g., [2]–[4]), event-triggered approaches are more responsive to changes in operating conditions and generally lead to lower overall network load and improved system robustness since they allow a more timely feedback of the potential changes in process operating conditions.

A close inspection of the available literature on this topic shows that while event-triggered control has been widely studied in the context of lumped parameter systems (e.g., see [5]– [7] for some results and references in this area), it has received only limited attention in the context of distributed parameter systems. This is an important gap that needs to be addressed given the fact that many important practical applications (e.g., transport-reaction processes and fluid flows) are characterized by spatial variations and are modeled by Partial Differential Equations (PDEs). Efforts aimed at the development of a framework for event-triggered control of spatially distributed systems have been initiated in a number of recent studies (e.g., [8]–[10]), where the focus has been on addressing various practical implementation issues such as performance optimization, the accommodation of control actuator faults and the handling of limited state measurements.

An important consideration in the implementation of the above approaches is the assumption that the communication medium is available whenever sensor-controller communication is needed. In other words, the communication medium is assumed to reliably transmit the broadcasted data from the sensor to the controller whenever needed. This assumption needs to be re-examined, especially in the context of wireless sensor and actuator networks where transmission failures, communication outages and environmental impacts are common and can interfere with the implementation of the event-triggered control strategy by precluding the execution of timely model updates. This consideration calls for an assessment of the event-triggered control system's robustness to failures in the communication medium, and the development of appropriate measures to help mitigate the impact of these failures on the stability and performance of the networked closed-loop system.

Motivated by these considerations, the objective of this work is to develop a model-based framework for the analysis and handling of sensor-controller communication failures in event-triggered networked control of a class of spatially-distributed processes. We focus on distributed processes whose dominant dynamics are finite-dimensional, but uncertain, with a finite number of spatially- distributed networked control actuators and measurement sensors. Initially, a model-based networked output-feedback controller is designed based on an approximate low-order model that captures the slow dynamics of the infinite-dimensional system. The controller is implemented using an event-triggered sensor-controller communication policy, which determines the times when the sensor data need to be transmitted to the controller to update the model state based on a suitably-designed stability threshold. An assessment of the networked control systems robustness to failures in the communication medium is carried out using an on-line forecasting approach that tracks the growth of the estimation errors during times of communication outages. An explicit characterization of the allowable down time that communication losses can be tolerated without compromising closed-loop stability is obtained in terms of the various control system design parameters, and used to devise various accommodation measures that enhance the networked closed-loop systems robustness against communication losses. A simulated diffusion-reaction process example is used to illustrate the developed approach.

References:

[1] Christofides, P.D., Davis, J., El-Farra, N.H., Clark, D., Harris, K., and Gipson, J. "Smart plant operations: Vision, progress and challenges," *AIChE J.*, 53(11), 2734–2741, 2007.

[2] Y. Sun and N. H. El-Farra, "Quasi-decentralized model-based net- worked control of process systems," *Comp. & Chem. Eng.*, vol. 32, pp. 2016–2029, 2008.

[3] E. Garcia, P. J. Antsaklis, and L. A. Montestruque, Model-Based Control of Networked Systems, Systems & Control: Foundations & Applications. Switzerland: Springer International Publishing, 2014.

[4] C. Zheng, M. Tippett, J. Bao, and J. Liu, "Dissipativity-based distributed model predictive control with low rate communication," *AIChE J.*, vol. 61, pp. 3288–3303, 2015.

[5] X. Wang and M. Lemmon, "Event-triggering in distributed networked control systems," *IEEE Trans. Automat. Contrr.*, vol. 56, no. 3, pp. 586–601, 2011.

[6] W. Heemels, M. Donkers, and A. Teel, "Periodic event-triggered control for linear systems," *IEEE Trans. Automat. Contr.*, vol. 58, no. 4, pp. 847–861, 2013.

[7] E. Garcia, Y. Cao, H. Yu, P. Antsaklis, and D. Casbeer, "Decentralized event-triggered cooperative control with limited communication," *International Journal of Control*, vol. 86, no. 9, pp. 1479–1488, 2013.

[8] Z. Yao and N. H. El-Farra, "Resource-aware model predictive control of spatially distributed processes using event-triggered communication," in *Proceedings of 52th IEEE Conference on Decision and Control*, Florence, Italy, 2013, pp. 3726–3731.

[9] D. Xue and N. H. El-Farra, "Networked event-triggered control of spatially distributed processes using a dual-mode communication strategy," in *Proceedings of 54th IEEE Conference on Decision and Control*, 2015, pp. 1899–1904.

[10] D. Xue and N. H. El-Farra, "Actuator fault-tolerant control of networked distributed processes with event-triggered sensor-controller communication," in *Proceedings of American Control Conference*, Boston, MA, 2016, pp. 1661–1666.

493385 Distributed Safeness Index-Based Predictive Control for Enhanced Process Operational Safety

Abstract:

Though industry has traditionally decoupled the designs of the process control and safety systems, recent work [1] has demonstrated that coordinating the control and safety systems through a common metric termed the Safeness Index can enhance process operational safety. The Safeness Index is a process-specific metric that is a function of the closed-loop state and therefore mathematically formalizes a systems approach to process safety [2]. It is developed from information on past process accidents, first-principles models, industrial safety studies, and process operating data. Thresholds on this index are used to set constraints in model predictive control and to trigger actions of the safety system. An economic model predictive control (EMPC) design with Lyapunov-based stability constraints and Safeness Index-based constraints (termed Safeness Index-based LEMPC) was developed in [1] with a centralized control architecture. Distributed control designs (e.g., sequential and iterative [3], [4]) can be beneficial for reducing the computation time of EMPC and are thus an attractive method for implementing Safeness Index-based LEMPC that may enhance its acceptance by industry.

In this work, we investigate the development of sequential and iterative distributed Safeness Index-based LEMPC designs. Due to the hard Safeness Index-based constraints that require the EMPC to maintain the process state within a region where the Safeness Index is below a given threshold, the distributed Safeness Index-based LEMPC design poses unique feasibility challenges that suggest that inputs should be partitioned such that those inputs that most affect the value of the Safeness Index are computed by one distributed controller. Feasibility considerations also affect the order in which the distributed controllers are recommended to be evaluated in a sequential distributed Safeness Index-based LEMPC design. Implementation strategies for both sequential and iterative distributed Safeness Index-based LEMPC are developed that are proven to maintain closed-loop stability in a bounded region of state-space and to always drive the closed-loop state into the region in state-space where the Safeness Index is below a threshold whenever it exits this region. A chemical process example is used to demonstrate the proposed control designs.

[1] Albalawi A, Durand H, Christofides PD. Process operational safety using model predictive control based on a process safeness index. Computers & Chemical Engineering. in press.

[2] Leveson NG, Stephanopoulos G. A system-theoretic, control-inspired view and approach to process safety. AIChE Journal. 2014;60:2-14.

[3] Liu J, Chen X, Muñoz de la Peña D, Christofides PD. Sequential and iterative architectures for distributed model predictive control of nonlinear process systems. AIChE Journal. 2010;56:2137-2149.

[4] Scattolini R. Architectures for distributed and hierarchical model predictive control - A review. Journal of Process Control. 2009;19:723-731.

483990 Two-Point Constraint Control of Water Quality in Distribution Networks

Abstract:

Providing potable drinking water with sufficient free residual chlorine (used as disinfectant worldwide) to prevent microbial regrowth and contaminant by-product formation is of utmost importance for municipal authorities. The work presented in this paper aims towards water quality control as it transports through water distribution networks. Maintaining free residual chlorine concentration levels in water is a critical task as chlorine reacts while it transports through the pipe network with incoming dissolved organic matter (DOM) in bulk flow as well as in biofilms present at the pipe surfaces. Presence of excessive chlorine in water causes the generation of carcinogenic disinfectant by-products such as tri-halo methane (THMs). On the other hand, lower chlorine levels in water results in microbial contamination. Hence, this is a two-point constraint control problem of maintaining free residual chlorine levels under the maximum and minimum bounds by suitably optimizing the disinfectant dosage at booster stations. However, transport delay, and complex interrelations present amongst the nodes in large water distribution network makes it difficult to design a global feedback control system. Therefore, in this work we have proposed to decentralize the water distribution system by clustering demand nodes of the network based on knowledge of interactions between manipulating and control variables using partial correlation analysis. In such multivariable control, due to the presence of interactions, the stability of system might get affected. Knowing the feasibility, stability and integrity of control structure by clustering variables in the above manner constitutes an important design aspect. To address this aspect, we have proposed efficient use of effective relative gain array analysis (ERGA) which accounts for transport delay also instead of regular steady state relative gain array analysis (RGA). The partitioning and control strategy presented in this work is based on concentration sensitivity matrix analysis by using PCA (Principle Component Analysis), Partial Correlation Analysis and ERGA stepwise.We also considered the results of partial correlation analysis to generate communication links which will be implemented for higher level coordinated decentralized control to account for dynamic, non-negligible interactions as disturbances to local control units. The proposed three step clustering-mapping strategy and mathematical formulation for two point control by optimized dosage is successfully verified for the steady state case on a prototype distribution example network 2 of EPANET (USEPA, 2000).

Relative Time-Averaged Gain Array

for Distributed Control Architecture Design

Presentation on 2017 AIChE Annual Meeting

November 1, 2017

<u>Wentao Tang</u>, Davood Babaei Pourkargar, Prodromos Daoutidis

Department of Chemical Engineering and Materials Science

University of Minnesota

1

Distributed Control

- Mass and energy integration – rule in modern chemical plants

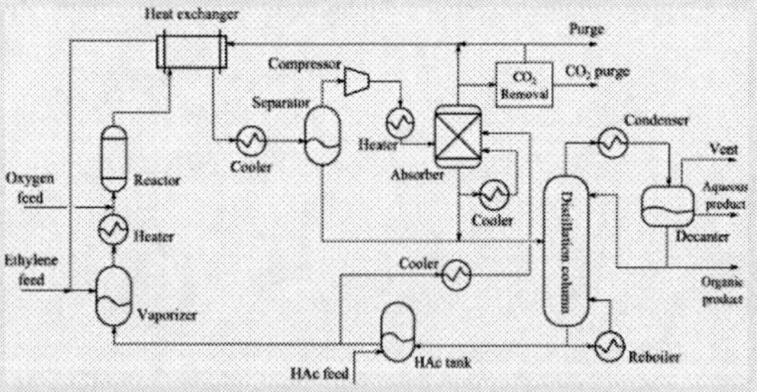

Vinyl acetate monomer process

Baldea & Daoutidis (2012). *Dynamics and nonlinear control of integrated process systems*. Cambridge University Press.
Luyben & Tyréus (1998). *Comput. Chem. Eng. 22*, 867-877.

- Distributed control

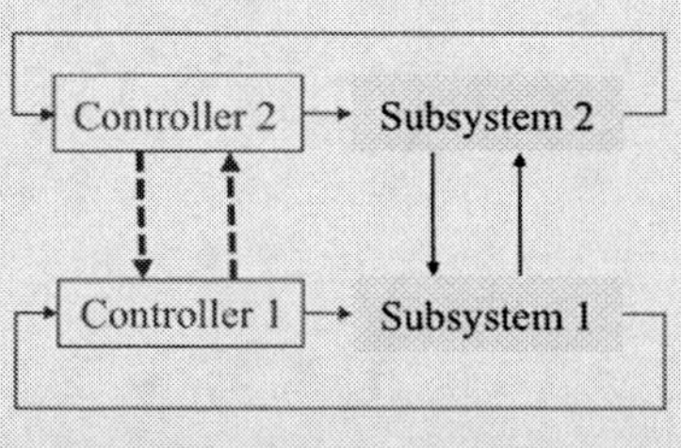

Distributed MPC – extensively studied for process control

Scattolini (2009). *J. Process Control 19*, 723-731.
Christofides, Scattolini, Muñoz de la Peña, & Liu (2013). *Comput. Chem. Eng., 51*, 21-41.

2

Decomposition

- Prerequisite for distributed control
- Performance of distributed control is affected by decompositions

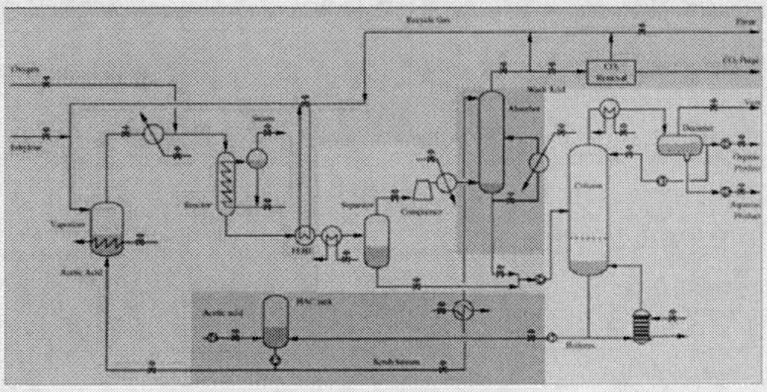

3

Input-output pairing

- Relative Gain Array (RGA)
- Numerous variants have been developed
- Widely applied in control of process units
- Based on infinite-time (static) response

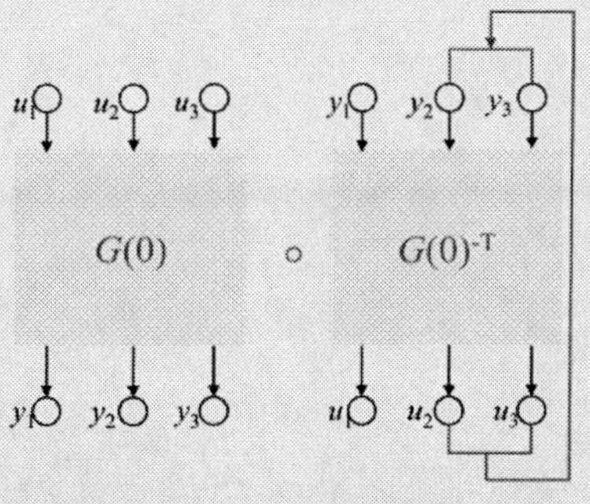

- Limitations
- Network topology is not considered
- Interactions are not minimized

McAvoy (1983). *Interaction analysis: Principles and applications*. ISA.
Khaki-Sedigh & Moaveni (2009). *Control configuration selection for multivariable plants*. Springer.

4

Community detection

- A major field study of network science
- Communities in networks – strongly connected inside, weakly connected between
- Community detection – modularity maximization
- Computationally efficient algorithms have been developed

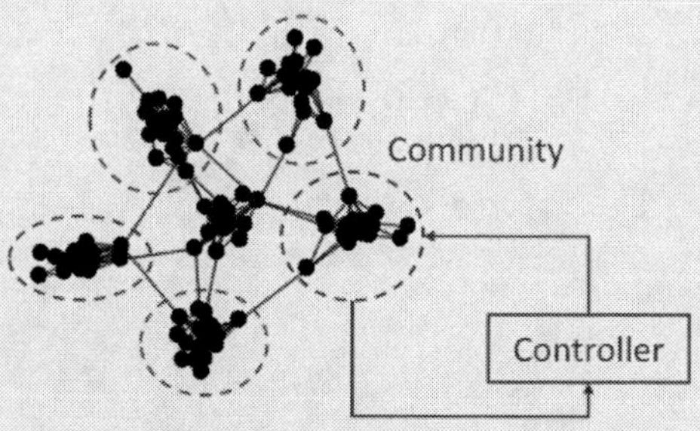

Exploitation of community structures for distributed control

Girvan & Newman (2002). *Proc. Nat. Acad. Sci. 99*, 7821-7826.
Newman & Girvan (2004). *Phys. Rev. E, 69*, 026113.
Barber, M. J. (2007). *Phys. Rev. E 76*, 066102.
Blondel, Guillaume, Lambiotte, & Lefebvre (2008). *J. Stat. Mech. Theor. Exp.*, P10008.

Community detection

- Applied in network representations of process systems

Bipartite graph

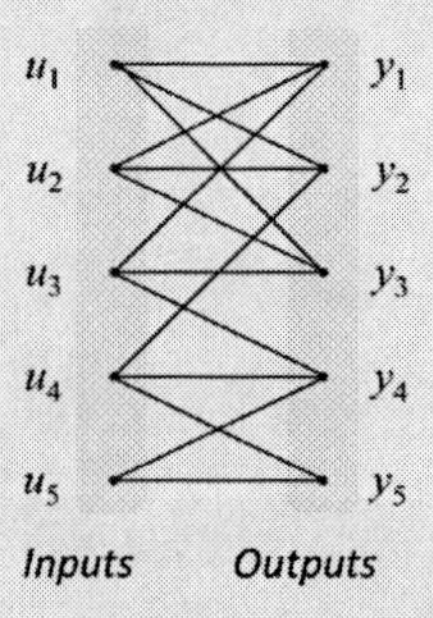

System digraph

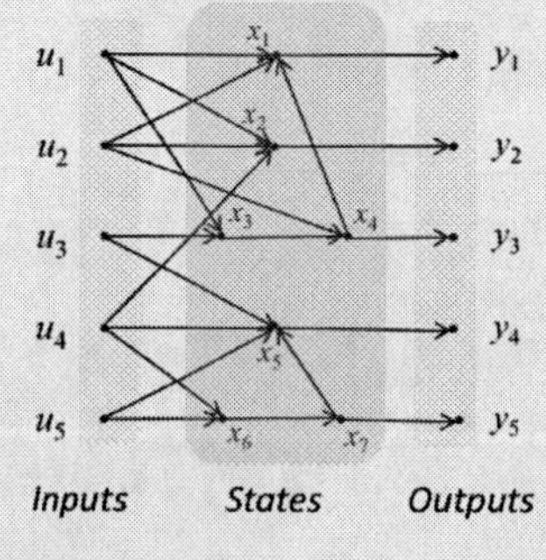

- Limitation

Connectivity does not capture response intensity except within an infinitesimal time scale

Heo & Daoutidis (2016). *AIChE J. 62*, 3177-3188.
Moharir, Kang, Daoutidis, & Almansoori (2017). *Comput. Chem. Eng. 106*, 532-543.
Tang & Daoutidis (2017). *J. Process Control*, submitted.
Jogwar & Daoutidis (2017). *Chem. Eng. Sci. 172*, 434-443.
Tang & Daoutidis (2017). *IFAC-PapersOnline 50*(1), 12015-12020.
Daoutidis, Tang, & Jogwar (2017). *Comput. Chem. Eng.*, in press.

Motivation

Finite time response

- Closed-loop systems evolve in finite time
- The characteristic time scale for closed-loop system is meaningful for analyzing input-output interactions

RGA analysis	←→	Community detection
Infinite time scale		Infinitesimal time scale

<u>This work</u>

- Interaction measure dependent on finite time scale
- Bipartite graph construction and community detection

7

Relative Time-Averaged Gain Array (RTAGA)

- $G(1/\tau)$ – Time-averaged gain array

Exponentially decaying weight function

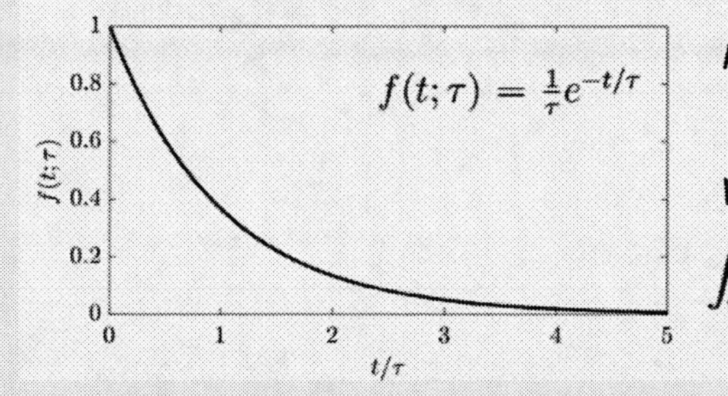

Response to step input

$$\bar{y}_i(s) = g_{ij}(s)\bar{u}_j(s) = \frac{g_{ij}(s)}{s}$$

Weighted average of response

$$\int_0^\infty y_i(t) f(t;\tau) dt = \frac{1}{\tau}\int_0^\infty y_i(t) e^{-t/\tau} dt = \frac{1}{\tau}\bar{y}_i\left(\frac{1}{\tau}\right)$$
$$= g_{ij}(1/\tau)$$

- Relative time-averaged gain array $\Lambda(1/\tau) = G(1/\tau) \circ G(1/\tau)^{-\mathsf{T}}$

8

Relative Time-Averaged Gain Array (RTAGA)

- Interpretation

$$\lambda_{ij}(1/\tau) = \frac{\text{Open-loop time averaged gain between } u_j \text{ and } y_i \text{ in time scale } \tau}{\text{Closed-loop time averaged gain between } u_j \text{ and } y_i \text{ in time scale } \tau}$$

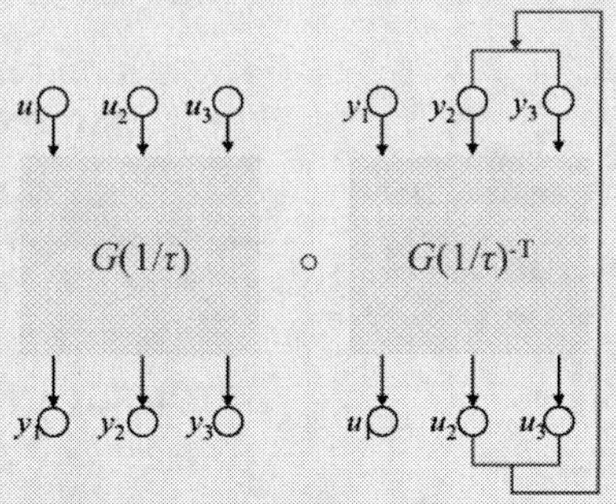

- Limits
- As $\tau \to \infty$, $\Lambda(1/\tau) \to$ RGA
- As $\tau \to 0$, $\Lambda(1/\tau) \to$ RSA (Relative Sensitivity Array)

Yin, X. & Liu, J. (2017). *AIChE J. 63*(4), 1226-1235.

9

Decomposition through community detection

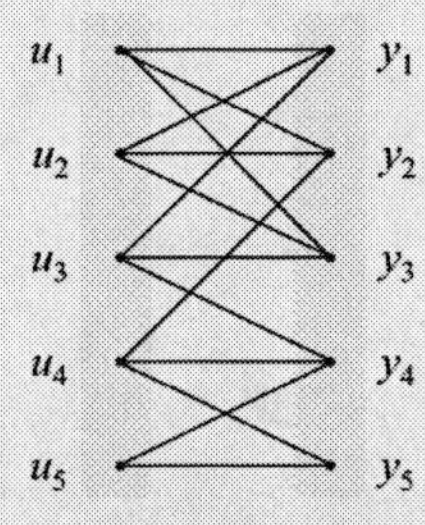

(Weighted) input-output bipartite graph

Edge weight between y_i and u_j

$$w_{ij} = \begin{cases} \lambda_{ij}, & 0 \le \lambda_{ij} \le 1 \\ 1/\lambda_{ij}, & \lambda_{ij} > 1 \\ 0, & \lambda_{ij} < 0 \end{cases}$$

"Closeness" between open-loop and closed-loop gains

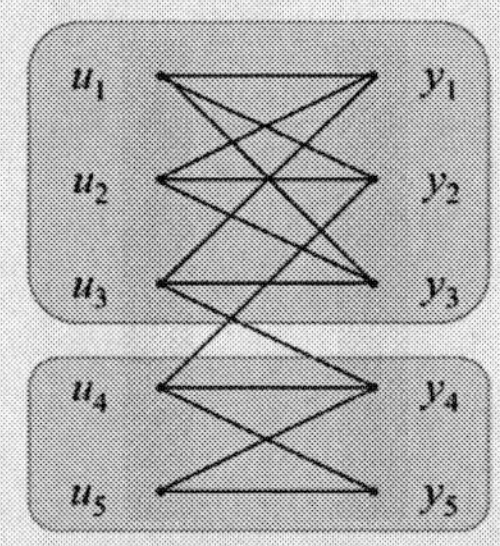

Community detection (dependent on τ)

- How to choose parameter τ?
- Identical decomposition for a perturbation in τ
- Input-output subsystems should be separated (distributed)

10

Case Study: Reactor-Separator Process

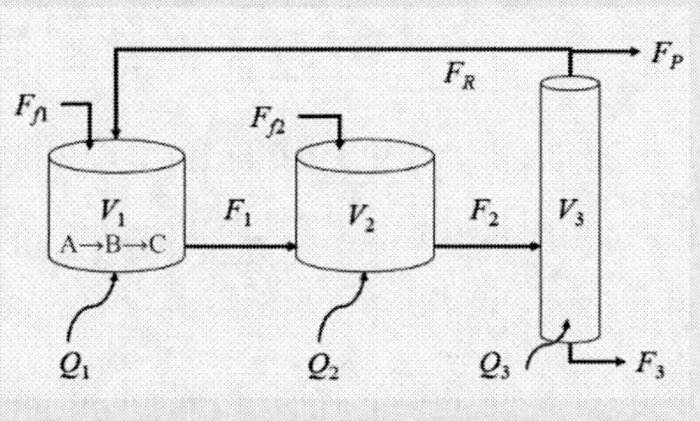

- Extensively studied for distributed MPC
- Intuitive decompositions
  - Unit-based partition
  - Physics-based partition (Q–T, F_f/F_R–x, F–V)

Inputs: $F_{f1}, F_{f2}, F_1, F_2, F_3, F_R, Q_1, Q_2, Q_3$
Outputs: $V_1, V_2, V_3, T_1, T_2, T_3, x_{B1}, x_{B2}, x_{B3}$

Liu, Muñoz de la Peña, & Christofides (2009). *AIChE J.* 55, 1171-1184.
Stewart, Venkat, Rawlings, Wright, & Pannocchia (2010). *Syst. Control Lett.* 59, 460-469.
Tippett & Bao (2013). *AIChE J.* 59, 787-804.

11

Decomposition using RTAGA

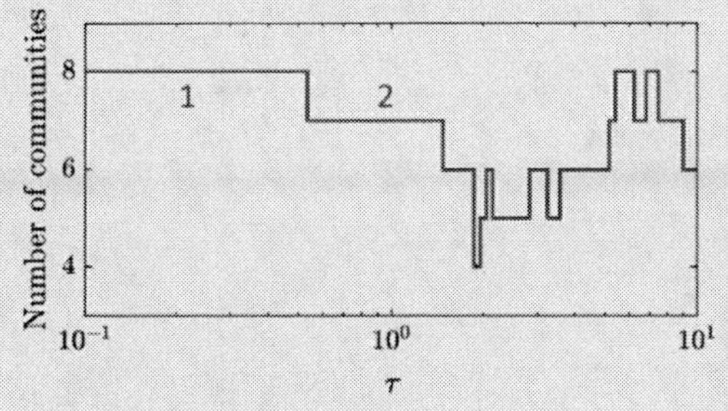

- Two highly similar decompositions are found.

τ	Subsyst.	Inputs	Outputs
0–0.5 h	I	F_{f1}, F_R	V_1, x_{B1}
	II	F_{f2}	x_{B2}
	III	F_1	V_2
	IV	F_2	x_{B3}
	V	F_3	V_3
	VI	Q_1	T_1
	VII	Q_2	T_2
	VIII	Q_3	T_3
0.6–1.5 h	I	F_{f1}, F_R	V_1, x_{B1}
	II	F_{f2}, F_1	V_2, x_{B2}
	III	F_2	x_{B3}
	IV	F_3	V_3
	V	Q_1	T_1
	VI	Q_2	T_2
	VII	Q_3	T_3

12

Distributed MPC

- Distributed optimization algorithm – non-cooperative, iterative, sequential quadratic programming (SQP)
- Communication between controllers – control and state profiles
- Sampling time $\Delta t = 1.5$ min
- Receding horizon $N = 15$
- Technical conditions – Matlab® 2015b on a 3.4 GHz Intel Core™ i7-6700 processor

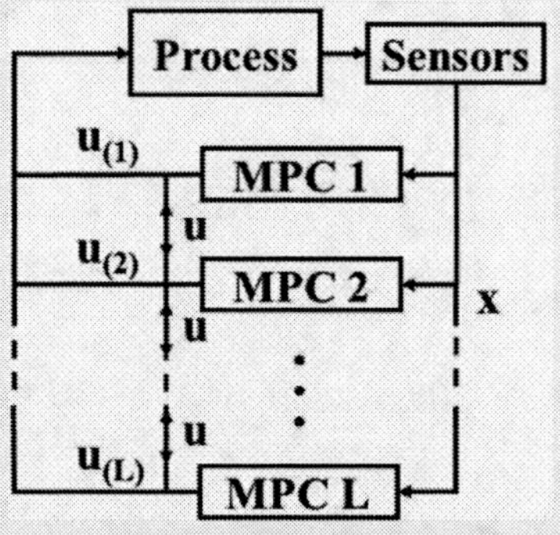

Babaei Pourkargar, Almansoori, & Daoutidis (2017). *Ind. Eng. Chem. Res. 56*, 9606-9616.
Babaei Pourkargar, Almansoori & Daoutidis (2017). *IFAC-PapersOnline 50*(1), 12452-12457.

13

Performance

- Closed-loop stable
- Conver

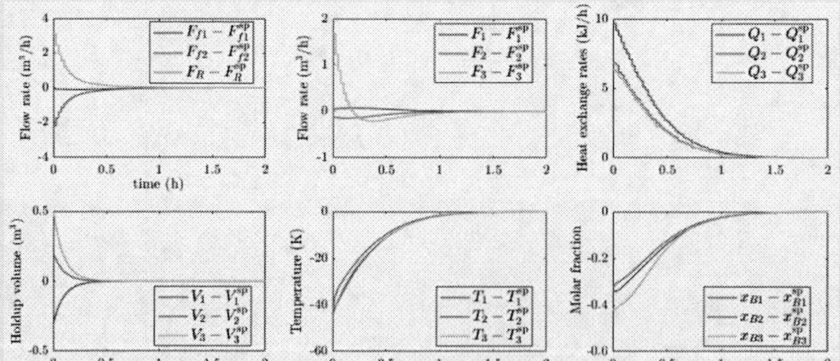

Simulation under decomposition 1

14

Performance

- Significant decrease in computational time (7.5%, 8.4% of centralized control)
- A certain extent of deterioration of control performance (209%, 181% of centralized control)
- Reason: RTAGA gives fine decompositions (8 and 7 subsystems)

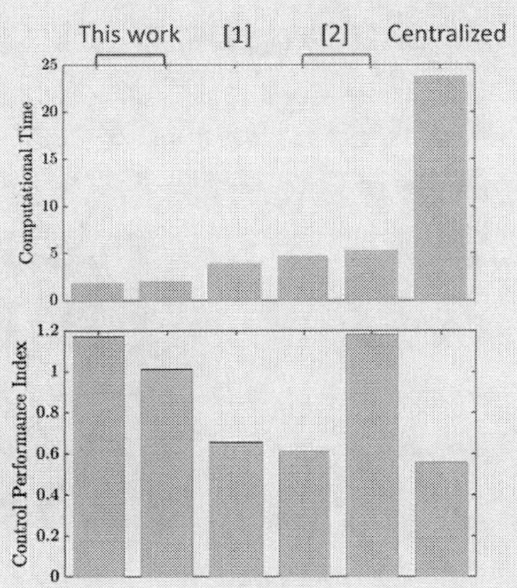

[1] Tang & Daoutidis (2017). *J. Process Control*, submitted.
[2] Jogwar & Daoutidis (2017). *Chem. Eng. Sci. 172*, 434-443.

15

Conclusion

- A systematic input-output partitioning method for distributed control is developed.
- Interaction measure – Relative time-averaged gain array
  - Finite-time response in closed-loop time scale
  - A middle ground between RGA and connectivity-based community detection
- Community detection is applied to generate input-output subsystems
- Case study proves the quality of the resulting decompositions
  - Fine decomposition that significantly decreases computational time

16

Acknowledgment

- Financial support from NSF-CBET
- The rest of my group

17

THANK YOU

18

500110 Innovative Design and Systematic Process Intensification Using Building Blocks

Abstract:

Process intensification addresses the development of techniques that leads to substantially smaller, cleaner, safer, and more energy-efficient technologies. It also enables the innovative design of novel multi-functional units (e.g., Eastman Chemical methyl acetate process that performs extractive distillation, reactive distillation, reaction and distillation within a single column [1,2]). Intensification methodologies have been developed in the past, several of which are described in [3-8]. In this work, we address the following design and intensification challenges for both steady-state and dynamic processes: (i) how to systematically select and combine multiple phenomena to create new opportunities and obtain "out-of-the-box" design solutions, (ii) how to systematically identify intensified equipment and, at the same time, generate process flowsheets containing these equipment, (iii) how to select optimal configurations of intensified equipment and process flowsheets, and (iv) how to incorporate process intensification into process synthesis and optimization without *a priori* postulation of intensification alternatives. To address these questions within a single framework, we depart from the classical unit operation-based representation of process units, flowsheets and superstructures, and propose a new representation using finite building blocks [9]. In this representation, a physiochemical phenomenon, a processing task or a function can be represented using a single block or using two neighboring blocks and their boundaries. Other representations also exist in literature, including the modular [10,11] and phenomena building blocks (PBBs) [6,7]. Here, we propose a novel and general process superstructure which is an ensemble of only building blocks in two dimensions. Each block is denoted by its coordinate within the superstructure. A block can have entering and outgoing streams in both horizontal and vertical directions. An assembly of blocks of the same type obtains a classical unit, while an assembly of blocks with different types results in an intensified unit. The block superstructure encompasses numerous process equipment and flowsheet alternatives, and can be used to design, synthesize and intensify a chemical process without *a priori* postulation of these alternatives. The overall problem is formulated using a single mixed-integer nonlinear optimization (MINLP) model.

For dynamic intensification of multi-functional and periodic/cyclic processes, we propose a similar block superstructure, where each block represents a specific material performing a specific task for a specific time step. The superstructure is specifically suitable for the design of integrated, multi-functional and modular adsorption-reaction systems embedded within a single column while exploiting the dynamic interactions and trade-offs between different phenomena. Again, from the same superstructure, one can systematically generate variants of adsorption-reaction systems, including single- or multi-layered packed adsorption columns, reactors, pressure/temperature swing adsorption (PSA/TSA) processes, pressure and temperature swing adsorption (PTSA) processes, sorption-enhanced reaction processes (SERP), simulated moving bed chromatography (SMB) processes, and reactive SMB processes. Using this method, various intensified and modular processes can be designed for the utilization of both conventional and unconventional resources including stranded natural gas, contaminated shale gas, fuel gas or landfill gas to produce sustainable fuels and chemicals [12].

References:

[1] Agreda, V.H., Partin, L.R. (1984). Reactive distillation process for the production of methyl acetate. U.S. Patent 4,435,595.

[2] Siirola, J. J., (1996). Industrial applications of chemical process synthesis. Advances in Chemical Engineering 23, 1–62.

[3] Stankiewicz, A. I., Moulijn, J. A. (2000). Process Intensification: Transforming Chemical Engineering. Chemical Engineering Progress 1, 22–34.

[4] Lutze, P., Gani, R., Woodley, J. M. (2010). Process intensification: a perspective on process synthesis. Chemical Engineering Processing: Process Intensification 49 (6), 547–558.

[5] Reay, D., Ramshaw, C., Harvey, A. (2013). Process Intensification: Engineering for Efficiency, Sustainability and Flexibility. Butterworth-Heinemann, Waltham, MA.

[6] Babi, D. K., Cruz, M. S., Gani, R. (2016). Fundamentals of process intensification: a process systems engineering view. In: Segovia-Hernández, J.G., Bonilla-Petriciolet, A. (Eds.), Process Intensification in Chemical Engineering., http://dx.doi.org/10.1007/978-3-319-28392-0 2.

[7] Tula, A. K., Babi, D. K., Bottlaender, J., Eden, M. R., Gani, R. (2017). A computer-aided software-tool for sustainable process synthesis-intensification. *Computers & Chemical Engineering*. http://doi.org/10.1016/j.compchemeng.2017.01.001.

[8] Baldea, M. (2015). From process integration to process intensification. *Computers & Chemical Engineering, 81*, 104-114.

[9] Demirel, S. E., Li, J., and Hasan, M. M. F., 2017. Systematic Process Intensification using Building Blocks, *Computers and Chemical Engineering*, http://dx.doi.org/10.1016/j.compchem eng.2017.01.04

[10] Papalexandri, K. P., Pistikopoulos, E. N. (1996). Generalized modular representation framework for process synthesis. *AIChE Journal, 42*(4), 1010-1032.

[11] Ismail, S. R., Proios, P., Pistikopoulos, E. N. (2001). Modular synthesis framework for combined separation/reaction systems. *AIChE Journal, 47*(3), 629-649.

[12] Iyer, S. S., Bajaj, I., Balasubramanian, P., Hasan, M. M. F. Modular Process Intensification of Carbon Capture and Conversion to Syngas. Submitted.

501702 Modeling and Simulation of a Hybrid Adsorptive-Membrane Reactor (HAMR) for Intensification of the Water-Gas Shift (WGS) Reaction Process

Abstract:

The Hybrid Adsorptive-Membrane Reactor (HAMR) is a dynamically operated process, which consists of a metal tube containing one or more tubular membranes. The HAMR's reaction section, surrounding the membrane tubes, contains catalyst and adsorbent material that accelerate reaction kinetics and adsorb some reaction products from the reacting mixture respectively. The HAMR's permeation section consists of the membrane tubes' interior, and physically removes some other reaction products from the reacting mixture. The HAMR is typically operated in reaction and regeneration modes, since the adsorbent meterial reaches its adsorption capacity after a period of operation, and thus needs to be regenerated.

In this work a mathematical model of HAMR is developed and simulated. The aim of this effort is to intensify the water gas shift (WGS) reaction process for hydrogen production. Indeed, the coupling of reaction and separation in the HAMR process can substantially improve reactant conversion, product selectivity, and can even yield near pure products. In addition, for reversible reactions such as the WGS, the HAMR process can be designed so as to deliver reactant conversions that exceed even equilibrium conversions attainable in conventional reactors. The HAMR system is composed of a, where some reaction products permeate. The proposed HAMR model quantifies velocity, species concentration, and temperature profiles throughout the reaction-adsorption, and permeation zones, by solving momentum/species/energy transport equations accounting for convection/reaction /diffusion/conduction mechanisms.

The developed model is used to intensify the Water Gas Shift Reactor (WGSR) Process. To this end, the performance of the traditional packed bed reactor is first quantified. Then parametric studies of the HAMR are carried out, so as to identify maximum intensification designs. These studies include a broad range of operating conditions and parameters (e.g. reactor operating temperature, catalyst and adsorbent weight to feed flowrate ratios, and others.

496925 Intensification of Chemical Processes Via Periodic Operation

Abstract:

Process intensification has emerged over the past few decades as a framework for increasing process performance by bringing chemical and physical phenomena into closer physical proximity. Intensification often transcends the traditional unit operations framework, with several conventional operations occurring in the same physical device. While this minimizes transport limitations and reduces equipment size, it can also lead to control challenges: intensified processes feature strong dynamic coupling between phenomena and have fewer control degrees of freedom than their conventional counterparts[1].

In this presentation, we describe a new means of operating intensified processes, focusing on deliberately imposing a transient (specifically, periodic) operating regime in order to improve average process performance. Additionally, we discuss periodic operation in terms of expanding the number of control degrees of freedom, in the sense that the amplitude and period of oscillation of a single manipulated variable can be set independently.

Using a prototype process featuring several coupled (and possibly competing) phenomena, we demonstrate theoretically that periodic operation can increase overall economic performance. We perform simulations to illustrate our theoretical findings, using both the canonical example described above, and the model of a dividing-wall column.

References

[1] Schembecker, G.; Tlatlik, S. Process Synthesis For Reactive Separations. Chem. Eng. Process. 2003, 42, 179-189.

498144 Intensification of the Hydride Vapor Phase Epitaxy Manufacturing Process for Solar Devices

Abstract:

Photovoltaics made of III-V compound semiconductors, especially in the form of multi-junctions, have demonstrated great conversion efficiencies [1]. However, their application is limited significantly by the expensive fabrication methods and materials. To address this challenge, hydride vapor phase epitaxy (HVPE) has regained research interest because of its intrinsically higher growth rate and lower precursor costs compared with other epitaxial growth methods. But the high growth rate and near-thermodynamic-equilibrium nature of this process also impose difficulties of producing large-area uniform wafer in a controlled way. The process needs to be geared into a more economic and manageable configuration to produce high quality products with stable properties at rapid throughput.

There are already a couple of practices applying the philosophy of process intensifications in this area. For example, through clever design, the compact horizontal reactor in [2] was capable of growing various III-V semiconductors (GaAs, InGaAs, GaInP, etc.) at high growth rates and depositing multilayers of those compounds in a controlled mode by manually operating transfer rod and shifting gas flow [2, 3]. Based on this reactor, an improved inline reactor was proposed [4], in which the growth method was adapted to sequentially deposit different layer of material on one wafer travelling through multiple reaction chambers separated by gas curtain. To further promote performance of process and product, we present novel alternatives to previous intensification techniques using new concepts in both equipment design and operation method. Computational modeling tools are used to verify design and operational strategies. We are expecting this work will not only aid the production of economically and industrially viable III-V solar cells, but also inspire more ideas to optimize other CVD processes that grow films with high cost-performance ratio.

[1] M. A. Green et al., "Solar cell efficiency tables (version 49)," Prog. Photovolt: Res. Appl., vol. 25, no. 1, pp. 3–13, Jan. 2017.
[2] K. L. Schulte, W. L. Rance, R. C. Reedy, A. J. Ptak, D. L. Young, and T. F. Kuech, "Controlled formation of GaAs pn junctions during hydride vapor phase epitaxy of GaAs," Journal of Crystal Growth, vol. 352, no. 1, pp. 253–257, Aug. 2012.
[3] K. L. Schulte et al., "Metalorganic vapor phase growth of quantum well structures on thick metamorphic buffer layers grown by hydride vapor phase epitaxy," Journal of Crystal Growth, vol. 370, pp. 293–298, May 2013.
[4] D. L. Young, A. J. Ptak, T. F. Kuech, K. Schulte, and J. D. Simon, "High throughput semiconductor deposition system," US20130309848 A1, 21-Nov-2013.

501295 Process Intensification of Reactive Separator Networks through the Ideas Conceptual Framework

Abstract:

A method to rigorously identify the performance limits of a reactive separator network is presented in this work. The quantification of the enhancement potential for a given technology can greatly benefit process intensification studies in the pursuit of radical improvements. The Infinite DimEnsionAl State-Space (IDEAS) conceptual framework is first reviewed and then shown to be capable of assessing the potential for process intensification of reactive separation processes. The IDEAS framework is employed to formulate an infinite linear program (ILP) that can synthesize optimal reactive separation networks, and establish rigorous tradeoffs between total network reactive holdup, and total network capacity. The proposed process intensification method is demonstrated on a case study involving the metathesis of 2-pentene through reactive distillation, an important chemical process for the oil industry. Total reactive holdup and total capacity, a surrogate for the amount of catalyst used and capital cost respectively, were selected as the systemâ€™s variables of interest for process intensification. Using a traditional reactor-followed-by-separation-system scheme as a baseline, an intensified IDEAS based reactive separation network can reduce the total reactive holdup and total capacity by 94.5% and 84.3% respectively. Results available in the literature were converted to the total holdup-total capacity space in the extent possible and compared with the IDEAS results. The comparison shows the existence of improvement potential, indicating that process intensification opportunities can be further explored in this reactive separation system.

490321 Intensified Reaction-Separation Schemes

Abstract:

Intensified reaction-separation schemes

Evangelia Koumaditi, John Woodley, Rafiqul Gani *Department of Chemical and Biochemical Engineering, Technical University of Denmark, DK-2800 Lyngby, Denmark*

The environmental and economic concerns are driving the chemical and biochemical industries to develop new, sustainable methods of production. Combining the reaction and the separation process in one unit, not only can it lead to cost and energy reductions for the overall process, but it can also have an effect on the reaction, in terms of yield. Motivated by the need to take advantage of the intensified process units, a method needs to be developed, which will allow the systematic generation and analysis of reactive separation configurations at an early stage in process synthesis. The objective of this work is the development of a method to design reaction-separation schemes, based on analysis of phase equilibrium of the involved chemical system.

Oxidation reactions are used as a generic system to highlight the conceptual basis of the design method. The selective oxidation of primary alcohols are examples of this generic system, which are used as case studies for the application of the design method. The properties of the pure components and the characteristics of phase equilibrium (azeotropes and liquid phase split) have been considered to identify new and novel intensified reaction-separation schemes. Also, the selective oxidation of alcohols to aldehydes is a challenging reaction, as aldehydes are relatively unstable products and also due to their use in the production of high-value products. Even though the reactions are considered to be bio-catalyzed by the enzyme Galactose Oxidase, the design method nevertheless is applicable to other similar systems and also not restricted to the use of biocatalysts only.

The specific examples highlighted in this work cover ternary systems involving alcohol + water + aldehyde, from which aldehyde is the product to be recovered, while alcohol and water are recycled back to the reactor. The first step of the method is the definition of the reaction system and data collection. The designs of the intensified separation schemes are based on the properties of pure components, as well as the properties of the mixture of the involved chemical system. The properties of pure components that are of interest are namely, the boiling point, the vapor pressure and the water solubility. After the pure component analysis is completed, the vapor-liquid and liquid-liquid equilibria are examined for each binary subsystem of the ternary mixture, employing available data or generated through model-based phase equilibrium predictions. The next step involves the classification of the systems according to their characteristics. The reaction-separation schemes are then determined by using phase equilibrium relations that favor the removal of the product from the reacting phase. Calculations of single-stage equilibrium based separations are performed in order to obtain the mass balance and the distribution of the compounds in the different phases. The last step involves validation tests to confirm the feasibility of the suggested configuration, by super-imposing the calculated mass balance lines on the predicted VLE and LLE plots. Success of the designed intensified reaction-separation scheme depends on the amount of the targeted compound removed from the reactor. Note that the separation and recovery of the product requires the use of solvents. Ionic liquids, because of their special properties have been considered as the needed solvents.

The presentation will highlight the results of the intensified reaction-separation schemes for a class of oxidation of primary alcohol reactions. Details of the design method together with the important concepts will also be highlighted.

499187 Process IntensificationÂ of Large-Scale Continuous Biobutanol Production via a Multi-Feed Bioreactor with *in situ* Gas Stripping

Abstract:

The economical production of alternative liquid fuels is vital to achieving energy independence from fossil fuel sources. While ethanol is often considered the primary candidate, butanol offers the advantages of higher energy density, lower vapor pressure, and lower volatility [1]. However, butanol is more toxic to cell cultures than ethanol, making it difficult to produce biologically at levels necessary for economic viability. Process intensification has been used in bioreactors by combining acetone, butanol, and ethanol (ABE) fermentation with separation phenomena, including pervaporation and gas stripping, to selectively remove the ABE products and overcome this limitation [2, 3]. Additionally, multi-feed inlet configurations have been investigated in continuous bioprocessing systems to limit inhibitory effects in bioreactors by manipulating the substrate concentration and dilution rate independently, limiting substrate concentration in the reactor without necessarily decreasing the reactor productivity [4]. The development of an intensified bioreactor with a multi-feed inlet configuration has the potential to overcome the toxicity limitation, increase process throughput, and produce butanol in a continuous, economically attractive manner that can compete with current bioethanol processes.

In this work, we examine the economic viability of large-scale, continuous production of biobutanol using *Clostridium acetobutylicum* and *in situ* gas stripping. First, the optimal control problem for a multi-feed intensified bioreactor is investigated using the flowrate of the stripping gas and inlet feeds as manipulated variable to maximize profitability. Batch reaction kinetics for the system, taken from Votruba et al., are extended to continuous processing in the formulation of the optimal control problem [5]. Solutions for the intensified bioreactor are compared to the economic viability of an optimally-controlled, continuous biobutanol reactor in the absence of gas stripping to emphasize the necessity of intensified process operations.

Once the optimal control policy for the bioreactor is determined, the process synthesis problem for large-scale biobutanol production from lignocellulosic biomass is explored by maximizing profitability. The synthesis problem considers using multiple bioreactors in parallel to process glucose made available from five options of cellulosic biomass, hybrid poplar, corn stover, sorghum, sugarcane bagasse, and switchgrass, in combination with four possible pretreatment methods, ammonia fiber expansion (AFEX), dilute sulfuric acid, sodium hydroxide, and liquid hot water. The use of lignin as a source of steam and salable electricity generation is also considered. The results of this study look to emphasize the importance of process intensification practices in the areas of bioprocess to develop new and otherwise economically impractical processes.

1. Liu, X.B., et al., *Enhancement of butanol tolerance and butanol yield in Clostridium acetobutylicum mutant NT642 obtained by nitrogen ion beam implantation.* Journal of Microbiology, 2012. **50**(6): p. 1024-1028.
2. Qureshi, N. and H.P. Blaschek, *Recovery of butanol from fermentation broth by gas stripping.* Renewable Energy, 2001. **22**(4): p. 557-564.
3. Li, J., et al., *Efficient production of acetone-butanol-ethanol (ABE) from cassava by a fermentation-pervaporation coupled process.* Bioresource Technology, 2014. **169**: p. 251-257.
4. Raftery, J.P., M.R. DeSessa, and M.N. Karim, *Economic improvement of continuous pharmaceutical production via the optimal control of a multifeed bioreactor.* Biotechnol Prog, 2017.
5. Votruba, J., B. Volesky, and L. Yerushalmi, *Mathematical-Model of a Batch Acetone Butanol Fermentation.* Biotechnology and Bioengineering, 1986. **28**(2): p. 247-255.

493035 Global Optimization of Nonconvex Problems with Convex-Transformable Intermediates

Abstract:

Current global optimization solvers rely on factorable programming techniques to construct convex relaxations of nonconvex optimizations problems [5–7]. These techniques iteratively decompose a nonconvex factorable function by introducing auxiliary variables for intermediate expressions, until each intermediate expression can be outer-approximated by a convex feasible set. While they are easy to automate, factorable programming techniques often result in large relaxation gaps.

Factorable relaxations of global optimization problems can be strengthened through the use of functional transformations. A particular example of this approach is the use of power and exponential transformations to convexify signomial terms [2–4]. A more general transformation scheme was recently proposed by Khajavirad et al. [1]. This technique exploits convex transformability of component functions of factorable programs, and is applicable to various classes of functional forms including signomials, products and ratios of convex and/or concave functions, and logarithmically-concave functions. As illustrated in [1], this transformation method often leads to relaxations which are tighter than those obtained by standard approaches.

Motivated by the potential of these new relaxations to enhance the performance of global solvers, in this paper, we introduce a new class of cutting planes for convex-transformable functions, and describe its implementation into the branch-and-reduce global solver BARON. The proposed implementation involves a recognition tool which identifies convex-transformable functions that are present in the problem, including those in intermediate expressions of factorable functions. To exploit local information regarding variable bounds for relaxation construction, as well as utilize the generated cutting planes for range reduction, we embed the proposed cutting plane generation technique at each node of the branch-and-reduce algorithm. An extensive computational study is carried out on problems selected from a collection of publicly available test sets. Results demonstrate that the generated cutting planes accelerate the convergence speed of the branch-and-reduce algorithm, by significantly reducing the average CPU time and number of nodes in the search tree.

References cited

[1] A. Khajavirad, J. J. Michalek, and N. V. Sahinidis. Relaxations of factorable functions with convex-transformable intermediates. Mathematical Programming, 144:107–140, 2014.

[2] H. Lu, H. Li, C. E. Gounaris, and Ch. A. Floudas. Convex relaxation for solving posynomial programs. Journal of Global Optimization, 46:147–154, 2010.

[3] A. Lundell and T. Westerlund. Convex underestimation strategies for signomial functions. Optimization Methods and Software, 24:505–522, 2009.

[4] C. D. Maranas and Ch. A. Floudas. Global optimization of generalized geometric programming. Computers & Chemical Engineering, 21:351–369, 1997.

[5] G. P. McCormick. Computability of global solutions to factorable nonconvex programs: Part I—Convex underestimating problems. Mathematical Programming, 10:147–175, 1976.

[6] H. D. Sherali and H. Wang. Global optimization of nonconvex factorable programming problems. Mathematical Programming, 89:459–478, 2001.

[7] M. Tawarmalani and N. V. Sahinidis. Global optimization of mixed-integer nonlinear programs: A theoretical and computational study. Mathematical Programming, 99:563–591, 2004.

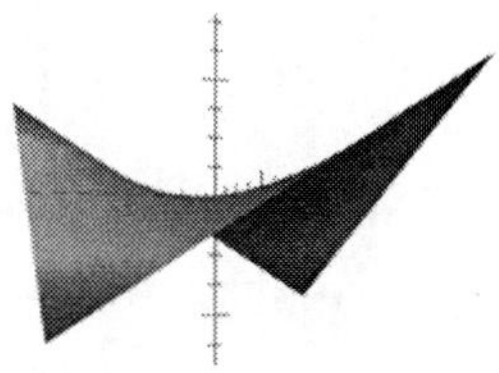

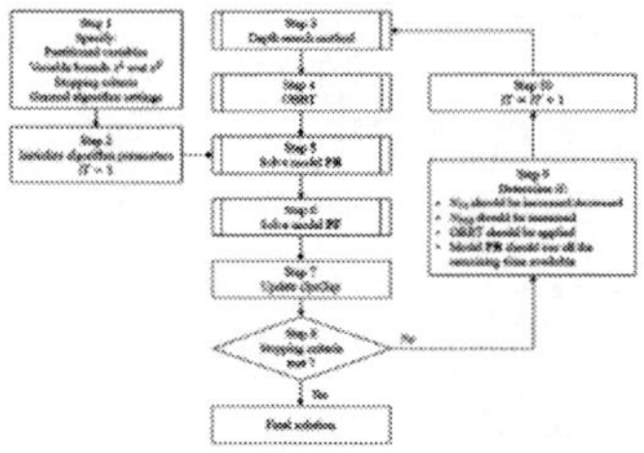

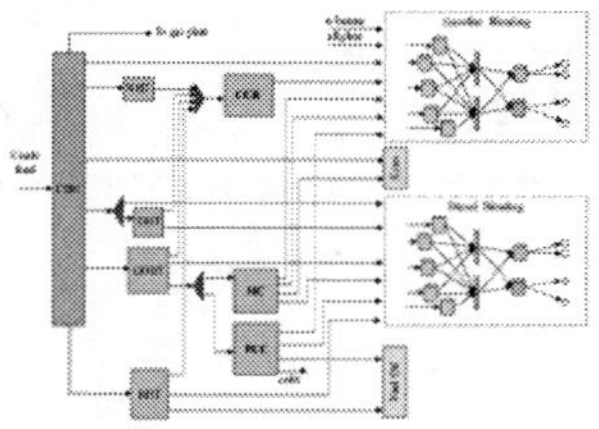

Deterministic Global Optimization Algorithm Using Piecewise Relaxations and Bound Tightening with Dynamic Partitioning

Pedro M. Castro (pmcastro@fc.ul.pt)

INVESTIGADOR FCT

Ciências ULisboa

Pedro Castillo Castillo
Vladimir Mahalec

McMaster University

MACC
McMaster Advanced Control Consortium

CMAF CIO
CENTRO DE MATEMÁTICA, APLICAÇÕES FUNDAMENTAIS E INVESTIGAÇÃO OPERACIONAL

Problem definition (MINLP)

McMaster University

- Mixed-integer bilinear program (MIQCP)

$$\boxed{\min} f_0(\boldsymbol{x},\boldsymbol{y})$$

$$f_q(\boldsymbol{x},\boldsymbol{y}) \leq 0 \quad q \in Q \backslash \{0\}$$

$$f_q(\boldsymbol{x},\boldsymbol{y}) = \sum_{(i,j)\in BL} a_{ijq} x_i x_j + B_q \boldsymbol{x} + C_q \boldsymbol{y} + d_q \qquad q \in Q \qquad \textbf{(P)}$$

$$0 \leq \boldsymbol{x}^L \leq \boldsymbol{x} \leq \boldsymbol{x}^U$$

$$\boldsymbol{x} \in \mathbb{R}^m, \boldsymbol{y} \in \{0,1\}^r$$

Relaxation gives lower bound f_0^R

$\boldsymbol{x}$ – continuous variables
$\boldsymbol{y}$ – binary variables
BL - (i,j) index set
$i \neq j$ – strictly bilinear problems
a_{ijq}, B_q, C_q, d_q- parameters

MILP relaxation of (P)

- Continuous variables w_{ij} replace bilinear term $x_i x_j$

$$f_0^R = \min f_0(\boldsymbol{x}, \boldsymbol{y})$$

$$f_q(\boldsymbol{x}, \boldsymbol{y}) \le 0 \quad q \in Q \backslash \{0\}$$

$$f_q(\boldsymbol{x}, \boldsymbol{y}) = \sum_{(i,j) \in BL} a_{ijq} w_{ij} + B_q \boldsymbol{x} + C_q \boldsymbol{y} + d_q \qquad q \in Q \qquad \textbf{(PR)}$$

$$w_{ij} \in \mathrm{WR}_{ij} \ (i,j) \in BL$$

$$0 \le \boldsymbol{x}^L \le \boldsymbol{x} \le \boldsymbol{x}^U$$

$$\boldsymbol{x} \in \mathbb{R}^m, \boldsymbol{y} \in \{0,1\}^r, \boldsymbol{w} \in \mathbb{R}^{|BL|}$$

- 3 alternatives for generating regions WR_{ij}
  - McCormick envelopes of bilinear function (convex hull)
  - Piecewise relaxations (non-convex)
    - Piecewise McCormick envelopes
    - Multiparametric disaggregation

Illustration piecewise relaxation techniques

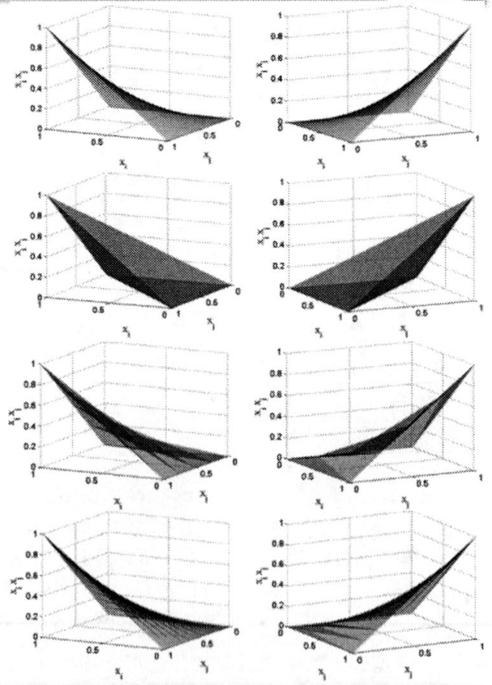

- Bilinear function $x_i x_j$ in $[0,1]^2$
- McCormick envelopes
- Piecewise McCormick envelopes $N = 4$
  - $N \in \mathbb{Z}^+$
- Multiparametric disaggregation $p = -1$ $(N = 10)$
  - Scales better than piecewise McCormick ($p \in \mathbb{Z}^-$, $N = 10^{-p}$)

Algorithms based on piecewise relaxations

McMaster University

- Standalone procedures
  - No need for spatial B&B (branches one variable at a time)

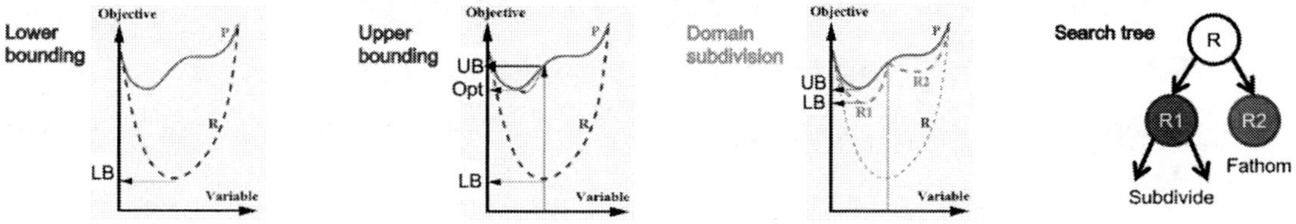

- Simultaneous partitioning of variables x_j
  - Relaxation gives lower bound $f_0^R \rightarrow f_0$ as $p \rightarrow -\infty$
  - Local solution of (**P**) gives upper bound f_0' (if feasible)
    - Add constraint $y = y^R$ to reduce (**P**) to an NLP (**PF**)
    - Solve (**PF**) with local solver using x^R as starting point
- Lower the gap by increasing number of partitions
  - Terminate when $optimality\ gap = (f_0' - f_0^R)/f_0^R \leq \varepsilon$
    - Solution (x', y') is globally optimal

How to make relaxation tighter?

McMaster University

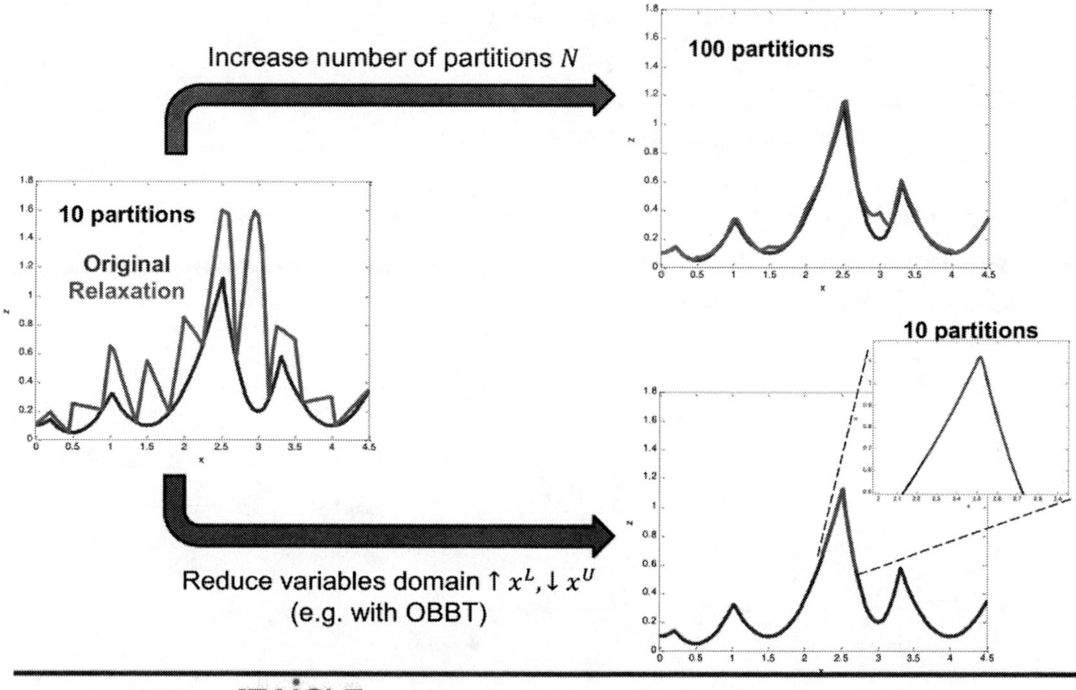

Tightening variable bounds x^L & x^U

McMaster University

- Optimality based bound tightening (OBBT)
  - Aim: $\uparrow x^L, \downarrow x^U$ for bilinear variables x_i & x_j
  - Focus on regions that improve upper bound f_0'
  - Mathematical problems $w_{ij} \in WR_{ij}$
    - Relaxation with McCormick envelopes (shown)
      - LP: $\boldsymbol{y} \in [0,1]^r$
      - MILP: $\boldsymbol{y} \in \{0,1\}^r$
    - Piecewise relaxations
      - MILP: original & new binaries
  - Applied in parallel

Computational time increasing ↓

Weaker bounds ↑

$$x_s^L := \min x_s \quad (x_s^U := \max x_s) \qquad \textbf{(PRB)}$$

subject to

$$\sum_{(i,j)\in BL} a_{ij0} w_{ij} + B_0 \boldsymbol{x} + C_0 \boldsymbol{y} + d_0 \le f_0'$$

$$\sum_{(i,j)\in BL} a_{ijq} w_{ij} + B_q \boldsymbol{x} + C_q \boldsymbol{y} + d_q \le 0 \quad \forall q \in Q \backslash \{0\}$$

$$\left.\begin{array}{l} w_{ij} \ge x_i \cdot x_j^L + x_i^L \cdot x_j - x_i^L \cdot x_j^L \\ w_{ij} \ge x_i \cdot x_j^U + x_i^U \cdot x_j - x_i^U \cdot x_j^U \\ w_{ij} \le x_i \cdot x_j^L + x_i^U \cdot x_j - x_i^U \cdot x_j^L \\ w_{ij} \le x_i \cdot x_j^U + x_i^L \cdot x_j - x_i^L \cdot x_j^U \end{array}\right\} \forall (i,j) \in BL \quad w_{ij} \in WR_{ij}$$

$$0 \le \boldsymbol{x}^L \le \boldsymbol{x} \le \boldsymbol{x}^U$$

$$\boldsymbol{x} \in \mathbb{R}^m, \boldsymbol{y} \in [0,1]^r$$

Sequential vs. parallel OBBT

McMaster University

- Results for McCormick ($N = 1$, LPs) and piecewise McCormick relaxation ($N = 2$, MILPs)
- Multiple OBBT cycles (return to 1st variable after going through all)
  - Impact in optimality gap becomes smaller
- Sequential implementation only slightly tighter
- Parallel implementation faster with increase in problem size
  - Only practical option for $N > 1$

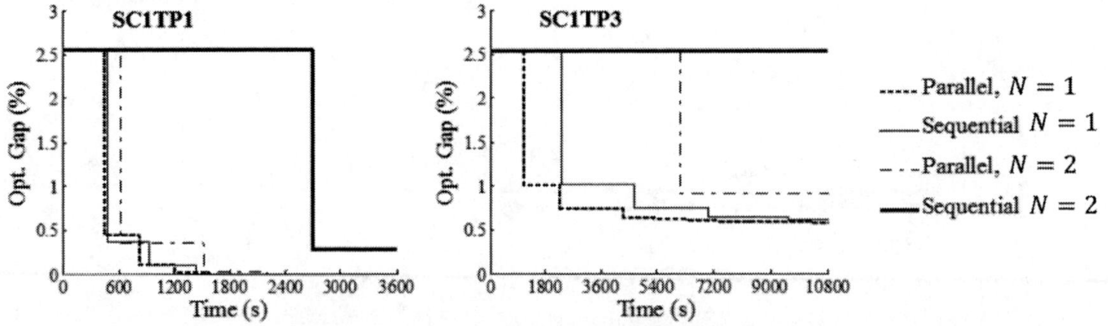

New algorithm features dynamic partitioning

- Tightening of x^L & x^U makes (**PR**) easier to solve
  - ↑ N_{PR} to try to improve lower bound
    - Backtrack in next iteration if jump in complexity is too high
- Improve average domain reduction (ADR) in OBBT
  - Cope with decrease in efficiency in later cycles by ↑ N_{PRB}
- Other key features
  - 2-stage MILP-NLP decomposition
    - Multiple (**PF**) initialization points from (**PR**) solution pool
  - Parallelization of OBBT & upper bounding stages

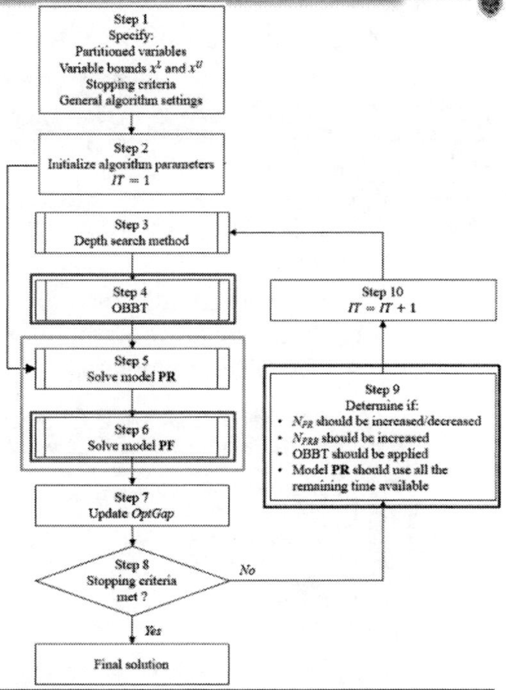

Benchmark problems

- Hydro energy system
  - Scheduling for day-ahead market
    - Binary variables identify hourly operation
    - Power generation is a bilinear function
    - Startup costs

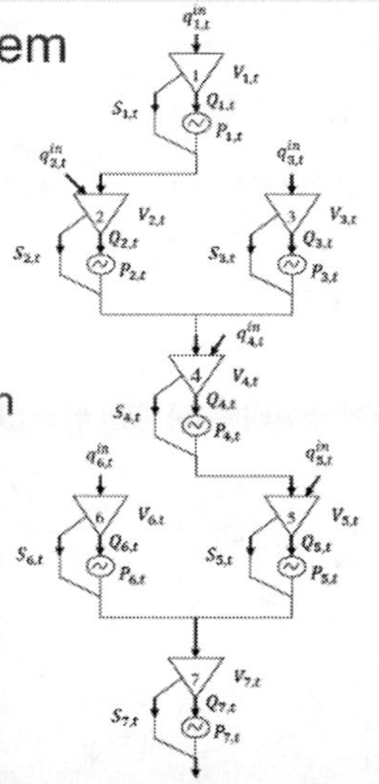

- Petroleum refinery
  - Weekly planning problem
    - Binary variables for active modes
    - Bilinear terms due to blending

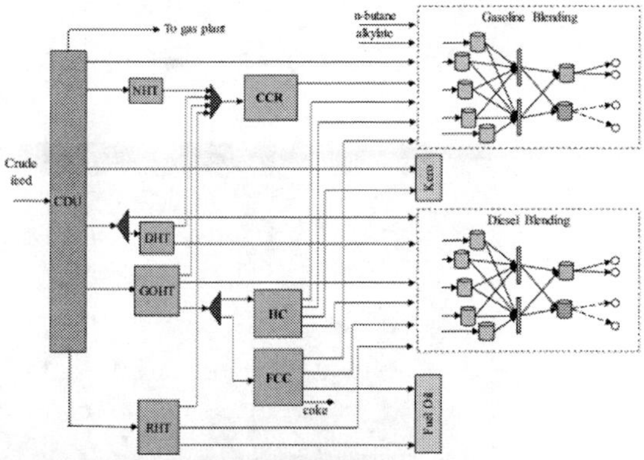

Benchmark problem	HYD2	HYD4	HYD7	SC1TP1-SC3TP1	SC1TP3-SC4TP3
Equations	573	1145	2003	1504	4526
Binary variables	96	192	336	12	36
Total variables	433	865	1513	1234	3716
Variables in bilinear terms	118	260	473	342	1132
Bilinear terms	192	384	672	476	1608

Computational results

McMaster University

- Comparison to different algorithms
  - Commercial global solvers BARON 16.5 & ANTIGONE 1.1, & our previous work

Algorithm/ Feature	Spatial B&B	Partitioning in relaxation step	OBBT calls	OBBT problems	Partitioning in OBBT step
Past (Hydro system)	Yes	Static ($N = 10$)	In every node of tree	LPs	N.A.
Past (refinery)	No	Dynamic (up)	Upon finding better solution	MILPs (McCormick envelopes)	N.A.
New algorithm	No	Dynamic (up/down)	While domain reduction is good	LPs or MILPs	Dynamic (up, $N \leq 7$)

- Highlights
  - New algorithm always returns lowest gap (4 solved to given tolerance ε)
  - Commercial solvers return one less suboptimal solution
  - ANTIGONE: 5 times faster in single period refinery problems (TP1)

		Optimality gap (%)				Wall time (s)				
Problem	Optimum	Literature	BARON	ANTIGONE	Algorithm	Literature	BARON	ANTIGONE	Algorithm	
HYD2	209,721.0	0.023	GO	0.536	GO	WTL	17417	WTL	5430	$\varepsilon = 0.0001\%$ WTL = 18,000 s
HYD4	371,811.8	0.585	1.359	1.323	0.465	WTL	WTL	WTL	WTL	
HYD7	744,963.7	1.369	2.285	1.816	1.260	WTL	WTL	WTL	WTL	
SC1TP1	55568.1	0.12	0.42	GO	GO	WTL	WTL	614	2973	$\varepsilon = 0.01\%$ WTL = 3,600 s
SC2TP1	49799.3	0.06	0.21	GO	GO	WTL	WTL	169	2062	
SC3TP1	53798.8	0.18	0.43	GO	GO	WTL	WTL	376	2900	
SC1TP3	55561.8	0.79	1.96	0.93	0.60	WTL	WTL	WTL	WTL	$\varepsilon = 0.01\%$ WTL = 10,800 s
SC2TP3	49878.5	0.27	1.36	0.30	0.26	WTL	WTL	WTL	WTL	
SC3TP3	53785.8	0.97	2.35	1.11	0.80	WTL	WTL	WTL	WTL	
SC4TP3	55323.3	0.66	2.39	0.79	0.54	WTL	WTL	WTL	WTL	

Detailed computational statistics

McMaster University

- Hydro problems benefit from piecewise relaxations
  - Last iteration of HYD2 with $N_{PR} = 1000$
- Refinery problems benefit from extensive OBBT
  - ADR > 78%
  - OBBT time >> PR+PF problems

Problem	HYD2	HYD4	HYD7	SC1TP1	SC2TP1	SC3TP1	SC1TP3	SC2TP3	SC3TP3	SC4TP3
Iterations	14	8	6	5	4	5	7	7	6	7
Time solving PR problems (s)	1140	12907	12386	571	424	553	940	864	900	938
Time solving PF problems (s)	155	86	69	63	46	58	143	108	107	130
Time in OBBT stage (s)	3902	3706	4418	2289	1554	2243	9665	9776	9750	9687
Time in depth search (s)	213	1301	1126	-	-	-	-	-	-	-
Instances of PF solved	203	111	134	160	107	116	256	166	218	273
Instances of PRB solved	2698	3610	3748	2630	1988	2630	13104	13094	10936	13104
ADR 1st iteration OBBT (%)	33.6	33.5	27.0	84.0	86.0	83.5	66.2	73.0	67.3	70.3
ADR final vs. initial bounds (%)	99.5	67.0	32.9	97.8	95.2	97.4	80.0	85.9	78.1	81.8
Final N_{PR}	1000	4	4	4	8	4	4	4	4	4
Final N_{PRB}	7	3	2	2	0	2	2	2	2	2

Stepwise decrease in optimality gap

McMaster University

- Major decrease in 1st iteration but takes time to go below commercial solvers (smoother profiles)
  - Sequence of procedures: OBBT, lower & upper bounding

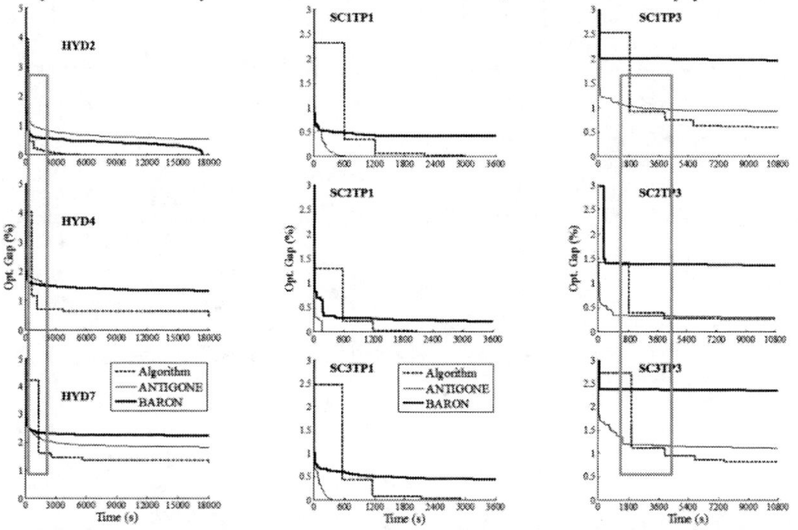

Conclusions

McMaster University

- New algorithm for MIQCPs without spatial B&B
  - Dynamic partitioning in piecewise relaxations
    - Switches between bilinear envelopes, piecewise McCormick & multiparametric disaggregation
    - Computing lower bound & reduce variables domain (OBBT)
  - Takes advantage of parallel computing
    - OBBT & generating upper bounds (solution pool of starting points)
- More problems solved to ε-global optimality compared to BARON & ANTIGONE
  - 10 real-life engineering problems from the literature
- Paper under review in Journal of Global Optimization
  - Special Issue Prof. Chris Floudas

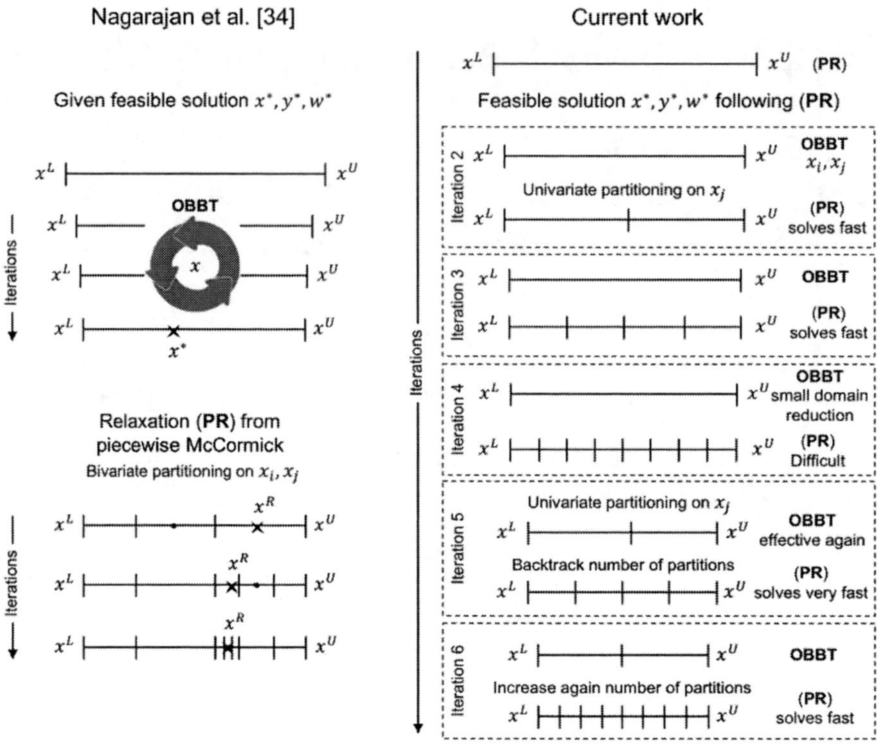

17AIChE
Annual Meeting, Minneapolis, MN

Solution from relaxation is good initialization

McMaster University

- Multiple starting points (e.g. CPLEX solution pool)

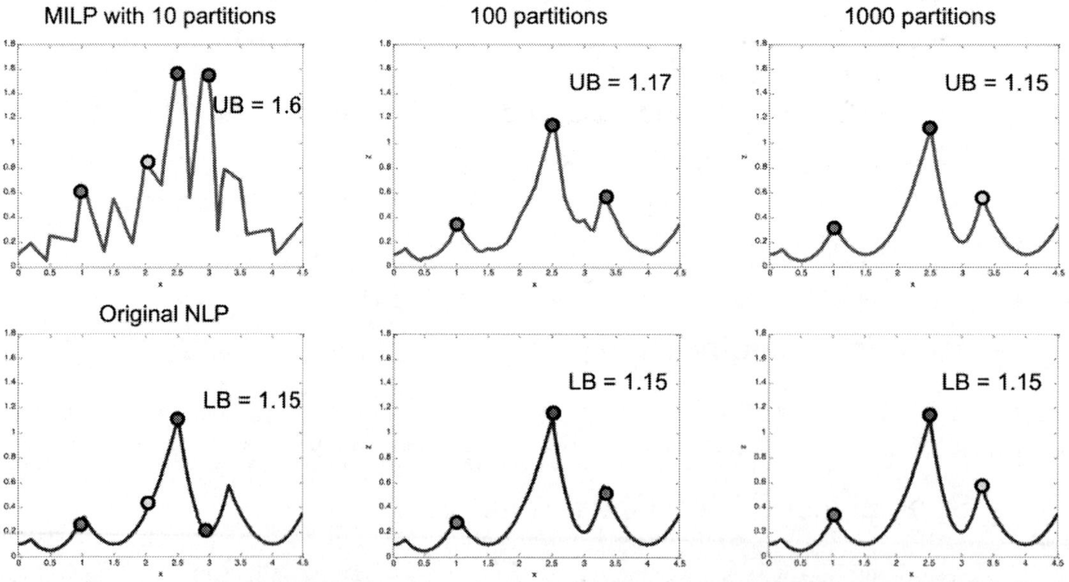

500211 A Level-Based Quadratic Outer Approximation Algorithm for Convex MINLP

Abstract:

A Level-Based Quadratic Outer Approximation Algorithm for Convex MINLP

Jan Kronqvist*, David E. Bernal**, Ignacio E. Grossmann**

*Chemical Engineering, Åbo Akademi University, Turku Finland, jan.kronqvist@abo.fi

**Chemical Engineering, Carnegie Mellon University, Pittsburgh PA

Mixed-integer Nonlinear programming (MINLP) deals with optimization problems containing both continuous and integer variables as well as linear and nonlinear functions. The integer variables make it possible to incorporate logic relations and discrete quantities in the optimization problem. The ability to incorporate logic decisions and nonlinear functions make it possible to accurately model complex systems. Therefore, there is a large number of applications in engineering and other fields, e.g., see [1] and [2]. An MINLP problem is usually regarded as convex, if an integer relaxation results in a convex nonlinear programming (NLP) problem [3]. There are several techniques for solving convex MINLP problems available, for example, the extended cutting plane (ECP) algorithm [4], extended supporting hyperplane (ESH) algorithm [5], outer approximation (OA) [6], generalized Benders decomposition (GDB) [7], and branch and bound techniques [8]. However, solving convex MINLP instances may still be a difficult task, as shown in a recent benchmark on convex MINLP solvers in [8].

OA can be considered as a decomposition technique, where the optimal solution is obtained by solving a sequence of mixed-integer linear programming (MILP) and convex NLP subproblems. A linear approximation of the MINLP problem is constructed and improved in each iteration in order to predict a non-decreasing sequence of lower bounds (assuming minimization problem). In each iteration, the integer variables are chosen as the minimizer of the linear approximation, and values for the continuous variables are obtained by solving a convex NLP problem where the integer variables are fixed. This problem yields an upper bound. When applied to a difficult MINLP problems, the normal OA technique may require a substantial number of iterations to obtain the optimal solution and verify optimality; i.e. converging the corresponding lower and upper bounds within a small tolerance.

Here we present a different approach for obtaining trial solutions, within an OA framework. The main idea is to use a different technique for obtaining trial values to the integer variables, and using these integer solutions in an OA algorithm. By utilizing second order derivatives of the nonlinear functions, we should be able to construct a better approximation of the MINLP problem, and thus, obtain better trial solutions. Instead of choosing the integer variables as the minimizer of a linear approximation, we propose to use a second order approximation of the Lagrangean function combined with the outer approximation cuts to obtain trial values for the integer variables.

In order to guarantee convergence, we cannot simply replace the linear objective of the MILP subproblems with a quadratic approximation of the Lagragean. However, by using ideas from the level method [9] we are able to guarantee convergence by introducing linear constraints that impose a specific improvement of the linearized objective function. These linear constraints restrict the search space into a specific level set of the linearized objective. Trial values for the integer variables are therefore obtained by minimizing a quadratic approximation the Lagrangean function such that all linear constraints and obtained cuts are satisfied. Thus, the algorithm uses second order information for obtaining trial solutions for the integer variables. Therefore, the algorithm should be able to find the optimal integer combination in fewer iterations. However, each iteration also becomes more computationally demanding compared to normal OA since it also requires the solution of a mixed-integer quadratic programming problem.

The level-based quadratic outer approximation algorithm will be presented in detail, including its theoretical properties. Furthermore, we present numerical comparisons against other convex MINLP techniques.

[1] Biegler, L. T., Grossmann, I. E., 2004. Retrospective on optimization. Computers & Chemical Engineering 28 (8), 1169–1192.

[2] Floudas, C. A., 2000. Deterministic Global Optimization, vol. 37 of Nonconvex Optimization and its Applications.

[3] Lee, J., Leyffer, S. (Eds.), 2011. Mixed Integer Nonlinear Programming. Vol. 154. Springer Science & Business Media.

[4] Westerlund, T., Petterson, F., 1995. An extended cutting plane method for solving convex MINLP problems. Computers and Chemical Engineering 19, S131–S136.

[5] Kronqvist, J., Lundell, A., Westerlund, T., 2015. The extended supporting hyperplane algorithm for convex mixed-integer nonlinear programming. Journal of Global Optimization 64 (2), 249–272.

[6] Duran, M. A., Grossmann, I. E., 1986. An outer-approximation algorithm for a class of mixed-integer nonlinear programs. Mathematical Programming 36 (3), 307–339.

[7] Geoffrion, A.M.: Generalized Benders decomposition. J. Optim. Theory Appl. **10**(4), 237–260 (1972)

[8] Dakin, R.J.: A tree-search algorithm for mixed integer programming problems. Comput. J. **8**(3), 250–255 (1965)

[9] Lemarechal, C., Nemirovskii, A., Nesterov, Y.: New variants of bundle methods. Mathematical programming **69**(1-3), 111-147 (1995)

Improved McCormick Relaxations in a Branch-and-Bound Framework[14]

Jaromił Najman and Alexander Mitsos

Process Systems Engineering
Aachener Verfahrenstechnik
RWTH Aachen University

[14] https://arxiv.org/abs/1710.09188

Aachener Verfahrenstechnik | RWTH AACHEN UNIVERSITY

Outline

Goal: Improve McCormick relaxations through subgradient propagation

Applications

P2X

- Sustainable fuel and chemicals from renewable electricity
- Design of efficient processes for production of Oxymethylene Ethers as Diesel substitutes
- Global optimization of chemical processes

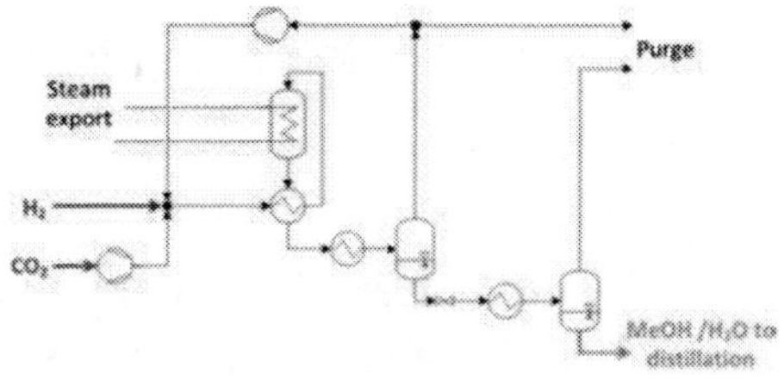

SynErgie

- Conceptual design of a cryogenic air separation unit
- High energy demand ($\approx$ 1% of industrial demand) due to compression
- Global optimization of physical processes

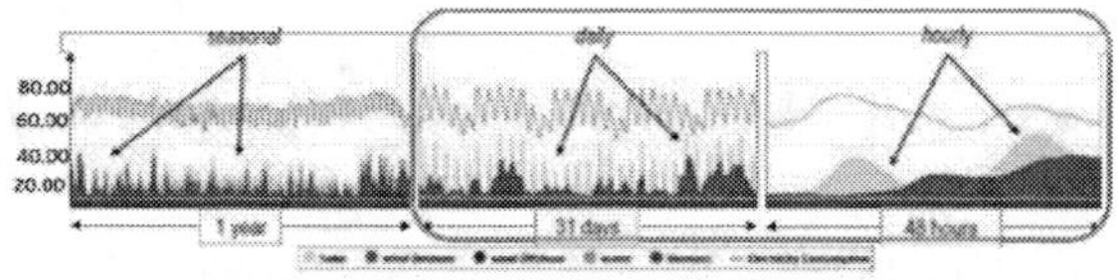

https://www.agora-energiewende.de/de/themen/-agothem-/Produkt/produkt/75/Agorameter/

Global Optimization Needs Tight Convex Relaxations I

Branch-and-bound (B&B) methods are state-of-the-art algorithms for deterministic global optimization

- Rely on favorable convergence order of the underlying relaxations (Du & Kearfott 1994 [6], Wechsung et al. 2014 [23], Kannan & Barton 2017 [8])

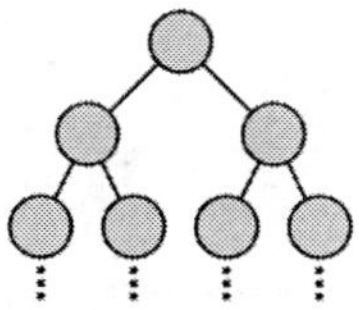

McCormick relaxations [10] may be constructed for a continuous factorable function

- Finite recursion of summation, multiplication, intrinsic functions and even procedures, $F_1 \circ f_1 + F_2 \circ f_2 \cdot F_3 \circ f_3$ (Mitsos et al. 2009 [12])
- Extended to multivariate outer functions (Tsoukalas & Mitsos 2014 [21])
- Used in the development of the auxiliary variable method (AVM) applied in state-of-the-art solvers (Smith & Pantelides 1993 [19], Tawarmalani & Sahinidis 2002 [20], Misener & Floudas 2014 [11])

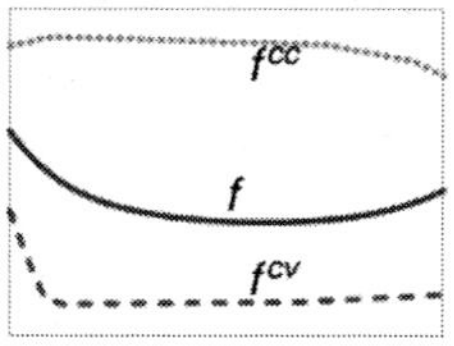

Global Optimization Needs Tight Convex Relaxations II

Domain and range reduction techniques improve quality of relaxations.

Many general tightening techniques have been developed:

- *Optimality-Based Bound Tightening* - solve additional optimization problems (Locatelli & Schoen 2013 [9])
- *Bound Propagation* - use constraints in order to tighten variable bounds (Brearley et al. 1975 [3])
- *Probing* - derive valid constraints from non-active ones (Tawarmalani & Sahinidis 2002 [20])
- ... and many more (Puranik & Sahinidis 2017 [15])

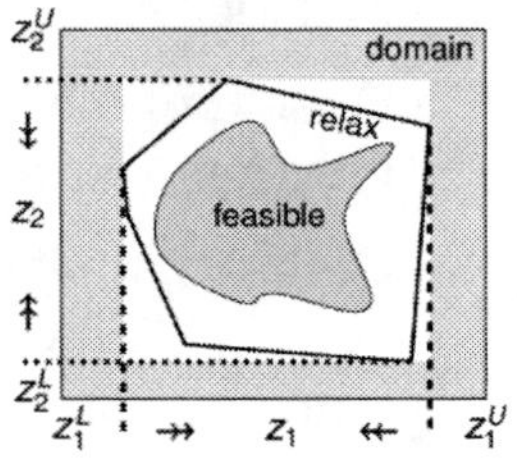

Most of the methods are (at least theoretically) applicable to McCormick relaxations

- Very few algorithms developed directly for McCormick relaxations (Wechsung et al. 2015 [24])
- We develop a new method for tighter McCormick relaxations

Factors of a Factorable Function

Let $g : Z \to \mathbb{R}, Z \in \mathbb{IR}^n$ be a (multivariate) *factorable* function consisting of factors f_j with $j \in \mathbb{F}$.
For each factor f_j, we have

- Convex and concave McCormick relaxations $f_j^{cv}(\bar{\mathbf{z}}), f_j^{cc}(\bar{\mathbf{z}})$ at $\bar{\mathbf{z}} \in Z$
- Convex and concave subgradients $\mathbf{s}_j^{cv}(\bar{\mathbf{z}}), \mathbf{s}_j^{cc}(\bar{\mathbf{z}})$ and corresponding affine functions $f^{cv,sub}(\bar{\mathbf{z}}, \mathbf{z}), f^{cc,sub}(\bar{\mathbf{z}}, \mathbf{z})$ at $\bar{\mathbf{z}} \in Z$
- Lower and upper bounds $I_{j,nat} = [f_j^L, f_j^U]$ on the range of f_j over Z

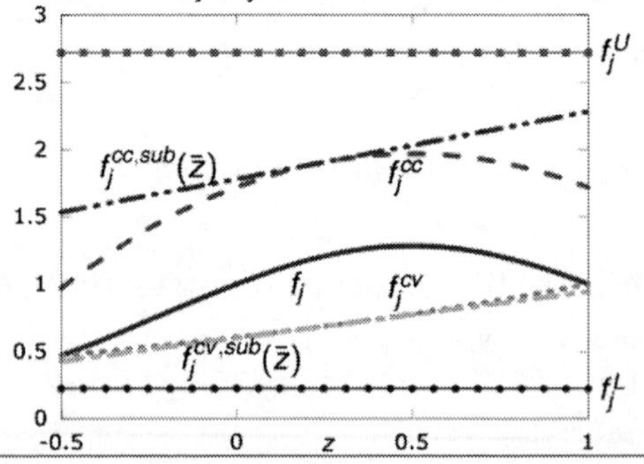

Computational Graph

A directed-acyclic-graph (DAG) of g is given by $G = (\mathbb{F}, E)$ ([17]).
Let $g(z) = (z - z^2) \cdot (z^3 - \exp(z))$ on $Z = [-0.5, 1]$.

- f_j depends on previous factors.

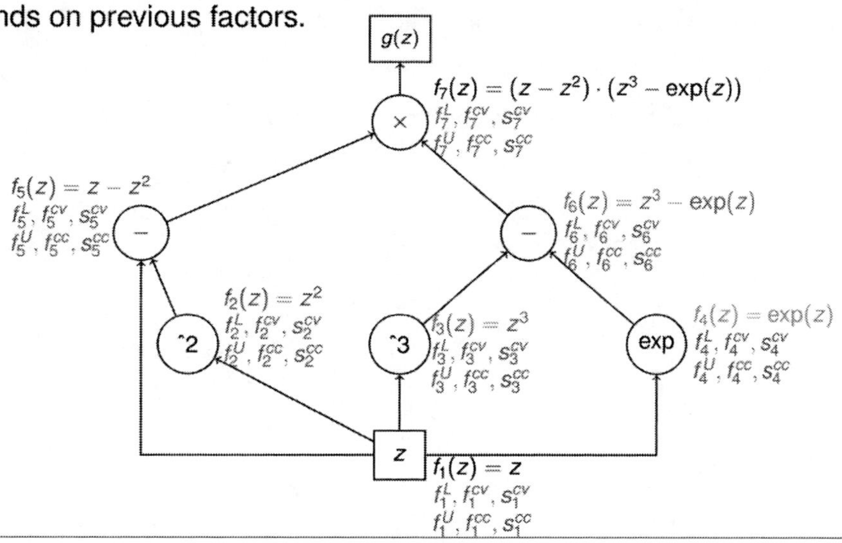

Basic Idea I

Consider $g(z) = (z - z^2) \cdot (z^3 - \exp(z))$ on $Z = [-0.5, 1]$

- In particular the factors $f_5(z) = z - z^2$ and $f_6(z) = z^3 - \exp(z)$.
- For factors f_1, f_2, f_3, f_4 envelopes and exact range bounds are known.

For f_5:

- Construct McCormick relaxations and interval bounds $I_{5,nat} = [-1.5, 1]$.
- Use subgradients at midpoint of Z to tighten $I_{5,nat}$ to $I_{5,alg} = [-0.75, 0.5625]$.

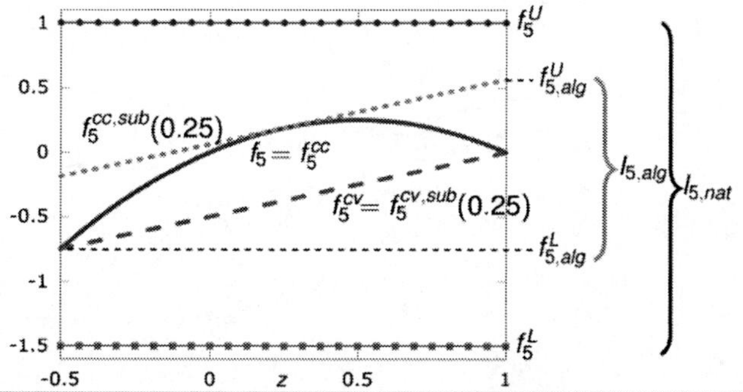

Basic Idea II

Consider $g(z) = (z - z^2) \cdot (z^3 - \exp(z))$ on $Z = [-0.5, 1]$
For $f_6(z) = z^3 - \exp(z)$:

- Construct McCormick relaxations and interval bounds $I_{6,nat} \approx [-2.843, 0.393]$.
- Use subgradients at midpoint of Z to tighten $I_{6,nat}$ to $I_{6,alg} \approx [-2.562, -0.446]$.

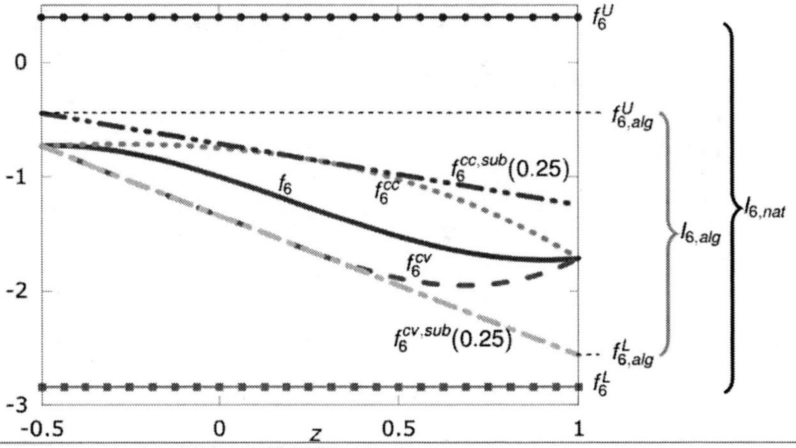

Basic Idea III

Construct McCormick relaxations for
$g(z) = f_7(z) = (z - z^2) \cdot (z^3 - \exp(z)) = f_5(z) \cdot f_6(z)$

- With *natural interval bounds* $I_{5,nat} \times I_{6,nat}$ to get $g_{nat}^{cv}, g_{nat}^{cc}$.
- With **improved** bounds $I_{5,alg} \times I_{6,alg}$ to get $g_{alg}^{cv}, g_{alg}^{cc}$.

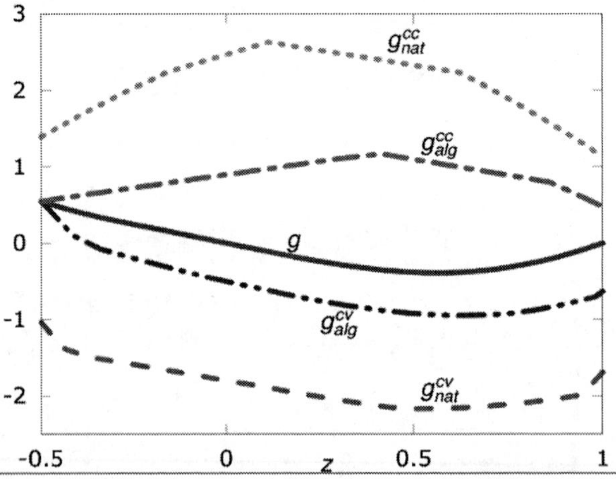

Formal Definition of the Method

$g : Z \to \mathbb{R}, Z = [\mathbf{z}^L, \mathbf{z}^U] \in \mathbb{IR}^n$, i =variable dimension, j =factor
k =iteration, N =max. number of iterations

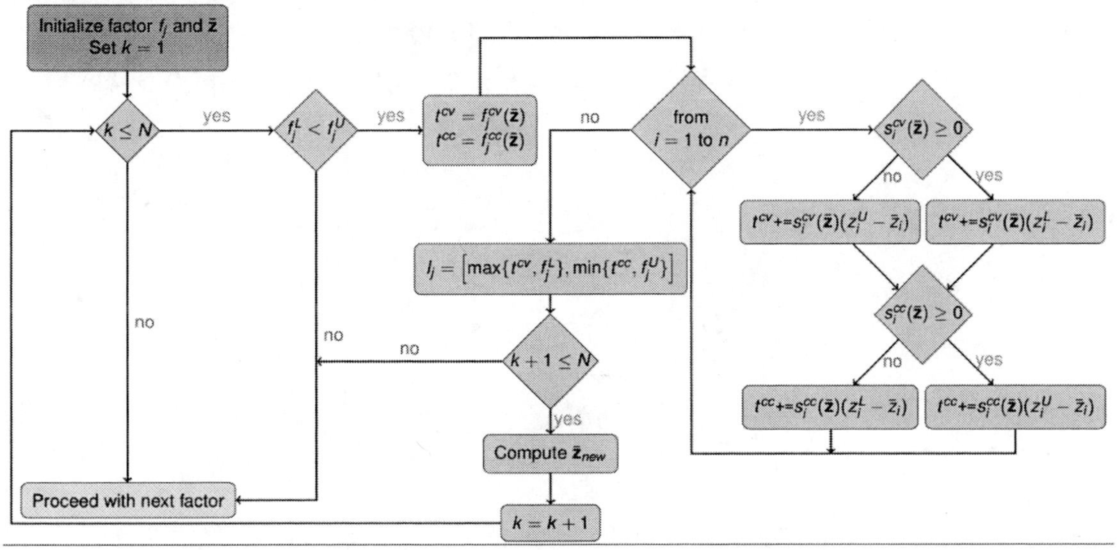

Choice of Linearization Point

Obtained range bounds are

- Valid and at best exact
- Not worse than a priori $f_j^L \leq f_{j,alg}^L \leq \min_{\mathbf{z} \in Z} f^{cv}(\mathbf{z})$ and $\max_{\mathbf{z} \in Z} f^{cc}(\mathbf{z}) \geq f_{j,alg}^U \geq f_j^U$

The choice of the initial point $\bar{\mathbf{z}}$ can be crucial

- $f(z) = \exp(z) - z^3$ on $Z = [-1, 1]$ with $\bar{z} = 1$ $\bar{z} = 0$

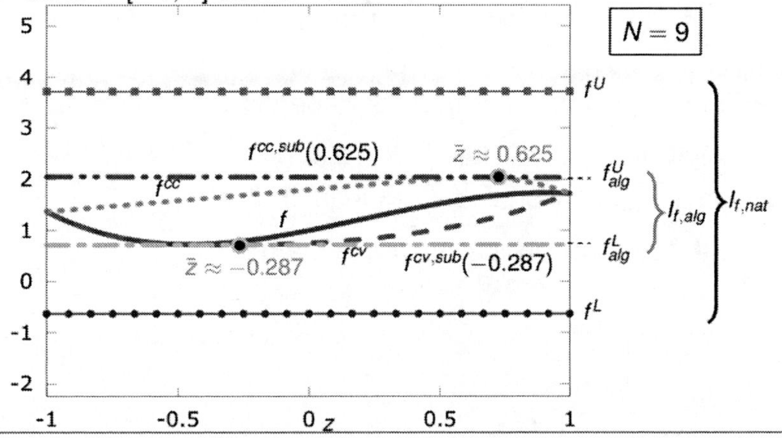

Other Interval Arithmetics

The method also works for more sophisticated interval arithmetics

- Quadratic Taylor form $g(c) + g'(c)(Z - c) + \frac{g''(Z)}{2}(Z - c)^2$
- $g : Z \to \mathbb{R}$ is twice differentiable, c is the middle point of Z

$$g(z) = (\log(z+1) - z^2) \cdot (\log(z+1) - \exp(z - 0.5)) \text{ on } Z = [-0.5, 1]$$

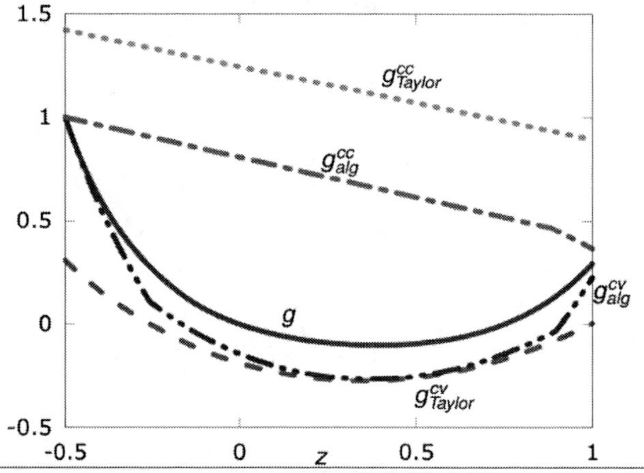

Approximate Scaling of Computational Cost

For $N = 1$

- Linear in the dimension of optimization variables $\mathbf{z} \in \mathbb{R}^n$, $\mathcal{O}(n)$ for factor f_j
- Comparable to computation of McCormick relaxations and subgradients
- No recalculation of factor f_j which depends on other factors
- Additional computational time is low

For $N > 1$

- Need to recalculate factor f_j which depends on $\mathcal{J}$ other factors
- This results in $\mathcal{O}(\mathcal{J}^N)$ computations (large for $\mathcal{J} \gg 1$ and $N > 1$)
- Additional computational time is clearly notable

Numerical experiments confirm the claims

Numerical Experiments: Computational Setup

We use our in-house global optimization solver MAiNG

- Upper bound by NLP Solver IPOPT (Wächter & Biegler 2006 [22])
- Automatic Differentiation via FADBAD++ [1]
- McCormick relaxations and subgradients via MC++ (Chachuat 2015 [4])
- Linearize the relaxations with subgradients and obtain lower bound through LP solver CPLEX [5]

Numerical examples:

- 13 small problems chosen arbitrarily from COCONUT library [18]
- 2 well-known functions (6-hump-camel, ursem-waves)
- 3 case studies of combined-cycle power plants (Bongartz & Mitsos 2017 [2])
- 2 case studies for parameter estimation (Mitsos et al. 2009 [12])

Utilization of our own library of functions in MC++ [13].

Numerical Results $N = 1$

abs&rel opt tol $\epsilon^{opt} = 10^{-4}$, abs&rel feas tol $\epsilon^{feas} = 10^{-6}$, max time$= 3600s$

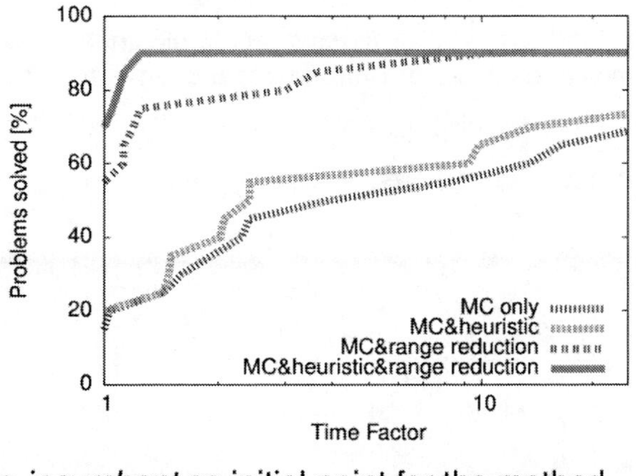

- Here we used the *incumbent* as initial point for the method
- Substantial speedup over McCormick relaxations with and without range reduction
- Range reduction (OBBT [7], DBBT [16]) shows drastic complementary speedup

Numerical Results $N > 1$

abs&rel opt tol $\epsilon^{opt} = 10^{-4}$, abs&rel feas tol $\epsilon^{feas} = 10^{-6}$, max time$= 3600s$

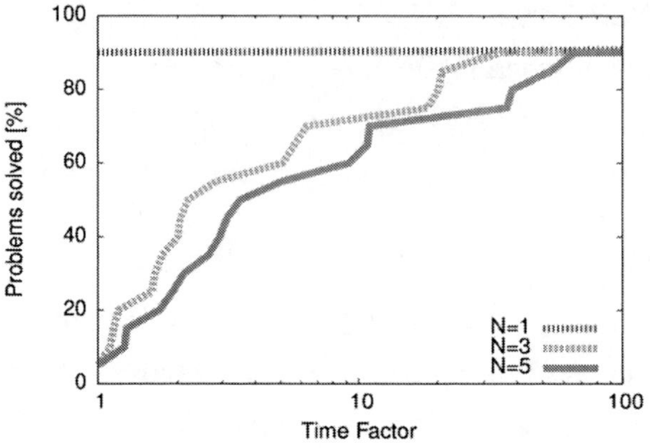

- Here only McCormick relaxations and the presented method were used
- We see an obvious loss in computational time for increasing N

Conclusions

- New method based on subgradient propagation for tightening of McCormick relaxations
  - Use subgradients of McCormick relaxations to construct linear estimators in each factor
  - Improve range bounds of each factor with the linear estimators
- Numerical results show very high potential for $N = 1$
- Additional iterations become expensive quickly

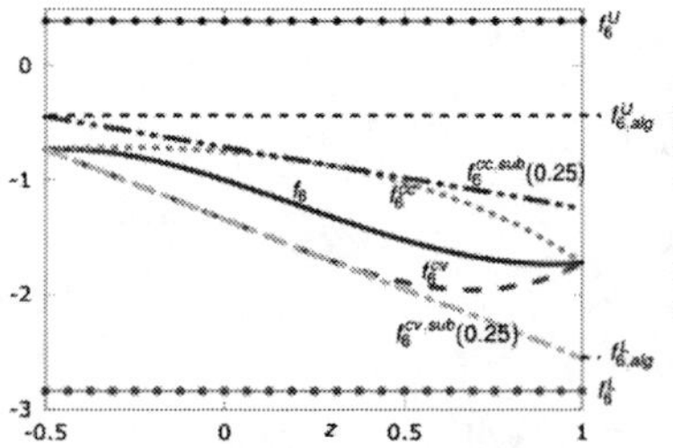

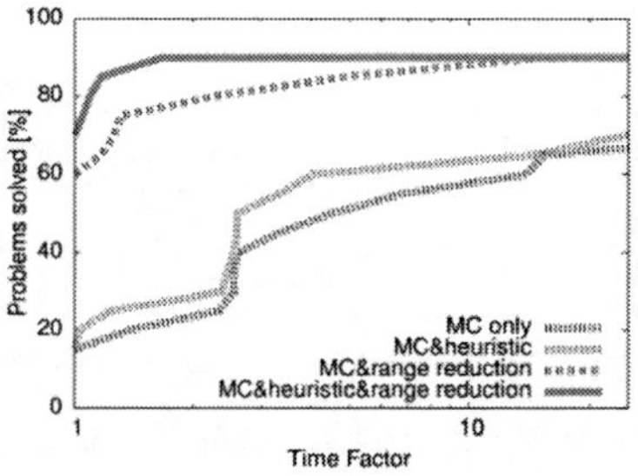

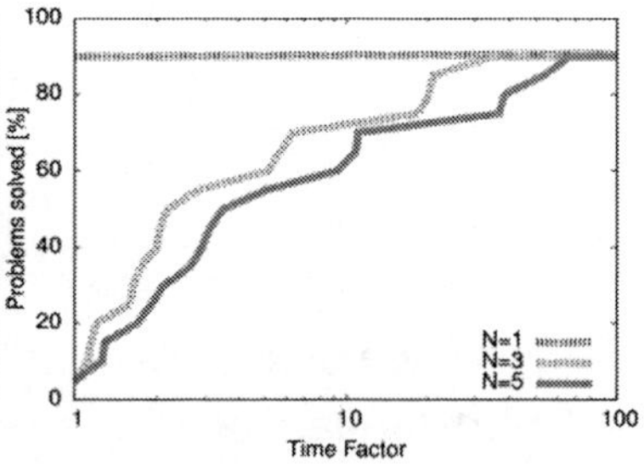

Acknowledgments

Thank you for your attention

- This project has received funding from the Deutsche Forschungsgemeinschaft (DFG, German Research Foundation) *Improved McCormick Relaxations for the efficient Global Optimization in the Space of Degrees of Freedom* MA 1188/34-1
- The authors gratefully acknowledge funding by the German Federal Ministry of Education and Research (BMBF) within the Kopernikus Project P2X: Flexible use of renewable resources exploration, validation and implementation of Power-to-X concepts.
- The authors gratefully acknowledge the financial support of the Kopernikus-project SynErgie by the Federal Ministry of Education and Research (BMBF) and the project supervision by the project management organization Projektträger Jülich (PtJ).
- Special thanks to Benoît Chachuat for providing MC++ [4]
- Dominik Bongartz for providing numerical case studies

References I

[1] C. Bendtsen and O. Staunin. *FADBAD++, a flexible C++ package for automatic differentiation. Version 2.1.* http://www.fadbad.com. Accessed 18 October 2016. 2012.

[2] Dominik Bongartz and Alexander Mitsos. "Deterministic global optimization of process flowsheets in a reduced space using McCormick relaxations". In: (2017). DOI: 10.1007/s10898-017-0547-4.

[3] A. L. Brearley, G. Mitra, and H. P. Williams. "Analysis of mathematical programming problems prior to applying the simplex algorithm". In: *Mathematical Programming* 8.1 (1975), pp. 54–83. ISSN: 1436-4646. DOI: 10.1007/BF01580428. URL: https://doi.org/10.1007/BF01580428.

[4] Benoît Chachuat et al. "Set-Theoretic Approaches in Analysis, Estimation and Control of Nonlinear Systems". In: *IFAC-PapersOnLine* 48.8 (2015). https://omega-icl.github.io/mcpp/(Accessed February 2017), pp. 981 –995. ISSN: 2405-8963. DOI: https://doi.org/10.1016/j.ifacol.2015.09.097. URL: http://www.sciencedirect.com/science/article/pii/S2405896315011787.

[5] International Business Machines Corporation: *IBM ILOG CPLEX v12.5. Armonk.* 2009.

[6] Kaisheng Du and R Baker Kearfott. "The cluster problem in multivariate global optimization". In: *Journal of Global Optimization* 5.3 (1994), pp. 253–265.

[7] Ambros M Gleixner et al. "Three enhancements for optimization-based bound tightening". In: *Journal of Global Optimization* 67.4 (2017), pp. 731–757.

[8] Rohit Kannan and Paul I. Barton. "The cluster problem in constrained global optimization". In: *Journal of Global Optimization* (2017). ISSN: 1573-2916. DOI: 10.1007/s10898-017-0531-z. URL: https://doi.org/10.1007/s10898-017-0531-z.

[9] Marco Locatelli and Fabio Schoen. *Global optimization: theory, algorithms, and applications.* SIAM, 2013.

[10] Garth P. McCormick. "Computability of global solutions to factorable nonconvex programs: Part I-Convex underestimating problems". English. In: *Mathematical Programming* 10 (1976), pp. 147–175. ISSN: 0025-5610.

References II

[11] Ruth Misener and ChristodoulosA. Floudas. "ANTIGONE: Algorithms for coNTinuous / Integer Global Optimization of Nonlinear Equations". English. In: *Journal of Global Optimization* 59.2-3 (2014), pp. 503–526. ISSN: 0925-5001. DOI: 10.1007/s10898-014-0166-2. URL: http://dx.doi.org/10.1007/s10898-014-0166-2.

[12] Alexander Mitsos, Benoît Chachuat, and Paul I Barton. "McCormick-based relaxations of algorithms". In: *SIAM Journal on Optimization* 20.2 (2009), pp. 573–601.

[13] Jaromił Najman and Alexander Mitsos. "Convergence Order of McCormick Relaxations of LMTD Function in Heat Exchanger Networks". In: *26th European Symposium on Computer Aided Process Engineering.* Ed. by Zdravko Kravanja and Milo Bogataj. Vol. 38. Computer Aided Chemical Engineering. Elsevier, 2016, pp. 1605 –1610. DOI: http://dx.doi.org/10.1016/B978-0-444-63428-3.50272-1. URL: http://www.sciencedirect.com/science/article/pii/B9780444634283502721.

[14] Jaromił Najman and Alexander Mitsos. "Tighter McCormick Relaxations through Subgradient Propagation". In: (2017). In Revision. URL: https://arxiv.org/abs/1710.09188.

[15] Yash Puranik and Nikolaos V Sahinidis. "Domain reduction techniques for global NLP and MINLP optimization". In: *Constraints* (2017), pp. 1–39.

[16] Hong S Ryoo and Nikolaos V Sahinidis. "Global optimization of nonconvex NLPs and MINLPs with applications in process design". In: *Computers & Chemical Engineering* 19.5 (1995), pp. 551–566.

[17] Hermann Schichl and Arnold Neumaier. "Interval analysis on directed acyclic graphs for global optimization". In: *Journal of Global Optimization* 33.4 (2005), pp. 541–562.

[18] Oleg Shcherbina et al. "Benchmarking Global Optimization and Constraint Satisfaction Codes". In: *Global Optimization and Constraint Satisfaction: First International Workshop on Global Constraint Optimization and Constraint Satisfaction, COCOS 2002, Valbonne-Sophia Antipolis, France, October 2002. Revised Selected Papers.* Berlin, Heidelberg: Springer Berlin Heidelberg, 2003, pp. 211–222. ISBN: 978-3-540-39901-8. DOI: 10.1007/978-3-540-39901-8_16. URL: http://dx.doi.org/10.1007/978-3-540-39901-8_16.

References III

[19] Edward MB Smith and Constantinos C Pantelides. "Global optimisation of nonconvex MINLPs". In: *Computers & Chemical Engineering* 21 (1997), pp. 791–796.

[20] Mohit Tawarmalani and Nikolaos V Sahinidis. *Convexification and global optimization in continuous and mixed-integer nonlinear programming: theory, algorithms, software, and applications.* Vol. 65. Springer Science & Business Media, 2002.

[21] Angelos Tsoukalas and Alexander Mitsos. "Multivariate McCormick Relaxations". In: *Journal of Global Optimization* 59 (2014), pp. 633–662.

[22] Andreas Wächter and Lorenz T. Biegler. "On the implementation of an interior-point filter line-search algorithm for large-scale nonlinear programming". In: *Mathematical Programming* 106.1 (2006), pp. 25–57. ISSN: 1436-4646. DOI: 10.1007/s10107-004-0559-y. URL: https://doi.org/10.1007/s10107-004-0559-y.

[23] Achim Wechsung, Spencer D. Schaber, and Paul I. Barton. "The cluster problem revisited". English. In: *Journal of Global Optimization* 58.3 (2014), pp. 429–438. ISSN: 0925-5001.

[24] Achim Wechsung et al. "Reverse propagation of McCormick relaxations". In: *Journal of Global Optimization* 63.1 (2015), pp. 1–36. ISSN: 1573-2916. DOI: 10.1007/s10898-015-0303-6. URL: https://doi.org/10.1007/s10898-015-0303-6.

494501 Scalable Global Algorithms for Stochastic Nonlinear Programming

Abstract:

Many large-scale nonlinear programs (NLPs) are block structured and weakly coupled [1,2] (i.e., the NLP is composed of block subproblems problems connected by few complicating variables). For example, a two-stage stochastic program has multiple scenarios connected by first stage variables. Problems arising from parameter estimation and machine learning can also fit in this category.

This presentation introduces a new global optimization algorithm for block structured NLPs that uses a specialized branch and bound strategy. For each node in the branch and bound scheme, a lower bound is constructed by relaxing coupling constraints between complicating variables and an upper bound is constructed by fixing the complicating variables. A key advantage of this approach is that both lower bounding and upper bounding problems can be decomposed into blocks that are solved in parallel to global and local optimality, respectively. Another key property of this approach is that, because the gap between the upper bound and lower bound converges to zero as the bounds of the coupling variables converges to zero, we only need to perform branching on the complicating variables. A similar approach has been recently proposed for stochastic MINLPs [5].

We show how the algorithm can be easily implemented in Julia and interfaced to modeling languages such as JuMP and PLASMO. Our implementation also contains typical global optimization algorithmic features such as convexification, outer approximation, feasibility-based bound tightening (FBBT), optimality-based bound tightening (OBBT), and local search. Numerical experiments are performed on a stochastic optimization formulation for controller tuning and a parameter estimation formulation for microbial community models. We compare the computational results with the commercial solver BARON [3] and with open-source solver SCIP [4], and we show that the new approach achieves significant speedups.

References:

[1] Kang, J., Cao, Y., Word, D.P. and Laird, C.D., 2014. An interior-point method for efficient solution of block-structured NLP problems using an implicit Schur-complement decomposition. Computers & Chemical Engineering, 71, pp.563-573.

[2] Kang, J., Chiang, N., Laird, C.D. and Zavala, V.M., 2015, December. Nonlinear programming strategies on high-performance computers. In Decision and Control (CDC), 2015 IEEE 54th Annual Conference on (pp. 4612-4620). IEEE.

[3] Tawarmalani, M. and Sahinidis, N.V., 2005. A polyhedral branch-and-cut approach to global optimization. Mathematical Programming, 103(2), pp.225-249.

[4] Maher, S.J., Fischer, T., Gally, T., Gamrath, G., Gleixner, A., Gottwald, R.L., Hendel, G., Koch, T., Lübbecke, M.E., Miltenberger, M. and Müller, B., 2017. The SCIP Optimization Suite 4.0.

[5] Li, Xiang, Asgeir Tomasgard, and Paul I. Barton. "Nonconvex generalized Benders decomposition for stochastic separable mixed-integer nonlinear programs." *Journal of optimization theory and applications* 151.3 (2011): 425.

496269 Analysis of the Alternating Direction Method of Multipliers for the Optimization of Distributed Nonconvex Systems

Abstract:

The alternating direction method of multipliers (ADMM) is an optimization method that has received a lot of attention for solving convex optimization problems. Some of this attention is because ADMM can be parallelized for problems with a certain separable structure. It is particularly suited to problems that have a decentralized or distributed nature.

However, very little analysis has been done on the theoretical properties and numerical performance of ADMM for the nonconvex problems of interest in chemical engineering applications. As a motivating example, consider the coordination or integration of various real-time optimization (RTO) applications across a refinery. Each RTO application can be a large nonconvex optimization problem with the aim of achieving economically optimal steady-state operation of a particular process unit or group of units. Of course these units, with separate RTO applications, are actually connected by material flows. Thus, the RTO problems can be connected by extra constraints in order to obtain a larger-scope optimization problem. Ideally, an optimization method for this larger, integrated optimization problem could take advantage of the existing infrastructure, and â€œsit on topâ€ of the models and software in use for the RTO applications. At least superficially, ADMM is a promising candidate, and numerical experiments with problems with similar structure show that ADMM does seem to work.

This work aims to support these numerical experiments and provide a convergence analysis of ADMM applied to nonconvex problems. Under fairly standard regularity assumptions on a local minimizer, we establish conditions under which ADMM does indeed converge to that minimum. These conditions include that the penalty parameter in the augmented Lagrangian be greater than a certain threshold. Bolstered by these theoretical guarantees, we propose putting ADMM in a trust region framework to improve its robustness and global convergence properties.

493722 Guaranteed Global Optimization of Expected-Value Minimization Problems with Continuous Ranom Variables

Abstract:

Stochastic programming is a powerful framework for determining optimal design and operational decisions under uncertainty, and is extensively used in manufacturing, chemical processing, power distribution, and wireless communication, to name only a few. In this talk, we propose a new algorithm for globally solving a specific class of stochastic programs with (i) nonconvex models, which arise in chemical processes, water and gas networks, AC power systems, etc.; (ii) continuously distributed random variables (RVs), which characterize uncertainty in, e.g., product demands, returns on investments, and renewable energy resources; (iii) expected-value objective functions; and (iv) no recourse decisions.

In the existing literature, such problems are most commonly addressed using sample-average approximation (SAA), which solves a deterministic approximation using a finite number of sampled scenarios. However, this often provides an unworkable compromise between accuracy and efficiency. Specifically, using too few scenarios can result in approximate solutions that are highly suboptimal or even infeasible. On the other hand, increasing the number of scenarios leads to high computational costs when computing sample-average approximations of the objective and constraints. Moreover, verifying the accuracy of a solution generally requires solving many repeats of the deterministic problem with independent samples, leading to an unmanageable computational burden for many problems of practical interest. Another common approach is the so-called stochastic branch-and-bound (B&B) algorithm. However, unlike standard B&B algorithms, the upper and lower bounds obtained in stochastic B&B are sample-based and are only valid in a probabilistic sense. Thus, stochastic B&B has a nonzero probability of fathoming optimal solutions.

To address these challenges, this talk presents a new Branch-and-Bound (B&B) algorithm that is guaranteed to converge finitely to epsilon-global solutions of nonconvex stochastic programs with continuous RVs, and does so with significantly higher efficiency than existing approaches. This is achieved using discrete approximations based on exhaustive partitions of the uncertainty space rather than on sampled scenarios. A key insight in our approach is that such approximations enable the construction of convergent convex relaxations that rigorously account for discretization error. In turn, this enables the computation of rigorous, deterministic upper and lower bounds that can be exploited within B&B to obtain global solutions with provable accuracy. To our knowledge, no existing approach can provide such deterministic bounds for nonconvex programs with continuous RVs. Moreover, by deterministically bounding the discretization error, it becomes possible to prove infeasibility or suboptimality of regions of the decision space using only coarse partitions of the uncertainty space. Using this insight, we achieve high efficiency by adaptively refining the uncertainty partition as the B&B search proceeds.

Compared to stochastic B&B, our B&B approach is advantageous because it uses deterministic rather than probabilistic upper and lower bounds. Thus, it is impossible to fathom a global solution. Compared to SAA, our approach is more rigorous because it provides epsilon-global solutions finitely, whereas SAA converges to the true solution with probability 1 only as the number of scenarios goes to infinity. Moreover, our approach can be significantly more efficient than SAA because we use a partition of the uncertainty space that is adaptively refined during B&B. In contrast, SAA must use a fixed number of scenarios during optimization, and must begin from scratch if additional scenarios are deemed necessary.

The advantages of our proposed B&B approach will be demonstrated through case studies in chemical reactor design and renewable energy systems.

494614 Go Now or Wait and See? - Optimal Investment Timing in National Power Systems

Abstract:

The reluctant deployment of low-carbon and zero-carbon power generation and energy storage technologies, perhaps besides wind and solar power technologies, are impeding the transition to power system decarbonisation. We put to test the notion that a strategy of waiting for a "unicorn technology" to become available could be more effective than deploying technologies which are commercial and viable today.
Especially in the context of Carbon Capture and Storage (CCS) equipped power plants, we are facing the question if it is economically and environmentally more sensible to "go now" or to "wait and see".

To answer such a question, we have expanded on previous work [1], in developing a mixed-integer linear formulation of a hybrid power generation capacity expansion and unit commitment model with a high level of technical detail. We apply a novel time clustering and profiling technique to compress hourly time-dependent data sets achieving an average error in system-level results of -1.7 % to 2.5 % compared to the full hourly time series over the planning horizon from 2015 to 2050. The national-scale model is implemented in different facets of temporal and spatial (dis-)aggregation; it considers up to 2000 units of 16 different power generation and storage technologies, including international interconnectors for electricity import and export, and grid-level energy storage. The Electricity System Optimisation framework (ESO-XEL) includes the consideration of endogenous learning for technology capital cost in a piece-wise linear fashion.

We compare a baseline scenario on the power system expansion of the United Kingdom (UK) with a scenario where a "unicorn technology" becomes available for deployment in 2035. This technology is parametrised as a zero-carbon emission, dispatchable, highly flexible power generation option at low capital and operational cost. We observe that the effect of such a technology becoming available reduces the deployment for nuclear, unabated gas-fired power generation, as well as intermittent renewables. Also, CCS-equipped power generation would reduce by 36 %, however, it remains a vital part of the least-cost capacity mix in 2050. Furthermore, its optimal investment timing remains unaffected by the availability of the future disruptive technology. Total system cost are reduced by 2 % by the deployment of the unicorn technology. More significantly however, without the deployment of CCS-equipped gas and coal-fired power plants total system cost by 2050 could be 44 % greater compared to the baseline scenario.

[1] C. F. Heuberger, I. Staffell, N. Shah, and N. Mac Dowell. Levelised Value of Electricity - A Systemic Approach to Technology Valuation. In 26th European Symposium on Computer Aided Process Engineering, volume 38, pages 721-726, 2016.

496660 Beyond Lcoe: Market-Based Design of Flexible Solar Thermal Systems

Abstract:

Fueled by diverse economic, social, and political factors, over 50% of electrical energy is expected to come from renewable sources by 2030 in states including California and New York. This poses new challenges for grid operators, as many renewable sources (e.g., solar, wind) are intermittent. New sources of flexibility are required to provide contingency and ensure generation always matches consumption.

Concentrated solar power (CSP) systems are a promising technology for providing both baseline and flexible energy to power grids. In brief, solar thermal energy is obtained by concentrating solar radiation into a heat transfer medium (e.g., Dowtherm A, molten salt, air) which ultimately used to drive a thermodynamic power cycle (e.g., Rankine, Brayton). Thermal energy storage (TES) is realized by installing insulated tanks. Thus, CSP systems with TES are dispatchable and able shift generation to times with intermittent solar irradiance. Current literature and Department of Energy evaluations focus on improving designs in order to reduce the so-called Levelized Cost of Electricity (LCOE) (i.e., the average electricity generation cost) [1]. We argue that this approach dramatically underutilizes the inherent flexibility of CSP systems, as it neglects the time value of electricity [2]. This is important because in many regions of the world electrical energy is traded in multiscale wholesale markets that exhibiti significant price fluctuations.

In this talk, we present a general framework to co-optimize design decisions (i.e., collector field and storage sizing) and operational decisions (i.e., generation schedules and market participation) for CSP systems [3,4]. We use the framework to compare different performance metrics, such as LCOE, power purchase agreements (PPA) with time-of-day (TOD) factors, and net present value (NPV) with market revenues. Using historical market and solar irradiation data for California, we compute optimal designs with 15 hours of storage and minimum LCOE of 147 $ / MWh. We show that the average value of the electricity delivered to the grid, computed from historic market prices, is only 31 $ / MWh, indicating an implicit incentive of 116 $ / MWh from the LCOE mechanism. As a reference, average market prices in CA range from 20 $ / MWh (in the early morning and solar noon) to 50 $ / MWh (in the late evening) [2]. Motivated by these results, we compute the necessary incentives to facilitate direct CSP participation in California energy markets. We then discuss the link between the financial metric and the optimal design. In summary, we find that LCOE favors larger storage systems (10+ hours), as the value of electricity is time-invariant. In contrast, market-based mechanisms favor smaller storage systems (4 to 6 hours) that can take advantage of the dominant evening price spikes and provide ancillary services throughout the rest of the day. In conclusion, we argue that the proposed new market-based design approach can provide CSP technologies with a better value proposition and differentiation from cheaper non-dispatchable photovoltaic technologies.

References:

[1] SunShot Initiative. https://energy.gov/eere/sunshot/sunshot-initiative

[2] Dowling, Zheng, Zavala (2017). Economic Assessment of Concentrated Solar Power Technologies: A Review. *Renewable and Sustainable Energy Reviews* 72, pg. 1019-1032.

[3] Dowling, Kumar, Zavala (2016). A Multi-Scale Optimization Framework for Electricity Market Participation. *Applied Energy* 190, pg. 147-164.

[4] Dowling, Dyreson, Miller, Zavala (2016). Economic Assessment and Optimal Operation of CSP System with TES in California Energy Markets. To Appear in *Proceedings for SolarPACES 2016*. Preprint available at zavalab.engr.wisc.edu.

500791 Enterprise-Wide Modular Process Intensification and Multi-Scale Optimization for Natural Gas Utilization

Abstract:

Natural gas will account for about 25% of the total energy consumption from primary sources by 2035.[1] It is also a clean fossil fuel source with abundant reserves. Though methane (CH_4) is the main component of natural gas, some of the natural gas reserves are uneconomical for exploitation due to high CO_2 contents sometimes up to 70% by volume.[2] CO_2 contents are also high for unconventional sources of methane such as biogas, coal bed methane, land fill gas etc. To meet specifications for transport through pipelines, the level of impurities in methane has to be less than 3â€"4 %. Existing standalone technologies for separation of methane from impurities involve separation, compression for pipeline transport and storage for consumption by the end user. The separation step usually via pressure swing adsorption (PSA) involves high pressure for sufficient adsorption, followed by vacuum depressurization for regeneration of the adsorbent, followed by re-pressurization for pipeline transportation. The successive pressure changes and the lack of integration between these steps makes the supply chain energy and capital intensive. Moreover, small and stranded sources of methane are not economical for pipeline transportation.

We have developed a modular one-step process[3] which intensifies the separation and storage of methane from sources in the same unit thereby eliminating the pressure changes involved in the supply chain. The technology is based on preferential adsorption of CH_4 using micro-porous materials such as zeolites. In this combined separation and storage (CSS) process, a mixture of $CH_4/CO_2/CH_4$ is fed to the column filled with zeolite, where CH_4 get separated and is stored in the column till the column is saturated while impurities leave the column from the other end. The column/modular unit saturated with CH_4 can then be used in applications such as vehicles and be sent back after use for refilling. The CSS process is described by a non-linear algebraic partial differential equation (NAPDE) model which incorporates mass and energy conservation equations, pressure drop correlations, driving force for mass transfer into the adsorbent and the adsorbent isotherm relating the equilibrium gas loading on the adsorbent material. To obtain optimal operating conditions for the process, the model is completely discretized in space and time into a system of non-linear algebraic equations whose size increases with the level of discretization.

To screen the existing zeolite materials along with process optimization, binary variables relating to the isotherm parameters are introduced into the model. We have also performed Grand Canonical Monte Carlo (GCMC) simulations for CH_4, CO_2 and N_2 on pure silica zeolites in the IZA-SC database to obtain the adsorption loading data at different pressures and temperatures. The isotherm model parameters are then obtained by solving the minimization problem of least-square error between the data and predicted model to global optimality. The simultaneous process optimization and materials screening problem thus becomes a mixed integer non-linear programming (MINLP) problem. We have solved the model to optimality to simultaneously select the best zeolite and the best design of the CSS process. Results indicate that the storage capacity of CH_4 on the material must be balanced with the requisite purity of the stored CH_4 by identifying the best material along with the corresponding optimal conditions, as the storage capacity and selectivity may not both be high enough for a given material. The multi-scale optimization framework developed herein is not only applied for simultaneous process optimization and materials screening but is also used to predict the values of optimal hypothetical isotherm parameters for the current application. This can assist experimental efforts to design best materials for gas separation and storage applications.

References:

(1) BP. *Energy Outlook 2017*; British Petroleum, **2017**.

(2) First, E. L.; Hasan, M. M. F.; Floudas, C. A. Discovery of Novel Zeolites for Natural Gas Purification through Combined Material Screening and Process Optimization. *AIChE J.* **2014**, *60* (5), 1767â€"1785.

(3) Iyer, S. S.; Hasan, M. M. F. A Novel Plug-and-Store Technology for Natural Gas Purification and Strage. In *CAMX 2015 - Composites and Advanced Materials Expo*; **2015**.

495136 Development of a Superstructure for Work and Heat Exchange Networks (WHENs)

Abstract:

Pinch Analysis is a well-known methodology in Process Systems Engineering for improving the energy efficiency of heat recovery processes. Up to recently, the focus has mainly been on heat integration, despite pressure-changing equipment often accounting for a significant portion of the total energy consumption in processes. Moreover, pressure manipulation of streams affects the heat sinks and sources in the process and thus the heat integration scheme. Consequently, the concept of Heat Integration can be expanded to Work and Heat Integration, in order to identify a further potential for enhanced energy recovery. An important principle in Pinch Analysis is the concept of Correct Integration (also referred to as Appropriate Placement). Enhanced heat recovery can be obtained by integrating various process units with the Heat Exchanger Networks (HENs). The correct integration of various equipment such as heat engines, heat pumps, reactors, evaporators and distillation columns is already well documented [1].

Compressors and expanders are two types of components that have not been extensively covered in Pinch Analysis. As these units involve both heat and work, the integration becomes rather more complex. Aspelund et al. [2] developed heuristic rules for the correct integration of pressure-changing equipment in subambient processes, where exergy instead of energy was used to combine the two modes of energy transfer (heat and work). The resulting heuristic rules became known as the Expanded Pinch Analysis and Design (ExPAnD) methodology. Later, Gundersen et al. [3] formulated the heuristics more precisely by saying that compression and expansion should start at the pinch. The heuristics have limited applicability however, and may result in suboptimal solutions if used without care. As a consequence, a series of theorems was developed for appropriate placement of compressors [4] and expanders [5] in above ambient processes and for compressors [7] and expanders [8] in sub-ambient processes. In addition to the theorems, the authors presented a graphical procedure for designing the Work and Heat Exchange Network. However, as temperatures as well as pressures must be taken into consideration when designing these networks, the manual procedure becomes tedious to solve even for small problems. Instead, mathematical optimization is required for studying large industrial sized problems.

Vikse et al. [8] discussed different superstructures for the optimization of WHENs. Of the models presented in that paper, only the superstructure by Uv [9] followed the rules of correct integration as determined by the theorems. On the other hand, the model required a sequential approach for which the sequence of integration may influence the solution. The model closely followed the manual procedure, and thus became quite involved for larger problems. Therefore, a new superstructure that takes into account the rules of correct integration and yet supports a simultaneous model approach must be developed. This work will develop a suitable superstructure for designing WHENs. The resulting models should be capable of solving larger and more complex problems. Furthermore, the superstructure should be versatile to easily adapt it for particular cases. The new superstructure will be compared with existing models, using several test problems of varying degree of complexity.

1. Smith, R. *Chemical process: design and integration*. 2nd edition, John Wiley & Sons, 2016.
2. Aspelund, A., D. O. Berstad, and T. Gundersen. "An extended pinch analysis and design procedure utilizing pressure based exergy for subambient cooling." *Applied Thermal Engineering* 27(16) (2007), 2633-2649
3. Gundersen, T., D. O. Berstad, and A. Aspelund. "Extending pinch analysis and process integration into pressure and fluid phase considerations." *Chemical Engineering Transactions* 18 (2009), 33-38.
4. Fu, C. and T. Gundersen. "Integrating compressors into heat exchanger networks above ambient temperature." *AIChE Journal* 61(11) (2015), 3770-3785.
5. Fu, C., and T. Gundersen. "Integrating expanders into heat exchanger networks above ambient temperature." *AIChE Journal* 61(10) (2015), 3404-3422.
6. Fu, C., and T. Gundersen. "Sub-ambient heat exchanger network design including compressors." *Chemical Engineering Science* 137 (2015), 631-645.
7. Fu, C., and T. Gundersen. "Sub-ambient heat exchanger network design including expanders." *Chemical Engineering Science* 138 (2015), 712-729.
8. Vikse M., C. Fu, P. I. Barton, and T. Gundersen. "Applying Mathematical Optimization for Work and Heat Exchange Networks (WHENs)." *Chemical Engineering Transactions*. Submitted for PRES 2017.
9. Uv, P.M. "Optimal design of heat exchanger networks with pressure changes." Master thesis, Norwegian University of Science and Technology (NTNU), Dep. of Industrial Economics and Technology Management, Trondheim, Norway, 2016.

497478 Multiresolution Modeling and Optimization of a Natural Gas Liquefaction Process Using Detailed Spiral-Wound Heat Exchanger Models

Abstract:

Liquefaction processing makes up around 52% of the cost of liquefied natural gas, and many research efforts have centered around improving the energy efficiency of the process [1]. Spiral-wound multistream heat exchangers, allowing for thermal contact between a refrigerant evaporating on the shell side and one or more hot streams circulating upwards on the tube side [2], are commonly selected for this process and are essential for tight energy integration.

The mathematical modeling of multistream heat exchangers (MHEXs) requires accurately predicting phase transitions and the corresponding changes in physical properties, and the MHEX models available in commonly used process simulators are typically limited to solving a set of energy balance equations with the purpose of determining the outlet conditions of one streams (all other inlet and outlet parameters must be specified). The heat duty is then usually discretized into finite enthalpy intervals and a stream sorting algorithm is used to establish the structure of the energy balance equations for the exchanger. While these formulations are acceptable in sequential-modular flowsheet simulation software, they pose considerable difficulties for equation-oriented process optimization.

Various works have shown how careful process design and optimization can significantly improve the economic performance of natural gas liquefaction, with several proposed approaches for multistream heat exchanger modeling and optimization [2] [3] [4]. In our previous work, we demonstrated a pseudo-transient process modeling approach for seamlessly incorporating unconventional or detailed mathematical models into process flowsheets [6] [7]. Furthermore, we demonstrated that describing different unit operations at different resolutions and length scales can be employed to increase the design degrees of freedom and to simultaneously identify the optimal (detailed) design of the respective units.

In this work, we apply our multiresolution flowsheeting approach to optimizing processes with a detailed spiral-wound MHEX model, capturing both thermal performance and the detailed geometry of the device. We employ semi-empirical correlations for pressure losses and heat transfer within the MHEX and use pseudo-transient modeling concepts to develop a robust simulation and optimization approach for the resulting heat exchanger models. We demonstrate the methodology by optimizing, at the flowsheet level, the PRICO® natural gas liquefaction process and comparing the results to previous solutions in literature.

References

[1] MIT Energy Initiative. The Future of Natural Gas: An Interdisciplinary MIT Study. Cambridge, MA: Massachusetts Institute of Technology, 2011.

[2] G Stephenson and L Wang. Dynamic simulation of liquefied natural gas processes. *Hydrocarbon Process Liquefied Nat Gas Dev*, 37-44, 2010.

[3] RS Kamath, LT Biegler, and IE Grossmann. Modeling multistream heat exchangers with and without phase changes for simultaneous optimization and heat integration. *AIChE Journal*, 58(1):190-204, 2012.

[4] MA Duran and IE Grossmann. Simultaneous optimization and heat integration of chemical processes. *AIChE Journal*, 32(1):123-138, 1986.

[5] RC Pattison and M Baldea. Multistream heat exchangers: equation-oriented modeling and flowsheet optimization. *AIChE Journal*, 61(6):1856-1866, 2015.

[6] RC Pattison, C Tsay, and M Baldea. Pseudo-transient models for multiscale, multiresolution simulation and optimization of intensified reaction/separation/recycle processes: Framework and a dimethyl ether production case study. *Comput. Chem. Eng.*, 2017. doi: 10.1016/j.compchemeng.2016.12.019

[7] C Tsay, RC Pattison, and M Baldea. Equation-oriented simulation and optimization of process flowsheets incorporating detailed spiral-wound multistream heat exchanger models. *AIChE Journal*, 2017. doi: 10.1002/aic.15705

496368 Optimal Design and Dynamic Modeling of Microtube Recuperators in an Indirect Supercritical Carbon Dioxide Recompression Brayton Power Cycle

Abstract:

Supercritical carbon dioxide (sCO_2) has strong potential as the working fluid for power generation cycles because of the potential for high efficiency, compact equipment, and lower costs. For instance, the highly-recuperated recompression Brayton cycle will use compact recuperators, rather than conventional shell-and-tube heat exchangers, resulting in a lower capital cost and footprint for the sCO_2 application. In the sCO_2 Brayton cycle, the overall thermal efficiency strongly depends on the design and performance of the recuperators because of the smaller cycle pressure ratio and the higher temperature of the turbine outlet stream exchanging heat with the cooler, compressed CO_2 entering the turbine inlet. Thus, the compact recuperator design is important. Furthermore, modeling the dynamic behavior is particularly important with regards to its impact within the cycle model for analyzing systems control and operation. It is also noted that the transport and thermodynamic properties of CO_2 vary considerably near its critical and pseudo-critical points, which makes most of the design approach and conventional thermal-hydraulic correlations based on the constant property assumption not suitable. Hence, instead of using conventional heat exchanger models available in commercial process modeling software, customized models must be developed.

With this motivation, a rigorous design model has been developed using Aspen Custom Modeler for high-temperature and low-temperature microtube recuperators in an indirect sCO_2 Brayton recompression cycle with improved steady-state economic characteristics and preferred dynamic performance characteristics. Additionally, a dynamic model has been developed to simulate the operating conditions spanning normal design point, to startup, shut down and other off-design conditions. This model is imported to a system-level dynamic model of the entire sCO_2 Brayton cycle to improve simulation accuracy when investigating the system dynamic behavior and developing advanced control strategies. In both models, the recuperators are discretized along the axes to handle the changing properties, and simulated with rigorous thermal-hydraulic correlations and commercially available microtube sizes. In this presentation, we will focus on the following aspects: (1) one-dimensional steady-state design approach of the recuperators in sCO_2 Brayton cycle, (2) design optimization for better economics along with superior dynamic characteristics by considering various objectives such as maximization of the heat transfer area, maximization of the compactness, minimization of the metal mass, minimization of the thermal residence time, by using successive quadratic programming algorithm, (3) sensitivity studies of different values of the design parameters, such as minimum temperature approach and maximum allowable pressure drop, which highly impact the thermal efficiency and the economic performance of the entire sCO_2 Brayton cycle, (4) two-dimensional dynamic model of the recuperators spanning both design and off-design conditions, and (5) re-design to obtain superior transient response of the recuperators.

500356 Locating Heat Exchangers in an EIP-Wide Heat Integration Network

Abstract:

Eco-industrial parks (EIP) are a cluster of manufacturing facilities located at proximity. They share and utilize each otherâ€™s resources, utilities, by-products and services for environmental, social and economic benefits, thus promoting sustainability [1]. Heating and cooling accounts for 73% of European Unionâ€™s total industrial energy consumption. Hence, there is a potential for inter-plant heat integration in such eco-industrial parks to conserve utilities [2]. However, it entails several challenges in terms of locating exchangers, confidentiality of streams, additional capital and operating expenses, and flexibility in the operation of individual plants [3]. To overcome these challenges, the prospect of locating the heat exchanger network (HEN) at a shared location was studied earlier [4]. This may not always be a feasible and an economical option, especially in case of space constraints or high land prices in locations such as Singapore. Hence, we propose an alternative approach where exchangers can be located at plant sites in enterprises.

In this work, we propose the idea of a distributed HEN, in which heat exchangers can be placed at some selected sites or participating plants. By allowing a heat exchanger to be located at a plant, we can eliminate the movement of streams that already belong to that plant. A plant can host an exchanger, only if the plant owns at least one of the exchanging streams. A stream may split into one or more substreams after entering a plant, but the substreams must merge again before leaving that plant. This is to minimize the cost of transporting multiple substreams and avoidance of operational issues. Also, we use the existing in-plant heaters/coolers at the exit of HEN. These heaters/coolers for the streams are retained or retrofitted to provide operational robustness. Existing in-plant movers (pumps or compressors) for the streams are retained or retrofitted to transport streams internally. One new mover is installed for transporting streams to other plants.

We develop a mixed integer non-linear programming (MINLP) model to synthesize the heat exchanger network for the above system. The model optimizes the HEN based on utility costs, exchanger and piping costs, and pumping expenses using a single objective function, Net Present Value (NPV). The superstructure-based formulation allows non-isothermal mixing in a stage [5, 6]. The model allows arbitrary cost expressions for exchangers and pumps with both fixed and variable components and cost multipliers for material and installation considerations. The model gives the optimum location of each exchanger with the capital and operating expenses associated with it.

For the first case study from Nair, et al. [4], the centralised HEN has NPV = \$1.84 million and Internal Rate of Return (IROR) = 78.8 %. In contrast, the distributed HEN has NPV = \$2.39 million with IROR = 161.5 %. Thus, by optimally locating the exchangers, the distributed HEN increases NPV by 30 % and IROR by nearly 100 %. Only two streams flow out of their plants in the distributed HEN versus six in the centralized HEN. The lower transportation needs reduce CAPEX by 45.9 %, and increases NPV and IROR. However, the distributed HEN saves less energy (i.e. utility heating and cooling) than the centralized HEN. The restricted availability of streams at each plant has limited exchanges and reduced energy savings.

Thus, the distributed HEN model designs a HEN based on the optimal streams matching, heat exchanger locations, utilities costs and investments required for pipes, pumps, to maximize the NPV of an EIP-wide HEN. Locating exchangers at plants reduced the number of streams travelling, saving significant piping, pump, and transportation investment. The distributed HEN model can choose and have a mix of inter-plant and intra-plant heat exchange considering the potential utility cost savings, CAPEX and OPEX before engaging the enterprises. With heat exchanger and stream locations as a variable, the distance between the plants play a significant role in deciding the HEN. This highlights the importance of trading off CAPEX versus OPEX, and true costs rather that simple energy savings to give a better HEN. Thus, the model can provide an optimal collaboration between many plants of many enterprises.

[1] R. P. Côté and E. Cohen-Rosenthal, "Designing eco-industrial parks: a synthesis of some experiences," *Journal of Cleaner Production,* vol. 6, pp. 181-188, 1998.

[2] E. Commission, "An EU Strategy on Heating and Cooling," ed. Brussels, 2016.

[3] C.-L. Chen and C.-Y. Lin, "Design of Entire Energy System for Chemical Plants," *Industrial & Engineering Chemistry Research,* vol. 51, pp. 9980-9996, 2012/08/01 2012.

[4] S. K. Nair, Y. Guo, U. Mukherjee, I. A. Karimi, and A. Elkamel, "Shared and practical approach to conserve utilities in eco-industrial parks," *Computers & Chemical Engineering,* vol. 93, pp. 221-233, 2016.

[5] T. F. Yee and I. E. Grossmann, "Simultaneous optimization models for heat integration--II. Heat exchanger network synthesis," *Computers & Chemical Engineering,* vol. 14, pp. 1165-1184, 1990.

[6] K. F. Huang and I. A. Karimi, "Simultaneous synthesis approaches for cost-effective heat exchanger networks," *Chemical Engineering Science,* vol. 98, pp. 231-245, Jul 19 2013.

492929 Thermochemical and Catalytic-Upgrading System Design to Convert Biomass to Liquid-Fuel Using a Superstructure-Based Approach

Abstract:

Nowadays, liquid fuels have become an essential commodity for day-to-day living, and this reliance is expected to increase in the future with increasing population, and raising standards of living. With dwindling petroleum resources and raising environmental concerns, biomass is considered to be a sustainable feedstock for liquid fuels. While multiple approaches to convert biomass to liquid fuels exist, we focus on the design and optimization of a biorefinery based on a multistage thermal decomposition [1,2] strategy.

The general multistage thermal decomposition process comprises of either one, or two, or three thermal decomposition stages for biomass pretreatment, each at different temperatures. While the streams leaving each of these stages, in practice, contain a wide variety of species, here, for the purpose of tractability, we represent all species using a fixed set of model-compounds, namely, acetic acid, acetol, furan, furfural, levoglucosan, toluene, guaiacol, char, carbon-di-oxide and water. The composition of these model-compounds in each stream leaving the thermal decomposition stages is dependent on the number of thermal decomposition stages used during pretreatment, which accordingly influences the sequence of downstream catalytic-upgrading chemistries to be used. The catalytic-upgrading steps are needed to increase the carbon-chain-length of the molecules, and reduce/eliminate oxygen-content so as to finally obtain fuel-like molecules. In this work, we consider the following kinds of chemistries for catalytic-upgrading as they are applicable to a variety of organic-molecules: acylation, aldol-condensation, alkylation, hydroxy-alkylation, ketonization, hydrogenation, oxidation and hydrodeoxygenation. With so many options for both thermal decomposition and catalytic-upgrading, it is non-trivial to identify cost-effective pathways to convert biomass to liquid fuel. To make the process-layout decisions and design the biorefinery in a systematic manner, we use a superstructure-based approach where we develop a mixed-integer nonlinear programming (MINLP) model.

We first develop a novel superstructure for a torrefaction-based biorefinery. The superstructure consists of reactors that are linked by a network of connecting streams associated with heat exchangers for temperature-conditioning. Further, we devise connectivity rules between the reactors. Some of these rules eliminate cyclical-paths and symmetrical solutions. Thereafter, we present strategies to model the individual elements of the superstructure.

If a reactor is present in the final solution, it is assigned one of many possible tasks. Each task in the thermal decomposition section is associated with a specific thermal decomposition temperature, while a task in the catalytic upgrading section is associated with a specific catalyst carrying out a specific kind of chemistry (listed earlier). For example, if acylation is the task assigned to a catalytic upgrading reactor, then, all molecules in the inlet stream to this reactor amenable for acylation react. So, multiple reactions can take place inside a reactor. To model the reactivities of the various molecules for a specific task, we use conversion data from literature and experiments. We also develop new methods to handle parallel reactions that any molecule can participate inside a reactor, alongside identifying the limitting reactant species of a reaction, which is unknown because stream flows are unknown. Furthermore, for each reactor, we determine the heating/cooling requirement based on the extents of reactions that occur in them. A new approximate method is proposed for the calculation of heating and cooling duties for phase- and temperature-conditioning the streams that connect reactors. While the overall heating/cooling duties along with the biomass and hydrogen usage contribute to the overall operating costs, capital costs are governed by the reactors used and the tasks that are assigned to them.

Using this general framework, we identify cost-optimal pathways to covert biomass to liquid fuels based on multistage thermal decomposition, and study the sensitivity of the solutions to different parameters like biomass costs, hydrogen costs, and fuel-range carbon-yield.

References:

[1] M.J. Prins, K.J. Ptasinski, F.J.J.G. Janssen, More efficient biomass gasification via torrefaction, Energy. 31 (2006) 3458â€“3470.

[2] J.A. Herron, T.Vann, N.Duong, D.E.Resasco, S. Crossley, L.L. Lobban, C.T. Maravelias, A Systems-Level Roadmap for Biomass Thermal Fractionation and Catalytic Upgrading Strategies, Energy Technology. 5 (2017) 130â€“150.

493434 Kepler Workflow in a Cloud Infrastructure for Temperature Balancing in a Steam Methane Reformer Furnace Using a Computational Fluid Dynamics and a Data-Driven Optimization Approach

Abstract:

Steam methane reforming (SMR) is widely used in industrial-scale hydrogen production process because of its relatively low cost compared to other methods such as electrochemical process. Even though the industrial SMR process is relatively well optimized, it still has room for improvement in the energy efficiency where industry can save hundreds of thousands of dollars in operation cost. SMR is a net endothermic process which requires heat to be supplied by the furnace. This, in turn, drives the reaction of methane and super-heated steam at approximately 1100 K to produce hydrogen. Hot spots can cause critical failure, and dramatically reduce the lifespan of the tubes. Prolonged increase of temperature even by 20 K can decrease the lifespan of the tubes by half [1], [2]. Therefore, balancing the temperature gradients inside the furnace and across all tube walls is important.

This was the motivation for our previous work to develop a computational fluid dynamics (CFD) model [3], [4] and a balancing algorithm [5] for a top-fired furnace with 336 tubes and 96 burners. This procedure requires an expert knowledge in CFD and mathematical models to implement, but we want this to be used by non-experts such as plant operators. This is also the vision of the Department of Energy Smart Manufacturing Project where studies such as this one, needs to be implemented in a reusable fashion across the industry.

In this study, we are using ANSYS Fluent to create an accurate CFD model of the reformer and a reduced order model (ROM) of temperature distribution and burner rate distribution is used to balance the furnace. To create the ROM, we are using IPOPT and CPLEX software packages. To build a converged ROM, we composed a mini workflow within Ansys Fluent itself using the user defined function (UDF) feature of Fluent and integrated the code for ROM as a shared library. In this way Ansys Fluent will be called in a loop as many times as required until the results from ROM calculation and CFD calculation are within acceptable tolerance. A typical CFD calculation takes 1 to 3 days on 80 cores of distributed computing power on a high-performance computing cluster and a ROM calculation can be carried out within minutes on a single core. So, in a real-time furnace temperature balancing process, a ROM calculation is quicker and practical.

Since we are designing this procedure for the automatic operation by plant operators. We also created a scientific workflow using Kepler software [6] to integrate and run this model in a cloud computing environment. We tested this procedure on a computing platform where the input data from a plant is being received through a Historian [7]. We used this input to run CFD and ROM calculations. Outputs of the calculations containing updated optimized valve positions to adjust the fuel flow rate through the burners and predicted temperatures are sent back to the plant operator. In a typical scenario, a plant operator needs to just visit the user interface and start the workflow. The workflow will then automatically perform a series of file transfer and computation processes. The entire process appears as a black box to the operator as all steps inside the workflow are already composed by an expert in the field.

[1] Latham D. Masterâ€™s Thesis: Mathematical Modeling of an Industrial Steam Methane Reformer. Queenâ€™s University, 2008.

[2] Pantoleontos G, Kikkinides ES, Georgiadis MC. A heterogeneous dynamic model for the simulation and optimisation of the steam methane reforming reactor. International Journal of Hydrogen Energy. 2012;37:16346-16358.

[3] Lao L, Aguirre A, Tran A, Wu Z, Durand H, Christofides PD. CFD modeling and control of a steam methane reforming reactor. Chemical Engineering Science. 2016;148:78-92.

[4] Aguirre A, Tran A, Lao L, Durand H, Crose M, Christofides PD. CFD modeling of a pilot-scale steam methane reforming furnace. In: Kopanos G, Liu P, Georgiadis M (Eds.). Advances in Energy Systems Engineering. Springer, Switzerland, 2017, pp. 75-117.

[5] Tran A, Aguirre A, Crose M, Durand H, Christofides PD. Temperature balancing in steam methane reforming furnace via an integrated CFD/data-based optimization approach. Computers & Chemical Engineering. in press.

[6] Korambath P, Wang J, Kumar A, Hochstein L, Schott B, Graybill R, Baldea M, Davis J. Deploying Kepler workflows as services on a cloud infrastructure for smart manufacturing. Procedia Computer Science. 2014;29:2254-2259.

[7] Korambath P, Wang J, Kumar A, Davis J, Graybill R, Schott B, Baldea M. A smart manufacturing use case: Furnace temperature balancing in steam methane reforming process via Kepler workflows. Procedia Computer Science. 2016;80:680-689.

496021 A Multiparametric C.F.D. Analysis of Multiphase Annular Flows for Oil and Gas Drilling Applications

Abstract:

The increasing global energy demand has pushed the oil industry towards developing more intelligent and advanced methods of enhancing oil recovery even under more unfavourable technical and environmental conditions. The severity of many operational problems affecting the drilling and production of oil and gas wells is worsened by inaccessibility; hence remedial efforts must be implemented from afar [1]. One of these problems is ensuring efficient removal of formation rock cuttings with a suitable fluid, whose rheology is often complicated. Furthermore, pressure losses along the annular geometry involved, and decreased Rates of Penetration (ROP) due to accumulated drill cuttings downhole, constitute significant portions of the total energy to be supplied [2]. Thus, the application of sophisticated modelling techniques with credible elucidation of the phase distributions (solids, liquids and gas) [3] and prevalent flow regimes becomes essential, if adequate and economic design of a drilling program is desired [4].

The advent of Computational Fluid Dynamics (CFD) and the growth in the available computational power to support it have provided an unprecedented opportunity to simulate and understand complex real flows especially when experimental methods may be too demanding [3]. The application of CFD for the analysis, design and optimisation of drilling programs [5] has mostly focused on vertical and concentric annular geometries, with few studies focusing on the intricacies that evolve due to hole inclination and eccentric geometric configurations [6]. Another complication characterising the modelling process is the occurrence of multiple phases in the flow domain, which requires robust multiphase closure equations and appropriate fluid-particle tracking models to be incorporated with the principal flow equations [7]; thus, enhancing solvability and accurate prediction of the drilling variables of interest.

The choice of the multiphase flow tracking scheme in turn significantly depends on the governing particle driving force during flow (drag, lift or collision). Nonlinearity of these multiphase interactions yield a variety of flow phenomena which are modelled using two major approaches. These include; the Lagrangian tracking of computational particles coupled with the Eulerian flow description of the continuous phase and the Eulerian-Eulerian description in which the solid particles are represented as a random field in the Eulerian reference frame [5]. Accuracy of mathematical representation, consistency of accompanying closure models and numerical stability are the attributes that make these techniques very applicable. However, there are still opportunities for improvement in these models as far as the particle-particle interactions and limits on solid phase concentrations are concerned. The limitation of the Lagrangian-Eulerian approach in handling flow systems involving high solids concentration [2] (>12%) makes the Eulerian-Eulerian approach more suitable for this computational analysis.

The present study employs the tool of Computational Fluid Dynamics to simulate a two-phase Solid-Liquid (SL) flow in an annulus, by elucidating multiple state variable distributions (namely cuttings concentration, pressure drop profiles, axial fluid and solid velocities) as a function of several drilling parameters: hole eccentricity, inclination, ROP, drill-pipe rotation and fluid rheology. Anon-Newtonian (power law) fluid model with well described flow parameters [8] has been implemented, considering a uniform cuttings size distribution (3 mm). A commercial CFD software suite (ANSYS FLUENT 17.1) has been used: the descriptive and predictive potential of the CFD software has been confirmed on account of the reasonable agreement with previously published experimental data (a relative error of less than 7% is achieved), as illustrated by the corresponding sensitivity plots. This multi-parametric CFD analysis study of multiphase cutting transport during drilling applications has confirmed that fluid velocity, hole inclination and annular eccentricity are the most influential factors governing the cuttings transport efficiency.

LITERATURE REFERENCES

1. Chin, W. C. (2001). *Computational Rheology for Pipeline and Annular Flow: Non-Newtonian flow Modeling for Drilling and Production, and Flow Assurance Methods in Subsea Pipeline Design*, Gulf Professional Publishing.
2. Pereira, F. A. R., Ataide, C. H., & Barrozo, M. A. S. (2010). A CFD Approach using a discrete phase model for annular flow analysis. *Latin American applied research*, **40**(1), 53-60.
3. Abdulkadir, M. (2011). *Experimental and Computational Fluid Dynamics (CFD) studies of gas-liquid flow in bends*, PhD Thesis, University of Nottingham, UK.
4. Gerogiorgis, D. (2013). Isothermal CFD modeling of annular multiphase flows during Underbalanced Drilling (UBD) in oil reservoirs, paper presented at the 2013 AIChE Annual Meeting, San Francisco, CA, USA.
5. Subramaniam, S. (2013). Lagrangian–Eulerian methods for multiphase flows. *Progress in Energy and Combustion Science*, **39**(2), 215-245.
6. Li, Y., & Kuru, E. (2003). Numerical modelling of cuttings transport with foam in horizontal wells. *Journal of Canadian Petroleum Technology*, **42**(10) (PETSOC 03-10-06).
7. Ofei, T. N., Irawan, S., & Pao, W. (2014). CFD method for predicting annular pressure losses and cuttings concentration in eccentric horizontal wells. *Journal of Petroleum Engineering*, **486423**, 1-16.
8. Rooki, R., Ardejani, F. D., Moradzadeh, A., & Norouzi, M. (2014). Simulation of cuttings transport with foam in deviated wellbores using computational fluid dynamics. *Journal of Petroleum Exploration and Production Technology*, **4**(3), 263-273.

498294 Development of a 3D Computational Fluid Dynamics Model for Microbial Fuel Cells

Abstract:

10D00 Modeling and Computation in Energy and Environment

Development of a 3D Computational Fluid Dynamics Model for Microbial Fuel Cells

Xianhua Li, Robert Ferrari, Jaclyn Guglielmi, Zuyi (Jacky) Huang

Department of Chemical Engineering, Villanova University, PA, 19085

Abstract:

Wastewater is an important new source of substantial energy. The energy potential contained in wastewater and its biosolids exceeds by 10 times the energy used to treat it [1]. However, conventional wastewater treatment currently consumes about 3% of all electrical power produced [2] and costs more than $25 billion annually in the United States [3]. Therefore, cost-effective methods are in high demand for converting the chemical energy contained in wastewater into bioenergy to address the energy crisis nowadays. Microbial fuel cell (MFC) is a bio-electrochemical system that is capable of producing electricity from wastewater under ambient conditions through the biodegradation of organic compounds by exoelectrogenic microorganisms [4]. Specific species of bacteria such as Geobacter and Shewanella spp. can transfer electrons directly to the anode by membrane-bound cytochromes through anaerobic respiration [5]. The generated electron flow can be used by an external electrical circuit as power supply. In the meanwhile, protons migrate to the cathode and react with oxygen to form water. While bench-scale MFCs have been studied for decades, for the practical implementation of MFCs in wastewater treatment plants, it is still challenging to scale up MFCs [6]. Regarding the complex multi-layer structure, fuel composition and electrochemical reactions in MFCs, a robust computational model will be helpful to direct the scaling up of MFCs.

This study aims to investigate the performance and scalability of MFCs with a three-dimensional computational fluid dynamics (CFD) model. In the CFD model, dynamic electricity generation data that was used for model validation was obtained from experiments conducted in different scales of single-chamber air-cathode MFCs with working volumes of up to one liter. Based on existing one-dimensional mathematical models that have been published [7-9], an integrated and advanced three-dimensional model with complex geometry specifications has been developed for the first time in CFD. User defined functions have been developed and compiled to CFD to accurately analyze process properties such as mass transfer rates, diffusion rates, electrochemical reactions, biofilm formation, bacterial endogenous metabolism, internal and external ohmic resistance. As a result, the 3D CFD model can quantify and visualize the MFC performance by showing electric potential, electrode surface potential, current density and species concentration distribution in any designated cross section across a MFC. From the model, the influence of current density, substrate concentration and conductivity on the long-term dynamics of power generation and coulombic efficiency has been revealed, thus optimal operational conditions have been identified. The developed three-dimensional MFC CFD models will be of great benefit for predicting and evaluating the scaling up of MFCs in the future.

References:

[1] W. E. R. Foundation, "Energy management exploratory team report executive summary," March 23, 2011.

[2] B. E. L. a. K. Rabaey, "Conversion of wastes into bioelectricity and chemicals by using microbial electrochemical technologies," *Science,* vol. 337, pp. 686-690, 2012.

[3] R. R. Hong Liu, and Bruce E. Logan, "Production of electricity during wastewater treatment using a single chamber microbial fuel cell," *Environ. Sci. Technol.,* vol. 38, pp. 2281-2285, 2004.

[4] "Exoelectrogenic bacteria that power microbial fuel cells," *Nature Reviews Microbiology,* vol. 7, pp. 375-381, 2009.

[5] Liang Shi, David J. Richardson, S. N. K. Zheming Wang, J. M. Z. Kevin M. Rosso, and J. K. Fredrickson, "The roles of outer membrane cytochromes ofÂ shewanellaÂ andÂ geobacterÂ in extracellular electron transfer," *Environmental Microbiology Reports,* vol. 1, pp. 220â€"227, 2009.

[6] B. E. Logan, "Scaling up microbial fuel cells and other bioelectrochemical systems," *Appl Microbiol Biotechnol,* vol. 85, pp. 1665-71, Feb 2010.

[7] R. P. Pinto, B. Srinivasan, M. F. Manuel, and B. Tartakovsky, "A two-population bio-electrochemical model of a microbial fuel cell," *Bioresour Technol,* vol. 101, pp. 5256-65, Jul 2010.

[8] C. Picioreanu, I. M. Head, K. P. Katuri, M. C. van Loosdrecht, and K. Scott, "A computational model for biofilm-based microbial fuel cells," *Water Res,* vol. 41, pp. 2921-40, Jul 2007.

[9] C. Picioreanu, K. P. Katuri, M. C. M. van Loosdrecht, I. M. Head, and K. Scott, "Modelling microbial fuel cells with suspended cells and added electron transfer mediator," *Journal of Applied Electrochemistry,* vol. 40, pp. 151-162, 2009.

500050 Optimal Resource Expansion and Placement in Smart Grid Communications Networks

Abstract:

With concerns about climate change, the adoption of new renewable standards and accelerated cost reductions are driving sharp growth in renewable energy technologies. In particular, the number of distributed solar photovoltaic (PV) installations is growing rapidly. However, as distributed solar power becomes an increasing fraction of total energy generation, the electric power grid must ensure continued reliability and cost-effectiveness. The development of the Smart Grid, as a natural evolution of the electric power grid, seeks to incorporate new technologies in order to support these distributed generation sources.

New grid devices, such as synchrophasors and Smart Meters, will be built to provide the existing Supervisory Control and Data Acquisition (SCADA) systems with vital power flow information. These SCADA systems allow operators to monitor and control transmission and maintain operational efficiency. But since network deployments can involve thousands of devices, manual reconfiguration and tuning of these devices are impractical. Therefore, automatic placement and optimization of network topology is critical to providing a high-performing, reliable, secure, and scalable communication network in support of the Smart Grid evolution.

In this work, we develop a procedure to optimally place communication hardware within a distributed PV network, while taking advantage of existing communications technologies to greatly reduce costs and maintaining quality of service for consumers. We propose a mixed-integer linear programming approach for optimal placement of communication nodes to ensure communication between Photo-Voltaic (PV) inverters and transmission substations in a multi-hop network via relay nodes, aggregation points, and various ad-hoc communication systems. We also consider uncertainty in link parameters, such as packet loss, latency, and bandwidth. We show results that are robust to changes in economic considerations and future technological improvements.

501771 Quantification of Parameter Space Regions Consistent with System Models and Associated Experimental Data

Abstract:

CO_2 adsorption on Hydrotalcite is an increasing field of study, as CO_2 is emitted in a variety of energy producing reactions and an increasing need for reduction of greenhouse emissions. With an ever-increasing demand for energy, accurate predictive models are imperative in order for lab to plant scale up. The Linear Driving Force (LDF) model is used to predict the concentration of adsorbed CO2 at equilibrium with parameters inputs from an isotherm. Additionally, we present a novel Two Step Adsorption mechanism that employs several parameters.

It is with this in mind that we also present a method for determining parameters for a model from experimental data. These models are typically comprised of ordinary differential equations that cannot be analytically integrated. In such cases, local non-linear regression techniques are employed in the solution. There are however shortcomings in local non-linear regression. The first is that it fails to guarantee the global optimality of the solution. Additionally, it also fails to ascertain whether or not the proposed problem parameterization is appropriate for the available data. In this work, we have addressed these issues by proposing a novel method for parameter identification. This method will be applied to the already established LDF, and compared with the feasibility of the Two Step Model. This method is guaranteed to identify the global optimum of the non-linear regression problem. In addition, given a predetermined level of accuracy, it also has the ability to define a range of suitable values for each parameter.

500661 Design and Intensification of Integrated Carbon Capture and Conversion to Chemicals

Abstract:

To address the rising CO_2 levels in the atmosphere, research efforts are directed towards making widespread deployment of carbon capture, utilization and sequestration (CCUS) cost-effective. CO_2 utilization can be implemented to produce as much as 2 Gt/yr. of fuels and 200 Mt/yr. of chemicals.[1] Utilization of CO_2 to deliver high-value products such as fuels and chemicals is an attractive and promising option. Existing standalone capture and utilization technologies perform energy-intensive CO_2 capture from dilute streams such as flue gas and then convert the captured CO_2 to syngas which is again energy-intensive. To intensify the capture and utilization steps to reduce energy inputs and hence costs, we design and optimize modular and multi-functional processes for direct utilization of CO_2 from flue gas to value-added fuels and chemicals.

We propose one such intensified process exploiting an integrated adsorption-purge-reaction system[2], which operates cyclically involving CO_2 adsorption from flue gas in the first step, followed by purging due to methane rich feed in the second step. The desorption of CO_2 from the column is achieved by a concentration driving force difference brought about by introduction of a second feed with a composition very different from that of flue gas. The N_2 separated from the column is vented out, while the CH_4/CO_2 purge mixture becomes the feed for dry reforming of CO_2 subsequent reactor section. A variety of methane rich feeds such as natural gas, fuel gas, unused methane, biogas etc. can be used by the process. The novel aspect of the process is the elimination of pressure and temperature swings employed in every cycle in conventional standalone CO_2 capture and integrating it with reactor feed pre-mixing which results in reduction in the overall cost of CO_2 capture and utilization. The process is integrated with existing CO_2 capture and utilization technologies.

The entire process is simulated using high-fidelity non-linear algebraic partial differential equation (NAPDE) model to identify the key decision variables, process metrics and the interactions between the adsorption and reaction sections. While it is important to operate the process such that it meets constraints pertaining to greenhouse gas emissions and product quality, it is only possible to obtain the values of these constraints after performing computationally intensive simulations. This poses a challenge in identifying the feasible and optimal operating conditions of the process. A data-driven constrained grey-box optimization framework is employed to obtain feasible operating conditions and to proceed towards the optimal objective desired. Applying the optimization framework, we are able to achieve a maximum of 99.67% net overall CO_2 utilization while accounting for auxiliary CO_2 emissions at costs ranging from \$110-\$130 per ton of syngas. About 14.6% of the total CO_2 input to the process is captured with low cost â€œdirectlyâ € from flue gas while still maintaining 91% overall CO_2 utilization. The rest of the CO_2 input is sourced from existing capture plants integrated with the developed process.

Using the above design, intensification and optimization framework, we also develop a novel cyclic process for the direct utilization of CO_2 to methanol. This process again uses a two-step process and exploits synergies between CO_2 capture, CH_4/CO_2 mixing, syngas production and syngas conversion to methanol. In this presentation, we will further elucidate the techno-economic feasibility of the methanol process.

References:

(1) Quadrelli, E. A.; Centi, G.; Duplan, J.-L.; Perathoner, S. Carbon Dioxide Recycling: Emerging Large-Scale Technologies with Industrial Potential. *ChemSusChem* **2011**, *4* (9), 1194â€"1215.

(2) Iyer, S. S.; Bajaj, I.; Balasubramanian, P.; Hasan, M. M. F. Modular Process Intensification of Carbon Capture and Conversion to Syngas. *Submitted.*

489213 Response Surface Models for SET POINT Determination of Smokeless Steam and Air Assisted Flares

Abstract:

Electronic code of Federal Regulations Â§63.670 requires flare operators to specify smokeless design capacity for flares and operate flares with no visible emissions when regulated vent gasÂ is sent to flares with flow rate less than the smokeless design capacity. Alternatively, 96.5% combustion efficiency (CE) or 98% destruction efficiency (DE) must be achieved using site specific operating limits to demonstrate the compliance in lieu of threshold values of minimum combustion zone net heating value (NHV_{cz}) as 270 BTU/scf during any 15â€"minute average period, maximum flare tip velocity (V_{tip}) less than 400 ft/s and net heating value dilution parameter (NHV_{dil}) as 22 BTU/ft^2. Site specific operating limits include threshold values for NHV_{cz}, V_{tip} and NHV_{dil} as well as vent gas flow rate, vent gas composition and steam/ air assist flow rate. This study aims at developing robust response surface models to express CE and opacity as a function of operating variables and to determine the set points for steam/air and make up fuel under smokeless flaring conditions. Multivariate response surface models were developed for steam and air assisted flares burning propylene, propane, natural gas, methane, ethylene and typical refinery vent gas mixtures. In this study, soot emissions, visible emission scale, DE, and CE test data from 1983 to 2014 (including 1983/1984/1985 EPA, 2010 TCEQ, 2009/2010 Marathon Detroit/Texas City, and 2014 Carleton University) were analyzed. The input variables include measurable variables like velocity, lower heating value, flare tip diameter, vent gas composition (Ï€-bond, carbon to hydrogen molar ratio and molecular weight), and operating variables (NHV_{cz}, NHV_{dil}, steam/air assists, and fuel). Minitab and Polymath were used to develop models and solve them to determine set points. Response surface models developed for CE and opacity showed high coefficient of determination (R^2) values 0.90 and 0.94 respectively for steam assisted and 0.91 and 0.93 respectively for air assisted flare test data.

502995 NMPC of Semi-Batch Processes under Uncertainty using Pontryagin's Minimum Principle

Abstract:

Nonlinear model predictive control (NMPC) is an important tool to perform real-time optimization for batch and semi-batch processes. Direct methods are often the methods of choice to solve the corresponding optimal control problems, in particular for large-scale problems (Wächter and Biegler, 2006; Zavala and Biegler, 2009). However, the matrix factorizations associated with large prediction horizons can be computationally demanding (Cannon, 2004; Cannon et al.,2008)). In contrast, indirect methods can be competitive for smaller-scale problems. Furthermore, the interplay between states and co-states in the context of Pontryagin's Minimum Principle might turn out to be computationally quite efficient (Cannon et al.,2008; Aydin et al.,2017).

This work proposes to use a novel indirect solution technique within shrinking-horizon in the context of NMPC. In particular, the technique deals with path constraints via indirect adjoining, which allows meeting active path constraints explicitly at each iteration (Hartl et al., 1995). Uncertainties are handled by the introduction of time-varying backoff terms for the path constraints (Srinivasan et al., 2003; Shi et al., 2016). The resulting NMPC algorithm is applied to a two-phase semi-batch reactor for the hydroformylation of 1-dodecene in the presence of uncertainty, and its performance is compared to that of NMPC that uses a direct simultaneous optimization method (Hentschel et al., 2015). The results show that the proposed algorithm (i) can enforce feasible operation for different uncertainty realizations both within batch or from batch to batch, and (ii) it is faster than direct simultaneous NMPC, especially at the beginning of the batch. In addition, a modification of the PMP-based NMPC scheme is proposed that enforces active constraints via tracking and reduces the complexity of the optimization further.

References

Aydin, E., Bonvin, D., & Sundmacher, K. (2017). Dynamic optimization of constrained semi-batch processes using Pontryagin's minimum principle—An effective quasi-Newton approach. Computers & Chemical Engineering, 99, 135-144

Cannon, M. (2004). Efficient nonlinear model predictive control algorithms. Annual Reviews in Control, 28, 229-237.

Cannon, M., Liao, W., & Kouvaritakis, B. (2008). Efficient MPC optimization using Pontryagin's minimum principle. International Journal of Robust and Nonlinear Control, 18, 831-844.

Hartl, R. F., Sethi, S. P., & Vickson, R. G. (1995). A survey of the maximum principles for optimal control problems with state constraints. SIAM review, 37, 181-218.

Hentschel, B., Kiedorf, G., Gerlach, M., Hamel, C., Seidel-Morgenstern, A., Freund, H., & Sundmacher, K.

(2015). Model-based identification and experimental validation of the optimal reaction route for the hydroformylation of 1-dodecene. Industrial & Engineering Chemistry Research, 54, 1755-1765.

Shi, J., Biegler, L. T., Hamdan, I., & Wassick, J. (2016). Optimization of grade transitions in polyethylene solution polymerization process under uncertainty. Computers & Chemical Engineering, 95, 260-279.

Srinivasan, Bonvin, D., Visser, E., & Palanki, S. (2003). Dynamic optimization of batch processes: II. Role of measurements in handling uncertainty. Computers & Chemical Engineering, 27, 27-44.

Wächter, A., & Biegler, L. T. (2006). On the implementation of an interior-point filter line-search algorithm for large-scale nonlinear programming. Mathematical Programming, 106, 25-57.

Zavala, V. M., & Biegler, L. T. (2009). The advanced-step NMPC controller: Optimality, stability and robustness. Automatica, 45, 86-93.

495022 Sample-Free Stochastic Nonlinear Model Predictive Control

Abstract:

Model predictive control (MPC) has been successfully applied for high-performance control of a wide range of complex systems due to its ability to handle multivariable dynamics, input and state constraints, and multiple (possibly conflicting) objectives [1]. However, the accuracy of the model predictions directly impacts the MPC behavior such that uncertainties (e.g., unknown parameters, incorrect model structure, or exogenous disturbances) can lead to degradation in the performance of the overall closed-loop system. Although the receding-horizon implementation of MPC provides some degree of robustness to uncertainty, this relatively small robustness margin may be inadequate in many safety-critical and high-precision control applications.

Extensive work has been done to develop so-called robust MPC (RMPC) methods that typically design control inputs with respect to the worst-case values of the uncertainty [2,3]. In many engineering systems, uncertainties are generally of a stochastic nature. When probabilistic descriptions of uncertainties are available, i.e., probability density functions (pdfs), they can be explicitly incorporated into a MPC formulation with the aim to achieve robustness to uncertainties in a probabilistic sense. This notion has led to stochastic MPC (SMPC) that solves an optimal control problem in which the cost function and constraints are defined in terms of the state pdfs [4]. Thus, SMPC can systematically tradeoff performance and constraint satisfaction in terms of uncertainty distribution, which is becoming increasingly important as high-performance operation pushes the operation toward system constraints. In addition, defining the control objectives in terms of the *state distributions* is essential for asymmetric cost functions that make the tails of the pdfs essential for effective control.

This work addresses the SMPC problem for nonlinear systems (i.e., SNMPC) with uncertain parameters and stochastic exogenous disturbances for the case of incomplete state information (commonly referred to as "output feedback"). Uncertainty propagation poses a key challenge to SNMPC since there exists no closed-form expressions for the time evolution of the state pdf as a function of the control inputs for nonlinear systems. Sample-based methods for uncertainty propagation (e.g., Markov chain Monte Carlo, sequential Monte Carlo, or the scenario approach) have been utilized in the context of stochastic optimal control of nonlinear systems [5,6]. However, the computational complexity of most sample-based approaches can be prohibitive for the real-time implementation of SNMPC since many samples must be drawn to provide accurate representations of the state pdfs (i.e., the number of required samples generally scales exponentially in terms of the number of uncertainties). To avoid the latter computational challenges, this work presents a sample-free approach to SNMPC which relies on closed-form expressions for describing the (approximate) evolution of the moments of the state pdfs as explicit functions of the control inputs. In addition to its attractive computational features from nonlinear optimization standpoint, a "sample-free" moment-based approach to SNMPC is further motivated by the fact that SNMPC cannot guarantee chance constraint satisfaction and state distribution shaping on the closed-loop system [7]. In this work, generalized polynomial chaos (gPC) [8] is used as the foundation for deriving explicit expressions for the time-evolution of the moments of the state distribution as a function of the control inputs that are amenable to fast, online computations. A moment-based surrogate for the intractable (nonconvex) chance constraints are proposed in terms of the Mahalanobis distance. A moment-matching formulation of the Bayesian estimation problem is presented for joint state and parameter estimation in order to account for incomplete state knowledge (i.e., output feedback implementation of SNMPC). The parameters are adapted (jointly with the states) at each time instant to reduce model uncertainty based on newly obtained measurements. The closed-loop performance of this proposed sample-free SNMPC approach is demonstrated on a continuous stirred-tank reactor benchmark problem [9] and is compared to that of scenario-based and certainty-equivalence NMPC.

References

[1] M. Morari, J. H. Lee. Model predictive control: Past, present and future, *Computers & Chemical Engineering*, 23, 667–682, 1999.

[2] A. Bemporad, M. Morari. Robust model predictive control: A survey, In: *Robustness in Identification and Control*, Springer, London, pp. 207–226, 1999.

[3] W. Langson, I. Chryssochoos, S. V. Rakovic, D. Q. Mayne. Robust model predictive control using tubes, *Automatica*, 40, 125–133, 2004.

[4] A. Mesbah. Stochastic model predictive control: An overview and perspectives for future research, *IEEE Control Systems Magazine*, 36, 30–44, 2016.

[5] A. Lecchini-Visintini, W. Glover, J. Lygeros, J. M. Maciejowski, Monte Carlo optimization for conflict resolution in air traffic control. *IEEE Transactions on Intelligent Transportation Systems*, 7, 470–482, 2006.

[6] N. Kantas, J. M. Maciejowski, A. Lecchini-Visintini. Sequential Monte Carlo for model predictive control, In: *Nonlinear Model Predictive Control*, Springer, Berlin Heidelberg, pp. 263–273, 2009.

[7] J. A. Paulson, E. Harinath, L. C. Foguth, R. D. Braatz. Nonlinear model predictive control of systems with probabilistic time-invariant uncertainties, In: *Proceedings of the 5th IFAC Conference on Nonlinear Model Predictive Control*, pp. 16–25, 2015.

[8] D. Xiu, G. E. Karniadakis. The Wiener-Askey polynomial chaos for stochastic differential equations, *SIAM Journal of Scientific Computation*, 24, 619–644, 2002.

[9] K. U. Klatt, S. Engell. Gain-scheduling trajectory control of a continuous stirred tank reactor, *Computers & Chemical Engineering*, 22, 491–502, 1998.

495822 Real-Time Optimization Via Modifier Adaptation â€“ on Updating the Model Outputs

Abstract:

Optimal operation of chemical processes is key for meeting productivity, quality, safety and environmental objectives. Both model-based and data-driven schemes are used to compute optimal operating conditions [1]:

- The model-based techniques are intuitive and widespread, but they suffer from the effect of *plant-model mismatch*. For instance, an inaccurate plant model can lead to operating conditions that are not optimal for the plant and may violate constraints. Furthermore, even with an accurate model, the presence of disturbances typically leads to a drift of the optimal operating conditions, so that measurement-based adaptation is needed to maintain plant optimality.

- The data-driven optimization techniques rely on *measurements* to adjust the optimal inputs in real time. Consequently, real-time measurements are used to help reach plant optimality. This field, which is labeled real-time optimization (RTO), has received growing attention in recent years. RTO schemes can be of two types: explicit schemes solve a numerical optimization problem repeatedly at each iteration, while implicit schemes adjust the inputs on-line in a control-inspired manner [2].

We consider here model-based explicit RTO schemes that solve an updated optimization problem at each iteration. For example, the two-step approach uses (i) measurements to update the model parameters, and (ii) the updated model to compute optimal inputs [3]. It has also been proposed to update the model differently. Instead of adjusting the model parameters, input-affine correction terms can be added to the cost and constraint functions of the optimization problem so that the modified model shares the first-order optimality condition with the plant [4]. The main advantage of this approach, labeled modifier adaptation (MA), lies in its proven ability to converge to the plant optimum, even in the presence of structural plant-model mismatch [5]. The ability of the algorithm to converge to the plant optimum is linked to the quality of the model, the filtering used, and the initial operating point. Methods for ensuring global convergence tend to slow down convergence [6], which makes them unsuited for some practical applications. The present contribution proposes a different correction to the model that can both increase the convergence speed and enlarge the set of acceptable initial operating points in some applications.

Standard MA introduces input-affine corrections of the cost and constraint functions, since these functions are linked to the KKT optimality conditions. The proposal here is to update the model outputs y instead, thus leading to the approach labeled MA-y. We will demonstrate that this approach results in input-affine corrections of the cost and constraint functions as well. Consequently, MA-y also leads to plant optimality upon convergence, provided perfect output measurements and gradient estimates are available. Furthermore, it turns out that, in cases where the cost and constraint functions are nonlinear functions of the output variables, the input-affine corrections of the outputs also lead to partial correction of the model Hessians, which results in the aforementioned improvements in convergence properties. For example, it can be shown that, if the outputs are affine in the inputs u, MA-y converges to the plant

optimum in a *single* iteration, while the standard MA might take several iterations to do the same. The approach will be illustrated on the performance optimization of the Williams-Otto continuous stirred-tank reactor.

References

[1] G. Franois and D. Bonvin, Measurement-Based Real-Time Optimization of Chemical Processes, In S. Pushpavanam, editor, *Advances in Chemical Engineering*, Vol. 43, 1-50, Academic Press (2013).

[2] B. Chachuat, B. Srinivasan and D. Bonvin, Adaptation Strategies for Real-Time Optimization, *Comput. Chem. Engng*, 33(10), 1557-1567 (2009).

[3] T. E. Marlin and A. N. Hrymak, Real-Time Operations Optimization of Continuous Processes, In *AIChE Symposium Series - CPC-V*, Vol. 93, 156-164 (1997).

[4] A.G. Marchetti, B. Chachuat and D. Bonvin, Modifier-Adaptation Methodology for Real-Time Optimization, *Industrial & Engineering Chemistry Research*, 48(13), 6022-6033 (2009).

[5] A.G. Marchetti, G. Franois, T. Faulwasser and D. Bonvin, Modifier Adaptation for Real-Time Optimization -- Methods and Applications, *Processes*, 4(4), 55 (2016) doi:10.3390/pr4040055.

[6] A.G. Marchetti, T. Faulwasser and D. Bonvin, A Feasible-Side Globally Convergent Modifier-Adaptation Scheme, *J. of Process Control*, in press (2017).

497216 Nonlinear Model Predictive Control for Zone Tracking

Abstract:

Model predictive control (MPC) is the most widely applied advanced control technique in the process industry due to its many advantages such as optimally handling process constraints and interaction. In the traditional paradigm, MPC fulfills the objective of set-point tracking by penalizing the deviation of the predicted state and input trajectories to the desired set-point. In practice, it is often beneficial to design MPC that tracks a zone region instead of a set-point. The superiority of zone control over set-point control lies in its ability to flexibly handle multiple objectives so that the degrees of freedom in the controller can be assigned to address more important objectives. Successful applications of zone MPC have been reported in various areas such as diabetes treatment [1], control of building heating system [2], and control of the FCC process in oil refinery [3].

Despite the widely successful industrial applications, there has not been a systematic approach for design and analysis of MPC that tracks a zone region. On the one hand, reported industrial zone MPC designs are heuristic in nature and lack stability guarantee. On the other hand, existing zone MPC designs with guaranteed stability essentially use ad hoc approaches to convert zone control to set-point control (e.g., [4]). Several difficulties arise when it comes to the design and analysis of MPC for zone tracking. First of all, the admissibility of a zone target needs to be carefully examined. Since the target zone is often user-specified based on the economic metrics of the process, it is not necessarily positive invariant. One might be tempted to find the largest positive invariant set in the target zone and convert the zone tracking objective to tracking of its largest positive invariant subset. Unfortunately, finding and characterizing the maximal positive invariant set in the target zone is very difficult for generic nonlinear systems, if possible at all. Even if such a set can be found, the aforementioned approach may not be desirable and may give deteriorated transient performance as shown by our simulation study.

In this talk, we propose a general framework for nonlinear model predictive zone control. The proposed zone MPC tracks a target zone characterized by coupled system state and input. A control invariant subset of the target zone is incorporated in the proposed zone MPC as the terminal constraint to ensure closed-loop stability. We resort to LaSalle's invariance principle and develop invariance-like theorem which is suitable for stability analysis of zone control. It is proved that under the proposed zone MPC design, the system converges to the largest positive invariant subset of the target zone. In the stability analysis, we focus on the evolvement of the state-input pair (x(n), u(n)) instead of merely the state trajectory x(n). This provides a better picture and more accurate description of the system behavior. Further discussions are made on enlargement of the region of attraction by employing an auxiliary controller as well as handling a secondary economic objective via multi-objective optimization.

We use two simple simulation examples to demonstrate the superiority of zone control over set-point control and the efficacy of our zone MPC design. The first example involves a discrete nonlinear scalar system. From its bifurcation diagram, it is clear how relaxing a set-point target to a zone target allows for more admissible operations in addition to steady state operations such as periodic orbits and chaotic attractors. The second example is a linear system with constrained state and input. In the simulation we compared three different zone MPC configurations: (1) a zone MPC that tracks the target zone, (2) a zone MPC that tracks the largest positive invariant set in the zone, and (3) the proposed zone MPC. Very interestingly, the results are analogous to the comparison between set-point MPC and economic MPC [5] where case (1), (2), (3), corresponds to economic MPC without terminal constraint, set-point MPC, and economic MPC with terminal constraint.

References

[1] Grosman, B., Dassau, E., Zisser, H.C., Jovanovič, L. and Doyle III, F.J., 2010. Zone model predictive control: a strategy to minimize hyper-and hypoglycemic events. *Journal of diabetes science and technology*, 4(4), pp.961-975.

[2] Privara, S., Široký, J., Ferkl, L. and Cigler, J., 2011. Model predictive control of a building heating system: The first experience. *Energy and Buildings*, 43(2), pp.564-572.

[3] Zanin, A.C., De Gouvea, M.T. and Odloak, D., 2002. Integrating real-time optimization into the model predictive controller of the FCC system. *Control Engineering Practice*, 10(8), pp.819-831.

[4] Ferramosca, A., Limon, D., González, A.H., Odloak, D. and Camacho, E.F., 2010. MPC for tracking zone regions. *Journal of Process Control*, 20(4), pp.506-516.

[5] Diehl, M., Amrit, R. and Rawlings, J.B., 2011. A Lyapunov function for economic optimizing model predictive control. *IEEE Transactions on Automatic Control*, 56(3), pp.703-707.

491498 Fast Robust Nonlinear Model Predictive Control Based on Sensitivity Update

Abstract:

As an advanced optimal control technique, Model Predictive Control (MPC) has been widely used in process control industry mainly because it is naturally suited to deal with constraints and multiple-input-multiple-output [1]. Its nonlinear counterpart, nonlinear model predictive control (NMPC), is getting more attention because it accurately captures the nonlinear effect of process dynamics. However, the system performance deteriorates under the influence of uncertainty. The current robust NMPC techniques are often shy of applications due to either model conservativeness or formidable computational effort. To improve this situation, Lucia et al. [2] introduced a multistage nonlinear model predictive control framework to serve as a promising non-conservative robust NMPC scheme. Multistage NMPC models the uncertainty evolution explicitly with a truncated scenario tree whose structure exploits the degrees of freedom by allowing future control inputs to adjust to future available information.

Multistage NMPC outperforms standard NMPC by robustly satisfying constraints and lowering the objective cost on average [3]. However, the price of the performance advancement with a multistage scheme is paid with a significant increase of CPU seconds. In order to reduce computational time, the sensitivity-based fast NMPC update algorithm is used to solve the parametric nonlinear program (PNLP) so that the update time is almost negligible compared to iterative solution procedures [4]. Allowing for non-unique multipliers, the path-following algorithm [5] solves parametric QP sub-problems and a jump step to capture the change of active sets.

In this talk, we first examine the performance of multistage NMPC compared with standard NMPC and two other robust NMPC approaches (min-max NMPC and back-off). Min-max optimizes the worst-case scenario while satisfying constraints for all the scenarios. Robust NMPC with back-off constraints tightens the nominal constraints with back-off terms, which are calculated from Monte Carlo simulations. A nonlinear CSTR example with a setpoint tracking objective has been explored with two separate uncertain parameters: activation energy and inlet concentration. Under both cases, the performance of multistage NMPC is superior than standard NMPC and robust NMPC approaches. Also, multistage scheme of different number of robust horizons is investigated and results are discussed. In addition, several probability distributions of the uncertain parameter are considered, where multistage formulation shows a clear advantage over other approaches. Lastly, additional case studies demonstrate that the sensitivity-based update can generate a rapid solution of multistage NMPC problems.

References:

[1] Grne, Lars, and Jrgen Pannek. "Nonlinear Model Predictive Control: Theory and Algorithms." (2013).

[2] Lucia, Sergio, Tiago Finkler, and Sebastian Engell. "Multi-stage nonlinear model predictive control applied to a semi-batch polymerization reactor under uncertainty." *Journal of Process Control* 23.9 (2013):

1306-1319.

[3] Lucia, Sergio, and Sebastian Engell. "Potential and Limitations of Multi-stage Nonlinear Model Predictive Control." *IFAC-PapersOnLine* 48.8 (2015): 1015-1020.

[4] Wolf, Inga J., and Wolfgang Marquardt. "Fast NMPC schemes for regulatory and economic NMPC–A review." *Journal of Process Control* 44 (2016): 162-183.

Degenerate Parametric Optimization Problems." *SIAM Journal on Optimization* 27.1 (2017): 538-564.

493392 Fault-Tolerant Economic Model Predictive Control with Empirical Process Models

Abstract:

Actuator faults in the chemical process industries, which prevent the control signal from being able to impact the actuator output, are a problem that has been addressed through a variety of detection and fault-tolerant control methods in the literature (e.g., [1], [2]). Model-based fault-tolerant control designs are needed that can incorporate empirical process models because many industrial control designs are developed based on empirical models. Linear empirical models are common, and a variety of methods for identifying linear empirical models exist (e.g., subspace model identification methods [3]). Due to the widespread use of linear empirical models in advanced control designs such as model predictive control (MPC) in industry, development of a methodology for handling actuator faults in such controllers is necessary. This is particularly important since an empirical model developed from data corresponding to the case that all inputs are varying (i.e., no fault has occurred) may no longer be accurate after a fault prevents one of the inputs from varying because the data corresponding to the case that an actuator output is fixed was not incorporated in the data used to develop the original empirical model corresponding to the case without faults.

In this work, we develop an approach for accounting for actuator faults in MPC based on empirical models by utilizing an error-triggering procedure to initiate on-line updates of the empirical model based on significant prediction error due to the fault, assuming that it is known which actuator has experienced a fault. The methodology presented is developed in the context of economic model predictive control (EMPC) with empirical models [4] due to the potential of this control design to dictate time-varying process operation by computing a time-varying input trajectory, which may cause it to compute inputs that persistently excite the process dynamics so that routine process operating data can be utilized for the on-line model identification procedure. Specifically, the implementation strategy of the EMPC is updated after a fault to either incorporate the value of the faulty actuator output within the empirical model (reducing the number of decision variables) when the value at which the faulty actuator output is fixed is known, or, when the value of the faulty actuator output is not known, the EMPC continues to solve for all inputs but only implements those for the non-faulty actuators. A moving horizon error detector computes the relative prediction error in all process outputs throughout a prediction horizon and initiates model re-identification when this error exceeds a pre-specified threshold [5], where the new model identified after a fault has one less input since the faulty actuator output is no longer available to be adjusted by the EMPC. A chemical process example demonstrates the effectiveness of the proposed strategy at compensating for the actuator fault compared to utilizing a single empirical model throughout the time of operation despite the faults. It also demonstrates that the moving horizon error detector is effective at determining the necessity of updating the empirical model after a fault, so that the model is only changed within the control design when significant prediction error is detected.

[1] Lao L, Ellis M, Christofides PD. Proactive fault-tolerant model predictive control. AIChE Journal. 2013;59:2810-2820.

[2] Mhaskar P, Gani A, El-Farra NH, McFall C, Christofides PD, Davis JF. Integrated fault-detection and fault-tolerant control of process systems. AIChE Journal. 2006;52:2129-2148.

[3] Van Overschee P, De Moor B. Subspace Identification for Linear Systems: Theory, Implementation, Application; Kluwer Academic Publishers: Boston, Massachusetts, 1996.

[4] Alanqar A, Ellis M, Christofides PD. Economic model predictive control of nonlinear process systems using empirical models. AIChE Journal. 2015;61:816-830.

[5] Alanqar A, Durand H, Christofides PD. Error-triggered on-line model identification for model-based feedback control. AIChE Journal. 2017;63:949-966.

494243 Fault-Tolerant Model Predictive Control of Nonlinear Process Systems Using a Forecast-Triggered Communication Strategy

Abstract:

Model predictive control (MPC), also known as receding horizon control, refers to a class of optimization-based control algorithms that utilize an explicit process model to predict the future response of a plant. At each sampling time, a finite-horizon optimal control problem with a cost functional that captures the desired performance requirements is solved subject to state and control constraints, and a sequence of control actions over the optimization horizon is generated. The first part of the control inputs in the sequence is implemented on the plant, and the optimization problem is solved repeatedly at every sampling time. Originally developed to meet the specialized control needs of power plants and petroleum refineries, MPC technology can now be found in a wide variety of application areas including chemicals, food processing, automotive, and aerospace applications (e.g., see [1]). Motivated by the advantages of MPC, such as constraint handling capabilities, performance optimization, handling multi-variable interactions and ease of implementation, an extensive and growing body of research has been developed over the past few decades on the analysis, design and implementation of MPC, leading to a plethora of MPC formulations (e.g., see [2], [3] for some recent research directions and references in the field).

Over the past few decades, and with the increasing demand for system performance and stability, fault-tolerance capabilities have become a critical component of modern day control system, especially for safety-critical systems, such as chemical processes, where malfunctions in the control actuators, measurement sensors and process equipment can lead to instabilities and safety hazards (e.g., see [4], [5] for some results and references on fault-tolerant control) and there are increasing calls to achieve zero-incident plant operations [6]. MPC, as a conventional feedback controller design methodology, is also faced with the challenges of dealing with device faults and handling the resulting deterioration in the closed-loop stability and performance requirements. While various methods have been investigated and developed for the design and implementation of fault-tolerant MPC for both linear and nonlinear processes (e.g., see [7]-[11]), the majority of existing methods at this point have been developed within the conventional feedback control setting where the sensor-controller communication links are assumed to be essentially dedicated links with flawless information transfer.

With the advent of networked control systems, however, and the subsequent increased levels of integration of communication networks in the feedback loop, this conventional design setting needs to be re-examined to address the fundamental challenges introduced by the intrinsic limitations on the processing and transmission capabilities of the sensor-controller communication medium. Examples of efforts to address this problem within the MPC framework include the results in [12], [13] where resource-aware MPC formulations have been developed using techniques from event-based control to enforce closed-loop stability with reduced sensor-controller communication requirements. In these studies, however, the problem of incorporating fault-tolerance capabilities in the MPC design framework was not addressed.

Motivated by these considerations, this work presents a methodology of the design of fault-tolerant MPC system for constrained nonlinear process systems, subject to model uncertainties, control actuator faults and sensor-controller communication constraints. The co-presence of faults, control and communication resource constraints creates a conflict in the control design objectives where, on the one hand, increased levels of sensor-controller communication may be required for robust fault handling, but, on the other hand, such levels may be either undesirable or unattainable due to the intrinsic limitations on the communication medium. To resolve this conflict, a resource-aware Lyapuonv-based MPC system that achieves the fault-tolerant stabilization objective with reduced sensor-controller communication is designed. In this approach, the control action is computed by solving on-line a finite-horizon optimal control problem based on an uncertain model of the plant subject to appropriate Lyapuonv-based stability constraints. The stability constraints are designed to ensure the desired closed-loop stability and performance properties in the presence of faults, and an explicit characterization of the state space region where fault-tolerant stabilization is guaranteed is obtained in terms of the fault size, the choice of the controller design parameters and the size of the plant-model mismatch. To keep sensor-controller communication to a minimum, a forecast-triggered communication strategy is used to determine when communication should be suspended or restored over the network. In this strategy, an update of the model state in the predictive controller using the sensor measurements at a given sampling time is triggered only when the Lyapunov function is forecasted to breach a certain threshold over the next sampling interval. The update-triggering threshold is derived using Lyapunov techniques and is explicitly parameterized in terms of the fault and a suitable fault accommodation parameter. Based on this characterization, fault accommodation strategies that guarantee closed-loop stability while simultaneously optimizing control and communication system resources are devised. Finally, the developed MPC formulation is illustrated using a chemical process example.

References:

[1] Qin, S.J. and Badgwell, T.A. "A survey of industrial model predictive control technology," *Control Engineering Practice*, 11(7), 733–764, 2013.

[2] Rawlings, J.B. and Mayne, D.Q. "Model Predictive Control: Theory and Design," Nob Hill Publishing, Madison, WI, 2009.

[3] Ellis, M., Liu, J., and Christofides, P.D. "Economic Model Predictive Control: Theory, Formulations and Chemical Process Applications," Springer-Verlag, London, 2017.

[4] Blanke, M., Kinnaert, M., Lunze, J., and Staroswiecki, M. "Diagnosis and Fault-Tolerant Control," Springer, Berlin-Heidelberg, 2003.

[5] Zhang, Y. and Jiang, J. "Bibliographical review on reconfigurable fault-tolerant control systems," *Annu. Rev. in Contr.*, 32, 229–252, 2008.

[6] Christofides, P.D., Davis, J., El-Farra, N.H., Clark, D., Harris, K., and Gipson, J. "Smart plant operations: Vision, progress and challenges," *AIChE J.*, 53(11), 2734–2741, 2007.

[7] Mhaskar, P. "Robust model predictive control design for fault-tolerant control of process systems," *Ind. Eng. & Chem. Res.*, 45, 8565–8574, 2006.

[8] Dong, J., Verhaegen, M., and Holweg, E. "Closed-loop subspace predictive control for fault-tolerant MPC design," In *Proceedings of 17th IFAC World Congress*, 2008.

[9] Camacho, E.F., Alamo, T., and de la Pena, D.M. "Fault-tolerant model predictive control," In *Proceedings of IEEE Conference on Emerging Technologies and Factory Automation*, 1–8. Bilbao, Spain, 2010.

[10] Lao, L., Ellis, M., and Christofides, P.D. "Proactive fault-tolerant model predictive control," *AIChE J.*, 59, 2810–2820. 2013.

[11] Knudsen, B.R. "Proactive actuator fault-tolerance in economic MPC for nonlinear process plants," In Proceedings of *11th IFAC Symposium on Dynamics and Control of Process Systems*, 1097–1102. Trondheim, Norway, 2016.

[12] Hu, Y. and El-Farra, N.H. "Quasi-decentralized output feedback model predictive control of networked process systems with forecast-triggered communication," In *Proceedings of American Control Conference*, 2612–2617. Denver, CO, 2013.

[13] Hu, Y. and El-Farra, N.H. "Adaptive quasi-decentralized MPC of networked process systems," In *Distributed Model Predictive Control Made Easy*, volume 69, 209–223. J. M. Maestre and R. R. Negenborn (Eds.), Springer, Netherlands, 2014.

494583 Multi-Grid Schemes for Multi-Scale Predictive Control

Abstract:

We propose to use multi-grid computing schemes to design multi-scale model predictive control (MPC) architectures. This is motivated by the fact that the presence of multiple spatial and temporal scales makes centralized MPC intractable [1]. Decentralized control schemes seek to mitigate this complexity by solving problems over individual subdomains [1,2]. Achieving central optimality in these schemes is important because desirable properties of centralized MPC such as recursive feasibility, stability, and robustness can be inherited [2]. Diverse decentralized coordination schemes such as Gauss-Seidel [3], Lagrangian dual decomposition [4], or alternating direction multipliers [5,6] have been proposed in the past. However, complex coupling between subsystems can make coordination of decentralized controllers quite slow.
Our study analyzes Gauss-Seidel schemes and we study how information can be propagated through distant subsystems by exchanging state and dual information. For strictly convex quadratic programs, we establish conditions under which the Gauss-Seidel scheme is guaranteed to converge to the optimal central solution. In particular, we prove that the convergence rate is exponential in the number of iterations. Our work is motivated by a number of applications that involve multiple spatial and temporal scales. In these applications, partitioning on the entire domain can be nontrivial, because the coupling structure can be irregular (e.g., as in power grids). We show that our analysis framework is general and can be applied to these types of problems with arbitrary structures. The analysis reveals that the coupling structure between subsystems significantly affects the convergence and we show that our established conditions can be used to identify what is the most suitable partitioned structure. Our analysis also reveals that convergence is preserved when different coordination orders are used, indicating that we can derive coordination orders to maximize the computational scalability. For instance, a red-black coordination order enables parallel computation.
We also show that one can use a low-resolution centralized control formulation (i.e., defined over a coarser mesh) to aid the convergence of the decentralized Gauss-Seidel scheme [7]. We argue that this multi-grid strategy can be seen as a hierarchical MPC structure, in which a low-resolution supervisory controller mitigates low-frequency (global) disturbances and oversees a set of high-resolution decentralized MPC controllers that mitigate high-frequency (local) disturbances. We demonstrate the performance of this hierarchical MPC architecture using an inventory management problem that is operated over multiple temporal scales and using a power grid problem that covers a 2-D spatial domain. We believe that the results have particular significance on the application of MPC for systems with multiple scales, as they allow for the systematic design of hierarchical control structures.

Reference:
[1] R. Scattolini. Architectures for distributed and hierarchical model predictive control–a review. Journal of Process Control, 19(5):723–731, 2009.
[2] J. B. Rawlings and D. Mayne. *Model predictive control*: theory and design. Nob Hill Publishing, 2009.
[3] Zavala, V. M.. New Architectures for Hierarchical Predictive Control. *IFAC-PapersOnLine*, *49*(7), 43–48. 2016.
[4] Q. T. Dinh, C. Savorgnan, and M. Diehl. Combining lagrangian decomposition and excessive gap smoothing technique for solving large-scale separable convex optimization problems. Computational Optimization and Applications, 55(1):75–111, 2013.
[5] R. T. Rockafellar. Augmented lagrange multiplier functions and duality in nonconvex programming. SIAM Journal on Control, 12(2):268–285, 1974.
[6] Summers, T. H., & Lygeros, J. (2012). Distributed model predictive consensus via the Alternating Direction Method of Multipliers. *2012 50th Annual Allerton Conference on Communication, Control, and Computing (Allerton)*, (1), 79–84.
[7] W. Hackbusch, Multi-grid methods and applications. Springer Series in Computational Mathematics, 1985.

502509 Exploiting Structure in Direct Simultaneous Methods for Global Dynamic Optimization

Abstract:

To date, a majority of the research in deterministic global optimization for dynamic systems has focused on direct sequential methods, mainly direct single-shooting [1-4] and more recently direct multiple-shooting [5]. Such sequential methods rely on set-propagation techniques in order to enclose the reachable set of the embedded dynamic system, for instance using interval, ellipsoidal or polynomial model arithmetic, and typically perform a branch-and-bound search in the reduced space of the main decision variables. Despite much progress in recent years, their practical applicability is currently limited to problems with around 10 decision variables and small-scale dynamic systems. In contrast, direct simultaneous methods based on orthogonal collocations have been used extensively for local optimization, but seldom have they been investigated in a context of global optimization [6-7]. The simultaneous approach presents the inherent advantage that the objective and constraint functions in the discretized optimization problem are now factorable, so that in principle state-of-the-art global optimization solvers such as BARON or ANTIGONE can be applied directly. Of course, an attendant drawback is the large number of variables and constraints, which leads to prohibitive computational times but for very small problems.

The main objective of this work is to further investigate direct simultaneous methods in deterministic global optimization. Our main focus is on exploiting the underlying structure of the discretized NLP problem in order to expedite the branch-and-bound search. Specifically, the residuals from the discretization of the differential equations present a bordered-block diagonal structure, which we are exploiting to generate additional constraints between the decision variables and the state collocation variables using novel implicit equation bounders [8]. This leads to a tightening of the NLP relaxations and provides a means of being more selective with regards to the branching variable selection. We also consider various discretization schemes for the differential equations, including time-series expansion, Runge-Kutta formula, and orthogonal collocation. Comparisons are made between these improved direct simultaneous methods and existing direct sequential methods for a number of numerical case studies.

References:

[1] Papamichail, I., Adjiman, C.S.: A rigorous global optimization algorithm for problems with ordinary differential equations. Global Optim. 24:1-33, 2002

[2] Chachuat, B., Singer, A.B., Barton, P.I.: Global methods for dynamic optimization and mixed-integer dynamic optimization. Ind. Eng. Chem. Res. 45(25):8373-8392, 2006

[3] Lin, Y., Stadtherr, M.A.: Deterministic global optimization of nonlinear dynamic systems. AIChE J. 53(4):866-875, 2007

[4] Sahlodin, A.M.: Global optimization of dynamic process systems using complete search methods. Ph.D. thesis, McMaster University, ON, Canada, 2012

[5] Diedam, H.: Global optimal control using direct multiple shooting. Ph.D. thesis, Ruprech-Karls-Universität Heidelberg, Germany, 2015

[6] Esposito, W.R., Floudas, C.A.: Global optimization for the parameter estimation of differential-algebraic systems. Ind. Eng. Chem. Res. 39(5):1291-1310, 2000

[7] Flores Tlacuahuac, A., Terrazas, S., Biegler, L.T.: Global optimization of highly nonlinear dynamic systems. Ind. Eng. Chem. Res. 47(8):2643-2655, 2008

[8] Rajyaguru, J.: Rigorous numerical analysis with polynomial models: Applications to implicit and differential-algebraic equations. Ph.D. thesis, Imperial College London, UK, 2016

496280 Demand Response Operation of Air Separation Units Utilizing an Efficient MILP Modeling Framework

Abstract:

Deregulation, abundance and affordability of natural gas, and the increase of renewable-based electricity generation from wind and solar power has transformed U.S. energy markets. While highly beneficial in terms of the environment, the variability of solar and wind power coupled with the variable demand of consumers creates a challenge for grid operators. Consequently, increased reliance on renewables carries serious energy security risks, manifest in brown-outs or black-outs when supply cannot meet demand. Electricity storage is a natural way to synchronize renewable electricity generation with demand but remains cost-prohibitive. Managing demand rather than generation, referred to as demand response (DR), has emerged as the approach of choice for mitigating demand variability. Electricity-intensive air separation units (ASUs) are a promising industrial DR candidate—production can be increased during off-peak hours, and excess product stored (in either liquid or gas form) for use during peak demand when production rate is lowered.

DR operation of ASUs requires that the process dynamic characteristics and performance be included in production scheduling calculations to ensure that production rate changes (typically scheduled at hourly intervals) are feasible and do not violate product quality and operating safety constraints. However, the (first-principles) mathematical models that are needed to describe and optimize the DR behavior of industrial facilities cover multiple time scales and are inevitably large-scale, nonlinear, and stiff, posing computational challenges for simulation and optimization.

In this work, we address this problem by developing a computationally efficient optimal scheduling framework for ASUs operating under DR conditions. In particular, we propose using (nonlinear) low-order data-driven models of the scheduling-relevant dynamics of the plant [1], and present a set of reformulation and (exact) linearization techniques that allow us to cast the overall DR scheduling problem as Mixed Integer Linear Program (MILP).

We illustrate these concepts using an Air Separation Unit (ASU) capable of producing 50 tons of nitrogen per day. We show that DR operation can save over 3% in operating costs compared to a scenario where the production rate is kept constant during the day, a considerable savings in the context of an established industrial system that generates a commoditized product. Further, we demonstrate the computational efficiency of this framework and its potential for real-time implementation.

[1] R. C. Pattison, C. R. Touretzky, T. Johansson, I. Harjunkoski, and M. Baldea, "Optimal Process Operations in Fast-Changing Electricity Markets: Framework for Scheduling with Low-Order Dynamic Models and an Air Separation Application," *Ind. Eng. Chem. Res.*, vol. 55, no. 16, pp. 4562–4584, Apr. 2016.

499222 Automatic Sensitivity Analysis and Structure-Exploiting Nonlinear Programming for Dynamic Optimization

Abstract:

CasADi [1] is an open-source software framework for numerical optimization, dynamic simulation and algorithmic differentiation. Of special interest are optimization problems constrained by differential equations, i.e. optimal control problems, both in an online and in an offline context. CasADi is a general-purpose tool, available for C++, Python, MATLAB and Octave, and uses a lower level of abstraction than is what is associated with popular algebraic modeling languages such as AMPL, GAMS, JuMP or Pyomo. This in makes CasADi suitable to be used as a building block for domain-specific tools, and to address non-standard problem formulations, in particular in the context of optimal control. Since its inception in late 2009, CasADi has been used successfully for academic teaching as well as in applications from multiple fields, including process control, robotics and aerospace.

In this presentation we give a general introduction to the tool and discuss two of the key features of the tool which sets it apart from other optimization modeling tools; namely (1) efficient and fully automatic sensitivity analysis for initial-value problems in ordinary differential equations (ODE) and differential algebraic equations (DAE) and (2) structure-exploiting nonlinear programming.

CasADi automates ODE/DAE sensitivity analysis by working with a relatively rich problem formulation, namely that of an initial-value problem in DAE coupled to a terminal-value problem in DAE. Problems of this form can be efficiently be solved with efficient integrator schemes such as backward differential formulas (BDF) or implicit Runge-Kutta methods, implemented in popular software tools such as SUNDIALS. CasADi supports differentiation of such objects, to arbitrary order and using any combination of forward and adjoint sensitivity analysis, by automatically generating augmented ODE/DAE systems and then exploiting their structure during numerical integration.

The second feature we discuss is that of structure-exploiting nonlinear programming (NLP), particular in the context of simultanous methods for optimal control. While sparsity exploiting NLP solvers - such as IPOPT - often work well for very sparse problems arising from fully simultanous discretization approaches, some type of structure exploitation is often needed for hybrid discretization approaches. These hybrid discretization methods, e.g the direct multiple shooting method, can then efficiently be applied to higher-dimensional optimal control problems. CasADi supports structure-exploiting nonlinear programming by two different approaches, namely by an implementation of the lifted Newton method [2] and via an interface to the block-structure exploiting NLP solver blockSQP [3].

[1] J. Andersson. A general-purpose software framework for dynamic optimization, Ph. D. thesis. Faculty of Engineering, KU Leuven, Leuven, Belgium (2013)

[2] J. Albersmeyer and M. Diehl. The Lifted Newton Method and its Application in Optimization. SIAM Journal on Optimization 20, 3Â (2010), 1655â€"1684

[3] D. Janka, C. Kirches, S. Sager, A. Waechter. An SR1/BFGS SQP algorithm for nonconvex nonlinear programs with block-diagonal Hessian matrix.Â Mathematical Programming Computation 8, 4 (2016), 435-459.

501742 Dynamic Modelling of an LNG Storage Tank in a Regasification Terminal

Abstract:

Abstract

Liquefied natural gas (LNG) is a convenient medium for storage and transport of natural gas, by virtue of its relative compactness. On the downside, LNG storage is a cryogenic process, which inevitably suffers from heat leak. This in turn generates boil-off gas, which is conventionally recycled to the storage tank via an energy-intensive re-liquefaction process. Generation of boil-off gas is particularly intensive at regasification terminals carrying large LNG inventories, especially due to the presence of recirculation flows which maintain cryogenic temperature in loading, unloading, and transfer lines. The present work introduces a comprehensive dynamic model for LNG storage tanks at regasification terminals, a semi-analytical solution to the model, and several case studies on minimization of boil-off gas generation[1].

The necessity for the present model arises from the observation that LNG storage tank modelling almost invariably assume thermal and mass equilibrium between the vapor- and liquid-phases, in contradiction with measured temperature data from our industrial partner. A case in point is the influential work of Miana et al. and , wherein the vapor- and liquid-phases are treated as a flashing mixture at each time step[2]. This approximation is expected to be appropriate for marine cargo tanks, wherein surface wetting results in greatly increase heat- and mass-transfer area, but not for the relatively still vapor-liquid interface in land-based LNG storage tanks. A second concern relevant to the aforementioned work concerns the use of an arbitrary, constant boil-off rate, which is expected to vary circumstantially. Others works e.g. Migliore et al. have eliminated the use of constant boil-off rate[3], but retain the equilibrium assumption. Note also that both the temperature and the mass flow rate of the boil-off gas are needed to accurately estimate the compressor duty of the re-liquefaction process.

To date, we were only able to identify two studies which incorporate heat leak arising from recirculation flows. The first study, published by Querol et al., is a simple but comprehensive attempt at estimating heat leak into LNG storage tanks[4]. The steady-state model presented therein incorporates heat leaks from various heat-transfer areas and dissipation of shaft work, and forms an excellent framework for a more detailed dynamic model. The second study, published by Park et al., incorporates the same features, albeit using a dynamic process simulator[5]. Both studies appear to capture the main features of LNG storage tank modelling in regasification terminals, but lack the details on the heat- and mass-transfer processes, and thus possess limited predictive value.

The present study consolidates the best features of the aforementioned studies in the form of a single model which accounts for the sources of heat leak indicated by Querol et al. and the heat-transfer mechanics of Migliore et al., while also accounting for vapor-liquid disequilibrium via vapor-liquid interface heat transfer and the dynamic effects introduced by changing liquid level. The model possesses a neat partial solution in the form of a set of algebraic equations for the special case of nitrogen-free LNG. The model is subsequently used to study minimization of boil-off generation as a function of tank aspect ratio and recirculation line properties for a representative case study. The insights gleaned from this model should prove valuable in future designs of LNG regasification terminals.

References

1. Effendy S, Khan MS, Farooq S, Karimi IA. Dynamic Modelling and Optimization of an LNG Storage Tank in a Regasification Terminal with Semi-Analytical Solutions for N 2-Free LNG. *Computers & Chemical Engineering.* 2017.

2. Miana M, Del Hoyo R, Rodrigálvarez V, Valdés JR, Llorens R. Calculation Models for Prediction of Liquefied Natural Gas (LNG) Ageing During Ship Transportation. *Applied energy.* 2010;87(5):1687-1700.

3. Migliore C, Tubilleja C, Vesovic V. Weathering Prediction Model for Stored Liquefied Natural Gas (LNG). *J. Nat. Gas. Sci. Eng.* 2015;26:570-580.

4. Querol E, Gonzalez-Regueral B, García-Torrent J, García-Martínez M. Boil Off Gas (BOG) Management in Spanish Liquid Natural Gas (LNG) Terminals. *Applied energy.* 2010;87(11):3384-3392.

5. Park C, Lee C-J, Lim Y, Lee S, Han C. Optimization of Recirculation Operating in Liquefied Natural Gas Receiving Terminal. *Journal of the Taiwan Institute of Chemical Engineers.* 2010;41(4):482-491.

492571 Optimization of Dynamic Flux Balance Analysis Systems

Abstract:

Optimization of Dynamic Flux Balance Analysis Systems

Jose A. Gomez[a] and Paul I. Barton[a]

[a] *Process Systems Engineering Laboratory, Massachusetts Institute of Technology, Cambridge MA 02139, USA*

Bioprocesses involving microbial communities have widespread applications in the pharmaceutical, food and biofuels industries. These complex bioprocesses can be modeled accurately using dynamic flux balance analysis (DFBA) [1], [2], [3], which combines genome-scale metabolic network reconstructions with dynamic process models. DFBA models result in dynamic systems with linear programs (LPs) embedded [4]. These LPs are embedded because their right-hand side depends on the dynamic states and the dynamic states depend on the solution vector of the LP. Newly available simulators [4], [5] have made the reliable and efficient implementation of DFBA possible. This modeling framework has opened a new set of possibilities with regards to the discovery of novel bioprocesses [6], [7], [8] and the optimal design and control of bioprocesses. The optimal design and control of bioprocesses requires optimization algorithms for DFBA systems.

Optimization algorithms that rely on transforming the DFBA system into a Mathematical Program with Complementarity Constraints [9] or possibly high-index differential-algebraic equations system [10] have been published in literature. However, numerical difficulties and shortcomings of these approaches have also been reported [11]. Here, we use the simulation framework reported in [11], [4], [5] that transforms the LP embedded into a lexicographic LP (LLP) to address complications associated with nonunique and infeasible LP solutions.

The objectives of a LLP in standard form as a function of its right-hand side are piecewise linear functions [12], and therefore, nonsmooth. This source of nonsmoothness can be propagated to the parametric dependence of the final states of the dynamic system. Therefore, there exist some parameter values for which the Jacobian of the dynamic system may not exist. Computing elements of Clarkeâ€™s generalized Jacobian for complex nonsmooth functions is challenging [13], but can be done efficiently for piecewise differentiable functions, such as LLPs parameterized by their right-hand side, with lexicographic-directional (LD) derivatives [14].

This paper presents an algorithm for the optimization of DFBA systems. This algorithm relies on LD-derivatives to compute sensitivity information. It then uses bundle methods to solve the resulting nonsmooth optimization problem. An example illustrating the application of this algorithm is presented.

Keywords: Linear programming, generalized Jacobian, lexicographic differentiation, LD-derivative, lexicographic optimization, nonsmooth sensitivities, nonsmooth equation solving, flux balance analysis, dynamic flux balance analysis.

References

[1] A. Varma and B. Ø. Palsson, "Stoichiometric flux balance models quantitatively predict growth and metabolic by-product secretion in wild-type Escherichia coli W3110," *Applied and Environmental Microbiology,* vol. 60, no. 10, pp. 3724-3731, 1994.

[2] J. D. Orth, I. Thiele and B. Ø. Palsson, "What is flux balance analysis?," *Nature Biotechnology,* vol. 28, pp. 245-248, 2010.

[3] R. Mahadevan, J. Edwards and F. I. Doyle, "Dynamic flux balance analysis of diauxic growth in Escherichia coli.," *Biophysical Journal,* vol. 83, no. 3, pp. 1331-40, 2002.

[4] K. Höffner, S. M. Harwood and P. I. Barton, "A reliable simulator for dynamic flux balance analysis," *Biotechnology and Bioengineering,* vol. 110, no. 3, pp. 792-802, 2013.

[5] J. A. Gomez, K. Höffner and P. I. Barton, "DFBAlab: A fast and reliable MATLAB code for Dynamic Flux Balance Analysis," *BMC Bioinformatics,* vol. 15, p. 409, 2014.

[6] J. A. Gomez, K. Höffner and P. I. Barton, "From sugars to biodiesel using microalgae and yeast," *Green Chemistry,* vol. 18, no. 2, pp. 461-475, 2016.

[7] J. Chen, J. Gomez, K. Höffner, P. Barton and M. Henson, "Metabolic modeling of synthesis gas fermentation in bubble column reactors," *Biotechnology for Biofuels,* vol. 8, no. 89, 2015.

[8] J. Chen, J. Gomez, K. Höffner, P. Phalak, P. Barton and M. Henson, "Spatiotemporal modeling of microbial metabolism," *BMC Systems Biology,* vol. 10, no. 21, 2016.

[9] A. U. Raghunathan, J. R. Pérez-Correa, E. Agosin and L. T. Biegler, "Parameter estimation in metabolic flux balance models for batch fermentation - formulation and solution using differential variational inequalities," *Ann Oper Res,* vol. 148, pp. 251-270, 2006.

[10] U. Kaplan, M. Türkay, L. Biegler and B. Karasözen, "Modeling and simulation of metabolic networks for estimation of biomass accumulation parameters," *Discrete Applied Mathematics,* vol. 157, no. 10, pp. 2483-2493, 2009.

[11] S. Harwood, K. Höffner and P. Barton, "Efficient solution of ordinary differential equations with a parametric lexicographic linear program embedded," *Numerische Mathematik,* vol. 133, no. 4, pp. 623-653, 2016.

[12] D. Bertsimas and J. N. Tsitsiklis, Introduction to Linear Optimization, Nashua, NH: Athena Scientific, 1997.

[13] F. H. Clarke, Optimization and Nonsmooth Analysis, Philadelphia: Society for Industrial and Applied Mathematics, 1990.

[14] K. A. Khan and P. Barton, "A vector forward mode of automatic differentiation for generalized derivative evaluation," *Optimization Methods & Software,* vol. 30, no. 6, pp. 1185-1212, 2015.

[15] K. A. Khan and P. I. Barton, "Generalized Derivatives for Solutions of Parametric Ordinary Differential Equations with Non-differentiable Right-Hand Sides," *Journal of Optimization Theory and Applications,* vol. 163, no. 2, pp. 355-386, 2014.

492805 Parallel Cyclic Reduction Decomposition for Dynamic Optimization Problems

Abstract:

Direct transcription of dynamic optimization problems, with differential-algebraic equations discretized and written as algebraic constraints, can create very large nonlinear optimization problems. When this discretized optimization problem is solved with an NLP solver, such as IPOPT, the dominant computational cost often lies in solving the linear system that generates Newton steps for the KKT system. Computational cost and memory constraints for this linear system solution raise many challenges as the system size increases. On the other hand, the linear KKT system for our dynamic optimization problem is sparse and structured, and can be permuted to form a block tridiagonal matrix. This study explores a parallel decomposition strategy for block tridiagonal systems that is based on cyclic reduction (CR) factorization of the KKT matrix. The classical CR method has good observed performance, but its numerical stability properties are unproved for our KKT system. Alternately, Yalamov and Pavlov proposed a modified, full-space CR algorithm with proven stability bounds. Here we compare the performance of both methods on random test matrices. Stable behavior is observed in both cases, and the classical CR method has lower computational cost. This result is explained by analyzing the complexity of both methods. Finally, we discuss modifications to the CR decomposition that improve performance, and we apply the approach to four industrially relevant case studies. These include a moving horizon estimation of a CSTR, nonlinear model predictive control of a CSTR, grade transition for a polymerization process, and nonlinear model predictive control of a bubbling fluidized bed reactor. The latter problem is discretized in both space and time, resulting in up to 250000 variables and constraints. On this problem, a parallel speedup of a factor of four is observed when using eight processors.

493837 Dynamic Optimization of Polymerization Processes with Detailed Molecular Weight Distributions

Abstract:

Dynamic optimization can be used to find fast and economical grade transitions in polymerization processes. The high degree of nonlinearity and large size of these systems makes this a very challenging problem. Many end-use polymer properties on the molecular weight distribution (MWD). Conventionally, a Flory-Schulz distribution with a small-scale moment model has been used to represent the MWD, but it is valid only under steady state or pseudo steady state assumptions. Ideally, the full population balance model should be included in the optimization process so that an operator can control from one desired MWD to another desired MWD in minimum time. However, population balance models consist of a countable system of differential equations. Even if chain length is truncated at a suitable value, these problems are too large for existing dynamic optimization methodologies. In this talk, we discuss alternative representations of the MWD using orthogonal collocation methods[1,2]. In the collocation method, the population balance equations of polymer chains are satisfied exactly at the collocation points in the chain length domain. The resulting solution is used to reconstruct a piecewise polynomial representation of the entire MWD. There are several outstanding challenges compared to typical applications of collocation. First, the inherently discrete MWD is being approximated by smooth functions, and the collocation points must be chosen as integers. In addition, reactions are driven by finite difference relations rather than derivatives, which introduces numerical challenges[1]. We investigate the impacts of these potential sources of error and discuss potential adaptive approaches to regulate and reduce the error.

The MWD collocation method is demonstrated on an industrial high-density polyethylene (HDPE) slurry process with a continuous stirred-tank reactor. A detailed mathematical model is presented to track the behavior of the entire plant with liquid and vapor recycles[3]. We will develop and discuss the steady state optimization of the process to produce a product of a specified MWD. A Monte-Carlo simulation method is used to verify the accuracy of the collocation approach. In addition, a dynamic grade transition problem will be presented, where a direct transcription optimization approach is applied to solve this problem after fully discretizing state and control variables in time.

References:

1. Canu, P., and Ray, W.H. Discrete weighted residual methods applied to polymerization reactions. Computers and Chemical Engineering, 1991. Vol 15, issue 8, pp 549-564
2. Pontes, K.V., Embirucu, M., Maciel, R., Hartwich, A., Marquardt, W. Optimal process operation for the production of linear polyethylene resins with tailored molecular weight distribution. AIChE Journal, 2011. Vol 57, No 8. pp 2149-2163.
3. Zhang, C., Shao, Z., Chen, X., Yao, Z., Gu, X. and Biegler, L. T. Kinetic parameter estimation of HDPE slurry process from molecular weight distribution: Estimability analysis and multistep methodology. AIChE Journal, 2014. 60: 3442â€"3459.Â

494109 Improved Bounds on the Solutions of Nonlinear Dynamic Systems Using Centered-Form Differential Inequalities

Abstract:

In this talk, we will present and compare several new techniques for computing interval enclosures of the solutions of nonlinear ordinary differential equations (ODEs) subject to a range of parameters and initial conditions. Bounding algorithms of this type are broadly used for uncertainty propagation, set-based state estimation, fault detection, and the global solution of optimal control problems. However, despite significant recent progress, state-of-the-art ODE bounding methods still provide an unsatisfactory compromise between computational efficiency and accuracy for many systems of practical complexity. In particular, classical methods based on interval computations are very efficient but often produce bounds that are extremely conservative. In contrast, methods based on higher-order set enclosures such as polytopes, zonotopes, and Taylor models often provide highly accurate enclosures, but are much more computationally demanding. Together, these issues severely limit the use of existing bounding methods for online control applications, and are also primarily responsible for the inefficiency of current global dynamic optimization algorithms.

To address this problem, this talk will present several new approaches based on extensions of the classical theory of differential inequalities (DI), which furnishes simple interval enclosures at very low computational cost. One of the primary drawbacks of the standard DI method is that it computes bounds assuming that, at every point in time, the uncertain quantities (i.e., all model parameters and state variables) can vary independently of one another (because they are bounded in a Cartesian product of intervals). This greatly simplifies computations, but also leads to severe overestimation of the true solution set in many cases. However, this problem is not unique to dynamic systems – a simpler form of the problem also in occurs when using interval arithmetic to bound the range of a given function, where it is known as the 'dependency problem.' Moreover, several well known strategies have been developed that partially resolve the dependency problem using derivative information while largely maintaining the computational efficiency of standard interval arithmetic. These techniques are broadly known as 'centered forms' and include the mean-value form and slope forms of several varieties.

In this work, we extend these centered-forms to dynamic problems by combining them with the theory of differential inequalities in a novel way. The resulting bounding methods are most similar to the method of Chachuat and Villanueva (Comp. & Chem. Eng., 2012), which combines differential inequalities with Taylor models (a form of enclosure for the range of a function that is represented by a polynomial approximation with a rigorous bound on the approximation error). However, the proposed methods use only first order derivative (or slope) information, whereas Taylor models use at least second order information. Thus, the proposed methods are expected to be more computationally efficient in general. We will present detailed numerical comparisons for several different centered-form ODE bounding methods, along with comparisons to existing state-of-the-art bounding methods, on a test set of nonlinear models of chemical reaction

systems, mechanical systems, and aerospace systems. Our results indicate that a significantly better compromise between efficiency and accuracy can be achieved in many cases. Moreover, the use of centered-forms can also enhance the effectiveness of another bounds tightening technique recently developed by the authors based on the use of solution invariants. We conclude with a discussion of key problem features that appear to dictate the effectiveness of the new approaches.

501368 A Mathematical Programming Approach for Integrating Distributed Urban Energy Systems and Iot

Abstract:

A mathematical programming approach for integrating distributed urban energy systems and IoT

Nagore Sabio[*][1,2], Evgenia Mechleri[1] and Harvey Arellano-GarcÃa[1]

[1]Department of Chemical and Process Engineering, University of Surrey, Guildford, United Kingdom

[2]UCL Energy Institute, University College London, London, United Kingdom

Over fifty per cent of the worldâ€™s population now lives in urban areas and much of the future population growth will take place in cities [1]. The growing penetration of variable renewable energy sources (VREs) is increasing the need for flexibility in the energy service domain. In the UK, installed capacity of renewable energy has increased more than six times in the last 15 years. Although electricity still dominates renewable generation, renewable heat contribution has risen up to 20% share recently [2]. VREs are commonly connected to the distribution grid in addition to other distributed energy systems (DES), which increases the complexity of the distribution grid management. In this sense, the demand and business case for frameworks capable of integrating demand and supply and providing an optimal design and operation of urban energy systems is increasing.

DESs are a suite of on-site, grid-connected or stand-alone technology systems that can be sourced by different types of distributed energy resources, such as natural gas, biomass, solar energy, wind and waste heat amongst others. Such DES offer great advantages over centralized generation by offering end users a diversified fuel supply, higher efficiency and lower emissions. However, the integration of renewable energy resources (e.g. wind turbines, PV units) which depend completely on the unsteady weather condition increase the complexity of energy provision continuation within DESs. On the other hand, the energy demands of end-consumers always fluctuate hourly which may cause energy unbalance between supply and demand sides. Our framework brings weather data and sensor information into a virtual energy plant optimization model that connects supplier and consumer in order to optimize potential flexibility gaps arising from supply and demand mismatch. The system includes two optimization levels: one that allows for providing the optimal distributed energy system design, and a subsequent level that optimizes its operation. The problem is posed as a two-stage hybrid MILP optimization model combining flexibility analysis and optimal synthesis for integrating energy supply and demand where environmental information is added to each stage, informing the consumer and supplier of the carbon footprints of the options selected. The novelties of the present framework therefore are threefold. First, we propose a smart metering design that makes use of the current information technologies and communication network infrastructure, bringing IoT and mathematical programming at the urban consumer level. Second, our approach integrates electricity and heat demands within a single frameworks for urban DES optimization. And finally, we combine traditional mathematical programming approaches such as flexibility analysis, optimal network synthesis and information technologies within a single optimization framework combining IoT and urban DES.

[1] Kirstead , J. and Shah, N. (2013). *Urban Energy Systems â€" An integrated approach*. London: Routledge

[2] Digest if UK Energy Statistics (DUKES), Chapter 6: Renewable sources of energy. (2016). Department of Business, Energy and Industrial Strategy,UK Government. Retrieved 30 October 2016.

[3] Staffell, I., Brett D.J.L., Brandon, N.P. and Hawkes, A.D. (2015). Domestic microgeneration: renewable and distributed energy technologies, policies and economics. Oxford: Routledge.

---

[*]Corresponding autor: n.sabio@ucl.ac.uk

501493 Life Cycle Optimization of Shale Gas Supply Chains with Comprehensive Environmental Impacts and Modeling of Modular Plants

Abstract:

In recent decade, the rapid expansion of shale gas industry has spurred great interest in the research of sustainable shale gas energy systems [1, 2]. The mathematical programming approach has been proven as a powerful tool in the design and operations of shale gas energy systems [3-5], and some pioneering studies have successfully addressed both the economic and environmental concerns of shale gas in their models [6-10]. However, most existing studies on shale gas energy systems solely focus on certain midpoint environmental impact indicators, namely the GHG emissions and water consumption [11-14]. Despite the importance of these two environmental impact categories, other environmental impact categories being ignored, such as land occupation, ecotoxicity, resource depletion, etc., are worth considering as well [2]. Additionally, the emergence of new technologies/operations, known as the modular manufacturing, have created great opportunities in the sustainable design and optimization of shale gas energy systems [15]. Thus, there is an urgent need of research to develop corresponding models for the assessment of their potential economic and environmental benefits.

Motivated by these research gaps, we propose the first life cycle optimization framework that considers the full environmental impacts for the shale gas energy systems. In this modeling framework, an endpoint-oriented life cycle impact assessment (LCIA) method is adopted and further integrated into a functional-unit-based multiobjective life cycle optimization model. Specifically, a full life cycle inventory (LCI) is first established throughout the life cycle of shale gas. The LCI is then combined with the state-of-the-art LCIA approach ReCiPe [16], which comprises of eighteen midpoint impact categories (e.g., climate change, ozone depletion, terrestrial acidification, etc.) and three endpoint impact categories, namely damage to human health, damage to ecosystem diversity, and damage to resource availability. Next, we can integrate the decision variables in the shale gas supply chain with their full environmental impacts based on the LCI data and characterization factors provided by ReCiPe. In this model, both economic and environmental objectives are considered. The economic objective is to optimize the net present cost associated with producing one functional unit product from shale gas, and the environmental objective is formulated as a total environmental score characterized by the ReCiPe associated with one functional unit. Additionally, design decisions regarding both the conventional processing plants and modular LNG plants are considered. The presented model also captures the mobile feature of modular plants. With this integrated model, we expect to investigate the advantage of modular plants in terms of both economic and environmental performance compared with its conventional counterpart. The resulting model is formulated as a mixed-integer linear fractional program that cannot be solved efficiently by any off-the-shelf global optimizers. Therefore, a parametric algorithm is implemented to improve the computational efficiency [17].

To illustrate the application of proposed modeling framework and solution algorithm, a case study based on a real Marcellus shale gas supply chain with twelve shale sites is presented. A â€œwell-to-wireâ€ life cycle is considered with the functional unit defined as 1 MWh electricity generation. Major design decisions regarding the drilling schedule, water treatment technologies, locations and capacities of modular and conventional processing plants, and infrastructure constructions are all considered. Based on the optimization results, the total environmental score of the economic-oriented solution is 21% higher than that of environmental-oriented solution. However, the levelized cost of electricity (LCOE) generated in the economic-oriented solution is $59.8/MWh, 72% lower than that in the environmental-oriented solution. The adoption of modular plants is preferred when economics is the major driving force in decision-making. Nevertheless, the shale gas supply chain design with modular plants does not show obvious advantage in terms of mitigating the life cycle environmental impacts of shale gas.

References

[1] C. L. Weber and C. Clavin, "Life cycle carbon footprint of shale gas: Review of evidence and implications," *Environmental Science & Technology,* vol. 46, pp. 5688-5695, 2012.

[2] J. Gao and F. You, "Design and optimization of shale gas energy systems: Overview, research challenges, and future directions," *Computers & Chemical Engineering*. DOI: http://doi.org/10.1016/j.compchemeng.2017.01.032.

[3] O. J. Guerra, A. J. Calderón, L. G. Papageorgiou, J. J. Siirola, and G. V. Reklaitis, "An optimization framework for the integration of water management and shale gas supply chain design," *Computers & Chemical Engineering,* vol. 92, pp. 230-255, 2016.

[4] D. C. Cafaro and I. E. Grossmann, "Strategic planning, design, and development of the shale gas supply chain network," *AIChE Journal,* vol. 60, pp. 2122-2142, 2014.

[5] J. Gao and F. You, "Deciphering and handling uncertainty in shale gas supply chain design and optimization: Novel modeling framework and computationally efficient solution algorithm," *AIChE Journal,* vol. 61, pp. 3739-3755, 2015.

[6] J. Gao and F. You, "Shale gas supply chain design and operations toward better economic and life cycle environmental performance: MINLP model and global optimization algorithm," *ACS Sustainable Chemistry & Engineering,* vol. 3, pp. 1282-1291, 2015.

[7] T. V. Bartholomew and M. S. Mauter, "Multiobjective optimization model for minimizing cost and environmental impact in shale gas water and wastewater management," *ACS Sustainable Chemistry & Engineering,* vol. 4, pp. 3728-3735, 2016.

[8] L. F. Lira-Barragan, J. M. Ponce-Ortega, M. Serna-Gonzalez, and M. M. El-Halwagi, "Optimal reuse of flowback wastewater in hydraulic fracturing including seasonal and environmental constraints," *AIChE Journal,* vol. 62, pp. 1634-1645, 2016.

[9] J. Gao and F. You, "Game theory approach to optimal design of shale gas supply chains with consideration of economics and life cycle greenhouse gas emissions," *AIChE Journal*. DOI: 10.1002/aic.15605, 2017.

[10] C. He and F. You, "Toward more cost-effective and greener chemicals production from shale gas by integrating with bioethanol dehydration: Novel process design and simulation-based optimization," *AIChE Journal,* vol. 61, pp. 1209-1232, 2015.

[11] I. J. Laurenzi and G. R. Jersey, "Life cycle greenhouse gas emissions and freshwater consumption of Marcellus shale gas," *Environmental Science & Technology,* vol. 47, pp. 4896-4903, 2013.

[12] M. Jiang, C. T. Hendrickson, and J. M. VanBriesen, "Life Cycle Water Consumption and Wastewater Generation Impacts of a Marcellus Shale Gas Well," *Environmental Science & Technology,* vol. 48, pp. 1911-1920, 2014.

[13] M. Yang and F. You, "Comparative Techno-Economic and Environmental Analysis of Ethylene and Propylene Manufacturing from Wet Shale Gas and Naphtha," *Industrial & Engineering Chemistry Research,* vol. 56, pp. 4038-4051, 2017.

[14] C. He and F. You, "Deciphering the true life cycle environmental impacts and costs of the mega-scale shale gas-to-olefins projects in the United States," *Energy & Environmental Science,* vol. 9, pp. 820-840, 2016.

[15] S. H. Tan and P. I. Barton, "Optimal dynamic allocation of mobile plants to monetize associated or stranded natural gas, part I: Bakken shale play case study," *Energy,* vol. 93, Part 2, pp. 1581-1594, 2015.

[16] M. Goedkoop, R. Heijungs, M. Huijbregts, A. De Schryver, J. Struijs, and R. van Zelm, "ReCiPe 2008-A life cycle impact assessment method which comprises harmonised category indicators at the midpoint and the endpoint level," 2009.

[17] Z. Zhong and F. You, "Globally convergent exact and inexact parametric algorithms for solving large-scale mixed-integer fractional programs and applications in process systems engineering," *Computers & Chemical Engineering,* vol. 61, pp. 90-101, 2014.

501834 Unit Commitment Operation of Energy Storage Systems: Comparison of Multi-Stage Stochastic Programming and Empc

Abstract:

It is widely recognized that a major concern with renewable energy is the fact that wind and solar sources are non-dispatchable. That is, the power produced from renewable sources is dependent on environmental conditions and is likely uncorrelated with the power demand from load centers. While fossil based sources are dispatchable and currently have the ability to respond to the full range of consumer loads, the additional range imposed by renewable sources is expected to exceed the dispatch capability of these fossil plants at the point of 20% renewable power. Thus, many have advocated the use of massive energy storage systems to provide the additional level of dispatch capability required to maintain grid solvency.

Due to the uncertainty of consumer demand as well as that of renewable generation, the problem of optimal placement of these storage units within the grid along with the selection of equipment sizes must be formulated as a stochastic program. However, rather than being a fairly simple two-stage stochastic program, the dynamics imposed by the storage devices requires the formulation to be of the far more challenging multistage class. Previously, we have demonstrated that Economic Model Predictive Control (EMPC) can be used to approximate the scenario based multi-stage stochastic programming version of the economic dispatch problem. However, since operation of storage units is expected to be at a larger time scale, it is more appropriate to consider a Unit Commitment (UC) type framework. Within this framework the multi-stage, stochastic program will contain mixed integer variables, accentuating the need for a more computationally attractive operating policy. In this work, the head-to-head comparison between EMPC and the scenario based solution is extended to the UC framework in an effort to validate the adoption of more tractable EMPC-based methods for the optimal sizing and placement of storage units.

491368 Cooling Limitations in Power Plants: Optimal Multiperiod Design of Natural Draft Cooling Towers

Abstract:

Thermal plants, either regular ones based on fossil fuels, geothermal or Concentrated Solar Plants (CSP) [1], require cooling units to reject the heat from the exhaust. The use air results in the fact that the operation becomes complex due to the variability of its conditions over time, temperature and humidity. On top of the variability of the cooling agent, it is even more challenging to consider the operation of CSP plants, where, there is a large rage in the power produced over time due to seasonality. Martín and Martín [2] and Martín [3] evaluated the operation of such plants for wet and dry cooling systems. However, no detailed structural design for the wet cooling tower was carried out.

In this work we develop a multiperiod optimization formulation for the geometric design and monthly operation of cooling towers for power plants aiming at minimum water consumption. Coal based and CSP plants are considered. While the operation of both depends on the weather, CSP plants operation is also characterized by a non steady production of energy that also depends on atmospheric conditions. CSP and regular power plants show that extreme conditions do not allow fully operation of the plant and worst case design results in large unused capacity. Cooling towers are modelled using Mickley method to compute the mass transfer operation [4] together with geometric considerations to account ofr the pressure drop across the structure [5,6]. A multiperiod formulation is developed in GAMS for the optimal design and operation of such units over time. Furthermore, a rigorous CHEMCAD simulation of the power block based on data from a real power plant is also developed to evaluate the limits in power production as a result of the cooling capabilities in different climates based on an actual Power plant in Spain [7]. We use a hot climate as a reference for the location of CSP plants [2] and the same location and another one in a colder location to compare the design of the tower under different weather conditions for plants whose fuel or energy source is biogas, natural gas or coal.

The formulation shows that the driving force is limited in winter. Furthermore, the extreme temperatures of summer reduce the production capacity of the plant due to the heat transfer capacity. Colder climates require larger towers and show lower water consumption. Hotter climates need additional heat transfer area until the point when production must decrease by increasing the exhaust pressure of the low pressure turbine. Furthermore, a correlation for the power produced as a function of the air temperature and exhaust pressure of the turbine has been developed as a reference to evaluate the efficiency losses depending on the weather.

Acknowledgement

M. Vicente. La Robla power plant.La Robla, León, Spain.

References

[1] Hooman, K (2010) Dry cooling towers as condensers for geothermal power plants. Int. Comm. Heat Mass transf.

37, 1215, 1220

[2] Martín, L, Martín M. Optimal year-round operation of a Concentrated Solar Energy Plant in the South of Europe Applied Thermal Engineering 2013;. 59: 627-633.

[3] Martín, M (2015) Optimal annual operation of the dry cooling system of a Concentrated Solar Energy Plant in the South of Spain. Energy 84, 774-782

[4] Geankoplis, C.J. (1993) *Transport Processes* and *Unit Operations* (3rd Edition) (1993) Prentice Hall Upper Saddle River, New Jersey. U.S.A.

[5] EPA (1970) A method for predicting the performance of natural draft cooling towers Pacifict Northwestern Water Laboratory

[6] Kreith, F., Goswami, D.Y (2005) The CRC Handbook of Mechanical Engineering. 2nd Edition CRC Press Boca Ratón

[7] Martín, M. (2013) Design of a power plant with carbon capture (In Spanish). MEng Thesis University of Salamanca

493274 Dynamic Process Modeling for a 10 MWe Supercritical CO_2 Recompression Brayton Pilot Plant Design

Abstract:

Supercritical carbon dioxide (sCO2) Brayton power cycles offer significant benefits over conventional steam Rankine cycle power plants. Closed, indirect Brayton power cycles use an external thermal source (e.g., solar, nuclear or fossil) to heat compressed working fluid which is expanded in a turbine, generating electricity and exhaust heat for recuperation. Using sCO2 as the working fluid provides the potential for thermal efficiencies over 50 percent, more compact designs, lower costs, and reduced greenhouse gas emissions. In view of these promising advantages, the U.S. Department of Energy under its Supercritical Transformational Electric Power Program is working with industry and research partners to design, build, and operate a 10 MWe (megawatts electrical) sCO2 pilot plant test facility. The new facility will support the future commercialization of sCO2 Brayton cycle energy conversion systems by testing and demonstrating the potential energy efficiency and cost benefits of this technology.

In this work, a dynamic pressure-driven process model of a closed, indirect 10 MWe recuperated sCO2 recompression Brayton cycle integrated with natural gas-fired burner has been developed in Aspen Plus Dynamics. The dynamic model can predict off-design and part-load performance and provide time-dependent profiles of key outputs such as power generated and thermal efficiency as a function of compressor inlet temperature and pressure, turbine inlet temperature, bypass recompression fraction, and other key process variables. The dynamic model, assumptions, and specifications will be described for the key components and overall cycle. The presentation will also highlight transient responses and operational strategies for maximizing sCO2 cycle efficiency and net power during typical pilot plant disturbances and off-design testing.

496745 Systematic Integration of Carbon Capture, Utilization and Storage Technologies to Meet CO_2 Emission Reduction Targets

Abstract:

Fossil energy sources have dominated the energy sector for the past century. However, increasing GHG emissions and diminishing fossil fuel reserves expedited implementation of renewable energy conversion processes such as biomass processing and solar power plants. Although these low carbon alternatives will shape the future of energy sector, achieving the CO_2 emission reduction goals will not be possible without the deployment of carbon capture, utilization and storage systems (CCUS) at scale. CCUS systems can be integrated with many processes including natural gas processing, power plants, and biomass conversion. Coal-fired power plants are important due to being the largest emission source, while bioenergy with CCUS (BECCS) is presented as a key for the future of CO_2 mitigation owing to be the only system that has negative net carbon emission. Various synergistic integration options and their theoretical limits should be studied from process systems perspective [1]. For cases of sustainable biomass, to have a reasonable yield, biomass conversion should be augmented by a carbon-free energy source [2]. Recently proposed hydricity process is an example to supply H_2, from solar thermal energy [3]. For the case of energy crops, growing biomass is not the most efficient use of limited land area [4]. To determine the potential of different alternatives we have created a process superstructure (PS) of conversion processes and CCUS technologies. PS includes coal-fired power plants, biomass conversion processes such as gasification and pyrolysis; upgrading processes such as FT and methane production; power generation options such as combined and steam cycles. PS also includes various CCUS systems such as pre-, post- and oxy-combustion, CO_2-conversion and storage options. Simulations and optimization of the proposed PS are performed using an integrated Matlab and Aspen platform [5]. The operating conditions and topological structure of the integrated process are determined via sensitivity analysis and optimization using genetic algorithm in Matlab. The PS is used for techno-economic assessment for integration of various biomass conversion processes and CCUS technologies. The study provides synergistic process designs that maximize energetic and economic performance. The promising integrations, preferred application areas and the details of modeling approach are presented.

1 Gençer E, Agrawal R. Strategy to synthesize integrated solar energy coproduction processes with optimal process intensification. Case study: Efficient solar thermal hydrogen production. Computers & Chemical Engineering, 2017. Doi: 10.1016/j.compchemeng.2017.01.038.

2 Mallapragada DS, Tawarmalani M, Agrawal R. Synthesis of augmented biofuel processes using solar energy. AIChE J., 60(7), 2533-2545, 2014.

3 Gençer E, Mallapragada DS, Marechal F, Tawarmalani M, Agrawal R, Round-the-clock power supply and a sustainable economy via synergistic integration of solar thermal power and hydrogen processes. PNAS, 112 (52), 15821-15826, 2015.

4 Mallapragada DS, Singh N, Curteanu V, Agrawal R, Sun-to-fuel assessment of routes for fixing CO_2 as liquid fuel, I&EC, 52(14), 5136-5144, 2013.

5 Gençer E, Tawarmalani M, Agrawal R, Integrated solar thermal hydrogen and power coproduction process for continuous power supply and production of chemicals, CACE, 37, 2291-2296, 2015.

498667 Representative Energy Costs for Optimization of Industrial Process Design and Operations – Systematic Comparison of Clustering Methodologies

Abstract:

Representative energy costs for optimization of industrial process design and operations – Systematic comparison of clustering methodologies

Holger Teichgraeber[1] Adam Brandt[1]

[1]Department of Energy Resources Engineering, Stanford University, Stanford, California 94305-2220, USA

The optimization of process synthesis and design problems is concerned with the optimal design and operation of an industrial system consisting of several components and subcomponents. Often, these components consume or generate power. This optimization problem requires that operations scheduling be solved for multiple model years. Because this is computationally expensive, it is of interest to represent input data using as few days or weeks as possible. Input data clustering can help obtain practical solutions to problems that we are not able to solve for the full set of input data on reasonable timescales. Input data examined here include hourly or sub-hourly electricity prices, wind speed, and solar irradiation (one can think of other data as well such as hourly demand, hourly temperature etc.).

Both Brodrick et al. [1] and Bahl et al. [2] have proposed using k-means clustering for finding representative periods in the optimization of process synthesis and design problems. They measure the accuracy of representative periods based on the objective function (net present value or total annualized cost). A similar approach has been proposed by Nahmmacher et al. [3] and Merrick [4] for electricity generation extension planning models using hierarchical clustering. The process synthesis and design problems thus far considered solved large-scale optimization problems for which the objective function and solution structure for the full set of input data is unknown.

We formulate several example problems that capture characteristics of real world applications, but are solvable for a whole year of input data. We use these problems to systematically investigate which information is lost and kept in the operational stage of the optimization problem during input data clustering. We test different clustering methodologies systematically. This information has implications for the design decisions. For different problem types, different kind of information is important: Variability is important for storage arbitrage problems, whereas average cut-off price is important for power generation problems with operational on-off decision. We investigate these systematic relationships between clustering methodologies and problem type.

Amongst others, we formulate an electricity storage problem to optimally take advantage of energy price arbitrage, a scheduling problem to perform energy-intensive tasks in times of low electricity prices, and a gas turbine power generation problem.
In addition, we also use an oxy-combustion combined cycle design and operational optimization problem [5] as an example for a large scale optimization problem to investigate clustering effects.

We use these example problems to investigate which information loss and inaccuracy occurs due to clustering for each of these problems. Methods that are subject to investigation are clustering method, normalization, cluster centroid vs. closest day, and application to multi-timescale input data.

[1] Brodrick, Philip G., Charles A. Kang, Adam R. Brandt, and Louis J. Durlofsky. "Optimization of carbon-capture-enabled coal-gas-solar power generation." *Energy*79 (2015): 149-162.

[2] Bahl, Björn, Alexander Kümpel, Matthias Lampe, André Bardow. "Time-series aggregation for synthesis of distributed energy supply systems by bounding error in operational expenditure." *Computer Aided Chemical Engineering* 38 (2016): 793-798

[3] Nahmmacher, Paul, Eva Schmid, Lion Hirth, and Brigitte Knopf. "Carpe diem: A novel approach to select representative days for long-term power system modeling." *Energy*112 (2016): 430-442.

[4] Merrick, James H. "On representation of temporal variability in electricity capacity planning models." *Energy Economics*59 (2016): 261-274.

[5] Teichgraeber, Holger, Philip G. Brodrick, and Adam R. Brandt., "Optimal Design and Operation of a Semi-Closed Oxy-Combustion Combined Cycle Power Plant". *2016 American Institute of Chemical Engineers Annual Meeting.* San Francisco, CA. November 2016.

Finding the Optimal Path
by Thomas F. Edgar

AIChE Annual Conference
11/1/2017

ABSTRACT

There are many possible career paths a university professor can pursue, but not all of them turn out to be "optimal." In this presentation Tom Edgar traces the career of Jim Rawlings, Professor of Chemical Engineering at the University of Wisconsin, from the time he was a student in Professor Edgar's class at the University of Texas until the present time.

503760 Model Predictive Control and Estimation â€“ Towards Decision Making in the Cloud

Abstract:

Technological advancements in communication and computation have led to the widespread appearance and use of cloud computing and data management, with significant impact on many applications, spanning from logistics, marketing, energy networks, to traffic control and monitoring.

Even so that the use of cloud technologies for process control, monitoring and operation is still in its infancy, they bear large potential to improve process performance, individualize control loops, and shade light on process phenomena which are not well understood. As will be outlined in the talk, the combination of model predictive control, model based estimation techniques, as well as learning and adaptation strategies are ideally suited to fully exploit the potential of cloud computing. On one side the combination of estimation, learning, and control allows to identify and adapt towards unknown and changing process parameters and disturbances, while maintaining satisfaction of constraints. It enables inclusion of preview information, such as disturbances and set point changes, and allows to cope with the often unavoidable, inherently asynchronous operation due to network communication. Additionally, cloud supported estimation and control strategies enable to exploit formerly unseen computational power, without the need to significantly modify existing local control hardware. Cloud based approaches allow to service and monitor many processes at the same time, while facilitating central maintenance of the underlying algorithms. Model and data mining based analysis of the collected data from the repeated operation of one process, or multiple similar processes, furthermore allows to shade light on phenomena not well understood, or to identify reasons for process failure or degradation.

We outline the potential of cloud based model predictive control and estimation considering two example processes â €“ the operation of batch bioreactors, as well as age aware charging and monitoring strategies of Li-Ion batteries.

503757 O Set-Free Tracking: There and Back Again

Abstract:

Feedback is necessary to reduce the e ect of disturbances and to cope with unavoidable modeling errors. Nonetheless,the way in which feedback is used to achieve o set-free tracking in the presence of persistent errors or non-zero mean disturbances appears to be often a question of personal preference among possible di erent, sometimes ad-hoc, methods. As a matter of fact, this aspect of controller design is typically overlooked in academic papers but is often fundamental for successful implementation.

After a preliminary introduction to linear o set-free Model Predictive Control (MPC) design principles based on disturbance models, explaining how the integral action is achieved in spite of modeling errors, we address the following results.

1. We propose an observer-based Internal Model Control (IMC) struc-ture which extends the simple IMC design principles to integrating and unstable plants, showing the conditions for internal stability and o set-free property. A connection with the Youla-Kucera parameter-ization is also established as a special case.

2. We show that several known alternative o set-free MPC algorithms (using velocity models) are special cases of the general disturbance models.

3. We extend the concepts of o set-free estimation to the design an economic MPC algorithm that is able to cope with persistent errors while still achieving the optimal ultimate economic performance.

503752 Linking Stability and Robustness of Nonlinear MPC and Economic MPC to Properties of Optimization Subproblems:Building on the Rawlings Results

Abstract:

This talk presents and analyzes optimization problem formulations for NMPC and eNMPC, and extends optimality and regularity conditions that are needed to fulfill well-known stability and robustness properties developed by Rawlings and coworkers. Optimization strategies that rely on these formulations are synthesized and several illustrative NMPC and eNMPC examples are presented that demonstrate their benefits.

503683 Adventures in Model Predictive Control – Tales from a 30-Year Collaboration with Professor Jim Rawlings

Abstract:

Nearly thirty years ago I arrived at the University of Texas at Austin to begin graduate studies in chemical engineering. The true depth of my ignorance became apparent as soon as classes began. The Advanced Mathematics class, taught by Professor Rawlings, was particularly challenging. But Professor Rawlings had a way of explaining the issues with clear, simple examples that made it much easier to learn the complex concepts. I decided to take every graduate class that he was offering. Soon I was learning about H-infinity, Linear-Quadratic-Gaussian theory, Model Predictive Control, and optimal control of a hot-air balloon. This early experience as a student in Jim's classes eventually blossomed into a long-term collaboration on MPC. In this presentation I'll tell the story of our collaboration, with a focus on some of the challenging technical issues that we encountered along the way. I'll end with a summary of Jim's impact on the field and a description of key personality traits that I believe led to his success.

497783 Calculation of Free Energy Barriers for Attachment of Molecules during Crystal Growth and Nucleation

Abstract:

Physical properties of crystalline materials depend upon their molecular structure and shape. Following the adoption of Quality by Design approaches by the Food and Drug Administration, the theoretical and computational approaches are increasingly implemented to predict structure (polymorphs), shape and properties of the crystalline materials. The most common method to predict crystal structure is the lattice energy minimization technique, which calculates the configuration of molecules in the lattice and its energy using the state-of-the-art optimization algorithms. Whereas the theoretical methods to predict crystal shape relies on the calculation of growth rates using either the interfacial energy or the kink energy. Such methods for the crystal structure and shape prediction depend on the estimation of crystal energies and hence to fail to capture the underlying dynamics of nucleation and growth, where solute molecules continuously attach to and detach from the crystal surfaces in a supersaturated solution. In order to effectively model the dynamics of crystallization processes, it is essential to know the free energy barriers for the attachment and detachment of solute molecules. Here we propose a method to calculate free energy barriers for the attachment and detachment of molecules using the double-well potential method. This approach consists of finding the configuration of molecules in the solvation shell, followed by the construction of the double-well potential diagrams to calculate free energy barriers. Application of the proposed method to calculate growth rates and nucleation rates of Glutamic acid will be shown.

492558 Population Balance Modeling and Optimization of an Integrated Batch Crystallizer – Wet Mill System for Crystal Size Distribution Control

Abstract:

Batch crystallization is one of the most efficient and economic separation and purification technique of solid crystalline materials, which has particularly important role in the fine chemical and pharmaceutical industry. From the point of view of particle formation is also significant as during the process the crystal size distribution (CSD) can be adjusted, within a certain domain.

Through the crystal size is often the most critical property of the end-product (keeping in mind that the purity, yield and crystal form specifications must be respected), the control of crystallization process in many practical case consists on controlling the product CSD. In crystallization systems' engineering, from the perspective of product size, the general objective is the production of large, uniform crystals [1] in order to facilitate the downstream operations. However, the production of small, uniform crystals is also of great interest due to the increased specific surface and bioavailability.

Since the crystallization is governed by the simultaneously ongoing nucleation and growth, which are non-linear functions of concentration, controlling the relative rate of these generally raises complex control issues. Generally, model predictive control (MPC) is employed to overcome the process nonlinearity's while respecting various constraints. In contrast with the continuous tank systems, the control of batch operation is sometimes more challenging due to the fact that the initial and final system states are far from each other, during which transition the process nonlinearity's has more significant effects [2]. Out of the model based techniques, with the quick spread of process analytical technologies the model free control methods, such as the direct nucleation control (DNC) and supersaturation control (SSC) [3]. The batch crystallization control, regardless on the way of implementations, in most cases relies in the manipulation of applied temperature profile, which influences the crystallization kinetics through the supersaturation. In crystallization practice the attainable particle size domain is always delimited by the crystallization kinetics [4].

In order to widen the attainable crystal size domain, other manipulated variables affecting crystal size are also implied, such as using antisolvent or, more often, applying combined antisolvent-temperature variations for supersaturation generation. Out of the nucleation and growth, the secondary operations, like the agglomeration and growth are also promising sup-processes to be controlled from the point of view of crystal size control [5]. The seed addition is a very popular startup procedure of batch crystallizers, which also presents high potential of tailoring the product CSD. Growth rate modifier additives were also successfully applied to adjust the crystal size and shape during the crystallization.

If the CSD cannot be obtained directly in the crystallization, secondary operations, such as granulation and milling are required to adjust the size of crystals properly. This however is not an advantageous solution as involves extra operations in the production line and, moreover, the quality of crystals not reaches those that obtained in direct crystallization. The straightforward question arises if the crystallization and milling could be coupled with each other for the sake of production uniform crystals in lower size domains? There is a forward looking experimental work analyzing a continuous crystallizer-wet mill cascade system [6] which was also successfully extended to continuous wet mill based direct nucleation control for cooling crystallization [7]. Others attempted to model and simulate the attainable crystal size distribution in a very similar setup.

For the mathematical description of particulate systems the population balance modelling (PBM) is the widely accepted and applied approach [8]. Since the introduction of PBM's a large number of publications was addressed to this topic. The PBM's, from the most basic form of nucleation and growth assuming one crystal dimension have been extended to multidimensional cases and involving additional crystallization mechanisms such as fragmentation, agglomeration, crystal-solution heat transfer [9] *etc*. From process engineering perspective, the numerical solution of the PB equations (PBE) is the bottleneck of crystallization simulation and is often more challenging that the derivation of the PBM. Numerous methods have been addressed to PBE solution from the moment based methods, through Monte Carlo simulations and method of characteristics to discretization based techniques. To improve the real time applicability of full PBM based simulations, the solution time is usually improved by various techniques from using adaptive grid, through application of parallel and GPU calculations to coordinate system transformations.

Despite of the fact that PBM modelling was successfully applied for various crystallization systems and that using a recycle wet-mill can considerably widen the attainable crystal size domain of batch crystallizers a study for the optimization of such a promising integrated system has not been published yet. The first objective of this work is to develop a PB based model for a batch crystallizer-wet mill system, involving primary nucleation, growth and dissolution for the crystallizer and fragmentation and attrition for the wet mill and to solve the equations efficiently and accurately [10]. The second objective of the study is to analyze and optimize the system through numerical experimentation. The authors address the current study as a quality-by-design (QbD) approach driven deeper understanding effort of simultaneous role of system integration and hidden potential of process optimization.

The optimization results indicate that the target crystal size distribution can be successfully achieved in the cases when the pure crystallizer cannot even approximate the shape of target CSD. The optimized system in all cases presents similar operation pattern, which can be summarized in 4 point:

1) The crystallizer is cooled quickly to produces some crystals

2) The pump is turned on to transport the crystals to the external wet mill (the mill is not running yet)

3) The pump is turned off (the crystallizer and wet mill is decoupled from each other). During this, the crystallizer is heated up to the solubility temperature (complete dissolution occurs) while the wet mill operates with high rotation speed (the existing crystals are milled).

4) When the crystals are sufficiently milled, the system is started to be cooled in controlled manner while the seeds are added dynamically through the recirculation stream.

Consequently, the optimal operation of integrated crystallizer-external wet mill system utilizes the strengths of optimal dynamic seeding and in situ seed generation, making unnecessary the time and high accuracy demanding seed preparation step.

Nevertheless, the system model was not validated for a particular system, in which context the results are not valid for any practical crystallization. Although, real crystallization kinetics and a realistic breakage model was used. Indeed, if the nucleation, growth, dissolution and breakage parameters would be change the attainable crystal size domain would certainly be modified but the contribution of this work, the optimal trajectories of temperature, flowrate and wet mill rotation speed, would remain the same.

References

[1] A. Mersmann, *Crystallization Technology Handbook*. Marcel Dekker Inc., New York, Basel, 2001.

[2] Z. K. Nagy and R. D. Braatz, "Robust nonlinear model predictive control of batch processes," *AIChE J.*, vol. 49, no. 7, pp. 1776–1786, Jul. 2003.

[3] Z. K. Nagy, G. Fevotte, H. Kramer, and L. L. Simon, "Recent advances in the monitoring, modelling and control of crystallization systems," *Chem. Eng. Res. Des.*, vol. 91, no. 10, pp. 1903–1922, Oct. 2013.

[4] T. Vetter, C. L. Burcham, and M. F. Doherty, "Regions of attainable particle sizes in continuous and batch crystallization processes," *Chem. Eng. Sci.*, vol. 106, pp. 167–180, Mar. 2014.

[5] B. Szilágyi and B. G. Lakatos, "Model-based analysis of stirred cooling crystallizer of high aspect ratio crystals with linear and nonlinear breakage," *Comput. Chem. Eng.*, vol. 98, pp. 180–196, 2017.

[6] Y. Yang, L. Song, T. Gao, and Z. K. Nagy, "Integrated Upstream and Downstream Application of Wet Milling with Continuous Mixed Suspension Mixed Product Removal Crystallization," *Cryst. Growth Des.*, vol. 15, no. 12, pp. 5879–5885, 2015.

[7] Y. Yang, L. Song, Y. Zhang, and Z. K. Nagy, "Application of wet milling-based automated direct nucleation control in continuous cooling crystallization processes," *Ind. Eng. Chem. Res.*, vol. 55, no. 17, pp. 4987–4996, 2016.

[8] H. M. Hulburt and S. Katz, "Some problems in particle technology. A statistical mechanical formulation," *Chem. Eng. Sci.*, vol. 19, pp. 555–574, 1964.

[9] B. G. Lakatos and B. Szilagyi, "Modeling Crystallization from Solution with Heat Effects," *Cryst. Growth Des.*, vol. 15, no. 12, pp. 5726–5737, 2015.

[10] B. Szilágyi and Z. K. Nagy, "Graphical Processing Unit (GPU) Acceleration for Numerical Solution of Population Balance Models Using High Resolution Finite Volume Algorithm," *Comput. Chem. Eng.*, 2016.

500509 A Cellular Automata Approach for Simulation of Crystal Growth

Abstract:

Cellular Automata (CA) provides a computational platform for combining micro-scale fluctuation theory with macro-scale field models to predict discontinuities, sharp transitions and moving boundaries in solidification of impure systems. The CA system can consider solute and heat conservation equations subject to complex boundary conditions at the solid-liquid interface. Constitutional and curvature undercooling on the equilibrium interface tare easy to incorporate. These variables are needed to predict the boundary between stable crystal growth and the unstable dendritic growth that lead to defect formation in many applications. The stochastic procedures in place for nucleation and dendritic growth can also be applied without making strong assumptions about the underlying rate processes. The price one has to pay can be significant computational cost due to the detailed modeling of the micro-structure.

The aim of the current paper is to consider the application of CA to the Horizontal Ribbon Growth (HRG) process. HRG was proposed by Shockley for producing single crystalline silicon sheets [1]. It is a continuous process and it has potential to overcome the limitations of traditional batch-scale wafer processes based on Czochralski crystallization and wire sawing. In HRG, a thin silicon solid sheet is produced and extracted continuously from a molten silicon pool, minimizing in this way material losses. In order to achieve fast production speed, the pulling rate has to be determined in conjunction with proper cooling set up and the presence of impurities may introduce defects and other instabilities that must be controlled.

Many technical challenges have been encountered in the effort to stable and fast ribbon production in the HRG process. Experiments show that sudden dendritic growth occurs at the crystal front when the pulling speed exceeds a threshold which depends on a number of parameters in the system including the accumulation of impurities [2]. Experimental results have shown cases of non-smooth and unstable solid-liquid interface and sharp wedge has occurred [3].

In the current paper we develop the CA approach for thin film crystal growth and show how the models can be used to derive stability/instability boundaries and control strategies to suppress these. We show results that demonstrate computational tractability and we present parametric tests to show how different cooling conditions, impurity concentrations, and other process related parameters relate to process stability. A method to control the process is proposed and it shown how model results relate to experimental results reported in the open literature.

References:

1. William, S. (1962). *U.S. Patent No. 3,031,275*. Washington, DC: U.S. Patent and Trademark Office.
2. Kudo, B. (1980). Improvements in the horizontal ribbon growth technique for single crystal silicon. *Journal of Crystal Growth*, *50*(1), 247-259.
3. Kellerman, P., Kernan, B., Helenbrook, B. T., Sun, D., Sinclair, F., & Carlson, F. (2016). Floating Silicon Method single crystal ribbonâ€"observations and proposed limit cycle theory.Â *Journal of Crystal Growth*,Â *451*, 174-180.

Optimal Crystal Size Control using a Plug Flow Crystallizer Configuration with Recycle

Xiaodong Xu, Yuan Yuan, Stevan Dubljevic

Department of Chemical and Materials Engineering
University of Alberta, Edmonton, Canada

1Nov 2017

American Chemical Engineering
Conference
Minneapolis

Outline

A. Introduction

B. Modeling of Plug Flow Crystallizer with Recycle

C. Problem Statement

D. Steady State

E. Case Study and Simulation Study

F. Conclusion

Introduction

Batch process

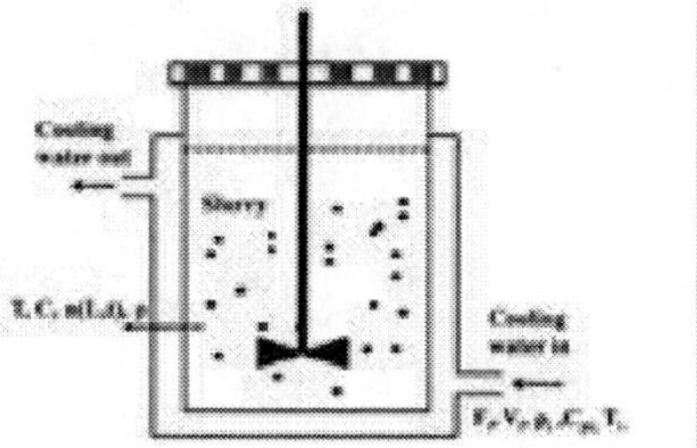

MSMPR

Plug flow process

- More manipulated variables: T, v_z;
- Control of crystal size and shape distribution;
- Produce narrow size and shape distribution.
- Contributions:
  - Myerson, A.S, 2002, Handbook of Ind. Crystal., B.-H. Elsevier Science, Boston
  - Debasis *et all*, 2006, Chem. Eng. Sci
  - Alvarez, A.J. *at all*, 2010, Cryst. Growth Des.
  - Nagy, Z.K and R.D. Braatz, 2012, Advanced and new directions in crystallization control, Ann. Rev. Chem. Bio. Eng.
  - Vetter *et. all.*, Chem. Eng. Sci. 2014

Introduction

PFC process with recycle

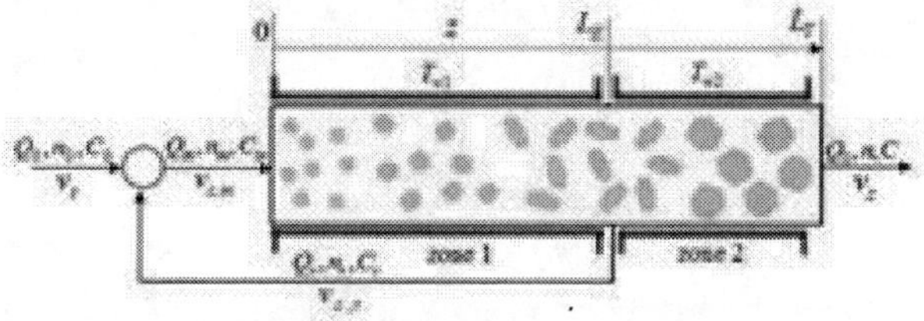

- $r = \frac{Q_r}{Q_0}$
- $v_{z,in} = (1+r)v_z$
- $C_{in} = \frac{rC(L_E)+C_0}{1+r}$
- $n_{in} = \frac{r}{1+r}n(L_E)$

Manipulated variables:

- Jacket temperature: T_{w1}, T_{w2}
- Flow velocity: v_z
- Extract position: L_E
- Recycle ratio: r

Mass balance

First part $0 \le z \le L_E$

Solution concentration:

$$\frac{\partial C_1}{\partial t} = -(1+r)v_z\frac{\partial C_1}{\partial z} - \rho_c \int_0^\infty G_v(\sigma_1)n_1(V,z,t)dV$$

$$C_1(0,t) = \frac{C_0 + rC_1(L_E,t)}{1+r}$$

$$C_1(z,0) = C_0$$

Second part $L_E \le z \le L_T$

Solution concentration:

$$\frac{\partial C_2}{\partial t} = -v_z\frac{\partial C_2}{\partial z} - \rho_c \int_0^\infty G_v(\sigma_2)n_2(V,z,t)dV$$

$$C_2(L_E,t) = C_1(L_E,t)$$

$$C_2(z,0) = C_0$$

- Solubility: $s(T) = 2.88\times10^{-4}T^3 - 1.65\times10^{-3}T^2 + 4.62\times10^{-2}T + 6.01\times10^{-1}$
- Supersaturation: $\sigma = \ln\left(\frac{C}{s}\right)$
- Growth rate: $G_v(\sigma) = 2G_{110}h_{110}h_{101} + G_{101}h_{110}^2$

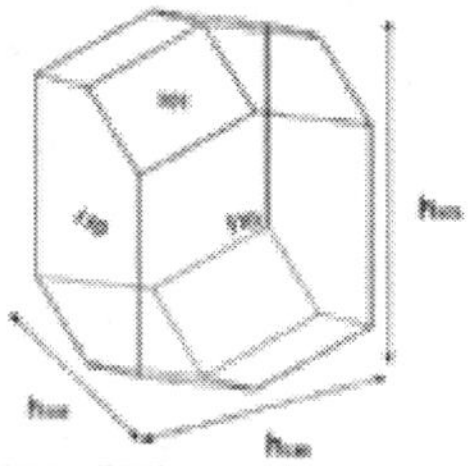

G. Cogni *et. all* Chemical Engineering Science, 138 (2015), 592–599

Energy balance model

First part $0 \le z \le L_E$

Temperature:

$$\frac{\partial T_1}{\partial t} = -(1+r)v_z\frac{\partial T_1}{\partial z} - \frac{\rho_c \Delta H_c}{\rho C_p}\int_0^\infty G_v(\sigma_1)n_1(V,z,t)dV - \frac{U_c A_c}{\rho C_p}(T_1 - T_{w1})$$

$$T_1(0,t) = \frac{T_0 + rT_1(L_E,t)}{1+r}$$

$$T_1(z,0) = T_0$$

Second part $L_E \le z \le L_T$

Temperature:

$$\frac{\partial T_2}{\partial t} = -v_z\frac{\partial T_2}{\partial z} - \frac{\rho_c \Delta H_c}{\rho C_p}\int_0^\infty G_v(\sigma_2)n_2(V,z,t)dV - \frac{U_c A_c}{\rho C_p}(T_2 - T_{w2})$$

$$T_2(L_E,t) = T_1(L_E,t)$$

$$T_2(z,0) = T_0$$

- Solubility: $s(T) = 2.88\times10^{-4}T^3 - 1.65\times10^{-3}T^2 + 4.62\times10^{-2}T + 6.01\times10^{-1}$
- Supersaturation: $\sigma = \ln\left(\frac{C}{s}\right)$
- Growth rate: $G_v(\sigma) = 2G_{110}h_{110}h_{101} + G_{101}h_{110}^2$
- Super-density: $\rho = 1 + C(z,t)$ [g/cm^3]

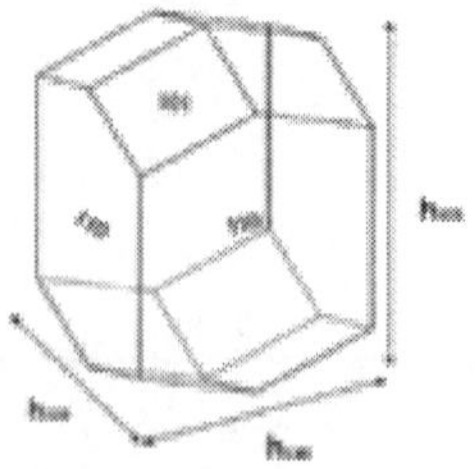

Population balance model

First part $0 \le z \le L_E$

NDF:

$$\frac{\partial n_1}{\partial t} = -(1+r)v_z\frac{\partial n_1}{\partial z}$$
$$-\frac{\partial(G_v(\sigma_1)n_1)}{\partial V}$$
$$-B_{seed}\delta(V-V_0, z-z_0)$$
$$n_1(V,0,t) = \frac{r}{1+r}n_1(V,L_E,t)$$
$$n_1(0,z,t) = 0$$
$$n_1(V,z,0) = 0$$

Second part $L_E \le z \le L_T$

NDF:

$$\frac{\partial n_2}{\partial t} = -v_z\frac{\partial n_2}{\partial z}$$
$$-\frac{\partial(G_v(\sigma_2)n_2)}{\partial V}$$
$$-B_{seed}\delta(V-V_0, z-z_0)$$
$$n_2(V,L_E,t) = n_1(V,L_E,t)$$
$$n_2(0,z,t) = 0$$
$$n_2(V,z,0) = 0$$

- Crystal volume: $V(z,t) = (h_{110}(z,t))^2 h_{101}(z,t)$
- Growth rate (110): $G_{110} = 0.1843\times\sigma^3 - 1.1699\times\sigma^3 + 2.8885\times\sigma^3 - 2.5616$
- Growth rate (101): $G_{101} = 0.1893\times\sigma^3 - 1.2264\times\sigma^3 + 2.9887\times\sigma^3 - 2.5348$
- Heights: $\frac{dh_{110}(z,t)}{dz} = \frac{G_{110}}{v_z}, \frac{dh_{101}(z,t)}{dz} = \frac{G_{101}}{v_z}$

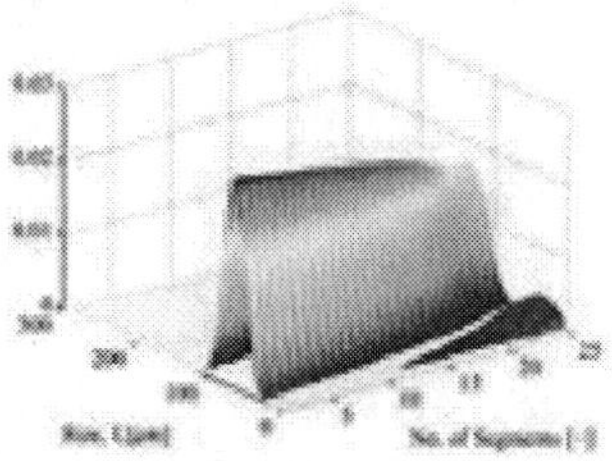

Problem statement

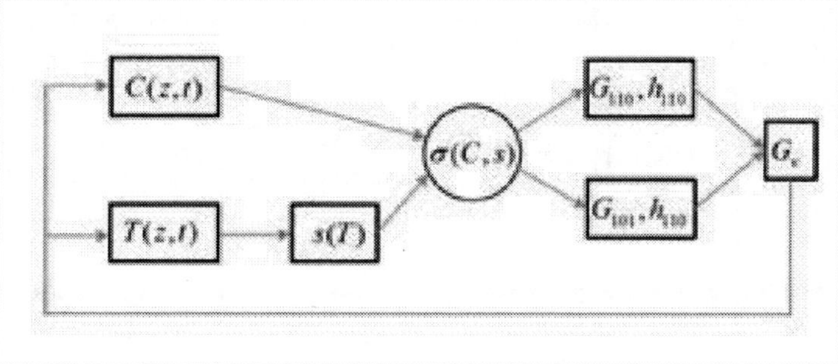

- Nonlinearity
- Coupling

- Design problem:

$$\min_{T_{w1},T_{w2},L_E} \gamma_1 L_T - \gamma_2\eta, \eta = \left[1 - \frac{C(L_T)}{C_{IN}}\right]\times 100$$

- Control problem:

$$\min_{T_{w1},T_{w2},r,L_E} \int_0^{t_f}\int_0^{L_T}\left[w_1(\alpha(z,t)-\alpha_{set})^2 + w_2(h_{110}(z,t)-h_{set})^2\right]dzdt$$

Method of moments

- Define the kth moment as:

$$M_k(z,t) = \int_0^\infty V^k n(V,z,t)dV$$

$M_0(z,t)$: Total number of crystals; $M_1(z,t)$: Total volume of crystals.

First part $z \in [0, L_E]$

$$\frac{\partial M_{0,1}}{\partial t} = -(1+r)v_z\frac{\partial M_{0,1}}{\partial z}$$
$$\frac{\partial M_{k,1}}{\partial t} = -(1+r)v_z\frac{\partial M_{k,1}}{\partial z} + kG_v(\sigma_1)M_{k-1,1}$$
$$M_{0,1}(0,t) = \frac{B_{seed}}{(1+r)^2 v_z} + \frac{r}{1+r}M_{0,1}(L_E,t)$$
$$M_{k,1}(0,t) = \frac{B_{seed}V_0^k}{(1+r)^2 v_z} + \frac{r}{1+r}M_{k,1}(L_E,t)$$

- B_{seed}: Seeding rate

Second part $z \in [L_E, L_T]$

$$\frac{\partial M_{0,2}}{\partial t} = -v_z\frac{\partial M_{0,2}}{\partial z}$$
$$\frac{\partial M_{k,2}}{\partial t} = -v_z\frac{\partial M_{k,2}}{\partial z} + kG_v(\sigma_2)M_{k-1,2}$$
$$M_{0,2}(L_E,t) = M_{0,1}(L_E,t)$$
$$M_{k,2}(L_E,t) = M_{k,1}(L_E,t)$$

- V_0: Seed size

Steady state model

- Moments

$$\frac{\partial M_{0,1}}{\partial z} = 0,\ \frac{\partial M_{k,1}}{\partial z} = \frac{kG_v(\sigma_1)}{(1+r)v_z}M_{k-1,1}$$
$$M_{0,1}(0) = \frac{B_{seed}}{(1+r)^2 v_z} + \frac{rM_{0,1}(L_E)}{1+r}$$
$$M_{k,1}(0) = \frac{B_{seed}V_0^k}{(1+r)^2 v_z} + \frac{rM_{k,1}(L_E)}{1+r}$$

$$\frac{\partial M_{0,2}}{\partial z} = 0,\ \frac{\partial M_{k,2}}{\partial z} = \frac{kG_v(\sigma_2)}{(1+r)v_z}M_{k-1,2}$$
$$M_{0,2}(L_E) = M_{0,1}(L_E)$$
$$M_{k,2}(L_E) = M_{k,1}(L_E)$$

- Concentration

$$\frac{\partial C_1}{\partial z} = \frac{\rho_c}{(1+r)v_z}G_v(\sigma_1)M_{0,1}$$
$$C_1(0) = \frac{C_0}{1+r} + \frac{r}{1+r}C_1(L_E)$$

$$\frac{\partial C_2}{\partial z} = \frac{\rho_c}{(1+r)v_z}G_v(\sigma_2)M_{0,2}$$
$$C_2(L_E) = C_1(L_E)$$

- Temperature

$$\frac{\partial T_1}{\partial z} = -\frac{\rho_c G_v(\sigma_1)M_{0,1}}{\rho C_p(1+r)v_z} - \frac{U_c A_c(T_1 - T_{w1})}{\rho C_p(1+r)v_z}$$
$$T_1(0) = \frac{T_0}{1+r} + \frac{r}{1+r}T_1(L_E)$$

$$\frac{\partial T_2}{\partial z} = -\frac{\rho_c G_v(\sigma_2)M_{0,2}}{\rho C_p v_z} - \frac{U_c A_c(T_2 - T_{w2})}{\rho C_p v_z}$$
$$T_2(L_E) = T_1(L_E)$$

Optimization for steady state model

$$\min_{T_{w1},T_{w2},vz} \int_{L_E}^{L_T} w_1(\alpha(z)-\alpha_{set})^2 + w_2(h_{110}(z)-h_{set})^2 dz$$

$s.t.$

$$G_{110} = 0.1843 \times \sigma^3 - 1.1699 \times \sigma^3 + 2.8885 \times \sigma^3 - 2.5616$$

$$G_{101} = 0.1893 \times \sigma^3 - 1.2264 \times \sigma^3 + 2.9887 \times \sigma^3 - 2.5348$$

$$G_v = 2G_{110}h_{110}h_{101} + G_{101}h_{110}^2$$

$$\frac{dh_{110}}{dz} = G_{110}, \frac{dh_{101}}{dz} = G_{101}, \alpha(z,t) \approx \frac{h_{110}(z,t)}{h_{101}(z,t)}$$

Manipulated variables:

- The superficial velocity v_z has to be a constant
- T_{w1} and T_{w2} can be spatially varying or constants along the space

Case study

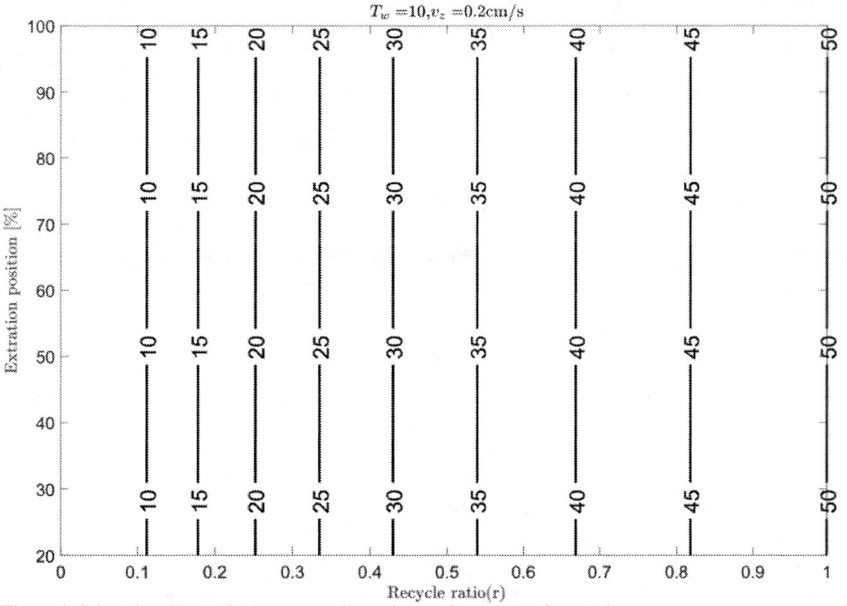

- The yield η is directly proportional to the recycle ratio r.

Case study

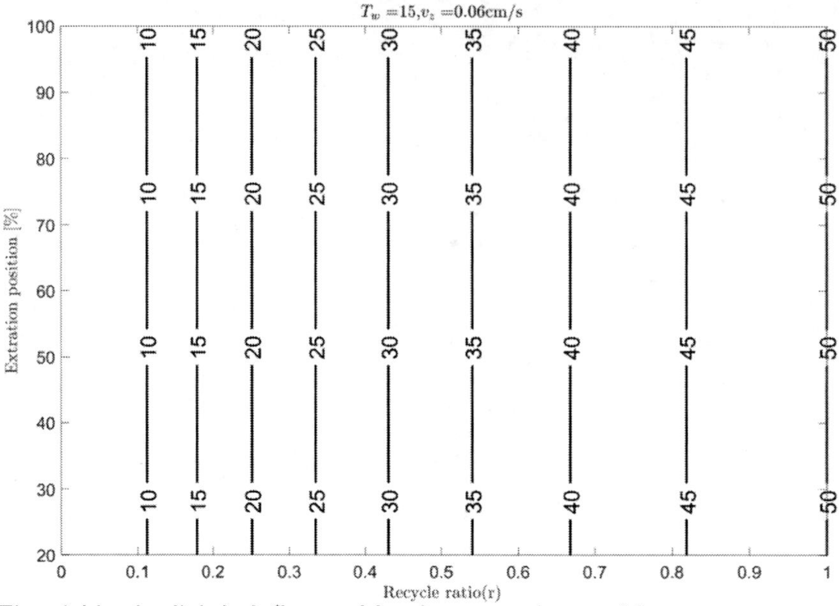

- The yield η is slightly influenced by the extraction position.

Case study

The shape of crystal

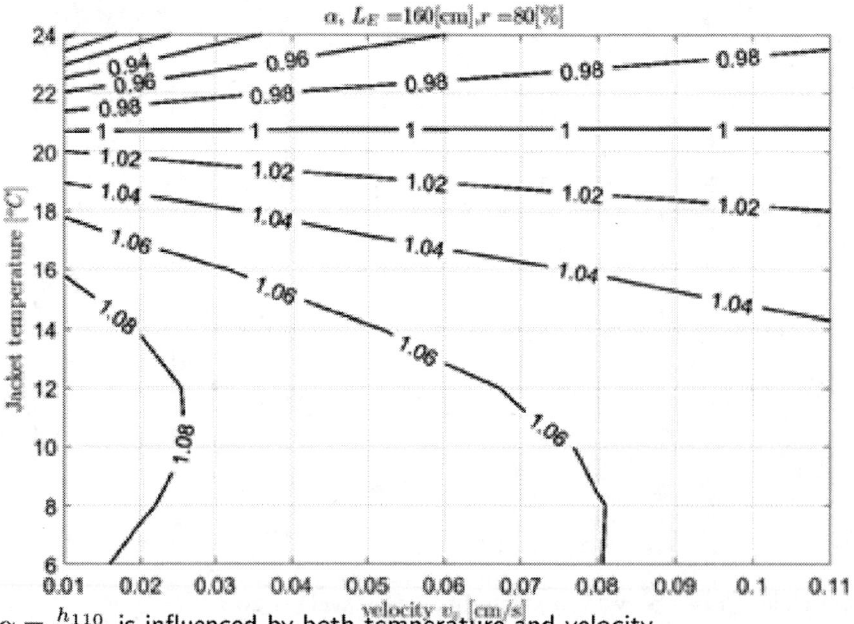

- $\alpha = \frac{h_{110}}{h_{101}}$ is influenced by both temperature and velocity.

Case study

The height of crystal (h_{110})

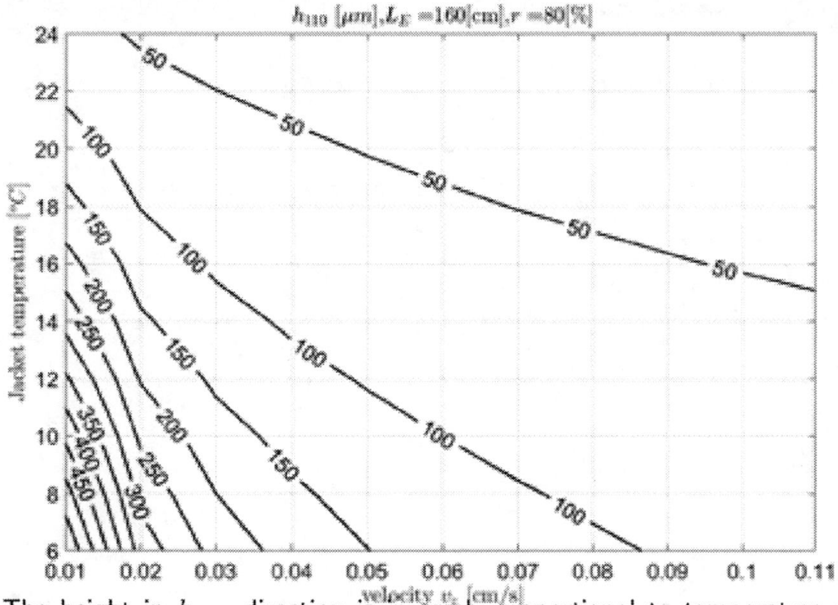

- The height in h_{110} direction is inversely proportional to temperature and flow velocity.

- The recycle ratio and the extraction position can be chosen to maintain appropriate yield.
- Temperature and superficial flow velocity are chosen to obtain desired crystals shape and size distribution.

Multivariable optimization problem

The resulting optimization problem:

$$\min_{T_{w1},T_{w2},v_z} \int_0^{t_f} \int_{L_E}^{L_T} w_1(\alpha(z,t)-\alpha_{set})^2 + w_2(h_{110}(z,t)-h_{set})^2 dzdt$$

$s.t.$

$$G_{110} = 0.1843 \times \sigma^3 - 1.1699 \times \sigma^3 + 2.8885 \times \sigma^3 - 2.5616$$

$$G_{101} = 0.1893 \times \sigma^3 - 1.2264 \times \sigma^3 + 2.9887 \times \sigma^3 - 2.5348$$

$$G_v = 2G_{110}h_{110}h_{101} + G_{101}h_{110}^2$$

$$\frac{dh_{110}}{dz} = G_{110}, \frac{dh_{101}}{dz} = G_{101}, \alpha(z,t) \approx \frac{h_{110}(z,t)}{h_{101}(z,t)}$$

$$\frac{\partial M_{0,1}}{\partial t} = -(1+r)v_z\frac{\partial M_{0,1}}{\partial z}, \frac{\partial M_{k,1}}{\partial t} = -(1+r)v_z\frac{\partial M_{k,1}}{\partial z} + kG_v(\sigma_1)M_{k-1,1}$$

$$\frac{\partial M_{0,2}}{\partial t} = -v_z\frac{\partial M_{0,2}}{\partial z}, \frac{\partial M_{k,2}}{\partial t} = -v_z\frac{\partial M_{k,2}}{\partial z} + kG_v(\sigma_2)M_{k-1,2}$$

$$\frac{\partial C_1}{\partial t} = -(1+r)v_z\frac{\partial C_1}{\partial z} - \rho_c G_v(\sigma_1)M_{0,1}$$

$$\frac{\partial C_2}{\partial t} = -v_z\frac{\partial C_2}{\partial z} - \rho_c G_v(\sigma_2)M_{0,2}$$

$$\frac{\partial T_1}{\partial t} = -(1+r)v_z\frac{\partial T_1}{\partial z} - \frac{\rho_c \Delta H_c G_v(\sigma_1)M_{0,1}}{\rho C_p} - \frac{U_c A_c}{\rho C_p}(T_1 - T_{w1})$$

$$\frac{\partial T_2}{\partial t} = -v_z\frac{\partial T_2}{\partial z} - \frac{\rho_c \Delta H_c G_v(\sigma_2)M_{0,2}}{\rho C_p} - \frac{U_c A_c}{\rho C_p}(T_2 - T_{w2})$$

Shape: $\alpha = 0.88$

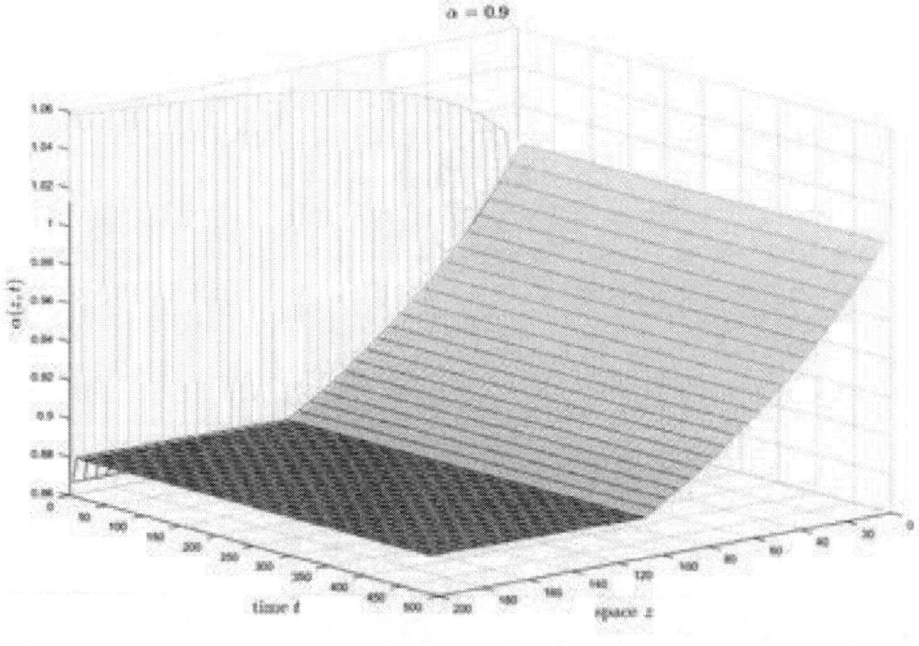

558

Height: $h_{110} = 130\mu m$

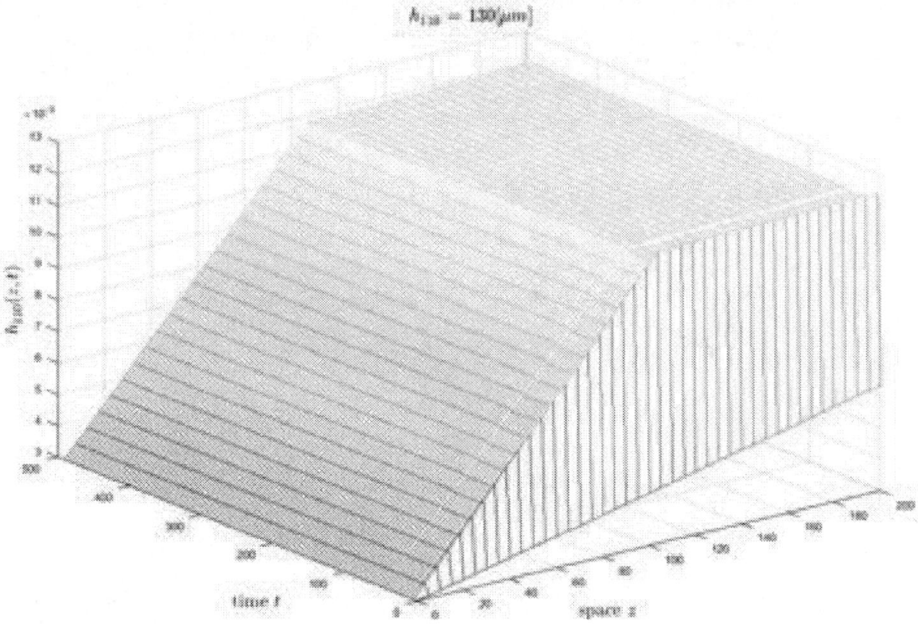

Temperature

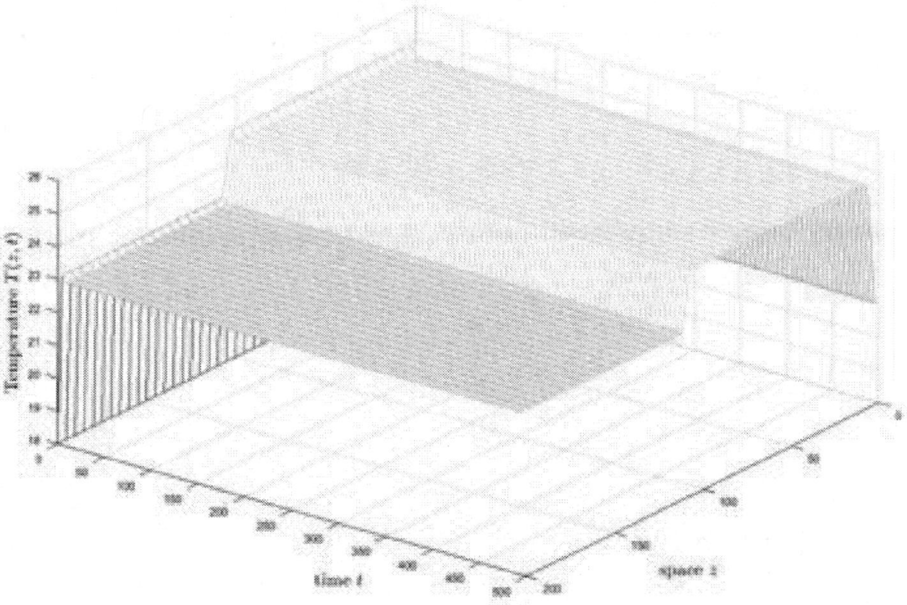

Concentration

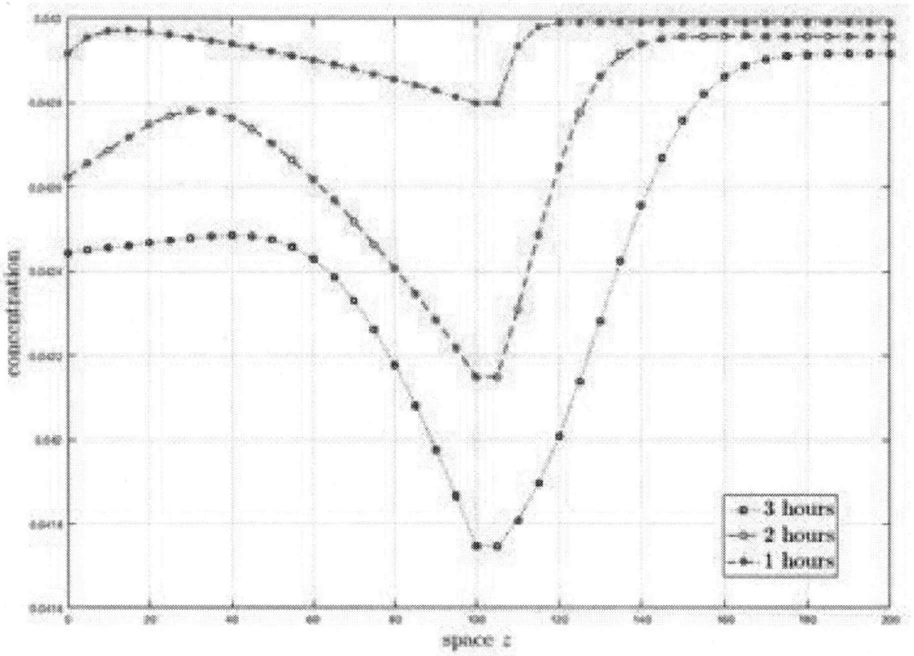

Conclusion:

- Model of PFC process with recycle and crystals growing from seeds
  - Mass balance
  - Energy balance
  - Population balance
- Reduced-order moments model: method of moment:

$$M_k(z,t) = \int_0^{\infty} V^k n(V,z,t)dV$$

- Manipulated variables: r,L_E, T_{w1}, T_{w2} and v_z.
  - yield
  - shape and size
- MOP for desired shape and size distributions

497753 Multiphysics Modeling and Simulation of Microfluidic Platforms for Screening of Pharmaceutical PolymorphsÂ

Abstract:

Polymorphs are the crystalline materials with different molecular arrangements and hence varying physicochemical properties such as bioavailability, chemical stability, and mechanical strength. Almost 60% of marketed oral drugs are produced in the solid form, whose quality is mostly determined by their polymorphic structure. Due to the regulatory issues, it is essential to develop a platform to effectively screen these polymorphs to identify stable and metastable forms of drugs. In the pharmaceutical industry, the commonly employed technique to screen polymorphs is a microtiter plate method which uses 96-well, 384-well, and 1586-well plates. Such high-throughput technique can evaluate up to 10,000 compounds per day. However, setbacks are often encountered during scale-up due to differences in the environmental conditions of a microtiter plate and a real crystallizer. Here we propose a design of a novel microfluidic crystallizer that can potentially overcome the drawbacks of the microtiter plate method by inducing hydrodynamically controlled nucleation and growth of single crystals. Our microfluidic platform can control the supersaturation at the stagnation point where single crystal can nucleate and grow. In this talk, I will present the results from the Multiphysics simulations of microfluidic crystallization and propose an optimal design to enable high-throughput screening of polymorphs.

497114 Motion-Based Multiple Object Tracking of Ultrasonic-Induced nucleation: A Case Study of L-Glutamic Acid

Abstract:

Motion-Based Multiple Object Tracking of Ultrasonic-Induced Nucleation: A Case Study of L-glutamic Acid

Zhenguo Gao [1,2] Dan Zhu [1,2] Yuanyi Wu [1] Sohrab Rohani [1]* Junbo Gong [2], Jingkang Wang [2]

1. The University of Western Ontario, Department of Chemical and Biochemical Engineering, London, Ontario N6A 5B9, Canada

2. Tianjin University, School of Chemical Engineering and Technology, Tianjin, 300072, P. R. China

Keywords: Ultrasonic irradiation, Nucleation, Motion-based multiple object tracking <>Introduction

Nucleation from solution has aroused great interest over the past decades. Scientists have considered atomistic levels on the order of 10^{-10} m and time scales in the order of 10^{-13} s to unravel the nucleation mechanism with the help of advanced instruments[1] . However, in industrial crystallization, there is still a limitation in monitoring nucleation at the initial stage of critical nucleus formation[2] . Existing measuring techniques used to monitor nucleation include the focused beam reflection measurement (FBRM), turbidity, ultrasonic velocity, electrical conductivity and light transmittance measurement of bulk solution[3] . The accuracy of the monitoring devices affects the measurement of nucleation kinetics. In order to improve the accuracy, progress has been made based on high-speed imaging, multivariate statistical process monitoring charts, etc[4,5] . Recently, motion-based multiple object tracking (MMOT) has been widely used in computer vision field. The MMOT method detects moving objects in a video stream, predicts their locations in the next frame, and records the object count automatically, which shows its potential application in monitoring the occurrence and disappearance of nuclei in a bulk solution.

External factors such as ultrasound, microwave, magnetic field and electric field have been studied to control crystal nucleation, which exhibit effectiveness in some cases[6,7] . Ultrasonic irradiation creates sequential compression and expansion of the solution, which leads to bubble formation and growth. Finally, the bubbles collapse and release energy, promoting nucleation within a short induction time and at a lower supersaturation level. Parameters, e.g. ultrasound power, frequency, with or without pulse, have significant influence on nucleation process and crystal qualities[8,9] . In this study, the MMOT was introduced in the crystallization area, for the first time, to monitor the nucleation of L-glutamic acid. Results showed that ultrasonic irradiation shortened the induction time dramatically. The MMOT-based device showed higher accuracy and precision compared to the FBRM technology as the bubbles were transparent in the MMOT technology, but considered as nuclei by the FBRM. <>Materials and Experiments

L-glutamic acid (LGA) was used as model compound. An ultrasonic processor was inserted into a double jacketed crystallizer to generate ultrasound. During nucleation process, a waterproof USB-based micro-camera (Magnification: 1-300X) was used to record a video stream to monitor the nucleation process as shown in Figure 1(a). A home-designed vial adaptor with a solution flow channel (Figure 1b) was printed by a 3D printer and connected to the end of the micro-camera to allow capture crystal images as the slurry flowed through the adaptor. Figure 2 shows the experimental setup. In the first step, clear solutions were prepared in the crystallizer at 80 . Afterwards, nucleation temperature was reached as soon as possible by switching the 3-way valve to the cooling refrigerated circulator. The temperature was kept constant to study the nucleation kinetics until the crystals appeared in the video stream. The images recorded by micro-camera were parsed by MMOT in MATLAB.

Figure 1. (a) Portable USB-based micro-camera setup; (b) 3D print vial adaptor.

Figure 2. **(a)** Schematic of experimental setup; **(b)** Experimental setup picture. Results and Discussions

Different conditions of solution concentration (30 g/L and 40 g/L) and nucleation temperature (35 and 45) were tested. The tracking model is solely based on motion, in which the background is subtracted and a Kalman filter is used to predict the assigned track's location in next frame. The nuclei count in MMOT was smoothed by a moving average (MA) model as shown in equations 1 and 2.

$$\text{MA} = \qquad = \qquad (1)$$

$$\text{MA}_{\text{current}} = \text{MA}_{\text{prev}} + \qquad (2)$$

Where c indicates the nuclei counts along the time series and MA_{prev} is the previous averaged value. The moving window n was optimized to be 30 which tightly fitted the nuclei counts. The time at which a continuous increase in the micro-camera counts was sensed was taken as the onset of nucleation. The induction time was calculated as the period between the time the ultrasound processor was switched on and the onset of nuclei occurrence. Under the ultrasonic irradiation at 40 g/L and 35 , Figure 3 and Figure 4 show the results at 38 s and 95 s of same process after the ultrasonic processor was started.

Figure 3. Detection results at 38 s with ultrasonic irradiation (14 W). **(a)** In-situ plot of nuclei counts versus time based on MMOT; **(b)** Plot of nuclei counts against time of the whole process. The blue line and black line are raw data and smoothed data. **(c)** Detected objects marked in an original video. **(d)** Detected objects marked in the video after background subtraction.

Figure 4. Detection results at 95 s with ultrasonic irradiation (14 W). **(a)** In-situ plot of nuclei counts versus time based on MMOT; **(b)** Plot of nuclei counts against time of the whole process. The blue line and black line are raw data and smoothed data. **(c)** Detected objects marked in an original video. **(d)** Detected objects marked in the video after background subtraction.

As shown in Figure 3c and 4c, only one object had been detected, which indicates no nucleation occurred before 38 s, while large quantities of objects had been detected after 95 s. A continuous increase of nuclei counts was detected after 72 s in Figure 4a, with the constant nucleation occurring in raw image (Figure 4c), which confirmed the onset of nucleation at 72 s. Comparing Figure 4a with Figure 3a, it can be concluded that the threshold of background noise is around 5, which can be ignored before the continuous increase of nuclei count. At least six experiments were repeated under the same condition and average results were reported in Table 1.

Table 1. Comparison of induction time and crystal form of LGA with/without ultrasonic irradiation.

Ultrasonic power (W)	Concentration (g/L)	Nucleation temperature ( )	S_0, a (C_0/C^*)	S_0, ß (C_0/C^*)	Induction time (s)	Standard deviation (s)	Crystal form
14	30	35	1.91	2.51	181.3	8.2209	a
14	40	35	2.54	3.35	77.8	4.0249	a
14	30	45	1.31	1.78	585.6	22.1088	a, ß
14	40	45	1.74	2.37	113.0	7.7136	a, ß
0	30	35	1.91	2.51	1824.3	100.8806	a
0	40	35	2.54	3.35	381.3	40.8084	a
0	30	45	1.31	1.78	3600.0	169.7056	ß
0	40	45	1.74	2.37	737.3	68.3002	ß

Note: Standard deviation was calculated by , and *X* means average of all (i=1, 2, 3..., n).

There is a sharp decrease in induction time compared with the nucleation kinetics without ultrasonic irradiation [10]. For a rising of nucleation temperature, the induction time is longer. Ultrasonic irradiation shows potential application to promote nucleation in crystallization process design. Experimental results show the measurement accuracy based on MMOT is much better than FBRM as the latter does not distinguish between the nuclei and the bubbles generated by ultrasound. However, limitation exists in MMOT method, which is similar to the PVM (Particle Vision and Measurement) technology, the detection limit of objects count is around 100 and the occurrence of multiple crystals can affect the accuracy. Conclusions

In this study, the MMOT method was, for the first time, introduced in the crystallization field to monitor nucleation process. Results show better accuracy compared with other methods such as the FBRM technology. The nuclei count measured by MMOT, which is different from the FBRM count, can be used as a parameter to control nucleation in designing crystallization processes. Based on the MMOT technology, potential applications can be developed to measure solubility, metastable zone width, particle size and even solution-mediated polymorphic transformation.
<>Reference

1. J. Zhang, P. Chen, B. Yuan, W. Ji, Z. Cheng, and X. Qiu, *Science*, 2013, **342**, 611–4.

2. A. S. Myerson, *Handbook of industry crystallization*, Butterworth-Heinemann, 2002.

3. C. Brandel and J. Horst, *Faraday discussions*, 2015, **179**, 199–214.

4. S. Bauerecker, P. Ulbig, V. Buch, L. Vrbka, and P. Jungwirth, *J Phys Chem C*, 2008, **112**, 7631–7636.

5. L. Simon, K. Oucherif, Z. Nagy, and K. Hungerbuhler, *Chem Eng Sci*, 2010, **65**, 4983–4995.

6. S. Jiang, University of Leeds, 2012.

7. D.-C. Yin, *Prog Cryst Growth Ch*, 2015, **61**, 1–26.

8. S. Bhangu, M. Ashokkumar, and J. Lee, *Cryst Growth Des*, 2016, **16**, 1934–1941.

9. M. Jiang, C. Papageorgiou, J. Waetzig, A. Hardy, M. Langston, and R. Braatz, *Cryst Growth Des*, 2015, **15**, 2486–2492.

10. Schöll, Vicum, Müller, and Mazzotti, *Chem Eng Technology*, 2006, **29**, 257–264.

487181 Distributed Fault Diagnosis for Networked Nonlinear Uncertain Systems

Abstract:

The last decades have witnessed increasing drive to economic efficiency leading to material and energy integration, and in turn, to increasing complexity. The complexity can be due to nonlinearities, uncertainties, strong interconnections between subsystems and high dimensionality (see e.g., [1]). Where possible, the resultant large scale process systems are often modeled as a set of interconnected subsystems and are termed networked systems. While advances in automation have led to process efficiencies, it has also driven the need to provide automated safeguards against actuator and sensor faults. Thus, monitoring schemes are required to avoid failures in the networked systems by diagnosing faulty equipments as soon as possible. Designing a centralized FDI scheme for this problem may not be the best solution due to computational limitations or reliability issues. In particular, a failure in a centralized FDI can lead to interruption in monitoring of all of the subsystems. If the interconnections are weak, a decentralized FDI scheme composed of an independent local FDI scheme for each subsystem can handle the situation. However, there are cases that interconnections between subsystems are not negligible, such as power systems or chemical plants, that results in poor performance of a decentralized FDI scheme. In this case, viable solution is designing an FDI scheme with distributed architecture. In the distributed architecture a local FDI scheme is designed for each subsystem while the local FDI schemes can communicate to exchange information. The exchanged information can be information about recently diagnosed faults in the other subsystems or any other information required by the other local FDI schemes.

These realizations have motivated the design of FDI schemes with distributed architecture for networked systems. There exist a plethora of research in designing fault detection and isolation framework for monitoring of engineering systems (see e.g., [2], [3], [4], [5], [6], [7], [8], [9], [10] and [11]). However, most of the existing results in the literature present a centralized FDI scheme that might not be applicable to networked systems due to computational capa- bilities, communication infrastructure and reliability issues as discussed above. There also exist several results for FDI design in networked systems. In [12], a distributed data based actuator fault identification scheme is presented for linear networked process systems. To this end, local FDI agents are designed for each subsystem using moving horizon estimator (MHE)s to identify the fault. In [13], a distributed actuators FDI scheme is proposed for a class of interconnected uncertain nonlinear systems. To this end, a local FDI framework is designed for each subsystem. In each local FDI scheme, first a fault detection scheme is designed which simply uses output estimation error as residual. Then, a bank of fault isolation estimators is designed using adaptive estimation techniques. In [14], a distributed actuators fault detection and isolation framework is presented for nonlinear uncertain large-scale discrete-time dynamical systems. To overcome the scalabil- ity issues, the large scale system under considerations is modeled as several interconnected subsystems. For each subsystem a local FDI scheme is designed using adaptive estimation techniques. Decisions made by the local fault diagnoser are gathered by a higher level global fault diagnoser to determine health of the overall system. A consensus-based estimator is designed to enhance the ability of FDI schemes in detection and isolation of faulty shared information between the subsystems. In [15], a distributed sensor fault diagnosis for a network of interconnected cyber-physicalsystem (CPS)s (see e.g., [16]) is presented. The FDI module consists of two layers. The first layer is built using local FDI schemes to isolate multiple local sensor faults. The detection is carried out using adaptive estimation techniques while the isolation is done using logic rules. The second layer is realized by applying a global decision logic designed to isolate multiple sensor faults that may propagate in the network. However, none of the existing results have addressed the problem of isolation of multiple actuator faults or simultaneous actuator and sensor faults in uncertain nonlinear networked systems. Also, the results addressing actuator faults are based on the assumption that full state measurements are available.

Motivated by the above considerations, in this work, we present a distributed fault diag- nosis framework for a generalized class of networked uncertain nonlinear systems. The idea is to design a bank of FDI schemes based on the robust FDI methodology presented in [11] in a distributed manner. In particular, the nature of the interconnection is first determined and appropriate FDI schemes are developed. To this end, at first boundedness of estima- tion error in the presence of uncertainty and exchange of information is established. Then a bank of residuals is designed for each subsystem using an appropriate subset of the available measurements (and associated state observers), to determine the expected behavior of the system and compare with the observed evolution, where each residual is sensitive to a subset of faults and insensitive to the rest, in the absence of uncertainty. To achieve FDI in the presence of uncertainty, thresholds are defined in a way that they account for the impact of the uncertainty on the estimation error and the prediction of the expected system behavior. In this way, each residual is still insensitive to a subset of faults in the presence of uncertainty and sensitive to the rest if the fault functionality satisfies a detectability condition. Then the detectability and isolability conditions for single and simultaneous faults are derived, when the detectability analysis establishes that the sensitive property of residuals is retained in the presence of uncertainty (see [11] for more on this). The local FDI schemes communi- cate to diagnose the faults in the shared interconnections between subsystems. Results are demonstrated using a networked CSTR example.

References:

[1] Xing-Gang Yan and Christopher Edwards. Robust decentralized actuator fault detection and estimation for large-scale systems using a sliding mode observer. International Journal of control, 81(4):591–606, 2008.

[2] Paul M Frank. Fault diagnosis in dynamic systems using analytical and knowledge-based redundancy: A survey and some new results. Automatica, 26(3):459–474, 1990.

[3] Mogens Blanke, Michel Kinnaert, Jan Lunze, Marcel Staroswiecki, and J Schr¨oder.

Diagnosis and fault-tolerant control, volume 2. Springer, 2006.

[4] Claudio De Persis and Alberto Isidori. A geometric approach to nonlinear fault detection and isolation. Automatic Control, IEEE Transactions on, 46(6):853–865, 2001.

[5] Xing-Gang Yan and Christopher Edwards. Nonlinear robust fault reconstruction and estimation using a sliding mode observer. Automatica, 43(9):1605–1614, 2007.

[6] Prashant Mhaskar, Charles McFall, Adiwinata Gani, Panagiotis D Christofides, and James F Davis. Isolation and handling of actuator faults in nonlinear systems. Auto- matica,44(1):53–62,2008.

[7] Xiaodong Zhang, Marios M Polycarpou, and Thomas Parisini. Fault diagnosis of a class of nonlinear uncertain systems with lipschitz nonlinearities using adaptive estimation. Automatica, 46(2):290–299, 2010.

[8] Xiaodong Zhang, Qi Zhang, and Niharika Sonti. Diagnosis of process faults and sensor faults in a class of nonlinear uncertain systems. Systems Engineering and Electronics, Journal of, 22(1):22–32, 2011.

[9] Miao Du and Prashant Mhaskar. Isolation and handling of sensor faults in nonlinear

systems. Automatica, 50(4):1066–1074, 2014.

[10] Miao Du, James Scott, and Prashant Mhaskar. Actuator and sensor fault isolation of nonlinear process systems. Chemical Engineering Science, 104:294–303, 2013.

[11] Hadi Shahnazari and Prashant Mhaskar. Simultaneous actuator and sensor fault isola- tion of nonlinear systems subject to uncertainty. In 2016 American Control Conference (ACC), pages 6857–6862. IEEE, 2016.

[12] Di Peng, Nael H El-Farra, Zhiqiang Geng, and Qunxiong Zhu. Distributed data-based fault identification and accommodation in networked process systems. Chemical Engi- neering Science, 136:88–105, 2015.

[13] Xiaodong Zhang and Qi Zhang. Distributed fault diagnosis in a class of interconnected nonlinear uncertain systems. International Journal of Control, 85(11):1644–1662, 2012.

[14] Riccardo MG Ferrari, Thomas Parisini, and Marios M Polycarpou. Distributed fault detection and isolation of large-scale discrete-time nonlinear systems: An adaptive ap- proximation approach. IEEE Transactions on Automatic Control, 57(2):275–290, 2012.

[15] Vasso Reppa, Marios M Polycarpou, and Christos G Panayiotou. Distributed sensor fault diagnosis for a network of interconnected cyberphysical systems. IEEE Transactions on Control of Network Systems, 2(1):11–23, 2015.

[16] Panos J Antsaklis, Bill Goodwine, Vijay Gupta, Michael J McCourt, Yue Wang, Po Wu, Meng Xia, Han Yu, and Feng Zhu. Control of cyberphysical systems using passivity and dissipativity based methods. European Journal of Control, 19(5):379–388, 2013.

489063 A Multi-Sensor Error Detection and Functional Redundancy Algorithm for Dynamic Systems

Abstract:

A Multi-Sensor Error Detection and Functional Redundancy Algorithm for Dynamic Systems

Jianyuan Feng [1], Iman Hajizadeh[1], Sediqeh Samadi[1], Mert Sevil[2], Nicole Frantz[2], Rachel Brandt[2], Caterina Lazaro[3], Zacharie Maloney[3], Xia Yu[4], Elizabeth Littlejohn[5], Laurie Quinn[6], Ali Cinar[1,2]

[1] Department of Chemical and Biological Engineering, Illinois Institute of Technology, Chicago, IL, USA

[2] Department of Biomedical Engineering, Illinois Institute of Technology, Chicago, IL, USA

[3] Department of Computer Engineering, Illinois Institute of Technology, Chicago, IL, USA

[4] Department of Control Theory and Control Engineering, Northeastern University, Shenyang, Liaoning China

[5] Department of Pediatrics, University of Chicago, Chicago, IL, USA

[6] College of Nursing, University of Illinois at Chicago, Chicago, IL, USA

Abstract:

Sensor errors degrade the performance of supervision and control systems. Sensor accuracy can be affected by many factors such as extreme working conditions, sensor deteriorations and interferences from other devices. Sensor error such as missing signals may disrupt system operation and signal bias may cause inaccurate manipulate variable calculation by the controller that may threaten the productivity and safety of the system. For a dynamic system, it may be difficult to distinguish between some sensor errors and real dynamic changes in the system. Many systems may have multiple sensors and the signals from those sensors may be correlated with each other. This offers the opportunity for using the redundancy to detect sensor errors and reconcile erroneous signals.

A hybrid online multi-sensor error detection and functional redundancy (SED&FR) system is developed to monitor the performance of multiple sensors and reconcile the erroneous signal values. The SED&FR system relies on two technologies, outlier-robust Kalman filter (ORKF) and a locally-weighted partial least squares (LW-PLS) regression model. The two methods have their unique way of using data, ORKF is comparing current signal samples with signal trace indicated by previous samples and LW-PLS is comparing previous samples with the samples from a database and use the most similar samples to build a model to predict current signal values. Sensor errors such as outliers can be detected more easily by ORKF. Long-duration sensor error detection and reconciliation can be handled by LW-PLS which relies on the prediction ability based on historical data sets. SED&FR leverages the advantages of automatic measurement error elimination with ORKF and data-driven prediction with LW-PLS.

The performance of this system is illustrated with a clinical case study which collects data from a continuous glucose monitoring (CGM) system and a wearable device that provides energy expenditure (EE) and Galvanic Skin Response (GSR) information from people with type 1 diabetes. More than 10,000 signal sets are used for testing. The results indicate that the proposed system can successfully detect most of the erroneous signals and substitute them with reasonable estimated values computed by functional redundancy system.

495005 Model-Based Fault Detection for Nonlinear Process Systems Using Multiparametric Programing for Parameter Estimation

Abstract:

Early detection and diagnosis of faults are essential to minimise the economic losses and fatal damage of the operators and the equipment. Several techniques have been developed to detect and localize the fault. With the aid of process models, estimation and decision methods, faults can be detected and monitored by process states, model parameters and characteristic quantities [1]. Model-based fault detection techniques can be categorised into observer-based, parity relation and parameter estimation methods. The observed-based uses the outputs of the system from the measurements by using some type of observer, and then constructs the residual by an output estimate error and the parity relation approach uses the parity check on the consistency of parity equation to generate residuals (parity vector). The inconsistency in the parity relations indicates the presence of faults. On the other hand, in the parameter estimation approach, the model parameters of the actual process are repeatedly estimated on-line and the results are compared with the reference model [2-7]. The key limitation however is that it requires solving an optimization problem online at regular time interval. The key issue with solving an online optimization problem is not only that it is time consuming, but also that the solution may not converge in a reasonable time.

Thus, this work presents a model based fault detection and diagnosis approach using parameter estimation for process systems using multiparametric programming. Multiparametric programming provides the optimization variables as an explicit function of the parameter [8]. In this work, a square system of parametric nonlinear algebraic equations is obtained by formulating optimality condition. These equations are then solved symbolically using Mathematica to obtain model parameters as an explicit function of measurement [9]. This allows computation of parameter estimates by simple function evaluation. The diagnosis of fault is carried out by monitoring the changes of the residual of model parameters. The estimated parameters should be close to observed parameters when no fault is present. Any substantial discrepancy between estimated model parameters and the observed model parameters may be interpreted as a fault, i.e., if the discrepancy goes above a certain threshold.

Case studies of fault detection and diagnosis for single stage evaporator system [10] and quadruple tank system [11] are presented. A number of faulty and fault free scenarios are considered to show the effectiveness of the presented approach of parameter estimate and faults can be detected by monitoring the residual of model parameters. The proposed approach successfully estimates the model parameters and detects the faults in the systems

[1] Isermann, R. Process Fault Detection Based on Modeling and Estimation Methods - A Survey. *Automatica* **1984**, *20* (4), 387.

[2] Dai, X.; Gao, Z. From Model, Signal to Knowledge: A Data-Driven Perspective of Fault Detection and Diagnosis. *IEEE Trans. Ind. Informatics* **2013**, *9* (4), 2226.

[3] Venkatasubramanian, V.; Rengaswamy, R.; Yin, K.; Kavuri, S. N. A Review of Process Fault Detection and Diagnosis Part I: Quantitative Model-Based Methods. *Comput. Chem. Eng.* **2003**, *27* (3), 293.

[4] Mhaskar, P.; McFall, C.; Gani, A.; Christofides, P. D.; Davis, J. F. Isolation and Handling of Actuator Faults in Nonlinear Systems. *Automatica* **2008**, *44* (1), 53.

[5] Du, M.; Mhaskar, P. Isolation and Handling of Sensor Faults in Nonlinear Systems. *Automatica* **2014**, *50* (4), 1066.

[6] Gertler, J. Fault Detection and Isolation Using Parity Relations. *Control Eng. Pract.* **1997**, *5* (5), 653.

[7] Isermann, R. Fault Diagnosis of Machines via Parameter Estimation and Knowledge processingâ€"Tutorial Paper. *Automatica* **1993**, *29* (4), 815.

[8] Pistikopoulos, E. N. Perspectives in multiparametric programming and explicit model predictive control. AIChE Journal **2009**, 55 (8), 1918-1925.

[9] Dua, V. Mixed Integer Polynomial Programming. *Comput. Chem. Eng.* **2015**, *72*, 387.

[10] Dalle Molle, D. T.; Himmelblau, D. M. Fault Detection in a Single-Stage Evaporator via Parameter Estimation Using the Kalman Filter. *Ind. Eng. Chem. Res.* **1987**, *26* (12), 2482.

[11] Johansson, K. H. The Quadruple-Tank Process: A Multivariable Laboratory Process with an Adjustable Zero. *IEEE Trans. Control Syst. Technol.* **2000**, *8* (3), 456.

493549 Event Detection and Estimation of Its Influence Based on Fuzzy Qualitative Representation of Measurements and Fuzzy Logic Estimator

Abstract:

Event Detection and Estimation of its Influence Based on Fuzzy Qualitative Representation of Measurements and Fuzzy Logic Estimator

Sediqeh Samadi *, Kamuran Turksoy **, Iman Hajizadeh *, Jianyuan Feng *, Mert Sevil **, Ali Cinar*

** Department of Chemical and Biological Engineering, Illinois Institute of Technology, Chicago, IL, USA (Corresponding author: cinar@iit.edu).*
*** Department of Biomedical Engineering, Illinois Institute of Technology, Chicago, IL, USA.*

Detection of underlying (temporal) events in chemical and biological systems is important in order to respond them by taking the appropriate control actions. The reflection of such events on the measured variables is employed to extract features for detection algorithms [1]. In this study, quantitative change of measured variable(s) over moving specified length time window (episode) is translated to the qualitative variables (shapes). Combination of first and second derivative (or difference) signs of every shape is distinct. We introduce fuzziness in qualitative variables where more than one shape can describe an episode. Fuzzy qualitative variables can be used to create the features for detection of the events. In addition to detection of the events, the estimation of their influence of is crucial for control purpose.

Using the proposed fuzzy qualitative representation of measurements, we define a detection feature named "increase of glucose trend index" to identify the occurrence of the meal. Detection of the meals based on continuously measured glucose is of interest for development of artificial pancreas (AP) control systems for people with type 1 diabetes mellitus (T1DM). The algorithm can be incorporated with the AP systems without manual meal announcements, or as a safety feature for people who may forget entering meal information manually. Using qualitative representation of the continuously measured glucose, a time period labeled as meal flag is identified. The criteria for activation and deactivation of meal flag described based on the numerical value of detection feature (increase of glucose trend index) computed every sampling time [2]. Following the detection of the occurrence of the meal, a fuzzy logic based estimator estimates the carbohydrate (CHO) content of the meal to determine the appropriate dose of insulin bolus for a meal. At every sampling time during active meal flag, a fuzzy system with two proposed input variables estimates the absorbed CHO and computes insulin meal bolus dose accordingly. The first input of fuzzy estimator is the ratio of glucose increase with respect to its increase at the detection time and the second input is the absorbed insulin modeled by a normalized finite impulse response (FIR) function [3].

To assess the performance of the algorithm, we used the academic version of UVa/Padova simulator [4] that contains 30 in silico patients (10 adults, 10 adolescents, 10 children). Basal insulin and CHO to insulin ratio for each patient are equal to their default values in the simulator. The proposed algorithm computes the time and dose of bolus insulin. For the evaluation data set, rates of sensitivity of meal detection are 87%, 94%, and 93% and false detection rates are 21%, 4%, and 3% for adults, adolescents and children, respectively. The absolute error in CHO estimation is 23.1%. The mean glucose concentration and the percentages in target range (70-180 mg/dl) are 135 mg/dl, 147 mg/dl, 145 mg/dl and 86.5%, 75%, 74.8% for adults, adolescents, and children, respectively.

[1] B. R. Bakshi and G. Stephanopoulos, "Representation of process trends-III. multiscale extraction of trends from process data," *Comput. Chem. Eng.*, vol. 18, no. 4, pp. 267–302, 1994.

[2] S. Samadi, K. Turksoy, I. Hajizadeh, J. Feng, M. Sevil, and A. Cinar, "Meal Detection and Carbohydrate Estimation Using Continuous Glucose Sensor Data ," *IEEE J. Biomed. Heal. Informatics*, doi: 10.1109/JBHI.2017.2677953, 2017.

[3] Q. Wang, P. Molenaar, S. Harsh, K. Freeman, J. Xie, C. Gold, M. Rovine, and J. Ulbrecht, "Personalized state-space modeling of glucose dynamics for Type 1 diabetes using continuously monitored glucose, insulin dose, and meal intake an extended Kalman filter approach," *J. Diabetes Sci. Technol.*, vol. 8, no. 2, pp. 331–345, 2014.

[4] C. D. Man, F. Micheletto, D. Lv, M. Breton, B. Kovatchev, and C. Cobelli, "The UVA/PADOVA type 1 diabetes simulator: new features.," *J. Diabetes Sci. Technol.*, vol. 8, no. 1, pp. 26–34, 2014.

497855 Fault Detection and Diagnosis of Continuous Processes Via Non-Linear Support Vector Machine Based Feature Selection

Abstract:

Advances in sensor and data collection technologies has valorized data-driven modeling approaches in process monitoring and fault detection in process systems engineering [1]. Today, machine learning and pattern recognition techniques play a significant role in attaining actionable insights and decision-making from the vast amounts of available process data by building accurate and robust data-driven models. One of the most popular machine learning techniques is Support Vector Machines (SVMs) [2-6] which allows the use of high dimensional feature sets for learning problems such as classification and regression. Yet reducing the dimensionality of the feature space in data-driven modeling, known as dimensionality reduction and feature selection, is still a key task in improving model accuracy as well as decreasing *a priori* data collection, which in turn yields enhanced efficiency in chemical processes.

In this work, we present the application of a novel non-linear (kernel-dependent) SVM-based feature selection algorithm [7-8] to process monitoring and fault detection of continuous processes. The developed methodology is derived from sensitivity analysis of the dual SVM objective and utilizes existing and novel greedy algorithms to rank features that also guides fault diagnosis. Specifically, we train two-class SVM models to detect known faults and use one-class support vector data descriptors to characterize normal operations. Here, the manipulated and measured variables of the process constitute the input feature space, and instances of normal and faulty operation yield training samples for our SVM models. The feature selection algorithm is used to improve the accuracy of fault detection models and perform fault diagnosis. We present results for the Tennessee Eastman process [9] as a case study and compare our approach to existing approaches for fault detection and diagnosis.

References

[1] Chiang, L. H., Braatz, R. D., & Russell, E. L. (2001). Fault Detection and Diagnosis in Industrial Systems. Springer Science & Business Media.

[2] Vapnik, V. (1995). The Nature of Statistical Learning Theory. Springer.

[3] Cortes, C., & Vapnik, V. (1995). Support-vector networks. Machine Learning, 20(3), 273-297.

[4] Schölkopf, B., Smola, A. J., Williamson, R. C., & Bartlett, P. L. (2000). New support vector algorithms. Neural Computation, 12(5), 1207-1245.

[5] Schölkopf, B., Platt, J. C., Shawe-Taylor, J., Smola, A. J., & Williamson, R. C. (2001). Estimating the support of a high-dimensional distribution. Neural Computation, 13(7), 1443-1471.

[6] Tax, D. M., & Duin, R. P. (2004). Support vector data description. Machine Learning, 54(1), 45-66.

[7] Guzman, Y.A., Kieslich, C.A., Floudas, C.A. (Submitted). A global optimization framework for feature selection with Support Vector Machines.

[8] Kieslich, C. A., Tamamis, P., Guzman, Y. A., Onel, M., Floudas, C. A. (2016). Highly Accurate Structure-Based Prediction of HIV-1 Coreceptor Usage Suggests Intermolecular Interactions Driving Tropism. PloS one,11(2), e0148974.

[9] Downs, J. J., & Vogel, E. F. (1993). A plant-wide industrial process control problem. Computers & chemical engineering, 17(3), 245-255.

499254 Optimal Test Design Framework for Model-Based Active Fault Detection and Isolation

Abstract:

The complexity and widely variable operating envelope of modern cyber-physical systems yields a need for advanced fault detection and isolation (FDI) algorithms. Efficient and accurate FDI tests can reduce maintenance cost, and improve system reliability and performance,[1,2] which are becoming increasingly significant in the aerospace and automotive industries. Recent advances in the field of active FDI[3,4] allow for better detectability and isolatability of faults by implementing advanced control logic for the Built-In Tests deploying FDI.[5,6] In this presentation, we will outline a comprehensive workflow that improves FDI robustness and accuracy through an optimal selection of input settings prior to FDI test execution.

Figure 1. Outline of framework for improving active FDI test designs

In the proposed workflow, model-based, active fault detection and isolation is accomplished by solving a set of optimization problems, constrained within the system operating space. Figure 1 presents an overview of this framework for test design selection. Models that accurately represent the system (especially the steady state and transient system responses to faults) are used to generate output forecasts with and without the anticipated fault(s) present. Modelling error and uncertainty in the form of environment variability are minimized, in terms of their impact on the FDI test. The improvement in fault identifiability is accomplished with two different optimal test design criteria, both using the output sensitivities, $\mathbf{Q}_i$, of i-th outputs at a predetermined number of sampling points with respect to the targeted faults and uncertainty, ξ. The first criterion designs tests satisfying D-optimality, which minimizes the correlation of all the elements in ξ.[7–9] The second approach maximizes the information with respect to the faults and minimizes the impact of uncertainty, using the so-called Ds-optimality criterion.[10] Sensitivities are compiled into modified formulations of the Fisher Information Matrix, $\mathbf{H}_\xi$, which is a function of the parameters representing faults and uncertainty in the system model and the test design vector. As shown in Eq.(1), $\mathbf{H}_\xi$ is partitioned into submatrices corresponding to $(\xi_1 \ldots \xi_f)$ and $(\xi_{f+1} \ldots \xi_u)$ that contain the correlations between faults (H_{ff}), uncertainty (H_{uu}), and that between faults and uncertainty, (H_{uf}, H_{fu}):

where s_{ij} is the estimated measurement variance between the i-th and j-th outputs. The optimal test design vector for each criterion is formulated for differential algebraic equations describing dynamic systems as shown in Eqs. (2) and (3) of Table 1.

Table 1. Mathematical formulation of the FDI test optimization for steady-state and dynamic test designs

Execution of the optimal FDI tests designed, leads to data that need to be analyzed in terms false alarm rates (FAR) and non-detection rates (NDR). First, the space of uncertain parameters and boundaries is explored to determine if similar output trajectories can be generated for faulty and fault-free systems at the chosen design of admissible inputs. The mathematical formulation of this search is based on practical identifiability of model parameters (not shown in this abstract). Thereafter, thresholds are designed to minimize the FARs and NDRs. We evaluate the residuals generated from an FDI test through a positive definite function (e.g. normalized root mean square). We iterate this calculation at various fault and uncertain parameter values, and then compare the results with fault thresholds to determine if the test satisfies FAR and NDR requirements. Threshold designs are analyzed in the form of receiver operating characteristic (ROC) plots. An example of a ROC curve is shown in Figure 2, where the optimal test design obtained using this framework significantly improves the overall quality of the FDI test in one of the systems studied.

Figure 2. Example ROC plot of a fouled heat exchanger at various fault detection thresholds for particulate fouling at nominal and optimal test designs.

Several case studies comparing the FDI effectiveness (in terms of FAR and NDR) at nominal and optimal test designs will be presented. The first case study will focus on built-in tests for a plate-fin heat exchanger from an aircraft environmental control system (ECS) that is prone to particulate fouling, which extends prior work on fouling quantification.[9] The second case study will also focus on particulate fouling in the integrated ECS, to illustrate the impact of increased system complexity of FDI, when multiple sources of uncertainty are present. False alarms and non-detections at nominal test conditions, and their improvement through the proposed framework will be presented. Lastly, the application of the proposed framework on the three-tank benchmark system will be presented. This analysis shows how the design of an active FDI test using the proposed framework compares to more computationally intensive methods proposed in the literature.

Acknowledgment

This work was sponsored by the UTC Institute for Advanced Systems Engineering (UTC-IASE) of the University of Connecticut and the United Technologies Corporation. Any opinions expressed herein are those of the authors and do not represent those of the sponsor. Help and guidance by Modelon and Modelon-AB are gratefully acknowledged.

References

1. Hwang, I., Kim, S., Kim, Y. & Seah, C. E. A Survey of Fault Detection, Isolation, and Reconfiguration Methods. *IEEE Transactions on Control Systems Technology* **18,** 636–653 (2010).

2. Isermann, R. Model-based fault-detection and diagnosis - Status and applications. *Annual Reviews in Control* **29,** 71–85 (2005).

3. Venkatasubramanian, V., Rengaswamy, R., Yin, K. & Kavuri, S. N. A review of process fault detection and diagnosis: Part I: Quantitative model-based methods. *Computers & Chemical Engineering* **27,** 293–311 (2003).

4. Venkatasubramanian, V. A review of process fault detection and diagnosis: Part III: Process history based methods. *Computers & chemical …* **27,** 293–311 (2003).

5. Henrik Niemann. A Setup for Active Fault Diagnosis. *Automatic Control, IEEE Transactions on* **51,** 1572–1578 (2006).

6. Gao, Z., Cecati, C. & Ding, S. X. A survey of fault diagnosis and fault-tolerant techniques-part II: Fault diagnosis with knowledge-based and hybrid/active approaches. *IEEE Transactions on Industrial Electronics* **62,** 3768–3774 (2015).

7. Han, L., Zhou, Z. & Bollas, G. M. Model-based analysis of chemical-looping combustion experiments. Part I: Structural identifiability of kinetic models for NiO reduction. *AIChE Journal* **62,** 2419–2431 (2016).

8. Han, L., Zhou, Z. & Bollas, G. M. Model-based analysis of chemical-looping combustion experiments. Part II: Optimal design of CH4 -NiO reduction experiments. *AIChE Journal* **62,** 2432–2446 (2016).

9. Palmer, K. A., Hale, W. T., Such, K. D., Shea, B. R. & Bollas, G. M. Optimal design of tests for heat exchanger fouling identification. *Applied Thermal Engineering* **95,** 382–393 (2016).

10. Bruwer, M.-J. & MacGregor, J. F. Robust multi-variable identification: Optimal experimental design with constraints. *Journal of Process Control* **16,** 581–600 (2006).

500900 Verification of Control Systems with Discrete and Continuous Dynamics

Abstract:

The need for automated and safe control of chemical processes requires the analysis of dynamical systems with interdependencies between continuous controls and discrete switches. The continuous variables result from the application of classical controllers, such as PID and feedforward and more advanced controllers such as MPC. The discrete dynamics can be due to batch operation, mode switches during start-up and shut down of continuous processes, operator intervention or hardware or software failures. Systems that have integrated discrete and continuous dynamics are said to be hybrid.

The design of hybrid control systems remains an unsolved and potentially unsolvable problem (as shown by Prof. P. Varaiya). The design and testing of process automation systems therefore remains heuristic in nature and very much dependent on the knowledge, experience and the creativity of the personnel involved. Many tools, such as HAZOP, LOPA, Fault Trees etc exist. But these are are not exhaustive and the potential for errors persists. Increased process automation demands that ever more complex chemical process control systems with hybrid dynamics must be evaluated and faults in the control logic continue to occur and cause accidents.

The aim of this work is to examine the application of model checking to aid the testing of process automation systems that integrate discrete logic and continuous controls. For a given set of specifications, the system dynamics are modeled by representing the desired behavior as transitions between discrete states. These specifications are written as computation tree logic (CTL) formulas, and model checking is applied to the discrete abstraction of the continuous system. Current software systems such as nuSMV is capable of handling systems with well over 10E20 states in reasonable time. So very large logic systems can be checked. It is the abstraction (over-approximation) of the continuous dynamics rather than the discrete controls themselves that lead to combinatorial explosion. Making the abstract coarse makes the problem tractable However, if the abstraction is too coarse, or the specifications are too broad, then model checking mail fail to verify the specifications even though the system is actually correct.

To address the concern of inadequate specifications, this work relates the discrete states like alarms, trips and interlock to the continuous variables they rely on. This allows the verifier to build CTL formulas for the specifications based on the values of continuous variables in discrete time. A CTL specification may look like AG(SAFE)

where SAFE is a discrete variable that depends on a combination of continuous and discrete variables. The specification checks whether SAFE is false at any point in the execution of the discrete model, either due to faulty logic, too coarse of a discrete abstraction, or inadequate specifications. If false returns then the system provides a path that lead to failure. It may be a faulty measurement, as in the BP Texas City explosion, a low level alarm that is stuck or some other event that makes the system unobservable or uncontrallable.

The problem with coarse over approximations is solved by applying the SAFE condition to material and energy balances. Checking such balances lead to a tractable problem since mass and energy balances can be represented as integrators and can be solved with explicit integration. This approach theroefore presents a first step towards developing a comprehensive methodology for checking automation code in the chemical process industry.

A Liquid Holding Tank model formulation is included to demonstrate application of the model-checking solver NuSMV, highlighting that

(1) A model can be faulty and yet adhere to all logic and operational specifications at the same time.

(2) A fault in logic, combined with instrumentation failure, may lead to a critical state.

(3) Combination of instruments and safety logics can be suggested to ensure reliability of a unit even in the case of multiple failures. The discrete state model is augmented with a continuous model in SIMULINK to verify the results of SMV.

The concept developed in this paper can be applied to test automation systems for chemical plants in a modular fashion. In some specific cases it will be possible to prove that the model satisfies the system specifications and that the specifications themselves are rich enough to account for all possible failures.

494094 Multi-Rate Sampled-Data Observer Design for Nonlinear Systems with Multiple Measurement Delays

Abstract:

The problem of nonlinear observer design has been intensively studied for systems under fast sampling [1-3], which can be potentially applied to estimation, process control, and fault detection and identification. Motivated by practical implementation needs, however, one of the biggest challenges is to design an observer for general multi-rate sampled-data systems with multiple measurement delays (e.g., chemical processes, biological systems, networked control systems). In the monitoring of polymerization processes, for instance, the slow-sampled measurements with significant output delay often provide important quality information on the products, and hence must be incorporated in an intelligent manner together with the fast-sampled measurements, to make the entire system observable as well as improve the estimation accuracy. Different sampling rates and different sizes of measurement delays, caused by sample preparation, analysis and calculation, need to be accommodated in the observer design framework. Furthermore, the presence of possible perturbations in the sampling schedule as well as in the size of measurement delay makes the real-time monitoring even more challenging.

The observer design for linear multi-rate systems was recently investigated in [4]. The proposed multi-rate observer adopts the idea in [5] of using a state predictor to approximate the inter-sample behavior, and moreover, multiple, asynchronous inter-sample predictors are used for the multi-rate system as a generalization. The proposed multi-rate observer is based on an available continuous-time Luenberger observer design coupled with asynchronous inter-sample predictors. Each predictor generates an estimate of a sampled output in between consecutive measurements, and will get reinitialized once the associated, most-recent measurement becomes available. Sufficient and explicit conditions in terms of maximum sampling period were established to guarantee exponential stability of the error dynamics. Although the presence of possible measurement delays was not considered, it would be straightforward to incorporate multiple output delays by using the algorithm of dead time compensation as will be mentioned later.

In this paper, a nonlinear multi-rate sampled-data observer design is developed as a starting point before taking multiple output delays into account. In the same spirit as in [4], the proposed multi-rate sampled-data observer is based on an available continuous-time design coupled with asynchronous inter-sample predictors for the sampled measurements [6]. The sampled-data system with the multi-rate observer forms a hybrid system and it is shown that the error dynamics of the overall system is input-to-output stable with respect to measurement errors by applying the vector small-gain theorem [6-7]. This sampled-data design also offers robustness with respect to perturbations in the sampling schedule. To handle the possible measurement delays, a dead time compensation algorithm is developed, which preserves the stability property of the multi-rate sampled-data observer, no matter how large the size of each delay is. Once a delayed measurement is obtained, the state estimates at its sampling instant can be re-calculated by using the multi-rate sampled-data observer. Then the process model can be utilized to predict the state at current time, given the delayed estimates, in the same spirit as in Smith-predictor methods. During the compensation, any available measurement can be treated as a delay-free output and the compensator will get reinitialized at every sampling instant where the measurement is available. This algorithm inherits all the nice properties (i.e., stability and robustness) of a multi-rate sampled-data observer. Moreover, it provides robustness with respect to the perturbations in the size of each measurement delay.

In the present work, we revisit the example of an industrial gas-phase polyethylene reactor in [4], which aims at monitoring the amount of active catalyst sites. The only fast-sampled output without delay is the temperature of the reactor. The gas concentrations in the reactor are measured on line by using gas chromatography, where the sampling normally occurs every 20 min with 8 min measurement delay. Cumulative melt index and density of polyethylene are measured off line in a quality control lab, where the sampling normally occurs every 40 min with 60 min measurement delay. However, perturbations in the sampling schedule as well as in the size of delay have been considered. A continuous-time nonlinear observer is first designed to cope with the nonlinearity of the process [3,8]. With the proposed framework of multi-rate multi-delay observer, we can obtain a continuous estimate of the unmeasured state with a satisfactory convergence rate. In addition, the output behavior between two consecutive samples can be estimated as well by using inter-sample predictors.

References:

[1] Krener, A. J., & Isidori, A. (1983). Linearization by output injection and nonlinear observers. *Systems & Control Letters*, 3(1), 47-52.

[2] Gauthier, J. P., Hammouri, H., & Othman, S. (1992). A simple observer for nonlinear systems applications to bioreactors. *IEEE Transactions on Automatic Control*, 37(6), 875-880.

[3] Kazantzis, N., & Kravaris, C. (1998). Nonlinear observer design using Lyapunovâ€™s auxiliary theorem. *Systems & Control Letters*, 34(5), 241-247.

[4] Ling, C., & Kravaris, C. (2017). Multi-rate observer design for process monitoring using asynchronous inter-sample output predictions. *AIChE Journal*. DOI: 10.1002/aic.15707.

[5] Karafyllis, I., & Kravaris, C. (2009). From continuous-time design to sampled-data design of observers. *IEEE Transactions on Automatic Control*, 54(9), 2169-2174.

[6] Ling, C., & Kravaris, C. Multi-rate sampled-data observers based on a continuous-time design. *Submitted to the 56th Conference on Decision and Control.*

[7] Karafyllis, I., & Jiang, Z. P. (2011). A vector small-gain theorem for general non-linear control systems. *IMA Journal of Mathematical Control and Information*, 28(3), 309-344.

[8] Kazantzis, N., Kravaris, C., & Wright, R. A. (2000). Nonlinear observer design for process monitoring. *Industrial & Engineering Chemistry Research*, 39(2), 408-419.

487195 Data Reconciliation with Inequality Constraints Induces Bias: A Cause for Concern?

Abstract:

1. Introduction

Data reconciliation comprises an important set of tools for data quality improvement. By removing always-present random measurement noise, one seeks to process the data so that the transformed data are an improved representation of the measured variables. Most often, this involves projection, filtering, or smoothing steps (e.g., Narasimhan & Jordache, 1999; Vachhani *et al.*, 2005, 2006). Data reconciliation can be based on mechanistic models (white box, Venkatasubramanian *et al.*, 2003a), empirical models (black box, Venkatasubramanian *et al.*, 2003b, or a combination of both (gray box). This work is focused on data reconciliation methods that rely on mechanistic process understanding (white box data reconciliation).

A majority of the data reconciliation literature is concerned with the application of equality constraints as a way to improve the accuracy of the available measurements. In this case, data reconciliation is interpreted geometrically as a projection to a plane or manifold. The application of linear and nonlinear equality constraints for data reconciliation has been studied in great detail already (Tamhane & Mah, 1985; Crowe, 1996; Veverka & Madron, 1997; Romagnoli & Sanchez, 1999).

The application of inequality constraints has been studied as well. An early example is found in Narasimhan & Harikumar (1993) where knowledge about the feasible range of values for a measured process state is accounted for. Vachhani *et al.* (2006) proposed a method for dynamic data reconciliation based on the unscented Kalman filter, thereby enabling the application of inequality constraints into an online data reconciliation method. More recently, special attention has been given to the incorporation of prior knowledge about function and signal shapes to optimally describe or reconcile time series data (Villez *et al.*, 2013; Villez & Habermacher, 2016; Derlon *et al.*, 2017; Vertis *et al.*, 2017; Masic et al, 2017; Srinivasan *et al.*, 2017). Most typical is that the data analysis is enhanced by incorporation of prior knowledge about the signs of the first and second derivative of the fitted function or the data series.

The application of inequality constraints offers several advantages. For instance, in Narasimhan & Harikumar (1993) and Srinivasan *et al.* (2017) the improved accuracy of the reconciled data is the principal motivation to consider inequality constraints. In Villez & Habermacher (2016), shape constraints lead to the formulation of a lack-of-fit statistic. Vertis *et al.* (2016) mainly motivate the application of shape constraints as a way to differentiate time series data without risk of noise amplification (Bhatt *et al.*, 2012) in turn enabling a reasonable guess of parameter values in kinetic models. A similar strategy to initialize parameter estimation is deployed in Masic *et al.* (2017).

The data reconciliation literature implicitly incorporates the idea that any prior knowledge should be applied during the data reconciliation process. This is based on the observation that application of knowledge-based equality and inequality constraints invariably improves the accuracy of the reconciled data as long as the applied constraints are correctly assumed for the data-generating process. The impact of using reconciled data, e.g. during model parameter estimation, has not been studied in detail yet. In this work it is demonstrated that data reconciliation with inequality constraints can induce significant bias during model parameter estimation.

2. Methods

The utility of data reconciliation is studied by means of a simple model parameter estimation problem. To this end, a batch experiment involving a single equilibrium reaction with two species, A and B, is repeatedly simulated. The simulated rate laws are linear in the consumed species concentrations for both directions of the reaction. Consequentially, the net reaction can be described by a kinetic rate law which exhibits two parameters: the reaction rate coefficient (k1=1 mol/L.h) and the equilibrium parameter (k2=0.2). The concentration of A is measured at regular intervals. The data collected in each experiment are first reconciled based on a monotonicity constraint and then used for parameter estimation. In parallel, the same data are also used for parameter estimation without any data reconciliation. In both cases, the initial process conditions and the kinetic rate law structure are assumed known. This complete procedure is repeated 10000 times to investigate the effects of data reconciliation in the parameter estimates.

3. Results

3.1 Simulation

Figure 1 displays the simulation results for a single batch experiment. One can see that the concentration of species A decreases with time from its initial value (1 mol/L) towards its equilibrium value (0.2 mol/L). Noisy measurements of the concentration of species A are obtained every 5 minutes during the 7-hour experiment.

Figure 1: A single batch experiment – Black line: simulated concentration; Blue circles: Measurements before data reconciliation; Red crosses: Measurements after data reconciliation.

3.2 Data reconciliation

Reconciled measurements are obtained by computing concentrations that are close as possible to the original measurements in the least-squares (WLS) sense while satisfying isotonicity (non-strictly increasing). This monotonicity constraint is implemented as a positivity constraint for the point-wise differences between consecutive concentration measurements. Thus, the data reconciliation problem is a convex quadratic program with linear inequality constraints. Figure 1 displays the reconciled concentration measurements of species A. One observes that the reconciled measurements are generally closer to their true values and that –as expected- they increase with time.

Figure 2 shows the cumulative density of the obtained root mean squared error (RMSE) with error defined as the deviation between a measurement and its true noise-free value. As one can see, data reconciliation improves the obtained accuracy.

Figure 2: Accuracy of the measurements – Empirical cumulative density of the RMSE before and after data reconciliation.

3.3 Parameter estimation

Both kinetic rate law parameters are estimated simultaneously with both the unreconciled and reconciled measurements and for each simulated experiment separately. To this end, the parameter estimates are adjusted so that the simulated concentrations are as close as possible to the (unreconciled or reconciled) measurements in the least-squares sense. The results of this can be seen in Figure 3. Individual parameter estimates are shown for the first 250 experiments. The results of the complete Monte Carlo simulation are summarized by the mean parameter vector and the variance-covariance matrix computed from all parameter estimates. Figure 3 shows the 99% confidence ellipsoids according to an assumed multivariate normal distribution for the parameter estimates. Most importantly, one can see that the application of data reconciliation leads to a shift of the parameter estimates away from their ground truth values. In contrast, fitting the model to the unreconciled data directly does not induce such an effect and delivers parameters that are practically unbiased, i.e. their average is very close to the true parameter values.

Figure 3: Accuracy of the parameter estimates – Blue circles: Parameter estimates with raw data; Blue dashed lines: mean and 99% confidence region for the parameter estimates with raw data; Red crosses: Parameter estimates obtained with reconciled data; Red full lines: mean and 99% confidence region for the parameter estimates obtained with reconciled data.

4. Conclusions

Data reconciliation with inequality constraints has been advocated as a generally applicable method to improve the accuracy of experimental data. In this study, it is shown with a simple simulation study that improved accuracy can come at a price when using the reconciled data for their ultimate purpose. Indeed, the parameter estimation study show that data reconciliation induces a biasing effect during parameter estimation. This is explained by the fact that (i) the reconciled data are distributed according to truncated multivariate normal distribution and (ii) the least-squares parameter estimation procedure assumes a multivariate normal distribution. So far, no data reconciliation method has been proposed that automatically accounts for this discrepancy. This, in part, explains why using the unreconciled data during parameter estimations leads to the best parameter estimates.

References

Bhatt N, Kerimoglu N, Amrhein M, Marquardt W, Bonvin D. *Chemical Engineering Science.* 2012;83:24-38.

Crowe C M. *Journal of Process Control.* 1996;6:89-98.

Derlon N, Thürlimann C M, Dürrenmatt D J, Villez K. *Water Research.* 2017;114:327-337.

Masic A, Srinivasan S, Billeter J, Bonvin D, Villez K. *Computers & Chemical Engineering.* 2017;99:96-105.

Narasimhan S, Harikumar P. *Computers & Chemical Engineering.* 1993;17:1115-1120.

Narasimhan S, Jordache C. Data reconciliation and gross error detection: An intelligent use of process data *Gulf Professional Publishing.* 1999.

Noorman H J, Romein B, Luyben K Ch A M, Heijnen J J. *Biotechnology and Bioengineering.* 1996;49:364-376.

Romagnoli J A, Sanchez M C. Data processing and reconciliation for chemical process operations (Vol. 2). *Academic Press* 1999.

Srinivasan S, Billeter J, Narasimhan S, Bonvin D. *Computers & Chemical Engineering.* 2017;101:44-58.

Tamhane A C, Mah R S. *Technometrics.* 1985;27:409-422.

Vachhani P, Rengaswamy R, Gangwal V, Narasimhan S. *AIChE Journal.* 2005;51:946-959.

Vachhani P, Narasimhan S, Rengaswamy R. *Journal of Process Control.* 2006;16:1075-1086.

Venkatasubramanian V, Rengaswamy R, Yin K, Kavuri S N. *Computers & Chemical Engineering.* 2003a;27:293-311.

Venkatasubramanian V, Rengaswamy R, Kavuri S N, Yin K. *Computers & Chemical Engineering.* 2003b;27:327-346.

Vertis C S, Oliveira N M C, Bernardo F P. *Computer Aided Chemical Engineering.* 2016;38:2121-2126.

Veverka V V, Madron F. Material and energy balancing in the process industries: From microscopic balances to large plants (Vol. 7). *Elsevier* 1997.

Villez K, Venkatasubramanian V, Rengaswamy R. *Computers & Chemical Engineering.* 2013;58:116-134.

Talks AIChE 2017

Stevan Dubljevic

November 23, 2017

AICHE 2017 MINNEAPOLIS Talk IV

Pipeline Big Data Analysis - Leak Localization

#4, 103F room Thursday, November 02, 2017 08:00 AM - 08:17 AM

Pipeline network in North America (Source Enbridge Inc)

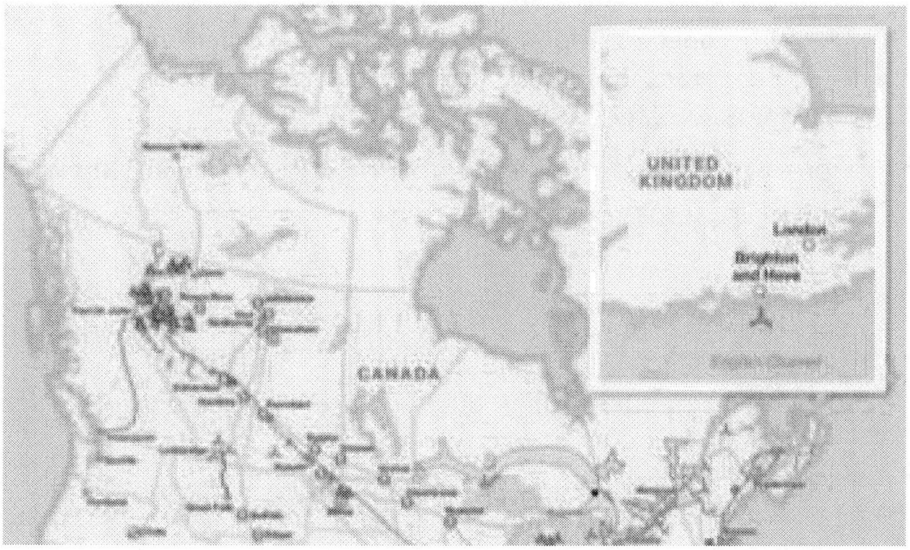

Pipeline Energy and Environmental Economics

- 55,377 km of gas pipelines
- 28,181 km of liquid pipelines
- Move 28% of crude oil produced North America and 23% of natural gas consumed in the United States
- Workforce of 15,400 people
- $27B+ in secured capital projects for growth
- No.12 on the 2016 Newsweek Green Rankings
- Environmental issues
- Insurance and leak prevention (pipeline integrity)

Pipeline inspection

- Inertial Measurement Unit
- External signal sink-source communication

Leak position localisation

- Measurements of IMU unit
  - Gyroscopes rotation angles $\omega_1, \omega_2, \omega_3$ - Euler angles
  - Linear translation motion a_x, a_y, a_z-acceleration
  - Magnetometer
- Error of integration prevents accurate localisation (known problem in aerospace engineering)

$$a_x = \frac{d^2x}{dt^2}$$

the meassurement error is accumulated by double integration of a_x signal

$$x(t) = \int\int a_x d\tau$$

Big Data Analysis

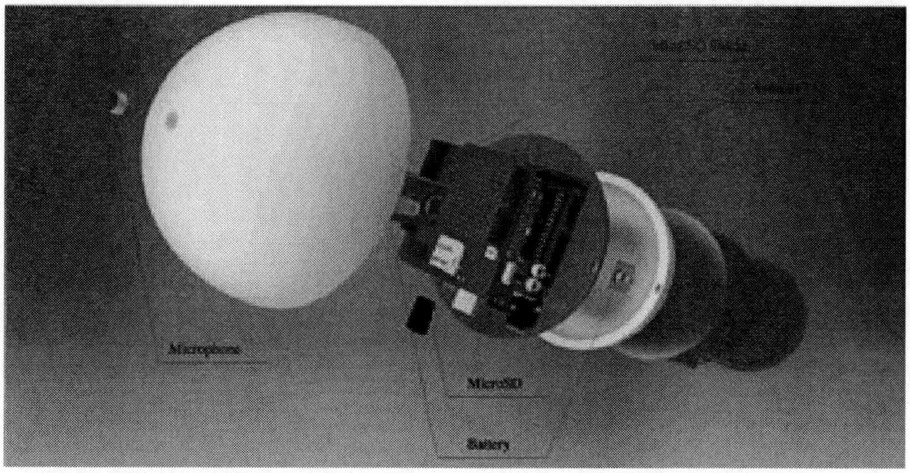

- Cheaper swimbot (87$)
- Large number of swimbot deployments (every batch of transfered fluid should have 2-3 swimbots)
- Large number of data for statistical processing and mining
  - Deep Neural Network
- Pipeline integrity and leak prevention

Big Data Processing

- Problem of integration drift is the main issue in all IMUs
- Algorithm is applied to fuse acceleration and and angular velocity to generate accurate positioning

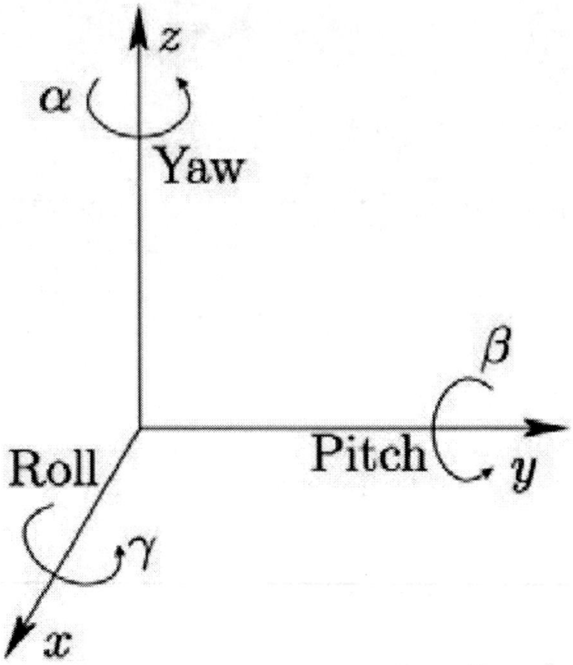

Model of Deep Neural Network

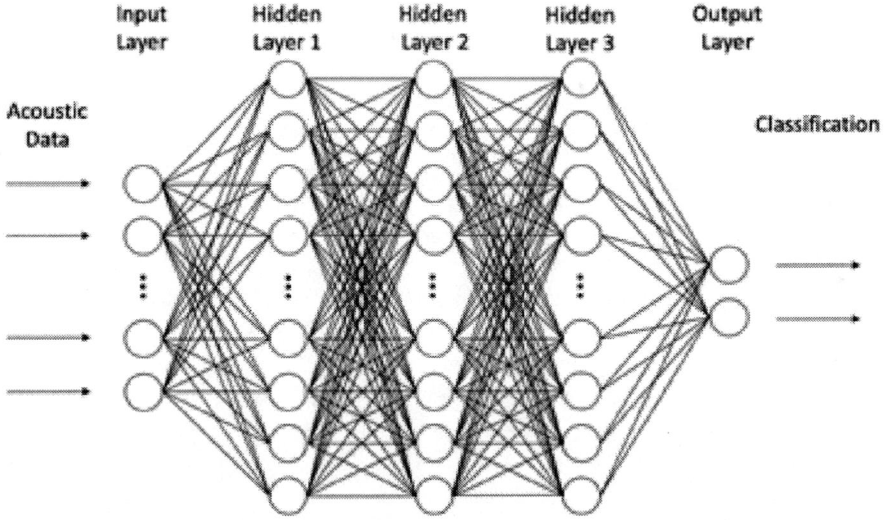

Experimental Test Environment

Experimental Results

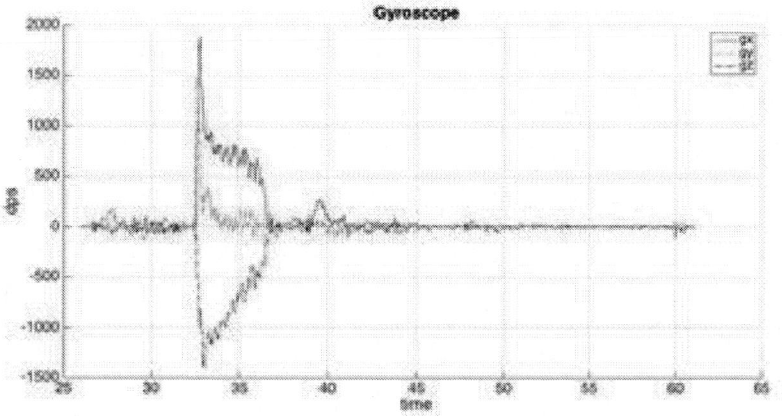

The experimental run takes 4-5 seconds in the recorded interval between 31 sec and 36 sec

Experimental Results

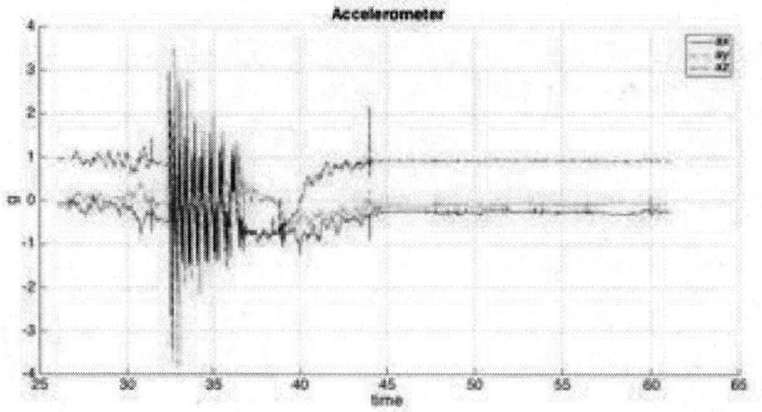

The experimental run takes 4-5 seconds in the recorded interval between 31 sec and 36 sec

Experimental Results

Position Calculated

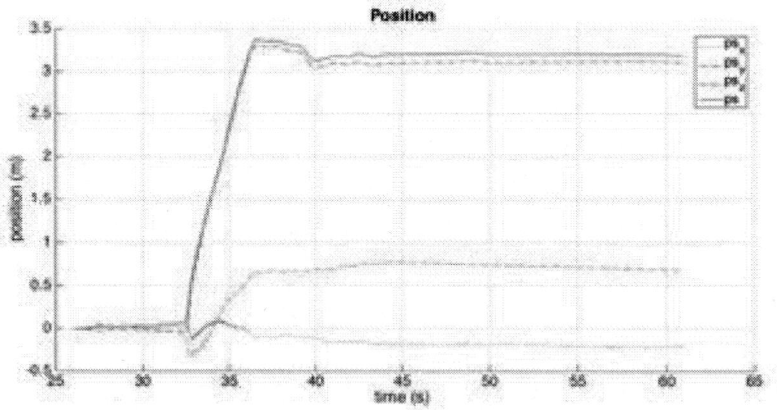

Pipe length is 3.22 m and relative error is 5.3%

Acustic data

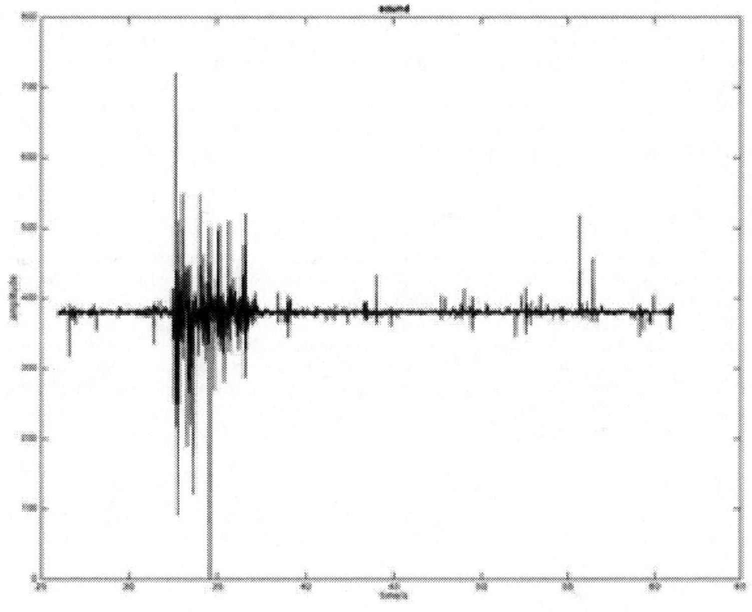

Summary

- Leak localization is explored
  - Issues of drift associated with IMU are addressed
  - Fusioon algorithm
  - Deep Neural Network
- Data exploration for large number of deployed swimbots

Future work:

- Bayesian inference
- Several agent swimbots with communication protocol

497644 Big Data Approach to Fault Detection, Diagnosis and Maintenance Optimization in Batch Processes

Abstract:

Batch process monitoring is an important task in order to attain a safe operability and minimize loss of productivity and profit, yet a challenging one. Characterized with a considerable number of interconnected variables, inherent non-stationarity, nonlinear response, batch-to-batch variability as well as finite duration, batch processes have become the focus of numerous researchers in the past few decades [1]. Today, statistical process monitoring (SPM) and data-driven process monitoring algorithms have been widely used in industry due to the advances in sensing technology which has facilitated enormous amounts of process data collection [2,3]. However, most novel techniques for fault detection and identification have focused on continuous processes, and the need of monitoring algorithm development for batch processes is evident.

In this work, we present a new data-driven framework for process monitoring and intervention in batch processes. Central to the framework are novel theoretical and algorithmic developments in nonlinear support vector machine-based feature selection [4, 5] which encapsulates highly nonlinear relationships between features, thus improving fault detection model accuracy and guiding fault diagnosis. We present results of a recent extensive benchmark, simulation dataset on Pensim model [6, 7] consisting 90,400 batches with numerous and diverse fault types, where we train time-specific models on the pre-aligned batch data trajectories in a moving horizon basis. The developed analysis framework aims for early detection of faulty batches and enables intervention to reduce loss of profit. Furthermore, simultaneous achievement of high process efficiency, safety and profitability is of utmost importance in modern process industries [8]. Therefore by using the insights attained on fault detection latency and preventative action from the built models, we will extend the discussion to the prediction of batch process duration as well as maintenance optimization of batch process operations [9, 10].

References

[1] Van Sprang, E. N., Ramaker, H. J., Westerhuis, J. A., Gurden, S. P., & Smilde, A. K. (2002). Critical evaluation of approaches for on-line batch process monitoring. Chemical Engineering Science, 57(18), 3979-3991.

[2] Joe Qin, S. (2003). Statistical process monitoring: basics and beyond.Journal of chemometrics, 17(8â€9), 480-502.

[3] Chiang, L. H., Braatz, R. D., & Russell, E. L. (2001). Fault Detection and Diagnosis in Industrial Systems. Springer Science & Business Media.

[4] Guzman, Y.A., Kieslich, C.A., Floudas, C.A. (Submitted). A global optimization framework for feature selection with Support Vector Machines.

[5] Kieslich, C. A., Tamamis, P., Guzman, Y. A., Onel, M., Floudas, C. A. (2016). Highly Accurate Structure-Based

Prediction of HIV-1 Coreceptor Usage Suggests Intermolecular Interactions Driving Tropism. PloS one,11(2), e0148974.

[6] Van Impe, J., Gins, G. (2015). An extensive reference dataset for fault detection and identification in batch processes. Chemometrics and Intelligent Laboratory Systems, 148, 20-31.

[7] Birol, G., Ündey, C., Cinar, A. (2002). A modular simulation package for fed-batch fermentation: penicillin production. Computers & Chemical Engineering, 26(11), 1553-1565.

[8] Cinar A., Palazoglu A., Kayihan F., (2006). Chemical Process Performance Evaluation. CRC Press.

[9] Pistikopoulos, E. N.; Vassiliadis, C. G.; Papageorgiou, L. G. (2000). Process design for maintainability: An optimization approach. Computers and Chemical Engineering. 24 (2-7), 203-208.

[10] Thomaidis, T. V.; Pistikopoulos, E. N. (2004). Criticality analysis of process systems. Annual Symposium on Reliability and Maintainability. pp 451-458.

499778 An Information Entropy Based Criterion for Variable Selection Performance Assessment

Abstract:

With ever-accelerating advancement of information, communication, sensing and characterization technologies, tremendous amount of data are generated and stored every day. Those so called "Big Data" are often extremely high-dimensional, contaminated by noise, and interspersed with a large number of irrelevant or redundant features, making it a challenging task to retrieve useful information from the data[1], [2]. Variable selection has been one of the practical approaches to reducing data dimensionality prior to data interpretation or modeling. Even for projection-based dimension reduction methods such as principal component analysis (PCA) and partial least squares (PLS), variable selection is often applied as a pre-processing step to further improve the modeling performance[3], [4]. In the last few years, many different variable selection methods have been reported. However, how to evaluate, in particular directly evaluate, the performance of variable selection methods has received limited attention. The commonly applied criteria to assess variable selection performance either indirectly measures the effects of variable selection, such as through prediction performance of a model, or require ground trough of variable relevancy, which is not available in practical applications.

To address this limitation, this paper presents an information entropy based consistency index (Ic) to directly evaluate the performance of variable selection method. The proposed method is based on the hypothesis that the same set of relevant variables would be selected when different training data sets are utilized to build a model. Therefore the proposed Ic index examines the consistency among variables being selected using different training data. The proposed Ic does not require any ground truth of variable relevancy, but can still make use of such information should it is available. Both simulated (with ground truth) and industrial (without ground truth) case studies are provide to demonstrate how Ic performs, which is compared with commonly used criteria. It is shown that the proposed index overcomes some of the limitations of existing indices, and the simulated case studies in this work show that Ic gave more objective assessments than the existing indices. The industrial case study shows that Ic is highly correlated with the performance of the resulted soft sensor, validating the need and benefits of directly assessing variable selection consistency.

References:

[1] J.-A. Ting, A. D'Souza, S. Vijayakumar, and S. Schaal, "Efficient learning and feature selection in high-dimensional regression," *Neural Comput.*, vol. 22, pp. 831–886, 2010.

[2] L. Comminges and A. S. Dalalyan, "Tight conditions for consistent variable selection in high dimensional nonparametric regression.," in *COLT*, 2011, pp. 187–206.

[3] Z. X. Wang, Q. He, and J. Wang, "Comparison of different variable selection methods for partial least squares soft sensor development," in *2014 American Control Conference*, 2014, pp. 3116–3121.

[4] Z. X. Wang, Q. P. He, and J. Wang, "Comparison of variable selection methods for PLS-based soft sensor modeling," *J. Process. Control.*, vol. 26, pp. 56–72, 2015.

499724 Health Monitoring of Multi-Scale Systems Using an Optimal Multi-Rate Wireless Sensor Network

Abstract:

Systems with multiple time scales where the fastest time scale can evolve in seconds or minutes whereas the slowest time scale can evolve over weeks or months are ubiquitous. Health monitoring of such systems is challenging since the traditional estimator-based approaches become computationally intractable. On the other hand, sensors that are embedded within the structures or are integral part of the equipment items are being increasingly investigated for health monitoring of critical equipment items. These sensors are often wireless. Measurements from wireless sensor networks can be noisy and suffer from communication constraints, packet dropouts, and synchronization errors. Due to the limitations in the communication bandwidth, only limited packets can be transmitted at any instant resulting in a multi-rate sensor network. In addition, random packet dropouts caused by unavoidable transmission error are unavoidable in wireless sensor networks. Furthermore, measurements can be out of sequence. There is hardly any work in the existing literature on process monitoring by using multi-rate measurements from a wireless sensor network that suffers from packet dropouts and synchronization errors.

In this work, an optimal sensor network is synthesized for a multi-scale system with widely varying time-scales considering that the measurements are available from a multi-rate wireless sensor network. It should be noted that in the existing literature, optimal sensor network synthesis for multi-rate systems has been mainly reported for linear systems.[1] In this work, algorithms are developed for optimal sensor placement and health monitoring of a multi-scale, nonlinear, differential-algebraic-equation system. The algorithm is tested on a smart refractory brick with embedded sensors. It is observed that the developed algorithms result in superior estimate of relevant variables for health monitoring that evolve over widely varying time scales from minutes to hours.

References

(1) Kadu, S. C.; Bhushan, M.; Gudi, R. Optimal sensor network design for multirate systems. ***J Process Control***, **2008**, *18*, 594-609.

497908 Learning Based Automated Identification of Nuisance and Correlated Alarms

Abstract:

Alarms systems are paramount for safe, efficient and reliable operation of modern industries. Ideally, alarms are suitably configured to provide an indication to process operators as soon as deviation from normal plant behavior is detected. The process operator can then take suitable remedial actions to ensure optimal plant behavior.

In practice, incorrectly configured alarm systems lead to problems including detection delays, alarm chattering, nuisance alarms and alarm flooding [1]. Due to the relative ease and low cost of measurement of process variables and collection of data, more alarms can be easily configured. However, this leads to the issue of generation of nuisance alarms which are alarms that require no response from the operator. Usually, an underlying fault in the plant propagates across several different units, leading to multiple alarms generated from different sources. In abnormal situations hundreds of alarms can arise in quick succession leading to operator overload. It is estimated that the US industry alone loses billions of dollars every year due to incorrect handling of alarms [2].

In this work, we present a method for the automatic identification of nuisance alarms and correlated alarms based on historically labelled alarm data. The method is designed to operate with an incoming stream of data â€“ newly available labelled data can be immediately used to update the machine learning based models. The method can be easily deployed over the cloud and scaled up to process alarm data at a very high frequency. The method is applied to network monitoring alarm systems spanning several sites. Results indicate the extremely high accuracy of our method for identification of nuisance and correlated alarms and demonstrate its utility in reduction of alarm flooding.

[1] B. R. Hollifield and E. Habibi, Alarm management: seven effective methods for optimum performance, ISA, 2007.

[2] W. Hu and A. W. Al-Dabbagh and T. Chen and S. L. Shah, Process Discovery of Operator Actions in Response to Univariate Alarms, IFAC-PapersOnLine, 49, 1026â€"1031, 2016.

500219 Dynamic Latent Variable Regression for Data Modeling and Monitoring

Abstract:

Multivariate process and quality measurements are often highly cross-correlated, but they are also often highly auto-correlated in most industrial processes due to dynamics. It is therefore critical to develop data modeling methods to efficiently capture these relations among process variables (**X**) and quality variables (**Y**). Partial least squares (PLS) and principal component analysis (PCA) serve as effective methods to capture cross-correlations, which have been widely used as basic methods to model the multivariate data [1-3], and used for multivariate statistical process monitoring [4-6].

For supervised learning or regression type of modeling, PLS is a dimensionality reduction algorithm that extracts latent variables (i.e., outer model projection) based on the maximum covariance criterion. However, the ultimate objective is to perform regression, as properly reflected in the inner modeling objectives. This discrepancy in the outer model and inner model objectives leads to many drawbacks. One of them is that the extracted scores of PLS may contain orthogonal variations, which are irrelevant to predict or monitor the quality variables. Due to the covariance objective in PLS, it usually requires multiple latent variables even to predict a single output variable [7].

Several subsequent works were proposed to overcome the above issue, including orthogonalized PLS [8] and concurrent PLS [9]. An alternative way is to use canonical correlation analysis (CCA) proposed by Hotelling [10]. CCA focuses only on extracting the multidimensional correlation between **X** and **Y**, and it performs better in prediction than PLS. One issue involved in CCA is that it pays no attention to the input variances, and it cannot exploit the input variance structure. Therefore, a concurrent CCA (CCCA) combineing CCA and PCA was proposed to exploit the variances and correlation in process-specific and quality-specific spaces [11].

Both PLS and CCA need additional processing to achieve good performance. Recently, Zhu and Qin [12] proposed a latent variable least squares (LVLS) method as an alternative method to exploit latent structured relations between **X** and **Y**. LVLS aims to minimize the prediction error between the input scores and the output scores, and it focuses on both the input variance structure and the prediction efficiency, which overcomes the drawbacks of PLS and CCA.

LVLS considers only static relations between **X** and **Y**. However, when the relationship between **X** and **Y** are dynamic, the static LVLS is not suitable for dynamic system modeling and subsequent process and quality monitoring. Several modified dynamic algorithms have been proposed for PLS and CCA to deal with dynamic processes [13-17]. Dong and Qin [17] developed a dynamic inner PLS (DiPLS) for dynamic system modelling, and it provides an explicit description for dynamic inner model and outer model.

In this paper, a dynamic inner LVLS (DiLVLS) algorithm is proposed to capture the dynamic relation between **X** and **Y** with a weighted combination of lagged process variables **X**. The objective of DiLVLS is to minimize the prediction error between the output scores and the weighted combination of lagged input scores, which builds the outer relation of

DiLVLS model. A regularization term is also included in the objective to overcome the collinearity problems. The consistent inner model is developed in inner modeling to describe the dynamic relations. After auto-correlation is extracted, the static LVLS model is then employed to exploit the static cross-correlations between the residuals of **X** and **Y**. The corresponding monitoring scheme is also developed for DiLVLS model. A synthetic case study and the Tennessee Eastman process are used to demonstrate the prediction and monitoring effectiveness of the proposed algorithm.

References

[1] Wold, S., Ruhe, A., Wold, H. and Dunn, III, W.J., 1984. The collinearity problem in linear regression. The partial least squares (PLS) approach to generalized inverses. SIAM Journal on Scientific and Statistical Computing, 5(3), pp.735-743.

[2] Geladi, P. and Kowalski, B.R., 1986. Partial least-squares regression: a tutorial. Analytica chimica acta, 185, pp.1-17.

[3] Höskuldsson, A., 1988. PLS regression methods. Journal of chemometrics, 2(3), pp.211-228.

[4] Nomikos, P. and MacGregor, J.F., 1995. Multivariate SPC charts for monitoring batch processes. Technometrics, 37(1), pp.41-59.

[5] Wise, B.M. and Gallagher, N.B., 1996. The process chemometrics approach to process monitoring and fault detection. Journal of Process Control, 6(6), pp.329-348.

[6] Joe Qin, S., 2003. Statistical process monitoring: basics and beyond. Journal of chemometrics, 17(8â€9), pp.480-502.

[7] Qin, S.J. and Dong, Y., DATA DISTILLATION, ANALYTICS, AND MACHINE LEARNING.

[8] Sun, L., Ji, S., Yu, S. and Ye, J., 2009, July. On the equivalence between canonical correlation analysis and orthonormalized partial least squares. In IJCAI (Vol. 9, pp. 1230-1235).

[9] Qin, S.J. and Zheng, Y., 2013. Quality-relevant and process-relevant fault monitoring with concurrent projection to latent structures. AIChE Journal, 59(2), pp.496-504.

[10] Hotelling, H., 1936. Relations between two sets of variates. Biometrika, 28(3/4), pp.321-377.

[11] Zhu, Q., Liu, Q. and Qin, S.J., 2016. Concurrent canonical correlation analysis modeling for quality-relevant monitoring. IFAC-PapersOnLine, 49(7), pp.1044-1049.

[12] Zhu, Q. and Qin, S.J., 2017. Latent variable least squares for process and quality modeling. Submitted to 56th IEEE Conference on Decision and Control.

[13] Kaspar, M.H. and Ray, W.H., 1993. Dynamic PLS modelling for process control. Chemical Engineering Science, 48(20), pp.3447-3461.

[14] Lakshminarayanan, S., Shah, S.L. and Nandakumar, K., 1997. Modeling and control of multivariable processes: Dynamic PLS approach. AIChE Journal, 43(9), pp.2307-2322.

[15] Li, G., Liu, B., Qin, S.J. and Zhou, D., 2011. Quality relevant data-driven modeling and monitoring of multivariate dynamic processes: The dynamic T-PLS approach. IEEE transactions on neural networks, 22(12), pp.2262-2271.

[16] Liu, Q., Zhu, Q., Qin, S.J. and Chai, T., 2017. Dynamic concurrent kernel CCA for strip-thickness relevant fault diagnosis of continuous annealing processes. Accepted by Journal of process control.

[17] Dong, Y. and Qin, S.J., 2015. Dynamic-Inner Partial Least Squares for Dynamic Data Modeling. IFAC-PapersOnLine, 48(8), pp.117-122.

493721 Bayesian Model Averaging for Estimating the Spatial Temperature Distribution in a Steam Methane Reformer

Abstract:

The steam methane reforming (SMR) process is a common method for commercial-scale hydrogen production [1]. At hydrogen manufacturing plants, the SMR process is typically carried out in a top-fired steam methane reformer that has two closed domains, i.e., the furnace side (in which fuel is burned in excess air to generate thermal energy that drives the SMR process) and the tube side (a collection of hundreds of nickel-based catalyst packed bed reactors, which are commonly referred to as reforming tubes). In top-fired reformers, a major factor that prevents reformers from being operated at the design capacity is the nonuniformity in outer tube wall temperature (OTWT) distributions (distributions of the outer reforming tube wall temperatures at a given distance from the reforming tube outlet) inside the reformer due to maximum temperature limitations of the material from which the reforming tubes are constructed. A number of references in the SMR literature have investigated furnace balancing methods (i.e., developing a fuel distribution for the burners that reduces the nonuniformity in the OTWT distribution at a given distance from the reforming tube outlet) in an attempt to improve the plant efficiency [2,3]. To determine an appropriate fuel distribution, the relationship between the mass flow rate of fuel through the burners and the outer tube wall temperatures at a specific distance from the reforming tube outlet must be modeled. First-principles models [4] and computational fluid dynamics (CFD) models [3] of the reformer may require significant computation time if solved repeatedly as part of an optimization problem that seeks to reduce nonuniformity in the OTWT distribution, and therefore would not be suitable for designing a real-time furnace balancing scheme that allows the fuel distribution to be adjusted to account for disturbances. Data-driven modeling is an appealing alternative as data-driven models are computationally inexpensive and can have reasonable accuracy. However, standard data-driven modeling procedures (i.e., subset selection) which could be used to derive a single best model relating the OTWT's and the mass flow rates through the burners may not be sufficient because they do not account for the fact that the single model selected is not necessarily the actual model [5]. If the model uncertainty is not accounted for when the model is used for predicting appropriate fuel distributions within the furnace balancing algorithm, there is a potential that the algorithm may select an inappropriate high-risk action (e.g., increasing a mass flow rate through a burner when it should not be increased due to temperature limitations of the reforming tubes).

Motivated by this, the present work focuses on developing a two-step data-driven modeling procedure that utilizes the Bayesian model averaging framework with non-informative prior density distributions on models and model-specific parameters, conditional autoregressive (CAR) modeling, information on the reformer geometry and knowledge of thermal radiative heat transfer to derive a hybrid data-driven model (i.e., a model based both on the geometry/physics of the reformer and also on the reformer operating data). Then, the performance on out-of-sample predictions of the model is demonstrated to be superior compared to those of the other models derived from standard subset modeling approaches (i.e., Lasso and Nonnegative Garrote). The proposed data-driven model also has the potential to be used for monitoring purposes (e.g., when a unit of an infrared camera system used for monitoring the outer reforming tube wall temperatures goes offline temporarily, the data-driven model can estimate the missing data based on the fuel

distribution).

[1] Ewan BCR, Allen RWK. A figure of merit assessment of the routes to hydrogen. International Journal of Hydrogen Energy. 2005;30:809–819.

[2] Kumar A, Baldea M, Edgar TF, Ezekoye OA. Smart manufacturing approach for efficient operation of industrial steam-methane reformers. Industrial & Engineering Chemistry Research. 2015; 54:4360–4370.

[3] Tran A, Aguirre A, Crose M, Durand H, Christofides PD. Temperature balancing in steam methane reforming furnace via an integrated CFD/data-based optimization approach. Computers & Chemical Engineering. in press.

[4] Latham D. Master's Thesis: Mathematical Modeling of an Industrial Steam Methane Reformer. Queen's University, 2008.

[5] Raftery AE, Madigan D, Hoeting JA. Bayesian model averaging for linear regression models. Journal of the American Statistical Association. 1997;92:179–191.

497985 Adaptive ARX Models for Non-Linear Chemical Processes: An Industrial Application

Abstract:

The ever-increasing capability of data collection and storage in chemical industry has provided us access to an immense amount of historical data. With open-loop or close-loop identification algorithms[1,2], dynamic predictive models are being developed to address both process and business related questions. However, a larger dataset does not always lead to a higher model accuracy. This is partly due to the fact that as the chemical processes usually evolve over time, the data collected over past years may not accurately reflect the current situation[3]. In addition to this, the intrinsic nonlinear behaviors of many chemical processes make the data analytics even more challenging.

In this paper, we discuss the utilization of a recursive identification algorithm[4] to analyze the historical data collected from time-evolving nonlinear industrial processes at Air Products. The identified dynamic models with properly selected forgetting factors were developed to be used for monitoring, optimizing and controlling Air Productsâ€™ assets[5].Â To verify the advantages of the proposed method, we compare the prediction accuracy of three models: an off-line ARX model, an adaptive ARX model and a set of two ARX models estimated at two main operating conditions. The set of two ARX models were found to be capable of accounting for the nonlinearity of the process, but not the time-evolving dynamics. Among the three models, the adaptive ARX model was found to achieve the smallest prediction error, which verifies that the proposed method is an effective option for modeling time-evolving nonlinear *real* industrial processes using manufacturing data.

Reference

1. Esmaili A, MacGregor JF, Taylor PA. Direct and two-step methods for closed-loop identification: a comparison of asymptotic and finite data set performance. *Journal of Process Control.* 2000;10(6):525-537.

2. Ljung L. *System identification.* Wiley Online Library; 1999.

3. Reis MS, Braatz RD, Chiang LH. Big Data: Challenges and Future Research Directions. *Chemical Engineering Progress.* 2016;112(3):46-50.

4. Ljung L, Söderström T. *Theory and practice of recursive identification.* Vol 5: JSTOR; 1983.

5. Kumar D, Chen Y, Esmaili A. Inclusion of Long-term Production Planning/Scheduling into Real-time Optimization. *Ifac Papersonline.* 2015;48(8):229-233.

496066 Performance Evaluation of Anomaly Diagnosis System Based on Adaptive Resonance Theory

Abstract:

Early detection of anomalies is crucial for maintaining high productivity at industrial plants. Clustering algorithms are promising methods for detecting anomalies. In particular, adaptive resonance theory (ART) can be used to make a powerful clustering algorithm [1], and anomaly detection systems based on ART have been installed in an industrial waste liquid treatment plant [2], thermal power plants [3], and distillation tower [4].

The anomaly detection systems of these studies were developed using a single model and a single algorithm. When applying ART to anomaly detection and identification problems, various models and algorithms can be used for detection and identification, and the model and algorithm selections affect the detection and identification performance. These effects, however, have not been evaluated yet.

In this study, we examined the performance of anomaly detection and identification of diagnosis systems based on ART by using the Tennessee Eastman process data. For anomaly detection, the effect of model division and multiple use of the ART model on the anomaly detection performance was evaluated. We examined a case in which an ART model was applied to the whole plant and a case in which an ART model was applied to each subsystem of the plant. We also evaluated how the performance of the system changes with the selection of vigilance parameters of the ART model. The performance of anomaly identification was examined by comparing the kinds of anomalies and the categories created by the ART models. The results of a case study indicate that model division can improve the anomaly detection performance and multiple use of ART models can reduce the false alarm rate. The results also show that anomaly diagnosis systems based on ART are applicable to anomaly identification.

References

1. Carpenter, G., and Grossberg, S., ART2: Self-Organization of stable category recognition codes for analog input patterns, Applied Optics, **26**(23), 4919-4930 (1987)

2. Yamashita, Y., A Clustered Class Distribution Approach for Process Monitoring and Fault Detection, Chemical Engineering Communications, **191**, 302-313 (2004)

3. Sekiai, T., Kusumi, N., Hori, Y., Shimizu, S., and Fukai, M., Auto tuning algorithm for vigilance parameter in the adaptive resonance theory model and its application to fault diagnosis system of thermal power plants, Proc. of the ASME 2011 Power Conference (POWER 2011) and the International Conference on Power Engineering 2011 (ICOPE-11), July 12â€"14, Denver, Colorado, USA, 227â€"234 (2011)

4. Hori, Y., Yamamoto, H., Suzuki, T., Okitsu, J., Nakamura, T., Maeda, T., Matsuo, T., Zabiri, H., Tufa, L., and Marappagounder, R., Online anomaly detection of distillation tower system using adaptive resonance theory,

Proceedings of 7th International Symposium on Design, Operation and Control of Chemical Processes (PSE Asia 2016), Tokyo (2016)

499674 Maintenance Optimization-Based Survival Analysis for Optimal and Safer Operation: Cooling Tower Case Study

Abstract:

In a process plant system, safe and reliable operations are highly sensitive to utilities such as power or steam, cooling water or nitrogen or instrument air. Most of these utilities play an important role as safety barriers. This is because a disturbance in their supply is likely to impact process operations downstream, reduce the production efficiency, lead to a sudden shut down or an unsafe condition. The main focus of this paper is to propose and evaluate a model for survival of a process system under an upset condition/s that relies on data-driven and model-based optimization approach utilizing resilience metrics[1]. These metrics integrate both technical (process parameters variations) and social (policy/regulations, human and organizational) factors [2,3].

In this work, we present a framework for survival evaluation using cooling tower operations as a case study. Poor or degraded performance of cooling towers is typically attributed to factors such as failure of gearbox, bearings, motor, driveshaft; blockage and fouling of piping; corrosion; or uneven water distribution *etc*[4]. Therefore, maintenance optimization of cooling towers with consideration of barriers robustness and any known/unknown uncertainties could improve the efficiencies of the process and overall safety and economics of the cooling tower. A methodology to select the best maintenance strategy for optimal and safer operation of the process system considering all three system components (plant/equipment, procedures and people) is developed. Three different types of maintenance alternatives considered as part of this analysis are: condition-based, predictive, and preventive [5,6] based on costâ€"benefit analysis which includes, the downtime cost, production loss, maintenance cost and safety costs. We describe a method to maximize the performance of safety barriers that will be utilized in case of a process-upset condition to prevent the escalation of the event into a catastrophic incident.

References

1. Jain, P., Pasman, H. J., Waldram, S. P., Rogers, W. J., & Mannan, M. S. (2016). Did we learn about risk control since Seveso? Yes, we surely did, but is it enough? An historical brief and problem analysis. Journal of Loss Prevention in the Process Industries.
2. Dinh, L. T., Pasman, H., Gao, X., & Mannan, M. S. (2012). Resilience engineering of industrial processes: principles and contributing factors. Journal of Loss Prevention in the Process Industries, 25(2), 233-241.
3. Jackson, S. (2009). Architecting resilient systems: Accident avoidance and survival and recovery from disruptions (Vol. 66). John Wiley & Sons.
4. Raghuvanshi, N. S., & Singh, A. (2014). Development of Maintenance Strategy to Improve Performance of Natural Draft Cooling Tower. International Journal of Scientific And Research Publications, 4(8), 1-7.
5. Vassiliadis, C. G., & Pistikopoulos, E. N. (2001). Maintenance scheduling and process optimization under uncertainty. Computers & Chemical Engineering, 25(2), 217-236.
6. Pistikopoulos, E. N., Vassiliadis, C. G., & Papageorgiou, L. G. (2000). Process design for maintainability: an optimization approach. Computers & Chemical Engineering, 24(2-7), 203-208.

494996 Scheduling and Feed Quality Optimization for Raw Materials in the Metals Industry

Abstract:

Scheduling and feed quality optimization for raw materials in the metals industry

Yingkai Song*, **Brenno C. Menezes***, **Pablo Garcia-Herreros§**, **Ignacio E. Grossmann***

*Department of Chemical Engineering. Carnegie Mellon University. Pittsburgh, PA 15213, USA.

§Research, Development & Innovation. Aurubis AG. Hamburg, Germany.

Abstract

We address the scheduling and feed quality optimization for processing of raw materials in non-ferrous metal industries. The model considers scheduled arrival of ships carrying solid concentrates, internal logistics, blending operations, and concentrate smelting. The raw materials are initially received at the port, where they are stored at warehouses and a first mixing step takes place. The mixed materials are sent to the processing facility through a river transportation system. Once the mixtures are received at the production facility, several conveyor belts deliver them to the smelter together with other non-concentrate materials. The proposed problem is similar to the classic crude oil scheduling optimization [1], although there are additional challenges to solve the logistics and quality problems that we describe below.

We develop a multi-period Mixed-Integer Non-Linear Programming (MINLP) model that includes several special features. First, special mixing rules are modeled for small ships and conveyor belts, which introduce non-convex bilinear terms. Second, variations in the concentration of the mixture are undesirable in the smelting process for operational stability. For this reason, penalties are considered for variations in the concentration; they provide a balance between profit and operational stability. In addition to upper bounds for concentrations in the final feed blend, which are common in crude oil scheduling problem, interdependency constraints based on individual element concentrations are considered in our model.

The large scale non-convex MINLP is prohibitively expensive to be solved by global optimization solvers. Furthermore, the non-convexities can make a local non-linear solver fail in finding feasible solutions or lead to suboptimal solution. Thus, we implement a decomposition scheme for the MINLP formulation representing the quantity-logic-quality phenomena [2,3]. We propose a two-stage solution approach that involves solving a Mixed-Integer Linear Problem (MILP) that relaxes the non-convex constraints, and a Non-Linear Problem (NLP) that fixes discrete decisions. Several solution strategies, such as rolling horizon, convex relaxations [4], and heuristics for finding feasible solutions have been implemented to avoid infeasibilities and to reduce the gap between the MILP and NLP problems. The examples show that the proposed solution strategy yields near optimal solutions with reasonable computational effort.

[1] Lee, H., Pinto, J., Grossmann, I., & Park, S. (1996). Mixed-Integer Linear Programming Model for Refinery Short-Term Scheduling of Crude Oil Unloading with Inventory Management. *Industrial & Engineering Chemistry Research*, *35*(5), 1630-1641.

[2] B.C. Menezes, J.D. Kelly, I.E. Grossmann, 2015, Phenomenological Decomposition Heuristic for Process Design Synthesis in the Oil-Refining Industry. *Comput Aided Chem Eng*, 37, 1877-1882.

[3] J.D. Kelly, B.C. Menezes, F. Engineer, I.E. Grossmann, 2017, Crude-oil Blend Scheduling Optimization of an Industrial-Sized Refinery: a Discrete-Time Benchmark. *Comput Aided Chem Eng*, In Press.

[4] Andrade, T., Ribas, G., & Oliveira, F. (2015). A Strategy Based on Convex Relaxation for Solving the Oil Refinery Operations Planning Problem. *Industrial & Engineering Chemistry Research*, *55*(1), 144-155

496337 Overcoming a Lack of Data in Decision Analysis

Abstract:

To remain competitive in the faces of challenging market forces, process operations continually improve and adapt. This includes optimizing the scheduling of turnarounds and equipment overhauls [1]. Even with the rise of Big Data in the process industries [2], there are still frequent situations where the required data is limited or in a form that makes the decision analysis a challenge. However, by treating uncertainty as a state of knowledge we can reframe our problem and develop a credible comparison methodology. The reframing allows us to overcome the lack of data and access the power of established tools such as lifetime analysis, physics of failure and simulation. Using these tools we can then quantify the differences and economic impact of different overhaul schedules and policies. This talk will present a case study developed from work on turbomachinery.

[1] Matteo Biondi, Guido Sand, Iiro Harjunkoski, Optimization of multipurpose process plant operations: A multi-time-scale maintenance and production scheduling approach, Computers & Chemical Engineering, Volume 99, 6 April 2017, Pages 325-339, ISSN 0098-1354, http://doi.org/10.1016/j.compchemeng.2017.01.007.

(http://www.sciencedirect.com/science/article/pii/S009813541730008X)

[2] J.F. MacGregor, M.J. Bruwer, I. Miletic, M. Cardin, Z. Liu, Latent Variable Models and Big Data in the Process Industries, IFAC-PapersOnLine, Volume 48, Issue 8, 2015, Pages 520-524, ISSN 2405-8963, http://dx.doi.org/10.1016/j.ifacol.2015.09.020.

(http://www.sciencedirect.com/science/article/pii/S2405896315011015)

492007 Mixed-Integer Nonlinear Programming Models for Line Pressure Optimization in Shale Gas Gathering Systems

Abstract:

Mixed-Integer Nonlinear Programming Models for Line Pressure Optimization in Shale Gas Gathering Systems

Markus G. Drouven[1] & Ignacio E. Grossmann[2]

Department of Chemical Engineering

Carnegie Mellon University

Pittsburgh, PA 15213, USA

[1]mdrouven@cmu.edu, [2]grossmann@cmu.edu

Abstract

Shale gas gathering systems are characterized by a few key elements, namely: existing and prospective well pads, existing and prospective gathering pipelines of varying sizes, one or several compressors and a coupling link to a large-diameter, long-distance transmission pipeline. On every existent pad one or more shale wells actively produce natural gas which is then fed into gathering lines at varying rates over time. The shale gas gathering system itself is typically operated at relatively low pressures, ranging from 50-200 psi. Low line pressures allow producers to extract the most gas from their shale wells, since they create a large differential between the reservoir pressure and the wellhead pressure. In other words, as the line pressure in a gathering system increases, overall production typically decreases (Lee & Wattenberger, 1996; Boyan & Ghalambor, 2014). Since the reverse statement is true as well, upstream producers generally prefer to operate their gathering systems at the lowest possible line pressure.

However, eventually the produced gas needs to be delivered to a transmission pipeline which will move the gas to major demand hubs. These transmission lines are operated by midstream companies at very high pressures between 900-1,200 psi in order to transport large quantities of natural gas over long distances. Therefore, it is up to the shale gas producer to overcome the pressure differential between the low-pressure gathering system and the high-pressure transmission line. This is typically accomplished through one or several compressors. Compressor stations allow upstream operators to produce the gas at low pressures, on the one hand, but still meet the transmission line's pressure delivery requirements on the other hand. Due to the significant pressure differential that needs to be overcome by the compressor, and the considerable volumes of gas that are processed, compression expenses can be a major cost factor in the operation of shale gas gathering systems. Consequently, upstream producers struggle to balance two conflicting objectives: a) operating their gathering systems at low pressures and thereby increasing gas production, and b) raising line pressures so as to minimize compression expenses.

In this work we present a multiperiod mixed-integer nonlinear programming model to address the line pressure optimization problem in shale gas gathering systems. The problem at hand can be stated as follows. Within an active development area, an upstream operator is actively producing natural gas from a set of existing shale wells into an existing gas gathering system. This pipeline system delivers the produced gas to a compressor station which feeds into a long-distance, wide-diameter, high-pressure transmission line. Within the foreseeable future the producer wishes to open up additional prospective wells to maximize the utilization of the available gas gathering capacity.

Our work is concerned with: a) determining the optimal schedule to turn prospective wells in-line, also referred to as the "turn-in-line (TIL) schedule", b) identifying the optimal pressure profile within the gas gathering network, and c) calculating the required compression power to deliver the gas into the interstate transmission network. The problem is complicated by the fact that as new wells are brought online, the production of previously producing wells is negatively affected. In other words, the increase in line pressure due to additional gas production curtails gas recovery from mature wells. This effect is particularly prominent due to the characteristically steep decline curves of new shale gas wells. Hence, the objective of this work is to determine the optimal "TIL" schedule, line pressure profile and compressor operation such that the net present value of the field development project is maximized.

We rely on a pressure-normalized decline curve model to quantify how line pressure variations impact the gas production of individual wells. The reservoir model itself is incorporated in a transmission optimization framework which rigorously evaluates pressure drops along pipeline segments. Moreover, we explicitly consider compression requirements to lift line pressure from gas gathering levels to setpoints dictated by transmission pipeline companies. Since the resulting optimization models are large-scale, nonlinear and nonconvex, we propose a solution procedure based on an efficient initialization strategy. Finally, we present a detailed case study and we show that the proposed optimization framework can be used effectively to manage line pressures in shale gas gathering systems by properly scheduling when, and how many, new wells are brought online.

References [1] Lee J, Wattenberger RA. Gas Reservoir Engineering. *Society of Petroleum Engineers*. **1996**. [2] Boyan G, Ghalambor A. Natural Gas Engineering Handbook. *Elsevier*. **2014**. [3] Drouven MG, Grossmann IE. Mixed-Integer Nonlinear Programming Models for Line Pressure Optimization in Shale Gas Gathering Systems. *Journal of Petroleum Science and Engineering*. Submitted for publication. **2017**.

494213 Optimal Startup of Hydrogen Plant Using Data-Driven Model

Abstract:

Hydrogen is one of the most extensively used chemical components, with wide range of applications in petroleum and chemical industries. An economic way to commercially produce hydrogen is to utilize steam methane reforming [1]. The process forms an intricate network of material and energy due presence of vast number of material and heat flows. Natural gas (NG) and superheated steam are fed to a reformer, consisting of catalyst tubes filled with nickel reforming catalyst. Majority of the hydrogen is produced in this reactor. The gas out of the reformer is then processed through a shift reactor to further produce hydrogen. Finally, the hydrogen flow is purified in a pressure swing adsorber (PSA), where high purity hydrogen is produced. This process presents various constraints and challenges during the start-up as well as nominal operation of the plant such as, the reformer exit temperature should be maintained at a desirable level. Further, the firebox pressure should not breach its lower and upper limits for safety. For instance, if the pressure is too low, the fire can be extinguished. If it is too high, it may impose safety hazards to facility and personnel [2].

A typical startup procedure is to implement a predefined recipe consisting of a series of events that take place over a period of time. Startup is typically categorized as cold startup and warm startup, based on the initial state of the plant from which the plant is required to be driven to nominal operation. The cold startup will be central to the discussion in this work, where the plant is started from completely shutdown conditions. A typical startup process involves ramping the natural gas fuel to reach a sufficient reformer exit temperature, circulation of natural gas feed and steam, regulation of nitrogen flow in the reactor tubes. It also includes making discrete decisions such as starting up the PSA, recirculating the tail gas from the PSA and ramping up to full capacity in several stages. Thus, the startup time is not fixed and varies based on the decisions taken at various stages. Existing startup recipes have been built heuristically from past successful startups, with the possibility of optimizing the startup procedure remaining unexplored.

One key recognition that we make is that due to the nature of startup/shutdown phase, they can be treated as a batch like operation with variable lengths, where the objective is to drive the process from one state to the other. For modeling and control of batch processes, a variety of a data-driven modeling techniques exist. One of the most widely used approach is partial least squares (PLS), which models the process in a projected latent space [3]. These models are essentially time-varying linear models, linearized around mean past trajectories, and therefore require the batches to be of same length, or to recognize an appropriate alignment variable. To account for these limitations, a multi-model approach was proposed in [4]. These models were based on the 'current measurements' of the process instead of the 'time'. These developments were followed by contributions in the area of integration of these data-driven models with the advanced control formulations [4-5]. More recently a subspace identification based batch control approach was proposed in [6] where a LTI state-space model of the batch process is estimated, and does not require the training batches to be of equal length. The batch subspace identification approach was recently utilized for data-driven modeling of the hydrogen plant [8]. The results demonstrated the capabilities of the approach in capturing the dynamics of a highly nonlinear plant through a LTI model.

Note that the development of a good model that can handle variable batch times paves the way for application of optimization algorithms that can improve not just the process economics along the startup/shutdown but also the operating time of the startup/shutdown procedure. Motivated by these considerations, in this work we address the problem of optimizing the startup procedure for the hydrogen production plant. From an optimization standpoint, the problem is quite different compared to existing application of data driven models for control of lumped parameter (such as in [6]) or distributed parameter process [7], where the objective was to reach a certain quality requirement in a given, predetermined batch length. The optimization problem for the present application is an optimal control problem where the objective is to reach the desired terminal state in minimum time, while adhering to various constraints, and also making discrete decisions along the way. To this end, first a high fidelity test bed model of the entire plant is developed in UniSim, capable of simulating the startup and shutdown phase, with appropriate adaptation of the plant SOP. Several simulated startups are performed to generate the training data. Then, an LTI data-driven model of the process using batch subspace identification [6-7] is identified. The identified data-driven model is subsequently utilized within an optimization framework to synthesize the startup SOP.

[1] JL Lynn and BH Bory. Kirk-othmer encyclopedia of chemical technology, 2000.

[2] Miao Du, Prashant Mhaskar, Yu Zhu, and Jesus Flores-Cerrillo (2014). Safe-Parking of a Hydrogen Production Unit. Industrial & Engineering Chemistry Research, 53 (19), 8147-8154.

[3] J. Flores-Cerrillo, J. F. MacGregor, Control of particle size distributions in emulsion semibatch polymerization using mid-course correction policies, Industrial and Engineering Chemistry Research 41 (7) (2002) 1805-1814.

[4] S. Aumi, B. Corbett, P. Mhaskar, and T. Clarke-Pringle. Data-based modeling and control of nylon-6, 6 batch polymerization. IEEE Transactions on Control Systems Technology, 21(1):94-106, Jan 2013.

[5] S. Aumi, P. Mhaskar, Integrating Data-Based Modeling and Nonlinear Control Tools For Batch Process Control, AIChE Journal 58 (2012) 2105-2119.

[6] B. Corbett, P. Mhaskar, Subspace identification for data-driven modelling and quality control of batch processes, AIChE Journal 62 (2016) 1581-1601.

[7] Abhinav Garg, P. Mhaskar, Subspace Identification Based Modeling and Control of Batch Particulate Processes, Industrial and Engineering Chemistry Research (2017), Submitted.

[8] Abhinav Garg, Brandon Corbett, Prashant Mhaskar, Gangshi Hu, Jesus Flores-Cerrillo, High Fidelity Model Development and Subspace Identification of a Hydrogen Plant Startup Dynamics, Computers and Chemical Engineering (2017), Submitted.

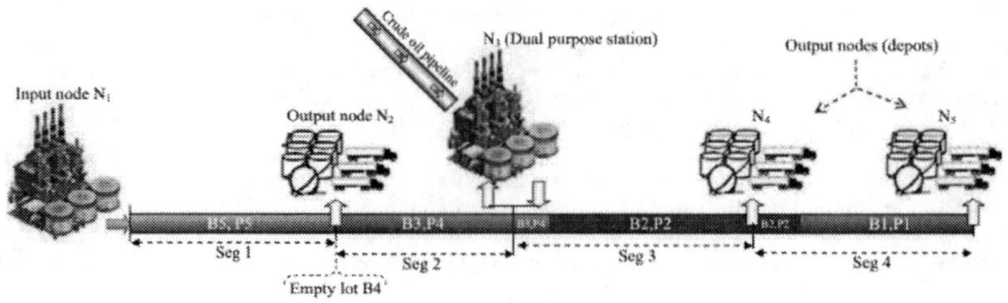

Product-Centric Continuous-Time Formulation for Straight Pipelines

Pedro M. Castro
(pmcastro@fc.ul.pt)

IF INVESTIGADOR FCT

Ciências ULisboa

U LISBOA UNIVERSIDADE DE LISBOA

Hossein Mostafaei

Azarbaijan Shahid Madani University

CMAF CIO
CENTRO DE MATEMÁTICA, APLICAÇÕES FUNDAMENTAIS E INVESTIGAÇÃO OPERACIONAL

Introduction

- Pipelines: most reliable and cost-effective way of transporting large volumes over long distances
- Petroleum supply chain
  - Crude oil
  - Refined products
- Pipeline systems
  - Straight pipelines
  - Multilevel treelike
  - Mesh structure
  - Reversible flow possible
- Main goal pipeline operation
  - Meet demand on time
  - Keep tanks inventory levels within limits

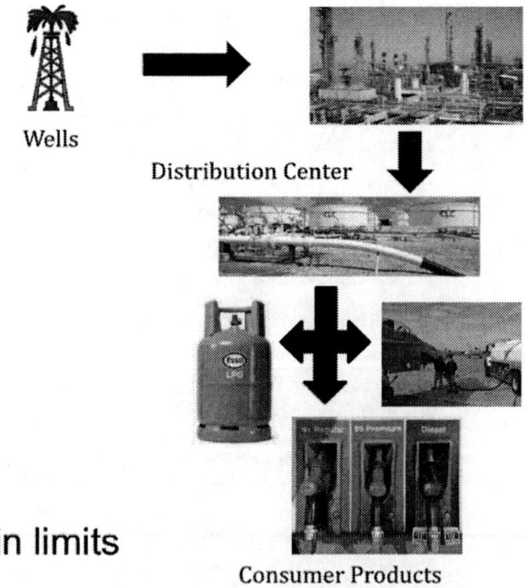

Problem statement

- Straight multiproduct (P) pipeline system
  - Nodes
    - Input (R), output (D) & dual purpose (DP)
    - Initial known volumes $v_{r,p}^{R,0}$, $v_{d,p}^{D,0}$ & $v_{dp,p}^{DP,0}$ (m^3)
  - Pipeline segments (S)
    - Located between consecutive nodes, of volume v_s^S
  - Other problem data
    - Bounds on pumping flowrates $\rho_{r,p}^{R,\min}$, $\rho_s^{S,\min}$ (m^3/h)
    - Product demand $f_{d,p}^{D,end}$ (end of time horizon)
- Find volumes & timing of injections/removals

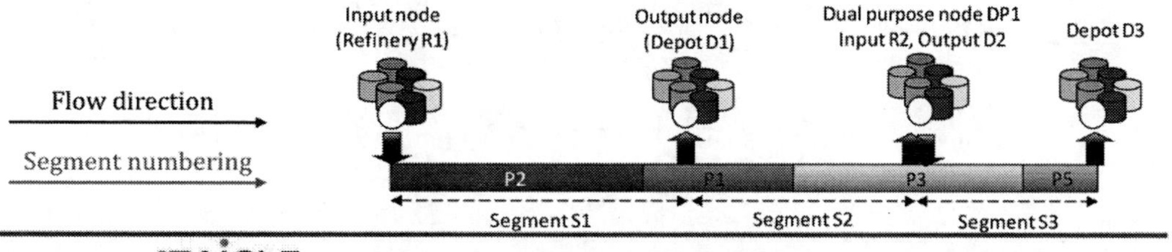

Continuous-time representation

- Single reference grid
  - Event points $t \in T$
  - Location T_t & slot duration L_t variables

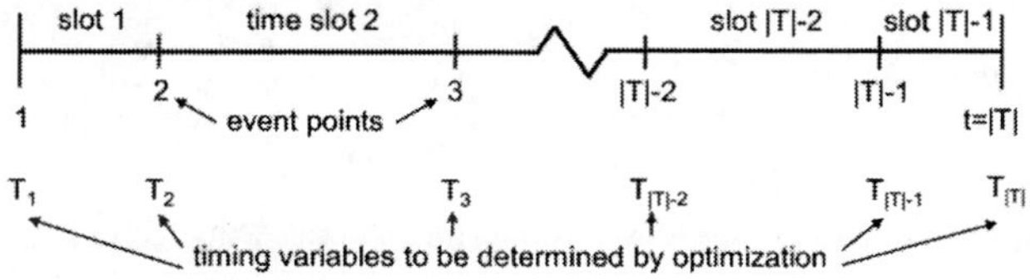

- Timing constraints
  - $T_{t+1} = T_t + L_t \; \forall t \neq |T|$
  - $T_{|T|} \leq h$ (time horizon)
  - Makespan minimization: $\min T_{|T|}$
- Important notes
  - $|T|$ affects solution quality (tuning parameter)
  - Need for iterative procedure to find global optimal solution
    - Single increments in $|T|$ until objective function stops improving

GDP model with exclusive disjunctions

- Refinery r pumping product p during slot t
  - $X^R_{r,p,t} = 1 \Rightarrow$ pumped volume & flowrate subject to bounds
  - $X^R_{r,p,t} = 0 \Rightarrow$ constraints relaxed

$$\bigvee_p \begin{bmatrix} X^R_{r,p,t} \\ f_r^{R,\min} \le F^R_{r,p,t} \le f_r^{R,\max} \\ \frac{F^R_{r,p,t}}{\rho_r^{R,\max}} \le L_t \le \frac{F^R_{r,p,t}}{\rho_r^{R,\min}} \end{bmatrix} \bigvee \begin{bmatrix} X^{R,no}_{r,t} \\ F^R_{r,p,t} = 0 \quad \forall p \\ L_t \le h \end{bmatrix} \quad \forall r, t \ne |T|$$

- Convex hull reformulation
  - $\sum_p X^R_{r,p,t} \le 1 \; \forall r, t \ne |T|$
  - $f_r^{R,\min} X^R_{r,p,t} \le F^R_{r,p,t} \le f_r^{R,\max} X^R_{r,p,t} \; \forall r, p, t \ne |T|$
  - $\frac{\sum_p F^R_{r,p,t}}{\rho_r^{R,\max}} \le L_t \le \frac{\sum_p F^R_{r,p,t}}{\rho_r^{R,\min}} + h \cdot \left(1 - \sum_p X^R_{r,p,t}\right) \forall r, t \ne |T|$

Segment-dependent coordinates

- Pick 2 out of 3 parameters for initial status
  - Volumes inside $v^{S,0}_{s,p}$
  - Left coordinate $lc^0_{s,p}$
  - Right coordinate $rc^0_{s,p}$
- General constraints
  - $RC_{s,p,t} = LC_{s,p,t} + V^S_{s,p,t}$
  - $\sum_p V^S_{s,p,t} = v^S_s \; \forall s, t$

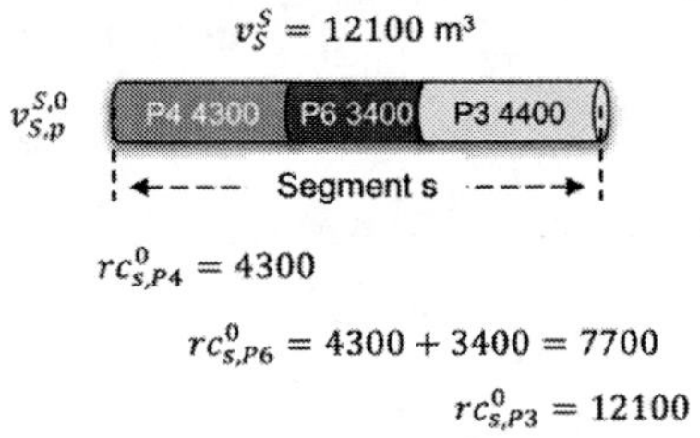

- Difficult to compute coordinates $\Leftarrow$ product reentrance
  - Need to reset coordinates ($ZC_{s,p,t}$) for products not inside ($\neg X^{S,is}_{s,p,t}$)

$$\begin{bmatrix} X^{S,is}_{s,p,t} \\ V^S_{s,p,t} \le v^S_s \\ ZC_{s,p,t} = 0 \end{bmatrix} \bigvee \begin{bmatrix} \neg X^{S,is}_{s,p,t} \\ V^S_{s,p,t} = 0 \\ ZC_{s,p,t} \le f_s^{S,\max} \end{bmatrix} \quad \forall s, p, t > 1$$

  - Not an issue in batch-centric models (Mostafaei & Castro '17)

How to allow product re-entry?

- System status just before t from variables $LC_{s,p,t-1}^{end}$, $RC_{s,p,t-1}^{end}$
  - $LC_{s,p,t}^{end} = LC_{s,p,t} + \sum_{p' \neq p} F_{s,p',t}^{S,in} \; \forall s,p,t \neq |T|$
  - $RC_{s,p,t}^{end} = RC_{s,p,t} + \sum_{p' \neq p} F_{s,p',t}^{S,out} \; \forall s,p,t \neq |T|$

  Coordinates increase from other products p' entering/leaving segment s regardless of p being inside
- Connection between just before t and t allows to reset coordinates
  - $LC_{s,p,t} = lc_{s,p}^{0}\big|_{t=1} + (LC_{s,p,t-1}^{end} - ZC_{s,p,t})\big|_{t>1} \; \forall s,p,t$
  - $RC_{s,p,t} = rc_{s,p}^{0}\big|_{t=1} + (RC_{s,p,t-1}^{end} - ZC_{s,p,t})\big|_{t>1} \; \forall s,p,t$

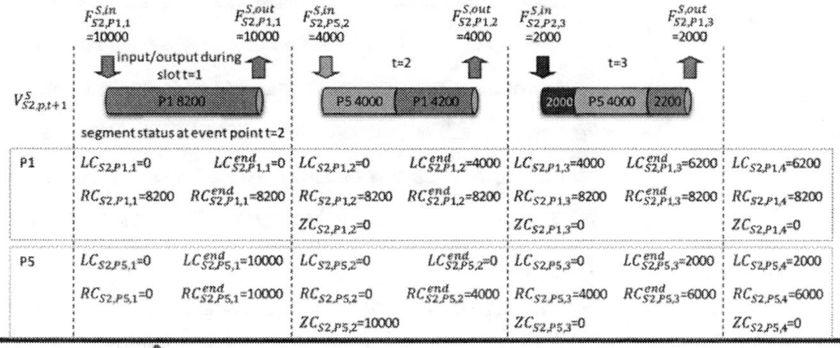

Alternative modeling approaches

- Resource-Task Network (RTN)
  - General formulation (Castro, '10)
  - Parameters from superstructure
  - Larger problem size (factor 3-5)

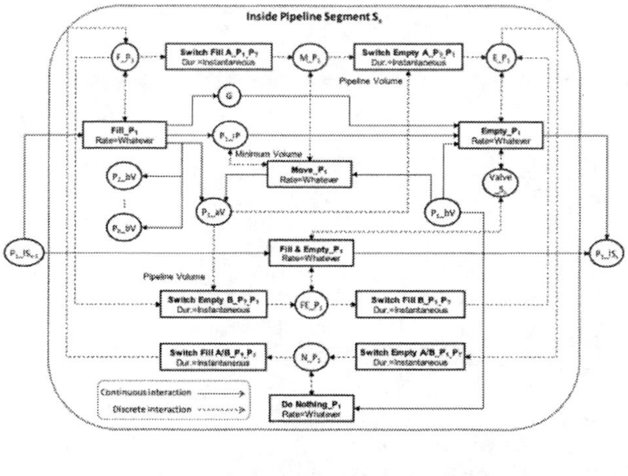

- Generalized Disjunctive Programming (GDP)
  - Disjunctions & logic propositions help derive complex constraints

$$\bigvee_{p} \begin{bmatrix} X_{s,p,t}^{S,in} \\ LC_{s,p,t} = 0 \\ f_s^{S,\min} \leq F_{s,p,t}^{S,in} \leq f_s^{S,\max} \\ \dfrac{F_{s,p,t}^{S,in}}{\rho_s^{S,\max}} \leq L_t \leq \dfrac{F_{s,p,t}^{S,in}}{\rho_s^{S,\min}} \end{bmatrix} \bigvee \begin{bmatrix} X_{s,t}^{S,no\ i} \\ LC_{s,p,t} \leq v_s^S \ \forall p \\ F_{s,p,t}^{S,in} = 0 \ \forall p \\ L_t \leq h \end{bmatrix} \forall s,t$$

$$\bigvee_{p} \begin{bmatrix} X_{s,p,t}^{S,out} \\ RC_{s,p,t} = v_s^S \\ f_s^{S,\min} \leq F_{s,p,t}^{S,out} \leq f_s^{S,\max} \end{bmatrix} \bigvee \begin{bmatrix} X_{s,t}^{S,no\ o} \\ RC_{s,p,t} \geq 0 \ \forall p \\ F_{s,p,t}^{S,out} = 0 \ \forall p \end{bmatrix} \forall s,t$$

  - Better computational performance

CPUs	RTN	GDP		RTN	GDP
Ex1	18000	7884	Ex5	16.8	1.85
Ex2	16320	111	Ex6	2.66	0.51
Ex4	110	6.22	Ex7	137	4.14

Forbidding product sequences $p \rightarrow p'$

- Logic proposition
  - $X^{S,in}_{s,p,t} \wedge_{t<t''<t'} X^{S,no\ i}_{s,t''} \Rightarrow \neg X^{S,in}_{s,p',t'}\ \forall s,p,p',t<t'<|T|$
    - Accounts for idle slots between products entry
- MILP constraint
  - $1+\sum_{p''}\sum_{t<t''<t'} X^{S,in}_{s,p'',t''} \geq X^{S,in}_{s,p,t} + X^{S,in}_{s,p',t'}\ \forall s,p,p',t<t'<|T|$

Forbidden sequences (within a segment):
P2-P3
P3-P2
P3-P4
P4-P3

Makespan=46.00 h
(44.2 h without)

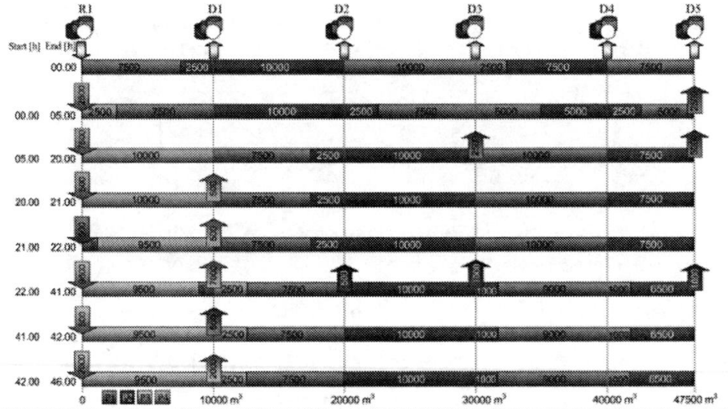

Computational results

- Previous batch-centric model (Mostafaei & Castro '17)
  - Capable of generating better solution for given $|T|$
    - Optimal solution with fewer events (Ex4, Ex6)
      - Multiple batches can enter/leave a segment per slot vs. one product
  - Unless initial decisions are poor (batch sequence, $|I|$)
- New product-centric model is better for Ex2
  - Infeasible for Ex3 ⇐ Single product batch inside segment
- Very tight formulations

	Batch centric			Product centric			Change		LP relaxation
Example	$\|T\|$	Makespan	CPUs	$\|T\|$	Makespan	CPUs	$\|T\|$	Makespan	Integrality gap
Ex0	4	**35.25**	0.16	4	**35.25**	0.10	0	0%	0%
Ex1	13	**197.44**	675	13	202.25	*18000*	0	+2.4%	6.4%
Ex2	10	134.72	4823	10	**125.00**	578	0	-7.2%	9.4%
Ex3	12	231.97	1786	11	Infeasible	0.09	-	-	0%
Ex4	7	**145.00**	1.22	8	**145.00**	6.22	+1	0	0%
Ex5	8	**158.33**	68.9	8	**158.33**	1.85	0	0	0%
Ex6	6	**114.29**	2.41	7	**114.29**	0.51	+1	0	0%

GAMS 24.6.1, CPLEX 12.6.3, parallel deterministic mode, up to 8 threads;
Intel i7-6400K (4.0 GHz), 16 GB RAM, SSD, Windows 10;
Relative optimality tolerance = 10^{-6}, maximum computational time = 18000 CPUs.

Best found solution for Ex2 ($|T| = 11$)

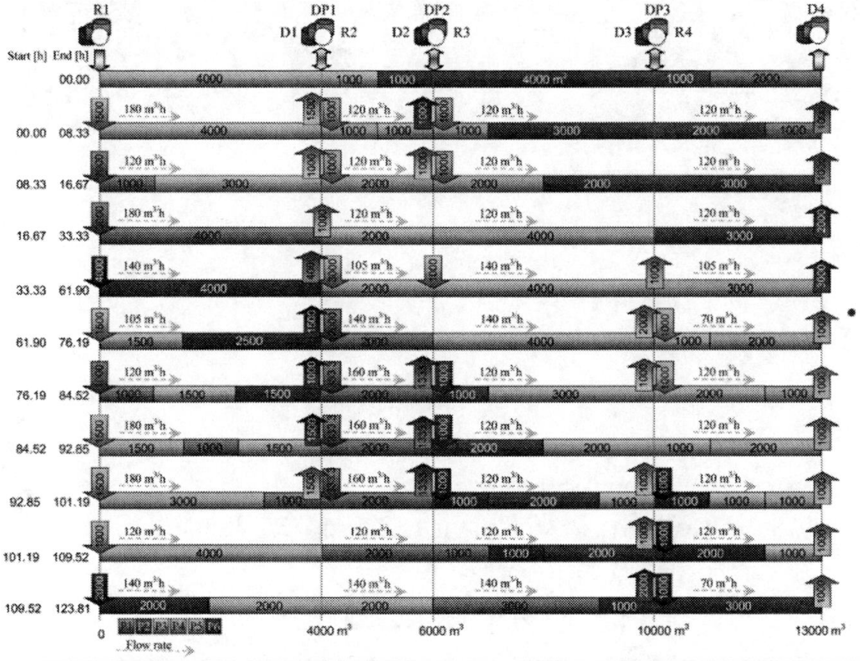

- Segments always operating
  - Varying flowrates
    - In time
    - Between segments
  - Makespan=123.81 h
    - 29511 CPUs
    - Would take $|I| = 16$ (batch-centric model)
      - Intractable problem

Conclusions

- New continuous-time formulation for straight pipelines
- MILP derived from Generalized Disjunctive Programming
  - Modular approach with nodes and segment elements
    - Single & dual purpose nodes
  - Convex hull reformulation $\Longrightarrow$ very tight formulation
- No overall best formulation
  - Batch-centric model
    - More tuning parameters, fewer events to generate schedule
  - New product-centric model
    - Can rigorously enforce forbidden product sequences
    - Easier to extend to other configurations
- See 733a (Planning & Scheduling I) to learn how to handle:
  - Tree-like & mesh structures, and systems with reversible flow
- Further details: CACE 104 (2017) 283-295

An optimization-based framework to define the probabilistic design space of pharmaceutical processes with model uncertainty

Richard(Shu) Xu
Advisor: Dr. Carl Laird
School of Chemical Engineering, Purdue University
Industrial advisor: Shankar Vaidyaraman, Salvador García-Muñoz
Lilly Research Laboratories, Eli Lilly & Company

Sandia National Laboratories is a multimission laboratory managed and operated by National Technology and Engineering Solutions of Sandia, LLC, a wholly owned subsidiary of Honeywell International, Inc., for the U.S. Department of Energy's National Nuclear Security Administration under contract DE-NA0003525.

Laird Research Group: http://allthingsoptimal.com

PURDUE CHEMICAL ENGINEERING

Outline

- Introduction
- Methodology
  - Design space determination
  - Flexibility index
  - Framework for probabilistic design space determination
- Steady state case study
  - Michael addition reaction
- Conclusions and future work
- Acknowledgment

PURDUE CHEMICAL ENGINEERING

Introduction

Quality by design (QbD)

- Quality by testing(QbT): assess product quality at the end of manufacturing stream
  - Disadvantage: time-consuming, economically inefficient, susceptible to process variability
- Quality by design(QbD): ensure product quality at the design stage
  - Advantage: efficient, manufacturing flexibility process robustness

PURDUE CHEMICAL ENGINEERING

Introduction

Design space

- Design space: the multidimensional combination of process parameters θ_p that have been demonstrated to provide assurance of quality (ICH Q8, 2009)

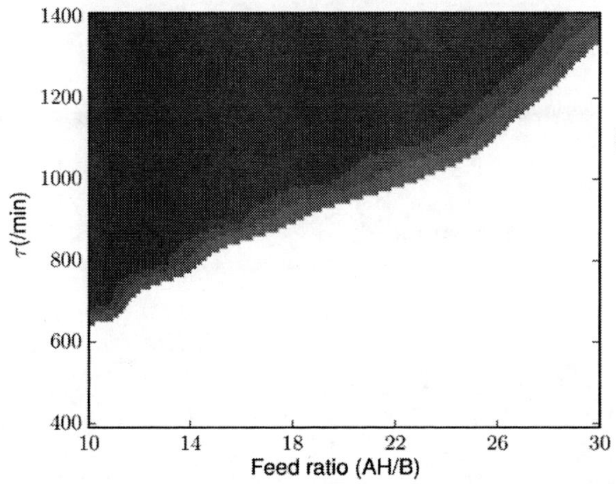

Demonstration of probabilistic design space (blue: $p \geq 0.85$ green: $0.7 \leq p < 0.85$ red: $0.5 \leq p < 0.7$ white: $p < 0.5$

PURDUE CHEMICAL ENGINEERING

Introduction

Design space

Advantages

- Offer operational flexibility for performance improvement
- Promote communication between industries and the regulator

Methodology

Physical model-based method

Monte Carlo method for design space

Assumption: continuous probabilistic design space

Computational cost

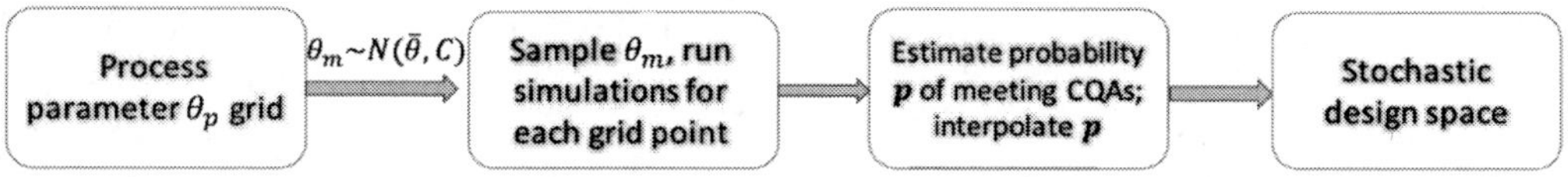

high

Optimization method for design space (1)

Assumption: continuous probabilistic design space; θ_m distribution shape

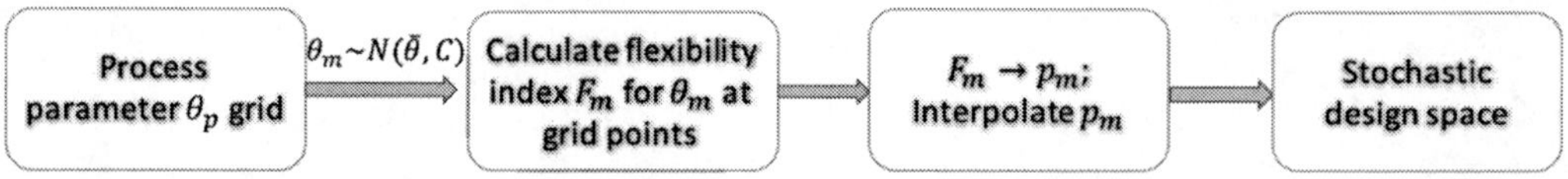

low

Optimization method for design space (2)

Assumption: continuous probabilistic design space; θ_m distribution shape with fixed p_m; feasible θ_p region; hyper-cubic design space

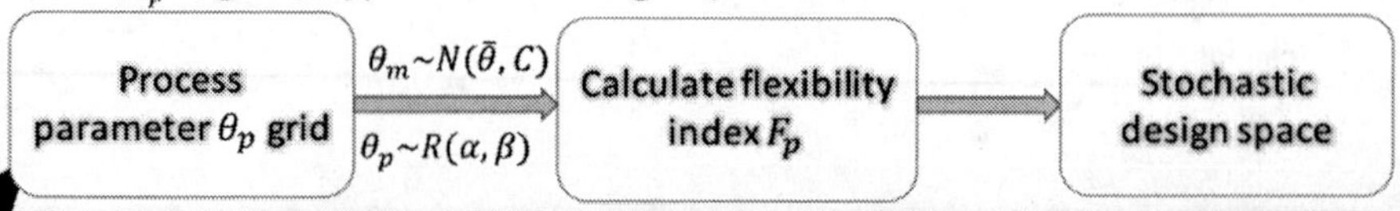

very low

Flexibility index

- **Flexibility index** find the largest hypercube within feasible region (Swaney and Grossmann, 1985a, b)
- Steady-state system
  - For a fixed design **d**, a steady-state system can be described by:

$$
\begin{aligned}
&h_l(\mathbf{d},\mathbf{x},\mathbf{z},\theta)=0 && l\in L \\
&g_k(\mathbf{d},\mathbf{x},\mathbf{z},\theta)\le 0 && k\in K \qquad (1)\\
&\mathbf{z}^L \le \mathbf{z} \le \mathbf{z}^U \\
&\theta^L \le \theta \le \theta^U
\end{aligned}
$$

x: state variables; z: control variables; θ: uncertain parameters

PURDUE CHEMICAL ENGINEERING

Flexibility index— steady state

Given a feasible nominal parameter value θ^N, the flexibility index is defined as the largest scaled value δ of the expected deviations $\Delta\theta^+, \Delta\theta^-$ without violating the feasible constraints:

$$
\begin{aligned}
F(\mathbf{d})= \quad & \max \delta \\
s.t. \quad & \chi(d)=\max_{\theta\in T}\min_{z}\max_{k\in K} g_k(\mathbf{d},\mathbf{x},\mathbf{z},\theta)\le 0 \\
& s.t.\ h_l(\mathbf{d},\mathbf{x},\mathbf{z},\theta)=0 \qquad l\in L \qquad (2)\\
& \quad \mathbf{z}^L \le \mathbf{z} \le \mathbf{z}^U \\
& \quad \theta^N-\delta\Delta\theta^- \le \theta \le \theta^N+\delta\Delta\theta^+ \\
& \quad \delta \ge 0
\end{aligned}
$$

PURDUE CHEMICAL ENGINEERING

Flexibility index — active constraint

$$\varphi(d,\theta^c) = \min_z \max_{k\in K} g_k(d,x,z,\theta) = 0 \xrightarrow[\text{strategy}]{\text{Active constraint}}$$

F is given by the smallest δ to the boundary $\varphi(d,\theta)=0$ (Floudas and Grossmann,1987)

$$\begin{aligned}
F(d) = \quad & \min \delta \\
s.t. \quad & s_k + g_k(\mathbf{d},\mathbf{x},\mathbf{z},\theta) = 0 \qquad k \in K \\
& h_l(\mathbf{d},\mathbf{x},\mathbf{z},\theta) = 0 \qquad l \in L \\
& \sum_k \lambda_k \frac{\partial g_k}{\partial x} + \sum_l \eta_l \frac{\partial h_l}{\partial x} = 0 \\
& \sum_k \lambda_k \frac{\partial g_k}{\partial z} + \sum_l \eta_l \frac{\partial h_l}{\partial z} = 0 \\
& \sum_k \lambda_k = 1 \\
& \sum_k y_k = n_z + 1 \\
& \lambda_k - y_k \le 0 \\
& s_k - U(1-y_k) \le 0 \\
& y_k \in \{0,1\} \\
& \delta, \lambda_k, s_k \ge 0, \\
& \theta^N - \delta\Delta\theta^- \le \theta \le \theta^N + \delta\Delta\theta^+
\end{aligned} \qquad (3)$$

$$\xrightarrow[\text{no control}]{n_z = 0}$$

$$\begin{aligned}
F(d) = \quad & \min \delta \\
s.t. \quad & s_k + g_k(\mathbf{d},\mathbf{x},\theta) = 0 \\
& h_l(\mathbf{d},\mathbf{x},\theta) = 0 \\
& \sum_k y_k = 1 \\
& s_k - U(1-y_k) \le 0 \\
& y_k \in \{0,1\} \\
& \delta, s_k \ge 0, \\
& \theta^N - \delta\Delta\theta^- \le \theta \le \theta^N + \delta\Delta\theta^+
\end{aligned} \qquad (4)$$

PURDUE CHEMICAL ENGINEERING

Constraints on uncertain model parameters

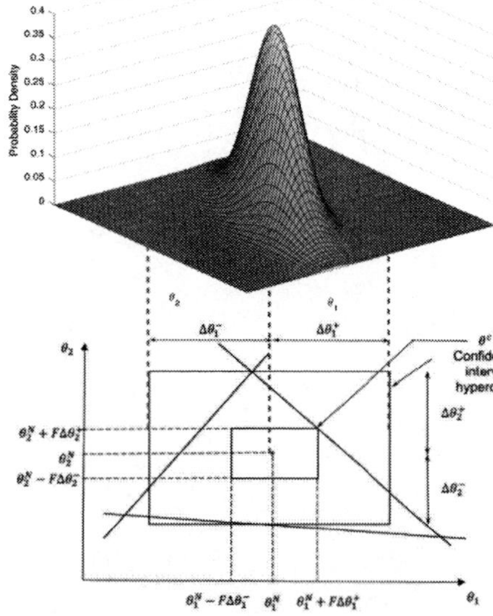

Geometrical representation of hyper-cubic T(F)

$$T(\delta_m) = \{\theta_m \mid \theta_m^N - \delta_m \Delta\theta_m^- \le \theta_m \le \theta_m^N + \delta_m \Delta\theta_m^+\}$$

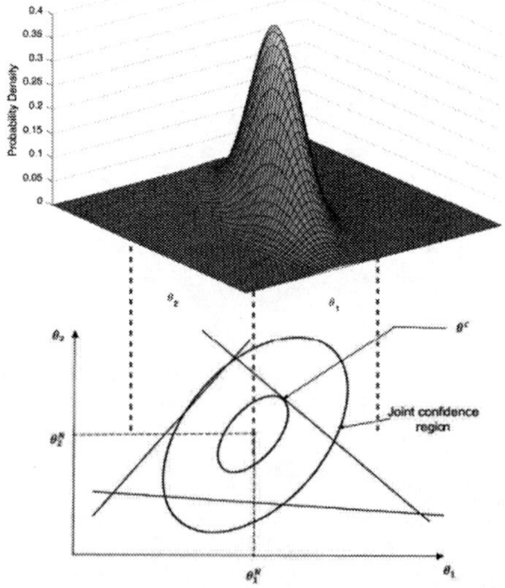

Geometrical representation of elliptical T(F)

$$T(\delta_m) = \left\{\theta_m \mid (\theta_m - \theta_m^N)^T \Sigma_{\theta_m}^{-1} (\theta_m - \theta_m^N) \le \delta_m\right\}$$

Geometrical representation of uncertainty space

PURDUE CHEMICAL ENGINEERING

Methodology — Flexibility index-based method

Optimization method (1)

Assumption: continuous probabilistic design space; θ_m distribution shape

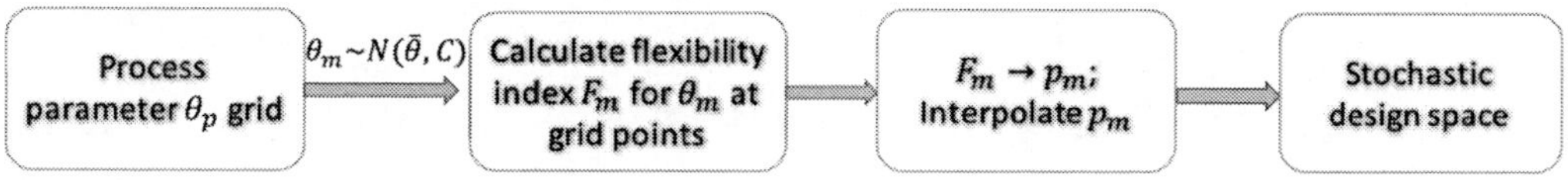

Optimization method (2)

Assumption: continuous probabilistic design space; θ_m distribution shape with fixed p_m; feasible θ_p region; hyper-cubic design space

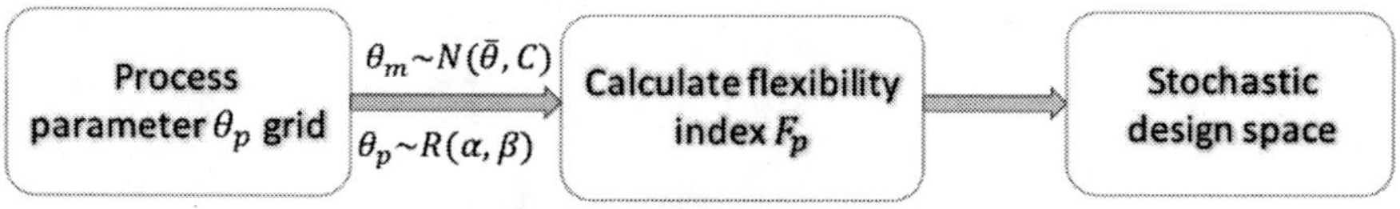

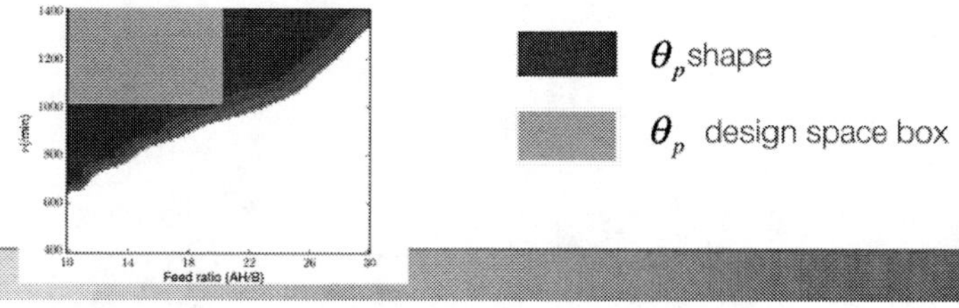

PURDUE CHEMICAL ENGINEERING

Case study

Our new framework is applied to an industrial case.
A series of experiments are performed:

- Monte Carlo simulation
- Flexibility index with a confidence interval hypercube
- Flexibility index with an ellipsoidal confidence region

PURDUE CHEMICAL ENGINEERING

Case study— Michael reaction

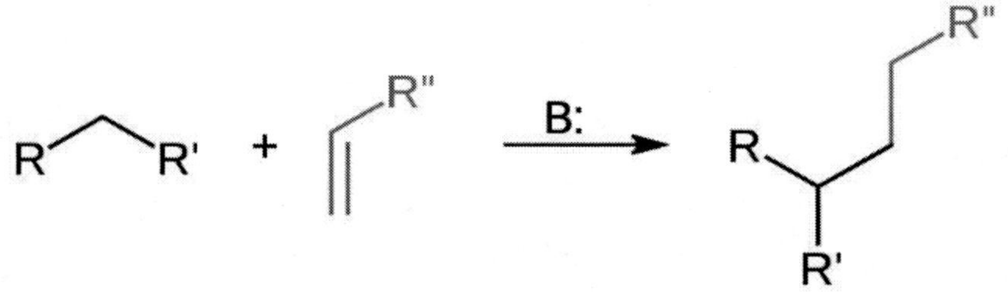

Michael Donor Michael Acceptor

PURDUE CHEMICAL ENGINEERING

Michael reaction

- The reaction kinetics of Michael addition can be described by:

$$
\begin{aligned}
&AH + B \xrightarrow{k_1} A^- + BH^+ \\
&A^- + C \xrightarrow{k_2} AC^- \\
&AC^- \xrightarrow{k_3} A^- + C \\
&AC^- + AH \xrightarrow{k_4} A^- + P \\
&AC^- + BH^+ \xrightarrow{k_5} P + B
\end{aligned}
\qquad (6)
$$

where AH(Michael Donor) and C(Michael Acceptor) are starting materials, B is a base, BH^+, A^- and AC^- are reaction intermediates, and P is the reaction product.

- The reaction rates r_i are defined as follows:

$$
\begin{aligned}
&r_1 = k_1 c_{AH} c_B \\
&r_2 = k_2 c_{A^-} c_C \\
&r_3 = k_3 c_{AC^-} \\
&r_4 = k_4 c_{AC^-} c_{AH} \\
&r_5 = k_5 c_{AC^-} c_{BH^+}
\end{aligned}
\qquad (7)
$$

PURDUE CHEMICAL ENGINEERING

Parameter uncertainty

- The reaction rate constants (k_i) are estimated based on experimental data, with variance-covariance matrix and optimal values shown by

$$Cov_{k_i} = \begin{bmatrix} 1.005 & -3.428\times10^{-4} & -1.006\times10^{-3} & -1.523\times10^{-3} & 2.718\times10^{-3} \\ -3.428\times10^{-4} & 0.412 & 7.951\times10^{-4} & -3.937\times10^{-3} & 2.364\times10^{-3} \\ -1.006\times10^{-3} & -7.951\times10^{-4} & 3.224\times10^{-3} & 1.466\times10^{-3} & -2.400\times10^{-3} \\ -1.523\times10^{-3} & -3.937\times10^{-3} & 1.466\times10^{-3} & 2.746\times10^{-3} & -4.102\times10^{-3} \\ 2.718\times10^{-3} & 2.364\times10^{-3} & -2.400\times10^{-3} & -4.102\times10^{-3} & 7.148\times10^{-3} \end{bmatrix} \quad (8)$$

Table 1 Uncertain reaction rate constants and deviations based on covariance matrix

	k_1	k_2	k_3	k_4	k_5
nominal value	49.7796	8.9316	1.3177	0.3109	3.8781
positive deviation($+3\sigma$)	3.007	1.926	0.170	0.157	0.254
negative deviation(-3σ)	-3.007	-1.926	-0.170	-0.157	-0.254

PURDUE CHEMICAL ENGINEERING

Steady state model

- Based on the mass balance of steady state CSTR:

$$0 = F_i^0 - F_i + V = \left(c_i^0 - c_i\right) + \tau \sum_j \nu_{ij} r_j \quad (9)$$

where F_i^0 and F_i are the inlet and outlet molar flow rates, ν_{ij} is the stoichiometric coefficient, r_j is reaction rate, and τ is the residence time.

We obtain the following equation:

$$\begin{aligned} & c_{AH}^0 - c_{AH} + \tau\left(-r_1 - r_4\right) = 0 \\ & c_B^0 - c_B + \tau\left(-r_1 + r_5\right) = 0 \\ & c_C^0 - c_C + \tau\left(-r_2 + r_3\right) = 0 \\ & c_{A^-}^0 - c_{A^-} + \tau\left(r_1 - r_2 + r_3 + r_4\right) = 0 \\ & c_{AC^-}^0 - c_{AC^-} + \tau\left(r_2 - r_3 - r_4 - r_5\right) = 0 \\ & c_{BH^+}^0 - c_{BH^+} + \tau\left(r_1 - r_5\right) = 0 \\ & c_P^0 - c_P + \tau\left(r_4 + r_5\right) = 0 \end{aligned} \quad (10)$$

where the initial concentrations $\{c_A^0, c_B^0, c_C^0, c_{BH+}^0, c_{A-}^0, c_{AC-}^0, c_P^0\}$ are set to be $\{0.3955, 0.3955/R, 0.25, 0, 0, 0, 0\}$ mol/L.

PURDUE CHEMICAL ENGINEERING

Problem formulation

- In this study, the design space is constructed by the molar ratio R between the feed of AH and B, the residence time τ .
- The individual flexibility ranges are given in Table 2.

Table 2 Discretization of initial guess of flexibility ranges

Parameters	Number of levels	Discrete points
$R_{AH\|B}$	21	[10,11,12,13,14,15,...,30]
τ (min)	11	[400,500,600,700,800,..., 1400]

- The feasibility of process operation is determined by the following constraint:

$$\frac{c_C^0 - c_C - c_{AC^-}}{c_C^0} \geq 0.9 \Rightarrow c_C + c_{AC^-} - 0.1 \times c_C^0 \leq 0 \quad (11)$$

$$c_{AC^-} \leq 0.002$$

which states that the conversion of C must be greater than 0.9, and the concentration of AC^- must be lower than 0.002 mol/L.

PURDUE CHEMICAL ENGINEERING

Monte-Carlo Simulation

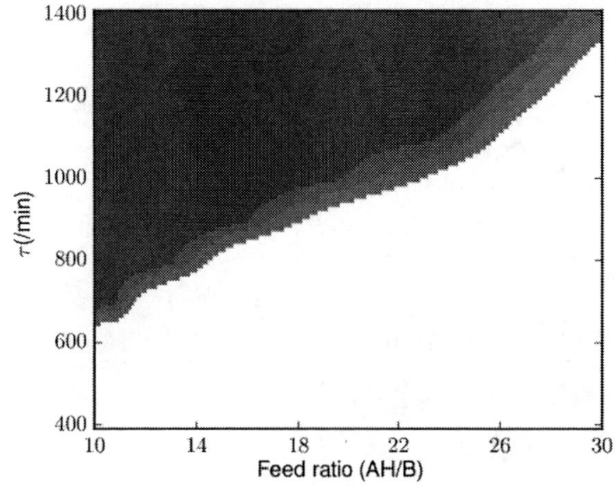

Michael reaction: probabilistic design space by Monte Carlo simulation (blue: $p \geq 0.85$ green: $0.7 \leq p < 0.85$ red: $0.5 \leq p < 0.7$ white: $p < 0.5$

(CPU time: 629.41 seconds, 200 repetitions, Parallel with 4 cores)

PURDUE CHEMICAL ENGINEERING

Flexibility index

- The formulation of the flexibility index problem is given by:

$$
\begin{aligned}
& F(d) = \min \delta \\
s.t. \quad & Eqs.(7),(10) \\
& s_1 + c_C + c_{AC^-} - 0.1c_C^0 = 0 \\
& s_2 + c_{AC^-} - 0.002 = 0 \\
& \sum_{i=1}^{2} y_i = n_z + 1 = 1 \\
& s_k - U(1 - y_k) \le 0 \\
& y_k \in \{0,1\} \\
& -\delta \times \sigma \le \theta - \theta^N \le \delta \times \sigma \quad \longrightarrow \text{confidence interval hypercube} \\
\text{or} \quad & (\theta - \theta^N)^T \Sigma_\theta^{-1} (\theta - \theta^N) \le \delta \quad \longrightarrow \text{ellipsoidal confidence region} \\
& \delta, s_k \ge 0; \theta = \{k_1, k_2, k_3, k_4, k_5\}
\end{aligned}
$$

where the initial concentrations $\{c_A^0, c_B^0, c_C^0, c_{BH^+}^0, c_{A^-}^0, c_{AC^-}^0, c_P^0\}$ are set to be $\{0.3955, 0.3955/R, 0.25, 0, 0, 0, 0\}$ mol/L.

PURDUE CHEMICAL ENGINEERING

Michael reaction: probabilistic design space by flexibility index

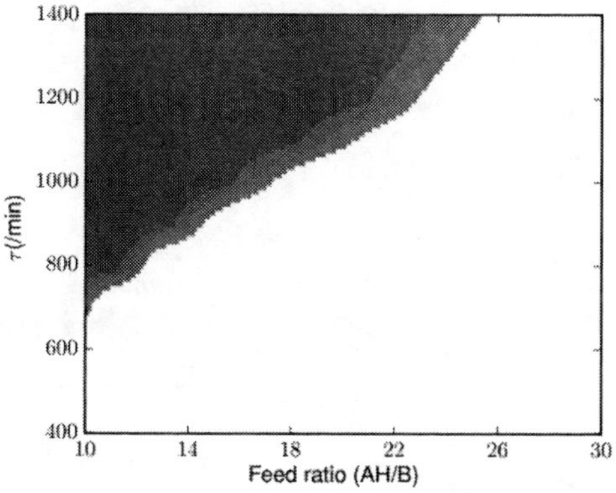

(a) confidence interval hypercube
(CPU: 5.153 sec; parallel with 4 core)

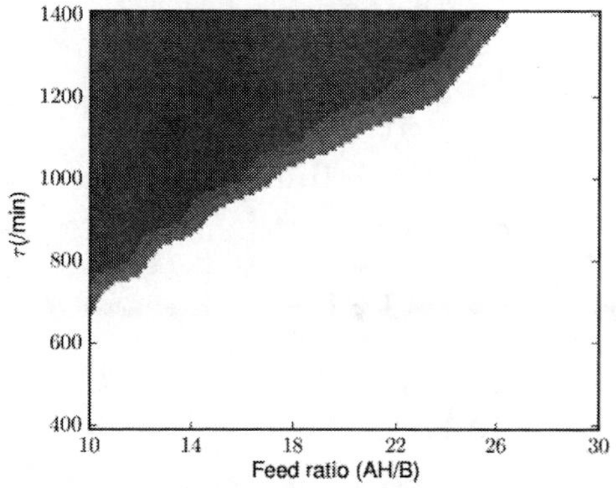

(b) ellipsoidal confidence region
(CPU 5.624 sec; parallel with 4 core)

Michael reaction case: probabilistic design space by flexibility index
(blue: $p \ge 0.85$ green: $0.7 \le p < 0.85$

red: $0.5 \le p < 0.7$ white: $p < 0.5$

PURDUE CHEMICAL ENGINEERING

CPU time comparison

- Environment: MacBook Pro with 4 cores; Parallel
- The Monte Carlo simulation costs **629.41 seconds** (200 repetitions)).
- The flexibility index with a confidence interval hypercube costs **5.153 seconds** .
- The flexibility index with an ellipsoidal confidence region costs **5.624 seconds** .

PURDUE CHEMICAL ENGINEERING

Hyper-cubic design space

- The hyper-cubic design space is determined by solving a flexibility index problem

$$
\begin{aligned}
& F(d) = \min \delta_p \\
s.t. \quad & Eqs.(7),(10) \\
& s_1 + c_C + c_{AC^-} - 0.1 c_C^0 = 0 \\
& s_2 + c_{AC^-} - 0.002 = 0 \\
& \sum_{i=1}^{2} y_i = n_z + 1 = 1 \\
& s_k - U(1 - y_k) \le 0 \\
& y_k \in \{0,1\} \\
& -\delta_m \times \sigma_m \le \theta_m - \theta_m^N \le \delta_m \times \sigma_m \longrightarrow \text{confidence interval hypercube} \quad \delta_m = 2.15 \\
\text{or} \quad & (\theta_m - \theta_m^N)^T \Sigma_{\theta_m}^{-1} (\theta_m - \theta_m^N) \le \delta_m \longrightarrow \text{ellipsoidal confidence region} \quad \delta_m = 8.12 \\
& \tau = 1400 - \Delta_\tau \delta_p \\
& R = 10.0 + \Delta_R \delta_p \\
& \delta_p, s_k \ge 0; \theta = \{k_1, k_2, k_3, k_4, k_5\}
\end{aligned}
$$

PURDUE CHEMICAL ENGINEERING

Michael reaction: Hyper-cubic design space

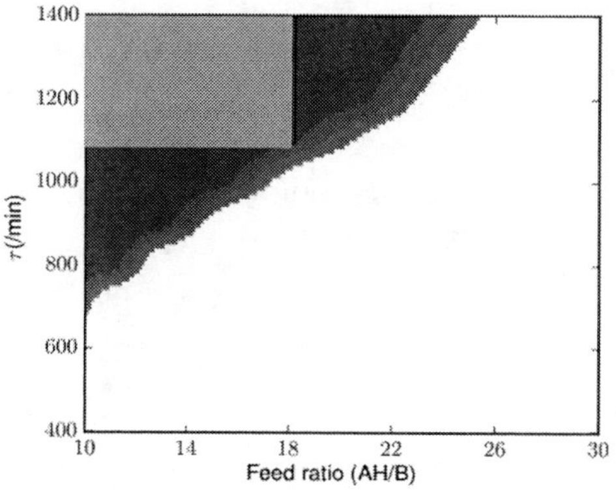

(a) confidence interval hypercube

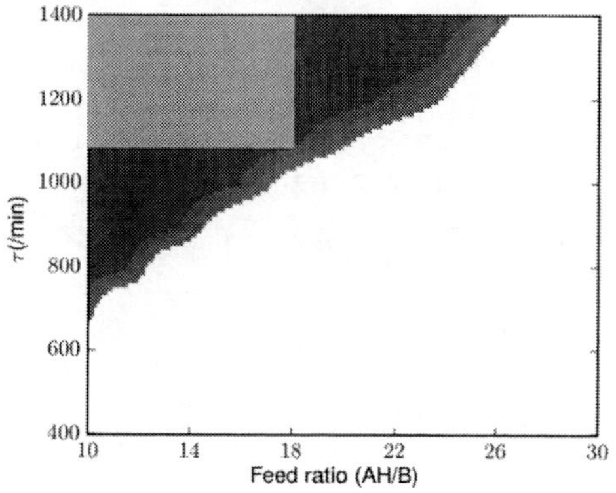

(b) ellipsoidal confidence region

Michael reaction: hyper-cubic design space

(blue: $p \geq 0.85$ green: $0.7 \leq p < 0.85$

red: $0.5 \leq p < 0.7$ white: $p < 0.5$

PURDUE CHEMICAL ENGINEERING

Conclusions and future work

- Area of design space: optimization < Monte Carlo
- Efficiency: optimization >> Monte Carlo
- Future work
  - ✦ Incorporate global optimization techniques and solve the flexibility index problem on a convex relaxation

PURDUE CHEMICAL ENGINEERING

Acknowledgment

- I would like to thank the help from Dr. Laird's group
- This work was supported by Eli Lilly and Company through the Lilly Research Award Program (LRAP)

PURDUE CHEMICAL ENGINEERING

References

- [1] Food and Drug Administration 2009. Guidance for Industry Q8 Pharmaceutical Development. Technical Report August, U.S. Department of Health and Human Services, Food and Drug Administration,Center for Drug Evaluation and Research (CDER), Rockville, MD.
- [2] Floudas, C. and I. Grossmann 1987. Synthesis of flexible heat exchanger net- works with uncertain flowrates and temperatures. *Computers & Chemical Engineering*, 11(4):319–336.
- [3] García-Muñoz, S., Luciani, C. V, Vaidyaraman, S., & Seibert, K. D. (2015). Definition of Design Spaces Using Mechanistic Models and Geometric Projections of Probability Maps. *Organic Process Research & Development*, **19**(8), 1012–1023.
- [4] Swaney, R. E. and I. E. Grossmann 1985a. An index for operational flexibility in chemical process design. Part I: Formulation and theory. *AIChE Journal*, 31(4):621–630.
- [5] Swaney, R. E. and I. E. Grossmann 1985b. An index for operational flexibility in chemical process design. Part II: Computational algorithms. *AIChE Journal*, 31(4):631–641.

PURDUE CHEMICAL ENGINEERING

496437 Supply Chain Design of Continuous Tablet Manufacturing Facilties Using Discrete Event Simulation

Abstract:

In the pharmaceutical industry, continuous manufacturing is becoming an attractable alternative to batch processing as the demand for small volume market specific products increases. However, the potential supply chain benefits of continuous manufacturing are difficult to quantify and vary significantly under different market conditions. In addition, to realize the full benefits of continuous manufacturing, changes to traditional production planning will need to be made. To address uncertainty, a discrete event simulation model was developed to inform the supply chain design of products and tablet manufacturing facilities. The outputs from the model define the product, market, and production planning conditions best suited for a continuous manufacturing or batch process platform. This ultimately enables a mechanism for identifying products and tablet manufacturing facilities that are a good fit for a continuous manufacturing platform as well as informing the production planning strategy.

496216 Assessment of Economic Potential of Furfural Platform

Abstract:

Furfural is identified as a potential bio-based building block by the department of energy. A suit of more than 80 chemicals that can be directly and indirectly derived from furfural have been identified. It is a daunting task to evaluate the economic potential of all of these furfural transformations. We, therefore, qualitatively screened and selected following seven chemicals that have near-term commercial potential: tetrahydrofuran (THF), tetrahydrofurfuryl alcohol (THFA), furfuryl alcohol (FA), maleic anhydride (MA), 2-methyltetrahydrofuran (MTHF), levulinic acid (LA), and 1,5-pentanediol (PDO). The minimum selling price (MSP) of furfural produced via acid dehydration of C5 sugar and the seven selected potential furfural derivatives were determined. The MSP of furfural produced in a plant with a capacity of 2000 metric ton per day of bone dry wood was estimated at $0. 78 per kg. This estimation indicates that the plant size of platform chemical production and the MSP are comparable to the proposed feasible size of lignocellulosic based biofuel production plant and the market price of super commodity chemicals, respectively. The estimated MSP ($/kg) of THF, THFA, FA, MA, MTHF, LA, and PDO are 1.47, 1.1, 0.94, 1.3, 1.54, 1.47, 1.28, respectively. The comparison of calculated MSP of furfural derivatives against the market prices has shown that all analyzed production processes are economically viable except the conversion of furfural to levulinic acid. The comparison of economic and modeling analysis of analyzed production pathways of furfural derivatives enable us to determine general rules of thumb necessary to screen early stage transformations of potential bio-based building blocks to various secondary commodity chemicals.

496231 Economic and Modeling Analysis of Product and Energy Driven Biorefineries Based on Thermal Deoxygenation Pathway

Abstract:

Biorefineries can be classified into product- and energy-driven biorefineries. The thermal deoxygenation pathway invented by University of Maine researchers offers flexibility in use of feedstock, which allow us to construct both product- and energy-driven biorefineries. In the product-driven biorefinary, lignocellulosic feedstock such as wood is converted primarily to furfural (a potential bio-based building block) along with co-products biofuel and biomaterials char and activated carbon. In the energy-driven biorefinary, cellulose rich feedstock such as cardboard is mainly used to produce biofuel along with the co-product levulinic acid (a potential bio-based building block). In this study, we performed economic and modeling analysis of product- and energy-driven biorefineries based on thermal deoxygenation pathway to determine the best type of a biorefinary in terms of economic returns. Greenfield and brownfield plant site scenarios were considered in the economic assessment of product- and energy-driven biorefineries. ASPEN plus simulation software was used to model each type of biorefinary. Capital and operating costs were estimated for a plant capacity of 2000 metric ton per day of bone dry biomass feedstock, using the simulated material and energy balances. The discounted cash flow analysis with an internal rate of return of 10% was employed to estimate the minimum selling price of bio-based products produced in both product- and energy-driven biorefineries. Finally, economic viability of producing activated carbon from char rather than burning it to produce energy was assessed. Determining which type of biorefinary is the best in terms of economic performance will enable us to effectively utilize limited resources for the development of economically viable biorefineries that eventually accelerate the growth of a bio-based economy.

496935 An Algorithm for Integrated Design of Organic Rankine Cycles

Abstract:

An algorithm for integrated design of Organic Rankine Cycles

Uku Erik Tropp, David H. Bowskill, Smitha Gopinath, George Jackson, Amparo Galindo, Claire S. Adjiman

Department of Chemical Engineering, Centre for Process Systems Engineering, Imperial College London

Organic Rankine Cycles (ORCs) are able to utilize waste heat or low quality heat to generate power. Thus, the use of ORCS can lead towards sustainable and environmentally friendly energy systems. The optimal operating conditions of the ORC depend on the specifications of the process (for e.g., temperature of the heat source), constraints of the system (for e.g., amount of coolant available) and the choice of working fluid. Optimal cycles may be designed by simultaneously choosing the fluid and the operating conditions, subject to the constraints. Such an integrated design of materials (molecules) and processes is also referred to as Computer Aided Molecular and Process design (CAMPD) [1]. In the ORC CAMPD, process decisions are usually continuous, whereas material (molecular) decisions are inherently discrete. The ORC working fluid (a pure-component is considered here) may be represented by the number of occurrences in the molecule of each functional group (in a given set of groups) and the connectivity between these occurring groups. Given a property prediction method, that translates the molecular composition of the fluid to physico-chemical properties of interest in the ORC model, the CAMPD problem may be formulated. The resulting design problem is a mixed integer nonlinear problem (MINLP) which is prone to non-convergence for a number of reasons.

The ORC process model (equality constraints of the design problem) consists of a number of non-convex coupled equations, while property prediction models tend to be highly nonlinear. Further, as the choice of working fluid changes, across optimizations, many molecules may be infeasible with respect to the process model, which could result in non-convergence of optimizations (especially if simulator-based optimization is used). For example, a chosen molecule may have a critical point that is lower than the range of allowable pressures and temperatures and can thus not be condensed into a liquid in the cycle. In other scenarios, a molecule with a boiling point which is higher than that of the hot source cannot be obtained as a vapour at the outlet of the evaporator. An attempt to evaluate the vapour properties of such a stream can result in numerical failure.

To address the challenge of non-convergence and improve computational performance, we extend an algorithm [2, 3] which was recently developed for CAMPD of separation systems. The algorithm builds on the concept of embedding screening of molecules within an iterative MINLP optimizer [4, 5], such as Outer-Approximation [6]. Within this algorithm, each molecule generated by the master problem passes through a sequence of tests, before the evaluation of the primal problem for that molecule. The aim of the tests is to identify *a priori* regions within the process domain that are infeasible for a given molecule. If the entire domain is infeasible for a molecule, the molecule may be eliminated without evaluation of the primal problem, and information on the violated constraints are added to the master

problem.Â For a molecule that passes the tests, the tests generate bounds on the feasible process domain for that molecule.

In this work, we develop tests specific to the design of ORC systems. The tests produce bounds on allowable operating pressures of the ORC for every feasible molecule. The tests also eliminate molecules whose phase behaviour (critical point, vapour pressure and so on) make them infeasible in the ORC. The algorithm is applied to solve three case studies of integrated ORC working fluid and process design. The SAFT-Î³ Mie [7] equation of state is used for prediction of thermodynamic properties of the working fluid. We find that the bounds obtained by the tests are useful in reducing the process domain and enable convergence of the primal problem. Without these bounds, the evaluation of the primal problem fails for certain molecules. The use of the algorithm enhances the robustness of the design problem, allowing one to obtain (locally) optimal solutions from a variety of starting points, which is useful as the problem is non-convex. Thus, the algorithm facilitates the use of rigorous process and property prediction models for the design of ORCs.

1 Adjiman, C. S., Galindo, A., and Jackson, G. (2014). Proceedings of the 8th International Conference on Foundations of Computer-Aided Process Design FOCAPD 2014. Elsevier

2 Gopinath S, Galindo A, Jackson G, Adjiman CS (2016). Proceedings of the 26th European Symposium on Computer Aided Process Engineering, edited by Kravanja Z, Computer Aided Chemical Engineering. Elsevier B.V.

3 Gopinath S, Jackson G, Galindo A, Adjiman CS (2016). AIChE Journal, 62:3484â€"3504

4 Buxton, A., Livingston, A. G., and Pistikopoulos, E. N. (1999). AIChE Journal, 45(4):817-843

5 Giovanoglou, A., Barlatier, J., Adjiman, C. S., Pistikopoulos, E. N., and Cordiner, J. L. (2003). AIChE Journal, 49(12):3095-3109

6 Duran, M. A. and Grossmann, I. E. (1986). Mathematical Programming, 36(3):307-339

7 Papaioannou, V., Lafitte, T., Aveñdano, C., Adjiman, C. S., Jackson, G., Müller, E. A., and Galindo, A. (2014). The Journal of Chemical Physics, 140(5)

497053 Integrated Computational Design of Cosmetic Products

Abstract:

Increased competition and globalized market dynamics have driven the accelerated growth of chemical industry [1]; consequently, different investigations have been directed in order to propose new resources in manufacture of high added-value products [2]. This work presents the implementation, in PYOMO®, of computational tool for integrated design of structured cosmetic products and its application in moisturizing lotion formulation.

The consumer-oriented products design requires a multiscale approximation, since the properties are closely related to molecular interactions that describe its microscopic and macroscopic structure [3]. Owing to the essential of multiscale observation approach, an integration of systemic procedures was performed in order to consider involved processes at different levels in product design; therefore, the tool was developed from an approximation defined by three modules:

- Surfactant Module: In charge of defining the surfactant (or mixture of surfactants) in order to specify molecular interactions that allows obtaining an appropriate surface tension and emulsion stability [4].
- Rheology Module: Defines the microstructure and macrostructure of emulsion from molecular scale with view to know its rheological properties [5].
- Consumer Module: Integrates a consumption preferences model based on sensory perception of customer and an economic model that aims at economic utility using market parameters [6].

In regard with previous ideas, this tool facilitates the design by projecting an appropriate formulation of cosmetic product, making it possible through an integrated optimization model raised to specify the variables of proposed modules, in quest of an optimal formulation, considering two objective functions according to user's requirements:

- Economic Function: Allows finding the product formulation and mass fractions per component focused on obtaining the maximum economic yield, taking into account the costs of manufacture, price and demand of the product.
- Preference function: Indicates the product formulation and mass fractions per component aimed at obtaining the maximum acceptance by consumer, according to established preferences parameters.

In conclusion, the implementation of this computational tool for integrated design of products is of great interest in the cosmetic industry; its importance lies in the optimization model that considers the interactions of proposed modules, for the purpose of generate a solid base in structured products formulation by simulations and thus reduce the resource spending in experimental design phases. Being a tool based on open source software with a flexible methodology brings with it certain advantages, it is not only possible to include new modules of interest according to the requirements of product design, but it also sets the basis for parallel developments of similar tools oriented to different Industries such as pharmaceuticals and food.

References

[1] B. V. Smith y M. G. Lerapepritou, «Integrative chemical product design strategies: Reflecting industry trends and challenges,» *Computers & Chemical Engineering,* vol. 34, nº 6, pp. 857-865, 2010.

[2] E. Cussler y G. Moggridge, Chemical Product Design, Cambridge: Cambridge University Press, 2011.

[3] R. Gani, «Chemical product design: challenges and opportunities,» *Computers & Chemical Engineering,* vol. 28, nº 12, p. 2441–2457, 2004.

[4] C. Wibowo y K. Ng, «Product-oriented process synthesis and development: Creams and pastes,» *AIChE Journal,* vol. 47, nº 12, pp. 2746 - 2767, 2001.

[5] M. Mattei, G. Kontogeorgis y R. Gani, «A comprehensive framework for surfactant selection and design for emulsion based chemical product design,» *Fluid Phase Equilibria,* vol. 362, nº 25, pp. 288 - 299, 2014.

[6] M. Bagajewicz, S. Hill, A. Robben, L. Heyde, M. Sanders, E. Sposato, C. Baade, S. Manora y J. Coradin, « Product Design in Price-Competitive Markets: A case study of a Skin Moisturizing Lotion,» *AIChE Journal,* vol. 57, nº 1, pp. 160 - 177, 2011.

497248 Modeling and Multi-Objective Optimization of an Industrial Ammonia Synthesis Process

Abstract:

Ammonia synthesis is a major important process in chemical industry. Ammonia application domain is very wide with major applications in manufacturing of agricultural fertilizers. Hence, it is particularly important for Canada where agricultural sector constitutes a significant part of national GDP. To keep up with continuously increasing demand, it is highly important to improve performance of existing facilities through process modeling, simulation and optimization. One of the way to achieve the goal is through mathematical modeling and optimization of operation and/or design parameters of a plant. Due to large scale of production, even minor improvements can result in significant profits or savings.

The conventional synthesis is based on Haber process - reaction of gaseous nitrogen and hydrogen over a solid catalyst to produce ammonia. The reaction is carried out at elevated temperature (400-500 °C) and pressure (80-230 atm.). The feedstock comes from a steam reforming unit at the same plant and is made of stoichiometric mixture of nitrogen and hydrogen with a minor portion of inert. The synthesis reaction occurs in one or few (usually, 1-4 beds) fixed catalytic beds. Since the process is exothermic, the vast amount of heat is generated and re-used in the process. In this work, a collaborative research was conducted with a large ammonia producing plant in Canada aiming to investigate possibilities of further improving ammonia production and heat duty in the ammonia converter. First-principle mathematical modeling of gas-solid catalytic reaction and heat transfer equipment was performed. A simple enough yet accurate model was developed to adequately simulate the ammonia converter. Main features of the model are one dimensional plug flow with Fick-type diffusion of reactant inside solid catalyst particle.

Importantly, kinetic parameters of catalyst are being the part of proprietary information, and hence, were not available. The model was first tuned with industrial data to provide estimate for catalyst kinetic parameters. For the proper estimation of kinetic parameters, a large set of industrial data was used covering time span over 12 years. Since data set is large and is full of information irrelevant to normal steady state operation, which included not only scheduled maintenance, but also short term disruptions, cold and hot start up and shutdown, sensor failures, etc., and hence, it was not evidently clear which data to use for the model validation. Hence, a data pre-treatment procedure based on data clustering around steady-states was performed to filter out relevant industrial data points. Thereafter, the model was fitted over several different time-dependent groups of data. It was discovered that even though catalyst life time for ammonia catalyst is large (> 10 years), the catalyst still experiences deactivation and it should be accounted at different stages of operation. Incorporating all the above, the model was validated over industrial data and showed good consistency with observed values. Finally, the validated model was used to perform multi-objective optimization with modified Non-Dominated Sorting Genetic Algorithm (NSGA-II). The primary goals were to maximize production of ammonia, minimize heat requirement for the reaction and minimize total catalyst loading in converter. Three optimization case studies were performed, where objectives were treated pair-wise to provide a Pareto optimal solution for converter operation.

In the first case, optimal values for three process parameters - namely, converter pressure, temperature of feed stream and ratio of quench stream to significantly reduce heat duty while keeping ammonia production above industrially acceptable level. In the second case, a different catalyst distribution among beds, was proposed to allow for higher ammonia production. Three cases with different distribution were obtained where total catalyst loading in the converter is lower, equal and higher than current. All three cases showed higher ammonia production compared to the base case. Additionally, a change in optimal conditions with respect to time-on-stream for the catalyst has been addressed and guidelines for the converter operation were provided like the first case.

497475 Generation of Sustainable Hybrid Process Flowsheets Using Procafd

Abstract:

The chemical industry is always looking for innovative process designs, which are both efficient and sustainable. To generate an optimal flowsheet, what is needed is a systematic way to identify the types of tasks-operations that need to be performed, the corresponding design of the operation-equipment, their configuration, mass-energy flows etc. Due to the fact that process synthesis problems are by nature combinatorial and open ended, a number of different solution approaches have been proposed. However, the solution for any synthesis-design problem is dependent on the search space of alternatives and the process performance criteria which in most cases are influenced by economic factors. This work focuses on the development of a computer aided tool (ProCAFD), capable of enumerating the entire feasible search space, analyze and determine the most sustainable process. ProCAFD is based on decomposition based method developed by Tula et al. (2015). On a higher level the workflow of ProCAFD can be divided into three stages: 1) Synthesis stage â€“ Here a hybrid approach based on group contribution is used to generate and rank the alternatives. In this stage, chemical process flowsheets are synthesized in the same way as atoms or groups of atoms are synthesized to form molecules in computer aided molecular design (CAMD) techniques. The building blocks in flowsheet synthesis are called process-groups, which represent a single or set of unit operations that are selected by employing a thermodynamic insights based method. These building blocks are then combined using connectivity rules to generate all the feasible flowsheet alternatives. The main advantage of representing the flowsheet with process-groups is that, the performance of the entire process can be evaluated from the contributions of the individual process-groups towards the selected flowsheet property (for example, energy consumed). Available flowsheet property models include energy consumption, carbon footprint, product recovery, product purity etc. In this way, the entire list of feasible chemical process flowsheets is quickly generated, screened and selected. 2) Analysis stage â€“ In this stage, comprehensive investigation consisting of sustainability, life cycle and economic analysis is performed to identify areas of improvement for the generated process alternative. The identified areas or process hotspots are translated into design targets that, if satisfied will guarantee process improvement. 3) Innovation stage â€“ In this stage different innovative design strategies are used to match the set design targets. Different design strategies include simultaneous process optimization and heat integration where determination of optimal operating conditions and design parameters is solved together with heat integration involving minimizing utility consumption. One other strategy ProCAFD uses is a generalized method based on Tula et al. (2016) to replace distillation columns in the selected design with a hybrid distillation-membrane system to reduce the operational cost substantially. Other strategies include process intensification, where the phenomena based intensification method developed by Babi et al. (2015) is employed to generate intensified processes.

In this work, the software implementation of the developed methodology will be highlighted through several examples involving chemical and biochemical processes. In all the cases studied, the developed software could generate innovative and more sustainable solutions very quickly.

References:

A. K. Tula, B. Befort, N. Garg, K. V. Camarda, R. Gani, Sustainable Process Design & Analysis of Hybrid Separations, Computers & Chemical Engineering (2016). http://dx.doi.org/10.1016/j.compchemeng.2016.11.031.

A. K. Tula, M. R. Eden, R. Gani, 2015, Process synthesis, design and analysis using a process-group contribution method, Computers & Chemical Engineering, 81, 245-259.

D. K. Babi, J. Holtbruegge, P. Lutze, A. Gorak, J.M. Woodley, R. Gani, 2015, Sustainable Process Synthesis-Intensification, Computers & Chemical Engineering, 81, 218-44.

499195 Total Site Integration As a Synthesis Tool to Select Biomass Valorization Paths and Schedule Multiple-Feedstock Operations

Abstract:

This paper presents a systematic approach to select and integrate product portfolios in multiple-feedstock biorefineries. Value chains include all potential routes from biomass to intermediate and end-products that may enter biorefineries. Once the problem requires minimum energy requirements to secure process sustainability, Total Site integration is called to minimize consumption of steam (heat transfer mean among processes). The use of Total Site Analysis to screen biorefinery paths has been recently presented by the authors (Kokossis et al., 2015) to detect promising process portfolios that optimally integrate each other in terms of energy use. The methodology introduced a modified transshipment model to evaluate energy synergies among candidate processes of value chains given the type and annual capacity of biomass feedstock. Once the majority of applications refer to lignocellulosic biorefineries, the analysis naturally extends to the assessment of changes in operation and efficiencies due to seasonal availability of each lignocellulosic biomass type. This paper extends previous concepts to assess the impacts of switching feedstocks on energy efficiencies and the selection of biorefinery portfolios.

Lignocellulosic biorefineries are based on the exploitation of the three fundamental chemical components: C5 and C6 sugars and lignin. Though products of value chains remain the same with changes in the type of lignocellulosic material – wheat straw, rice, poplar or birch, they all include sugars and lignin – changes appear in process capacities and energy consumption due to seasonal variations in capacities and yields in sugars and lignin for each biomass type. Accordingly, heat interactions and recovery among processes based on the valorization of each component change by switching feedstocks during the year.

The combinatorial process synthesis and integration problem is formulated by the mass balances along value chains (selection of paths) and energy balances at each interval of the proposed transshipment (evaluation of selected paths). The problem also assumes changes in biomass capacities at discrete time periods for each candidate type of biomass. Accordingly, the domain of flowrates along value chains is expanded by the indexes of biomass type and the time period that is applied. Large deviations in common chemicals flows and processes capacities at different periods are not industrially accepted. Likewise, the domain of heat flows in the transshipment model and utilities is also expanded by time and biomass index and limited for logic deviations. The MILP model combines selection of value chain paths with integration in transshipment model to evaluate energy savings at each period and thus, to synthesize the seasonal feedstocks plan and select the processes portfolio that minimize energy cost of the biorefinery site.

References

A.C. Kokossis, M. Tsakalova, K. Pyrgakis, Design of integrated biorefineries, Computers & Chemical

Engineering, 2015, 81, 40–56

496786 Grid-Level "Battery" Operation of Chemical Processes with Engagement in Short-Term Electricity Markets

Abstract:

The U.S. electric power generation sector is experiencing a transition from a set of concentrated resources to a more distributed system. This is in part driven by the deregulation of the electricity markets and the influx of renewables into the energy mix [1, 2]. Accompanying this transformation is the need for a robust grid, capable of handling variability on both the supply and demand sides. Mechanisms such as energy storage and demand response (DR) are being discussed and implemented by diverse stakeholders as a means to bridge instantaneous demand and supply mismatch and thus enhance grid operations.

Amongst such stakeholders are industrial consumers, who could potentially contribute significantly to peak load reduction by engaging in DR [3]. Industrial engagement in the operation of the power grid can occur over multiple time scales. Involvement in long-term (day-ahead) energy markets (LTM) can lead to considerable minimization of energy costs for the consumer subject to time-varying power tariffs [4]. To this end, industrial consumers typically reduce power use during peak-tariff hours by diminishing their production rates. Production is increased at off-peak hours, and the excess product is stored to meet demand during the subsequent peak period.

Participation in short-term markets (STMs) can be even more lucrative [5], but comes with some significant challenges. STMs are highly volatile, with conditions varying over time scales in the order of minutes (or even seconds). Since the dynamics of process systems are inherently slower, with time constants in the order of hours, altering production rates to capture fast changes in market conditions will imply a highly transient plant operation, with potentially abrupt changes in the production schedule (including e.g. schedules optimized to take advantage of the LTM conditions). As a consequence, it is imperative that, i) the production schedule be re-optimized to account for changes in STM conditions, and, further, ii) the process dynamics must be accounted for in the scheduling calculation [4] to guarantee that production schedules are feasible.

In our previous work [4], we addressed the issue of engaging chemical processes in LTM DR by proposing a novel scheduling framework that incorporates a representation of the process dynamics. The framework is predicated on the use of low-order representations of the scheduling-relevant dynamics of a process, which we refer to as "scale-bridging models" (SBMs) [6]. In this work, we extend this framework to account for participation in STMs. To this end, we formulate a two-tier optimization problem. In the top level problem, a base load profile is computed, that maximizes process profit for a fixed time horizon based on day ahead market (DAM) prices. The lower tier problem is solved over a shrinking time horizon, aimed at determining in every short-term time block any additionally available DR capacity for the process (which we define as the "DR headroom") and its cost relative to the base cost. Subsequently, process operators can offer the DR headroom for sale or auction in the open market. Thus, the process is acting as a grid-level "battery" in the STM, offering load reductions and earning revenue. Our approach accounts for any changes in the remaining long-term schedule, including potential future increases in production to compensate for the instantaneous decrease dictated by STM conditions. We illustrate our developments with a case study, showing that the potential for increased earning for chemical processes is significant.

References

1. Lacy, V., Matley, R., & Newcomb, J. (2012). Net Energy Metering, Zero Net Energy and the Distributed Energy Resource Future: Adapting Electric Utility Business Models for the 21st Century.
2. Davoli, F., Repett, M., Tornelli, C., Proserpio, G., & Cucchietti, F. (2012). Boosting energy efficiency through smart grids. International Telecommunication Union (ITU).
3. 2016 Assessment of Demand Response and Advanced Metering Staff Report. Federal Energy Regulatory Commission.
4. Pattison, R. C., Touretzky, C. R., Johansson, T., Harjunkoski, I., & Baldea, M. (2016). Optimal Process Operations in Fast-Changing Electricity Markets: Framework for Scheduling with Low-Order Dynamic Models and an Air Separation Application. Industrial & Engineering Chemistry Research, 55(16), 4562-4584.
5. Dowling, A. W., Kumar, R., & Zavala, V. M. (2017). A multi-scale optimization framework for electricity market participation. Applied Energy, 190, 147-164.
6. Pattison, R., Touretzky, C. R., Johansson, T., Baldea, M., & Harjunkoski, I. (2016). Moving horizon scheduling of an air separation unit under fast-changing energy prices. IFAC-PapersOnLine, 49(7), 681-686.

497645 A Decomposition Approach for the Integration of Scheduling and Model Predictive Control in Fast Changing Market Conditions

Abstract:

Globalization and intensive information exchange supported by new technologies have given rise to an environment with fast changing market conditions, which must be taken into account in order to achieve optimal process operation. An example of such conditions are electricity prices in real-time markets, which can change hourly (or faster) due to the change of electricity demand. Therefore, to maximize profit in electricity-intensive industries, changes in production rate and product grade must be scheduled with increased frequency.

Such frequent operational changes may result in processes operating in transient regimes, rather than in steady state conditions. This fact is contrary to the premise of most current scheduling methodologies, and emphasizes the necessity to account for the dynamic behavior of the system while making decisions at the scheduling level. Although the integrated scheduling and dynamic optimization problem has been addressed in the literature (Nystrom, Franke et al. 2005, Flores-Tlacuahuac and Grossmann 2006), few works considered the operation in transient regimes (Pattison, Touretzky et al. 2016), and even fewer take advantage of advanced control techniques in such integrated scheduling and control frameworks.

In this work, we propose a framework for the integration of scheduling and control for continuous processes operating in fast changing market conditions. The framework is based on an iterative procedure where problems at scheduling and control level are solved in each iteration. The first step is to identify the dynamic dependent variables and obtain algebraic approximations in terms of key decision variables (for example, production rate is dependent on production rate set points). The second step is to model the scheduling problem using the identified algebraic expressions. We use a discrete time representation of the system, and the number of slots is defined according to the frequency of change in parameters driving the costs of the system (for example, utility prices drive the costs of energy intensive industries). The problem is formulated as a mixed integer nonlinear problem. The scheduling problem defines the key decision variables (for example, production rate set points over time) in order to minimize costs. Since the scheduling problem is just an approximation of the real system, the costs predicted by the scheduling problem will provide a lower bound solution for the integrated problem.

The third step is to transfer the scheduling solution to the control level. At this level, we propose to use model predictive control in order to track the different set points over time while ensuring stability, robustness, safety and fast tracking. A closed loop simulation of the system is performed throughout the entire scheduling horizon, using the set points provided by the scheduling problem. The feasibility of the scheduling solution is verified and the actual cost of the system is calculated and compared to the scheduling cost. If the solution is infeasible, the sequence of set points defined in this iteration is cut from the scheduling problem, and the problem is resolved. If the solution is feasible, the scheduling cost associated to this sequence of set points is updated with the actual (simulated) cost, and the scheduling problem is resolved in the next iteration. The control problem sets an upper bound solution for the integrated problem, which is chosen at every iteration as the minimum simulated cost so far. The iterations proceed until the difference between upper and lower bounds are within a threshold. To illustrate the proposed approach, a case study involving the production of gaseous nitrogen in an air separation unit is presented.

Flores-Tlacuahuac, A. and I. E. Grossmann (2006). "Simultaneous cyclic scheduling and control of a multiproduct CSTR." Ind Eng Chem Res **45**(20): 15.

Nystrom, R. H., et al. (2005). "Production campaign planning including grade transition sequencing and dynamic optimization." Comput Chem Eng **29**: 17.

Pattison, R. C., et al. (2016). "Optimal Process Operations in Fast-Changing Electricity Markets: Framework for Scheduling with Low-Order Dynamic Models and an Air Separation Application." Ind Eng Chem Res **55**: 4562-4584.

498537 On the Design of an Online Scheduling Algorithm

Abstract:

Disruptions or arrival of new information can make the incumbent schedule suboptimal or even infeasible, thus motivating the need for online (re)scheduling (Subramanian et al., 2012). In this talk, we will investigate how to formulate the scheduling problem solved online (open-loop), so that it leads to high quality implemented (closed-loop) schedule. This is not trivial, because, even in the deterministic case, open-loop and closed-loop scheduling are two different problems (Gupta and Maravelias, 2016).

When uncertainty is present, online scheduling methods need not always explicitly model it; they can employ a nominal (deterministic) model, and still mitigate the effect of uncertainties by treating them as disturbances, and relying solely on recourse, through feedback (Gupta et al., 2016). Deterministic models are more computationally tractable, in real-time, than their stochastic counterparts, thus have a wider applicability. Furthermore, any stochastic modeling, if attempted, is always approximate which means that, eventually, stochastic optimization methods will also be faced with the challenge of mitigating uncertainties/disturbances. Therefore, the findings of this work are broadly applicable to all online scheduling methods.

There are two goals of this work. Our first goal is to study how demand uncertainty affects the quality of the implemented schedule. How does demand variability affect solution quality? Does the time at which uncertainty is observed affect solution quality? Do different demand patterns, characterized by, for example, the frequency of orders, pose any particular challenges? In addition, how do network characteristics (e.g. shared intermediate materials, availability of buffer production capacity), aggravate or diminish the above effect? We also study if and how horizon length and re-optimization frequency can be used to deal with demand uncertainty.

Our second goal is to develop a general method for the design of an online scheduling algorithm, that is, determining, specifically, but not limited to, the horizon length and the re-optimization frequency. In other words, we develop a method which, given an instance (e.g., network data and some demand uncertainty description), explicitly predicts the needed online scheduling parameters.

From our simulations, first, we find that demand uncertainty is an important consideration, and that it is indeed beneficial to invest in obtaining good forecasts. Second, counter-intuitively, this is more important for lightly loaded facilities than heavily loaded facilities. This is because the latter will run close to capacity irrespective of fluctuations in order sizes. Third, the choice of horizon is contingent on the mean load at which we are operating, and not the accuracy of the demand forecast, as has been sometimes covertly assumed. Fourth, re-optimization is very effective, especially, for under-loaded facilities, as it can help exploit uncommitted resources, which are not necessarily present under heavy loads. Interestingly, we show how the recommendations by Gupta and Maravelias (2016), who considered the deterministic problem, are consistent with the results of this study.

In accordance with goal 2, we first identify the key factors that affect the selection of online scheduling parameters. Some of these factors are production load, demand uncertainty, process network characteristics. Second, for each factor, we define an "auxiliary parameter" to quantify it. In some cases, defining this auxiliary parameter is straightforward (e.g., use standard deviation to describe demand variability) but in some cases it is far from trivial. For example, quantifying the load of a facility, subject to a demand pattern, depends on a series of factors and there is no formula to calculate it. We present procedures to calculate all these auxiliary parameters. Third, we present methods to determine the online scheduling parameters (e.g., horizon of optimization problem and re-optimization frequency) from the instance-specific auxiliary parameters. Finally, to test our framework, we present results for two networks that were not used during the development of the framework under different demand patterns.

To our knowledge, the work presented herein is the first systematic framework for the design of an online scheduling algorithm, and thus takes us one-step forward, towards synthesizing a general strategy to obtain high-quality closed-loop schedules.

References:

Subramanian, K.; Rawlings J.B.; Maravelias, C.T. "A state-space model for chemical production scheduling." *Computers & chemical engineering* 47 (2012): 97-110.

Gupta, D.; Maravelias, C.T. On deterministic online scheduling: Major considerations, paradoxes and remedies. *Computers & Chemical Engineering*, 94 (2016), 312–330.

Gupta, D.; Maravelias, C.T.; Wassick, J.M. From rescheduling to online scheduling. *Chemical Engineering Research and Design*, 116 (2016), 83-97

499626 Unified Treatment of Scheduling and Control: Same Story, Different Dynamics?

Abstract:

Traditionally, scheduling and control are viewed as two related but very different endeavors. For scheduling, the main decisions are typically discrete yes/no choices; the models capture only important discrete events but include many units; and, the objective is generally economic in some sense (minimize, e.g., cost or earliness). For control, the decisions are almost always continuous in nature; the models describe detailed temporal dynamics of the system but are local in scope; and, the objective function is artificially designed to drive the system to a predetermined setpoint. Despite these differences, both problems can be addressed by formulating a mathematical optimization and solving it repeatedly as new information is received. For control systems, re-optimization is necessary to correct for unmeasured disturbances or to respond to changes in setpoint, and these same general considerations also trigger rescheduling. While the timescales may be quite different (in terms of both the horizon of optimization and how frequently optimization is performed), this similarity begs the question of whether the two disciplines can be unified under a single mathematical treatment. In this presentation, we advance the idea that certain classes of scheduling and control problems are indeed two variants of the same overall problem that differ only in their respective system dynamics and decision space, and thus can be analyzed using a common set of tools.

To start, we revisit previous work that shows how a common scheduling model can be written in state-space form. This abstraction allows the model to be viewed as a dynamic system rather than as a set of combinatorial constraints. Next, we discuss results showing that, with suitable assumptions, the presence of discrete-valued control inputs does not affect the stability properties of model predictive control (MPC). Combining these ideas, we show that economically optimizing scheduling and control problems can both be viewed as cases of dynamic real-time optimization or economic MPC, which has important ramifications for closed-loop implementation. For example, without end constraints, dynamic economic optimization often exhibits the so-called "turnpike" property, whereby the system follows an optimal (or near-optimal) trajectory throughout most of the horizon, but at the end sharply deviates in pursuit of lower cost without regard for what the system will do after the prediction horizon. This behavior has two main effects: first, a given schedule remains useful for only part of its prediction horizon; and second, that it becomes very difficult to prove anything about closed-loop performance. However, by using a recursively feasible reference trajectory (e.g., a periodic solution) as an end constraint, it is possible to show that, in the nominal case, closed-loop cost will asymptotically be no worse than that of the reference trajectory. This and other types of terminal constraints can shrink the necessary prediction horizon, which allows optimization to be performed more rapidly, i.e., in real time instead of just when some specific upset triggers a reschedule. Thus, the decisions made in closed loop remain nearly optimal (as they are taken from the beginning of the prediction horizon, rather than the end) to achieve low closed-loop cost.

While it is important to understand the nominal properties of the closed-loop system, real systems always deviate from their models due to external disturbances, modeling errors, etc. Thus, we also examine what robustness properties can be applied to economic scheduling and control problems when the realized system behavior does not match the model's predictions. With a tracking objective, it can be shown that MPC is inherently robust (i.e., cannot be destabilized by small disturbances) even when discrete and continuous actuators are present. However, in economic contexts, these guarantees may no longer apply, and further, it is rare that scheduling environments are subject only to vanishingly small disturbances. Thus, using several small example systems, we examine whether nominal online optimization performs well when subject to disturbances despite having fewer theoretical guarantees. We then conclude with some directions for future research. The overall goal is to use insights from scheduling and control to advance a unified perspective that increases the applicability and improves the performance of closed-loop online dynamic optimization techniques.

Works Cited

1. Subramanian, K., Maravelias, C.T., Rawlings, J.B., 2012. *A state-space model for chemical production scheduling*. Comput. Chem. Eng. **47**, 97-160.

2. Rawlings, J.B., Risbeck, M.J., 2017. *Model predictive control with discrete actuators: Theory and application.* Automatica **78**, 258-265.

3. Gupta, D., Maravelias, C.T., 2016. *On deterministic scheduling: Major considerations, paradoxes, and remedies.* Comput. Chem. Eng. **94**, 312-330.

4. Angeli, D., Amrit, R., Rawlings, J.B., 2012. *On average performance and stability of economic model predictive control.* IEEE Trans. Auto. Cont. **57**, 1615-1626.

501744 Dynamic Scheduling to Maximize the Profitability of Air Separation Units

Abstract:

Dynamic Scheduling to Maximize the Profitability of Air Separation Units

Song Wang, Jian Zhang, and Qiang Xu

Dan F. Smith Department of Chemical Engineering
Lamar University, Beaumont, TX 77710, USA

Abstract

The Air Separation Units (ASUs) generate oxygen, nitrogen, and argon productions from air, which were widely used in many industries. Unlike other chemical processes, the cost of feedstocks of an ASU can be negligible because of the raw material of air is free; meanwhile, Besides, the energy supply of the entire process is totally in the charge of electricity power. ASUs are also characterized by fluctuating operating conditions to respond to changing product demands [1]. Although it is not much necessary to improve current ASU designs since there are quite proven techniques of cryogenic air separation design existing in industries; however, there are still significant potentials for improving their profitability in terms of optimal scheduling and operation ASU manufacturing.
Currently, many air separation plants are operated in a dynamic economic environment [2]. The electricity price could frequently fluctuate according to real time marketing (RTM) demands in different regions. In this paper, a general methodology based on dynamic simulation and scheduling of an ASU has been developed, which can optimize the ASU operation with respect to the electricity cost to achieve the best profit as well as highly efficiency. First, the methodology develops reliable dynamic simulation models under various production modes (capacities) and studies the dynamic transition performance among those production modes. Second, the optimal scheduling and transition will be conducted to maximize the plant profitability under the electricity cost pattern. Finally, the optimized ASU scheduling and operations will be fine turned and validated by rigorous dynamic simulations. The developed methodology can be further extended to help ASUs for the predictive and proactive smart manufacturing.

Keywords: Air Separation Unit, Production Scheduling, Maximum Profitability, Smart Manufacturing

References

[1] Rui, H; Victor M. Zavala; Lorenz T. Biegler *. Advanced Step Nonlinear Model Predictive Control for Air Separation Units. Journal of Process Control. www.elsevier.com/locate/jprocont.
[2] Li, T*; Thierry, R; and Marc, B. Production Scheduling of Air Separation Processes. Air Liquide, Newark, DE 19702; Air Liquide, Champigny-sur-Marne, France; Air Liquide, Madrid, Spain.

501802 Investigation of Closed-Loop Optimal Process Scheduling Policies

Abstract:

In the work of Lima et al., [1], the notion of inventory creep within a process scheduling context was introduced. In short, the inventory creep phenomenon is a gradual reduction of material inventory over time. To alleviate this myopic behavior, the authors of [1] advocate the use of larger prediction horizons, but quickly run into computational tractability issues. In our previous work [2], application of the Economic Model Predictive Control (EMPC) to a simple process scheduling problem (a single machine with two products), it was found that closed-loop performance (repeated rescheduling at each time-step) resulted in irregular performance as a function of horizon size. This is in contrast to most EMPC applications, where performance improves with each increase in horizon size [3, 4].

It has since been concluded that the open-loop EMPC problem does not converge as the horizon goes to infinity. In the current effort, we investigate the closed-loop EMPC problem using dynamic programming and the Hamilton-Jacobian-Bellman Equation (HJB). To avoid the curse of dimensionality a simple two state process scheduling example is used. Since the value function-results of the HJB depend on the objective function of the EMPC problem, a variety of (presumably equivalent) objective functions will be considered and compared.

[1] Lima, R. M.; Grossmann, I. E.; Jiao, Y. "Long-term scheduling of a single-unit multi-product continuous process to manufacture high performance glass." Comp. Chem. Eng. 2011, 35 (3), 554–574.

[2] Aleissa, Y.M., Inventory Creep Phenomenon in Process Scheduling, Master's Thesis, Illinois Institute of Technology (2016).

[3] Omell, B.P.; D.J. Chmielewski, "IGCC Power Plant Dispatch using Infinite-Horizon Economic Model Predictive Control" Ind. Eng. Chem. Res., 52(9) pp 3151-3164 (2013).

[4] Mendoza-Serrano, D.I.; D. J. Chmielewski, "Smart Grid Coordination in Building HVAC Systems: Computational Efficiency of Constrained ELOC " Sci. Tech. Built Env., 21, pp 812-823 (2015)

493345 Integration of Design, Scheduling and Control Under Uncertainty Via Model Based Multi-Parametric Programming

Abstract:

Multipurpose chemical processes are designed to meet the demands for multiple products or multiple grades of one product. Operability and profitability of these processes rely heavily on the decision making strategies in different time scales: short term (control), middle term (scheduling) and long term (design) strategies all of which contribute to capital and operating costs. Uncertainty also plays a key role at the different time scales, such as fluctuations in (i) the prices related to product, raw material, and energy, as well as (ii) raw material quality and specifications. The dynamics of the chemical processes are inherently affected by these uncertainties, necessitating the development of more robust and unified solutions. Simultaneous approaches for the decision making in the presence of uncertainties at these different layers are expected to deliver integrated and intensified processes with reduced cost and energy requirements [1] - a subject which has started to receive increasing attention in the open literature [2-7].

In this work, we present a unified framework to determine the long term optimal operation strategy of a process while accounting for the short term control and middle term scheduling decisions under process and market uncertainties. Based on (i) a single high-fidelity model, (ii) multi-parametric Rolling Horizon Optimization (mp-RHO) policies to readjust for the changing market structures and, (iii) multi-parametric Model Predictive Control (mp-MPC) for efficient set point tracking, we derive (i) design dependent and scheduling aware control strategies, and (ii) scheduling strategies that are design dependent and control aware. The framework is illustrated with three example problems of increasing complexity, (i) an isothermal continuous ow stirred tank reactor (CSTR) with multiple reactants and multiple products, (ii) two CSTRs in parallel to demonstrate resource allocation, and (iii) a system involving multiple residential heat and power (CHP) units.

[1] Pistikopoulos, E.N., Diangelakis, N.A., Towards the integration of process design, control and scheduling: Are we getting closer?, Computers and Chemical Engineering (2015), 91, pp.85-92.
[2] Patil, B.P., Maia, E., Ricardez-Sandoval, L.A., Integration of scheduling, design, and control of multiproduct chemical processes under uncertainty, AIChE Journal (2015), 61, pp.2456-2470.
[3] Terrazas-Moreno, S., Flores-Tlacuahuac, A., Grossmann, I.E., Simultaneous design, scheduling, and optimal control of a methyl-methacrylate continuous polymerization reactor, AIChE Journal (2008), 54, pp.3160-3170.
[4] Baldea, M., Harjunkoski, I., Integrated production scheduling and process control: A systematic review, Computers and Chemical Engineering (2014), 71, pp.377-390.
[5] Diangelakis, N.A., Burnak, B., Pistikopoulos, E.N., A multi-parametric programming approach for the simultaneous process scheduling and control - Application to a domestic cogeneration unit, Foundations of Computer Aided Process Operations/Chemical Process Control, 2017.
[6] Shi, H., You, F., A novel adaptive surrogate modeling-based algorithm for simultaneous optimization of sequential batch process scheduling and dynamic operations, AIChE Journal (2015), 61, pp. 4191-4209.
[7] Dias, L.S., Ierapetritou, M.G., Integration of scheduling and control under uncertainties: Review and challenges,

Chemical Engineering Research and Design (2016), 116, pp. 98-113.

500825 Superstructure-Based Process Synthesis of a Pre-Combustion Membrane CO_2 Capture System

Abstract:

Efficient improvement of carbon capture for fossil power generation is critical if timely mitigation of climate change is to be achieved. The state of the art pre-combustion capture approach is considered to be a solvent based Selexol system. To improve upon the Selexol based system for CO_2 capture, here we select an ionic liquid supported membrane based CO_2 separation system which has many opportunities to optimize process configuration and operation by varying membrane area, permeate recycle and compressor power. This work includes a flexible model of a counter-current CO_2-selective membrane module which serves as the basis for a computational framework for the large scale evaluation of this pre-combustion membrane-based carbon dioxide capture strategy.

This work employs a Mixed Integer Non-Linear Programming (MINLP) formulation to determine the optimal pre-combustion CO_2 membrane system configuration from a pre-defined set of potential process options. In MINLP, a set of binary state variables is distinguished from continuous state variables and characterizes the flowsheet topology. Binary variables in this formulation indicate whether a piece of equipment proposed in a superstructure is present in the optimal configuration. In general, this formulation is non-linear and non-convex presenting challenges for global optimization. The total problem formulation which characterizes the membrane system superstructure consists of $2.1\text{x}10^6$ potential process configurations characterized by 2,234 equations and 2,177 variables including 21 binary variables.

Results suggest that advanced membrane designs may be an improvement, albeit slight, over state of the art solvent (Selexol) based CO_2 separation systems. If projected improvements can be achieved in full, they can be expected to reduce required membrane area by 98% and compression power by 60% over current membrane performance. The COE of an advanced membrane based solution consequently drops by ~24% and becomes competitive with solvent based technology. However, the MINLP formulation synthesizes an optimal process configuration that would not be chosen by conventional heuristic design strategies for this type of application and results in an improvement of 7-8%. For a nominal 550MW plant, this represents an annual offset in required power generation costs of nearly $31M over the performance of conventional configurations.

497847 Screening for Economically Promising Bio-Based Chemicals

Abstract:

10A. Process Design: Innovation for Sustainability

Screening for Economically Promising Bio-Based Chemicals

Wenzhao Wu, Matthew Long, Jennifer Reed, Christos T. Maravelias*

Dept. of Chemical and Biological Engineering, University of Wisconsin-Madison, Madison, WI 53706

The last decade has seen tremendous progress in metabolic engineering and synthetic biology[1,2]. These advances enable the use of engineered microorganisms such as *Escherichia coli* and yeast for the production of chemicals that are currently derived mainly from fossil fuel feedstocks[3,4]. However, it is still unclear which chemicals have the highest economic potential, which is key in developing a sustainable chemical production system. To this end, we develop a framework for the identification of economically promising chemicals that can be produced biologically using microorganisms ("bio-based chemicals").

We first examine the US high-production-volume (HPV) chemicals[5], which are manufactured in or imported into the United States in amounts equal to or greater than 454 metric tons (MT) per year. HPV chemicals include all commodity chemicals and a portion of fine chemicals. We establish an HPV chemical database (3574 chemicals) by compiling several HPV lists published by the EPA over the past two decades. Then we intersect the HPV chemical database with the KEGG database, which includes most chemicals produced by characterized reactions in biological systems, and thus 613 overlapping chemicals are found. These chemicals are then imported into a genome-scale metabolic model we develop, and 170 chemicals are identified to be producible by microorganisms, often with the addition of heterologous reactions. These 170 chemicals are the complete pool of candidate targets for bio-based production. In addition, market volume and price data is collected for them.

We first use the following two screening criteria to quantify economic potential.

Criterion 1: market volume. The market volume (MT/year) should be greater than an expected production capacity.

Criterion 2: market value. The market value ($/year) should be large enough to attract investment and recover capital cost within an expected time horizon.

The difference between the price and the production cost also plays a key role in determining the economic potential of a product. The larger the difference between a chemical's selling price and its total production cost, the larger its economic potential. However, directly implementing a third criterion based on such a difference is challenging because the downstream separation cost is highly product-dependent and difficult to estimate. To this end, we use the "separation cost margin" as Criterion 3, which is the difference between the price and the upstream cost (including raw material supply cost and bio-conversion cost). We calculate the upstream cost using base cost data from the literature as well as the theoretical yield and productivity calculated using our metabolic model. For a specific product, if the margin is negative, then the product fails Criterion 3; if the margin is positive, then we compare it with the expected separation cost. In general, the separation cost depends on (1) the product properties, e.g., separation of extracellular insoluble products tends to be easier than that of extracellular soluble ones, especially when the product concentration is low; and (2) case-specific information, e.g., whether a known solvent exists for a particular product, and special product requirement such as colorlessness and dryness.

To account for these major separation cost drivers, we introduce a general downstream bio-separation superstructure optimization framework to provide guidance in the preliminary design of separation networks[6,7]. We then apply this framework and formulate superstructure models for different classes of bio-based chemicals[8] based on their properties such as localization (intracellular/extracellular), solubility (soluble/insoluble) and density with respect to water (light/heavy). Further, we identify the optimal separation processes for base cases and the key cost drivers (such as titer and technology performances). Finally, we generate sensitivity curves (e.g. Figure 1A) and "heat maps" (e.g. Figure 1B) to show how the overall cost changes with the variation in the key cost drivers, which can help us roughly estimate the cost of separation for different classes of products.

Figure 1. Cost estimates for intracellular insoluble products under varying parameters. (A) Sensitivity curve showing cost changes with the variation in titer; (B) "Heap map" showing the change of cost and the optimal cell harvesting technology with the variation in technology performances (centrifuge efficiency and filtration retention factor). The white lines show the threshold values where the optimal technology switches. The asterisk represents the base case.

As a *preliminary* result, we identify three chemicals as the most promising ones: glutaric acid, acrylamide and propanal, by setting a uniform separation cost margin threshold for all chemicals to be greater than 1.8 $/kg (an estimated margin for a case product - polyhydroxyalkanoate), a volume greater than 31751 MT/year (an estimated production capacity), and a market size greater than 200 million $/year (to recover capital cost in ~5 years supposing a 50% market share). More accurate results can be obtained by using the aforementioned bio-separation framework with a different separation cost margin.

References

[1] Gavrilescu, M. & Chisti, Y., 2005. Biotechnology- a sustainable alternative for chemical industry. *Biotechnology advances,* Volume 23, pp. 471-499.

[2] Bornscheuer, U. T. & Nielsen, A. T., 2015. Editorial overview: chemical biotechnology: interdisciplinary concepts for modern biotechnological production of biochemicals and biofuels. *Current Opinion in Biotechnology,* Volume 35, pp. 133-134.

[3] Wilson, S. A. & Roberts, S. C., 2014. Metabolic engineering approaches for production of biochemicals in food and medicinal plants. *Current Opinion in Biotechnology,* Volume 26, pp. 174-182.

[4] Zhang, X., Tervo, C. J. & Reed, J. L., 2016. Metabolic Assessment of E. coli as a Biofactory for Commercial Products. *Metabolic Engineering,* Volume 35, pp. 64-74.

[5] US EPA, 2004. High-production-volume (HPV) chemicals status and future directions of the HPV Challenge Program. *Office of Pollution Prevention and Toxics*, Washington DC.

[6] Wu, W., Yenkie, K.M. & Maravelias, C.T., 2017. A superstructure-based framework for bio-separation network synthesis. *Computers & Chemical Engineering*, Volume 96, pp. 1-17.

[7] Yenkie, K. M., Wu, W., Clark, R. L., Pfleger, B. F., Root, T. W., & Maravelias, C. T., 2017. A roadmap for the

synthesis of separation networks for the recovery of bio-based chemicals: Matching biological and process feasibility. *Biotechnology Advances*, Volume 34(8), pp. 1362-1383.

[8] Yenkie, K. M., Wu, W., Maravelias, C.T., 2017. Synthesis and analysis of separation networks for the recovery of intracellular chemicals generated from microbial-based conversions. *Biotechnology for Biofuels*

491541 Supercritical FluidÂ Carnot Cycle for Low Temperature Waste Heat Utilization

Abstract:

Supercritical Fluid Carnot Cycle for Low Temperature Waste Heat Utilization

Many chemical processes reject low-grade heat to the environment. While it is difficult to recover low temperature waste for work production, it would be useful to design processes that, in principle, could do this. In our presentation we describe computer simulations of a process for doing this using supercritical carbon dioxide as a working fluid for the power cycle. Given the low critical temperature of carbon dioxide and the desirability of keeping the working fluid in its one phase region, CO_2 is an ideal fluid for designing such a process. Using a temperature close to its critical point for the rejection leg of the Carnot cycle, we exploit the natural flattening of the critical isotherm in this region which leads to efficient work production in Carnot engines using small temperature driving forces between the input and rejection legs of the process. The concepts appear promising as a basis for designing experimental systems for recovering low- grade waste heat for conversion into work using supercritical CO2 as the power cycleÕs working fluid.

499181 Designing Techno-Ecological Synergies While Accounting for Ecosystem Dynamics

Abstract:

Industrial processes are designed from a superstructure of multiple production pathways and the most optimal design is selected based on multiple objectives. Design objectives usually include economic goals such as maximizing net present value while simultaneously reducing impact from processes by choosing raw materials that are least toxic and identifying pathways that result in minimum emissions. However, this method towards designing industrial systems fails to address the underlying issue that the current demand for natural resources and ecosystem goods and services exceeds the rate at which these services can be supplied by ecosystems. This difference between the ecosystem service demand and supply is largely unaccounted for during design of industrial systems and ignoring the role of supporting ecosystems and the limit on ecosystem service supply has been one of the major causes for ecosystem degradation.

Originating from the concept of `unit operations', this work expands the scope of traditional process design methodologies to include ecosystem services within the design framework, by treating ecological systems as unit operations, with the goal of building process flowcharts based on the unit operation concept of ecosystems. Previous work on the use of Techno-Ecological Synergy (TES) framework for the assessment and design of systems including industrial sites [1], residential [2] and agricultural systems [3] have shown the economic and environmental benefits of using this approach. However, none of these studies consider interactions between detailed process-level decisions and the design of supporting ecosystem services. In this presentation, we show that simultaneous selection of process design variables and ecological variables while accounting for the variability and dynamics of ecosystems results in robust systems that tend towards absolute sustainability while being economically feasible. Variability and dynamics of ecological systems arises due to changes in meteorological conditions and seasonal variations. We show that, such robust systems can be designed by identifying a combination of technological and ecological alternatives.

The overall system is designed to maximize the total net present value and maximize the net ecosystem services supply with the goal of achieving absolute sustainability. A multi-period optimization problem formulated as a nonlinear program (Mo-NLP) is solved and tradeoffs are identified between maximizing the net present value of the system and the net ecosystem service supply for multiple ecosystem services for multiple time periods in a year.

Application of this framework to a biofuel production system along with a forest and wetland ecosystem indicates that systems designed by accounting for ecosystem service supply results in novel designs that have a higher economic value, and higher sustainability index. Preliminary results for the multi-period optimization problem revealed that selecting a combination of technological and ecological options during certain seasons have a higher economic and environmental performance compared to a purely technological design. Thus, coupled TES systems are expected to have economic and environmental advantages over conventional systems design, mainly due to the reduction in external inputs by moving towards a closed loop production system.

References:

[1] Gopalakrishnan, V., Bakshi, B.R. and Ziv, G., 2016. Assessing the capacity of local ecosystems to meet industrial demand for ecosystem services. AIChE Journal, 62(9), pp.3319-3333.

[2] Urban, R.A. and Bakshi, B.R., 2013. Techno-ecological synergy as a path toward sustainability of a North American residential system. Environmental science & technology, 47(4), pp.1985-1993.

[3] Hanes, R.J, Gopalakrishnan, V., Bakshi, B.R. Synergies and trade-offs in renewable energy landscapes: Balancing energy production with economics and ecosystem services. Submitted. Applied Energy.

492835 Design and Dynamic Simulation of a Solar and Natural Gas Hybrid Power Plant to Investigate the Synergies of Hybridization

Abstract:

Limited fossil energy resources and the greenhouse gas effect on the environment is creating increasing pressure over the world to look for alternative energy resources that are sustainable and ecofriendly [1]. Solar energy is currently among the popular renewable energy resources due to its recent advancements in technology and economics. However, reliability of energy supply is considered among the major drawbacks for stand-alone solar systems, as solar energy is not available during nighttime and intermittent on cloudy days. Hybridizing solar systems with fossil fuel can resolve the issue of reliability and also reduce the infrastructure cost compared to a pure solar plant [2]. Concentrated solar power (CSP) system are viable for large scale power production [3]. Parabolic trough are widely used CSP technologies as they are easy to finance and their maturity level in the market [4]. Hybridizing solar power plant with natural gas requires minimal capital cost for infrastructure [5]. In this study, we have designed and modeled a hybrid solar and natural gas plant in comparison to a stand-alone solar plant with thermal energy storage. The novel features of the hybrid design include supplemental fuel firing to ensure that high superheat temperatures are always reached in order to ensure a high cycle efficiency and the ability to operate the plant flexibly between solar and fuel modes. Another novel contribution of this work is that the design is complete with a fully dynamic model to capture the effects of rapidly changing solar conditions and an active control system to ensure that plant automation can handle these fluctuations. The hybrid system demonstrates an increase in solar-to-electric efficiency of 15 to 26% when compared to the stand-alone system. The overall efficiency also increases from 15 to 34 % in the same comparison. The hybrid system also demonstrates better environmental performance than a stand-alone system that would otherwise rely on fossil fuel backup generators at grid scale.

Reference:

[1] G. Bakos and C. Tsechelidou, "Solar aided power generation of a 300 MW lignite fired power plant combined with line-focus parabolic trough collectors field," *Renew. Energy*, 2013.

[2] Y. Zhao, H. Hong, and H. Jin, "Evaluation criteria for enhanced solar–coal hybrid power plant performance," *Appl. Therm. Eng.*, 2014.

[3] H. Price *et al.*, "Advances in Parabolic Trough Solar Power Technology," *J. Sol. Energy Eng.*, vol. 124, no. 2, p. 109, 2002.

[4] J. Peterseim, A. Tadros, U. Hellwig, and S. White, "Increasing the efficiency of parabolic trough plants using thermal oil through external superheating with biomass," *Energy Convers.*, 2014.

[5] J. H. Peterseim, S. White, A. Tadros, and U. Hellwig, "Concentrated solar power hybrid plants, which

technologies are best suited for hybridisation?," *Renew. Energy*, vol. 57, pp. 520–532, 2013.

488553 The Design of Sustainable Carbon Dioxide Capture and Conversion Processes Considering Various Locations, Products and Routes

Abstract:

Due to climate change and growing concerns about global warming, methods of mitigating carbon dioxide emissions are important. The three most common approaches to mitigating emissions and reducing the concentration in the atmosphere are: improving the efficiency of processes, carbon dioxide capture and storage (CCS), and carbon dioxide capture and utilization (CCU). It is questionable, however, if they alone can satisfy the requirements; therefore, they need to be developed concurrently to provide an overall sustainable solution. Carbon dioxide utilization, which takes the captured carbon dioxide and reuses it, can provide an economic offset to the costs of capture and other new technologies in the formation of commercial products. Therefore, the sustainable design of carbon dioxide capture and utilization, especially conversion, technologies is important. To ensure this, a computer-aided, systematic framework for the synthesis-design of carbon dioxide utilization processes, inclusive of carbon dioxide capture, has been developed. This framework uses three stages, (i) process synthesis, (ii) process design and (iii) innovation, for the design of sustainable processes. In this work, the framework is implemented so that various scenarios - varying locations, prices or objective function â€" could be investigated. The results provide insight into what processes and products should be produced via carbon dioxide conversion and how much of emissions can be mitigated in this way.

In the first stage, a superstructure of 23 steps containing over 150 intervals is optimized to find the optimal route, amongst over 30 feasible routes, to take captured carbon dioxide to a product. To aid the optimization, the developed database contains over 20 reactions, and the production of over 10 products and byproducts (methanol, dimethyl ether, ethanol, methane, dimethyl carbonate, ethylene carbonate, ethylene glycol, propylene carbonate, propylene glycol, acetic acid and succinic acid) via over 30 feasible processing routes. Optimizations for seven different scenarios are performed, in terms of prices, demands, locations and objective, to determine the impact that carbon dioxide capture and conversion can have. The seven scenarios include, (i) optimization for a fixed amount of captured CO_2 considering operating costs and capital costs (covers 2 scenarios), (ii) optimization under price uncertainty with respect to products (1 scenario), (iii) optimization with price variability for utilities, chemicals and products, to reflect sensitivity and location variability (covers 2 scenarios), (iv) optimization with variable conversion (1 scenario), and (v) optimization considering a fixed demand of the products to determine the amount of CO_2 utilized for different paths (1 scenario). The results show that if the prices of certain products changed, such as if the price of dimethyl carbonate were doubled or the price of succinic acid were decreased by half, what to produce and how to produce it is greatly influenced. Changing the parameters â€" such as prices of utilities and materials or demands and availability of carbon dioxide sources - according to uncertainty and location, changes the optimal processing route. Considering seven scenarios, the optimal routes are found to produce succinic acid, dimethyl carbonate and methanol, amongst other value-added products.

After the optimal processing paths are found for the scenarios, they are simulated and analyzed in the second stage. A simulation library containing over 30 different capture and conversion processes to various products has been developed, including the optimal routes determined in the scenarios. In Stage 3, the production of dimethyl carbonate via different carbon dioxide utilization routes and the production of succinic acid are selected for further improvement as they are among the highest rated processes. For these cases, the targeted sustainability indicators can be improved; in the case of dimethyl carbonate production via ethylene carbonate, for example, the net carbon dioxide emissions (considering also energy production), operating costs and capital costs can be reduced by more than 10%. For already innovative processes, this is not always necessary. However, for processes which already exist or are not already carbon dioxide reducing, this can be used to find more sustainable processes.

This presentation will highlight the framework and the corresponding models, methods and tools, especially the database and simulation library. In addition, the application of this framework will be highlighted through a case study, which will encompass superstructure optimization of a large network of capture and utilization processes for all seven scenarios. The detailed design and analysis will be shown for the results of these scenarios. Finally, some conclusions about the role of capture and utilization as a bridge between current and future technology will be discussed.

491405 Optimal Integrated Plant for Waste to Biodiesel Production

Abstract:

Current developed societies generate large volumes of waste. The origin can vary from sludge from water treatment, urban waste, forest residues and manure mostly. Apart from the volume, their composition is complex and dangerous. Anaerobic digestion presents a technological option to stabilize the waste while generating added value in the form of methane and digestate that can be used as fertilizer. For decades, the aim of processing the waste has focused on power production [1]. However, biogas is an interesting carbon source, not only because of its content in methane, but also due to the amount of CO2 available. Dry reforming can be used to process this biogas into methanol [2]. A step further allows the production of biodiesel from waste. The production of biodiesel from algae requires nutrients for algae growing [3], that can be provided by the digestate. Next, the methanol produced from the biogas is used for the transesterification of the oil.

In this work we have optimized an integrated facility for the production of fully renewable biodiesel from waste via anaerobic digestion. On the one hand, biogas is produced from cattle and pig manure. After its purification, it is dry reformed to obtain syngas. Methanol is later synthesized from the syngas. On the other hand, the digestate remaining is used as nutrient to grow algae. The effect of the nutrient composition on the algae growth is considered. The algae are harvested and the oil is extracted using mechanical press and a solvent, that it is later recovered. The oil is transesterified using the methanol produced from the biogas. The problem is formulated as a large NLP for the optimal operating conditions, biogas composition and digestate nutrient content. Dry reforming, methanol and biodiesel synthesis reactors are modelled to account for the effect of the operating conditions on the yield. Distillation columns are modelled using short cut methods but including constraints on the reflux ratio [2,4], the operating pressures and temperatures from the literature and the rigorous simulation of the column in CHEMCAD [4].Optimal HEN and water networks are designed afterwards.

3110kt/y of manure are processed for the production of almost 90Mgal/y of biodiesel and 46.7 kt/y of methanol. The process uses part of the biogas to fuel the reformer but still needs 1.5 MJ/gal of FAME and 1.22 gal of water per galon of biodiesel. The production cost of biodiesel is competitive with previous work [2], 0.31€/gal, but the investment is higher due to the need to produce the biogas, 193 M€.

References

1.-León E., Martín, M. (2016) Optimal production of power in a combined cycle from manure based biogas Energ. Conv. Manag. 114 89–99

2.-Hernández, B, Martín, M. Optimal composition of the biogas for dry reforming in the Production of Methanol. IECR. 10.1021/acs.iecr.6b01044

3. -Xin, L., Hong_ying, H., Ke, G., Ying-xue, S. (2010) Effects of different nitrogen and phosphorus concentrations on the growth, nutrient uptake, and lipid accumulation of a freshwater microalga Scenedesmus sp Bioresource Technology 101 5494–5500

4.-Martín, M., Grossmann, I.E. (2012) Simultaneous optimization and heat integration for biodiesel production from cooking oil and algae. Ind. Eng. Chem Res. 51 (23), 7998–8014

497658 Advancing the Production of Olefins and Aromatics from Natural Gas Via Methanol: Chemical Looping for Syngas Generation

Abstract:

Due to technological advances in the shale gas industry, the supply of natural gas in the United States has greatly increased (37% since 2008[1]) and its price has significantly dropped ($2.52/MMBtu in 2016[2]). The shale gas boom has not only revitalized the U.S. economy, but also rejuvenated interests in C1 catalysis toward the production of chemicals typically derived from petroleum refining[3]. Olefins (ethylene, propylene, butane isomers, butadiene) and aromatics (benzene, toluene, xylenes) are examples of valuable chemicals in high demand.

As diesel continues to command a larger portion of the crude oil usage, less naphtha will be available for petrochemical production. Given its availability, low cost, and hydrogen-rich composition, natural gas is an attractive feedstock for the on-purpose production of chemicals[4]. Conversion of natural gas to a methanol intermediate for further upgrading to chemicals is a potential avenue for chemicals production and is a promising commercial candidate for natural gas based refineries. Currently, methanol is almost entirely produced from syngas.

However, a judicious utilization of natural gas resources necessitates a systematic analysis of all available processing routes. To this aim, a gas to olefins and aromatics (GTOA) superstructure has been developed[5,6] and includes several olefins and aromatics production technologies. Previous works investigated reforming as the primary method to convert natural gas to syngas and demonstrated the profitability of GTOA processes. Recently, the benefits of chemical looping[7,8] for syngas generation have been observed in lowering the breakeven oil prices for a gas to liquids (GTL) process[9]. In this work, we incorporate chemical looping into the GTOA superstructure as an alternative to reforming and study its impact on chemicals production.

Process synthesis of the GTOA superstructure forms a large-scale nonconvex mixed-integer nonlinear model (MINLP). A tailored deterministic global optimization branch-and-bound algorithm[10] is used to solve the MINLP and determine optimal natural gas refineries with the highest profit. Several case studies are investigated to compare chemical looping against reforming for natural gas conversion. The effect of different plant capacities and product ratios are explored as well. Key topological selection of process technologies will be discussed.

1. U.S. Energy Information Administration. "Monthly Energy Review – March 2017". Available at: http://www.eia.gov/totalenergy/data/monthly/pdf/mer.pdf. Accessed April 2017.

2. U.S. Energy Information Administration. "Henry Hub Natural Gas Spot Price". Available at:

http://www.eia.gov/dnav/ng/hist/rngwhhdm.htm. Accessed April 2017.

3. Mokrani T., Scurrel M. Gas conversion to liquid fuels and chemicals: the methanol route-catalysis and processes development. Catalysis Reviews 2009, 51, 1-145

4. Floudas, C. A.; Niziolek, A. M.; Onel, O.; Matthews, L. R. Multi-scale systems engineering for energy and the environment: Challenges and opportunities. AIChE Journal 2016, 62 (3), 602-623.

5. Onel, O.; Niziolek, A. M; Floudas, C. A Optimal Production of Light Olefins from Natural Gas via the Methanol Intermediate. Industrial & Engineering Chemistry Research 2016, 55 (11), 3043-3063.

6. Niziolek, A.M.; Onel, O.; Floudas, C.A. Production of Benzene, Toulene, and Xylenes from Natural Gas via Methanol: Process Synthesis and Global Optimization. AIChE Journal 2016, 62 (5), 1531-1556.

7. Luo, S.; Zeng, L.; Xu, D.; Kathe, M.; Chung, E.; Deshpande, N.; Qin, L.; Majumder, A.; Hsieh, T.-L.; Tong, A.; Sun, Z.; Fan, L.-S. Shale gas-to-syngas chemical looping process for stable shale gas conversion to high purity syngas with a H2:CO ratio of 2:1. Energy & Environmental Science 2014, 7 (12), 4104–4117.

8. de Diego, L. F.; Ortiz, M.; García-Labiano, F.; Adánez, J.; Abad, A.; Gayán, P. Hydrogen production by chemical-looping reforming in a circulating fluidized bed reactor using Ni-based oxygen carriers. Journal of Power Sources 2009, 192 (1), 27–34.

9. Tso, W.W.; Niziolek, A.M.; Onel, O.; Demirhan, C.D.; Floudas, C.A.; Pistikopoulos, E.N.; Enhancing Natural Gas-to-Liquids (GTL) Processes Through Chemical Looping for Syngas Production: Process Synthesis and Global Optimization. *In preparation*.

10. Baliban, R. C; Elia, J. A; Misener, R.; Floudas, C. A Global optimization of a MINLP process synthesis model for thermochemical based conversion of hybrid coal, biomass, and natural gas to liquid fuels. Computers & Chemical Engineering 2012, 42, 64-86.

497121 Minimizing the Economic Impact of Amine Scrubbing Using High Fidelity Modeling and Optimization

Abstract:

Minimizing the economic impact of amine scrubbing using high fidelity modeling and optimization

R.C. Pattison, C. Tsay, Y. Zhang, P.T. Frailie, G.T. Rochelle, and M. Baldea

McKetta Department of Chemical Engineering, The University of Texas at Austin, Austin, TX 78712

In an effort to minimize carbon emissions from existing power generation units, amine scrubbing technologies have garnered significant attention owing to the energetically favorable set of reactions between carbon dioxide and amine solvents. A large body of research has been dedicated to this field, with a primary focus on characterizing the reaction kinetics, mass transfer properties, and thermodynamic properties for various solvents [1-2]. Process modeling and simulation studies have also been carried out, with a focus on predicting the performance and cost of the absorption and stripping units [3]. However, relatively little research has been focused on the holistic economic analysis and optimization of the entire process using detailed unit models.

Owing to the complexity of these systems and the numerous design decisions that impact process performance, identifying the most economically compelling designs is extremely challenging, and calls for the use of modern optimization tools. These must be used to consider all the design decisions and constraints simultaneously, such that capital cost and energy use are minimized while simultaneously guaranteeing that the process meets required removal specifications and operational constraints.. While detailed rate-based mass transfer separation models are available in various software programs, the corresponding equations are typically highly nonlinear, coupled and stiff; they are difficult to solve (particularly when embedded within process flowsheets), and thus, not suitable for process-wide optimization.

In this presentation, we introduce a novel framework for plant-wide design optimization of amine-based carbon capture processes. We describe a set of experimentally-validated first-principles and empirical rate-based models of the absorption and stripping columns, along with empirical but detailed capital cost estimates for the process units. The numerical underpinnings of our approach are derived from the pseudo-transient continuation methods and the relaxation-based flowsheet optimization algorithms introduced in our previous work [4-5].

We demonstrate our framework with two extensive case studies, considering amine-based carbon capture plants having different carbon dioxide removal requirements (90% and 99%) from flue gas. The design optimization objective function is designed to minimize the present value (based on both capital and operating expenses) of the plant in the two cases while ensuring that the removal requirements are met and operational constraints are satisfied. The results show significant cost reduction when compared to a heuristics-guided design.

References

[1] Frailie, P.T.; Plaza, J.; Van Wagener, D.; Rochelle, G.T. Modeling piperazine thermodynamics. *Energy Procedia*, **2011**, 4, 35-42.

[2] Sherman, B.; Frailie, P.T.; Li, L.; Salta, N.; Rochelle, G.T. Thermodynamic and Kinetic Modeling of Piperazine/ 2-Methylpiperazine. *Energy Procedia*, **2014,** 63, 1243-1255.

[3] Frailie, P.T. *Modeling of carbon dioxide absorption/stripping by aqueous methyldiethanolamine/ piperazine*. Doctoral Dissertation, 2014.

[4] Pattison, R.C.; Baldea, M. Equation-oriented flowsheet simulation and optimization using pseudo-transient models. *AIChE J.*, **2014,** 60, 4104-4123.

[5] Pattison, R.C.; Tsay, C.; Baldea, M. Pseudo-transient models for multiscale, multiresolution simulation and optimization of intensified reaction/separation/recycle processes: Framework and a dimethyl ether production case study. *Comp. & Chem. Eng.*, **2017.**

IDAES
Institute for the Design of Advanced Energy Systems

Optimal model synthesis for solvent-based CO_2 capture systems by simultaneously using multi-scale data

Paul T. Akula[a], John Eslick[b], Debangsu Bhattacharyya[a], David Miller[b]

[a] *Department of Chemical and Biomedical Engineering, West Virginia University, Morgantown, WV*
[b] *National Energy Technology Laboratory, Pittsburgh, PA*

AIChE Annual Meeting, Minneapolis, MN
November 2nd, 2017

Outline

- **Background**
- **Model Development**
  - **Packed Column**
  - **Wetted- Wall Column (WWC)**
- **Parameter Estimation**
  - **Sequential**
  - **Simultaneous**
- **Challenges**
- **IDAES Platform**
- **Results**
- **Conclusion**

IDAES · NATIONAL ENERGY TECHNOLOGY LABORATORY · Sandia National Laboratories · Carnegie Mellon · West Virginia University · U.S. DEPARTMENT OF ENERGY

2

Solvent-Based CO_2 Capture System

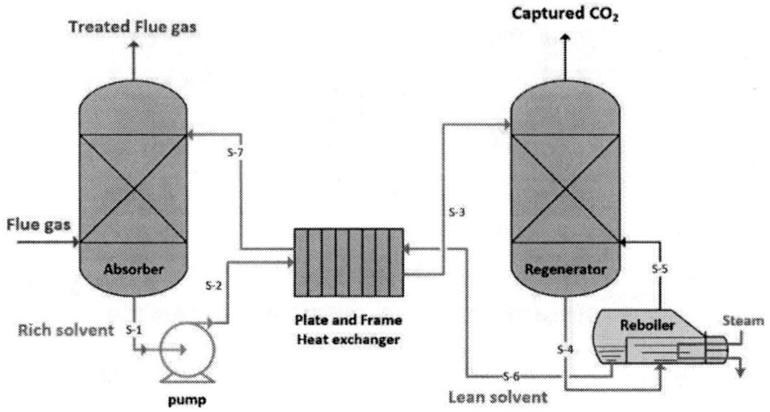

IDAES NATIONAL ENERGY TECHNOLOGY LABORATORY Sandia National Laboratories Carnegie Mellon West Virginia University U.S. DEPARTMENT OF ENERGY

Packed Column

Mass Balance

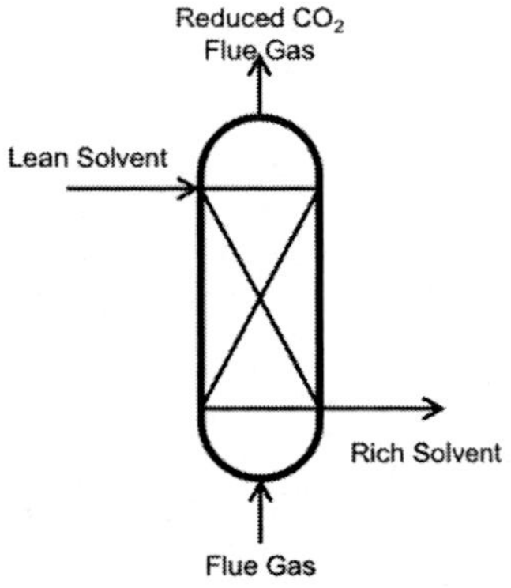

Vapor Phase

$$\varepsilon_V \frac{\partial C_{i,V}}{\partial t} = -u_V \frac{\partial C_{i,V}}{\partial z} - N_{i,V} \qquad i = 1 \ldots . n_{c^V}$$

Liquid Phase

$$\varepsilon_L \frac{\partial C_{i,L}}{\partial t} = u_L \frac{\partial C_{i,L}}{\partial z} + N_{i,L} \qquad i = 1 \ldots . n_{c^V}$$

Mass transfer rate

$$N_{i,V} = K_{V,i} a_e C_{tot,V} \left(y_i - y_i^{*,eq} \right) + N_{tot,V} \left(y_i \right) \qquad i = 1 \ldots . n_{c^V}$$

$$N_{i,V} = N_{i,L}$$

- 1D PDE-based model with plug flow
- Rate-based
- Enhancement factor approach
- NRTL model for liquid phase nonideality

IDAES NATIONAL ENERGY TECHNOLOGY LABORATORY Sandia National Laboratories Carnegie Mellon West Virginia University U.S. DEPARTMENT OF ENERGY

Packed Column

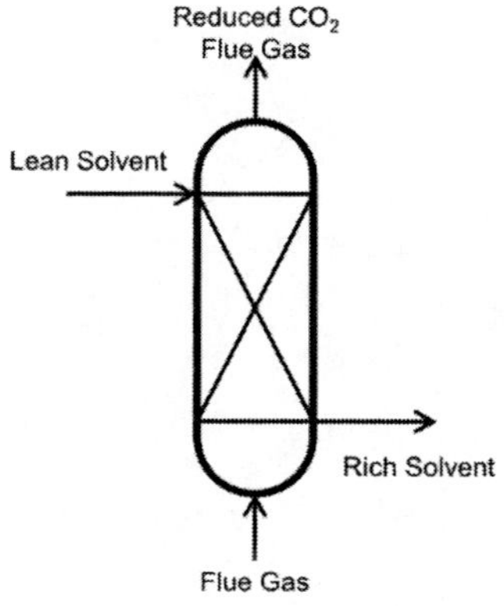

Energy Balance

Vapor Phase

$$\varepsilon_V C_{tot,V} C_{p,V} \frac{\partial T_V}{\partial t} = -u_V C_{tot,V} C_{p,V} \frac{\partial T_V}{\partial z} + Q_{h,V}$$

Liquid Phase

$$\varepsilon_L C_{tot,L} C_{p,L} \frac{\partial T_L}{\partial t} = u_L C_{tot,L} C_{p,L} \frac{\partial T_L}{\partial z} - Q_{h,L}$$

Heat transfer rate

$$Q_{h,V} = a_e h_{VL} (T_L - T_V)$$

$$Q_{h,L} = a_e h_{VL} (T_L - T_V) + N_{CO_2,V} \Delta H_{CO_2} - N_{H_2O,g} \Delta H_{H_2O}$$

Packed Column

Vapor Liquid Equilibrium

For MEA and H2O

$$y_i^* P = \gamma_i x_i^* p_i^s; \quad i = \text{MEA,H2O}$$

For CO2

$$y_i^* P = \gamma_i x_i^* \left(\frac{H_i}{\gamma_i^\infty} \right); \quad where\ \gamma_i^\infty = \lim_{x_i \to 0} \gamma_i\ for\ i = \text{CO2};$$

Henry's constant

$$\ln\left(\frac{H_i}{\gamma_i^\infty}\right) = \sum_s \ln\left(\frac{H_{is}}{\gamma_{is}^\infty}\right) \frac{x_s (V_{cs})^{2/3}}{\sum_{s'} x_{s'} (V_{cs'})^{2/3}} \qquad s = s' = \text{MEA,H2O}$$

Kinetics

For CO_2

$$\frac{1}{K_{V,CO_2}} = \frac{1}{k_{V,CO_2}} + \frac{m_{He,CO_2}}{k_{L,CO_2} E}$$

- Physical equilibrium
- Chemical equilibrium
- diffusivity

Enhancement factor

$$E_{abs} = \frac{\sqrt{k_{rx,CO_2} D_{L,CO_2}}}{k_{L,CO_2}}$$

$$E_{desp} = 1 + \frac{\left(D_{L,MEACOO^-} / D_{L,CO_2}\right) \sqrt{K} C^*_{MEA,L}}{\left[1 + 2 D_{L,MEACOO^-} \sqrt{\frac{K C^*_{CO_2}}{D_{L,MEA}}}\right] \left(\sqrt{C^*_{CO_2,L}} + \sqrt{C^{int}_{CO_2,L}}\right)}$$

Wetted-Wall Column (WWC)

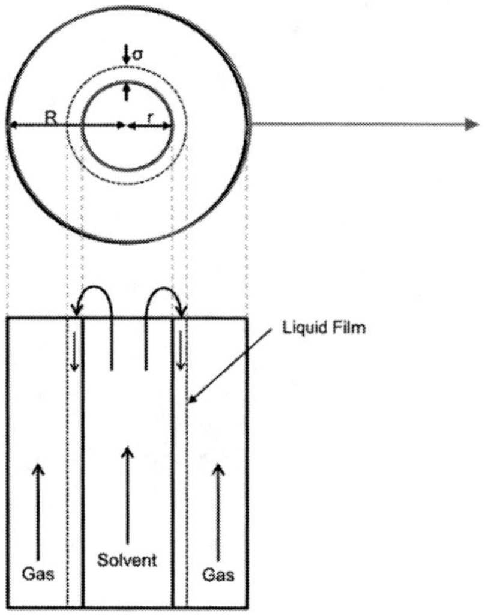

Equivalent Column diameter

$$D = 2\sqrt{(R+r)(R-r)}$$

Specific Interfacial Area

$$a_p = a_e = \frac{\text{interfacial area}}{\text{volume of annulus}} = \frac{2(r+\delta)}{(R^2 - r^2)}$$

Void fraction and hydraulic diameter

$$\varepsilon_P = 1, \quad d_h = \frac{4A_c}{P} = 2(R-r)$$

Liquid hold up

$$\varepsilon_L = \frac{\text{volume of liquid film}}{\text{volume of annulus}} = \frac{(r+\delta)^2 - r^2}{R^2 - r^2}$$

Average liquid velocity

$$u_L = \frac{\rho g \delta^2}{3\mu}$$

IDAES NATIONAL ENERGY TECHNOLOGY LABORATORY Sandia National Laboratories Carnegie Mellon West Virginia University U.S. DEPARTMENT OF ENERGY

Sub-Models Containing Estimated Parameters

Equilibrium constant

$$K_1 = \frac{C_{RNHCOO^-} C_{RNH_3^+}}{C^*_{CO_2,L} C^2_{RNH_2}}$$

$$K_2 = \frac{C_{RNH_3^+} C_{HCO_3^-}}{C^*_{CO_2,L} C_{RNH_2}}$$

Reaction Equilibrium Model
(WWC & Packed Column)

$$K_{eq,i} = \exp\left(a_i + \frac{b_i}{T} + c_i \ln T\right)$$

Mass transfer Model
(WWC & Packed Column)

$$k_{v,CO_2} = \frac{C_v}{\sqrt{\varepsilon_p - \varepsilon_L}} \frac{\sqrt{a}}{d_h} D_{v,CO_2} \left(\frac{u_v}{a\nu_v}\right)^{3/4} \left(\frac{\mu_v}{\rho_v D_{v,CO_2}}\right)^{1/3}$$

$$k_{L,CO_2} = 12^{1/6} C_L \left(\frac{u_L}{\varepsilon_L}\right)^{1/2} \left(\frac{D_{v,CO_2}}{d_h}\right)^{1/2}$$

Interfacial area Model
(Packed Column Only)

$$\left(\frac{a_e}{\beta_1}\right)^{\beta_2} = \frac{\rho_L}{\sigma_L} u_L^{\beta_3}$$

IDAES NATIONAL ENERGY TECHNOLOGY LABORATORY Sandia National Laboratories Carnegie Mellon West Virginia University U.S. DEPARTMENT OF ENERGY

Typical Approach To Parameter Estimation: Sequential

1st Step Optimization
(smaller scale apparatus or eqp.)

$$\min_{\theta_1}(y-\eta)^T\Sigma_1^{-1}(y-\eta)$$

s.t.

$$\dot{\eta}=f(\eta,u,\theta_1,\theta_2)$$
$$g(\eta,u,\theta_1,\theta_2)\le 0$$
$$u_{L1}\le u\le u_{U1}$$
$$y_{L1}\le y\le y_{U1}$$

Small scale (Wetted wall Column)

- Mass transfer
- Diffusivity
- Reaction kinetic Parameters

→

2nd Step Optimization
(larger scale apparatus or eqp.)

$$\min_{\theta_2}(y-\varphi)^T\Sigma_2^{-1}(y-\varphi)$$

s.t.

$$\dot{\varphi}=f(\varphi,u,\theta_1,\theta_2)$$
$$h(\varphi,u,\theta_1,\theta_2)\le 0$$
$$u_{L2}\le u\le u_{U2}$$
$$y_{L2}\le y\le y_{U2}$$

Large scale (Packed Column)

- Interfacial area
- Holdup
- Mass transfer
- Diffusivity
- Reaction kinetic Parameters

IDAES NETL National Energy Technology Laboratory Sandia National Laboratories Carnegie Mellon West Virginia University U.S. Department of Energy

9

Limitations of the Sequential Approach

- Mass Transfer and chemical reactions occur simultaneously.
- Experiment must isolate chemical reactions & mass transfer.
- Small scale data may not be representative of large scale equipment.

IDAES NETL National Energy Technology Laboratory Sandia National Laboratories Carnegie Mellon West Virginia University U.S. Department of Energy

10

Our Approach: Simultaneous

$$\min_{\theta_1,\theta_2}(y-\eta)^T\Sigma_1^{-1}(y-\eta)+(y-\varphi)^T\Sigma_2^{-1}(y-\varphi)$$

s.t.

$$\dot{\eta}=f(\eta,u,\theta_1,\theta_2)$$
$$\dot{\varphi}=f(\varphi,u,\theta_1,\theta_2)$$
$$h(\varphi,u,\theta_1,\theta_2)\leq 0$$
$$g(\eta,u,\theta_1,\theta_2)\leq 0$$
$$min(u_{L1},u_{L2})\leq u\leq max\,(u_{U2},u_{U2})$$
$$min(y_{L1},y_{L2})\leq y\leq max(y_{U2},y_{U2})$$

Data:
Small scale (Wetted wall Column)
Large scale (Packed Column)

IDAES NATIONAL ENERGY TECHNOLOGY LABORATORY Sandia National Laboratories Carnegie Mellon West Virginia University U.S. DEPARTMENT OF ENERGY

11

Simultaneous Parameter Estimation

$$\min_{\theta\in\{\theta_1,\theta_2...\}}\sum_{i=1}^{NG}w_i\left(\sum_{j=1}^{ND}(\varepsilon_{1,i,j}^T\sigma_{1,i,j}^{-2}\varepsilon_{1,i,j}+\varepsilon_{2,i,j}^T\sigma_{2,i,j}^{-2}\varepsilon_{2,i,j})\right)$$

subject to:

WWC model $\forall$i, $\forall$j
packed Column $\forall$i, $\forall$j
$\boldsymbol{\theta}^L\leq\boldsymbol{\theta}\leq\boldsymbol{\theta}^U$

where:
$\boldsymbol{\theta}$ = vector of estimated parameters
σ = standard deviation of data points
NG = number of data set
ND = number of data points in data set
w_i = weight for data set i

$$\varepsilon_{1,i}^T\sigma_{1,i}^{-2}\varepsilon_{1,i}=\frac{(CO_2flux_{wwc,i}^{exp}-CO_2flux_{wwc,i}^{cal})^2}{\sigma_{CO_2flux,wwc,i}^2}$$

$$\varepsilon_{2,i}^T\sigma_{2,i}^{-2}\varepsilon_{2,i}=\frac{(Loading_{abs,i}^{exp}-Loading_{abs,i}^{cal})^2}{\sigma_{loading,abs,i}^2}$$

Data

Bench scale - WWC
Pilot scale - NCCC

} Wide range of operating conditions {

- Liquid flowrate
- Gas flowrate
- Solvent loading
- Temperature
- Flue gas concentration

Data Sources
- National Carbon Capture Center (NCCC) Pilot Plant
- Bench Scale WWC[1]

[1]Experimental data retrieved from Dugas, R.E . (2009). *Carbon Dioxide Absorption, Desorption, and Diffusion in Aqueous Piperazine and Monoethanolamine.* (UT Austin Phd Thesis)

IDAES NATIONAL ENERGY TECHNOLOGY LABORATORY Sandia National Laboratories Carnegie Mellon West Virginia University U.S. DEPARTMENT OF ENERGY

12

Challenges

Size

- Number of equations (absorber ~15,000, WWC~6,000)
- Number of data points (absorber data from NCCC ~30 set, WWC data ~200 set)
- Total number of equation (simultaneously considered):~1,650,000

Nature

- Highly nonlinear equations
- Different scales (WWC height~9 cm and diameter~ 1.5 cm and NCCC absorber height ~20 m and diameter~ 0.64 m)

Capabilities Required

- Commercial simulators: primarily sequential
- Process <u>and</u> property sub-models
- Large # of equations
- Large amount of experimental data

Core Pyomo Capabilities Enabling IDAES

- **Hierarchical modeling** PYOMO ➡ **Property block & Unit models**
  - Blocks
  - External functions

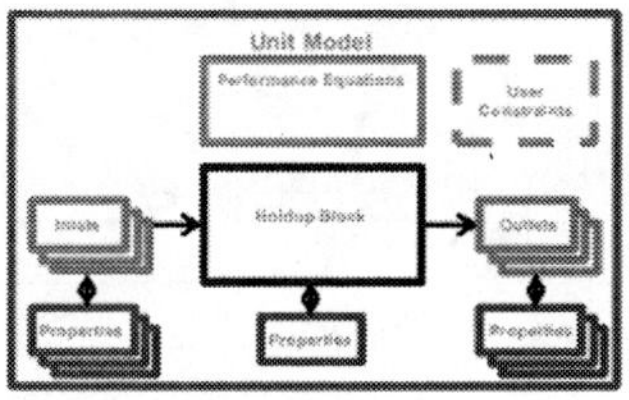

- **(Partial) differential equations** PYOMO DAE ➡ **Dynamic process models, optimal control**

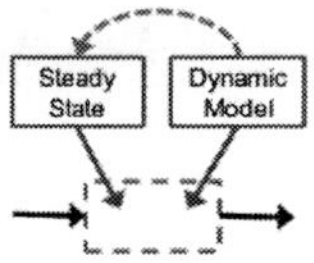

- **Stochastic programming** PySP ➡ **Uncertainty quantification**
- **Generalized Disjunctive Programming [Pyomo.GDP]** ➡ **Superstructure optimization**

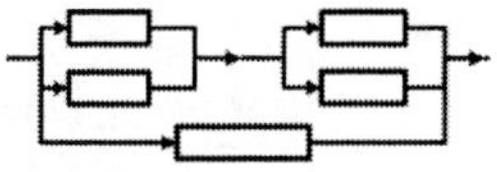

Pyomo: Extensible Object-oriented Algebraic Modeling in Python

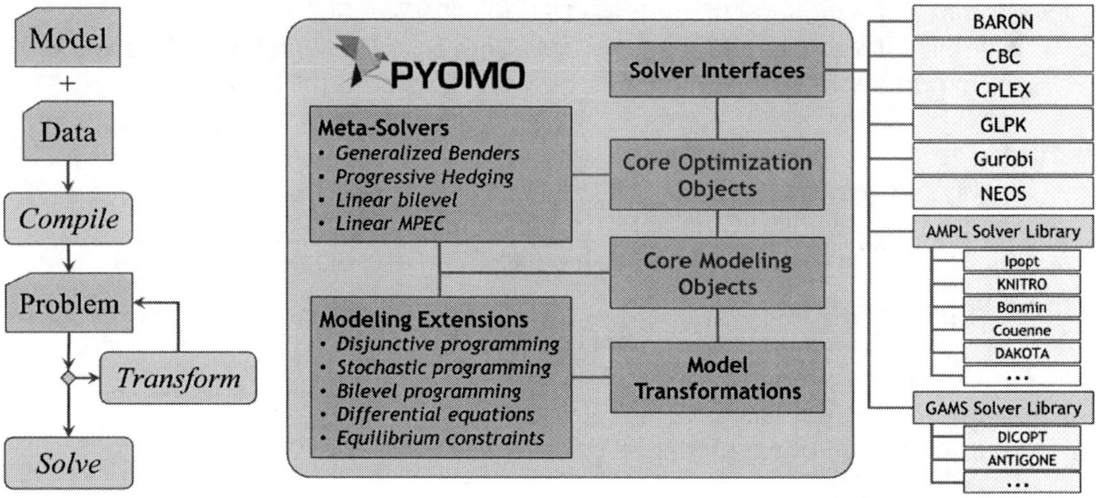

IDAES NATIONAL ENERGY TECHNOLOGY LABORATORY Sandia National Laboratories Carnegie Mellon West Virginia University U.S. DEPARTMENT OF ENERGY 15

Results

WWC & Packed Column

	Before opt	After opt
OBJECTIVE	**14.35**	**0.2058**

Model	Parameters	Before opt	After opt
Equilibrium reaction	a_1	198.4	200.0
	b_1	-1986	-1803
	c_1	-32.71	-32.98
	a_2	166.6	164.0
	b_2	-2102	-2200
	c_2	-26.51	-25.95
Mass transfer	Cv	0.357	0.1495
	CL	0.5	0.4903
Interfacial area	β_1	157.0	39.87
	β_2	8.621	23.95
	β_3	1.333	0.9962

IDAES NATIONAL ENERGY TECHNOLOGY LABORATORY Sandia National Laboratories Carnegie Mellon West Virginia University U.S. DEPARTMENT OF ENERGY 16

Results

Absorber column

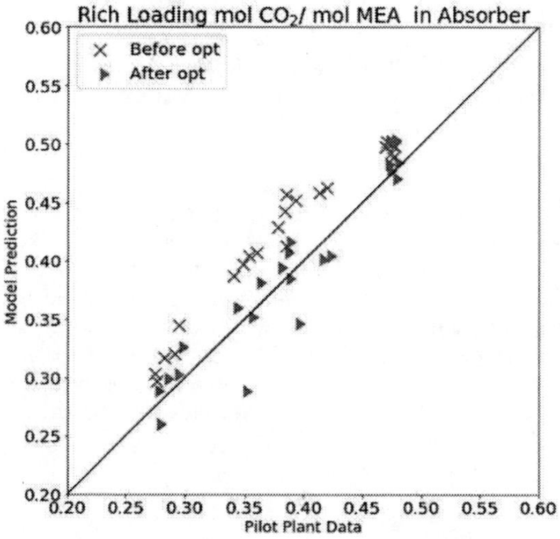

WWC

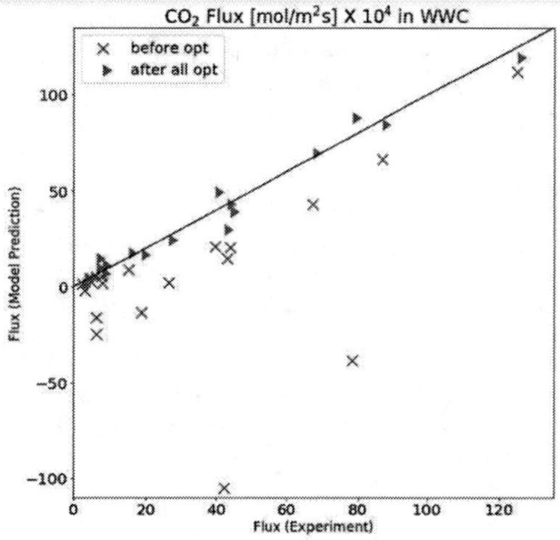

Conclusion

- Process models for packed bed and wetted wall columns
- IDAES framework demonstrated for parameter estimation
  - Complex process models
  - Multi-scale data
- Superior fitting

Continuing work:

- Larger subsets of model parameters with larger data sets
- Pyomo decomposition
  - Enable larger data sets and more complex models

ACKNOWLEDGEMENT & DISCLAIMER

Acknowledgement

This research was funded by the Lawrence Berkeley National Laboratory through Subcontract NO.7268587 as part of the U.S. DOE's Institute for the Design of Advanced Energy Systems (IDAES).

THANK YOU FOR YOUR ATTENTION

494599 The Value of Chemical Looping Combustion in Future Electricity Systems

Abstract:

Power generation via Chemical Looping Combustion (CLC) is inherently low in carbon dioxide emissions while maintaining the advantages of conventional thermal power plants such as dispatchability, flexibility, or firmness of capacity. The energy penalty incurred by the carbon dioxide removal process is performed at a fraction of other carbon capture techniques [1], such as post-combustion capture using amine solvents.

In this work, we quantify the role and value of three different CLC technologies and their integration into the future electricity system; CLC with natural gas coupled to a combined cycle (NGCC-CLC), integrated gasification combined cycle with CLC (IGCC-CLC) and CLC with in situ gasification and combustion of solid fuel (iG-CLC). We develop a national-scale capacity expansion and unit commitment model based on a mixed-integer linear program (MILP), minimising total system cost while limiting carbon emission, and maintaining security, and operability requirements [2]. Detailed operational constraints (e.g., start-up, ramp rates, minimum stable generation) on an hourly time discretisation enable the analysis of individual power plant performance characteristics and an assessment of their system-level impact.

We investigate the effects of the integration of CLC power plants into a 2030 and 2050 electricity system of the United Kingdom on the optimal capacity mix, the operational and system cost, and dispatch patterns. All three CLC power generation technologies prove valuable to the electricity system, as their deployment reduces system-wide investment and operational cost. Power generation from intermittent renewables becomes less economical as CLC technologies become available. Consequently, the requirements for carbon-intense reserve capacity, energy storage and overseas electric interconnection reduces. NGCC-CLC operating at virtually zero-carbon dioxide emission and moderate operation cost compares most favourably to IGCC-CLC and iG-CLC. The most cost-effective operational strategy for CLC power plants is in a steady manner, however, their ability for flexible operation can complement intermittent renewable power generation.

Additionally, we present the sensitivities to capital and operating expenditure and efficiency. A reduction in the efficiency of NGCC-CLC power plants from 46 % to 42 % reduces their annual average utilisation rate by only 1 %. For iG-CLC, the sensitivity to the cost of the oxygen carrier material is small; a variation of -50 % to +70 % causes a deviation in the average utilisation factor of -2 % to 3 %. For iG-CLC and IGCC-CLC, a 20-40 % increase in capital cost can reduce optimal capacity deployment levels considerably. NGCC-CLC deployment is more robust against investment cost increase. The results emphasises the high value of CLC power generation in a carbon-constrained system and can indicate and guide a prioritisation of future research efforts.

[1] C. Ekstrom, F. Schwendig, O. Biede, F. Franco, and G. Haupt. Technoeconomic evaluations and benchmarking of pre-combustion CO_2 capture and oxy-fuel processes developed in the European ENCAP project. Energy Procedia,

1(1):4233-4240, 2009.

[2] C. F. Heuberger, I. Staffell, N. Shah, and N. Mac Dowell. Levelised Value of Electricity - A Systemic Approach to Technology Valuation. In 26th European Symposium on Computer Aided Process Engineering, volume 38, pages 721-726, 2016.

498958 Superstructure-Based Optimization of Membrane-Based Carbon Capture Systems

Abstract:

This work aims to optimize the plant design, process configuration and operating conditions of a post-combustion membrane-based CO_2 capture plant. A new superstructure-based mathematical model is presented, modeling discrete and continuous decision variables to minimize the cost of electricity (COE). The proposed approach tackles a major issue in carbon capture and storage (CCS) for the power generation industry, involving the simultaneous optimization of process configurations and operating conditions (gas temperature, pressure ratio, sweep feed, etc.).

The development of efficient and cost-effective CCS systems is important to mitigate the CO_2 emission in fossil fuel-based power generation. Available technologies for post-combustion carbon capture include liquid-solvents, gas-permeation, and solid sorbent systems. Solvent and Sorbent â€" based systems can considered mature technologies for CCS, while membrane systems are generally less well-studied.

This work focuses on membrane-based systems for CCS applied to gas permeation, where the membranes with high CO_2 selectivity separate the CO_2 from other components in the flue gas. The main driving force of the process is the partial pressure between the feed and the permeate sides. Membrane systems often consist of multiple stages of membranes and compressors with intercooling. Most of the applications consider membrane models discretizing the membrane area and solving the separation problem for each membrane-section. This often results in a large-scale model typically analyzed through process simulation-based frameworks (ASPEN, gPROMs, etc.)[1] developed primarily for accurately modeling performance as opposed to optimization. Therefore, the number of configurations that can be optimized by conventional simulation-based frameworks is limited.

Significant efforts have been made for developing multi-stage systems, where previous studies focused on optimizing the number of stages and their operation[2,3], instead of analyzing advanced process configurations[4]. Advanced process configurations can lead to important savings in operating and capital costs. This work analyzes advanced process configurations while considering rigorous process models in a mathematical optimization framework.

A set of streams, and multiple processes (membrane stages, mixers, splitters, pumps and compression units with intercooling) is considered. Discrete design decisions determine the plant layout by selecting the optimal number of stages (where each stage consists of a membrane, compressor, pump, and heat exchanger units) and the process configuration. At the operating level, continuous decision variables involve the optimal flows, temperatures, concentrations, pressure ratio, and sweep feed. The proposed model is formulated as a mixed integer nonlinear programming model that minimized the cost of electricity to derive an optimal design and operation of a post-combustion carbon capture plant. The optimization of several process configurations led to lower CCS costs.

[1] Eslick J. C., Tong C., Lee A., Dowling A. W., Mebane D. S. (2015). Optimization under uncertainty with rigorous

process models. Proceedings AIChE Annual Meeting 2015.

[2] Hasan M. M., Baliban R. C., Elia J. A., Floudas C. A. (2012). Modeling, Simulation, and Optimization of Postcombustion CO_2 Capture for Variable Feed Concentration and Flow Rate. 1. Chemical Absorption and Membrane Processes. Industrial and Engineering Chemistry Research. 51, 15642-15664.

[3] Arias A. M., Mussati M. C., Mores P. L., Scenna N. J. (2016). Optimization of multi-stage membrane systems for CO_2 capture from flue gas. International Journal of Greenhouse Gas Control. 53, 371-390.

[4] Merkel T. C., Lin H., Wei X., Baker R. (2010). Power plant post-combustion carbon dioxide capture: An opportunity for membranes. Journal of Membrane Science. 359, 126-139.

Imperial College London

Bio-energy with carbon capture and storage (BECCS): Opportunities for efficiency improvement

Mai Bui[a,b], **Mathilde Fajardy**[a,b], **Niall Mac Dowell**[a,b]

[a]Centre for Environmental Policy, Imperial College London, London SW7 1NA UK
[b]Centre for Process Systems Engineering, Imperial College London, London SW7 2AZ UK

cep

Centre for Process Systems Engineering

2nd November 2017

EPSRC
Engineering and Physical Sciences Research Council

Imperial College London

Meeting the below 2°C and 1.5°C COP21 targets

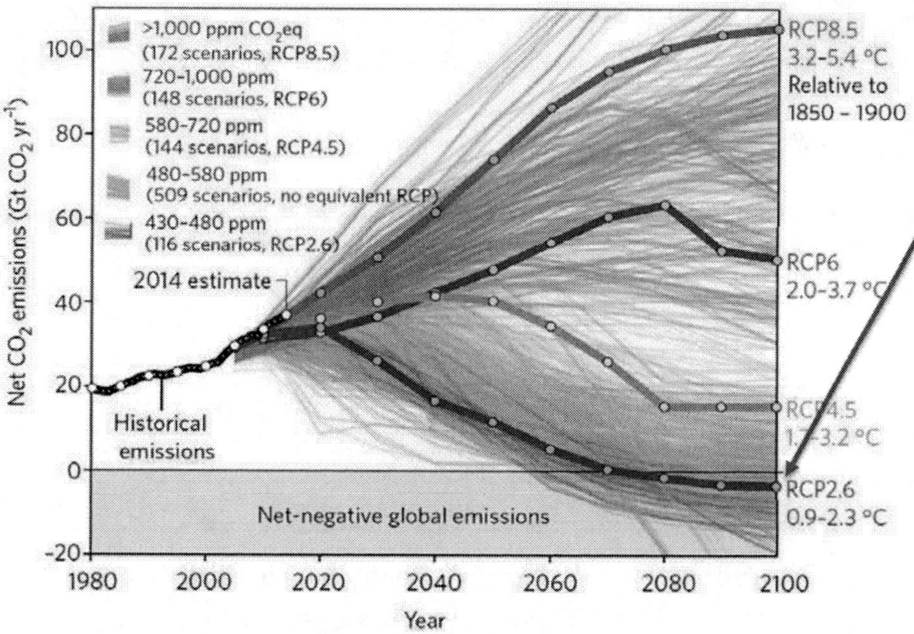

- Mitigation costs 138% more without CCS
- Without CCS, <50% of IAM models solve the problem
- Most feasible solutions use BioEnergy + CCS substantially

→ BECCS is a key negative emissions technology

IPCC Fifth Assessment Synthesis Report November 2014, Fuss, S., et al. (2014). Nature Climate Change, 4(10), 850–853

2

Imperial College London

Negative emissions technology: BECCS

Atmospheric CO_2 is captured and permanently stored → Negative CO_2 balance

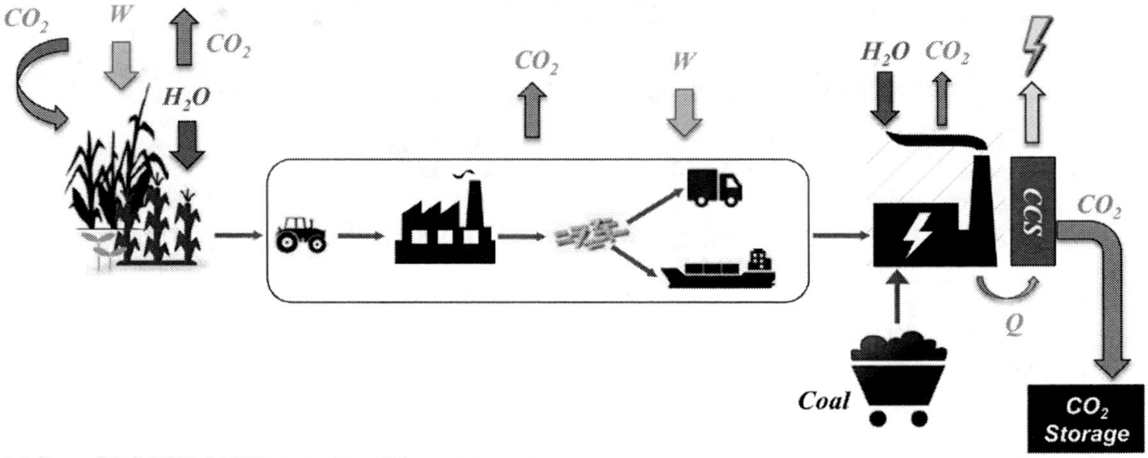

Reference: Fajardy, M. & Mac Dowell, N. (2017). Can BECCS deliver sustainable and resource efficient negative emissions? Energy and Environmental Science, 10 (6), 1389-1426.

3

Imperial College London

Fuel properties and composition

Property	Coal	Wood	Herbaceous
Ash content (wt% dry)	8.5 – 10.9	0.4 – 4.0	5 – 7.5
Moisture content (wt% wet)	6 – 10	5 – 60	15 – 25
Lower heating value, LHV (MJ/kg)	26.0 – 28.3	18.4 – 20.0	17.1 – 17.5
C (wt% dry)	76 – 87	47 – 52	45 – 47
H (wt% dry)	3.5 – 5.5	5.8 – 6.7	5.7 – 6.0
O (wt% dry)	2.8 – 11.3	38 – 46	40 – 46
Cl (wt% dry)	< 0.1	0.01 – 0.05	0.09 – 0.97
S (wt% dry)	0.5 – 3.1	0.02 – 0.10	0.05 – 0.2
N (wt% dry)	0.8 – 1.5	0.1 – 0.8	0.4 – 1.04

Efficiency penalty due to: Biomass (moisture & low heating value)
+ CO_2 capture process requires heat

4

Imperial College London

Fuel properties and composition

Property	Coal	Wood	Herbaceous
Ash content (wt% dry)	8.5 – 10.9	0.4 – 4.0	5 – 7.5
Moisture content (wt% wet)	6 – 10	5 – 60	15 – 25
Lower heating value, LHV (MJ/kg)	26.0 – 28.3	18.4 – 20.0	17.1 – 17.5
C (wt% dry)	76 – 87	47 – 52	45 – 47
H (wt% dry)	[illegible]	[illegible]	5.7 – 6.0
O (wt% dry)	[illegible]	[illegible] 46	40 – 46
Cl (wt% dry)	< 0.1	0.01 – 0.05	0.09 – 0.97
S (wt% dry)	0.5 – 3.1	0.02 – 0.10	0.05 – 0.2
N (wt% dry)	0.8 – 1.5	0.1 – 0.8	0.4 – 1.04

SO_x and NO_x emissions

Reference: Bui, M., Fajardy, M. & Mac Dowell, N. (2017). Bio-Energy with CCS (BECCS) performance evaluation: Efficiency enhancement and emissions reduction. Applied Energy, 195, 289–302.

5

Imperial College London

Heating and condensation of gas phase

Solids and liquids formed at various stages

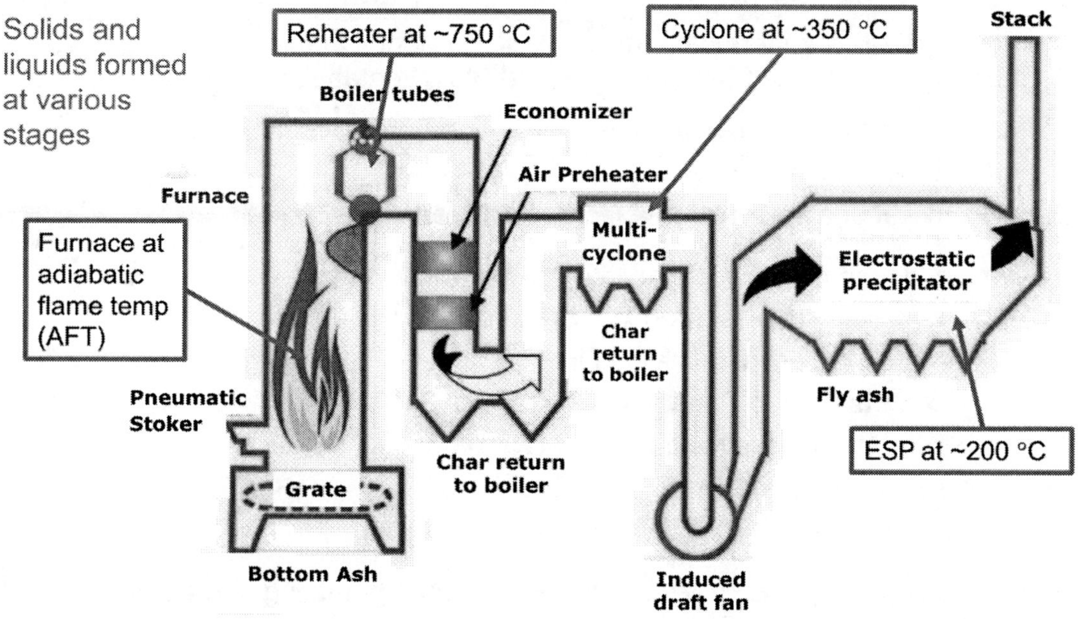

Image: Overend, Ralph P. Direct combustion of biomass, Encyclopedia of Life Support Systems

6

Imperial College
London

Flue gas condensation in a power plant

Unit	Temperature (°C)	Solid/liquid removed (%)	Solid/liquid carried to next step (%)
Furnace[1]	AFT	80	20
Reheater[2]	750	15 (fouling)	85
Multi-cyclone[3]	350	80	20
ESP[4]	200	99	1

[1] Biomass combustion ash is typically 80% bottom ash, 20% fly ash (Jensen et al., 2013; Hansen et al., 2014)

[2] Biomass ash fouling at ~750 °C based on T_{15} (varies with Cl & K content) (Mansikkasalo et al., 2014)

[3] Multi-cyclones: separation efficiency of 80-95% (US EPA, 2016)

[4] ESP: separation efficiency of 99-99.9% (US EPA, 2016)

References: Jensen, P. A., *et al.* (2013). Characterization and quantification of deposits build up and removal in straw suspension fired boilers. Energinet.dk project no. 7217

Hansen, S. B., *et al.* (2014). Deposit probe measurements in large biomass-fired grate boilers and pulverized-fuel boilers. Energy & Fuels, 28 (6), 3539–3555.

Mansikkasalo, J., *et al.* (2014). High power generation from recovery boilers, what are the limits? [Online]. 2014 International Chemical Recovery Conference, Finland.

USEPA (2016). Air Pollution Control Technology Fact Sheets. United States Environmental Protection Agency (USEPA) Technology Transfer Network.

7

Imperial College
London

Automation of FactSage: Macros

FactSage 7.0 is a thermo-chemical modelling tool – accesses thermodynamic data from the FACT and SGTE databases

Macros can be used to automate a large number of simulations:

- Incremental temp increase
- Removal of a particular phases from the reaction

Designed macros used to simulate the following:

- Combustion Process: (200°C to AFT)
- Flue gas condensation divided into:
    i. Furnace to Reheater (AFT to 750°C)
    ii. Reheater to Multi-cyclone (750°C to 400°C)
    iii. Multi-cyclone to ESP (400°C to 200°C)
    iv. ESP to Stack (200°C to 120°C)

Can observe flue gas composition changes during heating and cooling of the flue gas

8

Imperial College London

Flue gas combustion-condensation graph

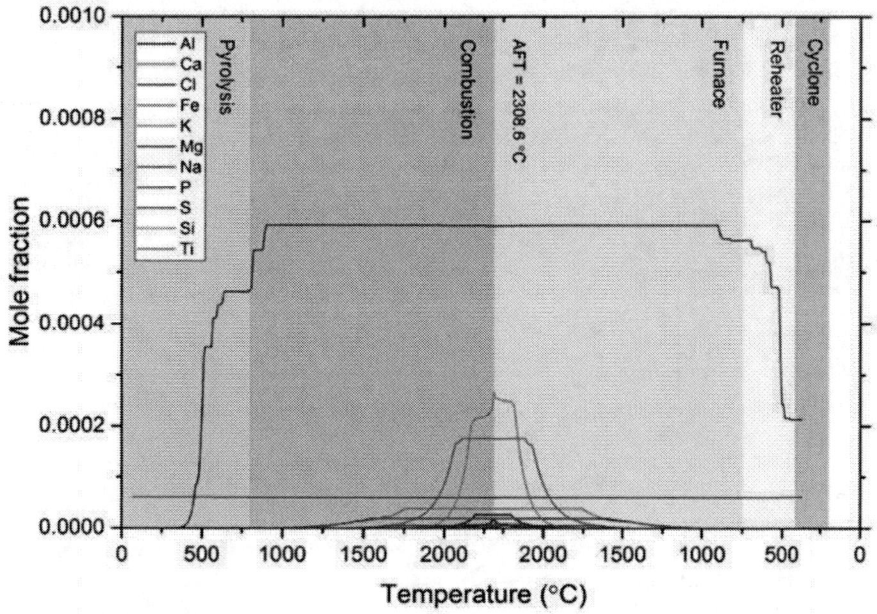

9

Imperial College London

Energy flow in a thermal power plant

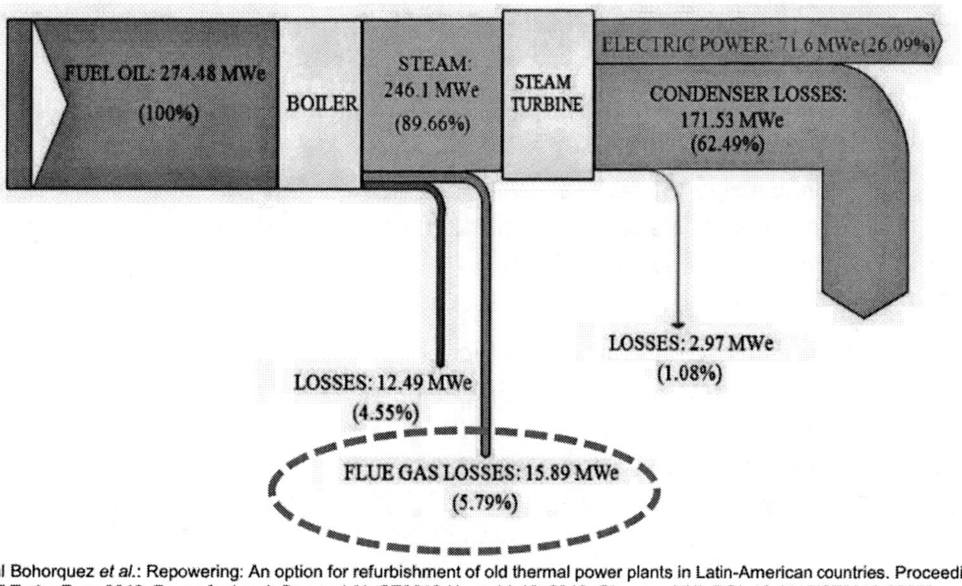

Irrazabal Bohorquez *et al.*: Repowering: An option for refurbishment of old thermal power plants in Latin-American countries. Proceedings of ASME Turbo Expo 2010: Power for Land, Sea and Air GT2010 (June 14-18, 2010, Glasgow, UK). DOI: 10.1115/GT2010-23058
http://www.sankey-diagrams.com/tag/power-plant/

10

Imperial College London

Flue gas heat recovery to improve power plant efficiency

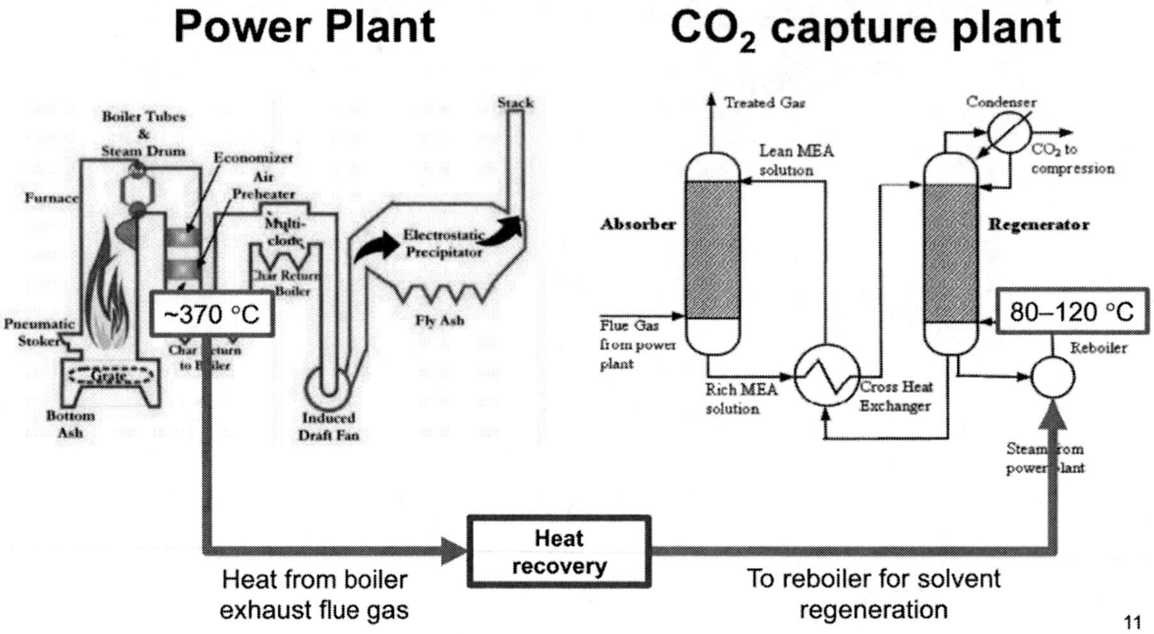

11

Imperial College London

BECCS performance evaluation study: Objectives

- Assess the performance of a BECCS system using a 500 MW ultra-supercritical pulverized fuel power plant with 90% CO_2 capture.
- The effects of the following factors were evaluated:
  i. Coal types: high sulphur 2.51 wt% and medium sulphur 0.86 wt%
  ii. Biomass type (wheat straw and clean wood chips)
  iii. Variable moisture content of the biomass (5%, 16%, 50%)
  iv. Biomass co-firing proportion %
  v. Solvent selection (MEA, Cansolv, biphasic solvents)
- Key performance indicators include:
  - Combustion adiabatic flame temperature (AFT)
  - Pollutant emissions e.g. SO_X and NO_X
  - Recoverable heat from the flue gas
  - Energy efficiency
  - Carbon intensity

12

Co-firing scenarios modelled in FactSage

Scenario	Coal type	Biomass type
A	Medium sulphur coal	Wheat straw 5% moisture
B	Medium sulphur coal	Wheat straw 16% moisture
C	Medium sulphur coal	Wood chip 5% moisture
D	Medium sulphur coal	Wood chip 50% moisture
E	High sulphur coal	Wheat straw 5% moisture
F	High sulphur coal	Wheat straw 16% moisture
G	High sulphur coal	Wood chip 5% moisture
H	High sulphur coal	Wood chip 50% moisture

Biomass Co-firing: Adiabatic Flame Temperature (AFT)

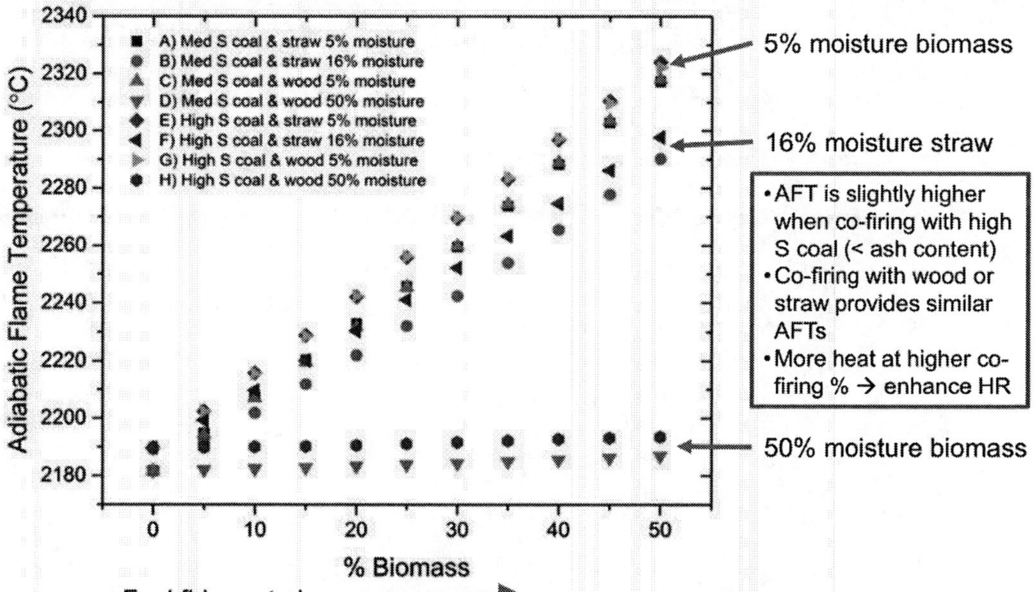

Fuel firing rate increases →

- AFT is slightly higher when co-firing with high S coal (< ash content)
- Co-firing with wood or straw provides similar AFTs
- More heat at higher co-firing % → enhance HR

Reference: Bui, M., Fajardy, M. & Mac Dowell, N. (2017). Applied Energy, 195, 289–302.

Imperial College
London

Coal co-firing with biomass : SO_x emissions

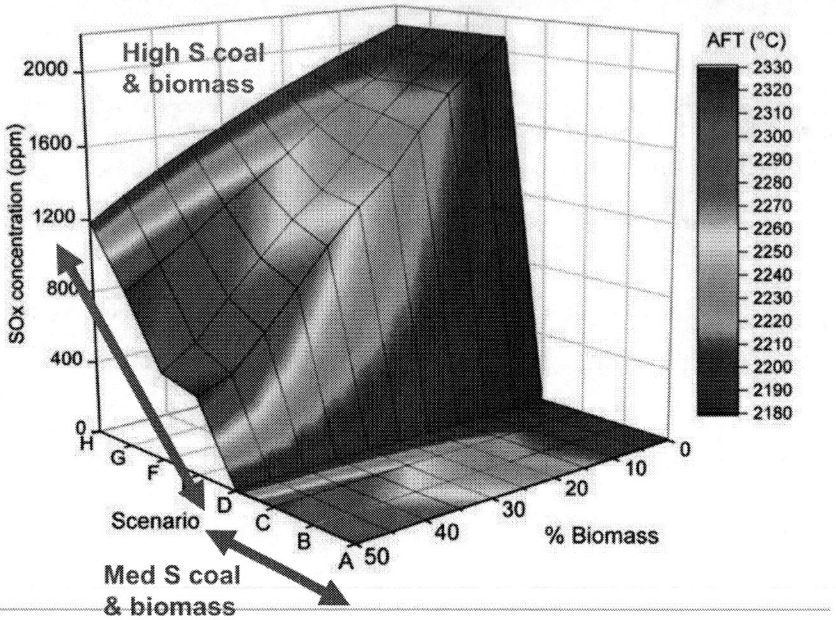

Reference: Bui, M., Fajardy, M. & Mac Dowell, N. (2017). Applied Energy, 195, 289–302.

15

Imperial College
London

Coal co-firing with biomass : SO_x emissions

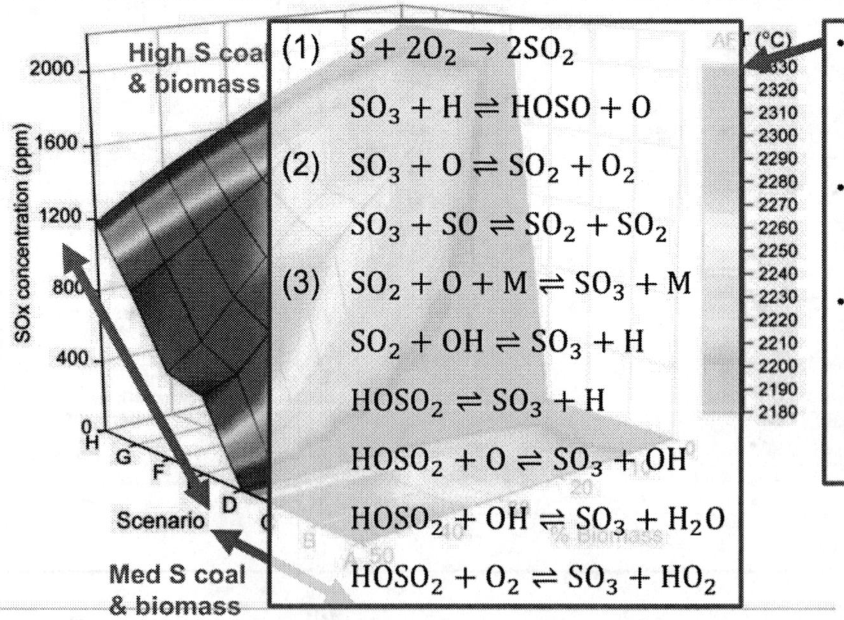

(1) $S + 2O_2 \rightarrow 2SO_2$

$SO_3 + H \rightleftharpoons HOSO + O$

(2) $SO_3 + O \rightleftharpoons SO_2 + O_2$

$SO_3 + SO \rightleftharpoons SO_2 + SO_2$

(3) $SO_2 + O + M \rightleftharpoons SO_3 + M$

$SO_2 + OH \rightleftharpoons SO_3 + H$

$HOSO_2 \rightleftharpoons SO_3 + H$

$HOSO_2 + O \rightleftharpoons SO_3 + OH$

$HOSO_2 + OH \rightleftharpoons SO_3 + H_2O$

$HOSO_2 + O_2 \rightleftharpoons SO_3 + HO_2$

- SO_X formation mechanism involves equilibrium reactions
- SO_2 formation thermodynamically favoured
- Temperature ↓ : Equilibrium shifts towards SO_3 formation (iron oxide can catalyse reaction 2)

Reference: Bui, M., Fajardy, M. & Mac Dowell, N. (2017). Applied Energy, 195, 289–302.

16

Imperial College London

Coal co-firing with biomass : SO_x emissions

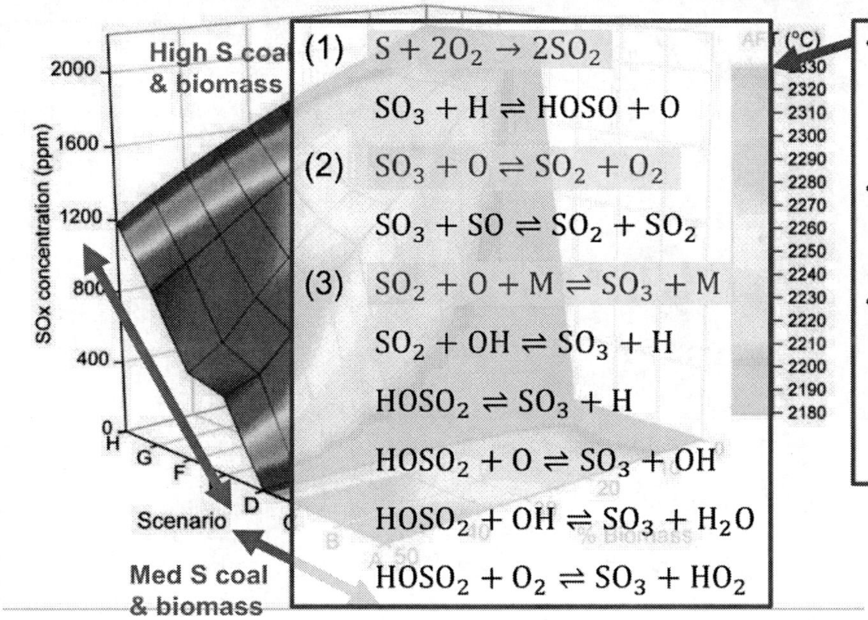

(1) $S + 2O_2 \rightarrow 2SO_2$

$SO_3 + H \rightleftharpoons HOSO + O$

(2) $SO_3 + O \rightleftharpoons SO_2 + O_2$

$SO_3 + SO \rightleftharpoons SO_2 + SO_2$

(3) $SO_2 + O + M \rightleftharpoons SO_3 + M$

$SO_2 + OH \rightleftharpoons SO_3 + H$

$HOSO_2 \rightleftharpoons SO_3 + H$

$HOSO_2 + O \rightleftharpoons SO_3 + OH$

$HOSO_2 + OH \rightleftharpoons SO_3 + H_2O$

$HOSO_2 + O_2 \rightleftharpoons SO_3 + HO_2$

- SO_X formation mechanism involves equilibrium reactions
- SO_2 formation thermodynamically favoured
- Temperature ↓ : Equilibrium shifts towards SO_3 formation (iron oxide can catalyse reaction 2)

Reference: Bui, M., Fajardy, M. & Mac Dowell, N. (2017). Applied Energy, 195, 289–302.

17

Imperial College London

Coal co-firing with biomass : SO_x emissions

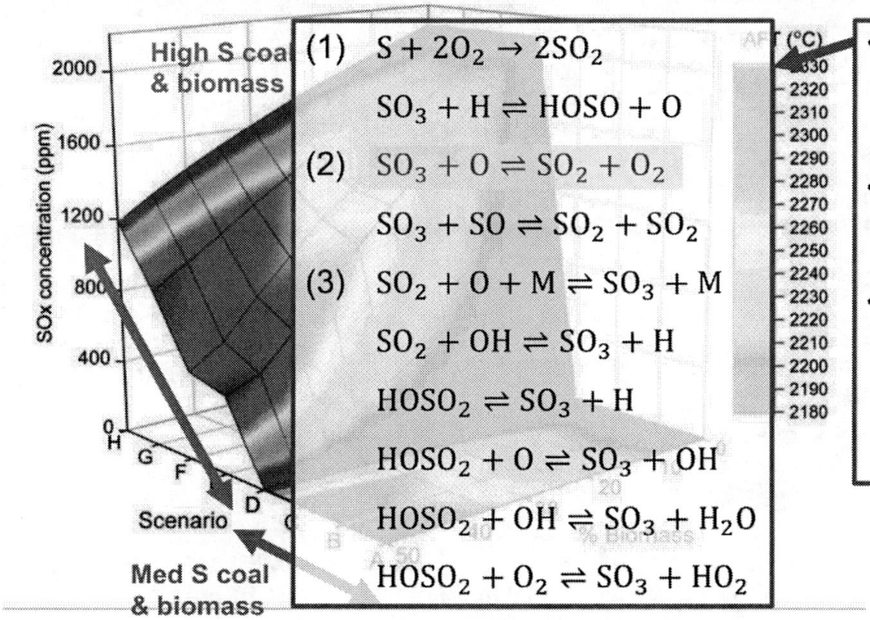

(1) $S + 2O_2 \rightarrow 2SO_2$

$SO_3 + H \rightleftharpoons HOSO + O$

(2) $SO_3 + O \rightleftharpoons SO_2 + O_2$

$SO_3 + SO \rightleftharpoons SO_2 + SO_2$

(3) $SO_2 + O + M \rightleftharpoons SO_3 + M$

$SO_2 + OH \rightleftharpoons SO_3 + H$

$HOSO_2 \rightleftharpoons SO_3 + H$

$HOSO_2 + O \rightleftharpoons SO_3 + OH$

$HOSO_2 + OH \rightleftharpoons SO_3 + H_2O$

$HOSO_2 + O_2 \rightleftharpoons SO_3 + HO_2$

- SO_X formation mechanism involves equilibrium reactions
- SO_2 formation thermodynamically favoured
- Temperature ↓ : Equilibrium shifts towards SO_3 formation (iron oxide can catalyse reaction 2)

Reference: Bui, M., Fajardy, M. & Mac Dowell, N. (2017). Applied Energy, 195, 289–302.

18

Imperial College
London

Coal co-firing with biomass : SO_x emissions

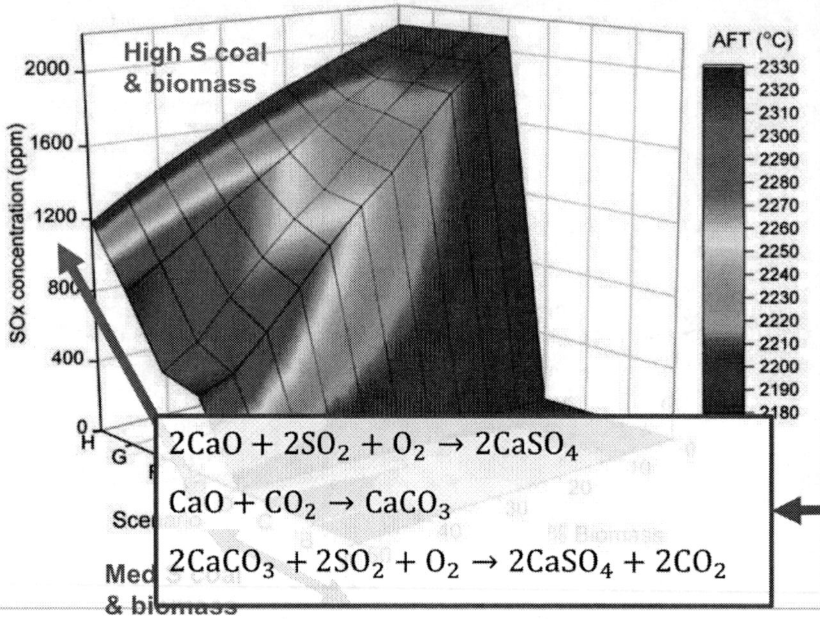

$$2CaO + 2SO_2 + O_2 \rightarrow 2CaSO_4$$

$$CaO + CO_2 \rightarrow CaCO_3$$

$$2CaCO_3 + 2SO_2 + O_2 \rightarrow 2CaSO_4 + 2CO_2$$

- SO_X formation mechanism involves equilibrium reactions
- SO_2 formation thermodynamically favoured
- Temperature ↓ : Equilibrium shifts towards SO_3 formation (iron oxide can catalyse reaction 2)
- Alkali oxides in ash capture SO_2 and SO_3

Reference: Bui, M., Fajardy, M. & Mac Dowell, N. (2017). Applied Energy, 195, 289–302.

19

Imperial College
London

High sulphur coal & biomass: SO_x emissions

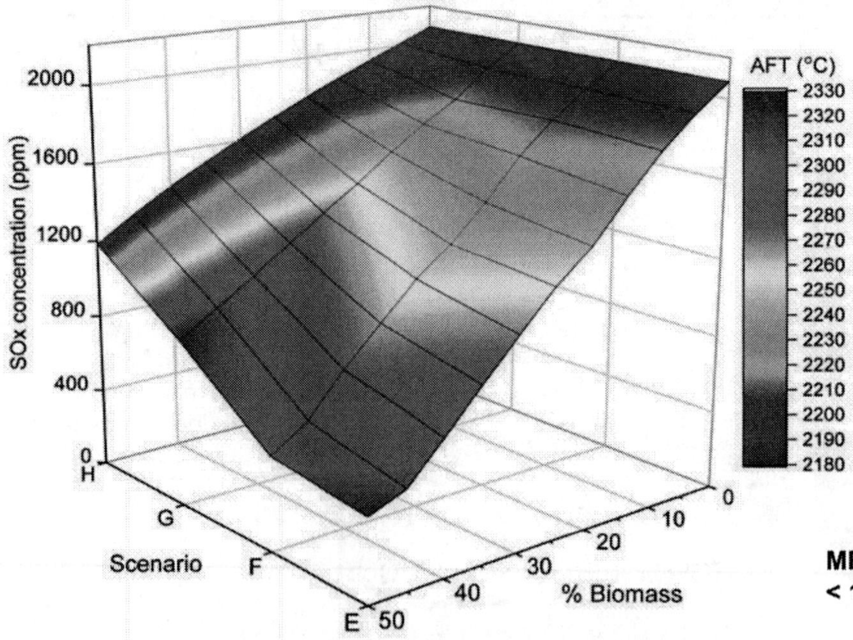

- SO_X decreases linearly with % biomass due to decreased S content

MEA tolerance to SO_X < 10 ppm

Reference: Bui, M., Fajardy, M. & Mac Dowell, N. (2017). Applied Energy, 195, 289–302.

20

Imperial College
London

Medium sulphur coal & biomass: SO_x emissions

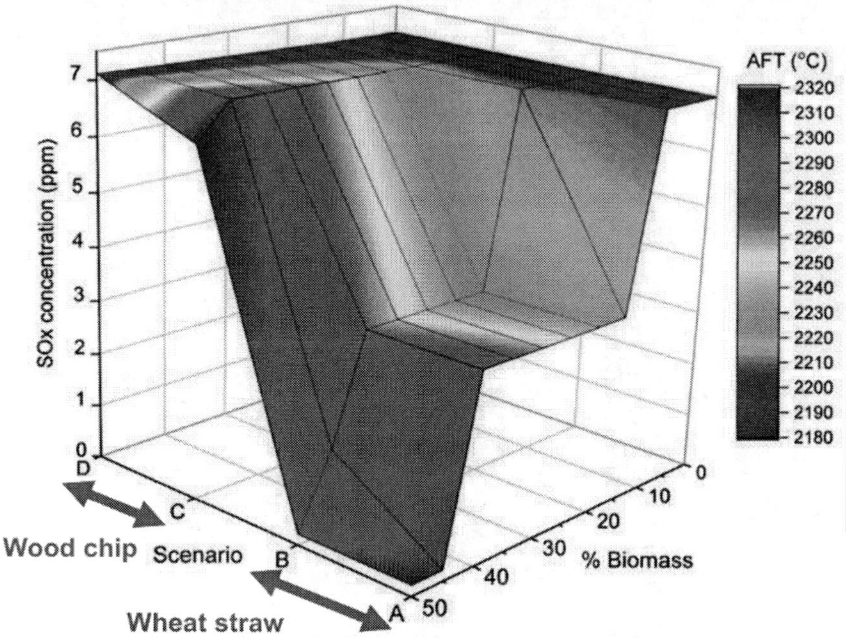

- FactSage results show that CaO, MgO, Na_2O, K_2O captures SO_X
- In FactSage, Al_2O_3, Fe_2O_3 & MnO_2 also involved in SO_X ↓
- Wheat straw has higher sulphur content than the wood chip (0.1 wt% Vs 0.05 wt%)
- Wheat straw has more ash than wood chip (5.5 dry wt% Vs 2 dry wt%)
- Alkali oxides in ash → reduce SO_X emissions

MEA tolerance to SO_X < 10 ppm

Reference: Bui, M., Fajardy, M. & Mac Dowell, N. (2017). Applied Energy, 195, 289–302.

21

Imperial College
London

BECCS: Recoverable heat from the flue gas

Analysis of scenario F (high S coal and 16% moisture wheat straw)

Solvent	HD (MJ/t_{CO2})	Reg T (°C)
MEA	3600	120
Cansolv	2300	120
Biphasic	2000	80

Data for biphasic solvent systems published by Ye *et al.* (2015) and Raynal *et al.* (2011)

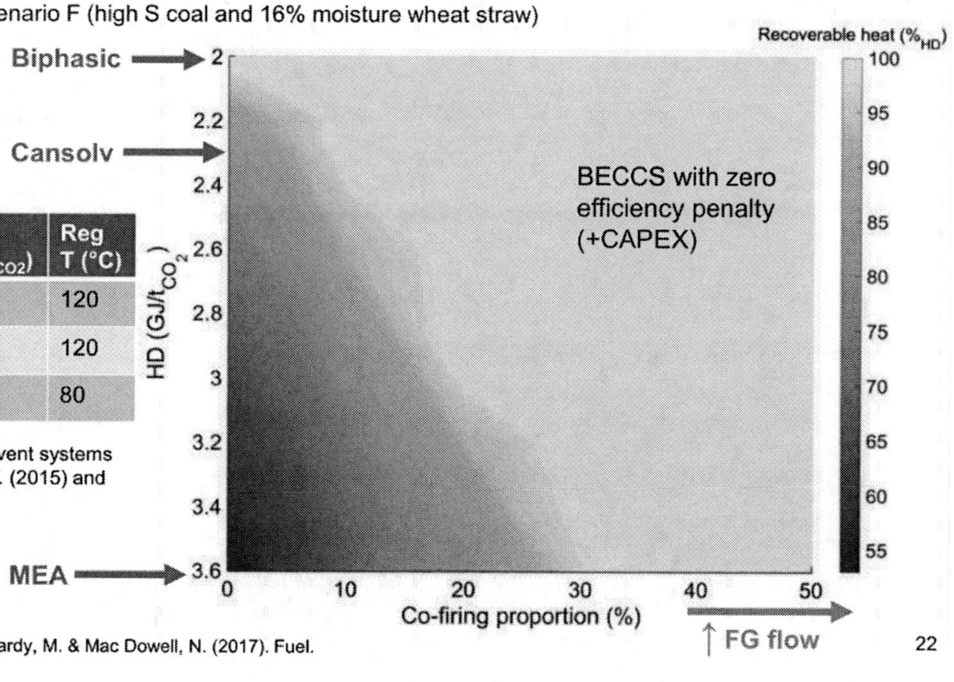

Reference: Bui, M., Fajardy, M. & Mac Dowell, N. (2017). Fuel.

22

Imperial College London

BECCS power plant efficiency improvement

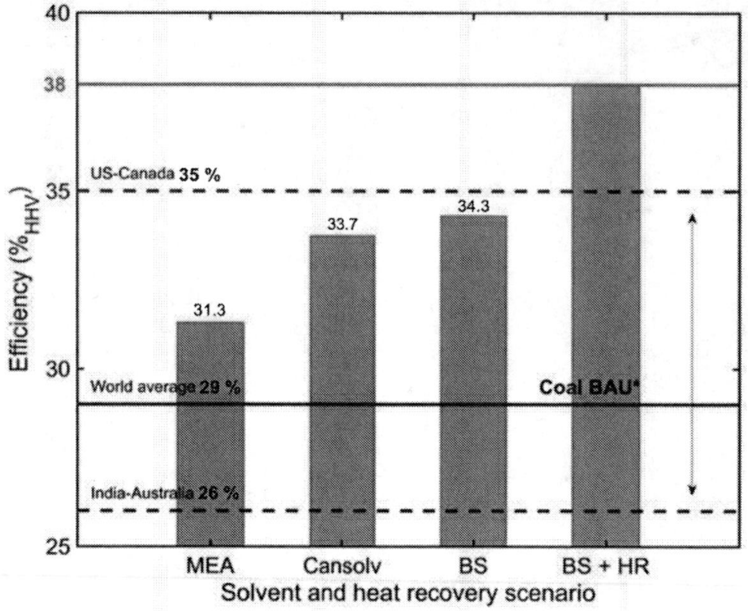

- Efficiency for a 500 MW BECCS system **co-firing 50% biomass**
- Efficiency can be significantly improved by using (i) high performance solvents, (ii) heat recovery

BS = biphasic solvent

HR = heat recovery

BAU = Business As Usual (average efficiencies around the world)

Reference: Bui, M., Fajardy, M. & Mac Dowell, N. (2017). Bio-Energy with carbon capture and storage (BECCS): Opportunities for performance improvement. Fuel.

23

Imperial College London

BECCS carbon intensity

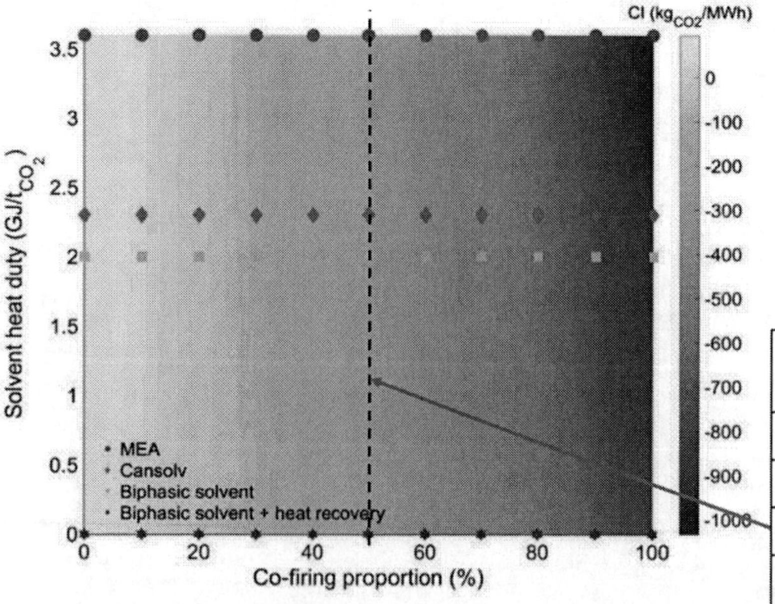

As system efficiency increases, less fuel is burned per MWh produced → less CO_2 is captured per MWh of electricity generated

At 50% biomass co-firing and 90% CO_2 capture:

System	Efficiency %HHV	Carbon intensity kg CO_2/MWh
MEA	31.3	-296
Cansolv	33.7	-275
BS	34.3	-271
BS + HR	38.0	-244

Reference: Bui, M., Fajardy, M. & Mac Dowell, N. (2017). Fuel.

24

Imperial College
London

CO_2 mitigation potential of a 500 MW BECCS system

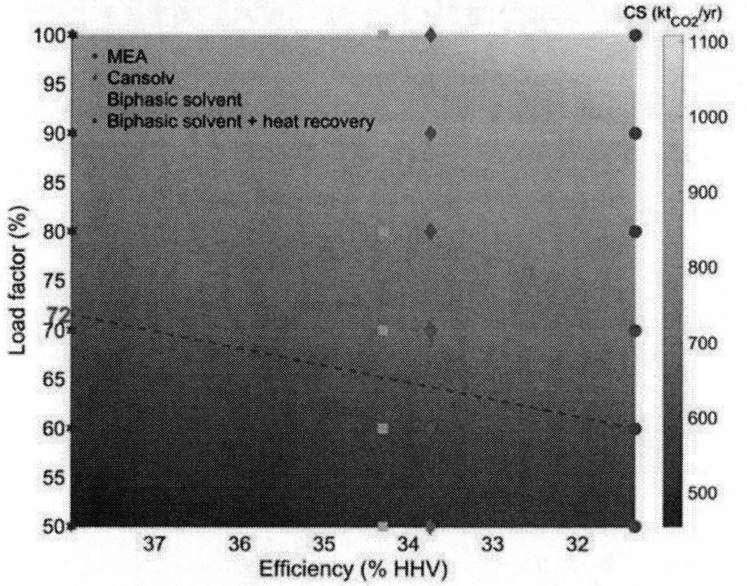

- Power plant load factor is a function of power plant efficiency.
- Improving efficiency reduces marginal costs of electricity generation, enabling operation at higher load factor (Mac Dowell & Shah, 2015; Mac Dowell & Staffell, 2016)
- The efficiency of a BECCS system will have an impact on its dispatch rate within an electricity market
  → Influences CO_2 mitigation potential

Objective: (i) more efficient at electricity generation, or (ii) more carbon negative & capture CO_2?

Reference: Bui, M., Fajardy, M. & Mac Dowell, N. (2017). Applied Energy, 195, 289–302.

25

Imperial College
London

Effect of biomass carbon footprint (BCF) and co-firing proportion on the system carbon intensity (CI)

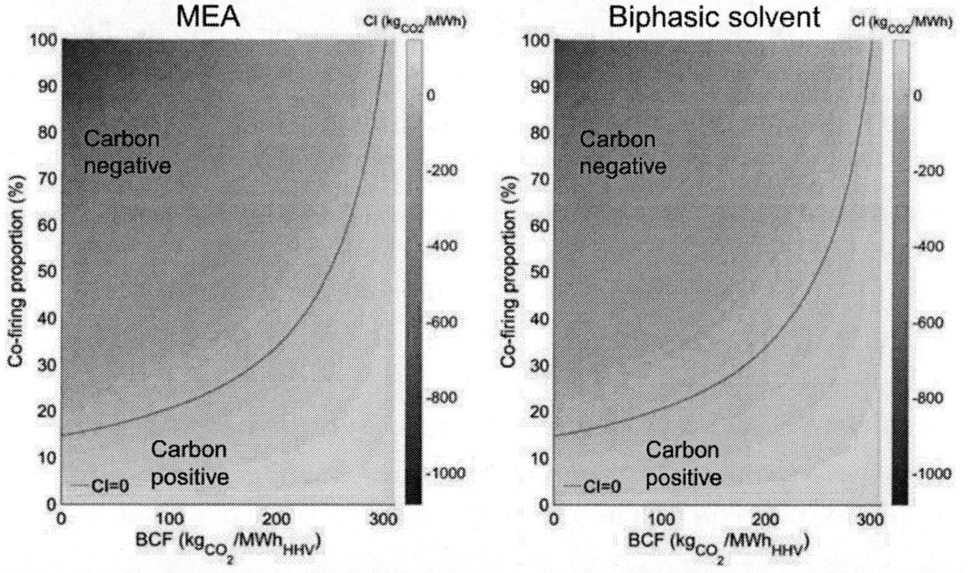

Bui, M., Fajardy, M. & Mac Dowell, N. (2017) Fuel.

Low efficiency MEA system is more carbon positive above the CI = 0 limit & more carbon negative below this limit. To ensure the BECCS system is actually carbon negative, the impact of biomass carbon footprint (BCF) must be considered.

Imperial College London

Conclusion: BECSS Performance Matrix

Performance parameter	Ash content	Sulphur content	Moisture content	HHV	Biomass co-firing %	Solvent heat duty
NO_X ↓	dependent on combustion conditions					
SO_X ↓	↑	↓	↓		↑	
AFT ↑	↓		↓		↑	
Heat recovery ↑	↓		↓	↓*	↑	↓
Efficiency ↑	↓		↓	↑↓	↓	↓
CO_2 negativity ↑	↑		↑	↑	↑	↑
Exhaust gas flow rate ↑				↓	↑	

Enhance efficiency: high performance solvents (low heat duty) and higher heat recovery (higher AFT and flue gas flow rate)
Greater carbon negativity: lower efficiency systems (burn more biomass fuel, thus more CO_2 is captured and permanently stored), low biomass carbon footprint.
Minimise pollutant emissions: Increasing biomass co-firing proportion significantly reduces SO_X emissions, fuel selection (low S, metal oxides), combustion control is required to regulate NO_X emissions.

27

Imperial College London

Conclusion

- BECCS has the potential provide negative emissions, whilst providing reliable firm power generation capacity.
- Opportunities to improve energy efficiency of BECCS include enhancing heat recovery and using high performance solvents for CO_2 capture (*e.g.*, biphasic).
- Power generation efficiency increased from $31\%_{HHV}$ (MEA system) to $38\%_{HHV}$ (biphasic system with heat recovery).
- Important to consider the complex trade-offs between carbon intensity, power plant efficiency and biomass carbon footprint.
- Inefficient BECCS → removes more CO_2 from the atmosphere.
- However, inefficient systems might have lower dispatch rate compared to high efficiency systems.

Bui, M., Fajardy, M. & Mac Dowell, N. (2017). Applied Energy, 195, 289–302.
Bui, M., Fajardy, M. & Mac Dowell, N. (2017). Energy Procedia, 114 (Supplement C), 6010-6020.
Bui, M., Fajardy, M. & Mac Dowell, N. (2017). Bio-Energy with carbon capture and storage (BECCS): Opportunities for performance improvement. Fuel.

28

493767 A Process Synthesis Approach to Natural Gas Liquefaction

Abstract:

A new approach to the design of natural gas liquefaction processes, which could significantly reduce their energy consumption, is proposed. A graphical technique based on exergy and enthalpy is used to determine optimal paths for refrigeration systems. The analysis reveals that keeping refrigeration cycles in the two-phase region, including compression and heat removal, is the most efficient way of supplying the work required to remove heat from sub-ambient processes. The approach is validated using an Aspen Plus simulation of a natural gas liquefaction process with pure refrigerant in cascade; it is shown that work efficiency higher than 90% can be achieved when the process fluid as well as the working fluids in the cascade cycles are kept in the two phase region at every stage.

500732 Observer, Filters and Moving Horizon Estimator Design for Linear Transport-Reaction Distributed Parameter Systems

Abstract:

Fundamental conservation laws in process control, manufacturing, transport systems are present by distributed parameter systems (DPS) models. The burden of such a general representation is that DPS models take form of partial differential equations containing higher order derivatives in space and time. The complexity of dealing with a partial differential equations (PDEs) in the case of chemical engineering relevant linear PDE models lies in necessity of modellers to account for model spatial characteristics by an approximation of underlying model through some spatial approximation arriving to a finite dimensional model representation amenable for subsequent control, observer and/or monitoring device design [1], [2].

This work provides foundation for systematic development of modelling framework for a design of **observer, filters and/or moving horizon estimator** for linear DPS system which which is accomplished ***without application of any spatial approximation and/or order reduction***. Namely, we will presented a general observer and filter design methodologies which follow the seminal works of Luenberger (Luenberger observer [3]) and Kalman (Kalman Filter [4]). In particular, we will link the existing well known observer and/or filters design concepts to the infinite dimensional PDE settings and point the differences or analogies with the finite dimensional designs. Results of this novel approach are applied to the DPS emerging from the chemical transport-reaction systems varying from the convection dominated models of a plug flow reactor to diffusion dominated models of an axial dispersion reactor. In particular, the discrete model of a distributed parameter system is obtained by using energy preserving Cayley-Tustin discretization [5]. *This ultimately leads to discrete (in time and continuous in space) DPS models which are low dimensional, energy preserving and do not dissipate numerically (symplectic)*. In particular, infinite dimensional discrete setting is suitable to filter and moving horizon estimator design which can account for optimality and naturally present constraints in the system. Finally, the design performances are assessed by numerical simulation with application to different linear transport-reaction distributed parameter systems which are endowed with process and measurement noise.

[1] Ray, W. Harmon, Advanced Process Control, McGraw-Hill Inc.,USA, 1980

[2] Curtain, Ruth F and Zwart, Hans, An introduction to infinite-dimensional linear systems theory, Springer, 1995.

[3] David Luenberger, An Introduction to Observes, IEEE Transactions on Automatic Control, vol, 16, no 6, 1971.

[4] R. E. Kalman, A new approach to linear filtering and prediction theory, J. of Basic Engrg, Trans. ASME, Series D, vol 82, pp-34-45, 1960.

[5] V. Havu, J. Malinen, The Cayley transform as a time discretization scheme, Numerical Functional Analysis and Optimization 28 (7-8) (2007) 825-851.

501035 Model Parameterization through Data-Mining

Abstract:

Introduction: The detailed evolution of complex systems often involves disparity of scales in time and space. Low-dimensional representation of such complex dynamics is usually achievable through systematic model reduction approaches. Most of the time, the low-dimensional models can be parameterized by a small number of (nonlinear combinations of) parameters [1]. Therefore, one should find the reduced number of identifiable and effective parameters for the low-order model. In the present study, we employ a data-driven approach to identify parameters which are affecting the modelâ€™s outcome globally. The approach is explained and validated through a sequence of examples, including Michaelis-Menten type reaction networks.

Methodology: Principal Component Analysis (PCA) is the well-established approach to find parameters which are spanning the low-dimensional (linear) subspace of the system, or better say the good global reduced coordinates (parameters). However, it is known that in physical and chemical kinetics such a linear low-dimensional subspace is a poor description of the system and nonlinear embedding approaches should be considered. Diffusion maps (Dmap), as one of nonlinear model reduction techniques have been repeatedly applied to find low-dimensional, nonlinear manifolds underlying high-dimensional datasets [1] and offers the potential of finding â€˜goodâ€™ global reduced coordinates [2].

By applying diffusion distance, Dmaps offers more meaning-full distance (metric) than Euclidean metric specifically in nonlinear manifolds. However the metric and kernel in Dmap can be modified further to organize the intrinsic parameters even better than its original formulation. For example, if the observed data on parameter space are locating on neighborhood of level-sets of a function then one can deduce that similarity between two data on the level set is more than similarity of one of them with other point somewhere else in the parameter space but with the same metric. This issue is addressed and solution is proposed in this work, lead us to find identifiable, effective global parameterization of low order model, while the problem set-up can be treated as a black box which we have only access on the observed output. The solution can be considered as combination of parameterization and sensitivity analysis while the numerical cost is considerably lower. The method is applied on datasets organized in the form of an nsamples'nvariables matrix which can be an output of numerical simulation or experiments.

Results: In the first example a dataset was found by Michaelis-Menten enzyme kinetics which the parameterization and low-dimensional model are well-established analytically. Then the methods is applied on the filtered highly resolved Euler-Lagrange gas-particle simulation to identify optimal set of filtered variables and sub-grid correlations, seeking constitutive models for two-phase flow setup.

Summary: We use a modification of Diffusion Maps, a manifold learning technique, that employs an output-inspired metric in the input (i.e. parameter) space of a dynamic model. This approach allows us to identify the number and type of relevant parameter combinations, and to identify singular as well as regular perturbation regimes without formulas, in a purely data-driven manner.

References:

[1] Coifman, Ronald R., and Stéphane Lafon. "Diffusion maps." Applied and computational harmonic analysis 21.1 (2006): 5-30.

[2] Singer, Amit, Radek Erban, Ioannis G. Kevrekidis, and Ronald R. Coifman. "Detecting intrinsic slow variables in stochastic dynamical systems by anisotropic diffusion maps. " Proceedings of the National Academy of Sciences 106, no. 38 (2009): 16090-16095.

493474 Run-to-Run Control of PECVD Systems: Application to a Multiscale CFD Model of Amorphous Silicon Deposition

Abstract:

Uniform deposition of thin film layers remains a challenge in silicon processing industries due to the lack if in situ measurements and drift in the plasma composition caused by fouling [1],[2]. Specifically, in plasma-enhanced chemical vapor deposition (PECVD) of amorphous silicon (*a*-Si:H), growth rate non-uniformities greater than 20% across the surface of the wafer are common [3], which may lead to inefficient solar cells and microelectronics of poor quality. Process operators typically avoid this issue by preconditioning the reaction chamber and manually adjusting deposition conditions such that thickness non-uniformities may be reduced. Recently, Crose et al. developed a run-to-run based, offline control scheme which has been demonstrated to reduce the offset in product thickness from 5% to less than 1% within ten batches of operation [3]. While these results are promising, the multiscale model used relies on a first principles approach to solving the continuous mass, momentum and energy balances; however, recent advances in parallel computation strategies have made possible the application of multiscale computational fluid dynamics (CFD) to the modeling of PECVD reactors [4].

Consequently, we propose the application of the newly developed run-to-run control strategy to the simulated deposition of amorphous silicon (*a*-Si:H) thin films using a transient, multiscale CFD model. The two-dimensional axisymmetric model recently developed by Crose et al. [4] has been shown to accurately capture the complex behavior of the PECVD reactor and to reproduce experimentally observed non-uniformities with respect to the film thickness and porosity, and will be used throughout this work. A computationally efficient parallel processing scheme allows for the application of the run-to-run algorithm to 40 serial batch simulations which are shown to reduce the product offset from the 300 nm thickness set-point. Additional considerations are made with respect to the porosity and hydrogen content of the amorphous silicon product.

[1] Armaou A, Christofides PD. Plasma enhanced chemical vapor deposition: modeling and control. Chemical Engineering Science. 1999;54:3305-3314.

[2] Gabriel O, Kirner S, Klick M, Stannowski B, Schlatmann R. Plasma monitoring and PECVD process control in thin film silicon-based solar cell manufacturing. EPJ Photovoltaics. 2014;55202:1-9.

[3] Crose M, Kwon JSI, Tran A, Christofides PD. Multiscale modeling and run-to-run control of PECVD of thin film solar cells. Renewable Energy. 2017;100:129-140.

[4] Crose M, Tran A, Christofides PD. Multiscale computational fluid dynamics: methodology and application to PECVD of thin film solar cells. Coatings. 2017; 7 (2) , 22.

494229 Dynamic Actuator Scheduling in Networked Distributed Processes Using a Receding-Horizon Optimization Approach

Abstract:

One of the fundamental problems that has received significant attention in the context of control of distributed parameter systems is the optimal placement and scheduling of control actuators. The realization that the selection of the control actuator locations in a spatially-distributed control system has a significant influence not only on the controllability and stabilizability of the infinite-dimensional system but also on the achievable closed-loop performance has motivated numerous research studies on this problem. The literature on this topic is quite extensive. A few examples of works in this area include the development of a methodology for the computation of optimal locations of point control actuators for non- linear feedback controllers in transport-reaction processes described by quasi-linear parabolic PDEs [1], the use of the notions of spatial and modal controllability to formulate the optimal actuator placement for linear parabolic PDEs [2], the development of control actuator scheduling policies for fault-tolerant control [3] and constrained stabilization of parabolic PDEs [4], and the development of integrated controller and actuator switching policies aimed at improving closed-loop performance and compensation of spatiotemporally varying disturbances [5].

A close inspection of the available results, however, shows that the emphasis of existing formulations is placed on optimizing the closed-loop system performance, with respect to the response speed and the control effort, without considering the cost associated with the sensor-controller communication. This cost is an increasingly important consideration in the design of feedback control systems for processes controlled over resource-constrained communication networks. Sensor and control systems that are accessed over bandwidth-limited communication channels are commonplace in many modern industrial systems due to the appealing economic savings and operational flexibility that networked control systems provide. For these systems to fulfill their promise, however, a balance must always be maintained between the extent of network utilization, on the one hand, and the achievable closed-loop performance, on the other. Keeping communication costs to a minimum favors reduced network utilization whereas maximizing closed-loop performance requires the opposite.

While research on networked control systems has been extensive (see [6], [7] for some surveys of results in this area), results on the analysis and design of networked control systems for spatially distributed systems have been more limited by comparison. A number of efforts have been made in recent years to address this problem by leveraging tools and techniques from infinite-dimensional systems, model reduction and model-based control (e.g., see [8]–[10]). The focus of these studies has been on ensuring closed-loop stability with reduced sensor-controller communication and explicitly characterizing the closed-loop stability properties in terms of the different control and communication design parameters. While the resulting characterizations establish a direct linkage between the placement of the control actuators and the level of network utilization, the problem of how to optimally select or schedule the control actuators in a way that simultaneously optimizes the closed-loop performance and communication costs has not been considered in previous works.

Motivated by these considerations, we present in this work a methodology for the integrated dynamic scheduling of control actuators and communication in a class of networked distributed processes using an optimization-based formulation. The objective is to simultaneously optimize the control system performance and minimize sensor-controller communication costs. We focus on processes modeled by uncertain PDEs with intrinsically low-order dominant dynamics. Based on an approximate low-order model that captures the dominant dynamics of the infinite-dimensional system, a model-based state feedback controller is initially designed and its closed-loop stability properties are explicitly characterized in terms of the model update period and the feasible control actuator locations. Based on this characterization, a finite-horizon optimization problem is formulated and solved on-line to determine the control actuator locations and update rates that simultaneously optimize closed-loop performance and sensor-controller network resource utilization. To this end, an objective function that includes suitable penalties on the response speed, the control effort and the model update frequency is formulated subject to appropriate closed-loop stability constraints. The optimization problem is solved in a receding horizon fashion resulting in an integrated dynamic scheduling policy that varies the control actuator placement and communication rate and allow the process to cope with its uncertain operating environment. The developed methodology is illustrated and evaluated through an application to a simulated diffusion-reaction process example.

References:

[1] C. Antoniades and P. D. Christofides, "Computation of optimal actuator locations for nonlinear controllers in transport-reaction processes," *Comp. & Chem. Eng.*, vol. 24, pp. 577–583, 2000.

[2] A. Armaou and M. Demetriou, "Optimal actuator/sensor placement for linear parabolic PDEs using spatial H2 norm," *Chem. Eng. Sci.*, vol. 61, pp. 7351–7367, 2006.

[3] N. H. El-Farra and S. Ghantasala, "Actuator fault isolation and reconfiguration in transport-reaction processes," *AIChE J.*, vol. 53, pp. 1518–1537, 2007.

[4] S. Ghantasala and N. H. El-Farra, "Stability-based actuator scheduling in distributed processes with control and communication constraints," in *Proceedings of 17th Mediterranean Conference on Control and Automation*, pp. 492–497, 2009.

[5] M. Demetriou and N. Kazantzis, "A new actuator activation policy for performance enhancement of controlled diffusion processes," *Automatica*, vol. 40, pp. 415–421, 2004.

[6] J. P. Hespanha, P. Naghshtabrizi, and Y. Xu, "A survey of recent results in networked control systems," *Proceedings of the IEEE*, vol. 95, no. 1, pp. 138–162, 2007.

[7] Y. Ke-You and X. Li-Hua, "Survey of recent progress in networked control systems," *Acta Automatica Sinica*, vol. 39, pp. 101–117, 2013.

[8] Y. Sun, S. Ghantasala, and N. H. El-Farra, "Networked control of spatially distributed processes with sensor-controller communication constraints," in *Proceedings of American Control Conference*, St. Louis, MO, pp. 2489–2494, 2009.

[9] Z. Yao and N. H. El-Farra, "Resource-aware model predictive control of spatially distributed processes using event-triggered communication," in *Proceedings of 52nd IEEE Conference on Decision and Control*, Florence, Italy, pp. 3726–3731, 2013.

[10] D. Xue and N. H. El-Farra, "Output feedback-based event-triggered control of distributed processes with communication constraints," in Proceedings of *55th IEEE Conference on Decision and Control*, Las Vegas, NV, pp. 4296–4301, 2016.

498118 Modified Equation-Free Control of Distributed Parameter Systems with Model Mismatch

Abstract:

Many chemical and material industry processes exhibit spatial variation of the state variables because of diffusion, convection and reaction. These processes can be mathematically described by partial differential equations based on conservation equations. This kind of system is called distributed parameter system (DPS). To achieve economic profit and improve the product quality, controlling DPS is important. Since the infinite dimensional state space dynamics of DPS makes controller design nontrivial, a standard way to tackle the issue is to construct a reduced order model (ROM) using Galerkinâ€™s method. Then controller and observer can be designed based on the ROM with less computational cost. However, most of the established methods require a mathematical model of the system.

To relax the requirement on the mathematical model, in [1] we proposed an equation-free method to control systems when the knowledge of the governing law is unavailable or incomplete and the actuator effect is known. This method was developed based on the feature of discrete empirical interpolation method (DEIM) [2] that the selection of the interpolation indices can limit the growth of the error of the approximation by DEIM. A set of snapshots (observation of the system) is assumed available. Estimation of the dynamics can be made using these previous observations and continuous measurement on limited number of locations.

In this presentation, we improve the accuracy of the estimation of the dynamics when the system dynamics no longer match previous observations. A correction term is added to reduce the error of the estimation of the system dynamics. The difference between the expected value of the state and the actual value is used to update the correction term. The modified version is applied to a diffusion reaction process.

REFERENCES

[1] Yang, M., Armaou, A. On the design of equation-free controllers for dissipative PDEs via DEIM, American Control Conference, 2017.

[2] Saifon Chaturantabut and DC Sorensen. Nonlinear model reduction via discrete empirical interpolation. SIAM Journal on Scientific Computing, 32(5):2737â€“2764, 2010.

498373 Local Dynamic Mode Decomposition with Control: Itâ€™s Application to Model Predictive Control of Hydraulic Fracturing

Abstract:

More often than not, a chemical process can be accurately described using a mathematical model which is generally comprised of nonlinear partial differential equations (PDE). In order to accurately capture the dynamics of systems described by PDEs, a large number of state snapshots are required and this makes it computationally expensive to design online control strategies. Dynamic Mode decomposition with control (DMDc) is an effective model reduction technique that has been developed by the fluids community to analyze observational data arising from high-dimensional dynamic systems with external inputs applied to the system [1]. It is a data driven, equation free framework that approximates the underlying dynamics from snapshot measurements alone. Computationally, DMDc assumes a linear model that best represents the underlying dynamics, even if those dynamics stem from a nonlinear process. Although it might seem equivocal describing a nonlinear system by superposition of modes whose dynamics are governed by the corresponding eigenvalues, DMDc can be considered a numerical approximation to Koopman spectral analysis providing theoretical justification for characterizing nonlinear systems [2].

However, for a highly nonlinear system, the assumption of linear relation might not work well especially, only if limited number of measurements are available to characterize the data. Additionally, it fails to capture the local dynamics when the process parameters change with space and time. Intuitively, for a system with nonlinear dynamics, this assumption of linear relation can still be applicable when we consider shorter periods of time (i.e., time-clustered data). Based on these observations, in order to capture the local dynamics of a complex nonlinear system more effectively, the embedded coherent structures must be tailored to the local behavior of every portion of the solution trajectory [3].

In this work, we extend the concepts of DMDc to better capture the local dynamics associated with highly nonlinear processes and develop local reduced-order models that accurately describe the fully-resolved data. In this context, we first partition the data into clusters using a Mixed Integer Nonlinear Programming based optimization algorithm, the Global Optimum Search, which incorporates an added feature of predicting the optimal number of clusters [4]. Next, we compute the reduced-order models that are tailored to capture local system dynamics by applying DMDc within each cluster. The developed models can be subsequently used to compute approximate solutions to the original high-dimensional system and to design feedback control systems. We successfully applied the proposed local model reduction method to design a model predictive controller (MPC) of a hydraulic fracturing process, which is characterized by a system of nonlinear highly-coupled PDEs with time-dependent spatial domains, to achieve uniform proppant concentration at the end of pumping [5]. We also compared the performance of MPC based on the reduced-order models developed by the proposed local DMDc method and the Multivariable Output Error State Space (MOESP) identification method (of same order). The results showed that the proposed method outperforms the MOESP by approximating the nonlinear system more accurately.

Literature cited:

[1] Proctor, J.L., Brunton, S.L., Kutz, J.N. Dynamic mode decomposition with control. *SIAM J. Appl. Dyn. Syst.* 2016;15:142-161.

[2] Rowley, C.W., Mezic, I., Bagheri, S., Schlatter, P., Henningson, D.S. Spectral analysis of nonlinear flows. *J. Fluid Mech.* 2009;641:115-127.

[3] Amsallem D, J. Zahr M., Farhat C. Nonlinear model order reduction based on local reduced-order bases. *Int. J. Numer. Methods Eng.* 2012;92:891-916.

[4] Tan M. P., Broach J. R., Floudas C. A. A novel clustering approach and prediction of optimal

number of clusters: global optimum search with enhanced positioning. *J. Glob. Optim.* 2007;39:323-346.

[5] Economides, M.J., Nolte, K.G., 2000. Reservoir stimulation. Chichester, Wiley.

493450 Three-Dimensional Multiscale CFD Modeling for PECVD of Amorphous Silicon Thin Films

Abstract:

Continuous strides have been made in the multiscale modeling of plasma-enhanced chemical vapor deposition (PECVD) with applications to the manufacturing of silicon thin films for use in the photovoltaic and microelectronics industries [1],[2],[3],[4]. Due to difficulties associated with continuous, in situ measurements during the deposition process, accurate modeling of thin film deposition is necessary in order to improve product quality and to cut down on manufacturing costs. Recently, Crose et al. proposed a novel multiscale computational fluid dynamics (CFD) model, which combined a macroscopic CFD domain with a microscopic surface domain [5]. While this model was successful in capturing the behavior of the PECVD reactor with regard to the non-uniform deposition of *a*-Si:H films and in providing a basis for multiscale CFD modeling, the two-dimensional axisymmetric nature of the model means that some features of the true three-dimensional system remain unexplored. Specifically, the showerhead holes that provide process gas to the plasma region, a key feature when considering the uniformity of thin film products, are not possible to represent using two-dimensional (2D) axisymmetric models.

Motivated by the above considerations, the 2D axisymmetric framework previously developed is extended to the three dimensional (3D) spatial domain. Using a 3D render of a typical chambered, parallel-plate PECVD reactor, a CFD model is proposed which is capable of reproducing both accurate plasma chemistry and fluid flow into the reaction zone through the showerhead region. Additionally, a detailed kinetic Monte Carlo (kMC) algorithm developed previously [3] is applied to simulate the the surface of the wafer in order to capture both the exchange of mass and energy, as well as the character of the *a*-Si:H thin-film. Given the computationally demanding nature of the transient simulations, a parallel computation strategy is applied which allows for the discretization of both the macroscopic CFD volume and the microscopic kMC algorithm. The outlined multiscale model is applied to the deposition of 300 nm thick *a*-Si:H films revealing significant non-uniformities in the thickness and porosity of the thin-film product. Additionally, comparison to the previously mentioned 2D axisymmetric model provides insight to the utility of reduced dimensional approaches.

[1] Collins D, Strojwas A, White D. A CFD model for the PECVD of silicon nitride. IEEE Transactions on Semiconductor Manufacturing. 1994;7:176â€"183.

[2] Da Silva A, Morimoto N. Gas flow simulation in a PECVD reactor. In Proceedings of the 2002 International Conference on Computational Nanoscience and Nanotechnology, San Juan, Puerto Rico, 22â€"25, 2002.

[3] Crose M, Kwon JSI, Nayhouse M, Ni D, Christofides PD. Multiscale modeling and operation of PECVD of thin film solar cells. Chemical Engineering Science. 2015;136:50-61.

[4] Armaou A, Christofides PD. Control and optimization of multiscale process systems. Computers and Chemical Engineering. 2006;30:1670-1686.

[5] Crose M, Tran A, Christofides PD. Multiscale computational fluid dynamics: methodology and application to PECVD of thin film solar cells. Coatings. 2017; 7 (2), 22.

498381 A Model Reduction Approach for Mechanistic Biochemical Network Modeling

Abstract:

Cell signaling proteins function through their modular structures, which contain site-specific features, such as binding domains and phosphorylation motifs. To mechanistically understand signaling pathway systems, it is important to develop models incorporating these site-specific features of the signaling proteins. Nevertheless, modeling a signaling network system incorporating such molecular details could be computationally challenging. A key challenge is to deal with the complexity of many possible combinations of site-specific binding and transformations of the molecules. This property, known as combinatorial complexity, can lead to a very large number of species and reactions in a network. This challenge has been partly addressed in a recent technique called the rule-based modeling approach (RBM). However, for systems with high degree of combinatorial complexity, the RBM can become computationally prohibitive. In this work, we present a model reduction framework that can systematically reduce the state-space dimension of a biochemical network model with combinatorial complexity. We demonstrate this framework by developing a highly mechanistic rule-based model of insulin growth factor receptor (IGF1R) signaling in HeLa cells. The model accounts for the detailed membrane-proximal events that lead to IGF1R crosslinking and activation by the extracellular insulin growth factor (IGF1), IGF1R phosphorylation, and protein recruitment in phosphorylated IGF1R. The model in its unreduced form represents a large reaction network system that is computationally intractable. By applying our model reduction technique, we show a dramatic reduction in the network size of the model. We use the reduced model to analyze IGF1-induced temporal and steady-state autophosphoryaltion of six tyrosine sites in the intracellular domain of IGF1R. We also analyze receptor recruitment of 16 HeLa cell-specific signaling proteins that directly compete for the six IGF1R autophosphorylation site binding.

496310 Economic Nonlinear Model Predictive Control of Continuous Pharmaceuticals Manufacturing Processes

Abstract:

Continuous manufacturing of pharmaceuticals (CMP) is being explored as an alternative to traditional batch-wise production. This shift holds great promise to increase production efficiency, enable smaller production facilities, minimize waste, energy consumption, and raw material use and to enable drug quality monitoring on a continuous basis. However, CMP may imply short operational campaigns, with long transient phases (start-up and shutdown), constituting up to 30% of the entire time horizon [1]. This significantly hampers the plant's economic viability.

In previous work the problem of offline economic open-loop optimal control was investigated on toy problems [2] as well as on more realistic plant-wide models [3]. The problem formulation involves a hybrid (discrete-continuous) dynamical system, where switching between on/off-spec production modes occurs at discrete times, which are themselves optimization variables. These modes' order is predetermined, according to the turnpike property of optimal dynamic systems, allowing to perform correct sensitivity analysis, to be used in gradient based optimization algorithms. The dynamic optimization approach maximizes the accumulated on-spec production directly over the entire time horizon. In this sense the process being controlled has properties of both continuous and batch processes.

When considering online control, previous work has focused mainly on controlling a predetermined steady state operation, using traditional PI controllers [4] or MPC [5], without considering the economics of the overall production. Here, on the contrary, we focus on real-time control strategies that optimize an economic objective function. Some of the challenges this approach is facing are: 1. The dynamic optimization involves nonsmooth and often hybrid behavior, which results in high computational cost of the sensitivity analysis. 2. Solving the optimal control profile for shorter times than the actual campaign time may result in different (sub-optimal) solutions.

In this contribution we study the implementation of non-linear model predictive control (NMPC) schemes to operate such a plant. An economic-NMPC approach is investigated for online optimal operation of the manufacturing process. We compare the results of a bench-scale ideal receding horizon economic-NMPC approach to a hierarchical approach, where the control is conducted in two layers: 1. On the top we have a real-time dynamic optimization layer that considers the entire time horizon (campaign time) 2. Below we apply NMPC controllers using simplified models and shorter horizons. We also make comparison to an open-loop optimal control. Choosing the control variables and dynamic set-points will also be discussed.

1. S. Mascia et al., "End-to-End Continuous Manufacturing of Pharmaceuticals: Integrated Synthesis, Purication, and Final Dosage Formation", Angew Chem. Int. Ed. Engl., 2013, 52(47),12359-63.

2. A. Sahlodin and P.I. Barton, "Optimal Campaign Continuous Manufacturing", Ind. Eng. Chem. Res., 2015, 54(45), 11344-59.

3. M. Shoham Patrascu and P.I. Barton, "Modeling and Dynamic Optimization of an End-to-End Continuous Pharmaceuticals Manufacturing Plant", in preparation.

4. R. Lakerveld et al., "The Application of an Automated Control Strategy for an Integrated Continuous Pharmaceutical Pilot Plant.", Org. Process Res. Dev., 2015, 19, 1088-1100.

5. A. Mesbah et al., "Plant-wide Model Predictive Control for a Continuos Pharmaceutical Process", In American Control Conference (ACC), 2015, pp. 4301-4307.

496699 A Scheduling Perspective on the Monetary Value of Improving Process Control

Abstract:

Advanced process control (APC) strategies such as Model Predictive Control (MPC) have become the standard control technique in the process industries due to their unique capability of dealing with complex interactions, process nonlinearities, and operational constraints [1]. The use of APC offers many advantages, such as an increase in throughput, process stability, yield, or a reduction in energy consumption, waste, raw material, and costs. However, the implementation of a novel APC technology can be expensive due to the costs of manpower, hardware, software, and production loss due to installation downtime. While there are well-established metrics for the evaluation of the closed-loop performance of a control technology (either in-silico or once implemented on a physical system), a strong case for its implementation can only be established by quantitatively defining the associated monetary benefits. The question of how to assess the economic impact of any control system has therefore been a concern of the process systems community since the inception of the field [2,3].

During steady-state operation, an effective control system minimizes the variability of the controlled variables, such that the process can be operated close to the constraints on those states, which normally translates to higher profits. The most common approach for quantifying the monetary value of any proposed control system has therefore been through performance functions that define profit as a function of the distance between a process value and its operational limit [3]. In transient operation, an effective control system is one that minimizes the transition time between operating points, and limited work has been done on quantifying the monetary value of control systems in the context of transient operation.

In todayâ€™s fast-changing markets, transient operation has become ever-more relevant, and thus there have been many efforts towards integrating the dynamic process control calculations within process scheduling frameworks. This integration has been shown to yield significant economic benefits [4]. However, the problem of how to quantify the monetary value of a particular control technology across the horizon of a production schedule remains open.

In this work, we posit that the monetary value of a particular control system can be evaluated by defining control performance metrics and quantifying the change in a production scheduling profit objective value as the performance metric varies. We arrive at an explicit expression that is applicable to any production scheduling problem, and illustrate our methodology for several case studies.

References

[1] S. J. Qin, T. A. Badgwell. A survey of industrial model predictive control technology. *Control Engineering Practice* 11 (2003) 733-764.

[2] M. Bauer, I. K. Craig. Economic assessment of advanced process control â€“ A survey and framework. *Journal of Process Control* 18 (2008) 2-18.

[3] T. F. Edgar. Control and operations: when does controllability equal profitability? *Computers and Chemical Engineering* 29 (2004) 41-49.

[4] M. Baldea, I. Harjunkoski. Integrated production scheduling and process control: A systematic review. *Computers and Chemical Engineering* 71 (2014) 377-390.

497717 A Computationally Efficient Approach to Economic Model Predictive Control Via Carleman Approximation

Abstract:

A Computationally Efficient Approach to Economic Model Predictive Control via Carleman Approximation

Yizhou Fang and Antonios Armaou, the Pennsylvania State University, University Park, PA

Proposed Session: Systems and Process Control (10B) 10B00 Economics and Process Control

Abstract:

In recent years, Economic Model Predictive Control (EMPC) has gained popularity in chemical and petrochemical industries. The primary difference of EMPC from traditional MPC is that EMPC is directly formulated to maximize the economic profits. However, one of the remaining major challenges is the heavy burden in computation. The economically optimal cost functions representing the economic performances, unlike traditional tracking MPC, are usually non-quadratic or even non-convex. That may require significant amount of computation and puts a heavy load on the optimizer to identify the economically optimal control inputs. If the optimizer fails to converge to the solution fast enough, the delay in sending the control signals, or sending non-optimal control signals to the system may degrade its closed loop performance, or even cause potential stability issues.

The approach we propose to address this problem is built upon two foundations: the theory of Carleman approximation and the theory of bilinear control systems. Following a Two-Tier approximation, we use high order polynomial states to capture the nonlinearity of the original dynamic process. With an extended bilinear expression and assuming piecewise constant manipulated inputs, we then analytically predict future state trajectories and economic performance. We also analytically calculate the sensitivity of the economic cost function to the manipulated inputs. The sensitivity information significantly increases optimizer performance and reduces the number of iterations. Hence, the computational effort in solving the EMPC problem is sufficiently reduced.

As an application example, we consider a CSTR where ethylene is oxidized by air in a catalytic environment to produce ethylene oxide. The economic performance of the reactor is characterized by the time-averaged yield of the product. We simulate the system under nominal conditions, under system noise and under model mismatches. Compared with the â€œstandardâ€ EMPC method via numerical simulation, we demonstrate that the proposed method is applicable and leads to significant computational effort savings.

Key words: Economic Model Predictive Control, Carleman Approximation, Computational Efficiency

Reference:

1. Fang Y, Armaou A. Nonlinear Model Predictive Control using a bilinear Carleman linearization-based formulation for chemical processes. 2015 Am Control Conf. 2015;1:5629-5634. doi:10.1109/ACC.2015.7172221.

2. Fang Y, Armaou A. Carleman approximation based quasi-analytic model predictive control for nonlinear systems. AIChE J. 2016;62(11):3915-3929. doi:10.1002/aic.15298.

3. Fang Y, Armaou A. A Formulation of Advanced-step Bilinear Carleman Approximation-based Nonlinear Model Predictive Control. Proc 55th IEEE Conf Decis Control. 2016:4027-4032.

4. Fang Y, Armaou A. A Computationally Efficient Approach of Economic Model Predictive Control via Carleman Approximation. Proc 56th IEEE Conf Decis Control. 2017:submitted.

498353 Handling Economic and Practical Considerations in Feedback Control of Hydraulic Fracturing in Ultra-Low Permeability Reservoirs

Abstract:

Typically, the term shale oil refers to natural oil trapped in rock of low porosity (2% or less) and ultra-low permeability (0.01 to 0.0001 md or even less) [1]. In practice, the value and even the meaning of reservoir permeability is questionable. Therefore, the use of traditional fracture design approaches such as Unified Fracture Design, which makes use of rock formation properties to obtain the optimal fracture geometry, is not the method of choice for shale formations [2, 3].

In ultra-low permeability formations, horizontal wells are typically drilled and for each well, multiple hydraulic fractures are generated for enhanced recovery of oil and gas. Based on the assumption that the propped fracture has practically infinite conductivity, in this work, we will use a section based method [3] to calculate the optimal number of horizontal wells, number of fractures, the length of the fracture and the drainage area aspect ratio that maximize the overall productivity of the well-fracture system. Economic considerations are represented in the form of fixed total length of all the fractures that can be created from the available resources. With a reasonable additional technical constraint (maximum length of fracture that can be placed repeatedly and reliably from any point of the horizontal well) there is a unique solution to the optimization problem.

Once the fracture design parameters are determined, it is also important to achieve uniform proppant concentration at the end of pumping, because it is directly related to the overall efficiency of hydraulic fracturing processes [4]. From a control engineering viewpoint, hydraulic fracturing treatments have been traditionally operated in an open-loop manner. However, such an open-loop operation may lead to poor performance if there is a plant-model mismatch and uncertainties in the process model parameters. While some attempts to employ model-based control schemes have been made [5], they did not consider economics as well as practical and safety considerations in the design of pumping schedules. Motivated by these considerations, we initially focus on the development of a first-principle model of a hydraulic fracturing process. Second, a novel numerical scheme is developed to efficiently solve the coupled partial differential equations (PDEs) defined over a time-dependent spatial domain. Third, a reduced-order model is constructed and used to design a Kalman filter to accurately estimate unmeasurable variables such as proppant concentration inside the fracture. Lastly, model predictive control (MPC) theory is applied for the design of a feedback control system to achieve uniform proppant concentration across the fracture and to realize the optimal fracture geometry obtained by the open loop optimization. Within the controller, we explicitly take into account the practical considerations such as the actuator limitations and safety issues. We demonstrate that the proposed control scheme is able to generate the required fracture extent and the spatial concentration profile within which is closer to the target compared to other state-of-the-art pumping schedules.

References:

[1] Nikolaou, M. (2013). Computer-aided process engineering in oil and gas production. *Computers & Chemical Engineering*, *51*, 96-101.

[2] Economides, M.J., Oligney, R.E., Valko, P., 2002. Unified fracture design. Orsa Press.

[3] Liu, S., & ValkÓ, P. P. (2017). Optimization of spacing and penetration ratio for infinite conductivity fractures in unconventional reservoirs- A sectional based approach. *Society of petroleum engineers* (SPE-186107-PA).

[4] Yang, S., Siddhamshetty, P., Kwon, J.S. (2017), Optimal pumping schedule design to achieve a uniform proppant concentration level in hydraulic fracturing. *Compt. Chem. Eng.*, 101, 138-147.

[5] Gu, Q., & Hoo, K. A. (2015). Model-based closed-loop control of the hydraulic fracturing process. *Industrial & Engineering Chemistry Research*, *54*(5), 1585-1594.

499883 Integrating RTO with Stabilizing Economic MPC

Abstract:

Integrating RTO with stabilizing economic MPC
Douglas A. Allan and James B. Rawlings
April 17, 2017

Model predictive control (MPC) is an advanced control algorithm that combines a system model with an objective function and numerical optimization to provide improved control performance. Traditionally, the objective functions used in MPC have been tracking objective functions that penalize deviations from a target steady-state combination of system states and inputs. Recently, a new form of MPC that optimizes other performance criteria, termed economic MPC has been proposed (Diehl, Amrit, and Rawlings, 2011).

This type of MPC has the possibility of directly optimizing important performance criteria, such as maximizing plant profit or minimizing pollution or carbon dioxide emissions, rather than optimizing a surrogate tracking cost. However, in general this type of MPC does not stabilize any particular steady state unless the system and stage cost satisfy a certain dissipation inequality (Angeli, Amrit, and Rawlings, 2012); if this inequality is not satisfied, then (under a reachability condition) it is more profitable to operate the system periodically (Muller, Angeli, and Allgower, 2015). However, industrial systems are often not designed to operate in such a fashion. Periodic operation may cause excess fatigue on equipment, and periodic operation may meet resistance from process operations personnel.

There are conditions under which economic MPC is known to be stabilizing. In particular, if a problem has a linear model and convex objective function, economic MPC stabilizes the optimal steady state under mild assumptions (Diehl et al., 2011). Typically industrial dynamic models are linear, but profit functions typically have both linear and bilinear terms. We propose a method to obtain a local convex approximation of the profit function from the real-time optimization (RTO) layer and use it in Economic MPC. We compare this technique to both the current industrial practice of RTO combined with tracking MPC and fully nonlinear economic MPC in a case study.

References

D. Angeli, R. Amrit, and J. B. Rawlings. On average performance and stability of economic model predictive control. IEEE Trans. Auto. Cont., 57(7):1615-1626, 2012.

M. Diehl, R. Amrit, and J. B. Rawlings. A Lyapunov function for economic optimizing

model predictive control. IEEE Trans. Auto. Cont., 56(3):703-707, 2011.

M. Muller, D. Angeli, and F. Allgower. On necessity and robustness of dissipativity in economic model predictive control. Automatic Control, IEEE Transactions on, 60(6): 1671-1676, June 2015. ISSN 0018-9286. doi: 10.1109/TAC.2014.2361193.
1

490364 Combining Self-Optimizing Control and Extremum-Seeking Control â€“ Applied to Ammonia Reactor Case Study

Abstract:

In this work, we consider self-optimizing control and extremum seeking control in the context of Real time optimization. Often real-time optimization (RTO) uses a process model to determine the optimal setpoints by solving an optimization problem online. Online computations are often expensive and is susceptible to plant-model mismatch. Alternatively, there are methods that avoid repeated numerical optimization by simply transforming the optimization problem into a feedback control problem. Extremum seeking control, Necessary conditions of optimality (NCO) tracking and self-optimizing control belong to such feedback-based methods.

Self-optimizing control uses a system model to solve an offline optimization problem for expected disturbances. This translates the economic objectives into control objectives by finding an appropriate set of controlled variables, which when kept at a constant setpoint gives acceptable loss [1]. This eliminates the need for an expensive online optimizer to re-optimize when disturbances occur and converts the slow optimization problem into a fast control problem. However, the optimal setpoint may change for large deviations from the nominal optimal point because of nonlinearity in the process. One of the main advantages of self-optimizing control is that it can handle unmeasured disturbances, but cannot handle unmodelled disturbances.

Extremum seeking control on the other hand is a class of model free methods that optimizes the system based on measurements. However, due to the time scale separation requirements, extremum-seeking control causes the convergence to be very slow [2]. Extremum seeking control is also often affected by disturbances. The effect of disturbance is clearly motivated in [3], with a suggested modification for measured disturbances. The effect of unmeasured disturbances, however, remains unsolved.

The self-optimizing control and extremum seeking control methods have been developed relatively independently since 2000. Self-optimizing control combined with NCO tracking was proposed in [4]. However, the simple finite difference method used in to obtain the gradient made it difficult to make the NCO tracking work in [4]. In this work, we propose the combination of extremum seeking control and self-optimizing control in a hierarchical fashion to exploit the synergy between the two methods. The self-optimizing control layer reacts fast to changes in unmeasured but modelled disturbances thereby keeping the process in the near optimal region, and extremum-seeking control accounts for the unmodelled disturbances and fine-tunes the optimal setpoint to further minimize the loss from the self-optimizing control. By combining self-optimizing control and extremum seeking control, we can handle a wider class of disturbances.

The proposed method is tested on an ammonia reactor case study from [5]. The self-optimal variables are identified using the exact local method [6] and the corresponding setpoints were provided by an extremum-seeking scheme as proposed in [7]. Simulation results show the performance improvements gained from the proposed method compared with just a self-optimizing control or extremum-seeking control. References

[1] Skogestad, S. (2000). Plantwide control: the search for the self-optimizing control structure. Journal of Process control, 10(5), 487{507.

[2] Nesic, D. (2009). Extremum seeking control: Convergence analysis. In Control Conference (ECC), 2009 European, 1702-1715.

[3] Krishnamoorthy, D., Pavlov, A., and Li, Q. (2016). Robust extremum seeking control with application to gas lifted oil wells. IFAC-PapersOnLine, 49(13), 205-210.

[4] Jäschke J, Skogestad S. NCO tracking and self-optimizing control in the context of real-time optimization. Journal of Process Control. 2011 Dec 31;21(10):1407-16.

[5] Morud, J. and Skogestad, S. (1998). Analysis of Instability in an Industrial Ammonia Reactor. AIChe Journal, 44(4), 888-895.

[6] Halvorsen, I.J., Skogestad, S., Morud, J.C., and Alstad, V. (2003). Optimal selection of controlled variables. Industrial & Engineering Chemistry Research, 42(14), 3273-3284.

[7] Hunnekens, B., Haring, M., van de Wouw, N., and Nijmeijer,H. (2014). A dither-free extremum-seeking control approach using 1st-order least squares fits for gradient estimation. In 53rd IEEE Conference on Decision and Control, 2679-2684.

493482 Combining First-Principles and Empirical Modeling for Computation Time Reduction of Economic Model Predictive Control

Abstract:

For large-scale nonlinear process systems, the computation time required to solve the optimization problem of an economic model predictive controller or EMPC [1] (a control design that computes control actions that optimize process economics), may be large due to the need to solve a nonlinear program with tens or hundreds of optimization variables. For processes requiring a sampling time shorter than the time to solve such a problem due to the time scale on which the process dynamics evolve, this significant computation time may prohibit the use of EMPC for a given process. One method of reducing the computation time of EMPC is by utilizing linear [2] or nonlinear [3] empirical models for making state predictions in the EMPC compared to using a first-principles model. These empirical models may have lower computation times than first-principles models because they may have less terms to evaluate during numerical integration of the model, they may be less stiff than the first-principles model, or, in the case of linear empirical models, they have an analytic solution that can be taken advantage of when computing state predictions within the EMPC. Prior works that have examined EMPC with empirical models [2,3,4] assume that the empirical model is used throughout the entire prediction horizon. However, due to the increased accuracy of a first-principles model, it may be more economically beneficial to utilize a first-principles model for making state predictions in EMPC, when such a model is available, rather than an empirical model.

Motivated by this, the present work investigates the trade-off between computation time and model accuracy for an EMPC design that utilizes a first-principles model for the first several sampling periods of the prediction horizon (including the first sampling period since the control actions computed for the first sampling period are implemented on the process according to the receding horizon implementation strategy of EMPC) and an empirical model for the remaining sampling periods (for computation time reduction). The number of sampling periods for which the first-principles model is used, the sampling period length, and the prediction horizon length are evaluated with respect to their impact on closed-loop economic performance and computation time. The use of the error-triggered on-line empirical model update strategy from [4] is investigated for updating the empirical model used within the EMPC with the combined first-principles/empirical model as the closed-loop state moves throughout state-space. Conditions that guarantee closed-loop stability and feasibility of the EMPC with the combined first-principles/empirical model are developed. In addition, the impact of changing the empirical model partway through a sampling period (i.e., allowing the EMPC to start using the empirical model after a non-integer multiple of a sampling period has passed) is investigated both from a closed-loop stability and feasibility standpoint and from a practical computation time and economic performance standpoint. A chemical process example compares the computation time and economic performance of a nonlinear process under the proposed combined first-principles/empirical EMPC design with the computation time and performance under both an EMPC utilizing only a first-principles model and an EMPC using only an empirical model.

[1] Ellis M, Durand H, Christofides PD. A tutorial review of economic model predictive control methods. Journal of Process Control. 2014;24:1156-1178.

[2] Alanqar A, Ellis M, Christofides PD. Economic model predictive control of nonlinear process systems using empirical models. AIChE Journal. 2015;61:816-830.

[3] Alanqar A, Durand H, Christofides PD. On identification of well-conditioned nonlinear systems: Application to economic model predictive control of nonlinear processes. AIChE Journal. 2015;61:3353-3373.

[4] Alanqar A, Durand H, Christofides PD. Error-triggered on-line model identification for model-based feedback control. AIChE Journal. 2016;63:949-966.

493378 Economic Performance Improvement forÂ Lyapunov-Based Economic Model Predictive ControlÂ Using Disturbance Probability Distributions

Abstract:

Lyapunov-based economic model predictive control (LEMPC) [1] is an economic model predictive control (EMPC) [2] scheme that can dictate two modes of operation. In the first mode, LEMPC can impose time-varying operation through a state constraint while maintaining the state in state-space within the stability region of a nonlinear process operated under a Lyapunov-based controller. In the second mode of operation, LEMPC utilizes a contractive constraint to drive the closed-loop state closer to the origin (in the sense that it is driven into Lyapunov level sets defined by a smaller upper bound on the Lyapunov function) even in the presence of uncertainty. Due to the guaranteed closed-loop stability properties of this LEMPC paradigm, it has been used over the past years to promote various control purposes such as process safety [3] and production management [4]. However, the constraints that define the two modes of operation of the LEMPC to ensure that the stability region is a forward invariant set for the closed-loop process under LEMPC in the presence of disturbances are activated based on the location of the process state in state-space (i.e., the contractive constraint is activated whenever the measurement of the closed-loop state is outside a subset of the stability region at a sampling time). This subset of the stability region is chosen such that even if the worst-case upper bound on the disturbance was reached throughout the entire sampling period and the process is operating in a time-varying fashion (the first mode of operation), the state would still be within the stability region at the end of the sampling period. This is a highly conservative approach for designing the region in which time-varying operation is allowable, given that the worst-case upper bound on the disturbance is unlikely to be obtained throughout an entire sampling period under normal operation of a chemical process. Therefore, the economic performance of the LEMPC design may be significantly improved if the subset of the stability region that dictates the activation of the first or second mode of operation is chosen in a less conservative way.

Motivated by the above considerations, we propose an LEMPC design in which the region in state-space within which time-varying operation is allowable is expanded compared to the traditional LEMPC design by accounting for the probability distribution of the disturbance magnitude (rather than assuming that the upper bound occurs throughout the entire sampling period). Specifically, the region in which time-varying operation is allowed is chosen such that if the disturbance at any given time throughout a sampling period takes a value within its bounds with a probability P according to a disturbance probability distribution obtained from routine process operating data, then the closed-loop state is maintained within the stability region throughout the sampling period. The role that the probability distribution shape plays in the size of the region of time-varying operation is evaluated, and the closed-loop stability and feasibility properties of this less conservative LEMPC design are investigated. Practical suggestions are made for obtaining a disturbance magnitude probability distribution and for designing the LEMPC when the probability distribution itself has uncertainty since it is obtained from limited process data. A chemical process example is used to evaluate the economic benefits of designing the LEMPC with a less conservative state constraint activating the first or second mode of operation compared to the traditional LEMPC design methodology.

[1] Heidarinejad M, Liu J, Christofides PD. Economic model predictive control of nonlinear process systems using Lyapunov techniques. AIChE Journal. 2012;58:855-870.

[2] Ellis M, Durand H, Christofides PD. A tutorial review of economic model predictive control methods. Journal of Process Control. 2014;24:1156-1178.

[3] Alanqar A, Durand H, Albalawi F, Christofides PD. An economic model predictive control approach to integrated production management and process operation. AIChE Journal; 2017; in press.

[4] Albalawi F, Alanqar A, Durand H, Christofides, PD. A feedback control framework for safe and economically-optimal operation of nonlinear processes. AIChE Journal. 2016;62:2391-2409.

500230 Economic Optimization of Large-Scale Embedded Battery Applications

Abstract:

Commercial buildings are responsible for 20% of the total U.S. energy consumption. In particular, large-scale heating, ventilation, and air conditioning (HVAC) systems account for a large portion of these energy expenditures. Due to the lack of optimal control algorithms in buildings (Afram & Janabi-Sharifi, 2014), these systems have become a focus for both academic and industrial research. Many papers have been published in this area showing that significant cost savings can be attained by capitalizing on the time-varying nature of electricity prices, including (Oldewurtel et al., 2010; Mendoza-Serrano & Chmielewski, 2012). Savings are achieved using various forms of energy storage to shift the energy load from peak hours when prices are high to off-peak hours when prices are low (Touretzky & Baldea, 2016; Avci, Erkoc, Rahmani, & Asfour, 2013). Several forms of storage can be used including passive thermal energy storage (e.g., mass of the buildings), active thermal energy storage (e.g., insulated chilled water tanks), and electricity storage (e.g., batteries).

Model predictive control (MPC) is an advanced control method that is well-suited for load shifting due to its ability to forecast and optimize. MPC relies on a model of the system to predict the process variables based on the actions taken by the controller (Rawlings & Mayne, 2009). An optimization problem is then solved using this model to achieve desired control objectives. MPC has been highly successful with thousands of applications in the chemical and petroleum industries alone (Qin & Badgwell, 2003). Economic MPC is one form of MPC in which the objective is to minimize cost or maximize profit. MPC can also handle both continuous decisions such as temperature setpoints and discrete decisions such as when to turn equipment on and off (Rawlings & Risbeck, 2016). Hence, heuristics are no longer needed to make discrete decisions.

Large-scale embedded battery applications are the primary focus of this work. As the cost of batteries decreases, they can be embedded directly into any piece of HVAC equipment that draws power, including roof-top units (RTUs), air-handler units (AHUs), variable refrigerant flow (VRF) units, fans, pumps, and compressors. Large-scale applications, such as university campuses and industrial complexes, may contains hundreds of buildings and thousands of zones. Each building can have a multiplicity of RTUs, VRFs, and/or AHUs. A centralized MPC formulation for these cases is neither feasible nor desirable, so a decomposition is needed. Distributed control can be used for this purpose. Managing the peak demand over the entire campus is also key since most pricing structures include a peak demand charge. The distributed control system design must account for this peak demand charge.

In this talk, we propose a hierarchical control architecture using MPC for the economic optimization of large-scale commercial HVAC systems with embedded batteries. The high-level problem serves as a coordinator for the distributed controllers in each building or subsystem. The aforementioned issues are addressed in this decomposition. We conclude with a large-scale simulation study to demonstrate the savings potential using embedded batteries with the proposed control strategy.

References

Afram, A., & Janabi-Sharifi, F. (2014, Feb). Theory and applications of HVAC control systems—A review of model predictive control (MPC). Build. Environ., 72, 343–355.

Avci, M., Erkoc, M., Rahmani, A., & Asfour, S. (2013, May). Model predictive HVAC load control in buildings using real-time electricity pricing. Energ. Buildings, 60, 199–209.

Mendoza-Serrano, D. I., & Chmielewski, D. J. (2012). HVAC control using infinite-horizon economic MPC. In Decision and control (cdc), 2012 ieee 51st annual conference on (pp. 6963–6968).

Oldewurtel, F., Parisio, A., Jones, C. N., Morari, M., Gyalistras, D., Gwerder, M., ... Wirth, K. (2010). Energy efficient building climate control using stochastic model predictive control and weather predictions. In American Control Conference (ACC), 2010 (pp. 5100–5105). Baltimore, MD, USA.

Qin, S. J., & Badgwell, T. A. (2003). A survey of industrial model predictive control technology. Control Eng. Pract., 11(7), 733-764.

Rawlings, J. B., & Mayne, D. Q. (2009). Model predictive control: Theory and design. Madison, WI: Nob Hill Publishing. (576 pages, ISBN 978-0-9759377-0-9)

Rawlings, J. B., & Risbeck, M. J. (2016). Model predictive control with discrete actuators: Theory and application. Accepted to Automatica.

Touretzky, C. R., & Baldea, M. (2016). A hierarchical scheduling and control strategy for thermal energy storage systems. Energ. Buildings, 110, 94–107.

494522 Stochastic Model Predictive Control for Battery Systems

Abstract:

Battery storage systems are flexible assets that can be used to provide frequency regulation for ISOs and to modulate loads for energy-intensive facilities to aid utilities (e.g. buildings or manufacturing) [1]. This flexibility is becoming increasingly valuable as more intermittent renewable power is injected into the grid. In particular, ISOs have reported an increased demand for frequency regulation services to mitigate fast renewable power fluctuations and stranded power [2]. Battery storage systems can be used to mitigate these issues and can be strategically placed in the network to maximize system-wide flexibility. Various studies have shown the economic benefits of using stationary battery systems to provide frequency regulation to the grid and to save cost by reducing demand charges in buildings and microgrids (via load peak shaving) [3,4,5]. These studies assume perfect knowledge of markets and loads. In real-time settings, however, diverse uncertainties can hinder battery performance and revenue potential. While stochastic optimization models for these types of applications have also been recently explored [6], an important general limitation of these studies is that they do not evaluate the economic potential of stochastic formulations against deterministic counterparts. Moreover, as in general stochastic optimization applications, uncertainty modeling and scenario generation from operational data remains a challenge. This is particularly relevant in battery systems performing simultaneous frequency regulation and demand peak shaving, because load uncertainty limits the amount of frequency regulation revenue that can be obtained to the ISO and can induce large demand charges if regulation participation potential is overestimated.

In this work we propose a stochastic model predictive control (MPC) framework to determine optimal participation strategies for stationary battery systems in ISO frequency regulation markets while simultaneously mitigating demand charges from a local utility associated to a set of attached loads that need to be modulated. The proposed framework solves a two-stage stochastic program that maximizes the expected revenue over a receding horizon and considers uncertainty of the modulated load. We propose to use a Ledoit-Wolf covariance estimator [7] to generate load scenarios from limited historical data and to capture short- and long-term load correlations. We use the framework to study the flexibility and economic benefits provided by a 1 MW battery system attached to a load from a collection of buildings. We use the proposed framework to study the benefits of stochastic MPC policies compared to those obtained with deterministic MPC and perfect information MPC strategies. We also study the effect of the prediction horizon length and of peak demand carryover prices on the performance of the MPC policies.

Using real load data of a typical university campus and price data from PJM, we find that stochastic MPC can significantly outperform its deterministic counterpart and improves the value of the battery. Our simulations illustrate that stochastic MPC can mitigate large demand charges and better manage frequency regulation commitments. Notably, stochastic MPC can recover 88% of the ideal value of the battery obtained with operations under perfect information, while deterministic MPC can only recover 79%. Moreover, we show that stochastic MPC can be used to modulate revenue volatility and with this mitigate risk. We have also found that the length of the prediction horizon and scaling of the demand

charge significantly affect economic performance.

References:

[1] A. Oudalov, D. Chartouni, C. Ohler, and G. Linhofer, "Value Analysis of Battery Energy Storage Applications in Power Systems," in 2006 IEEE PES Power Systems Conference and Exposition, pp. 2206–2211, 2006.

[2] K. Kim, F. Yang, V. M. Zavala, and A. A. Chien, "Data Centers as Dispatchable Loads to Harness Stranded Power," IEEE Transactions on Sustainable Energy, vol. 3029, no. c, pp. 1–1, 2016.

[3] A. Oudalov, R. Cherkaoui, and A. Beguin, "Sizing and optimal operation of battery energy storage system for peak shaving application," in 2007 IEEE Lausanne POWERTECH, Proceedings, pp. 621–625, 2007.

[4] Y. Wang, B. Wang, C. C. Chu, H. Pota, and R. Gadh, "Energy management for a commercial building microgrid with stationary and mobile battery storage," Energy and Buildings, vol. 116, pp. 141–150, 2016.

[5] A. Lucas and S. Chondrogiannis, "Smart grid energy storage controller for frequency regulation and peak shaving, using a vanadium redox flow battery," International Journal of Electrical Power and Energy Systems, vol. 80, pp. 26–36, 2016.

Vehicles," IEEE, 2012.

[7] O. Ledoit and M. Wolf, "A well-conditioned estimator for large-dimensional covariance matrices," Journal of Multivariate Analysis, vol. 88, no. 2, pp. 365–411, 2004.

482127 High Fidelity Dynamic Simulation of Integrated Process, Control and Electrical Systems

Abstract:

High fidelity dynamic simulation of integrated process, control and electrical systems

The practice of dynamic simulation involves the use of software programs that model the time dependent behavior of complex process systems using rigorous thermodynamic and first principles models that adhere to laws of conservation of momentum, mass and energy. The first dynamic process simulations were developed considering only the process and equipment involved so their functioning could be understood better to improve their design. As the use of such simulations broadened, the need to include process control behavior and interactions became a critical component. Having process simulations and controls together in one platform opened up a plethora of opportunities that ranged from designing and checking out controls to using the integrated system to train operators and evaluate what-if scenarios on the process as a whole. Further advancements in software technology have helped take these systems to a point today where a plant operator would find it difficult to differentiate between the experience on a Training Simulator from that in the real control room. The impact such systems have on improving plant availability and safety are well known. Clearly, over the past two decades, the methodology has gained wide acceptance as an effective tool in applications throughout various stages of a plant lifecycle from concept evaluation to design to commissioning, startup and operations.

The proposed paper discusses adding a third dimension to the technology by including electrical systems to the process and controls model. Electric drives are gaining more popularity today because of the various advantages it offers in terms of low maintenance, reduced emissions, ease of operation, lower lifecycle costs and increased availability. In addition they also offer improved flexibility of operations through the ability to control output power through variable frequency drives and they completely eliminate the impact of ambient temperature on equipment performance vis-a-vis a Gas Turbine driver. Industrial electric drives come in various sizes ranging from a few KWs to in excess of 60MW. Plants typically have in-house captive power plants to produce the power to drive these or in some cases they rely on public utility grids. As process plants are being driven towards increasing energy efficiency, the industry is also seeing an increase in the number of cogeneration plants being constructed. As such, large amounts of electric power generation and consumption have become key components of most large scale process plants. The process and electrical systems differ in their characteristics inherently as the time constants involved in the electrical system is typically much smaller compared to the ones on the process side. The systems however impact each other significantly to warrant analyzing them together. As an example, start-up or loss of a large driver could induce frequency and voltage fluctuations in the grid which could potentially impact other users and hence the overall stability of operations. Similarly loss of partial power generation or steam production, if not managed properly, could cascade into severe instabilities and even loss of entire operations.

The paper discusses the approach, benefits and value associated with integration of a first principles based dynamic model of electric power generation, distribution and consumption with a similar high fidelity dynamic model of the process and control systems. Example case studies will be presented to highlight the points the author intends to make.

495701 A Supervisory Predictive Control System for Solar-Load Balancing: Application to Building Energy Management

Abstract:

Energy consumption is typically divided among four major sectors, including residential, commercial, industrial, and transportation sectors. Non-industrial energy usage comprises a significant part of the total energy budget in the United States. For example, in 2016 20% of total energy use was due to the residential sector, while another 18% of end use energy was by the commercial sector [1]. As a result, even small improvements in residential and commercial building energy efficiency, if widely adopted, hold the potential for significant impact. It is now largely recognized that addressing energy use in buildings can reduce total fossil fuel consumption and associated greenhouse gas emissions. Benefits such as reduced building operational energy costs have prompted growing interest among policy makers, the technical community, and the general public in addressing building energy issues and investigating solutions for decreasing building energy consumption (e.g., [2]-[9]). Reducing existing building energy consumption consists of two synergistic approaches. One is to reduce the need for energy through implementation of energy efficiency measures, and the other is to offset the remaining building energy needs through use of on-site renewable energy systems. The latter approach also holds the promise of reducing transmission and distribution losses associated with the electricity generated by power plants.

In practice, the most easily accessible and logistically feasible renewable energy source for buildings is rooftop solar. Consequently, local solar deployment at both residential and commercial sites has grown in recent years as local users seek to save on their electric utility bills while contributing less to non-renewable energy sources. This increased renewable energy penetration into the existing grid infrastructure is necessary for sustainability reasons, although with it new problems have emerged. For the budding years of residential and commercial on-site solar generation the connection with the grid was assumed to give the user access to an infinite entity where any amount of electricity could be sold to or bought from at any time. While on small scales this is accurate, as any excess or shortage a single user would reasonably generate could easily be handled by the grid, when this paradigm is expanded to all grid users this treatment of the grid connection is no longer a viable option.

The current usage of this technology is that when on-site solar generation outpaces local demand (or when it is financially advantageous to sell power due to time of use), the user could sell electricity back to the grid, and when the reverse is true electricity could be bought. While this proves an attractive option - as it provides additional source of income for users installing on-site generation capabilities - it is not deployable at larger scales. One can see how if a substantial amount of users are trying to sell electricity at similar times then this is problematic from a supply demand standpoint.

As more solar generation from these users enters a geographic area, this typically leads to a reduction in demand from the generation side during daylight hours. However, due to the nature of solar generation, the availability is not consistent. Most buildings with on-site generation do not take into account the varying solar load when implementing building controls. For commercial users this is especially problematic due to a demand charge being implemented for higher use electric bills, where a fee is paid for each kilowatt-hour used in the peak fifteen minute usage period over the past billing cycle. It can be imagined that if a cloud passes over the solar array of a building this could sharply increase the electricity needed from the grid, potentially resulting in the monthly peak demand. The result of this oversight is that the demand peaks for such buildings can be unnecessarily high. This is non-ideal for both the commercial grid user and the grid itself, where the grid user pays a higher bill and the grid has to supply high loads for short periods of time which is cumbersome from an operational standpoint.

Motivated by these considerations, we focus in this work on the scalability problem from a local consumer user standpoint. The reasoning being that it would be advantageous for each building on the grid to act in such a way to minimize its own grid impact independent of other grid users. This would allow a larger number of local users to integrate local cost saving renewable energy, while at the same time reducing the collective demand on the grid, all without the need for any inter-user communication or coordination. To this end, an optimization-based supervisory predictive control system is developed to leverage flexible building loads throughout the day while taking into account solar forecasting to reduce the cost of electricity to the user, as well as ease the load to the grid of peaks resulting from the on-site solar generation. From the user standpoint, this will result in a monthly reduction in demand charges as well as a reduction in electricity cost with solar installation, and in turn the grid supplies a more consistent load to the user resulting in longevity of equipment and higher operational efficiency resulting in a win-win scenario.

The developed supervisory model predictive control (MPC) system is tested on a simulated three-story 150,000 square foot office building located in Merced, California, with dimmable lights, rooftop fixed-tilt solar, electric vehicle charging stations, HVAC system, and simulated miscellaneous loads. The predictive controller is implemented on the building with various objective functions in an attempt to find a computationally agreeable function that reasonably minimizes the grid impact from a practical point of view. Three functions are selected for the test cases to minimize over the MPC horizon, including the net building load, the peak building load, and a weighted sum of both the peak and net building loads. Case studies are carried out with both perfect and imperfect forecasting of the available solar energy. For the perfect forecasting case, historical data of solar irradiance are used for both the building solar generation and the MPC solar availability forecast. In the imperfect forecasting case, the historical solar data are used for the building solar generation while a clear sky model is used for the MPC solar availability forecast. The clear sky model data are corrected in real simulation time via a moving-horizon estimation scheme that uses the historical data over a past horizon to estimate a solar ``fault" value to correct the clear sky model output over some future horizon [10]. The building energy and demand costs are compared between the cases with both perfect and imperfect forecasting, and with reference to a rules-based base case for context. Through deployment of a framework for building energy management taking into account the grid impact and cost from the local user's standpoint, this paves the way for a more substantial utilization of solar energy on a large scale.

References:

[1] “Energy information administration energy review 2016.” https://www.eia.gov/consumption/.

[2] F. Oldewurtel, A. Parisio, C. N. Jones, D. Gyalistras, M. Gwerder, V. Stauch, B. Lehmann, and M. Morari, “Use of model predictive control and weather forecasts for energy efficient building climate control,” *Energy and Buildings*, vol. 45, pp. 15 – 27, 2012.

[3] T. Salsbury, P. Mhaskar and S. Z. Qin, “Predictive control methods to improve energy efficiency and reduce demand in buildings,” *Comp. & Chem. Eng.*, vol. 51, pp. 77-85, 2013.

[4] V. M. Zavala, “Real-time optimization strategies for building systems,” *Industrial & Engineering Chemistry Research*, vol. 52, no. 9, pp. 3137–3150, 2013.

[5] P. Zhao, S. Suryanarayanan, and M. G. Simoes, “An energy management system for building structures using a multi-agent decision-making control methodology,” *IEEE Transactions on Industry Applications*, vol. 49, pp. 322-330, 2013.

[6] Skogestad, S. and P. Mhaskar (Eds.), Energy Efficient Buildings, *Special Issue of Journal of Process Control*, vol. 24, 2014.

[7] P. H. Shaikh, N. B. M. Nor, P. Nallagownden, I. Elamvazuthi, and T. Ibrahim, "A review on optimized control systems for building energy and comfort management of smart sustainable buildings," *Renewable and Sustainable Energy Reviews*, vol. 34, pp. 409 – 429, 2014.

[8] Y. M. Lee, R. Horesh, and L. Liberti, "Simulation and optimization of energy efficient operation of HVAC system as demand response with distributed energy resources," in *2015 Winter Simulation Conference (WSC)*, pp. 991–999, Dec 2015.

[9] X. Wang, A. Palazoglu, and N. H. El-Farra, "Operational optimization and demand response of hybrid renewable energy systems," *Applied Energy*, vol. 143, pp. 324-335, 2015.

[10] J. T. Allen and N. H. El-Farra, "A model-based framework for fault estimation and accommodation applied to distributed energy resources," *Renewable Energy*, vol. 100, pp. 35-43, 2017.

492808 Maximizing the Output of a Solar and Natural Gas Hybrid Power Plant Using Real-Time Optimization

Abstract:

Due to limited supply of fossil fuels and growing concern of the impact of greenhouse gases on the environment, the world is looking for alternative fuels, which are sustainable and environmentally friendly. Solar energy technologies are considered among the leading renewable energy sources due to recent technological advancement and improved economics. Because intermittency of supply is a problem in solar energy systems, they can be hybridized with dispatchable fuel sources to ensure supply when the sun is not available [1]. The importance of hybridizing renewable energy systems is increasing due to the fact that it improves the economics, efficiency and reliability [2]. Hybridizing these energy sources also provides the advantage of sharing infrastructure that further reduces the capital cost. This work focuses on using real-time optimization (RTO) for a 100 MW_e hybrid plant so that the system utilizes its flexibility to ensure peak performance. This operational strategy allows the plant to read in the current environmental conditions, such as the amount of solar energy available in real time, in order to rigorously determine the plant's optimal set points. Because the plant is continuously optimized as conditions change, it is able to fully leverage its degrees of freedom, most notably the ability to direct solar heat to different parts of the plant, depending on the current level of radiation. Findings show that the system is synergistic in that, by operating in hybrid mode and applying RTO, the system makes much more efficient use of the available solar energy.

Reference:

[1] K. Powell and T. Edgar, "Modeling and control of a solar thermal power plant with thermal energy storage," *Chem. Eng. Sci.*, 2012.

[2] M. Ghofrani and N. Hosseini, "Optimizing Hybrid Renewable Energy Systems: A Review," *intechopen.com*.

[3] V. Reddy, S. Kaushik, and S. Tyagi, "Exergetic analysis of solar concentrator aided natural gas fired combined cycle power plant," *Renew. Energy*, 2012.

[4] J. Servert, G. S. Miguel, and D. Lopez, "Hybrid solar-biomass plants for power generation; technical and economic assessment," *Glob. Nest J*, 2011.

499089 Dynamic Simulation and Optimization of Power Plants Operating at Transient Electricity Demand and Carbon Footprint Constraints

Abstract:

Modern power plant designs are required to provide better operational flexibility and improved performance, and adapt to the rapidly evolving energy infrastructure.[1–3] Dynamic simulation and optimization can be useful tools in analyzing the dynamic performance of power plants in response to seasonal and daily fluctuations in power demand and in the assessment of the value proposition and feasibility of new programs focused on renewable energy. Moreover, dynamic optimization offers the potential of reducing operating cost, improving exergy efficiency, meeting environmental requirements and reducing CO_2 emissions.

Figure 1: Simplified diagram of the steam side system of the 605 MW subcritical-pressure coal-fired power plant with control system.

In this work, a high-fidelity dynamic model was developed, for a 605 MW coal-fired, subcritical power plant with regenerative reheat cycle, as shown in Figure 1. Conventional regulatory controllers were incorporated in the plant model. The model was validated against static, full-load data from a fossil-fueled power plant reported in the literature.[4] This validated plant model was used as a test-bed for dynamic simulations, in which the power plant operation was studied under transient power demand. The performance in closed-loop operation in response to step changes in coal load was assessed, showing that the plant model is stable and reliable for sudden changes, and the control system is able to maintain the controlled variables at their set points. Advanced power plant-level control architectures, including controllers for coal load, feed water load, and preheated air load, were used to adjust the power plant operations, in response to time-varying power demand. A supervisory control architecture was designed for the purpose of optimizing the net efficiency of the power plant operating under steady state and dynamically, in response to realistic load changes, employed using power demand from the New England ISO data. The value proposition of static and dynamic optimization of the plant operating at variable load was assessed. An improvement in time-averaged efficiency of up to 1.95% points was shown feasible with corresponding savings in coal consumption of 184.7 tons/day and carbon footprint decrease of 0.0352 kg/kWh. In summary, this presentation will show the complete work flow of data collection, model development and validation, control tuning, dynamic optimization formulation and solution, and supervisory control architecture for a coal-fired subcritical power plant, focusing on the potential efficiency, energy and carbon footprint benefits of model-based approaches.

References

1. Chen, C., Han, L. & Bollas, G. M. Dynamic Simulation of Fixed-Bed Chemical-Looping Combustion Reactors Integrated in Combined Cycle Power Plants. *Energy Technol.* **4,** 1209–1220 (2016).

2. Chen, C., Zhou, Z. & Bollas, G. M. Dynamic Modeling, Simulation and Optimization of a Subcritical Steam Power Plant. Part I: Plant Model and Regulatory Control. *Energy Convers. Manag.* (unpublished) (2017).

3. Chen, C., Zhou, Z. & Bollas, G. M. Dynamic Modeling, Simulation and Optimization of a Subcritical Steam Power Plant. Part II: Dynamic Optimization under Time-Varying Power Load. *Energy Convers. Manag.* (unpublished) (2017).

4. *Combustion Fossil Power: A Reference Book on Fuel Burning and Steam Generation.* (Combustion Engineering, 1991).

Acknowledgements

This material is based upon work supported by the National Science Foundation under Grant No. 1054718. CC gratefully acknowledges support by the GE Graduate Fellowship for Innovation and helpful advice and guidance from Alstom Power.

494085 Comparison in Dynamic Response of Energy-Storing Cryogenic and Chemical Absorption Carbon Capture Systems to Electricity Demand

Abstract:

Projections of energy production from different resources shows that fossil fuel will remain the primary energy source by 2040 [1]. Concerns over CO2 emission from the combustion of fossil fuels have necessitated the technology development to mitigate the CO2 emission from these resources. Several technologies are developed for carbon capture such as physical and chemical absorption, oxyfuel combustion, pre-combustion, membrane and cryogenic separation [2]. In this study, the dynamic performance of two of these carbon capture systems is compared in response to volatile electricity prices and demand. The systems in consideration are amine-based chemical absorption and cryogenic carbon capture (CCC). Amine-based chemical absorption is a mature technology in gas processing units and is under investigation and commercialization for the power industry. The CCC process is a novel technology that requires less energy in comparison to many carbon capture systems including amine-based chemical absorption. This system captures CO2 from power plant flue gas by cooling it to the desublimation temperature of CO2. Solid CO2 is then separated from the remaining gases by filtration that is followed by its liquefaction. To liquefy CO2, the heat integration inside the carbon capture plant it utilized, resulting in significantly lower energy consumption in comparison to other capture systems. The CCC process is able to capture 99% of CO2 from the flue gas as well as other pollutants such as SOx, NO2, mercury, and particulate matter. Additionally, the CCC process is able to store energy in large-scale and in the form of liquefied refrigerant that can be used in peak hours, resulting in a more stable power grid [3-11].

This investigation considers the integrated system of a coal-fired power generation unit and the above mentioned capture systems. The energy-storing versions of both capture systems are considered in this analysis, which allows for more flexibility toward volatile electricity prices. Unlike many of the previous analyses that consider a constant electricity demand, the goal in this analysis is to compare the performance of both carbon capture systems in response to the dynamic electricity demand while maximizing the operational profit. Inability of both systems in meeting the electricity demand is severely penalized. Thus, meeting the power demand is given the higher priority in comparison to maximization of the profit. The dynamic model of the CCC process is developed previously by the authors and is used in this analysis [3-7]. The model from [12-16] is also adopted for the analysis of the amine-based carbon capture. Some modifications are needed, however, to develop a common basis for comparison. The dynamic models of both systems are developed in the GAMS modeling language and are run for 8 days of simulation time on the NEOS Server [17]. Nonlinear solvers such as KNITRO, CONOPT, and IPOPT are used in analysis to solve the models. A 2014 electricity demand profile from a residential area in San Diego, CA, with a maximum demand of 2000 MW is adopted in this study. A common assumption in comparing the carbon capture systems is 90% capture rate. Thus, 90% capture rate is also assumed in this study while a penalty of $50/tonne is applied for the emission of remaining CO2 to the atmosphere. Wind power is also utilized in meeting the total electricity demand and results in a more sustainable power production. In both systems, it is observed that they are capable of meeting the dynamic electricity demand throughout the simulation horizon while priority is given to using the wind power in meeting the demand. The remaining power requirement is met from coal. It is also observed that the CCC process consumes less total energy than the amine system to capture the same amount of CO2. Additionally, the operational profit of running the CCC process is significantly higher than the amine system. The lower energy consumption of the CCC process, higher operational profit, and the large-scale energy storage capability of it suggest that the CCC process could be a promising system for large-scale integration with the power grid which helps in stabilizing the grid.

[1] https://www.eia.gov/outlooks/aeo/pdf/0383(2017).pdf

[2] Jensen, M., 2015. â€œEnergy Processes Enabled by Cryogenic Carbon Capture.â€ PhD thesis, Brigham Young University, February. 63, 66, 67, 75, 77, 95

[3] S. M. Safdarnejad, J. D. Hedengren, L. L. Baxter, L. Kennington, Investigating the impact of cryogenic carbon capture on the performance of power plants, Proceedings of the American Control Conference (ACC),Chicago, IL, 2015.

[4] S. M. Safdarnejad, J. D. Hedengren, L. L. Baxter, Plant-level dynamic optimization of cryogenic carbon capture with conventional and renewable power sources, Applied Energy 149 (2015) 354-366.

[5] S. M. Safdarnejad, J. D. Hedengren, N. R. Lewis, E. L. Haseltine, Initialization strategies for optimization of dynamic systems, Computers & Chemical Engineering 78 (2015) 39-50..

[6] S. M. Safdarnejad, J. D. Hedengren, L. L. Baxter, Effect of Cryogenic Carbon Capture (CCC) on Smart Power Grids, Proceedings of the American Institute of Chemical Engineers (AIChE), Austin, TX, 2015.

[7] S. M. Safdarnejad, J. D. Hedengren, L. L. Baxter, Dynamic Optimization of a Hybrid System of Energy-Storing Cryogenic Carbon Capture and a Baseline Power Generation Unit, Applied Energy

[8] E. Ebrahimzadeh, P. Wilding, D. Frankman, F. Fazlollahi and L. L. Baxter, "Theoretical and experimental analysis of dynamic heat exchanger: Retrofit configuration." Energy 96 (2016) 545-560

[9] E. Ebrahimzadeh, P. Wilding, D. Frankman, F. Fazlollahi and L. L. Baxter, "Theoretical and experimental analysis of dynamic plate heat exchanger: Non-retrofit configuration." Applied Thermal Engineering 93 (2016) 1006-1019

[10] F. Fazlollahi, A. Bown, S. Saeidi, E. Ebrahimzadeh and L. L. Baxter, "Transient natural gas liquefaction process comparison-dynamic heat exchanger under transient changes in flow." Applied Thermal Engineering 109 (2016) 775-788

[11] F. Fazlollahi, A. Bown, E. Ebrahimzadeh and L. L. Baxter, "Transient natural gas liquefaction and its application to CCC-ES (energy storage with cryogenic carbon capture (TM))." Energy 103 (2016) 369-384

[12] S. M. Cohen, G. T. Rochelle, M. E. Webber, Optimizing post-combustion CO2 capture in response to volatile electricity prices. International Journal of Greenhouse Gas Control, 8, 180-195.

[13] S. M. Cohen, G. T. Rochelle, M. E. Webber, Turning CO2 capture on & off in response to electric grid demand: A baseline analysis of emissions and economics. ASME Journal of Energy Resources Technology, 132 (2010b)

[14] S. M. Safdarnejad, Developing Modeling, Optimization, and Advanced Process Control Frameworks for Improving the Performance of Transient Energyâ€"Intensive Applications, PhD Dissertation, Brigham Young University, 2016

[15] S. M. Cohen, A Techno-economic Plant- and Grid-Level Assessment of Flexible CO2 Capture, PhD Dissertation, The University of Texas at Austin, 2012

[16] S. M. Cohen, The implications of flexible CO2 capture on the ERCOT electric grid. Master's thesis, The University of Texas at Austin, 2009

[17] J. Czyzyk, M. P. Mesnier, J. J. Moré, The NEOS Server. IEEE Journal on Computational Science and Engineering 5(3), 1998, 68-75.

495622 Implications of Heat Integration in Energy Savings during Heat Treating of Steel

Abstract:

In the United States, metal processing is the fourth largest contributor to energy consumption in the industrial sector with annual demands of around 2 quadrillion BTU (quads) [1]. For a particular steel processing plant, 65% to 85% of the energy is spent as fuel required to operate furnaces.

Steel slabs are typically forged into the desired shape and the resulting workpieces are heat treated for hardness and strength. During heat treating, the finished or pre-finished parts are indirectly heated in a furnace in an inert atmosphere to prevent surface oxidation and then quenched in water, oil or brine. Heat treating is an energy-intensive process and its economics can be improved by optimizing furnace design and furnace operations, in particular fuel use.

In this talk we focus on a currently operating heat treating furnace at an industrial partner, with the goal of increasing the energy efficiency by model-based system reconfiguration and optimization. The two-scale model [2] developed in our previous work is used to simulate the furnace operation. The model predicts the macro-scale furnace energy usage and the part temperature distribution and micro-scale grain size distribution as a function of processing time. We observe that about 40% of the energy input to the system is lost via burner exhaust gases. Hence, we propose to achieve energy savings by installing a recuperator, i.e., a feed-effluent heat exchanger that captures heat from the burner exhaust gases to preheat burner inlet air-fuel mix. We study the steady-state and open-loop dynamic implications of implementing the recuperator. Further savings can be achieved by identifying the optimal operating conditions of the furnace that both reduce the energy consumption and satisfy the product quality. To this end, we proceed by identifying surrogate models of the (steady-state) system behavior, which we then use to solve a pseudo-steady state energy optimization problem. The furnace temperature profile obtained from the optimization is tracked using a set of distributed temperature controllers. The proposed scheme results in considerable energy savings, and we demonstrate that such savings are available both in the nominal (steady-state) case and in the presence of transient disturbances.

REFERENCES

[1] V Viswanathan, R Davies, and J Holbery. Opportunity analysis for recovering energy from industrial waste heat and emissions. Pacific Northwest National Laboratory, 2005

[2] HS Ganesh, TF Edgar, and M Baldea. Model based control of product toughness and energy minimization of an austenitization furnace. Journal of Process Control (submitted), 2017.

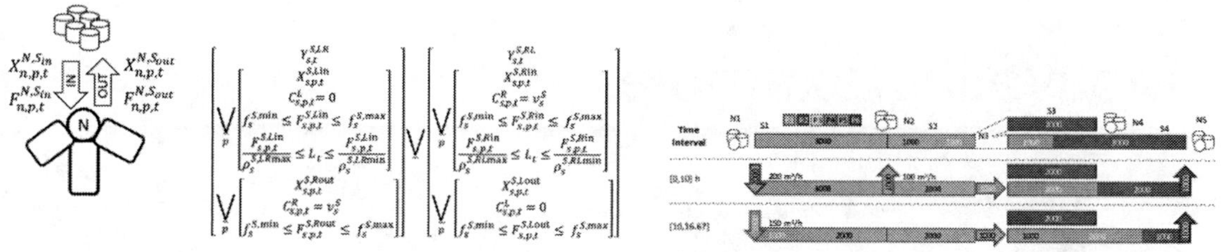

Optimal Scheduling of Multiproduct Liquid Pipelines with Reversible Flow

Pedro M. Castro (pmcastro@fc.ul.pt)
http://www.researcherid.com/rid/C-3642-2008

Associate Researcher
CMAFCIO
Faculdade de Ciências
Universidade de Lisboa

INVESTIGADOR FCT

Ciências ULisboa

LISBOA UNIVERSIDADE DE LISBOA

CMAF CIO CENTRO DE MATEMÁTICA, APLICAÇÕES FUNDAMENTAIS E INVESTIGAÇÃO OPERACIONAL

Motivating example

- Central node connects 3 dual purpose terminal nodes
- Each terminal node supplies one product & demands the other two
  - Operating flowrate $\rho_s^S \in [20,200]$ m^3/h
- What is the optimal makespan?

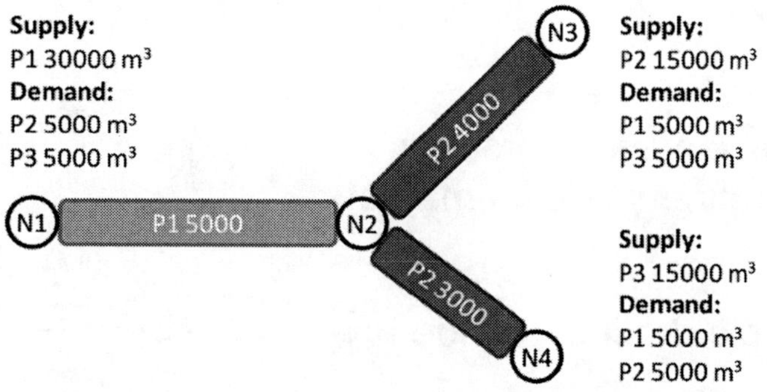

Motivating example

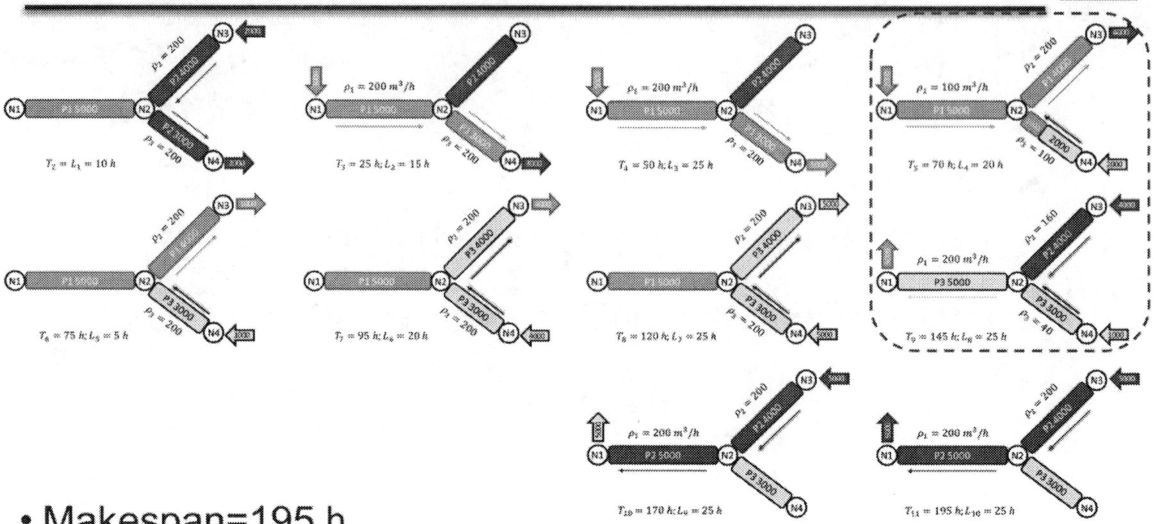

- Makespan=195 h
  - All segments are idle at some point in time
  - Two active input nodes simultaneously send material to other node
  - P2 in S2 backtracks to N3, later returns to fulfill demand in N1

Building a pipeline network

- Nodes

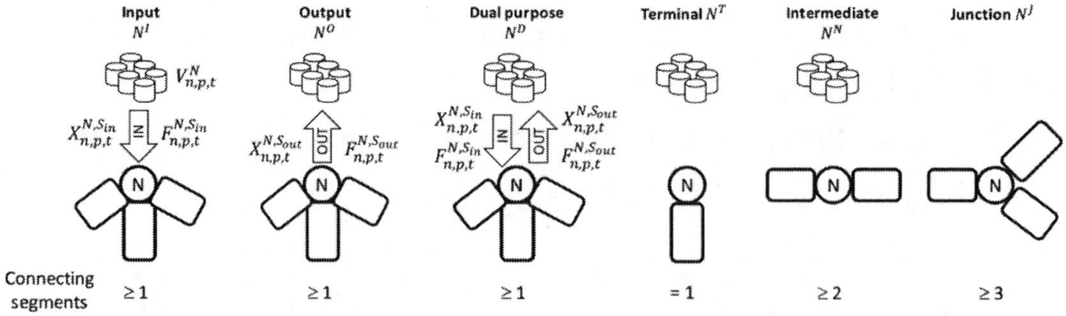

  - Storage system interacts with refinery & local markets

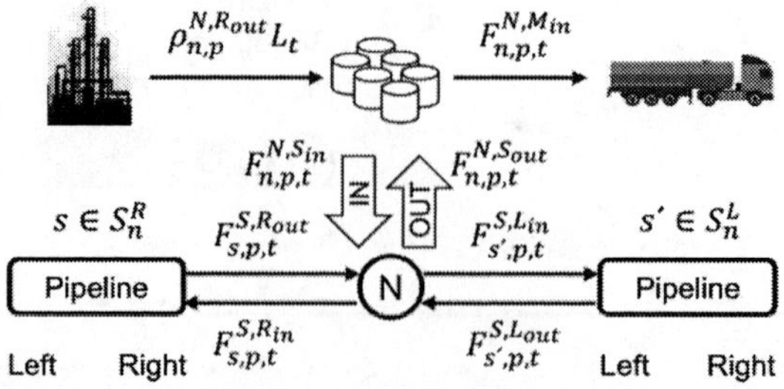

- Segments
  - Multiple products inside $V^S_{s,p,t}$

Mathematical formulation (MILP)

- Continuous-time representation with single grid

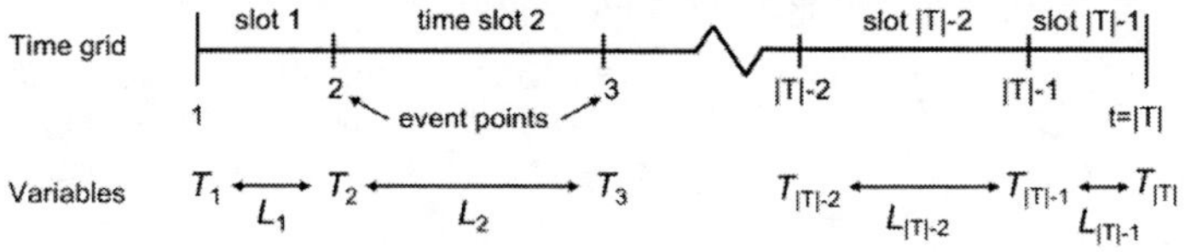

- Derived from Generalized Disjunctive Programming
  - Eqs. only with continuous variables
    - Mass balances: $V_{n,p,t}^{N} = v_{n,p}^{N,0}\big|_{t=1} + V_{n,p,t-1}^{N} + F_{n,p,t-1}^{N,Sout} - F_{n,p,t-1}^{N,Sin} + \rho_{n,p}^{N,Rout} L_t - F_{n,p,t-1}^{N,Min}\ \forall n,p,t$
    - Enforcing bounds: $F_{n,p,t}^{N,Min} \leq \rho_{n,p}^{N,Mmax} L_t\ \forall n,p,t \qquad v_{n,p}^{N,\min} \leq V_{n,p,t}^{N} \leq v_{n,p}^{N,\max}\ \forall n,p,t$
  - Only binary variables
    - Logic propositions $X_{s,p,t}^{S,L} \Longrightarrow \neg X_{s,p',t}^{S,Lin}\ \forall s \in S^{LR}, p, p', t \qquad X_{s,p,t}^{S,R} \Longrightarrow \neg X_{s,p',t}^{S,Rin}\ \forall s \in S^{RL}, p, p', t$
  - Both types
    - Disjunctions

$$\bigvee_{p \in P_s} \begin{bmatrix} X_{s,p,t}^{S,L} \\ V_{s,p,t}^{S} \geq f_s^{S,Smin} \\ C_{s,p,t}^{L} = 0 \end{bmatrix} \forall s \in S^{LR}, t \qquad \bigvee_{p \in P_s} \begin{bmatrix} X_{s,p,t}^{S,R} \\ V_{s,p,t}^{S} \geq f_s^{S,Smin} \\ C_{s,p,t}^{R} = v_s^{S} \end{bmatrix} \forall s \in S^{RL}, t$$

Segments with reversible flow

- Three alternative operating modes

$$\begin{bmatrix} Y_{s,t}^{S,LR} \\ \bigvee_{p} \begin{bmatrix} X_{s,p,t}^{S,Lin} \\ \boxed{C_{s,p,t}^{L} = 0} \\ f_s^{S,\min} \leq F_{s,p,t}^{S,Lin} \leq f_s^{S,\max} \\ \dfrac{F_{s,p,t}^{S,Lin}}{\rho_s^{S,LR\max}} \leq L_t \leq \dfrac{F_{s,p,t}^{S,Lin}}{\rho_s^{S,LR\min}} \end{bmatrix} \\ \bigvee_{p} \begin{bmatrix} X_{s,p,t}^{S,Rout} \\ \boxed{C_{s,p,t}^{R} = v_s^{S}} \\ f_s^{S,\min} \leq F_{s,p,t}^{S,Rout} \leq f_s^{S,\max} \end{bmatrix} \end{bmatrix} \bigvee \begin{bmatrix} Y_{s,t}^{S,RL} \\ \bigvee_{p} \begin{bmatrix} X_{s,p,t}^{S,Rin} \\ \boxed{C_{s,p,t}^{R} = v_s^{S}} \\ f_s^{S,\min} \leq F_{s,p,t}^{S,Rin} \leq f_s^{S,\max} \\ \dfrac{F_{s,p,t}^{S,Rin}}{\rho_s^{S,RL\max}} \leq L_t \leq \dfrac{F_{s,p,t}^{S,Rin}}{\rho_s^{S,RL\min}} \end{bmatrix} \\ \bigvee_{p} \begin{bmatrix} X_{s,p,t}^{S,Lout} \\ \boxed{C_{s,p,t}^{L} = 0} \\ f_s^{S,\min} \leq F_{s,p,t}^{S,Lout} \leq f_s^{S,\max} \end{bmatrix} \end{bmatrix} \bigvee \begin{bmatrix} Y_{s,t}^{S,idle} \\ C_{s,p,t}^{L} \leq v_s^{S}\ \forall p \\ C_{s,p,t}^{R} \geq 0\ \forall p \\ F_{s,p,t}^{S,Lin} = 0\ \forall p \\ F_{s,p,t}^{S,Rout} = 0\ \forall p \\ F_{s,p,t}^{S,Rin} = 0\ \forall p \\ F_{s,p,t}^{S,Lout} = 0\ \forall p \\ L_t \leq h \end{bmatrix} \forall s,t$$

Decision of product to enter/leave the segment depends on its left or right coordinate

Allowing for product re-entrance

- Required changes compared to unidirectional flow
  - Coordinate variables $C^{L,\text{end}}_{s,p,t}$, $C^{R,\text{end}}_{s,p,t}$, $C^{Z}_{s,p,t}$ can take negative values
    - $C^{L,\text{end}}_{s,p,t} = C^{L}_{s,p,t} + \sum_{p' \neq p}(F^{S,L\text{in}}_{s,p',t} - F^{S,L\text{out}}_{s,p',t})\ \forall s,p,t$
    - $C^{R,\text{end}}_{s,p,t} = C^{R}_{s,p,t} + \sum_{p' \neq p}(F^{S,R\text{out}}_{s,p',t} - F^{S,R\text{in}}_{s,p',t})\ \forall s,p,t$

    Left/right coordinate increases/decreases from other products p' entering/leaving on the left/right side regardless of p being inside

    - $C^{L}_{s,p,t} = c^{L,0}_{s,p}\big|_{t=1} + C^{L,\text{end}}_{s,p,t-1} - C^{Z}_{s,p,t}\ \forall s,p,t$
    - $C^{R}_{s,p,t} = c^{R,0}_{s,p}\big|_{t=1} + C^{R,\text{end}}_{s,p,t-1} - C^{Z}_{s,p,t}\ \forall s,p,t$
  - Reset variables $C^{Z}_{s,p,t} \neq 0$ only if p not inside
    - Also applies to $t = 1$ to decide first flow direction

$$\begin{bmatrix} X^{S,is}_{s,p,t} \\ V^{S}_{s,p,t} \leq v^{S}_{s} \\ C^{Z}_{s,p,t} = 0 \end{bmatrix} \underline{\bigvee} \begin{bmatrix} \neg X^{S,is}_{s,p,t} \\ V^{S}_{s,p,t} = 0 \\ -f^{S,\max}_{s} \leq C^{Z}_{s,p,t} \leq f^{S,\max}_{s} \end{bmatrix} \forall s,p,t$$

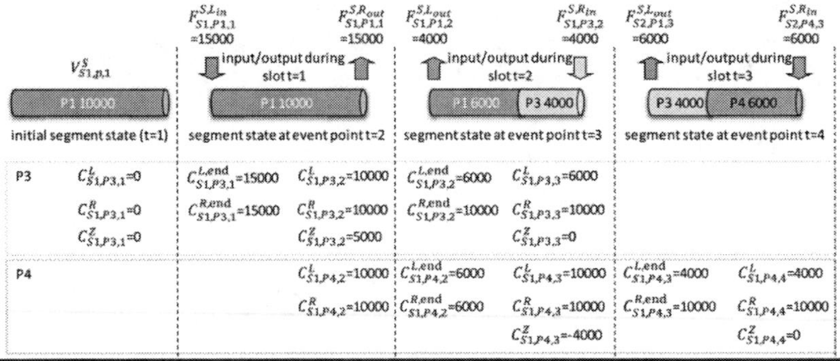

Benchmark problems

Example	Source	Structure	Flow Direction	Market Supply	Makespan (h) Reported	This work	Reduction
Motivating	This work	Tree-like	Reversible	No	-	195	-
Ex1	Cafaro & Cerdá (2014), Ex. 1	Straight	Reversible	No	129.00[a]	168.00	-30.2%
Ex2	Cafaro & Cerdá (2014), Ex. 2	Straight	Reversible	Yes	327.50[a]	310.32	5.2%
Ex3	Cafaro & Cerdá (2014), Ex. 3	Straight	Reversible	Yes	325.71[a]	310.00	5.1%
Ex4	Mostafaei et al. (2015), Ex. 1	Tree-like	Unidirectional	No	154.17	138.33	10.3%
Ex5	Mostafaei et al. (2015), Ex. 2	Tree-like	Unidirectional	No	179.98	162.00	10.0%
Ex6	Cafaro & Cerdá (2012), Ex1	Mesh	Unidirectional	Yes	179.00[a]	194.29	-8.5%

GAMS/CPLEX 12.6.3
i7-6400K (4.0 GHz)
16 GB RAM
relative tolerance=10^{-6}
8 parallel threads
18,000 CPUs

- Main purpose is model validation
  - Variety of structures & directions of flow
  - Makespan from previous works just an indication
    - Obtained for cost minimization (term in objective for Ex1-Ex3, Ex6)
  - Makespan can be reduced significantly except for:
    - Ex1: lower performance due to single product delivery per time slot
    - Ex6: misleading results, issues in model by Cafaro & Cerdá (2012)

Computational results

- Very challenging problems
  - First feasible solution only after several minutes of computation
  - Unable to prove global optimality (except Ex4)
    - Tuning parameter: $|T|$ (# slots in time grid)
  - ↑ integrality gap in reversible flow problems

Example	$\|T\|$	Integrality gap	Binaries	Variables	Equations
Motivating	10	102%	792	1910	2818
	11	92%	879	2116	3122
Ex1	11	207%	1019	2535	3891
	12	153%	1120	2782	4271
Ex2	16	21%	574	1680	2219
	17	19%	612	1790	2364
Ex3	14	19%	1100	2784	3992
	15	19%	1184	2994	4293
Ex4	11	2.5%	1102	3414	5392
	12	2.5%	1210	3742	5912
Ex5	11	12%	2142	6574	10443
	12	13%	2352	7206	11451
Ex6	12	21%	1329	4009	6252

Example	$\|T\|$	Makespan (h)	CPUs
Motivating	8	Infeasible	28.1
	9	210	551
	10	205	2152
	11	195	15454
	12	195	18000
Ex1	10	Infeasible	475
	11	204	10356
	12	168	18000
Ex2	11	Infeasible	50.9
	12	355	111
	13	345	286
	14	335.625	591
	15	319.405	5852
	16	314.652	10209
	17	310.323	18000
Ex3	12	Infeasible	406
	13	323.33	1975
	14	310	11789
	15	310	18000
Ex4	10	Infeasible	306
	11	138.333	704
	12	138.333	11004
Ex5	10	Infeasible	1285
	11	162	3126
	12	164.29	18000
Ex6	11	Infeasible	581
	12	206.429	18000
	13	no solution	18000

Optimal schedule (Ex3)

$\rho_s^{S,LR} \in [100,600]$ m³/h

$\rho_s^{S,RL} \in [50,400]$ m³/h

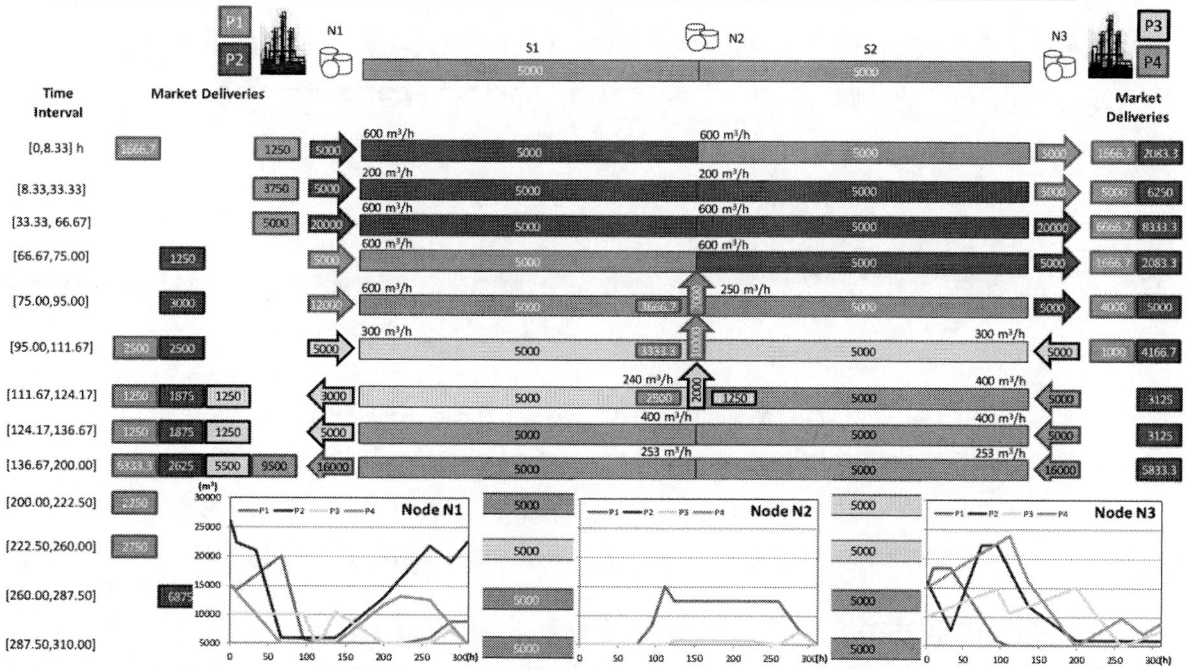

Ex6

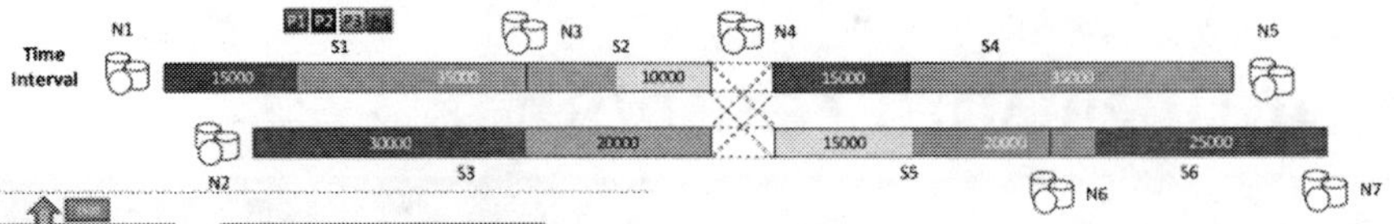

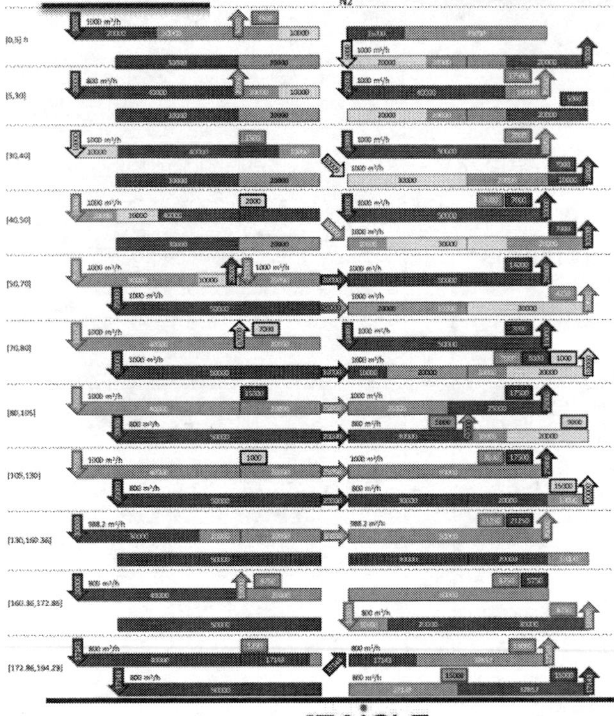

- Model by Cafaro & Cerdá ('12)
  - Constraints not rigorously enforced
    - Flowrate bounds in S2 & S6 (part of lines after nodes N3 & N6)
    - 4150 m³ short of demand to N5
  - Better classified as planning model

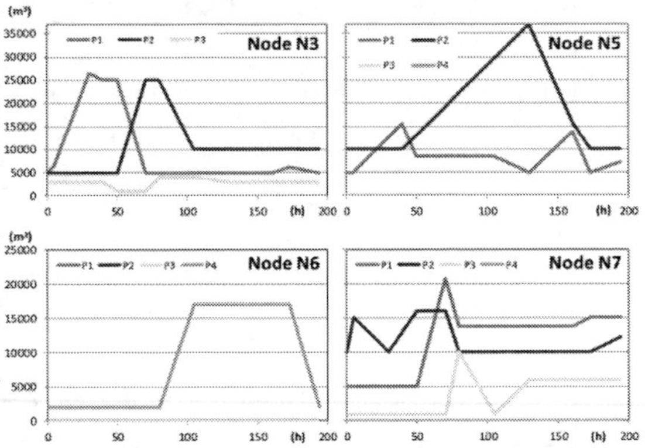

Conclusions

- Most general formulation for integrated scheduling of pipeline systems and associated storage tanks
  - Network of segments connecting different types of nodes
    - Can handle multilevel tree-like structures
  - Allows movement of products in both directions
  - Forbidden product sequences rigorously enforced on entry
- Known limitations
  - Only one batch per product inside a segment
    - May compromise feasibility when dealing with long segments
  - Single product entering/leaving a segment per time slot
    - Performance vs. batch-centric models (unidirectional flow)
- Further details: IECR 56 (2017) 9638-9656

492458 Combining the Strengths of Continuous and Discrete Time Representations: A General Solution Refinement Method for Discrete-Time MIP Models

Abstract:

Production planning and scheduling of batch processes have received considerable attention in the process systems engineering community in the last 20 years (Harjunkoski *et al.*, 2014). Mainly, two different time representation have been adopted for the mathematical programming models developed: discrete and continuous. Both representations have pros and cons (Floudas and Lin, 2004; Sundaramoorthy and Maravelias, 2011). Specifically, discrete-time formulations are tighter and can readily account for limited shared resources and intermediate shipments/deliveries, but suffer from large model size and inaccuracy of the solution. On the other hand, continuous-time formulations are more accurate and can easily model variable processing time, but exhibit large solution times when the number of periods increases, and cannot be readily extended to linearly account for inventory and resource costs. Although models may benefit from the advantages of the time representation adopted, the short comings that follow are unavoidable.

In this work, we propose a solution method that allows to harness the advantages of both modeling approaches while overcoming their limitations. Specifically, we discuss a â€œrefinementâ€ method for discrete-time MIP models, applicable to all types of production environments. In this method, we solve the problem in two stages: solve for an approximated solution using discrete-time formulation in the first stage, which is refined to a more accurate solution in the second stage using a continuous-time LP formulation. The information transfer from the discrete-time solution (i.e. task assignment and sequencing) to the continuous-time model is achieved through a mapping algorithm we propose. The continuous-time model adopts material-specific grid to ensure material balances and unit-specific grids to ensure unit utilization constraints, while additional utility-specific grids can be adopted, if needed, for shared utilities. The discrete-time model used in the first stage allows us to obtain a good solution faster, as well as consider limited shared resources with time-varying resource availability and costs, as well as intermediate shipments/deliveries. The continuous-time model used in the second stage refines the solution to enhance solution accuracy as well as model variable processing time, while maintaining the feasibility of the solution. We investigate three objective functions, namely minimization of makespan/cost and maximization of production.

Furthermore, since the optimal solution obtained by the discrete-time model in the first stage may not be the optimal for the continuous-time model, we propose a method that allow us to generate a solution pool during the first stage. By collecting non-degenerate solutions, we increase the chance of obtaining the optimal solution during the second stage, which is extremely fast because it involves the solution of LPs.

Finally, we perform a thorough computational study to show that discrete-time models with the proposed solution refinement method are computationally superior when compared to standalone discrete- and continuous-time models. In terms of solution quality, the proposed method is shown to obtain the optimal solution in the majority of instances. The proposed method enable us to solve instances that were considered intractable with continuous-time models, while

accounting for process features that only continuous-time models can handle efficiently (e.g., variable processing times).

Reference

1. Floudas, C. A. and Lin, X. X. (2004) Continuous-time versus discrete-time approaches for scheduling of chemical processes: a review. *Computers & Chemical Engineering*. **28**, 2109-2129.
2. Harjunkoski, I., Maravelias, C. T., Bongers, P., Castro, P. M., Engell, S., Grossmann, I. E., Hooker, J., Mendez, C., Sand, G. and Wassick, J. (2014) Scope for industrial applications of production scheduling models and solution methods. *Computers & Chemical Engineering*. **62**, 161-193.
3. Sundaramoorthy, A. and Maravelias, C. T. (2011) Computational Study of Network-Based Mixed-Integer Programming Approaches for Chemical Production Scheduling. *Industrial & Engineering Chemistry Research*. **50**, 5023-5040.

497698 A New Approach for Scheduling of Operations in Scientific Services Facilities Via Multi-Commodity Flow

Abstract:

During the past two decades, the short-term scheduling of batch and continuous processes have been extensively discussed. Various models, using different time formulations such as discrete time [1,2], global event point time [3,4], unit-specific event point time [5-7] and time slots [8], as well as different process representations, such as State-task Network [1] (STN) and resource task network (RTN) have been presented. Unit-specific event time formulations are the most efficient of all other formulations as they formulate problems with less variable and constraints and as a result less computational time is required [9]. Their limitation however, is that an iterative procedure is required in order to define the optimal number of events as well as in some cases the total number of events that a task is allowed to span. Additionally, the majority of the proposed models fail to solve medium-term and large-term scheduling problems.

A scientific service facility examines the physical and chemical properties of a great number of samples from different clients. The examination of such properties is done by a number of machines or manpower, which have considerable large capacity and as a result samples from more than one customers can be processed simultaneously. However, in all models presented in the literature, there is the constraint that only one task can be processed in a unit during a specific time slot. Therefore, existing formulations cannot be implemented in order to optimize the scheduling of such facilities. Only recently, Patil et. al [10] and Lagzi et al. [11] presented models for optimizing the scheduling of a scientific services facility. Patil et. al. [10] presented a model based on discrete time formulation. For large-scale problems, they used the rolling horizon technique, where samples that were not able to be processed in the current scheduling horizon, were scheduled to be processed during the next scheduling horizon. Lagzi et al. [11] proposed a model based on global event time formulation for the same problem. The results presented in their work proved that the proposed multitasking formulation proposed is superior than single-tasking models, as a better solution was received. However, the proposed formulation failed to provide an optimal schedule in reasonable amount of time.

In this work, a new model based on unit-specific event time representation, for scheduling of scientific services facilities was proposed. the novelty of this formulation is that an iterative procedure for defining the maximum number of events that a task is allowed to be processed is not required. Instead, it is defined whether it is allowed one task to span in more event point. Furthermore, the proposed model can handle the feature that more than one task can be processed in a unit during a specific time slot. By solving a number of random small, medium and large-scale examples generated for the purpose of this work, the proposed formulation formulates problems with much fewer binary variables and constrains and generates same or better results in significantly less computational time than the global event time formulation of Lagzi et. al. [11].

References

[1] Kondili E., Pantelides C. C., Sargent R. W. H. A general algorithm for short-term scheduling of batch operations-I MILP formulation, *Computers & Chemical Engineering,* 1993;17(2);211-227

[2] Dedopoulos I. T., Shah N., Optimal sort-term scheduling of maintenance and production for multipurpose plants, *Industrial and Engineering Chemistry Research,* 1995;34(1);192-201

[3] Castro P., Barbosa-Póvoa A. P. F. D., Matos H. An improved RTN continuous-time formulation for the short-term scheduling of multipurpose batch plants, *Industrial and Engineering Chemistry research,* 2001;40(9);2059-2068

[4] Zhang X., Sargent R., W. H. The optimal operation of mixed production facilities-A general formulation and some approaches for the solution, *Computers and Chemical Engineering,* 1996;20(6-7);897-904

[5] Ierapetritou M., G. Floudas C., A. Effective continuous-time formulation for short-term scheduling. 1. Multipurpose batch processes, *Industrial & Engineering Chemistry,* 1998;37(11);4341-4359

[6] Ierapetritou M., G. Floudas C., Effective Continuous-Time Formulation for Short-Term Scheduling 2. Continuous and Semicontinuous Processes, *Industrial & Engineering Chemistry,* 1998;37(11);4360-4374

[7] Shaik M. A., Floudas C. A. Novel Unified Modeling Approach for Short-Term Scheduling, *Industrial & Engineering Chemistry Research,* 2009;48(6);2947-2964

[8] Sundaramoorthy A., Karimi I. A. A simpler better slot-based continuous-time formulation for short-term scheduling in multipurpose batch plants, *Chemical engineering science,* 2005;60(10);2679-2702

[9] Floudas C., A. Lin X. Mixed Integer Linear Programming in Process Scheduling Modeling, Algorithms, and Applications, *Annals of operation research,* 2005;139(1);131-162

[10] Patil B. P., Fukasawa R., Ricardez-Sandoval L. A. Scheduling of operations in a large-scale Scientific services facility via multicommodity flow and an optimization-based algorithm, *Industrial & Engineering Chemistry,* 2015;54(5);1628-1639

[11] Lagzi S., Fukasawa R., Ricardez-Sandoval L. A multitasking continuous time formulation for short-term scheduling of operations in multipurpose plants, *Computers & Chemical Engineering,* 2017;97;135-146

488422 An Integrated Chemical Site Planning and Scheduling Framework---Model and Algorithm

Abstract:

A wide range of math programming frameworks are available for planning and scheduling of multiproduct batch and continuous processes chemical plants, notably state-task network (STN) [1,2], resource-task network (RTN) [3], and corresponding variants thereof---[4], [5], and [6], to name a few. There are also superstructure and flowsheet-based frameworks that incorporate appropriate connectivity aspects as well as proposed recently by [7] and [8]. However, such unified frameworks for site-wide modeling of integrated chemical sites are either limited in scope or were developed as a part of an application, and thus, are not necessarily general---[9], [10], [11], and [12], for instance. In this work, we propose a modeling framework for enterprise-wide planning of integrated chemical sites that can suitably be applied for long-term as well as medium-term strategic and tactical planning. We also propose solution strategies to tackle such planning problems via a mixed-integer linear programming (MILP) model.

In particular, the model framework simultaneously captures (1) different discrete and continuous operating modes of production units; (2) storage facilities including tanks, reservoirs, pipelines, rail cars, and barges and associated time delay and rate constraints; (3) handling of redundancy of utilities such as steam through smoothing and symmetry breaking methods; (4) production ramping constraints, recycles, and residence time delays; (5) maintenance and turnaround outages and downtimes; (6) reliability of production units including unplanned events and yield and selectivity degradation; (7) transport, supply and demand constraints; and (8) economics. The model is generalized to flexibly handle time-dependent and uncertainty-dependent data with multiple time scales. We discuss algorithmic aspects such as heuristics and relaxations, preprocessing techniques, and possible reformulations for different model components. We provide computational results on a large industrial-size test case.

References:

1. E. Kondili, C. C. Pantelides, and R. W. H. Sargent. A general algorithm for short-term batch operations. 1. MILP formulation. Computers and Chemical Engineering, 17, 211—217. 1993.
2. N. Shah, C. C. Pantelides, and R. W. H. Sargent. Optimal periodic scheduling of multipurpose batch plants. Annals of Operations Research, 42, 193—228. 1993.
3. C. C. Pantelides. Unified frameworks for optimal processing planning and scheduling. Proceedings of the 2nd conference on the Foundations of Computer Aided Process Operations, page 253. Cache Publications, New York. 1994.
4. M. G. Ierapetritou and C. A. Floudas. Effective continuous-time formulation for short-term scheduling. 1. Multipurpose batch processes. Industrial and Engineering Chemistry Research, 37 (11), 4341—4359. 1998.
5. C. T. Maravelias and I. E. Grossmann. New general continuous-time state-task network formulation for short-term scheduling of multipurpose batch plants. Industrial and Engineering Chemistry Research, 42 (13), 3056—3074. 2003.
6. P. M. Castro and I. E. Grossmann. New continuous-time MILP model for the short-term scheduling of multistage batch plants. Industrial & Engineering Chemistry Research, 44 (24), 9175—9190. 2005.
7. D. Zyngier and J. D. Kelly. UOPSS: a new paradigm for modeling production planning & scheduling system. In proceedings of the 22nd European Symposium on Computer Aided Process Engineering, London, June 17—20, 2012.
8. W. Wu, C. A. Henao, and C. T. Maravelias. A superstructure representation, generation, and modeling framework for chemical process synthesis. AIChE Journal, 62(9), 3199—3214. 2016.
9. J. M. Wassick. Enterprise-wide optimization in an integrated chemical complex. Computers and Chemical Engineering, 33(12), 1950—1963. 2009.
10. S. Mitra, I. E. Grossmann, J. M. Pinto, and N. Arora. Integration of strategic and operational decision-making for continuous power-intensive processes. In proceedings of the 22nd European Symposium on Computer Aided Process Engineering, London. Computer Aided Chemical Engineering 30, 232—236. 2012.
11. S. Terrazas-Moreno, I. E. Grossmann, J. M. Wassick, and S. J. Bury. Optimal design of reliable integrated chemical production sites. Computers and Chemical Engineering, 34(12), 1919—1936. 2010.
12. B. Sharda and S. J. Bury. A discrete event simulation model for reliability modeling of a chemical plant. In proceedings of the 2008 Winter Simulation Conference, Miami, December 7—10, 2008.

492547 Scheduling of Distributed Chemical and Renewable Power Production with Regulated Energy Exchange

Abstract:

The coupling of distributed renewable power and chemical production can help to provide a use for stranded renewable energy resources in rural areas, particularly if the chemicals produced will be used in the same rural areas. However, such systems will still require interaction with the power grid due to the stochastic nature of renewable production. It is well documented that introducing additional renewable energy into the power market results in a required increase in the ramp rates of fossil fuel generators and a risk of over or underproduction due to the stochasticity. One way to mitigate these challenges is to introduce a regulatory market structure which obligates renewable energy producers to explicitly limit the uncertainty and variability in their load by providing a day-ahead commitment. Adjustable parameters can be used to specify how close consecutive commitments must be to each other (mitigating variability), and how close the realized energy exchange must be to the commitment value (mitigating uncertainty). Our previous work has applied this regulatory structure to a system where an office building is powered by renewable energy and shown that meeting commitments solely by adjusting power-producing or storage units (i.e. microturbines or batteries) was quite expensive for renewable energy producers [1].

In this work, we apply our regulatory energy exchange framework to a system where ammonia production is powered by wind energy, using electrolysis to generate hydrogen from water. Because the power load for this system is a chemical process, there exists some inherent flexibility in the load itself which can be optimized by scheduling. For example, ammonia production can be scaled up or down, and electrolyzers can be turned on or off. The scheduling problem itself is formulated as a mixed integer linear program that minimizes operating cost each hour using a 48-hour receding horizon. Unit operating levels can change every hour, with the exception of the ammonia reactor, which can change at most every 4 hours and must obey constraints in ramp rates. Power exchange commitments are made 24 hours in advance, and we show that the flexibility in hydrogen, nitrogen, and ammonia production can be used to avoid violating these commitments and paying penalties. In addition, we show that the cost of regulating the energy exchange is minimal and that the variability and uncertainty of energy exchange is significantly reduced.

We extend our work by analyzing how the power exchange variability and operating cost change as various key design parameters are varied. We note that a tradeoff results from changing some design parameters; for example, an ammonia plant which is too large will require more purchasing of power, while one that is too small will make it difficult to meet power exchange commitments. Other design parameters will have diminishing returns, such as the number of electrolyzers installed or the amount of gas storage available. We also analyze the effect of location on operating cost and establish that locations with a higher wind capacity factor can enable larger ammonia plants relative to the installed wind capacity. This analysis enables better long term planning on where and how to build renewable ammonia plants, and can be applied into a larger supply chain optimization problem.

[1] Zachar, M. and Daoutidis, P. Microgrid/macrogrid energy exchange: A novel market structure and stochastic scheduling. IEEE Trans. Smart Grid, 8(2017), pp. 178-189.

498299 Robust Refinery Planning Under Exogenous and Endogenous Uncertainty

Abstract:

This paper proposes a modeling and optimization framework for supporting the decision-making of optimal feed acquisition in refinery planning while simultaneously considering both exogenous and endogenous sources of uncertainty. The presented approach consists of two distinct decision stages that are separated in terms of time-scales: (a) a strategic stage in which we determine the optimal feedstock slate (such as crudes or chemicals feeds) to purchase for the long-term horizon (using a contract approach - stage 1), and (b) a tactical stage in which we determine the optimal feedstocks to purchase on a short-term horizon basis (utilizing spot market purchase opportunities - stage 2). These stages are evaluated at separate time intervals throughout the planning horizon. Unlike existing approaches such as stochastic programming or multi-stage recourse, that often require complex decomposition strategies, the proposed algorithm allows hedging against future risk while maintaining computationally-tractable problem instances using the principles of Monte-Carlo simulation and discreet event simulation. Specifically, the uncertainty considered in this work can be (a) exogenous, such as supply, demand and pricing, as well as (b) endogenous, such as asset availability that is directly calculated from reliability and maintenance considerations, that are typically the outcome of discrete-event simulation. In principle, the proposed two-stage approach can be used with both exogenous and/or endogenous uncertainties for both short and long term decision making horizons. In this work, we utilize the use of chance-constrained optimization for the determination of robust crudes (price-resilient) for contract purchase in stage 1, and perform a breakeven analysis under equipment reliability and availability uncertainty in stage 2. We apply this strategy to an industrial case study involving a refinery model in order to assess tangible economic benefits and evaluate the effectiveness of the proposed algorithm. The numerical results presented here indicate that this systematic methodology provides significant economic benefits and it far outperforms the deterministic approach, leading to an increase in the expected refinery profit of more than \$20 million, on average, per year.

501945 LP Reformulation to Approximate Non-Convex Blending in MILP Scheduling Problems Using Factors

Abstract:

Blend scheduling operations in chemical process industries, as crude-oil refineries and metal processing facilities, involves logistics and quality problems in large scale non-convex models. However, these problems are prohibitively expensive to be solved in full space Mixed-Integer Non-Linear Programming (MINLP) approaches since the initial steps of the binary variables relaxation in a Non-Linear Problem (NLP). To overcome this, the main strategy found in the literature is to decompose the quantity-logic-quality phenomena into a two-stage solution [1,2,3]. First, scheduling assignments in quantity-logic relationships are solved in a Mixed-Integer Linear Problem (MILP) by neglecting the non-convex blending constraints also referred to as *pooling*. Then, a quantity-quality NLP problem is solved with fixed discrete decisions. The major drawback from this partition in MILP and NLP models is the possible distance or gap between their objective functions. In addition, infeasibilities in the NLP solution may arise by setting up undesired discrete decisions in the MILP.

We develop a Linear Programing (LP) reformulation to proxy or approximate NLP quality constraints (from the blending of streams) in the MILP stage using parameters or coefficients of qualities, what we call *factors*. In this approach, factors are augmented equality balance constraints for each property or intensive variable ($factor = f_i$) between raw material sources (x_i) and product sinks, using slacks (for <= constraints) or surpluses (for >= constraints) to guarantee the balance and replacing the blended material by the desired specification (f_s) in the formula: $\mathrm{sum}(i,f_i{*}x_i) = f_s{*}\mathrm{sum}(i,x_i) + (f_{slack}$ or $f_{surplus})$. This potentially reduces the MILP-NLP gap as more quality information is programmatically transferred to the logistics optimization. If the quality information is neglected in the MILP problem, several degenerated binary solutions will be found, all with the same MILP objective for different sets of sources to sinks assignment, which unfortunately may substantially increase the time to reduce the MILP-NLP gap.

A similar proposition for calculating overall contaminant flow balances is applied in water network systems to significantly improve the strength of the lower bound in global optimum models [4,5] though this is considered as an NLP by the bilinear terms of blending of water stream flows with different contaminant concentrations. In our proposition, the concentrations are parameters instead of variables, therefore this approach is valid only for the mixing point between raw materials and products whereby properties or other intensive variables cannot propagate throughout the network. Although analogous reformulation is commonplace in non-linear planning problems by simplifying or aggregating the real topology of process networks, it can be included in scheduling problems under certain circumstances as in product blend-shops with pseudo-constant concentration of the raw materials and also in the preparation of feed processing of crude-oils to oil-refineries.

[1] H. Lee, J. Pinto, I.E. Grossmann, S. Park, 1996, Mixed-Integer Linear Programming Model for Refinery Short-Term Scheduling of Crude Oil Unloading with Inventory Management. *Industrial & Engineering Chemistry Research*, *35*(5), 1630-1641.

[2] Z. Jia, M. Ierapetritou, J.D. Kelly, 2003, Refinery Short-Term Scheduling Using Continuous Time Formulation: Crude-Oil Operations. *Industrial & Engineering Chemistry Research*, 42, 3085-3097.

[3] J.D. Kelly, B.C. Menezes, F. Engineer, I.E. Grossmann, 2017, Crude-oil Blend Scheduling Optimization of an Industrial-Sized Refinery: a Discrete-Time Benchmark. In FOCAPO, Tucson, AR.

[4] R. Karuppiah, I.E. Grossmann, 2006, Global Optimization for the Synthesis of Integrated Water Systems in Chemical Processes. *Computers Aided Chemical Engineering*, 30, 650-673.

[5] E. Ahmetovic, I.E. Grossmann, 2011, Global Optimization for the Synthesis of Integrated Water Systems in Chemical Processes. *AICHE Journal*, 57(2), 434-457.

496470 Linear Surrogate Dynamical Models for Embedding Process Dynamics in Optimal Production Scheduling Calculations

Abstract:

On-demand manufacturing and fast changing electricity markets have created a need for production schedule changes to occur over time scales (hours or minutes) comparable to the process time constants. For these production schedules to be feasible for plant operation, the dynamics of the plant must be accounted for in scheduling calculations. Embedding dynamic information in scheduling calculations is challenging due to the multiple time-scales involved—short term setpoint execution must be balanced with long term scheduling decisions.

Detailed models describing the dynamics of industrial processes are almost invariably nonlinear and highly dimensional, making it challenging to solve the associated optimal scheduling problems (which are in a mixed-integer nonlinear program form) in a practical amount of time. Motivated by this, in this paper we present a novel approach to formulating and solving optimal scheduling problems under dynamic constraints, predicated on developing a Mixed Integer Linear Program (MILP) formulation amenable to real-time solution.

Specifically, in our previous work[1], we have shown that (nonlinear) scheduling-relevant low-order models of the process dynamic can be derived [2] using routine operating data. In this work, we focus on two classes of such models (Hammerstein-Wiener (HW) and Finite Step Response (FSR)). We propose a novel framework for reformulating HW models as a set of linear equations comprising both continuous and integer variables. Further, we demonstrate that for specific choices of the model nonlinearity, this linearization is exact. We compare various linearization techniques and discuss scaling and computational cost. In addition, comparisons between linearized and discretized HW models and inherently linear and discrete FSR models are drawn.

On this basis, we study the formulation and solution strategies for the production scheduling problem, including adaptive sampling, relaxation, and decomposition. A polymerization reactor and an air separation unit are utilized to demonstrate the theoretical concepts developed, showing considerable improvements in computation time compared to previous works [1].

[1] R. C. Pattison, C. R. Touretzky, T. Johansson, I. Harjunkoski, and M. Baldea, "Optimal Process Operations in Fast-Changing Electricity Markets: Framework for Scheduling with Low-Order Dynamic Models and an Air Separation Application," *Ind. Eng. Chem. Res.*, vol. 55, no. 16, pp. 4562–4584, Apr. 2016.

[2] S. A. Billings, *Nonlinear system identification : NARMAX methods in the time, frequency, and spatio-temporal domains*. Chichester, West Sussex: John Wiley & Sons, 2013.

486849 Machine Learning for Autonomous Crystal Structure Identification

Abstract:

We present a machine learning technique to discover relevant structures from molecular simulation snapshots or particle tracking data. Unlike other popular methods for structural identification, our technique requires no a priori description of the target structures. Instead, we use nonlinear manifold learning to infer structural relationships between particles according to their local topology. This graph-based approach yields unbiased structural information which allows us to quantify the crystalline character of particles near defects, grain boundaries, and interfaces. We demonstrate the method by identifying relevant structural features in a simulation of colloidal crystallization, some of which are missed by standard techniques.

499380 Intelligent, Autonomous Exploration of Self Assembly Simulation Parameter Space

Abstract:

Computational "experiments" using coarse-grained models are immensely popular and have helped us understand the behavior of systems with length and time scales ranging from those of atoms to proteins. For exploratory studies in particular, researchers often spend large amounts of computational resources simulating not-so-interesting behavior of these models, such as large regions of parameter space that form trivial structures or kinetically arrested states. It is usually difficult to know which areas in parameter space should be sampled or how systems should be annealed in order to yield interesting or desirable phase behavior. Without the guidance of prior simulations, this leads to an iterative "trial-and-error" approach in many exploratory studies. Here we discuss strategies to incorporate machine learning into the experimental design loop in order to optimize computational time spent simulating interesting phenomena in the area of self assembly.

498035 Efficient Phase Diagram Sampling By Active Learning

Abstract:

We propose and implement an active learning-based efficient phase diagram sampling method, which can save upwards of an order of magnitude of state point exploration time. This adaptive sampling method is inspired by advances in machine learning research. It works by adaptively calculating the potentially most informative state points by considering their proximity to explored state points and the estimated phase boundary location. It incorporates little computational overhead and is easily generalizable to cases of high state space dimensionality. Batch sampling is also supported to take advantage of modern parallel test settings, such as supercomputing clusters. We demonstrate its use, and compare its results to traditional fixed-grid sampling. This proposed method can assist our research community by saving time and computational resources.

485440 A Path Entropy-Based Approach to Predict Transition Rates from Limited Information

Abstract:

We are interested inferring rate processes on networks. In particular, given a network's topology, the stationary populations on its nodes, and a few global dynamical observables, can we infer all the transition rates between nodes? We draw inferences using the principle of maximum path entropy. We maximize the entropy of the distribution over all possible paths a dynamical system can take subject to stationary and dynamical constraints. The present work leads to a particularly important analytical result: namely, that when the network is constrained only by a mean jump rate, the rate matrix is given by a square-root dependence of the rate, kab $(\pi b/\pi a)1/2$, on πa and πb, the stationary-state populations at nodes a and b. This leads to a fast way to estimate all of the microscopic rates in the system. As an illustration, we show that the method accurately predicts transition probabilities among the metastable states of a small peptides as well as proteins at equilibrium.

499861 Iterative Manifold Extension for Efficient Discovery of Transition Pathways

Abstract:

Our understanding of high-dimensional dynamical systems, including many found in molecular modeling, can often be improved through effective free energy surfaces (FES). One of the main challenges is to find the right coarse variables upon which to construct our FES when exploration on the manifold that defines it is computationally restricted by high transition barriers. We present an algorithm to more efficiently cross transition barriers and discover new stable states by relying on the smoothness of the manifold in the original space. We first establish the effective low-dimensionality from a relatively short initial simulation burst, which gives us a local picture of the underlying manifold. After identifying the boundary of the explored region, we can extrapolate geometrically outwards to find new initial conditions to run new simulation bursts, relying on the time scale separation to correct for extrapolation errors that result in off-manifold initializations. This approach biases the simulator to cross transition barriers on the effective FES, iteratively expanding our knowledge about the underlying manifold. We demonstrate our approach on several illustrative systems.

495537 Learning Free Energy Landscapes Using Artificial Neural Networks

Abstract:

The use of adaptive biasing potential methods to estimate free energies of relevant chemical processes from molecular simulation have gained popularity in recent years. Although metadynamics and its well-tempered variant remain the most widely used, new methods such as basis function sampling, variationally enhanced sampling and Greenâ€™s function sampling have been developed which improve upon some of the shortcomings of metadynamics. In particular, the use of a basis expansion to represent the biasing potential overcomes the limitations of a Gaussian-shaped kernel and offer more flexibility in representing the often nuanced and highly nonlinear free energy landscapes. Though said approaches can outperform traditional metadynamics under many circumstances, they remain limited by the functional form or basis set chosen to represent the underlying free energy surface (FES) of interest. Without *a priori* knowledge of the FES, the choice of parameters, such as the number of terms in a basis expansion, can significantly affect the accuracy and convergence behavior of the algorithm. FESs that have sharp curvature can also introduce characteristic oscillations due to the Runge or Gibbs phenomenon. This limitation can lead to oscillations which lead to dramatic under-sampling and a non-convergent bias.

Here we develop a powerful method wherein artificial neural networks (ANNs) are used to obtain the adaptive biasing potential, and thus learn free energy landscapes. As ANNs typically represent a form of *supervised learning*, we develop an iterative scheme which refines an unbiased estimator of a system's partition function. We demonstrate that this method is capable of rapidly adapting to complex free energy landscapes and is not prone to boundary or oscillation problems. The method offers a substantial degree of flexibility to the end-user in specifying the network architecture when the topological features of the FES of interest are not known. Importantly, because the bias learned by the ANN obtains the best continuous approximation of the free energy, we see a dramatic improvement in convergence, especially for poorly sampled states over currently available and broadly used techniques.

501192 QM/ML: A Hybrid Quantum-Mechanics/Machine-Learning Scheme

Abstract:

Hybrid quantum-mechanics / molecular-mechanics (QM/MM) simulations are popular tools for the simulation of extended atomistic systems, in which the atoms in a core region of interest are treated with a QM calculator and the surrounding atoms are treated with an empirical potential. Recently, a number of atomistic machine-learning (ML) tools have emerged that provide functional forms capable of reproducing the output of more expensive electronic-structure calculations. Such ML tools are intriguing candidates for the MM calculator in QM/MM schemes, and have some obvious advantages such as the avoidance of mis-matches (e.g., between lattice constants, mechanical responses, or liquid structures) between the QM and MM calculators, and the removal of the need for the researcher to gain expertise in the MM literature. However, accurate ML models generally require a substantial amount of training data that anticipates atomic configurations that will be encountered during the course of a simulation, provided on the basis of QM methods. Here, we present a simple quantum-mechanics / machine-learning algorithm created with the goal of minimizing the amount of QM training data that needs to be anticipated and generated. Importantly, by re-training the ML calculator to the new QM information at each step, we can identically cancel the overlapping QM and MM energy and forces in a subtractive scheme, producing a simple algorithm for treating boundary interactions between the regions. The approach is analogous to the subtractive framework of the QM/MM methodology, but is not dependent on empirical parameterization, and allows the simulation of arbitrary atomic configurations while keeping QM precision, within certain physical limitations. In this talk, we outline the basic theory and algorithm and also demonstrate the application of QM/ML on the basis of simple examples.

500646 Automation of an Energy Renormalization Approach for the Temperature Transferable Coarse-Graining of Glass-Forming Polymers

Abstract:

A key objective of the Materials Genome Initiative (MGI) for soft materials is to create a coherent framework to perform coarse-grained (CG) simulations in a streamlined and efficient manner to reduce the time and cost of novel material discovery and design. Such an effort involves both advancement in coarse-graining theory, and development of appropriate machinery to give effective practical implementation. Recently, we proposed a coarse-graining strategy for glass forming (GF) liquids based on an energy renormalization (ER) scheme in which cohesive interactions are rescaled using the altered configurational entropy [1,2]. This approach yields a temperature dependent coarse-grained potential with simple form that preserves atomistic dynamics and GF activation energy from the high-temperature melt state to the low-temperature glassy state.

In this talk, we describe the implementation of the above algorithm in a computational â€œworkbenchâ€ we are developing to provide an integrated computational and data environment to support the development of coarse-grained force-fields using quantitative, progressive methods. The workbench environment has three essential elements: a modular program structure that supports the addition of new functionality through Python scripting ; a hierarchical data structure which enables unified representation of materials at different levels of granularity; an ontology based database environment. The starting point for the procedure is a set of all-atom (AA) simulation data at state points that span from the melt state to the glassy regime that are stored in the database as reference data. Starting with this upstream data, we then develop a workflow using the workbench environment to automated the parametrization of the ER potential using multiple properties derived from the AA simulation set. The theory and procedure are illustrated using a number of examples.

References

1. Wenjie Xia, Jake Song, Cheol Jeong, David D. Hsu, , Frederick R. Phelan Jr., Jack F. Douglas, and Sinan Keten, "Energy-Renormalization for Achieving Temperature Transferable Coarse-Graining of Polymer Dynamics," submitted to Proceedings of the National Academy of Sciences (PNAS) 2017.

2. Jake Song, David D. Hsu, Kenneth R. Shull, Frederick R. Phelan Jr., Jack F. Douglas, Wenjie Xia, Sinan Keten, "Energy Renormalization Method for the Coarse-Graining of Polymer Viscoelasticity," in preparation, (2017).

501681 Developing Theory and Data-Driven Benchmarks for General Coarse-Grained Mapping Operators

Abstract:

Coare-graining (CG) is a dimensionality reduction technique used to more efficiently simulate multiscale systems. CG methods rely on grouping similar particles together (mapping) and then redefining the potential energy in terms of the reduced particle number. There is a mature theory about generating potentials for CG models. There is not a rigorous theory for generating mappings from the fine-grain (FG) system to the CG system. This missing component is essential because past CG work shows that many mappings lead to homogeneous, weakly interacting, gas-like CG models. Identifying a mapping operator that preserves relevant physics is usually the most heuristic but essential step in developing CG models. The number of possible mappings is exponential with respect to the number of atoms, which is especially a problem since CG mapping are desired when the number of atoms is large.

In this talk I will outline a new graphical rigorous theory for CG mappings with techniques from the field of computer vision. Due to the ill-defined problem of evaluating specific mapping operators, the talk will emphasize data-driven benchmarks for general mapping strategies. Resulting mappings for solute-solvent and liquids will be explored.

495174 New Coarse-Grained Models of Hydrocarbons

Abstract:

New coarse-grained models for hydrocarbons are developed by utilizing particle swarm optimization (PSO) method. In this model, the bonded potentials are based on all-atom simulations of alkane, while the non-bonded potentials are optimized by the PSO method to reproduce experimental properties. Specifically, all-atom simulation trajectories were utilized to estimate the average bond length and angle. The PSO method was employed to optimize the 12-6 Lennard-Jones Potentials. New CG force-field parameters were optimized to reproduce the experimental properties of hydrocarbons including density (*Ï*), self-diffusion coefficient (*D*), heat of vaporization (*H*v) and surface tension (*Î³*). The potentials are transferable over a range of hydrocarbons with different chain lengths .

494168 Constrained Best Subset Selection Methodology for the Regression of Helmholtz Energy Equations

Abstract:

Accurate thermodynamic properties are important for the development of new processes and technologies and simulating an optimized flowsheet. Unfortunately, equations of state typically used, such as Peng Robinson and Soave-Redlich Kwong, are inaccurate in critical regions to be able to provide accurate models. A recent development in equations of state uses first principles to determine one unifying equation explicit in Helmholtz energy to overcome these inaccuracies. These equations have been developed for over 100 pure substances and have been expanded to mixtures [1, 2, 3, 4].

A major challenge in the development of Helmholtz energy equations is the fitting of large datasets containing correlated data to the main unifying equation [3]. The current fitting procedures depend on cycling between linear and nonlinear regression techniques that are restricted to equality constraints and result in local solutions requiring multiple starts or an experienced user manually selecting an initial starting point [4]. Advancing these techniques to eliminate the need for multistart heuristics and enforce inequality constraints on the resulting models would allow us to control the thermodynamic behavior, improve extrapolation behavior, and maintain thermodynamically feasible limitations on the regressed equation while simultaneously fitting the data [5].

In order to address these challenges associated with fitting data to Helmholtz energy equations, we have developed a new model-building procedure from thermodynamic data. The main idea is to apply best subset selection to allow for the controlling of thermodynamic slopes with inequality constraints while fitting all the data simultaneously. A bank of terms is chosen to best represent the different phase regions and thermodynamic behavior. In order to avoid overfitting, our procedure systematically selects a subset of these terms to optimally fit the multiple thermodynamic property data sets. We rely on a global optimization solver [6] to find an optimal solution that optimizes the fit as well as the number of terms in the model according to an information criterion.

The proposed approach was used to fit a new Helmholtz energy equation for toluene. The resulting models were validated against results from the literature [7]. With the same data and bank of terms, the new methodology results in accuracy within the range of the experimental error of the thermodynamic properties incorporated. Our methodology can be applied to other pure and pseudo-pure chemical data sets to automatically generate new Helmholtz energy equations for use in simulations and optimization.

References cited

[1] Lemmon, E. W.; Huber, M. L.; McLinden, M. O. NIST Standard Reference Database 23: Reference Fluid Thermodynamic and Transport Properties-REFPROP, Version 9.1, National Institute of Standards and Technology. 2013; https://www.nist.gov/srd/refprop.

[2] Span, R. Multiparameter equations of state: An accurate source of thermodynamic property data; Springer-Verlag, 2000.

[3] Span, R.; Wagner, W.; Lemmon, E. W.; Jacobsen, R. T. Multiparameter equations of state - recent trends and future challenges. Fluid Phase Equilibria 2001, 183-184, 1-20.

[4] Wagner, W.; PruÃŸ, A. The IAPWS formulation 1995 for the thermodynamic properties of ordinary water substance for general and scientific use. *J. Phys. Che. Ref.*Data 2002, 387.

[5] Lemmon, E. W.; Jacobsen, R. T. A New Functional Form and New Fitting Techniques for Equations of State with Application to Pentafluoroethane (HFC-125). J. Phys. Chem. Ref. Data 2005, 34, 69â˘'108.

[6] Tawarmalani, M.;Sahinidis, N. V. Global optimization of mixed-integer nonlinear programs: A theoretical and computational study, Mathematical Programming, 99, 563-591, 2004.

[7] E. W. Lemmon. Correlations short fundamental equations of state for 20 industrial fluids. J. Chem. Eng. Data, (51):785{850, 2006.

501374 Identifying Equilibrated Simulation Trajectories with Artificial Neural Networks

Abstract:

Determining which microstates to include in an ensemble average is a ubiquitous simulation task whose reproducibility is hampered by being underspecified. For a single simulation, the task is not so bad: Including or omitting one microstate from a sufficiently large set won't change the answer, so heuristics like "The potential energy is higher here, so we'll cut those states out" will do. The problem arises when thousands of simulation trajectories need to be analyzed, demanding efficient heuristics that are automatable and deterministic. Artificial neural networks are one type of machine learning algorithm that provides a reproducible path for applying pattern recognition heuristics. Here we use the TensorFlow machine learning library to train an artificial neural network on the problem of distinguishing "equilibrated" from "not equilibrated" hypothetical observation sequences. We generate training populations and test populations of normally-distributed observation sequences with embedded linear and exponential correlations. We train a two-neuron artificial network to distinguish the correlated and uncorrelated sequences. We characterize the predictive capabilities of the network as a function of training population size, the number of points in each sequence, and the number of training rounds performed. We find diminishing returns on predictive accuracy with sample sizes over 512 points, training populations of 8000 sequences, and 2000 training rounds, which takes about three minutes to train. We find that this simple network is powerful enough for over 98% accuracy in identifying exponentially-decaying observation sequences in microseconds. These observations suggest neural networks as an effective tool for automating this and other heuristic bottlenecks in high-throughput simulation pipelines.

493486 Enhanced Furnace Balancing Scheme Via an Integrated Computational Fluid Dynamics/Data-Based Optimization Approach

Abstract:

The steam methane reforming (SMR) process is highly endothermic and is widely used for commercial-scale production of hydrogen from superheated steam and methane in nickel-based catalyst packed bed reactors [1]. It is typically carried out in a top-fired steam methane reformer that has two closed domains (i.e., the furnace side in which fuel is burned in excess air to generate thermal energy to drive the SMR process, and the tube side in which the SMR process takes place to produce hydrogen). Reformers are often energy-intensive such that the efficiency of hydrogen plants depends on the reformer efficiency, but the plant efficiency is often decreased by nonuniformity in the outer reforming tube wall temperature (OTWT) distributions (i.e., the distributions of the radially-averaged tube outer wall temperatures at a fixed height within the reformer) [2,3,4]. Specifically, when there is a large discrepancy between the maximum and minimum temperature for the distribution at a given reformer height but the average temperature of the distribution is close to the design value of the reforming tube wall material to maximize production of hydrogen from the endothermic SMR process, the outer wall temperature values of some reforming tubes would likely be higher than the design value. This would shorten the reformer service life and put the hydrogen plant at risk of having substantial capital and production losses due to creep rupture of the reforming tubes [3,4,5]. An intuitive strategy to resolve this problem is to reduce the total mass flow rate through the burners to reduce the heat generated in the furnace side and thus to reduce the maximum outer reforming tube wall temperature below the design value. However, this also decreases the methane conversion and hydrogen production rate, and thus should not be performed in an ad hoc fashion, but should instead be part of an optimization-based approach for furnace balancing (reducing the nonuniformity in the OTWT distribution at a given height within the reformer by adjusting the distribution of flow rates through the burners).

Motivated by this, the present work focuses on developing an advanced furnace balancing scheme via an integrated computational fluid dynamics/data-based optimization approach that optimizes not only the distribution of burner mass flow rates but also the total mass flow rate through the burners such that the degree of temperature nonuniformity inside the reformer is minimized, while the maximum outer reforming tube wall temperature is kept strictly below the design temperature of the wall material, and the hydrogen production rate is maximized without reducing the reformer service life. The furnace balancing scheme employs computational fluid dynamics modeling to generate data that is used in identifying a data-driven model for the optimization-based furnace balancing scheme and for validating the data-driven model. The furnace balancing optimization problem simultaneously solves for the optimal total mass flow rate through the burners and the individual mass flow rates through each burner. The robustness of the furnace balancing scheme to disturbances is compared with the robustness of a furnace balancing scheme developed in [4] that assumes a constant total mass flow rate. The data-driven model development follows statistical principles to allow it to make reasonable out-of-sample predictions of the OTWT distribution for various mass flow rate distributions and also for various total mass flow rates. The ability to adjust the total mass flow rate for the advanced furnace balancing scheme is of special interest for the hydrogen manufacturing industry as it can potentially lead to substantial savings in the re-tubing cost of the reformer.

[1] Ewan BCR, Allen RWK. A figure of merit assessment of the routes to hydrogen. International Journal of Hydrogen Energy. 2005;30:809â€"819.

[2] Dunn AJ, Yustos J, Mujtaba IM. Modelling and simulation of a top-fired primary steam reformer using gPROMS. Developments in Chemical Engineering and Mineral Processing. 2002;10:77â€"87.

[3] Kumar A, Baldea M, Edgar TF, Ezekoye OA. Smart manufacturing approach for efficient operation of industrial steam-methane reformers. Industrial & Engineering Chemistry Research. 2015;54:4360â€"4370.

[4] Tran A, Aguirre A, Crose M, Durand H, Christofides PD. Temperature balancing in steam methane reforming furnace via an integrated CFD/data-based optimization approach. Computers & Chemical Engineering. in press.

[5] Pantoleontos G, Kikkinides ES, Georgiadis MC. A heterogeneous dynamic model for the simulation and optimisation of the steam methane reforming reactor. International Journal of Hydrogen Energy. 2012;37:16346-16358.

497398 Stochastic Programming Approach Vs. Estimator-Based Approach for Sensor Network Design for Maximizing Efficiency

Abstract:

The measurement technology plays a key role in achieving efficient operation of the process plants without violation of the environmental and operational constraints. Inaccuracies in the measurements of the controlled variables can lead to inefficient operation (or even unsafe operation under extreme conditions) of the plant. Inaccuracies in the measurements of other variables that are used for monitoring purposes can lead to undesirable conditions and can damage equipment and reduce their life. However, it may be difficult, if not impossible to measure a number of variables or the measurements may be noisy, inaccurate, or unreliable. Furthermore, the measurement technology for measuring certain variables may be costly or the measurement is available after undesired delay. It may be possible to estimate these variables accurately by optimally located sensors. For a large-scale process plant, it is difficult to decide which variables should be measured for maximizing the plant efficiency while satisfying desired estimation accuracy in other variables. With this motivation, the focus of the current work is on development of sensor placement (SP) algorithms to obtain the numbers, locations, and types of sensors for a large-scale process with the estimator-based control system.

Two SP algorithms are developed and investigated. In one algorithm (Paul et al., 2015, 2016), dynamics in the process efficiency loss due to the estimator-based control system that receives measurements from a candidate sensor network are explicitly accounted for. For a large-scale process with large number of candidate sensor locations, this approach leads to a computationally expensive mixed integer nonlinear programming problem. A number of novel approaches is developed by modifying the SP algorithm, problem formulations, and computational approaches to make the solution tractable. In another algorithm, the estimation error is accounted for in terms of probability distributions and therefore, a stochastic programming approach is used to solve the SP problem (Sen et al., 2016). A novel algorithm called BONUS to solve the problem (Diwekar, 2015).

The developed SP algorithms are implemented in an acid gas removal (AGR) unit as part of an integrated gasification combined cycle (IGCC) power plant with pre-combustion CO
model is developed in Aspen Engineering Suite. The model has several thousand states and hundreds of candidate sensor locations. In this presentation, we will compare and contrast these two SP algorithms by evaluating the efficiency loss of the optimal sensor network synthesized by each of these algorithms along with their computational performance.

References:

1. Paul et al., Sensor Network Design for Maximizing Process Efficiency: An Algorithm and Its Application, AIChE Journal, V. 61, pp 461-476, 2015.
2. Paul P, Bhattacharyya D, Turton R, Zitney S "Dynamic Model-Based Sensor Network Design Algorithm for System Efficiency Maximization", Computers & Chemical Engineering, 89, 27-40, 2016
3. Diwekar U. and A. David, BONUS Algorithm for Large Scale Stochastic Nonlinear Programming Problems, Springer, 2015.
4. Sen P., K. Sen, and U. Diwekar, A multi-objective optimization approach to optimal sensor location problem in IGCC power plants, Applied Energy 181, pp 527–539, 2016.

495329 Optimal Scheduling of a Microgrid on a Steam-Assisted Gravity Drainage (SAGD) Facility

Abstract:

Steam-Assisted Gravity Drainage (SAGD) is a thermal recovery technique, which largely accounts for the extraction of the 1.7 trillion barrels of heavy and extra heavy oil (bitumen) present in the oil sands of Western Canada. Due to the injection of superheated steam in the oil sands formation, at the end of the production, substantial amount of waste heat is left behind in the reservoir. In this work, we explore the opportunity of utilization of this waste heat to produce electricity using a steam turbine operating on an Organic Rankine Cycle (ORC). The premise being that the electricity produced could then meet the load demands of other operational SAGD facilities, thus making the SAGD facilities less dependent on a macrogrid, more profitable, and leave a lower carbon footprint in the process. However, preliminary research has shown that the waste heat left behind in the reservoir, is by itself, incapable of sustaining the operations of an entire SAGD facility. This issue can be overcome by incorporating other renewable sources of energy generation working synergistically with the SAGD waste heat based steam turbine. This leads to a challenging scheduling problem since one has to make optimal decisions in the volatile electricity and natural gas market in the face of inherent intermittency in renewable generation and local demand.

In this work, we focus on the dispatch of a microgrid consisting of a bi-directional connection to the macrogrid, gas turbine, wind turbine, steam turbine operating on an ORC and a battery bank. We propose and formulate a Kelly Criterion (KC) based optimization problem to supply electricity (from renewables and waste heat) and make profitable decisions in the volatile electricity and natural gas pricing system. The KC is mainly used in economics to hike the return on investment and is built on the principle of maximization of the growth rate based on successive investments or bets as a fraction of the total investible capital. The technique does not have any explicit dependence on the starting capital, and provides the optimal betting or investment fraction for maximal expected return on investment, and in turn, capital growth. For this study, the starting capital is analogous to the amount of electricity generated, and the KC based optimization technique provides the optimal strategy to distribute that electricity generated as a function of the electricity and natural gas pricing markets. The electricity price was forecasted using a three-layered neural network working on a Nonlinear Autoregressive time series system identification technique with Exogenous inputs (NLARX), while the natural gas price was estimated based on a Gaussian distribution centered around the current natural gas price.

Since the KC technique does not depend on the amount of electricity generated, therefore, an accurate forecast of the parameters influencing the generators, such as wind for wind turbines, is not required. The optimization study was made possible by a bidirectional communication between MATLAB (the price forecaster) and General Algebraic Modeling System (GAMS) (the optimizer) software. The results indicate that the KC based optimization technique for microgrid scheduling outperforms the standard Dynamic-Real Time Optimization technique.

493099 Design and Implementation of MPC Strategies for Supercritical Pulverized Coal-Fired Power Plant Cycling with Carbon Capture

Abstract:

In recent years, with increasing energy demand and focus on greenhouse gas emissions, concerns on whether coal should be used as a primary energy source are heightened. According to a reference case study for power generation from the years 2000 to 2040, coal-fired power plants would still account for more than 30% in future power generation [1]. However, as renewable energy sources increasingly penetrate into the grid, coal-fired power plants which were designed to operate at base-load conditions would need to cycle their load. This cycling includes ramping the power generation rate up/down for load-following, as well as shut down/start up according to load demand. Previous literature studies have shown the impact of wind generation when integrated into gas and coal generation [2]. In particular, a sudden decrease in the wind power generation would lead to increased emissions and efficiency reduction from the coal-fired power generation due to cycling. To systematically achieve the optimal profiles for cycling the power plant, the development of advanced control algorithms is needed.

In this work, the design and implementation of linear and nonlinear model predictive control (MPC) strategies are performed for base-load and cycling scenarios. The application system is a supercritical pulverized coal-fired (SCPC) power plant with post-combustion carbon capture. The linear MPC control strategy is based on the dynamic matrix control (DMC) method. For this method, a multiple-input-multiple-output dynamic matrix is obtained by performing step response tests in the SCPC plant. The proposed nonlinear MPC strategy is an extension of NLMPC, which is based on the direct transcription method [3, 4]. For the NLMPC implementation, the nonlinear programming (NLP) solver IPOPT is employed, which is an efficient interior point-based large-scale nonlinear optimization algorithm [5]. Also, an open-source automatic differentiation package ADOL-C is used for improving the accuracy and calculation speed of system derivatives [6]. For the extended nonlinear MPC strategy, sequential quadratic programming (SQP) algorithms are analyzed for potential computational improvements when solving the posed NLP problem related to the large-scale power plant application.

In this presentation, a number of scenarios for the SCPC power plant will be discussed including: (i) trajectory tracking associated with power generation demands according to cycling starting from the original power generation setpoint; (ii) disturbance rejection for maintaining the carbon capture rate. In this case, the power generation should be kept at the original setpoint while potential disturbances affect the system, such as the carbon content in the coal is changed; and (iii) combination of both setpoint tracking and disturbance rejection cases. Preliminary results on the implementation of the DMC strategy for the carbon capture sub-system show that the controller can successfully keep the carbon capture rate at the specified 90% setpoint when the fluegas flowrate entering the system acts as step and ramp disturbances. Results on the closed-loop responses for different scenarios related to the SCPC plant will be compared and analyzed considering the advanced linear and nonlinear MPC strategies.

References:

1. Annual energy outlook 2015 with projection to 2040. Contract No.: DOE/EIA-0383(2015).
2. Bentek Energy LLC. Wind power, and unintended consequences in the Colorado energy market. Evergreen, CO: Bentek Energy, LLC, 2010.
3. F.V. Lima, R. Amrit, M. Tsapatsis, and P. Daoutidis. Nonlinear model predictive control of IGCC plants with membrane reactors for carbon capture. In Proceedings of the American Control Conference. Washington, DC, June 17-19 2013.
4. F.V. Lima, Xin He, R. Amrit and P. Daoutidis. Advanced control strategies for IGCC plants with membrane reactors for CO_2 capture. Process systems and materials for CO_2 capture: modeling, design, control and integration, A.I. Papadopoulos and P. Seferlis (eds.), Wiley, 2017.
5. A. Wächter, L.T. Biegler. On the implementation of an interior-point filter line-search algorithm for large-scale nonlinear programming. Mathematical programming. 2006;106(1):25-57.
6. A. Walther, A. Griewank. Getting started with ADOL-C. Combinatorial Scientific Computing. 2012:181-202.

495197 Fast-Nonlinear Model Predictive Control Implementation with Open-Source Tools

Abstract:

New modeling platforms and state of the art optimization solvers combined allow to build algorithms, where mathematical optimization is the basis, with relative low difficulty. For instance, Nonlinear Model Predictive Control (NMPC) that requires the solution of several parametric nonlinear optimization problems.

In this context, the resulting NLPs must be constrained by a DAE model of the plant. This can be approached in several ways of which the direct transcription method is possibly the most efficient for large DAE models. This implies that the DAE model of the plant is fully discretized with a high order implicit method like collocation, and then solved simultaneously during the optimization. The resulting NLP problem will have a large and sparse structure, which can be solved with Ipopt.

We have selected Pyomo as our modelling platform and solver interphase, because it has been founded on python, and therefore allows access to other packages and high-level programming capabilities. Consequently, the NMPC implementation allows for the controller input calculation and feedback from a plant model in the same platform. This; however, comes with a computational cost which can be circumvented by using sensitivity based NMPC.

The sensitivity NMPC relies on the first order curvature information of the optimality conditions that can be used to calculate the change of the solution of the optimization problem given a perturbation of the initial set of parameters in a much faster way than solving the NLP from the beginning. Then, the controller input can be calculated for a predicted state of the plant before the real state is known, and when the real state is known the solution can be updated via sensitivity. This strategy keeps most of the stability and robustness of the original NMPC while being faster. However, sometimes the sensitivity calculation will be not reliable if the matrix associated with the linearized optimality conditions is close to singularity; therefore, we propose re-scaling and restructuring if necessary, so the calculations become more robust.

A large problem was then used as example of the application of the framework. This was the bubbling fluidized bed reactor for CO2 capture which is partial differential algebraic model that consists of 11 PDEs, some of which contain both time and space differential terms, and around 80 algebraic equations that include boundary conditions, correlations for the hydrodynamics, and thermodynamic properties. The original optimal control problem for this model takes more than 100 seconds to solve, which makes the real time NMPC implementation infeasible. Nevertheless, by implementing the previously described sensitivity based NMPC, as well as further model reduction, the controller input can be calculated in a reasonable time frame.

Finally, as part of the ongoing work, we are developing the framework for state estimation using the Moving Horizon Estimation (MHE) approach with the same tools described before. These involve calculation of the arrival cost and advanced step approaches and will demonstrate the ability of open-source tools to combine optimization and modeling

for large problems.

500959 Robust Stabilization of a Two-Stage Continuous Anaerobic Bioreactor

Abstract:

Continuous anaerobic digestion processes cover a wide range of applications in biochemical engineering and environmental technology, such as the anaerobic water treatment and biogas production.Â The biogas produced via this process, which consists of mainly methane and CO_2, serves as an important alternative renewable energy source. Compared with other biochemical production processes from biomass, the anaerobic digestion process with production of biogas has shown its benefits of robustness to a wide variety of substrates, potential incorporation with carbon capture, minimal safety requirements and low cost. Meanwhile, due to the constantly changing microbial environment and the systemâ€™s marginally stable property at moderately high dilution rates, there are major challenges in the dynamic operation and control of the process.

Traditional operation of anaerobic digestion involves a single stage, where a large number of microbial populations with varying characteristics grow in the same reactor. To enhance the system stability and product quality, the two-stage process, consisting of two continuous bioreactors connected in series, with different reactions occurring in separate stages, has been proposed in the literature [1]. In this way, it is possible to achieve: 1) higher yield, 2) higher methane concentration in the output gas, and 3) a more flexible gas production [2][3]. However, the stable region around the operating point is still narrow when it is operated at moderately high dilution rates to maximize product yield and purity. Consequently, it is likely to lead to the washout of biomass when disturbances occur. Therefore, a controller is needed to enlarge the size of the stability region around such optimal operating conditions. There is a large body of literature on control algorithms for stabilization of traditional one-stage processes, e.g. [4][5][6]. However, there are currently no research results on the control of two-stage anaerobic processes.

In this work, we study the problem of robust stabilization of a two-stage anaerobic bioreactor. The steady state properties of the system are first investigated to show that the need for control is due to the fact that the optimal steady state is extremely close to an unstable steady state and the stability region is in need of enlargement. The control problem is rigorously tackled based on a three state approximate dynamic model, and later applied to a more detailed seven state model to test its performance. With dilution rate as an input and methane production rate and VFAs concentration as outputs, a constant-yield nonlinear controller first derived based on intuitive considerations. A Lyapunov function approach is used to prove the stability of the closed-loop system over the entire positive orthant. The robustness properties of the control law are also studied. The control law is proved to be robust to both errors in the kinetic model parameters and disturbance in the inlet feed concentration, in the sense of preserving the stability region of the system. Performance of the control system can be even enhanced if a feedforward measurement of the inlet feed concentration is incorporated in the control law.

References:

1. Pohland, F. G., and S. Ghosh. "Developments in anaerobic stabilization of organic wastes-the two-phase concept." Environmental letters 1.4 (1971): 255-266.

2. Parawira, Wilson, et al. "Energy production from agricultural residues: high methane yields in pilot-scale two-stage anaerobic digestion." Biomass and Bioenergy 32.1 (2008): 44-50.

3. Salomoni, C., et al. "Enhanced methane production in a two-phase anaerobic digestion plant, after CO_2 capture and addition to organic wastes." Bioresource technology 102.11 (2011): 6443-6448.

4. Syrou, L., et al. "Robust global stabilization of continuous bioreactors." 7th International Conference on Dynamics and Control of Process Systems (DYCOPS 7), Boston, MA. 2004.

5. Savoglidis, Georgios, and Costas Kravaris. "Constant-yield control of continuous bioreactors." Chemical engineering journal 228 (2013): 1234-1247.

6. Gouze, Jean-Luc, and Gonzalo Robledo. "Feedback control for competition models with mortality in the chemostat." Decision and Control, 2006 45th IEEE Conference on. IEEE, 2006.

Optimal Operation and Control Policy of a Solar-Thermal District Heating System

Xiaodong Xu, Yuan Yuan, Stevan Dubljevic

Department of Chemical and Materials Engineering
University of Alberta, Edmonton, Canada

October 29 - November 3, 2017

2017 AIChE Annual Meeting
Minneapolis, MN, USA

Outline

A. System Description

B. System Models

C. Main Results

D. Simulation Study

F. Conclusion

System Description

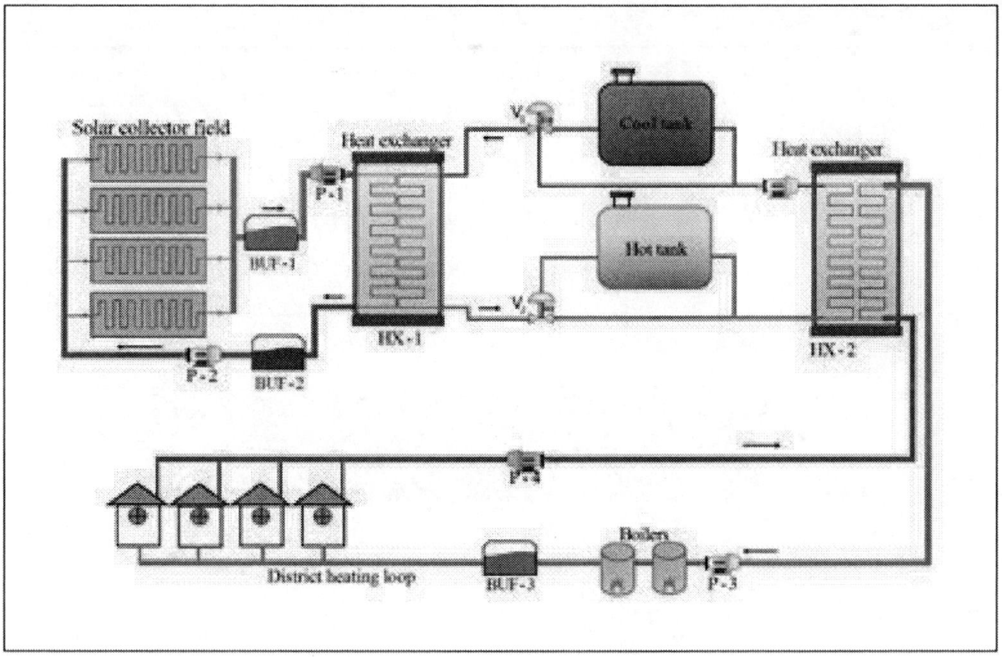

Solar Collector Field

$$\rho_m C_m A_m \frac{\partial T_{sol,m}}{\partial t}(t,\zeta) = \mu_0 GI(t) - D_m \pi h_l (T_{sol,m}(t,\zeta) - T_a(t)) - D_f \pi h_t (T_{sol,m}(t,\zeta) - T_{sol,f}(t,\zeta))$$

$$\rho_f C_f A_f \frac{\partial T_{sol,f}}{\partial t}(t,\zeta) = -\rho_f C_f v_{sol,f}(t) \frac{\partial T_{sol,f}}{\partial \zeta}(t,\zeta) + D_f \pi h_t (T_{sol,m}(t,\zeta) - T_{sol,f}(t,\zeta))$$

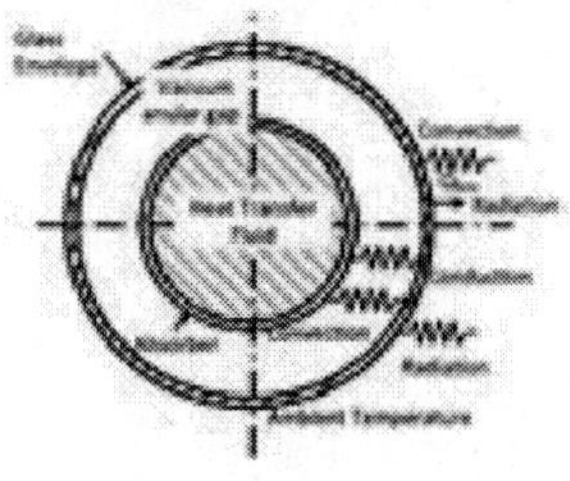

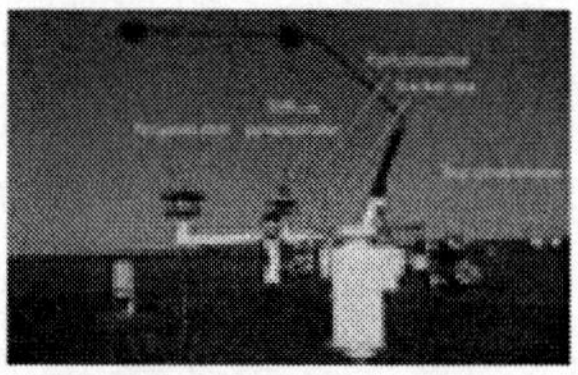

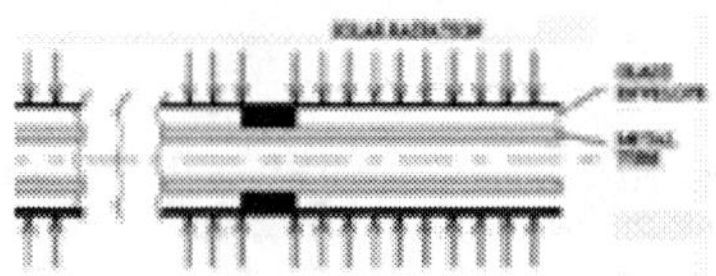

Heat Exchanger Model

$$\rho_f C_f A_{F_1} \frac{\partial T_{F_1}}{\partial t}(t,\zeta) = \rho_f C_f v_{F_1}(t) \frac{\partial T_{F_1}}{\partial \zeta}(t,\zeta) - k\varphi\left(T_{F_1}(t,\zeta) - T_{F_2}(t,\zeta)\right)$$

$$\rho_w C_w A_{F_2} \frac{\partial T_{F_2}}{\partial t}(t,\zeta) = -\rho_w C_w v_{F_2}(t) \frac{\partial T_{F_2}}{\partial \zeta}(t,\zeta) + k\varphi\left(T_{F_1}(t,\zeta) - T_{F_2}(t,\zeta)\right)$$

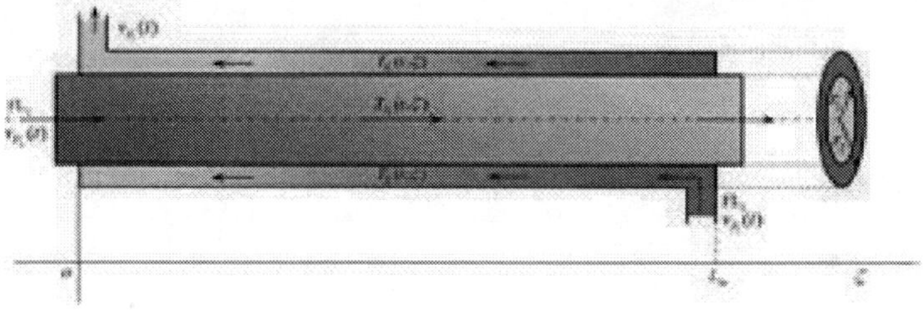

Solar Collector with Storage System

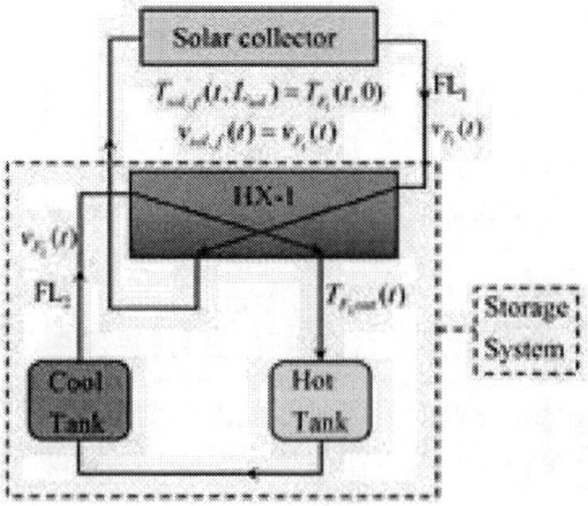

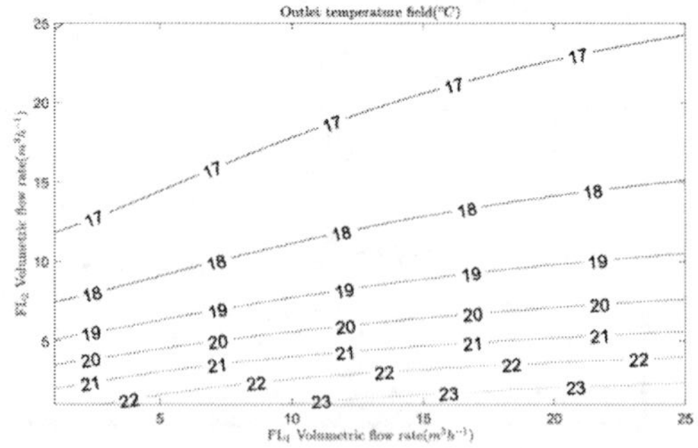

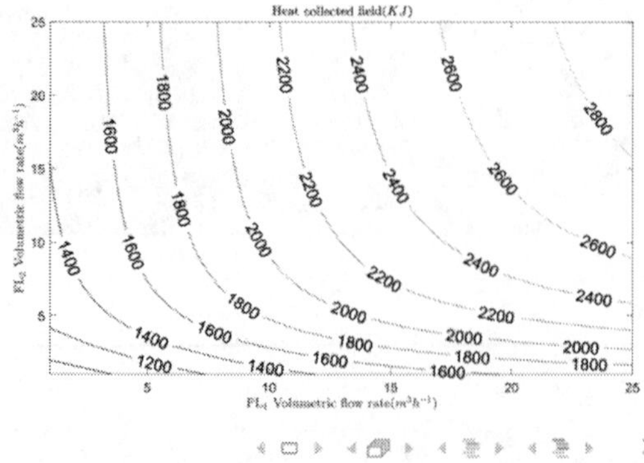

District Heating Loop System (Boiler Gas System)

$$\rho_w C_w V_{bol} \frac{dT_{bol}}{dt}(t) = \rho_w C_w v_{bol}(t)(T_{bol,in}(t) - T_{bol}(t)) + H_R(t) h_{fg}(Reg(t) T_{gas}(t) - T_{bol}(t))$$

$$\rho_w C_w A_{dis} \frac{\partial T_{dis}}{\partial t}(t,\zeta) = -\rho_w C_w v_{dis} \frac{\partial T_{dis}}{\partial \zeta}(t,\zeta), \zeta \in [0, L_{dis}]$$

$$T_{dis}(t,0) = T_{bol}(t)$$

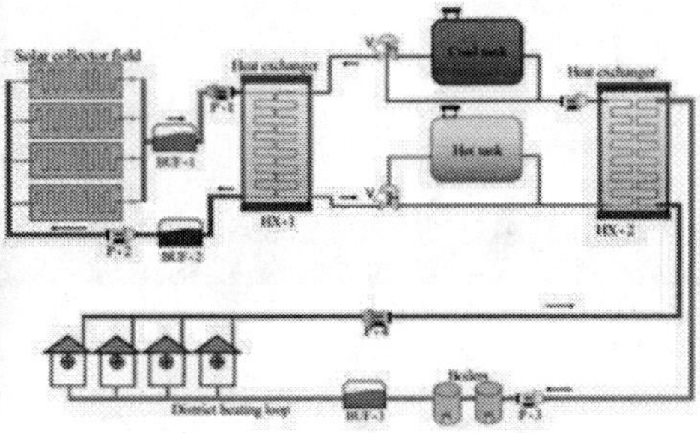

Boiler Gas System

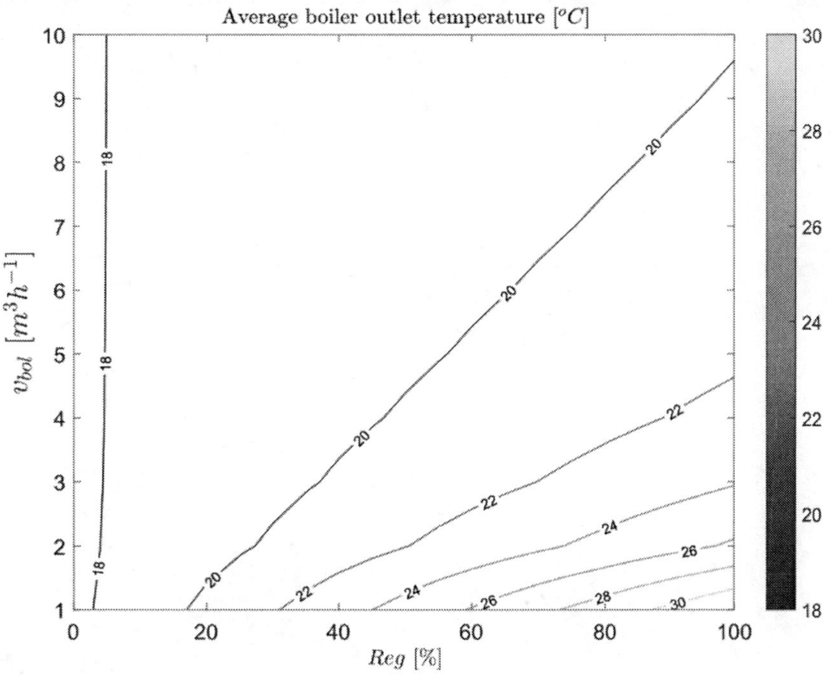

Observer

Observer Design for Solar Collector

Boundary state observer:

$$\rho_m C_m A_m \frac{\partial \hat{T}_{sol,m}}{\partial t}(t,\zeta) = \mu_0 G I(t) - D_m \pi h_l(\hat{T}_{sol,m}(t,\zeta) - T_a(t)) - D_f \pi h_t(\hat{T}_{sol,m}(t,\zeta) - \hat{T}_{sol,f}(t,\zeta))$$

$$\rho_f C_f A_f \frac{\partial \hat{T}_{sol,f}}{\partial t}(t,\zeta) = -\rho_f C_f v_{sol,f}(t) \frac{\partial \hat{T}_{sol,f}}{\partial \zeta}(t,\zeta) + D_f \pi h_t(\hat{T}_{sol,m}(t,\zeta) - \hat{T}_{sol,f}(t,\zeta))$$

$$\hat{T}_{sol,f}(t,0) = T_{sol,fin}(t) - \kappa \Delta T_{sol,f}(t, L_{sol})$$

$$\Delta T_{sol,f}(t,\zeta) = \hat{T}_{sol,f}(t,\zeta) - T_{sol,f}(t,\zeta)$$

Theorem

If $\kappa^2 \leq e^{-\mu}$ where μ is a tuning positive parameter, the proposed boundary state observer exponentially converges, i.e., the observer error system is exponentially stable.

Observer Design for Solar Collector

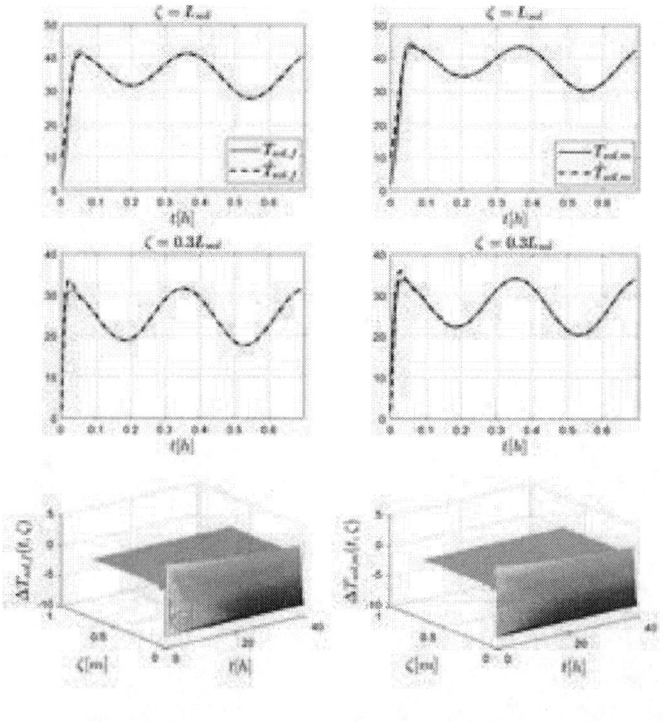

Multivariable Optimization Problem

$$\min_{v_{F_1}(t), v_{F_2}(t)} J_H\left(t, T_{F_2}, t_f, T_{des}^{F_2}\right)$$

s.t. Solar Collector Model

Heat Exchanger Model

$$J_H\left(t, T_{F_2}, t_f, T_{des}^{F_2}\right) = \omega_1 J_{F_2}\left(t, T_{F_2}, t_f, T_{des}^{F_2}\right) - \omega_2 H_{tran}(t, t_f)$$

$$J_{F_2}\left(t, T_{F_2}, t_f, T_{des}^{F_2}\right) = s_1 \int_t^{t+t_f} \left(T_{F_2 out}(\tau) - T_{des}^{F_2}(\tau)\right)^2 d\tau + s_2 \int_t^{t+t_f} \left(\frac{dv_{F_1}(\tau)}{d\tau}\right)^2 d\tau + s_1\left(T_{F_2 out}(t+t_f) - T_{des}^{F_2}(t+t_f)\right)^2$$

$$H_{tran}(t, t_f) = \int_t^{t+t_f} C_w \left(T_{F_2 out}(\tau) - T_{F_2 in}(\tau)\right)\rho_w v_{F_2}(\tau) d\tau + \int_0^{L_{he}} \rho_w A_{F_2} C_w \left(T_{F_2}(t_f, \zeta) - T_{F_2}(0, \zeta)\right) d\zeta$$

Simulation Study

The resulting input $v_1(t)$ and $v_2(t)$

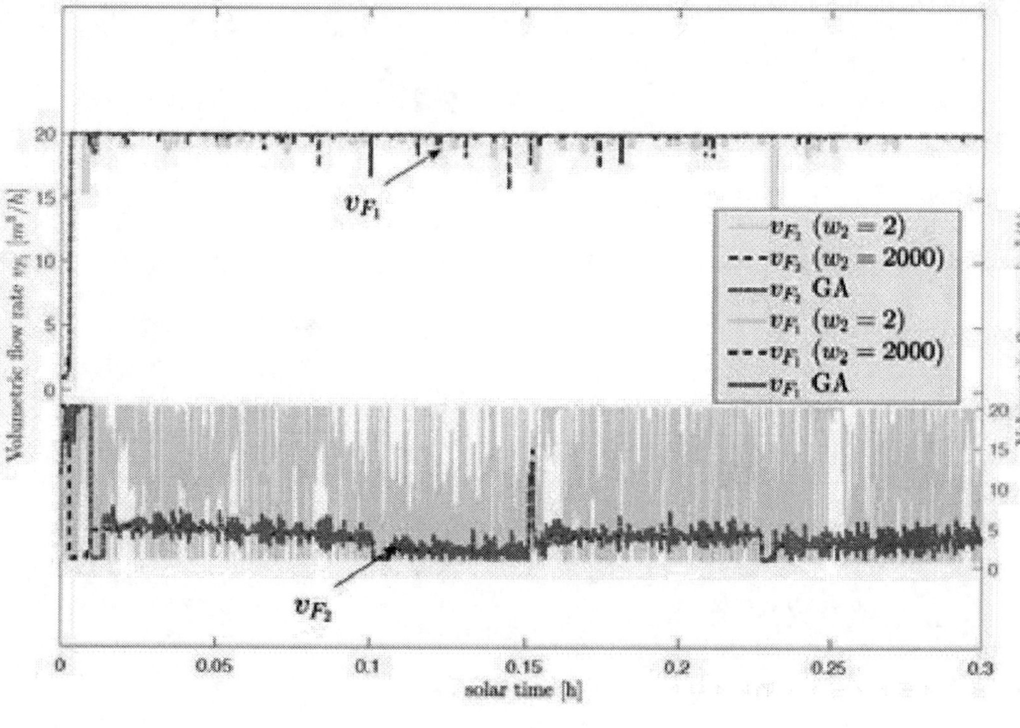

Simulation Study

The resulting tracking performance

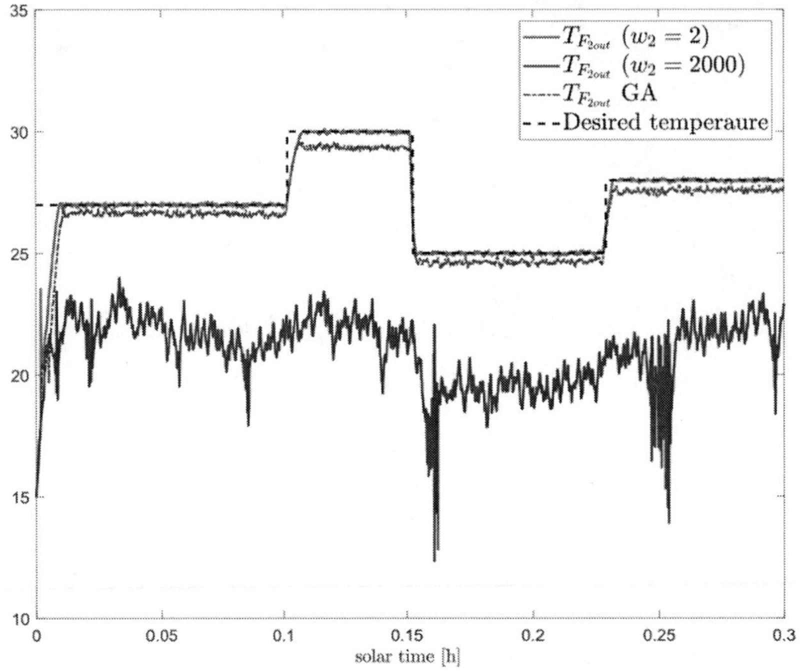

Simulation study

The transferred (stored) energy

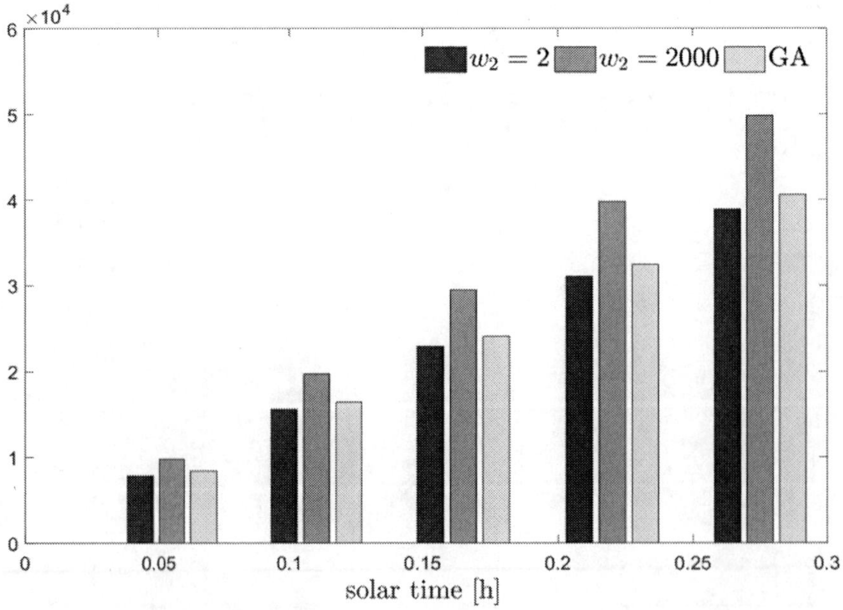

Servo-control of Boiler-Heating System

Step 1. Apply Internal Model Principle based servo-control approach to generate a sequence of desired input $T^*_{bol}(t)$ which can be regarded as the set-point for the boiler outlet temperature $T_{bol}(t)$ so that $\lim_{t\to\infty} \left(T_{dis}(t, L_{dis}) - T^*_{dis}(t)\right) = 0$.

Step 2. Apply the optimal control to drive the boiler-gas system so that the boiler temperature $T_{bol}(t)$ tracks the desired temperature $T^*_{bol}(t)$ obtained in **Step 1**.

Servo-control of district heating system (step 1)

$$\dot{w}(t) = Sw(t), \quad T^*_{dis}(t) = Fw(t)$$

$$\begin{aligned} &\frac{v_{dis}}{A_{dis}} \frac{dm^T}{d\varsigma}(\varsigma) + m^T(\varsigma)S = 0 \\ &m^T(L_{dis}) = F \\ &T^*_{bol}(t) = m^T_w w(t) \end{aligned}$$

$\sigma(S) = 0 \cup \{iw_k\}, k = 1, 2 \cdots, n-1$: modelling of steplike and sinusoidal exogenous signals

Outline System Description System Models Observer Design Conclusion

Servo-control of Boiler-Heating System

Step 1. Apply Internal Model Principle based servo-control approach to generate a sequence of desired input $T^*_{bol}(t)$ which can be regarded as the set-point for the boiler outlet temperature $T_{bol}(t)$ so that $\lim_{t\to\infty} \left(T_{dis}(t, L_{dis}) - T^*_{dis}(t)\right) = 0$.

Step 2. Apply the optimal control to drive the boiler-gas system so that the boiler temperature $T_{bol}(t)$ tracks the desired temperature $T^*_{bol}(t)$ obtained in **Step 1**.

Optimal tracking control of the boiler system (step 2)

$$\begin{aligned} \min_{v_{bol}, Reg} \quad & J_{bol} = \int_t^{t+t_f} \left(T_{bol}(\tau) - T^*_{bol}(\tau)\right)^2 d\tau \\ \text{s.t.} \quad & \text{Boiler gas system} \\ & 0 \leq Reg \leq 1 \\ & 0 \leq v_{bol} \leq 10 \end{aligned}$$

$$\begin{aligned} J_{bol} = & \int_t^{t+t_f} \left(T_{bol}(\tau) - T^*_{bol}(\tau)\right)^2 d\tau \\ & + \int_t^{t+t_s} \left(\frac{dv_{bol}}{d\tau}(\tau)\right)^2 d\tau + \int_t^{t+t_s} \left(\frac{dReg}{d\tau}(\tau)\right)^2 d\tau \end{aligned}$$

The desired boiler temperature

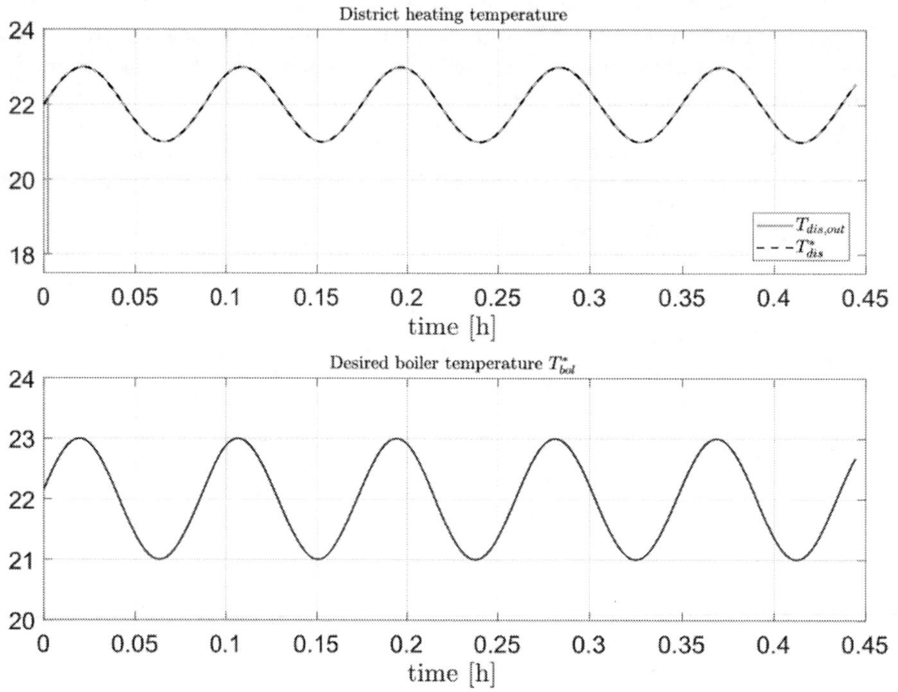

The resulting boiler flow velocity $v_{vol}(t)$ and regulator value Reg

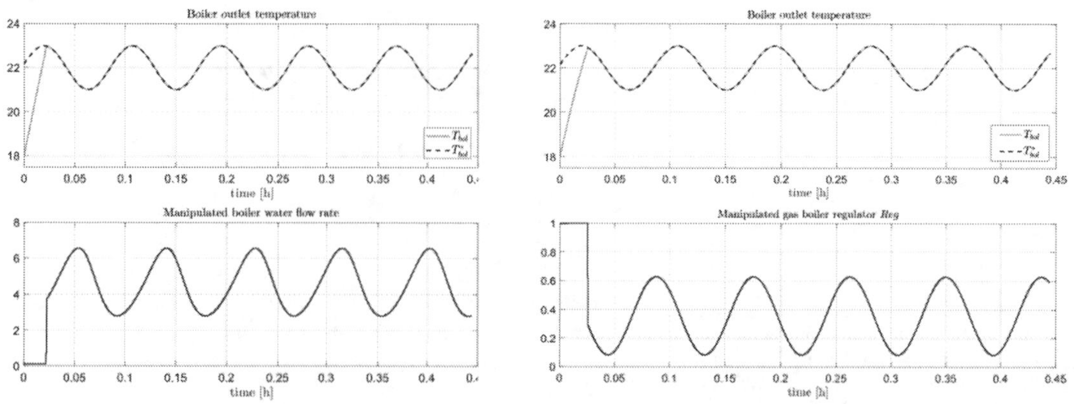

Temperature evolution and profile in DHL-system

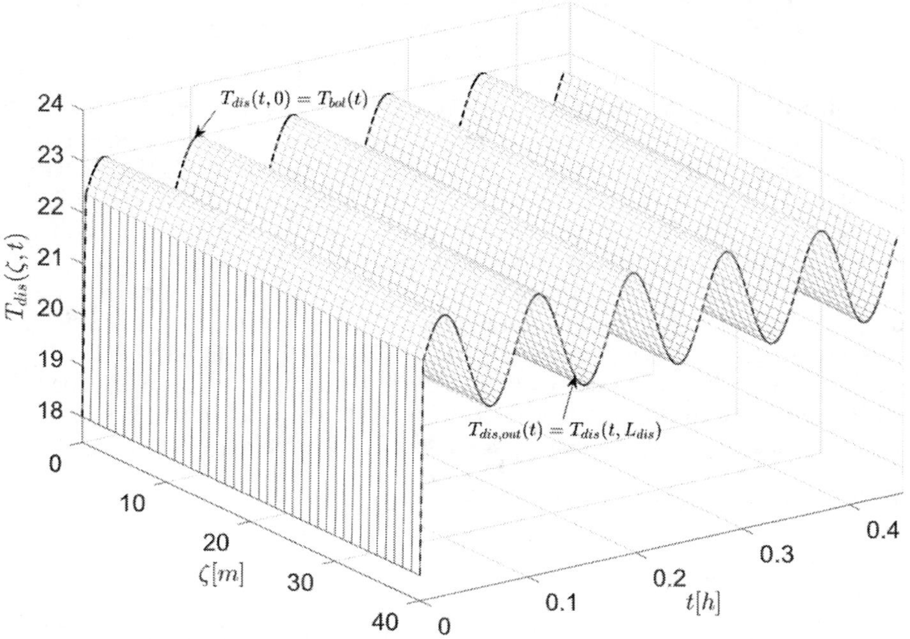

Conclusion:

- Model of Solar Thermal District Heating System
  - Solar collector field
  - Heat exchanger
  - Boiler-heating system
- Multivariable Optimization Problem
  - Solar collector filed
  - Heat exchanger
  - Boiler-gas system
- Internal Model Principle
  - District heating loop system

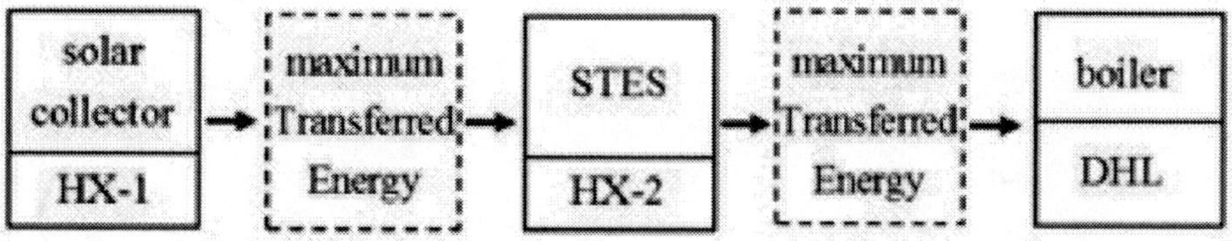

496108 Control System Design for Small Size Isolated Concentrated Solar Power Generation Units

Abstract:

Electricity generated by small size concentrated solar power (CSP) driven Rankine cycle (RC) is an increasingly explored alternative for powering isolated homes / small communities. In these units, solar energy is captured to raise the temperature of a heat transfer fluid (HTF). The hot HTF drives a low temperature Rankine cycle to generate electricity for the household(s). Since the solar flux follows the diurnal transient cycle along with high variability in the power consumed, very large inventory of the HTF is needed to store the solar energy and supply heat to the Rankine cycle as per power demand, including when the solar flux is zero at night. There then is no steady state for the process and it must be designed for inherently transient operation over a very large operating window. A well designed control system that provides robust operation for such a large operating window is then essential. In particular, it should effectively handle process constraints such as maximum/minimum HTF tank level limits etc. This paper evaluates robust control system design for a small size isolated CSP-RC power generation unit. A robust control strategy with appropriate overrides for handling process constraints is developed and tested using rigorous dynamic simulations. The closed loop dynamic simulation is also used for sizing the necessary HTF inventory.

495149 Strategic Planning of Oil Sands SAGD Drainage Area Development Under Uncertainty

Abstract:

During the lifecycle of an oilfield project, development is the most critical phase due to intensive investments required for capital and operating costs. Determining whether it is economic to develop an oilfield or not and finding the best schedule for field development are the main concerns in this field of study. Albertaâ€™s oil sands are the third largest proven global reserves of oil with current estimates of 26.6 billion cubic meters of crude oil. The majority of the oil sand deposits in Alberta Canada can only be extracted using in situ methods, mainly the Steam Assisted Gravity Drainage (SAGD). SAGD operations use super-heated steam to decrease the viscosity of the bitumen in the reservoir, such that bitumen can easily flow and be pumped up to the surface. A SAGD project normally targets the development of a number of drainage areas consisting of multiple injector and producer well pairs. Drainage area placement and development planning are crucial in a SAGD project.

Current practices of commercial placement of SAGD wells largely rely on reservoir characterization and engineering judgement. Optimization methods for the placement of SAGD drainage areas are relatively new in academia. Kumar investigated a space packing optimization method for generating a compact and non-overlapping arrangement of drainage areas [1]. Geometric transformations of global rotation, global translation and column translation were used by the space packing algorithm to maximize the recoverable bitumen. Restrictions such as non-placement of surface pads over surface restrictions or non-placement of drainage areas in thief zones were considered by incorporating a penalty function in the objective function. Manchuk and Deutsch used an adaptive grid search algorithm to determine the optimal arrangement of drainage areas [2]. The optimization algorithm determined the positions and orientations of the surface pads and drainage areas over a reservoir area to economically maximize the recoverable bitumen.

In this work, an optimization framework for planning the development of SAGD drainage areas under uncertainty is presented. The proposed framework includes the following major elements. First, an algorithm for compact drainage areas arrangement towards maximizing the amount of extractable bitumen is discussed. Second, a mixed integer optimization model is developed to arrange the multiperiod development plan of the drainage areas with consideration of capital and steam allocation restrictions [3]. Third, uncertainties in crude oil price and reservoir property are investigated based on a multistage stochastic programming model. The proposed method is applied to a case study with multiple drainage areas. The results demonstrate that the method can effectively generate a good drainage area layout and an economically optimal development plan that maximizes the net present value under crude oil price and reservoir uncertainty.

References:

[1] A. Kumar. Optimal drainage area and surface pad positioning for SAGD development. Thesis, University of Alberta, Canada. 2011.

[2] J.G. Manchuk and C.V. Deutsch. Optimization of Drainage-Area Configurations To Maximize Recovery From SAGD Operations. Journal of Canadian Petroleum Technology, 2013, 52, 233-242.

[3] H. Shahandeh, S. Rahim, Z. Li. Strategic Optimization of the Oil Sands Development with SAGD: Drainage Area Arrangement and Development Planning. Journal of Petroleum Science and Engineering. 2016, 137, 172-184.

488418 A Multistage Stochastic Programming Approach to Long-Term Electricity Procurement for Large Industrial Consumers

Abstract:

Due to the increasing volatility in electricity price, managing the procurement of electricity in power-intensive business has become a major challenge. Large industrial electricity consumers often enter into long-term bilateral contracts with favorable rates; however, such power contracts typically require the consumers to commit themselves to the amount that they are going to purchase months in advance. The vast majority of existing works on electricity procurement assume that the electricity demand is given, which implies that a separate production planning problem has already been solved, from which the electricity demand is determined. As shown in recent works on industrial demand side management (Zhang & Grossmann, 2016), this sequential approach is likely to be suboptimal.

In this work, we optimize long-term electricity procurement and production planning simultaneously, which poses two additional challenges: (1) how to account for time-sensitive electricity prices that change on an hourly basis while optimizing over a planning horizon that may span multiple months or years; (2) how to consider uncertainty in product demand, which can cause major disruptions in the production and electricity procurement plans. In order to solve this problem, we propose a multiscale multistage stochastic programming model in which a one-year planning horizon is divided into seasons, with each season represented by two characteristic weeks; also, each season corresponds to a stage at which the demand for that season is revealed. When applied to real-world industrial problems, the proposed approach results in model instances with tens of millions of variables and constraints. In order to solve these large-scale instances, we apply a decomposition algorithm based on the concept of progressive hedging.

We emphasize the importance of computing the value of stochastic solution (VSS) when evaluating the results of the model. The VSS was developed for two-stage stochastic programming and has rarely been applied to the multistage case. In this work, we apply the definition of VSS for multistage problems (VMSS) proposed by Escudero et al. (2007) and also extend it to define a VSS for applying a two-stage approximation of the problem (VTSS). When applied to an illustrative example, the results show that high VMSS can be achieved if the level of uncertainty is high, yet the VMSS are only marginally higher than the VTSS. Similar results have been observed when applying the proposed framework to an industrial air separation case.

References

Escudero, L. F., Garín, A., María, M., & Pérez, G. (2007). The value of the stochastic solution in multistage problems. *Top*, *15*, 48–64.

Zhang, Q., & Grossmann, I. E. (2016). Enterprise-wide optimization for industrial demand side management: Fundamentals, advances, and perspectives. *Chemical Engineering Research and Design*, *116*, 114–131.

502678 Multi-Stage Stochastic Programming Models for Pharmaceutical Clinical Trial Planning

Abstract:

Pharmaceutical industry is a global business with over one trillion U.S. dollars per year market with extensive supply chains throughout the world [1]. A potential drug identified at discovery stage goes through pre-clinical testing. The goal of these laboratory and animal model studies is to understand how the drug works and assess its safety. The clinical trials aim to demonstrate the safety and efficacy of the potential drug, and are designed with and carried out under strict guidelines and supervision of regulatory bodies. If a drug successfully completes the clinical trials and is approved by the regulatory bodies, the drug is manufactured and distributed to the market. Pharmaceutical manufacturers are under pressure to improve the efficiency of the pharmaceutical R&D pipeline, partially because the patent protections of a number of significant brand-name drugs will soon expire [2].

Managing the scheduling and planning of clinical trials is one of the efficient ways to reduce the cost of developing new drugs. There are three phases in clinical trials. The goal of Phase I clinical trials is to assess the safety of the drug and to understand how it is metabolized in the body. Phase II clinical trials are used to evaluate the drug's effectiveness and short-term side effects on a limited number of target patient volunteers. Phase III clinical trials aim to assess the benefit-risk ratio of the drug using a large number of target patient volunteers (in the order of thousands) [3]. The clinical trial planning is complicated due to its highly stochastic nature: the pharmaceutical companies do not know which drugs will successfully complete clinical trials a priori. The outcomes of clinical trials significantly influence drug development plan, the investments, and overall profits. An effective approach to solve clinical trial planning problem employs multi-stage stochastic programming (MSSP).

Colvin and Maravelias [4] developed a MSSP model for clinical trial planning. They explicitly incorporated the impact of endogenous uncertainty [5], i.e., the outcome of a clinical trial is only revealed once that clinical trial is completed, to their model. In a later study, they exploit the structure of the problem to reduce the number of scenarios, and extend their model to account for resource planning by introducing outsourcing decisions [6]. developed a sample average approximation (SAA) based algorithm to generate candidate solutions for the MSSP [7]. Colvin and Maravelias [8] introduce a number of theoretical properties, which reduce the problem size and improve the tightness of the formulation. Then, they develop a novel branch and cut algorithm to solve the resulting problem efficiently.

Motivated by above, we propose two new MSSP formulations for pharmaceutical clinical trial planning problem. Givens are: (1) a set of candidate drugs ($i \in I$)that should go through a set of clinical trials ($j \in J$), (2) the length of the planning horizon, which is discretized into equal time periods $t = 1, 2, 3...T$ (period t starts at time t-1 and ends at time t), (3) the lengths, resource requirement, and costs associated each clinical trial for each drug, and (4) potential revenue of each drug if it successfully completes all clinical trials. The decisions are which clinical trials of drugs to carry out, and when to start the clinical trials. The objective is to maximize the expected net present value (ENPV) [4]. The clinical-trial outcome uncertainty of each drug is defined by parameter, whose outcome space is {I-fail, II-fail, III-fail, III-pass}. The scenarios for the MSSP model are constructed as Cartesian product of uncertain parameter outcomes, and, the total number of scenarios for a problem with |I| drugs is $|S|= 4^{|I|}$.

The first formulation, which will be referred to as CM1, uses two binary decision variables, $X_{i,j,p,s}$ and $Y_{i,j,t,s}$. The first binary variable, $X_{i,j,p,s}$, tracks when a drug starts a clinical trial. Formally, it is equal to 1 if drug *i* starts clinical trial *j* at time period *p* in scenario *s*. The second binary variable, $Y_{i,j,t,s}$, tracks when a clinical trial is completed, and is equal to 1 if drug *i* completes clinical trial j at time period *t* in scenario *s*. The constraints ensure that (1) drug *i* starts clinical trial *j* only once, (2) drug *i* finishes clinical trial *j* only once, (3) clinical trial *j*+1 for each drug *i* cannot be completed before its previous trial *j*, and (4) the utilized resources at any given time period remains within the available resources. The second formulation, which will be referred to as CM2, introduces one more binary decision variable, $W_{i,j,t,p,s}$, which tracks both start and end time periods of drug *i* clinical trial *j*. This decision variable, is equal to 1 if (drug, clinical trial) pair *(i, j)* starts at time period p and end at time period *t*. The new binary variable satisfies the following $W_{i,j,t,p,s} \Longleftrightarrow X_{i,j,p,s} \wedge Y_{i,j,t,s}$.

We applied both formulations to solve 60 different instances [9] of clinical trial planning problem. For comparison, all instances were also solved using the MSSP formulation of [4], which will be referred to as CM3 here. All models were implemented in Pyomo [10], and solved using CPLEX 12.6.3 on Auburn University Hopper Cluster. All problems were solved to 0.1% optimality gap with all three formulations. In general, the solution times of CM3 were the longest, and the solution times of CM2 the shortest. For example, CM2 only took 8114 CPUs to solve the six-product case, while CM1 12991 CPUs, and CM3 108746 CPUs. Similar improvements were observed for all problems solved, and the differences in solutions times are more pronounced for larger instances of the problem. Furthermore, most problems with all formulations were solved at the root node. For instances where CPLEX branched, CM2 consistently required fewer branches than CM1 and CM3 to close the optimality gap, suggesting that the additional binary variable provided CPLEX a more efficient branching variable.

Acknowledgements

This work was completed in part with resources provided by the Auburn University Hopper Cluster. The authors would like to acknowledge financial support from the NSF CAREER Award #1623417.

References

1. PhRMA, 2016 Biopharmaceutical Research Industry Profile, Pharmaceutical Research and Manufacturers of America. 2016, PhRMA: Washington, DC.

2. IMS, Institute global use of medicines: outlook through 2016. Future Prescr, 2013. 14(1): p. 2-10.

3. FDA. FDA Drug Approval Process - US Food and Drug Administration. 2017 [cited 2017 30 January 2017]; Available from: http://www.fda.gov/downloads/Drugs/ResourcesForYou/Consumers/UCM284393.pdf.

4. Colvin, M. and C.T. Maravelias, A stochastic programming approach for clinical trial planning in new drug development. Computers & Chemical Engineering, 2008. 32(11): p. 2626-2642.

5. Goel, V. and I.E. Grossmann, A Class of stochastic programs with decision dependent uncertainty. Mathematical Programming, 2006. 108(2/3): p. 355-394.

6. Colvin, M. and C.T. Maravelias, Scheduling of testing tasks and resource planning in new product development using stochastic programming. Computers & Chemical Engineering, 2009. 33(5): p. 964-976.

7. Solak, S., et al., Optimization of R&D project portfolios under endogenous uncertainty. European Journal of Operational Research, 2010. 207(1): p. 420-433.

8. Colvin, M. and C.T. Maravelias, Modeling methods and a branch and cut algorithm for pharmaceutical

clinical trial planning using stochastic programming. European Journal of Operational Research, 2010. 203(1): p. 205-215.

9. Christian, B. and S. Cremaschi, Variants to a knapsack decomposition heuristic for solving R&D pipeline management problems. Computers & Chemical Engineering, 2017. 96: p. 18-32.

10. Hart, W.E., et al., Pyomo–optimization modeling in python. Vol. 67. 2012: Springer Science & Business Media.

492470 Two-Stage Stochastic Programming with Chance Constraints for Refinery Optimization

Abstract:

Maximizing profit while meeting product quality specifications is always the primary goal of refining industry in a highly competitive market with increasingly stringent environmental regulations. Noting that the properties of feedstocks and operating conditions in fact are subject to uncertainties, any decisions based on the nominal parameters can be infeasible in reality and result in a substantial loss. Hence, a robust decision-making system is highly necessary for refining industry. To this end, the entire operations of refinery is usually modeled by two-stage stochastic programming (SP) with a number of sampled scenarios [1]. By solving this formulation to global optimum, we may find a robust strategy to direct real operations.

However, in practice, some constraints at Stage-II in SP can be softened since recovery operations can be employed if an infeasible scenario occurs at Stage-II. This motivates us consider a SP with probabilistic constraints. For this new SP, at Stage-I, the decision of upstream operations is made to ensure that final products meet quality specification with high probability. At Stage-II, the uncertainty is realized and the optimal blending recipe is decided accordingly. If the real scenario is infeasible to the normal operations, a special operation will be triggered to recover the quality of products. To make this formulation more tractable, we consider a linear model of refinery operations and assume that uncertain parameters satisfy normal distribution. This novel two-stage decision process will lead to a mixed-integer linear programming (MILP) with sampled scenarios. Here we allow the recovery operations in 5% scenarios at most, if they are infeasible under the decision made at Stage-I. In other scenarios, we have to use normal operations. There are two difficulties for applying this model. First, the number of binary variables in MILP is proportional to the number of scenarios. On the one hand, more scenarios implies more accurate solution to the real process; on the other hand, the resulting formulation becomes more computationally demanding. Second, the feasibility of unsampled scenarios are not guaranteed.

To overcome these two limitations, we first improve the existing Benders decomposition method [2] to obtain the global optimal solution of scenario-based chance-constrained programming much more rapidly. Second, given the optimal solution of Stage-I based on sampled scenarios, we integrate second-order cone programming and linear decision-rule to evaluate its feasibility. We can show that if there exists a linear decision-rule such that the proposed second-order cone programming is feasible, then the Stage-I solution satisfies the probabilistic constraints no matter how many scenarios are considered in proposed MILP.

The presented scheme is able to handle a large number of uncertain parameters in the model and provide a global optimal solution promptly. The performance and effectiveness of this approach are demonstrated by optimizing a simplified refining process model under uncertainties.

Reference

[1] Yang Y, Barton PI. Integrated crude selection and refinery optimization under uncertainty. AIChE Journal. 2016;62:1038-1053.

[2] Liu X, Kucukyavuz S, Luedtke J. Decomposition algorithms for two-stage chance constrained programs. Mathematical Programming. 2016;157:219-243.

490564 Adjustable Robust Optimization for Multi-Tasking Scheduling with Reprocessing of Imperfect Tasks

Abstract:

The optimal scheduling of operations has progressively become more challenging due to the ever-increasing complexity of production networks and the agile competition that necessitates the transition to more versatile production schemes [1]. Therefore, a variety of short-term, multi-purpose process scheduling models have been proposed in the literature [2]–[4] that are pertinent to numerous real-life production environments. One of the cases that up until recently had not been investigated pertains to multi-tasking, where multiple tasks are being processed simultaneously by the same processing unit.

Multi-tasking scheduling is vital for the analytical services industry (ASI), where samples originating from different orders must undergo a number of processing steps (tasks) that are typically order-specific. These tasks are executed by processing units that are capable of handling samples from different orders, giving rise to a multi-tasking scheduling problem. Recent works have treated this as a multi-commodity flow problem with a discrete time grid [5], and later as a special case of multi-purpose environment [6] using a modified continuous time, slot-based formulation [7]. In both cases, however, parameter uncertainty was not considered.

Uncertainty is of paramount importance in process scheduling, since optimization under nominal conditions can lead to suboptimal, or even infeasible solutions, in view of the actual realized values of the uncertain parameters [8]. Some of the main sources of uncertainty, also relevant to ASI, are variations in raw materials specifications, or fluctuations in operating conditions of the processing units that may result to imperfect products requiring reprocessing to meet the standards.

To address reprocessing of imperfect tasks, we present a systematic way of modeling it via the introduction of recycle streams in conjunction with uncertain variability in the associated production yields. To that end, we introduce a State-Task-Network representation for multi-tasking environments with recycles. This representation also allows the utilization of any multi-purpose scheduling model with only minor adaptation of the material balance and state level related constraints. Using a global event based model [9], we show the application of the multi-stage Adjustable Robust Optimization framework [10] and its ability to overcome the limitation of traditional Robust Optimization by obtaining solutions that maintain their feasibility, despite the existence of uncertainty in the production yields of imperfect tasks. Finally, we present a comprehensive computational study across multi-tasking scheduling benchmark problems for various levels of uncertainty, which allows us to assess both the quality of the robust solutions as well as the computational effort required to obtain those.

References:

[1] I. Harjunkoski, C. T. Maravelias, P. Bongers, P. M. Castro, S. Engell, I. E. Grossmann, J. Hooker, C. A. Méndez, G. Sand, and J. M. Wassick, "Scope for industrial applications of production scheduling models and solution methods," Comput. Chem. Eng., vol. 62, pp. 161–193, 2014.

[2] C. A. Méndez, J. Cerdá, I. E. Grossmann, I. Harjunkoski, and M. Fahl, "State-of-the-art review of optimization methods for short-term scheduling of batch processes," Comput. Chem. Eng., vol. 30, no. 6–7, pp. 913–946, 2006.

[3] C. A. Floudas and X. Lin, "Mixed Integer Linear Programming in Process Scheduling: Modeling, Algorithms, and Applications," Ann. Oper. Res., vol. 139, no. 1, pp. 131–162, 2005.

[4] M. Baldea and I. Harjunkoski, "Integrated production scheduling and process control: A systematic review," Comput. Chem. Eng., vol. 71, pp. 377–390, 2014.

[5] B. P. Patil, R. Fukasawa, and L. A. Ricardez-Sandoval, "Scheduling of Operations in a Large-Scale Scientific Services Facility via Multicommodity Flow and an Optimization-Based Algorithm," Ind. Eng. Chem. Res., vol. 54, no. 5, pp. 1628–1639, 2015.

[6] S. Lagzi, R. Fukasawa, and L. Ricardez-Sandoval, "A multitasking continuous time formulation for short-term scheduling of operations in multipurpose plants," Comput. Chem. Eng., vol. 97, pp. 135–146, 2017.

[7] S. Gupta and I. A. Karimi, "An Improved MILP Formulation for Scheduling Multiproduct, Multistage Batch Plants," Ind. Eng. Chem. Res., vol. 42, no. 11, pp. 2365–2380, 2003.

[8] Z. Li and M. G. Ierapetritou, "Process scheduling under uncertainty: Review and challenges," Comput. Chem. Eng., vol. 32, no. 4–5, pp. 715–727, 2008.

[9] P. M. Castro, A. P. Barbosa-po, H. A. Matos, and A. Q. Novais, "Simple Continuous-Time Formulation for Short-Term Scheduling of Batch and Continuous Processes," Ind. Eng. Chem. Res., vol. 43, no. 1, pp. 105–118, 2004.

[10] N. H. Lappas and C. E. Gounaris, "Multi-stage adjustable robust optimization for process scheduling under uncertainty," AIChE J., vol. 62, no. 5, pp. 1646–1667, 2016.

493740 A Data-Driven Multistage Adaptive Robust Optimization Framework for Planning and Scheduling Under Uncertainty

Abstract:

Robust optimization has gained increasing popularity because it provides a good balance between solution quality, feasibility and computational tractability [1]. Robust optimization methods can be roughly classified into three categories, namely static robust optimization, two-stage adaptive robust optimization (ARO), and multistage ARO. Two-stage ARO grants more flexibility than static robust optimization by introducing recourse decisions, and typically generates less conservative solutions [2-5]. To overcome the limitation of two-stage structures, multistage ARO is recently proposed for non-anticipative sequential decision making processes [6, 7]. Notably, most of the existing robust optimization approaches assume that uncertainty sets are defined or given *a priori* and fall short of fully utilizing uncertainty data. Consequently, these approaches hedge against the uncertainty realization regardless of its occurrence probability, and could generate conservative solutions. The abundance of uncertainty data and recent advances in machine learning hold the promise to address the conservatism issue, and call for research efforts on data-driven robust optimization [8, 9].

To the best of our knowledge, data-driven approach for multistage ARO has not been considered in the existing literature. Therefore, this paper aims to fill the knowledge gap and take the first step of developing a data-driven multistage ARO framework. In this work, we propose a novel data-driven multistage ARO framework that leverages robust statistics and machine learning techniques for multistage ARO methods. To extract an accurate probability distribution from uncertainty data that could be possibly large-scale and messy, we introduce a nonparametric robust kernel density estimation (RKDE) [10], which is obtained via a kernelized iteratively re-weighted least squares (KIRWLS) algorithm. Based on the information extracted from uncertainty data, we further propose a data-driven uncertainty set for multistage ARO using quantile functions and uncertainty budgets to organically integrate uncertainty data information with the multistage ARO framework. The quantile functions provide the bounds of uncertain parameters based on cumulative distribution functions. Moreover, uncertainty budgets are employed to adjust the level of conservatism. We apply affine decision rules for recourse variables [11], and further develop a data-driven multistage adaptive robust counterpart in a general form to facilitate the solution of the aforementioned ARO problem. Moreover, we introduce an approach to capture correlations between uncertain parameters in the data-driven uncertainty set using covariance matrix [12]. We further develop a corresponding data-driven multistage adaptive robust counterpart that addresses not only the correlations between different uncertain parameters, but also the temporal correlations between different stages.

To illustrate the applicability of the proposed framework, two typical applications in process operations are presented. One application is the multi-period strategic planning of process networks [13], and the other one is the multipurpose batch process scheduling [14, 15]. In a case study on process planning, our proposed approach generates 23.9% more net present value (NPV) than that of the multistage ARO with a box uncertainty set. Compared with multistage ARO using KDE, the proposed method returns 22.6% more NPV. In a case study on batch process scheduling, the proposed approach returns 31.5% and 7.6% more profits than those of multistage ARO with a box uncertainty set and a KDE based uncertainty set, respectively.

References

[1] D. Bertsimas, D. B. Brown, and C. Caramanis, "Theory and applications of robust optimization," SIAM Review, vol. 53, pp. 464-501, 2011.

[2] H. Shi and F. You, "A computational framework and solution algorithms for two-stage adaptive robust scheduling of batch manufacturing processes under uncertainty," AIChE Journal, vol. 62, pp. 687-703, 2016.

[3] J. Gong, D. J. Garcia, and F. You, "Unraveling Optimal Biomass Processing Routes from Bioconversion Product and Process Networks under Uncertainty: An Adaptive Robust Optimization Approach," ACS Sustainable Chemistry & Engineering, vol. 4, pp. 3160-3173, 2016.

[4] J. Gong and F. You, "Optimal processing network design under uncertainty for producing fuels and value-added bioproducts from microalgae: Two-stage adaptive robust mixed integer fractional programming model and computationally efficient solution algorithm," AIChE Journal, vol. 63, pp. 582-600, 2017.

[5] D. Yue and F. You, "Optimal supply chain design and operations under multi-scale uncertainties: Nested stochastic robust optimization modeling framework and solution algorithm," AIChE Journal, vol. 62, pp. 3041-3055, 2016.

[6] E. Delage and D. A. Iancu, "Robust multistage decision making," in The Oper Res Rev. , ed, 2015, pp. 20-46.

[7] N. H. Lappas and C. E. Gounaris, "Multi-stage adjustable robust optimization for process scheduling under uncertainty," AIChE Journal, vol. 62, pp. 1646-1667, 2016.

[8] D. Bertsimas, V. Gupta, and N. Kallus, "Data-driven robust optimization," Mathematical Programming, 2017. DOI:10.1007/s10107-017-1125-8

[9] C. Ning and F. You, "Data-driven adaptive nested robust optimization: General modeling framework and efficient computational algorithm for decision making under uncertainty," AIChE Journal, 2017. DOI: 10.1002/aic.15717

[10] J. Kim and C. D. Scott, "Robust kernel density estimation," Journal of Machine Learning Research, vol. 13, pp. 2529-2565, 2012.

[11] A. Ben-Tal, A. Goryashko, E. Guslitzer, and A. Nemirovski, "Adjustable robust solutions of uncertain linear programs," Mathematical Programming, vol. 99, pp. 351-376, 2004.

[12] Y. Yuan, Z. Li, and B. Huang, "Robust optimization under correlated uncertainty: Formulations and computational study," Computers & Chemical Engineering, vol. 85, pp. 58-71, 2016.

[13] F. You and I. E. Grossmann, "Stochastic inventory management for tactical process planning under uncertainties: MINLP models and algorithms," AIChE Journal, vol. 57, pp. 1250-1277, 2011.

[14] Y. Chu, J. M. Wassick, and F. You, "Efficient scheduling method of complex batch processes with general network structure via agent-based modeling," AIChE Journal, vol. 59, pp. 2884-2906, 2013.

[15] J. M. Wassick, A. Agarwal, N. Akiya, J. Ferrio, S. Bury, and F. You, "Addressing the operational challenges in the development, manufacture, and supply of advanced materials and performance products," Computers & Chemical Engineering, vol. 47, pp. 157-169, 2012.

492483 Discrete-Time Mixed-Integer Programming Models for Simultaneous Batching and Scheduling in Sequential Environments

Abstract:

Scheduling problems in sequential environments are among the most common in batch manufacturing facilities. In general, a sequential production environments has a series of stages in which batches have to go through without being mixed, split or recycled. Among problems in sequential environments, multistage batch plants are the facilities that have received the most attention (Harjunkoski *et al.*, 2014). However, only a limited number of studies is available for, the more general, (sequential) multipurpose batch plants. Furthermore, most of the existing approaches for scheduling in sequential environments are based on several restrictive assumptions. Specifically, batching decisions are assumed to be decoupled from scheduling decisions, often leading to suboptimal solutions. Another simplifying assumption commonly made is that unlimited intermediate storage is available. Although models that overcome these limitations have been proposed for multistage batch plants (Liu and Karimi, 2008; Sundaramoorthy *et al.*, 2009), no models addressing all these limitations are available for multipurpose facilities.

In this work, we develop two novel discrete-time mixed integer programming models for simultaneous batching and scheduling in multipurpose facilities with storage constraints. The proposed models adopt two different modeling approaches. The first is based on explicit labeling of the batches of an order. Although each batch is labeled and scheduled individually, they are linked through the order satisfaction constraint, as well as the unit utilization constraint. The second is based on identifying possible unit routings for each order and the corresponding batch size intervals. After batch intervals are identified, the problem becomes an assignment problem where batches of a given order are assigned to different batch size intervals to meet demand. Since it is not trivial to identify unit routings and batch size intervals, we develop an algorithm which only need to be run once for each facility. In addition, we propose extensions for both models that allow us to consider limited shared utilities (both fixed and time-varying) and storage with capacity limits.

We provide illustrative examples to showcase the impact of simultaneously considering (1) batching decisions, (2) limited shared resources, and (3) storage with capacity limits on the final schedule. Finally, we carry out a computational study to understand how instance characteristics, such as expected number of batches per order and uniformity in unit capacities, impact the effectiveness of the proposed models. We show that due to the different modeling approaches adopted, their performance shows a clear trend with respect to instance characteristics. Hence, carefully selecting the model allows us to effectively solve large-scale instances.

Reference

1. Harjunkoski, I., Maravelias, C. T., Bongers, P., Castro, P. M., Engell, S., Grossmann, I. E., Hooker, J., Mendez, C., Sand, G. and Wassick, J. (2014) Scope for industrial applications of production scheduling models and solution methods. *Computers & Chemical Engineering*. **62**, 161-193.
2. Liu, Y. and Karimi, I. A. (2008) Scheduling multistage batch plants with parallel units and no interstage storage. *Computers & Chemical Engineering*. **32**, 671-693.
3. Sundaramoorthy, A., Maravelias, C. T. and Prasad, P. (2009) Scheduling of Multistage Batch Processes under Utility Constraints. *Industrial & Engineering Chemistry Research*. **48**, 6050-6058.

AUTHOR INDEX

AUTHOR INDEX

AUTHOR INDEX

AUTHOR INDEX

AUTHOR INDEX

AUTHOR INDEX

AUTHOR INDEX

AUTHOR INDEX

AUTHOR INDEX